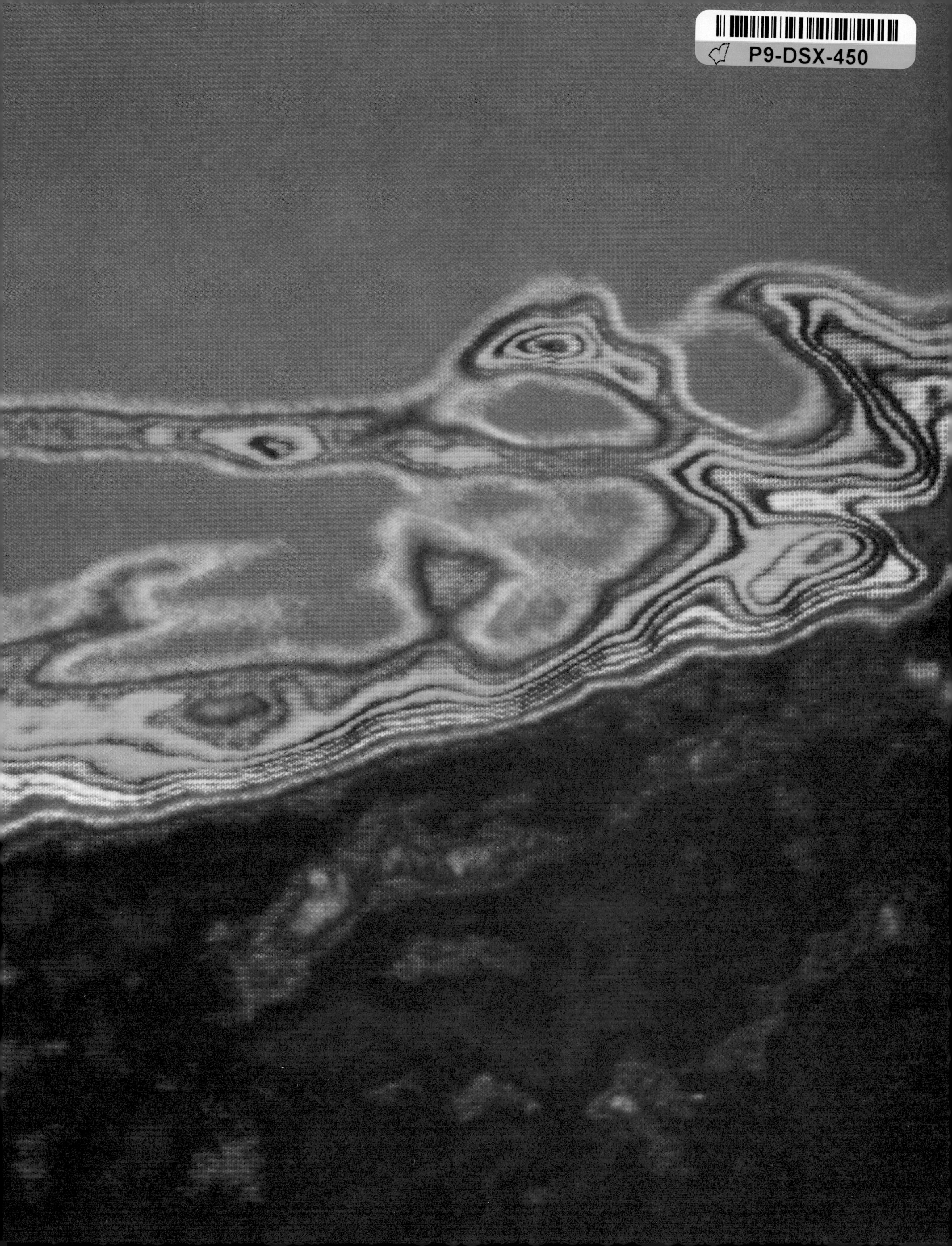

Larousse Astronomy

Contributors

Roger-Maurice Bonnet, *research director of the French National Center for Scientific Research, director of scientific programs, European Space Agency*

Pierre Bourge, *founder of the French Astronomy Association and the journal* Ciel et Espace

Françoise Combes-Bottaro, *junior lecturer at the Physics Laboratory of the École Normale Supérieure and at the Department of Radioastronomy of the Paris-Meudon Observatory*

Pierre Encrenaz, *senior astronomer at the Paris Observatory, director of the Millimetric Radioastronomy Laboratory of the École Normale Supérieure and of the Paris-Meudon Observatory*

Renaud Foy, *research supervisor at the Paris-Meudon Observatory*

Lucienne Gouguenheim, *professor at the University of Paris XI, director of the Nançay Radioastronomy Center*

James Lequeux, *director of the Marseilles Observatory*

Anny-Chantal Levasseur-Regourd, *professor at the University of Paris VI, researcher at the Aeronomy Department of the French National Center for Scientific Research*

Robert Lucas, *professor of radioastronomy at the Medical Scientific University of Grenoble*

François Raulin, *junior lecturer at the Environmental Physics and Chemistry Laboratory of the University of Paris XII*

Hubert Reeves, *research director of the French National Center for Scientific Research (Saclay Nuclear Studies Center, Astrophysics Division)*

Gérard Toupance, *senior lecturer at the Environmental Physics and Chemistry Laboratory of the University of Paris XII*

Larousse Astronomy

Editor-in-Chief:
Philippe de la Cotardière
Edited by **Mark R. Morris**, Professor of Astronomy, UCLA (Los Angeles, California)

Facts On File Publications
New York, New York • Oxford, England

ASTRONOMY

New Edition with updated text

First published in 1987 in the United States by

Facts on File, Inc.
460 Park Avenue South
New York, New York 10016

Library of Congress Cataloging in Publication Data
Astronomie. English
Astronomie.

Includes index.
1. Astronomy. I. La Cotardiére, Philippe de.
QB43.2.A8713 1985 520 84-28770
ISBN 0-8160-1219-9

Photo selection by Françoise Guillot
Drawings by Alain Doussineau, Frank Bouttevin, Alain Vial
Layout by Juan Cousino based on a design by F. Longuépée and H. Serres-Cousiné
Translation by Susan Wald

Printed in Italy

9 8 7 6 5 4 3 2 1

Contents

Introduction

By Philippe de la Cotardière

In the past twenty years, direct exploration of the solar system by spacecraft, the launching of instruments that can extend observation to the various radio wavelengths that are blocked out by the earth's atmosphere, and the installation on Earth of a number of high-performance visual and radio telescopes have led to the spectacular growth of astronomy.

Widely publicized by the media, the great expansion in our knowledge of nearby planets has attracted particular public attention. But the growth in knowledge has been no less substantial with regard to the stars, interstellar space and galaxies. The traditional observation of the starry sky led to a vision of the universe as a system ruled by order and calm. Recent studies in wavelengths invisible to the naked eye—ultraviolet, X-ray, gamma ray—have swept aside that vision and established an image of a cataclysmic universe.

Since most existing works on astronomy are now out of date, we have tried to fill the void. This work aims to present a contemporary and complete panorama of the universe—incorporating the most recent discoveries—for everyone who, while not expert in astronomy, is interested in the development of that science.

Without disregarding the substantial progress made in recent years in the study of the extrasolar universe, we have deliberately placed the solar system in the forefront: first, because the sun—being the only star whose surface and atmosphere may be observed in detail—is a prototype, so to speak, on which specialists test some of their theories regarding stellar structure or evolution; second, because the past decade has emphasized the exploration of the planets, and it seemed to us that the public, while kept informed by the press of the progress of space missions, remained generally unaware of the results obtained by those missions.

In the last chapter, we wished to raise the issue of extraterrestrial life and the possibilities of contact with any civilizations on other planets. This is a fascinating question that various astrophysicists and biologists around the world are actively exploring.

Beyond the wealth of information it contains, in which we hope readers will find answers to some of their questions regarding the universe, this book will fully achieve its goal if it becomes, for those who browse through it, a gateway to dreams—a passport for a voyage toward infinity.

Aerial view of the Pic du Midi Observatory in southern France. Famous for the quality of the solar, lunar and planetary observations that have been done there over the past several decades, this institution now is also concerned with the study of distant objects, with the help of its new 2-meter (79-inch) telescope (tower and cupola visible at far left). Photo: O.P.M.T.

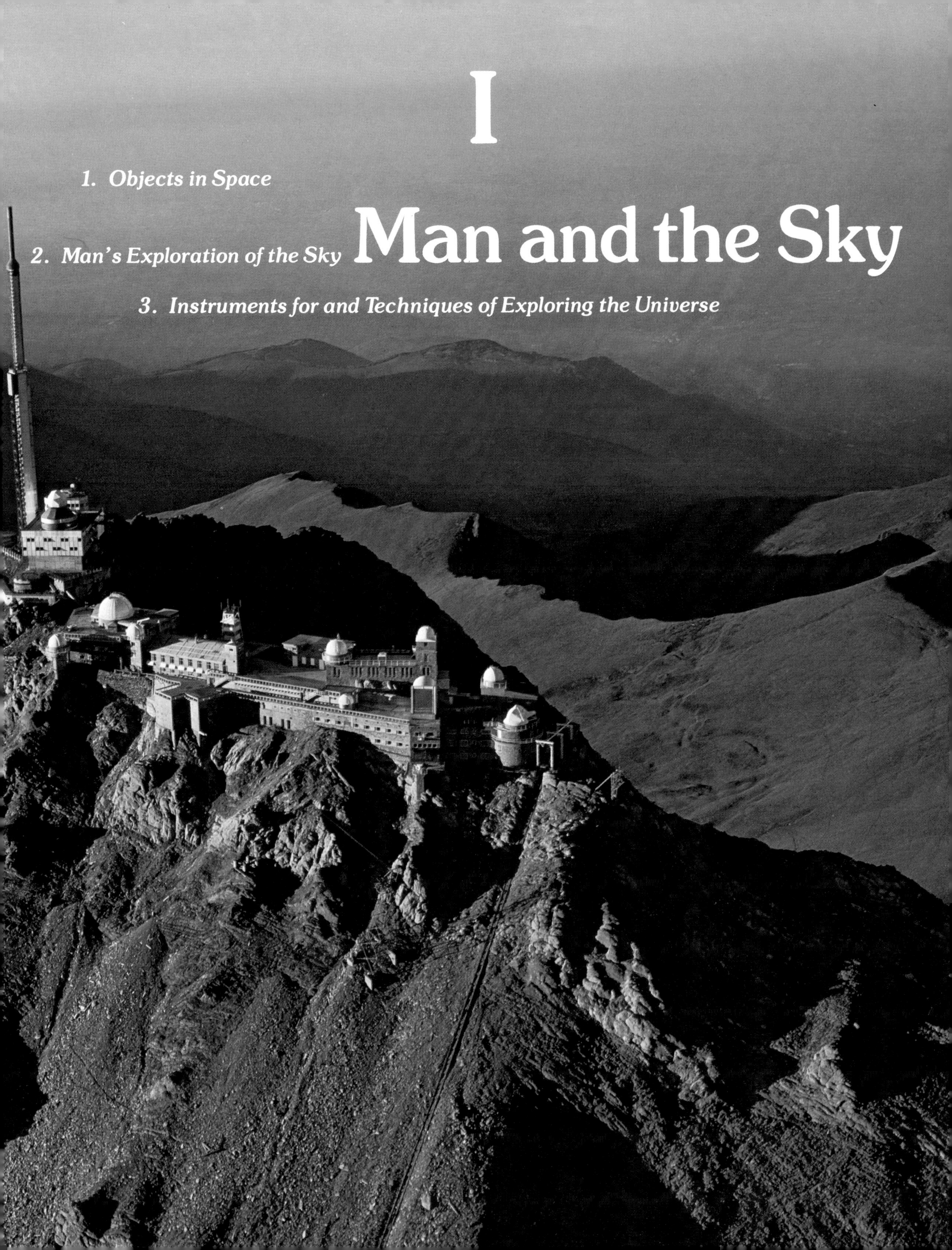

I

Man and the Sky

1. Objects in Space

2. Man's Exploration of the Sky

3. Instruments for and Techniques of Exploring the Universe

1. Objects in Space

When we confront the heavens, everything at once attracts our gaze and awakens our intellectual curiosity. Above us, as though driven by a slow, relentless circular motion, each in turn gives way to the other—the dazzling sphere of the Sun, the Moon with its changing shapes, the thousands upon thousands of stars—like so many glittering points that light up one by one after darkness falls.

***The Earth viewed from space:** For millennia, human beings have gazed with admiration on the other worlds of the universe. Today, with the help of spacecraft, we can also discover the face of our planet and appreciate its beauty. Photo: ESA.*

And then there are the special or unusual phenomena—comets with strange features, eclipses of the sun and moon, and also the sudden and transitory appearances of certain brilliant meteors. These are some of the spectacles that at times are bound to make a vivid, strong impression. From this visual attraction, in the earliest epochs of thinking humanity, from this sometimes admiring, sometimes fearful contemplation, astronomy was born—perhaps the most exciting of all branches of science.

Originally, knowledge of astronomy was based solely on unaided human observation. But while this method of investigation informs us effectively in various respects, it can also lead us to err. To begin with, there is one illusion that can be dispelled only through rational thought—that of the "celestial sphere" over which all the stars appear to be scattered and grouped, as though they were candles on a curved ceiling, but extending to an immeasurable height. Such an impression can be explained once we know that our eyes are incapable of discerning distance variations by way of depth sensation beyond relatively nearby limits, and so they place all the stars on the same plane, regardless of the prodigious and very uneven distances at which they actually lie.

Nevertheless, false though it may be, the celestial sphere is evident to our senses. It is as though we were in the center of a perfect sphere of infinite radius. Hence, when we speak of the "sky," it must be understood to mean the apparent panorama rather than the real distribution of stars in space that gives rise to it, owing to the incapacity of our eyesight and the effects of perspective.

Thus the starry sky is nothing but an image. The atmosphere that surrounds us and is essential to the maintenance of life, is a gaseous layer extending above the surface of the Earth and matching the curvature of the globe; and its arched shape is made materially visible by the cloud layers that, although spread out at different levels, in their totality create a genuine ceiling, extending in perspective all around us toward the horizon. The condition of the atmosphere also determines the degree of the stars' visibility, since they are necessarily seen through the more or less transparent layer lying between them and human eyes.

The meteorologists' sky, with its traditional shade of blue, is therefore very different from that of the astronomers. But while they should not be confused, we will see that they must not be completely disassociated, either, for the atmosphere plays a crucial role in the production of various astronomical phenomena.

The astronomers' sky, the successive pictures of which are so many visible translations of the universe, has often been compared to the pages of a book lying open to the eyes of all. But one must still know how to read this book, for to many, it is written in an unknown language. Yet while it requires painstaking study that obviously demands attention and thought, it can provide indescribable joys as one gradually becomes capable first of spelling it out, then of understanding the subject and then of trying to unravel its secrets. At the end, the mind eager for knowledge will find this reading as fascinating as a novel. But—one will object—how can you weigh dramatic action or events against a spectacle that, in spite of its imposing splendor, always remains the same except for a few periodic variations? Here again, appearances are deceiving. For while the universe may appear unchanging at the very instant we contemplate it, what formidable motions are occurring in it without interruption! What magnitudes in length and duration, what transformations and cataclysms are revealed to us, staggering the imagination and arousing the most prodigious amazement! If human beings cannot grasp all of this at first glance, it is because they are too small on the scale of such events; their senses are imperfect and their lives too short. But then, for those who become aware of it, knowledge of these awesome facts will seem more compelling than anything. However, we can only understand what happens and what we witness on condition that this idea is solidly ingrained: the Earth on which we live is a celestial body just as much as the others; like them, and with them, it is plunged into the midst of space. Within that space, our world is driven by various forces, and the spectacles that continually unfold before our eyes result from the combination of those forces with those of the celestial bodies that appear everywhere in the vastness.

The Diversity of Celestial Objects

Of all the stars, that which is most prominent to our eyes is the Sun. It is a ball of bright gas, hot in and of itself, a powerful, incandescent furnace with an intensity to which nothing on Earth can be compared, but fortunately far enough away to give us light and heat without "baking" us.

The Sun's powerful illumination enables us to perceive other objects rotating around it, which do not shine by themselves but whose surfaces are lit by the Sun—the planets, to which family the Earth belongs. Some planets themselves possess satellites that accompany them in their journey around the Sun; such is the case, specifically, of the Earth, which has a single natural satellite, the Moon.

Because of their great distance, which reduces their apparent diameter to an infinitesimal number, the planets appear to be indistinguishable from the stars, a term that in everyday language refers to all the glittering points in the sky without distinction. But by a bit of close observation, in the absence of knowledge, one can detect the difference. The real stars are arranged in unchanging patterns making up the constellations—such as the Big Dipper—that some people like to gaze at or recognize while out walking on a clear night. One the other hand, despite their stellar appearance, the planets—most of which appear to us as brighter than the stars, even the brightest ones—change position fairly rapidly relative to the stars. That is a result of their movement around the Sun and their relative proximity compared to the stars, whose distances are so great that they can be considered fixed landmarks on an infinite backdrop. The stars are suns more or less comparable to our own—that is, immense balls of incandescent gas emitting a colossal flux of light and heat. But they are so far away that they are reduced to mere shining points in the sky.

A mere glance at the celestial sphere is sufficient to notice the variety in the brightnesses of stars, from the most brilliant ones to those that the eye can barely make out. Yet we must realize that we are seeing only a very modest sample of the celestial host. The instruments of modern astronomy have revealed the existence of billions upon billions of stars we cannot possibly detect with the naked eye.

As we perceive them scattered over the expanse of the sky, the stars appear closer together in some places, as though belonging to groups arranged so that they form certain patterns and dominate our view to some extent. Thus figures appear to take shape. Some have irregular outlines, while others show a kind of geometrical regularity that is sometimes remarkable. These are the *constellations*. But these groupings are only apparent. Their varied shapes result solely from the effects of perspective. Where we think we see several adjoining stars, in reality they are spread out at very unequal depths depending on the direction of our line of sight.

The General Motions of Celestial Bodies and the Phenomena to Which They Give Rise

If we ignore the even, sustained motion with which the stars seem to march tirelessly by, the first impression we get is that the spectacle of the heavens is a majestic symbol of calm and immobility. In fact, close observation shows that from one night to the next, the Moon changes position relative to the stars; even a few hours are sufficient to note its displacement if it is close to a bright star that can be used as a reference point. Similarly, within a few days or a few weeks we see the planets occupy different positions among the constellations. These examples point to the movement of heavenly bodies in space. But many more could be mentioned. Indeed, not a single star is motionless. All, in fact, are propelled at very great speeds, though their movement may be generally imperceptible to us because of their great distance from us.

Our closest neighbor, the Moon, revolves around the Earth at a speed of about 1 kilometer per second. Pluto, the slowest planet (because it is the farthest from the Sun), follows its orbit at 4.7 kilometers (2.92 miles) per second, and Mer-

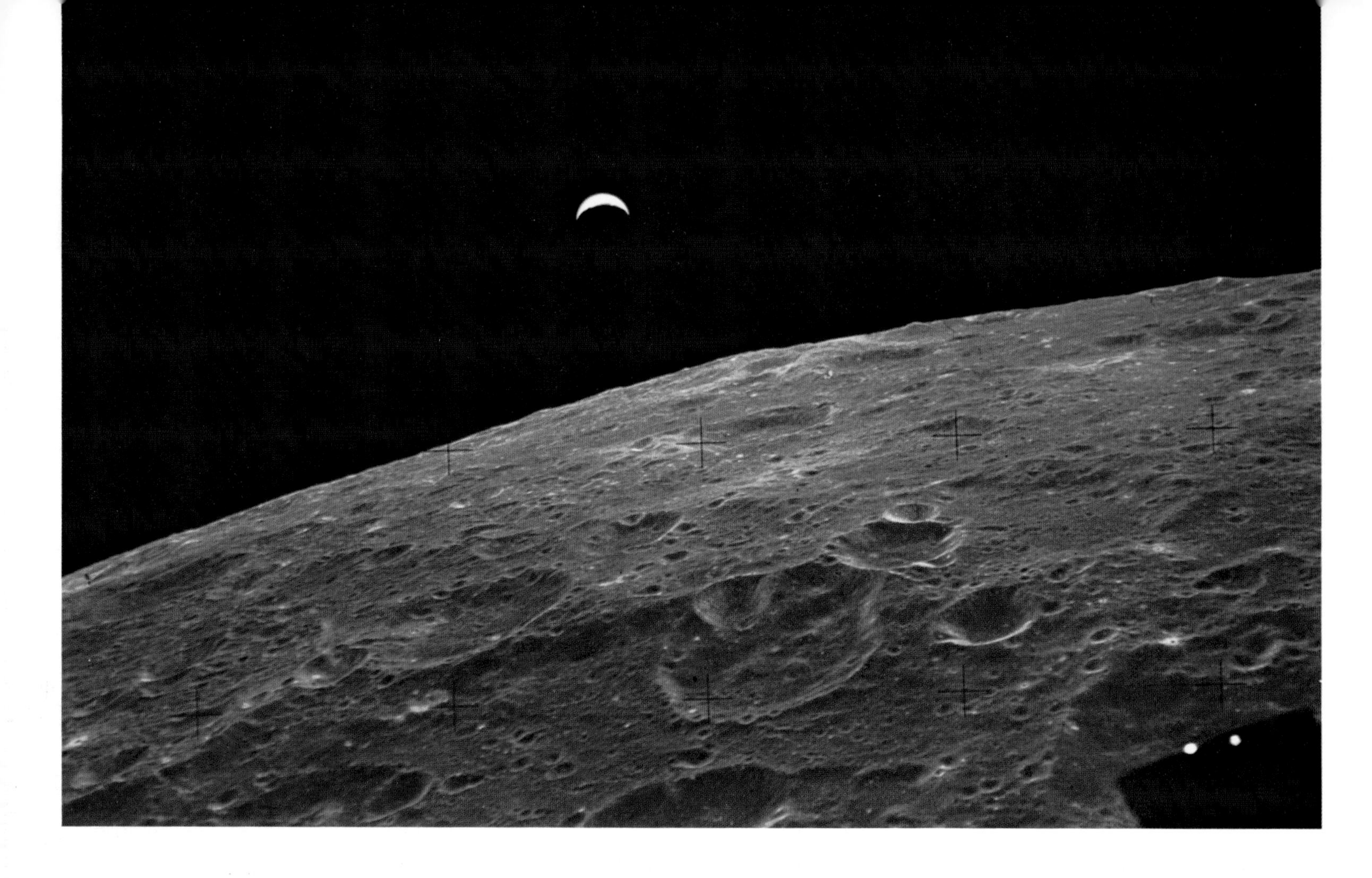

cury, the swiftest, at 47.9 kilometers (29.77 miles) per second. The stars attain much higher speeds: 100, 200 even 300 kilometers (62, 124, 186 miles) per second. As for the galaxies, the most distant sidereal formations we know of, they move at fantastic speeds measured in thousands or tens of thousands of kilometers per second!

Yet what do we observe? That the Moon visibly drops to the horizon in a few hours, whereas the motion of the planets becomes perceptible only after a substantially longer time, and that the stars seem to remain practically motionless. For as long as human beings have been pondering the sky, the relative positions of the stars outlining the constellations have not undergone any apparent change. It is only since the nineteenth century that extremely precise measurements have made it possible to detect the stars' own motions.

These appearances are caused by the differing distances of the bodies we are able to see. The Moon, which moves at a relatively modest speed, appears fast because it is very close. The stars, on the other hand, despite their much greater speed, seem motionless because they are very far away.

Celestial Distances

It is difficult to picture the scale of the distances between celestial bodies. In our daily lives we are used to measuring distances in miles and kilometers, and journeys of several thousand kilometers, like a trip "around the world," already seem great to us. They are, however, quite insignificant compared with celestial distances. At most they can be compared with the distances of some satellites from the planets they are attached to. But these exceptions notwithstanding, it would take millions, hundreds of millions, billions of kilometers and even more, laid end to end!

Hence, in most cases, one must try to make the immense distances of the celestial bodies more imaginable. The closest ones to us are first the Moon, then certain planets, and the Sun, which they revolve around as our globe does. The Earth belongs to a group, the solar system, which extends over a radius of about 6 billion kilometers (3.73 billion miles) to the most distant planet we know of at this time. However great this number seems to us, it is nil, so to speak, next to those that express the distances of the majority of stars. Thus, better than a nearly unintelligible string of numbers, a physical fact enables us to measure these distances—the time that light takes to travel to them.

Light waves are propelled at a speed of about 300,000 kilometers (186,451 miles) per second. They cover the distance between us and the Moon in 1.25 seconds; between us and the Sun in a little more than eight minutes; and between the Sun and Pluto, the farthest planet in the solar system, in about six hours.

Yet this is nothing compared with the distances between us and the stars, since they are measured not in hours or days but in years and centuries! A light wave starting from the nearest of these

The Moon: *Long considered inaccessible, but nevertheless very close to the Earth, it was the principal target of the first spacecraft. Photo: NASA.*

other suns would not reach us until four years and three months later. Then there are others for which the travel time would be lengthened to 10, 20, 50, 100 or 1,000 years or more. Finally, light from those stars that have been identified at the outermost reaches attainable with current methods of investigation would not reach us for several billion years! Nevertheless, we must admit that journeys of such staggering durations may be just as disconcerting to our understanding of the the distances involved as a string of kilometers expressed in unreadable numbers. Whatever the case may be, we now use a unit to simplify the numbers that equals the distance traveled by light in a year. This unit, called a *light-year*, is equal to 9.461 trillion kilometers (5.88 trillion miles). Astronomers also rely on other units of measurement, which will be defined in a later part of this book.

The Visible Universe and Our Place in It

In conclusion, we are on a planet that, along with others, revolves around the Sun. This celestial family constitutes a system whose dimensions we have indicated. Beyond the limits of this system, and immensely far away in all directions, the stars—those other suns—shine in the depths of space, as isolated from each other as they are from the Sun; but as a result of perspective we see them somewhat clustered, forming what appear to be groupings—first, the constellations, somewhat spread out and dispersed, then others pulled together into a mass that makes up the bright clouds of the Milky Way.

From the standpoint of the real distances between them, which are so great, the stars appear to have no relation to one another. But that is not at all the case. All the stars we see make up a localized entity in space. This prodigious grouping, which we call a *galaxy*, is a roughly lens-shaped cluster, where the stars are distributed fairly unevenly, gathered here and there into larger or smaller concentrations. Our galaxy is an entity revolving slowly around an axis perpendicular to the plane of the lens—a replica, approximately, of what we saw in the solar system.

The Sun is one of the stars in the galaxy, but it is every bit as unremarkable as any individual lost in an immense crowd, for it is neither the biggest nor the smallest, and it occupies no special position. Sharing the same destiny as its innumerable counterparts, the Sun in turn is propelled across space, trailing its procession of planets at great speed.

And just as a great vacuum extends from the solar system until it comes in contact with other stars, so the galaxy is surrounded by an immensity of empty space separating it from similar concentrations. From some of these galaxies light takes hundreds of millions, indeed billions of years to reach us, whereas it crosses our entire galaxy in some 100,000 years. We can thus approximately judge the limited size of our galaxy compared to the space existing between it and its counterparts.

Such are the major outlines of the visible universe.

A characteristic feature of the stellar universe: *The "North American" nebula—whose shape resembles that of the continent—in a star-studded region of the heavens. Photo copyright © 1959 by California Institute of Technology and Carnegie Institution of Washington.*

2. Man's Exploration of the Sky

Babylonian cuneiform tablet *with a depiction of the world. Photo: British Museum.*

From earliest antiquity the sky has been the subject of close observation. The walls of certain prehistoric cave dwellings are decorated with the familiar drawings of various constellations, and the oldest documents that have come down to us (Assyrian and Babylonian tablets in particular) show that the first civilizations, some 5,000 years ago, already possessed rudimentary knowledge of astronomy.

Why should it surprise us, since the sky's rhythms are imposed on life on Earth, that the Sun's regular motion, the Moon's cycle of phases, the slow daily journey of the stars on the celestial vault, and phenomena as striking as the sudden appearance of comets or eclipses of the Sun and Moon should have attracted man's attention and awakened his curiosity at a very early stage?

In the beginning, sky-watching was certainly motivated by human fears in the face of natural events that stubbornly disrupted the regular order of things. To primitive man, tempests, storms,

Detail from Egyptian papyrus of Nespakashuti, 21st dynasty (Louvre). Reclining, the god Geb represents the Earth. Arched above him, the goddess Nout stands for the sky. Between the two, the sun god, Ra, sails on his barge. Photo: Louvre.

flood rains, and darkness accompanying an eclipse of the Sun seemed to be expressions of heavenly wrath. The large-scale meteorological and astronomical events for which the sky provides a backdrop, and the obviously essential role that the Sun plays in all earthly activity led to the conferring of a personal, individual and characteristic power on the stars, to seeing in them the expression of supreme forces and deities, favorable or harmful, whose influences it was important to try to predict, whose wrath must be deterred and whose good will must be secured through offerings and sacrifices. This explains astronomy in its first forms and for a long time it was closely associated with astrology and religious beliefs.

But the needs of daily life are also what prompted man to raise his eyes skyward from remotest times. The Sun's motion gave him the first clock, lending a rhythm to his daily activities with the alternation of days and nights. The phases of the Moon, on which early calendars were based, enabled him to count the days and measure long periods. The seasonal appearance or disappearance of certain stars signaled to early farmers the return of favorable times for planting or harvesting, while travelers and sailors learned to find their way by observing the position of the Sun by day, and by night, that of the stars, always visible to the north.

Primitive Astronomy in Chaldea, Egypt and China

Chaldea, Egypt and China are the three oldest centers of development of astronomy for which evidence has been found.

The Chaldeans, about 4,000 years before Christ, lived in the plains of Mesopotamia between the Tigris and Euphrates rivers, but their civilization spread widely thereafter. Their history is often confused with that of the Babylonians, of whom, along with the Assyrians, they were sometimes allies and sometimes enemies. It is traditional to consider them the founders of astronomy; at least they appear to have been the first in the ancient Western world to engage in a systematic study of the heavens. Favored with unusually good atmospheric conditions, they acquired through observation a remarkable knowledge of the apparent motion of the Sun, the Moon and the planets. From remotest times, they divided the stars into constellations. Ancient tablets mention 17 constellations in the road of Anou (the strip around the equator), 23 in the road of Enlil (the boreal, or northern zone) and 12 in the road of Ea (the austral, or southern zone)—Anou, Enlil and Ea being, respectively, the gods of heaven, earth and water. Monuments from the twelfth century B.C. bear constellations we can recognize as Taurus, Leo, Scorpio and Capricorn. A tablet from the time of King Assurbanipal (650 B.C.) mentions 12 zodiacal divisions, and an unusual tablet dating from the seventh year of the reign of Cambyses II (523 B.C.) bears the 12 signs of the zodiac with the names they retained in Babylon up to our era, with each sign divided into three segments of 10 degrees. (The Chaldeans knew the division of the circle into 360 degrees, and the sexagesimal division of the degree, the minute and the second they used in a duodecimal numbering system.) Other tablets show how some very bright stars were used as a reference for locating others. The Chaldeans also appear to have discovered the cycle of eclipses of the Sun and Moon (*saros*); at least they managed, with a greater or lesser degree of success, to predict in an empirical way, without any idea of the laws behind them, those events that were especially feared because they were accompanied by the disappearance of one of the two most venerated celestial bodies.

In fact, the Chaldeans believed that human destiny was inscribed in the heavens, and the desire to predict the future was what drove their astronomy forward. Observation of the planets was a privilege of the priestly class, one of whose essential functions was the establishment of a calendar (based on the phases of the Moon), but who were also responsible for predicting events of major importance to the state, based on their examination of the sky.

The astronomical knowledge of the Egyptians was much more limited, it appears, despite what has often been written on this subject. To be sure, the pyramids are lined up with the points of the compass—to within a few degrees for the oldest ones (such as the pyramid of Djoser at Sakkarah), and to within only a few tenths of a degree for the newer ones (the pyramids of Gizeh, specifically). This testifies to the fact that those who built them had remarkable powers of observation for the time. However, some of the supposedly amazing connections between their dimensions that some authors say exist and that supposedly point to a high degree of scientific knowledge are the height of fantasy.

As farmers whose activity was governed by the seasons, the Egyptians were mainly interested in developing a calendar adapted to their needs, in which changes of season constituted fixed intervals. Having observed, some 3,000 years before our era, that the heliacal rise (the appearance in the eastern sky just before the Sun) of Sirius, the brightest star, slightly preceded the annual flooding of the Nile, they paid special attention to this event and thus discovered the approximate length of the year, the period at the end of which the Sun returns to the same position relative to the stars; they were, it seems, the first to adopt the 365-day year. But they were interested neither in the motion of the planets (which they could nonetheless distinguish from the stars) nor in the prediction of eclipses, and they made few measurements.

In ancient China as in Mesopotamia, astronomy was essentially religious and astrological. Still, it is hard to establish a chronological survey of knowledge in China, since the Chinese emperor, disgruntled with his astrologers, had all the astronomy books burned in 213 B.C. The oldest texts dealing with astronomy that have been found go back no farther than the ninth century B.C., but their content permits the assumption that close attention had been paid to the sky in the Far East since a much more distant period.

The Chinese understood very early that the motion of the celestial sphere must take place about an imaginary axis, one of whose end points was located in the northern part of the sky. They then traced the circle of the celestial equator perpendicular to this axis. The concept of the ecliptic was not introduced until much later, in an incomplete way, as the course followed by the Sun and planets in traversing the equator. Measurements were relative to the equator; the tilt of the Earth's axis with respect to the celestial equator was not appreciated until the end of the first century A.D. The heavens were divided into 28 unequal sections spaced along the equator and converging toward the pole; the equator was thus divided into 28 sectors. Twenty-four principal stars that were easy to identify were used to measure time and locate other bodies on the celestial sphere. Toward the end of the fourth century B.C., the Chinese made the year commence at the vernal equinox (spring) (solstices and equinoxes were identified by observing four carefully selected stars) and perfected their first astronomical calendar. They also forced the solar and lunar years to coincide by inserting seven extra lunar months within a cycle of 18 years.

However, despite the astronomical knowledge acquired through patient and repeated observation of the sky, neither in China, Egypt nor Mesopotamia do we find an attempt at a rational explanation of the universe. Their cosmologies, permeated with mythology, are naïve and reflect only immediate impressions—the Earth is flat, circular like the horizon or rectangular like the land (Egypt); it is surrounded by a river or ocean on which it floats, and over it lies a hollow dome forming the sky, where the chariots (or barges) of the gods travel back and forth. Sometimes it is supported by columns or pillars (China, Egypt); beneath it, or in subterranean caves, lie Hades and the realm of the dead.

The Greek Contribution

LITTLE BY LITTLE, HOWEVER, given the unvarying cyclical nature of many celestial phenomena, mankind became aware of the existence of a world where the caprice of the elements and the unpredictability of supernatural forces was unknown but that was governed by laws capable of being discovered. This was the beginning of astronomy's emergence as a science.

Greece was the birthplace of this transition. From the sixth century B.C. onward, a scientific rationalism that rejected magical and supernatural interpretations of events gradually took hold. To be sure, cosmologies remained arbitrary but were no longer limited, at least, to mere observation. On the contrary, there was an attempt to come up with a correct description of the universe. The nature of the planets was pondered, efforts were made to explain their motion, and consideration was given to their distance and dimension.

The Ionian School

Greek science began with the philosophers of the Ionian school, Thales (c. 624–c. 548 B.C.), Anaximander (c. 610–c. 547 B.C.), Anaximenas (c. 550–c. 480 B.C.) and Anaxagoras (499–428 B.C.). Convinced that the universe was capable of being understood and governed by simple laws, they were the first to raise the problem of a primordial element on which all others depend. This was the sign that mankind was already beginning to reject ready-made explanations based on superstition or mythology and was seeking to understand what the universe is made of and how it began. Thales and his followers believed that any material element is living, and vice versa. It was thus sufficient to discover the fundamental element of matter to be able to explain all of nature. But their notions of astronomy remained simplistic, for they made the mistake of lumping together celestial and meteorological phenomena. Although Thales continued to believe that the Earth floated on the ocean, his intuition told him that it was illuminated by the Sun, and he made himself famous by successfully predicting certain eclipses, such as the solar eclipse that occurred in 585 B.C. and ended the war between the Medes and the Lydians. Above all, in the course of his travels he collected the knowledge and observations amassed by the Chaldeans and Egyptians and carried them back to his country. Anaximander was the first to state that the Earth is isolated in space; Anaxagoras had the brilliant hunch that the Moon and planets are stones in motion, not bright in and of themselves but getting their light from the Sun; he was thus able to explain correctly lunar eclipses as the result of the Moon moving into the Earth's shadow.

The Pythagoreans

At the end of the sixth century B.C., in face of the political upheavals that were rocking the Ionian Peninsula, some Greek thinkers sought refuge in northern Italy. There, in Crotona, Pythagoras (c. 572–c. 500 B.C.) founded a new school of philosophy that taught that numbers with their properties are the basis of all things. If the Pythagorean school did not completely neglect the meteorological concerns of the Ionian philosophers, it at least added to them a concern for understanding the mechanics of celestial motion. Pythagoras remains a legendary figure in large part, and it is quite hard to separate his ideas from those of his followers. Did he perhaps have an inkling of the Earth's real shape? In any event, it was among the Pythagoreans that the idea first arose, at the beginning of the fifth century B.C., that the Earth was round, without any experimental basis but for simple reasons of geometric harmony and perfection. Only a century later, Aristotle listed pertinent arguments in favor of this hypothesis: the fact that the masts of ships in the distance become visible before their hulls, the fact that the field of visible stars varies with the latitude of the observation point, and the rounded shape of the Earth's shadow on the Moon during lunar eclipses. Moreover, the Pythagoreans unraveled the seemingly complicated course of the Sun around the Earth. Its motion was broken down into two rotations: one, the diurnal rotation, from east to west; the

Astronomy and Megaliths

Stone monuments, or megaliths, are the most notable and common vestiges of prehistoric Europe. They are chiefly found in Brittany, Scotland, the South of England and Scandinavia. Through the use of carbon-14 dating, it has been determined that the oldest ones go back about 6,000 years—i.e., to the Neolithic period—while the newer ones were built in the Bronze Age, about 3,000 years ago.

European megaliths are classified into four main kinds: menhirs (single upright stones), groups of menhirs forming linear arrangements or rings (cromlechs), funeral chambers (dolmens) and megalithic temples.

One of the many enigmas surrounding these monuments is their purpose. It has been determined that the dolmens correspond to gravesites, but we still do not know what the menhir alignments or rings were used for. Some of them seem to be placed according to astronomical criteria and perhaps served as observatories from which the Sun and Moon were sighted at given times of the year.

This hypothesis has been offered with respect to the Carnac alignments in the Morbihan (France). In 1805, J. Cambry linked the Breton menhirs to Sun-worshiping. In about 1874, H. du Cleuziou, in a short description of the Carnac alignments for the École des Beaux-Arts, suggested for the first time that the stones were arranged in ac-

The megalithic monument of Stonehenge, *near Salisbury (England), was built in several stages between 2800 and 1500 B.C. It contains, in the center, a group of raised stones arranged in a horseshoe, surrounded by a circle of smaller, scattered stones, and a circle of megaliths 5 meters (16.5 feet) high and 2 meters (6.6 feet) wide, connected by lintels; on the periphery, a double circle of single stones and holes. It may have been an observatory used in establishing a calendar and in the prediction of eclipses—different alignments make it possible to spot the position of the Sun and Moon when rising and setting, in midsummer and midwinter, because of the four stones (here labeled A, B, C, D) arranged in a rectangle around the circle of 56 holes called the Aubrey circle. Due to the swing of its orbit, the Moon rises and sets each day at a slightly different point on the horizon, following a cycle of 18.6 years, which the megaliths woiuld also have revealed. Photo: Seraillier-Rapho.*

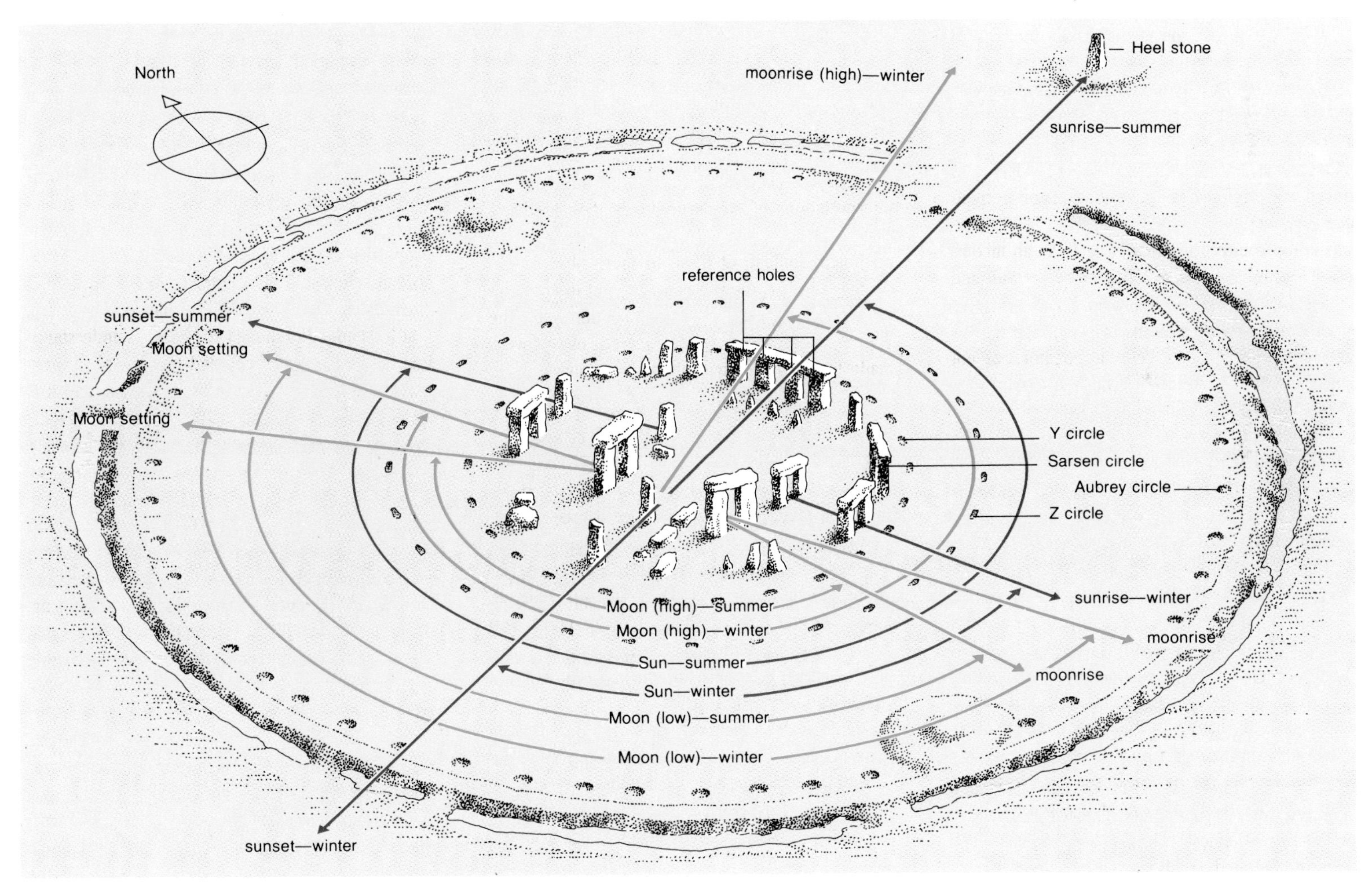

cordance with sunrises and sunsets at the solstices and equinoxes. This theory was further developed in 1888 and 1897 by the native Morbihan archaeologist F. Gaillard, who directed the state's acquisition of monuments and their first restoration. The theory was further refined by P. Grossin (1898) and especially A. Devoir (1899). The latter, in collaboration with Sir Norman Lockyer (1906–09), then developed a more extensive solstitial theory encompassing the groups of megaliths to the north and west of Finistère. For Morbihan, the archivist R. Merlet took very precise measurements with a theodolite and pushed the solstitial theory to its fullest extent.

According to the British astronomer Alexander Thom, a large number of megalithic monuments played a role in the observation of the phases and movements of the Moon. In particular, the Carnac alignments may have been part of a kind of huge observatory designed to watch risings and settings of the Sun and Moon in order to predict eclipses. The focal point of this group might have been the great broken menhir of Locmariaquer, which would have served as a sighting device for a large number of distant targets from lookout points several kilometers away.

An astronomical function also has been ascribed to the famous Stonehenge megalith in the South of England. According to a hypothesis advanced in 1901 by Lockyer but revitalized in 1963 by Gerald Hawkins of the University of Boston and also defended by Fred Hoyle and Alexander Thom, the monument made it possible to predict eclipses of the Sun and Moon.

The various theories ascribing astronomical functions to the megaliths nonetheless remain extremely controversial.

other taking place in a year's time, from west to east, in a special big circle to which the name *ecliptic* was given.

The Moon and the five planets visible to the naked eye have more complex movements. But the Greeks, aware of simple geometric combinations, were quickly convinced that the planets' courses—despite their apparent caprices—could be explained by combinations of circular and uniform motions. No doubt it was observation of the steady, unchanging diurnal movement that prompted this conclusion. In any case, the necessity of circular motion dominated all of ancient and medieval astronomy until Kepler.

Philolaeus (c. 450–c. 400 B.C.) worked out the first cosmological model, properly speaking. He taught that at the center of the universe is a motionless ball of fire—the abode of the gods and the principle governing the motion of all the planets. The planets make circular trajectories around the central fire. At the boundaries of the universe, and all around it, is another igneous region, the supreme fire. Between the central fire and the supreme fire, space encompasses three concentric realms—sky, earth and Olympus, which are found in that order as one goes toward the periphery. All elements within Olympus are in a state of perfect purity. It is where the "fixed" stars are. From Olympus to the sky, as one travels around the world, is a succession of the five planets; the Sun, which is transparent and illuminated by the central fire; and the Moon, lit up by the Sun. The planets move around the central fire in a single plane from east to west. The sky lies between the Moon and the central fire, and the Earth makes its revolution from east to west in a different plane than the planets. This revolution is made in such a way that the Earth always presents the antipodes (the point on the opposite side of the Earth) of Greece to the central fire so that the fire remains invisible.

The universe thus contains nine bodies in all: the sphere of fixed stars, the Sun, the Moon and the Earth. But convinced that numbers are the first cause of all phenomena, the Pythagoreans attributed a special value to the number 10. Thus Philolaeus postulated the existence of a tenth body, the anti-Earth, closer to the central fire than the Earth and similarly revolving around the fire, facing it always on the same side. Its motion is such that the inhabitants of Earth can never see it. Its existence explains why, in a given place, lunar eclipses are seen more often than solar ones.

Though it may amuse us, make us smile, this model at least has the merit of teaching that the Earth travels in space without any special status, like the other planets. It represents a first step toward a less anthropocentric view of the world.

The Platonic School; Aristotle

In the fourth century B.C., Plato (428–348 B.C.) outlined his vision of the world in his *Dialogues,* essentially in *Timaeus*. His ideas were inspired by his own metaphysics, but also by his Pythagorean predecessors. He imagined an infinite space in which floats a limited, closed and spherical universe. Emptiness is banished from it. The Earth reigns at the center of the world. Between it and the confines of the universe lies ether. The planets pursue their course according to immutable laws. The universe is divided into nine contiguous and concentric regions. The outermost one is the supreme orb, the abode of the fixed stars. It rotates uniformly from east to west around the axis of the world. Its motion is combined with the individual rotations of the lesser orbs. Thus the diurnal (daily) motion of all celestial objects is explained. The lesser orbs, on which Saturn, Jupiter, Mars, Mercury, Venus, the Sun and the Moon are ranged in succession, revolve uniformly around an axis perpendicular to the ecliptic, each at a fixed speed. The last, motionless orb contains the Earth at its center.

However, at the time, astronomical observations had made considerable progress. The stationary and retrograde motions of the planets were already known (see "The Solar System," page 65), as well as other apparently whimsical aspects of their motion. But Plato's model, while tempting, left these incongruities unexplained. Plato laid down a categorical imperative: *Save the phenomena*—in other words, explain the observed movements by means of calculations, but only on the basis of combinations of uniform circular motions, the only ones, he believed, that could account for the incorruptible and eternal essences of the planets. Astronomers spent twenty centuries in this effort, gradually complicating the celestial machinery to the utmost.

Eudoxas of Cnidas (c. 406–c. 355 B.C.) was the first to try to come up with a more precise description of planetary motion within the Platonic framework. He believed that the fixed stars are located on a solid sphere whose rotation from east to west takes place in one sidereal day. The motion of each wandering planet (Sun, Moon, planets) is due to a mechanism that is peculiar to it and separate from other mechanisms; it results from the combination of motions of several concentric spheres that are contiguous to each other. The number of spheres and the details of their rotation vary from one planet to another. All of these spheres, except for the innermost one, which carries the planet at its equator, are invisible and can be revealed only through geometric logic.

Not long after, the persisting inconsistencies between theory and observations, particularly the unequal length of the seasons, led the astronomer Callippos to complicate Eudoxas's system by introducing new spheres called "compensating spheres." Then the system was again revised by Aristotle, who gave it its final form. The number of spheres, initially 26, rose to 55. Thus remolded, the theory of homocentric spheres became the first of the physical cosmologies.

However, while it satisfies the demands of Plato as to the circularity of motion, it remains powerless to explain certain phenomena revealed by observation, such as the considerable differences in the brightness of Mars and Venus, the variations of the Moon's apparent diameter, the lack of uniformity of the Sun's apparent motion along the ecliptic, and so on. Indeed, the theory implicitly

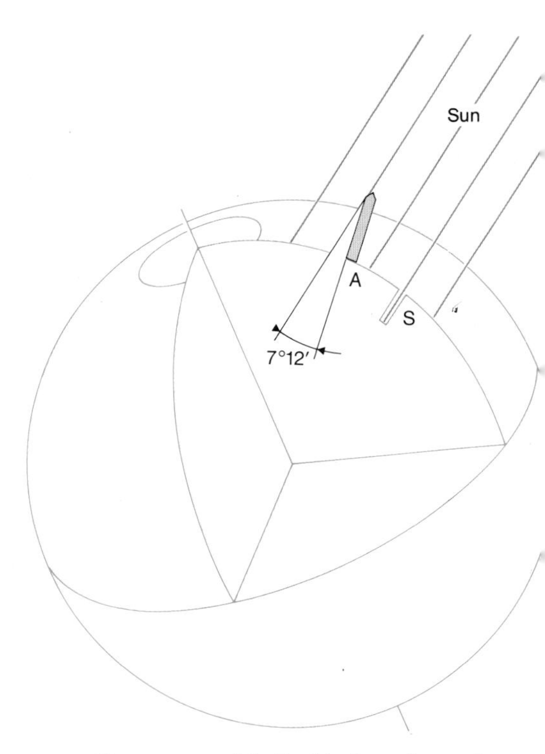

Measurement of the Earth's circumference by Eratosthenes. *At Syene (S), at noon on the day of the summer solstice, the Sun casts no shadow and lights up the bottom of a well: it is at its zenith. The same day, in Alexandria (A), located farther north, the shadow of an obelisk shows that the Sun makes an angle of about 7 degrees (or one-fiftieth of the circumference) with the zenith. This angle represents the difference in latitude between the two cities. Now, the distance from one city to the other, measured on the ground by land surveying, is 5,000 stadia (about 800 kilometers [480 miles]). From this the Earth's circumference can be deduced: 250,000 stadia, or about 40,000 kilometers (24,000 miles).*

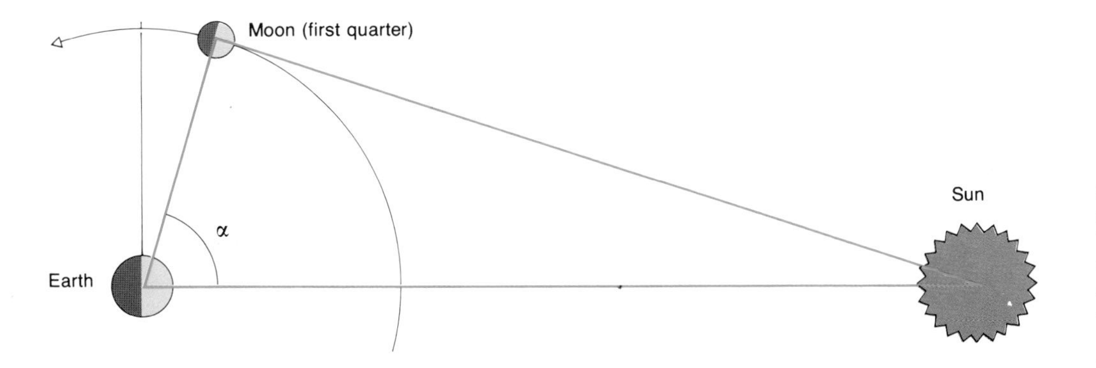

Comparison by Aristarchus of distances of the Moon and Sun. *When the Moon is in the first or last quarter, the angle of the Sun, Moon and Earth equals 90 degrees. By measuring the angle of the Moon, Earth and Sun (alpha), one can deduce the relative distances of the Sun and the Moon.*

assumes that the distances between the Earth and the Sun, the Moon and the planets are invariable, which is far from true.

Conscious of the difficulties it raised, Heraclitus (388–315 B.C.) rejected the doctrine of homocentric spheres and put forward new hypotheses. He explains diurnal motion as the Earth's rotation from west to east in approximately one day, making the distinction between the solar day and the sidereal day. Furthermore, in order to account for the apparent motion of Mercury and Venus, which always remain fairly close to the Sun, he conceived of a partial heliocentrism: He assumed that the Sun makes an annual revolution around the Earth and imagined that Mercury and Venus, while included in this motion, make smaller circles in the same direction around the Sun. This clever model explains the apparent observations very well.

The Alexandrian School

As a result of the conquests of Alexander the Great, the Greek culture and language penetrated the East, and scientific activity was concentrated in the southern Mediterranean basin, particularly in Alexandria. There, for four centuries, astronomy underwent a remarkable development, with respect to both theory and observation.

The first representative of the Alexandrian school is Aristarchus of Samos, at the beginning of the third century B.C. By measuring the time it takes the Moon to cross the Earth's shadow during a lunar eclipse, he was able to calculate the ratio of the lunar radius to the Earth's radius (about one to three). He also tried to compare the distances of the Moon and Sun by an ingenious trigonometric method, but he came up with the wrong result for lack of a sufficiently precise sighting instrument. Above all, he seems to us a remarkable precursor of Copernicus; he imagined the Sun as motionless at the center of the sphere of stars, and the Earth driven by a dual rotation on its own axis (in one day) and around the Sun (in one year). Moreover, he was the first, it seems, to emphasize the considerable distance between the Earth and the sphere of fixed stars. Violently fought, his heliocentric model fell quickly into oblivion, for he was accused of departing from the principles of Aristotle's laws of motion. These laws gave "proof" of the Earth's motionlessness that were unanimously convincing to the philosophers.

Another Alexandrian scholar, Eratosthenes (c. 275–c. 195 B.C.), was the first to establish the length of the Earth's circumference, using a method as noteworthy for its simplicity as for the precision of its result. He also determined the obliquity of the ecliptic—i.e., the inclination of the plane of the Sun's apparent annual trajectory (the ecliptic) relative to the plane of the celestial equator.

As the accuracy of observations increased, the inadequacies of the theory of homocentric spheres became more and more apparent, and it lost its last defenders. It was replaced by a system based on eccentrics and epicycles. To explain the nonuniformity of the Sun's apparent motion, it was assumed that the ecliptic is eccentric relative to the Earth—that is, the Earth is displaced from its center. As for the motion of the planets, it was interpreted by introducing the notion of two circles: The first, called the *deferent,* centers on the Earth, which is assumed to be the center of the

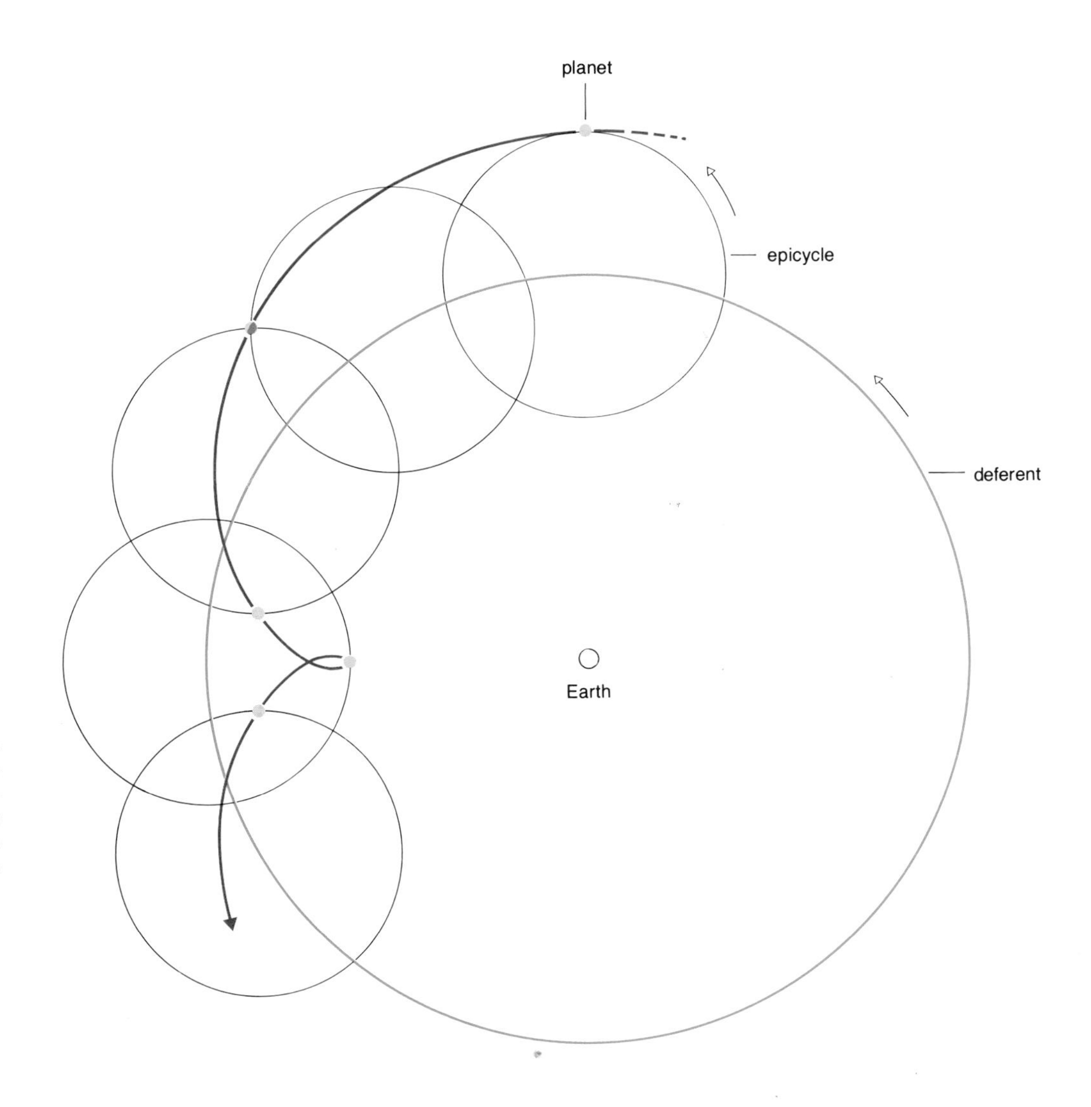

Epicycle and deferent. *To explain the irregularities of the apparent motion of the planets in the sky, Ptolemy imagined that each planet travels with a uniform motion, in a direct sense, around a small circle, the epicycle, the center of which also moves on a much bigger circle, the deferent, with the Earth in the middle.*

universe; the deferent lies in the same plane as the ecliptic. A point describes this circle in a period of time equal to the length of the planet's revolution in the zodiac. This point is the center of a second circle, the *epicycle,* which is thus pulled around the deferent. The epicycle carries the planet and makes a complete turn in the time interval that brings the planet, the Sun and the Earth back to the same relative positions. The combination of these movements makes it possible to reconstruct the seemingly capricious courses of the planets.

These mechanisms—the eccentric on one hand, the deferent and epicycle on the other—seem to be different in nature but are in fact equivalent. In the second century B.C., Hipparchus put the final touches on them and brought them to a previously unknown state of perfection. Even more than his forebears, Hipparchus emerges as an unrivaled observer and theoretician, which is why he is regarded today as the greatest astronomer of antiquity and the founder of astrometry.

Among his many works, which have come down to us thanks to Ptolemy (see further), we find the exact measurement of the length of time of the Moon's orbit and the angle of the lunar orbit to the ecliptic, the distance between the Earth and the Moon, and tables showing the motions of the Sun and Moon, permitting, among other things, a correct prediction of eclipses, etc. In 134 B.C., the sudden appearance of a temporary star in the constellation Scorpio prompted Hipparchus to measure accurately the coordinates of the stars in the sky, so as to be able to detect any future unexpected appearance, as well as to demonstrate the possible motions of stars relative to one another. He thus produced a catalog of 1,025 stars, grouping them for the first time into six "sizes" according to their brightness. The observations he made in order to prepare this catalog further led him to the discovery of the *precession of equinoxes* (see "The Earth," p. 131—that is, the very slow displacement of the vernal point on the ecliptic relative to the stars—which he introduces into the theoretical description of celestial movements: He was the first to distinguish between the *tropical year* (the interval at the end of which the Sun returns to the same position relative to the equinoxes—the seasonal year), and the *sidereal year,* the length of time the Sun takes to return to the same position among the stars.

Hipparchus's contribution was, moreover, not limited to astronomy. He was also the father of trigonometry and even of spherical trigonometry, the inventor of stereographic projection and author of the first scientific method for determining longitudes.

Ptolemy's system. *Print taken from* ***Harmonia macrocosmica*** *by Andrea Cellarius, 17th century (Bibliothèque Nationale, Paris). Photo: Y. Bressy.*

After him, Greek astronomy produced no important result for nearly 300 years. It was not until the second century, with Ptolemy, that it regained its luster—and then gave off its final sparks.

A keen observer, Ptolemy invented several new astronomical instruments and discovered the change of position of the celestial pole among the stars as a correlative of the precession of the equinoxes, as well as the unevenness of lunar motion known as *evection* (see "The Moon," page 140). He was also the first to recognize clearly the effect of *astronomical refraction*—that is, the fact that the stars appear to be elevated near the horizon relative to stars at higher altitudes, a phenomenon that had scarcely been perceived prior to his time.

But above all, as heir to the entire Greek scientific and philosophical tradition, Ptolemy carried on and completed the work of his predecessors. In his *Mathematical Collection,* which has come down to us in its Arabic translation as the *Almagest,* he set forth the entire astronomical science of his time. This treatise became the astronomers' bible up to the 16th century. It includes in particular the famous geocentric model of the world, which held sway until Copernicus. At the center of the universe, the Earth stands motionless. All around it are the various celestial spheres on which the Moon, Sun, planets and stars revolve; with the eighth sphere, to which the stars are attached, the universe ends. The motion of the planets is explained as a combination of epicycles and deferents. Ptolemy also introduced a few innovations, such as the equant, an extra complication necessary to account for the observations better.

As a culmination of the efforts of an entire line of astronomers, this system is in keeping with its ambitions. It does not pretend to describe a reality that cannot be known, but is merely a mechanical representation based on the observations of the time and the principles of Aristotle's physics.

Ptolemy's system reflects both the strengths and the weaknesses of the Greek scientific thought that produced it. Because they believed in a permanent order underlying the changing world of observation, the Greeks created scientific theory and gave science the features we know it by. Under their impetus, astronomy was transformed into a science based on precise, rigorous observations. But in remaining too abstract, Greek scientific thought could not make the connection between *mathematics* and *physics,* between *calculation* and *experiment,* which is now a condition of progress in the sciences.

The decline of the Greeks marked the beginning of a long period of stagnation for the sciences in general and astronomy in particular. The Romans, essentially concerned with technique, disdained pure science. Inclined to superstition, they were much more interested in astrology than in astronomy. Broadly speaking, they made no original contribution to the scientific disciplines; they produced only compendiums, such as Pliny's famous *Natural History.* In astronomy their sole contribution was the Julian calendar. It was the Arabs who collected the Greek legacy; having substantially enriched it with their own experience, it passed on to the Western Christian world.

Arab Astronomy

Greek knowledge of astronomy came to the Arabs by way of India, where Greek thought had penetrated as a result of the conquests of Alexander the Great. Baghdad in the ninth and 10th centuries and Cairo in the 10th, 11th and 12th centuries were flourishing centers of astronomy. Greek books were translated and studied there. But the Arabs also did original work: They improved instruments (particularly the astrolabe) and made numerous observations. These essentially served to develop tables of the movement of the stars and planets for astrological or nautical purposes (determining latitude and longitude). Al-Battānī (858–929) reobserved the Sun's positions; came up with a better estimate of the tropical year; corrected the value of the precession constant found by Ptolemy; determined the obliquity of the ecliptic; and succeeded in developing a new theory for determining the conditions necessary for the visibility of the new moon, a fact of supreme importance to the Islamic religion. He also made such accurate observations of lunar eclipses that the 18th-century astronomer Dunthorne still referred to them. Abū al-Wafā' (940–97?) perfected the theory of the Moon. His pupil Ibn Yūnus (979–1009) made many observations of the Sun, Moon and planets and drew up tables of their movement (the Hākīmite tables, 1007). Al-Bīrūnī (973–1048) built various instruments of observation, studied sunsets and solar eclipses, produced tables of the declination of the stars and so on. 'Umar Khayyām in 1079 established a calendar at least as accurate as all those that preceded the Gregorian reform.

In 1258, Baghdad fell to the Mongols, but astronomical research did not come to a halt. On the contrary, the Mongols set up new observation centers. Houlagou had a huge observatory built at Meragah in Azerbaijan for the scholar Nassir Eddin Thousi and his collaborators. In the 15th century, Ulug Beg (1393–1449), grandson of the famous conqueror Tīmūr Lang (Tamerlane), built an observatory in Samarkand with a massive quarter-circle meridian (40.2 meters in radius), with which he made remarkably accurate observations for the time: He drew tables of the Sun and planets, and in particular established a catalog of 1,018 stars, giving their coordinates not only in degrees but also in minutes for the first time.

By the 10th century, Arab science began to penetrate to the West by way of Spain. Cordova and Toledo in particular became major astronomy centers. For a long time, latitudes were based on the Toledo meridian; the Toledan tables, and later the Alphonsian tables (published in 1252 at the behest of King Alphonse X of Castille) enjoyed great fame. Beginning in the 12th century, the original works of the Islamic scholars and their translations of Greek works began to circulate in Europe in Latin translations.

While the Islamic contribution did not include great discoveries, it at least proved extremely valuable in the field of observation and astronomical calculations. In particular, the Arabs were the first

Planispheric astrolabe dating from 1216 *(P. Dupuy Museum in Toulouse). This instrument, much used in the Middle Ages, has two main parts: a stationary disk called the tympan, with another disk called the spider, much of which has been cut out, pivoting around it. On the tympan are engraved the projections of circles from the celestial sphere by which the positions of the planets are determined. The position of the horizon relative to the visible celestial pole varies with the latitude of the observation site; therefore, a special tympan is needed for each latitude (the photograph shows several, which can be attached to the instrument). The spider represents a projection of the celestial sphere, reduced to the principal stars (whose positions are marked by ridges) in the zodiac and the ecliptic. By swiveling it around the tympan, one reproduces the apparent motion of the sky relative to the Earth. Therefore, one can instantly obtain a graphic solution to astrological or astronomical problems such as the position of the Sun at a given moment, stages of the rising and setting of a planet, etc. Photo: Lauros-Giraudon.*

to apply trigonometry to astronomy systematically. Their influence can still be seen today in certain astronomical terms ("zenith," "nadir") and in the names of many bright stars (Aldebaran, Altair, Betelgeuse, Deneb, Mizar, etc.).

Nicolas Copernicus *(1473–1543). Painting on wood, Pomeranian school, 16th century (Museum of Torun, Poland). Photo: Lauros-Giraudon.*

The West: From Obscurantism to a Rebirth of Science

In the Latin West, the penetration of Greek and Arab knowledge ended a long period of scientific obscurantism. Revealed in their Latin translations, the works of the Greek philosophers—Aristotle's in particular—elicited much comment from the 13th century onward. Still, it was not until the 15th century that a genuine rebirth occurred, facilitated by, among other things, the invention of printing, which helped to spread knowledge. At that time, Germany and Austria, in particular, became bustling centers of astronomical learning. Georg Peuerbach (1423–1461) produced a theory of the planets. His pupil Johann Müller (1436–76), better known by the name of Regiomontanus, set up an observatory in Nuremberg in 1471, an instrument factory and a print shop, enabling him to make many observations and to publish ephemera. Finally, in Belgium, Nicolas de Cusa (1401–64) adopted and defended Aristarchus's hypothesis, though he unfortunately did not succeed in bolstering it with the decisive arguments capable of convincing his contemporaries.

The Copernican Revolution

The disintegration of the ancient cosmology did not begin until the 16th century, with the Polish monk Nicolas Copernicus (1473–1543). In a work entitled *De revolutionibus orbium coelestium* (On the Revolution of Heavenly Bodies), published in 1543, the year he died, Copernicus developed a heliocentric model of the world that marked the advent of the modern conception of the universe: All the planets revolve around the Sun in orbits whose dimensions are infinitely small in regard to the distances to the stars; the Earth is just a planet like the rest, rotating on its own axis every 24 hours and completing a revolution around the Sun in one year. The Earth's rotation explains the apparent diurnal motion of the Sun; the Earth's revolution, the change of seasons. To be sure, Copernicus retained many of the aesthetic and philosophical prejudices of his predecessors—like them, he believed the universe was round, orbits circular and movements uniform, and as these hypotheses did not sufficiently account for his observations, he was compelled to reintroduce eccentrics and epicycles into his model, after having rejected those of Ptolemy. He even adopted Aristotle's notion of the orbs, or mobile material spheres, and while he placed the Sun at the center of his universe, it was mainly because he felt such a splendid body deserved the place of honor! The line of reasoning he developed in favor of his system is often incorrect and abounds in unscientific assertions borrowed from the ancients. In fact, for the men of his time, the superiority of his system over that of the ancients could not yet appear obvious, since no decisive argument had been offered in particular to prove that the Earth moved.

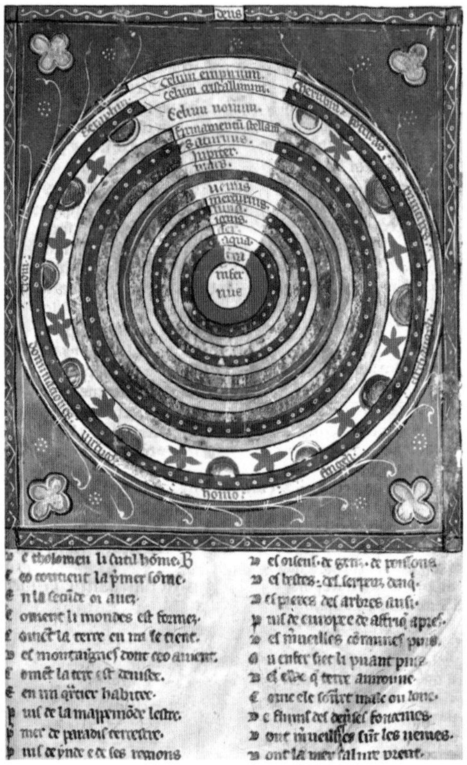

The universe *as it was pictured in the Middle Ages, after the rediscovery of Ptolemy's works. Illumination by Gauthier de Metz (Bibliothèque Nationale, Paris). Bibliothèque Nationale photograph.*

The fact remains, nonetheless, that in freeing astronomy from the hypothesis of the immobility of the Earth and in substituting for the authority of the ancients the principle of subordination to the facts as the source of all knowledge, Copernicus's work marks an essential turning point in the history of ideas and scientific progress. For science as well as philosophy, Copernicus ushered in a new era. After him, human thought, reined in for centuries by a false conception of the universe, was finally liberated, and the basis established on which to build a rational scientific structure. Still, it took another century until his system was finally

Astronomy in the PreColumbian Civilizations

The preColumbian civilizations that flourished in Central America between the beginning of the Christian era and the Middle Ages attained quite a remarkable level of astronomical knowledge, as evidenced by the inscriptions on the steles they left behind. In particular, the Aztecs and Mayans, for whom the Sun was the greatest and most fearsome of gods, were very close observers of the sky for religious reasons. They could recognize the planets; find the length of the year, seasons and lunar month with amazing accuracy; predict eclipses; determine the average period of Venus's synodic revolution, etc. Although their instruments were very rudimentary, their measurements nonetheless achieved an astonishing degree of accuracy due to meticulous and repeated observation. The Mayans showed a strong interest in the measurement of time. Struck by the observation that the same phenomena were repeated at regular intervals, they thought that the same events must occur on the same date. The importance they attributed to dates is shown by the inscription on all their monuments of the date their projects were begun and finished. They used two calendars: one, a solar calendar, the ***haab***, lasting 18 months and 20 days, plus 5 "unlucky" days; the other, a ritual calendar, the ***tzolkin***,

accepted. It encountered strong resistance from theologians, who refused to accept that the Earth moved. One of his most ardent defenders, Giordano Bruno—who even dared to conceive of an infinite universe with a center everywhere and a circumference nowhere, and the existence of an infinite multitude of inhabited worlds—was sent to the stake by the Inquisition in 1600.

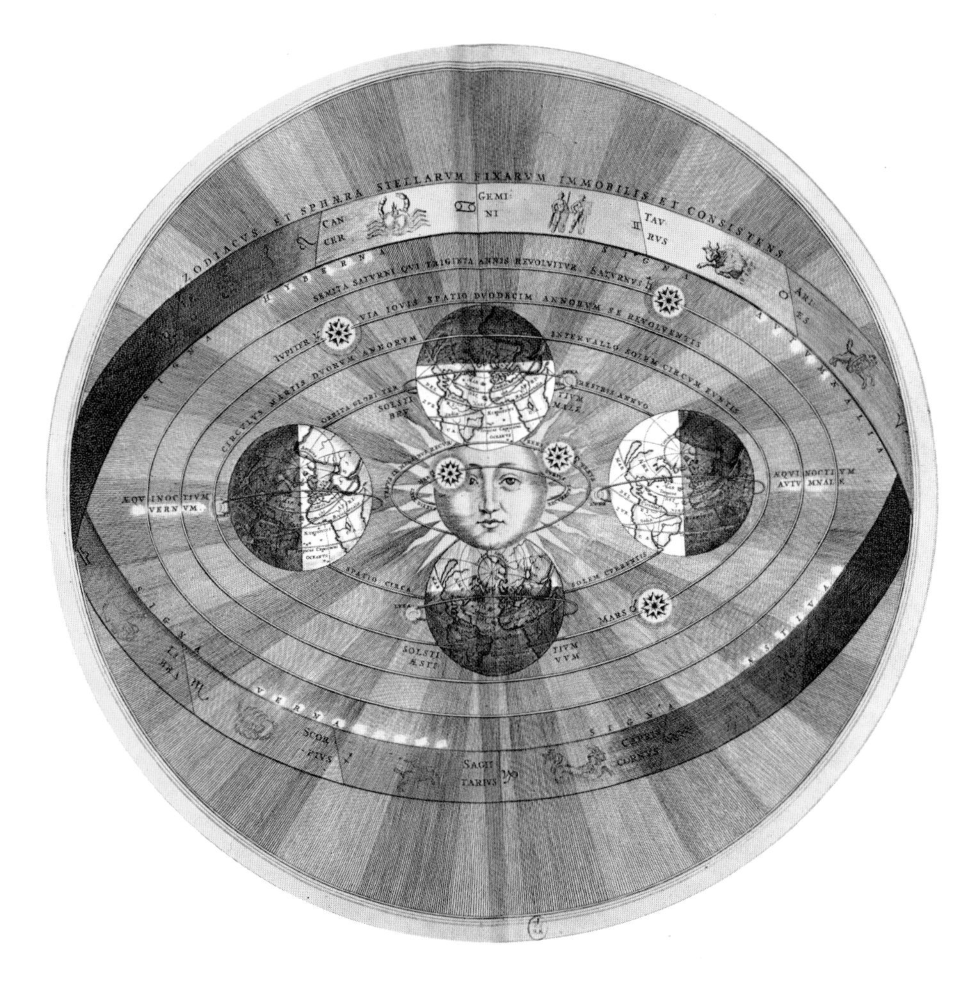

***The system of Copernicus.** Print taken from **Harmonia macrocosmica** by Andrea Cellarius, 17th century (Bibliothèque Nationale, Paris). Photo: Y. Bressy.*

Tycho Brahe and Kepler

The adoption by scientists of the Copernican model in the 17th century owes much to the work of two astronomers with complementary talents: an unparalleled observer, the Dane Tycho Brahe (1564–1601), and a remarkable mathematician, the German Johannes Kepler (1571–1630).

Attracted to astronomy from early youth, Tycho Brahe made a definite commitment to observing the heavens as a result of the appearance of a new star in 1572—in fact a supernova, as brilliant as Venus but that decreased in brilliance and finally disappeared after 18 months. It filled him with amazement, for it contradicted the theory of the immutability of the heavens that had been accepted since the time of Aristotle. Brahe was able to show, however, by way of accurate measurements, that the object was indeed a star, as distant as the "sphere of fixed stars" and not some unexpected meteor within the planetary system. This spectacular event also gave him the idea—as with Hipparchus 18 centuries earlier—of compiling an accurate catalog of stars that would make it possible to note immediately any subsequent change in the celestial arch. He realized this ambition thanks to the generosity of King Frederick II of Denmark, who in 1576 gave him the island of Hveen, near Copenhagen, and funds for the construction of an observatory on the island. Brahe built Uranieborg—the "palace of Uranus"—there and equipped it with the largest and most accurate instruments of the age. There, for 20 years, he compiled thousands of observations of the Sun,

***Remnants of a Mayan observatory:** Caracol, in Chichén Itzá, in Yucatán (Mexico). Photo: A. Monclème.*

containing 20 periods of 13 days, each referred to by a special name preceded by a number from 1 to 13. The Aztecs used similar calendars. In fact, the Mayan calendar amounted to an indefinite series of days placed in an arbitrary order, without regard to astronomical phenomena. The combination of the *tzolkin* and the *haab* caused the repetition of this sequence in an identical fashion every 18,890 days. Thus the Mayans, and the peoples who borrowed their calendar, placed great importance on this cosmic cycle, equal to 52 *haabs* or 73 *tzolkins*.

The beginning of the Mayan era probably does not correspond to an historical or astronomical event but rather to a mythical event. It took place about 3,400 years before the dates inscribed on the oldest steles that have been found, but we have not yet been able to establish a precise correlation between our chronology and that of the Mayans; at present, dates in Mayan history can be estimated at best to within 1½ centuries.

While they were able to describe celestial movements accurately, the Mayans did not, like the Greeks, try to develop cosmological systems. The Mayans did not possess the wheel, and it is hard to imagine how they could have pictured a cyclical motion.

of the world although the planets revolved around the Sun.

When he died in 1601, Tycho Brahe left his priceless observation logs and the task of carrying on his work to his pupil Kepler. Kepler was a confirmed supporter of Copernicus, and in his first book, (*Mysterium cosmographicum*), published in 1596, he explained why Ptolemy's system must give way to the Copernican model of the world. Endowed, however, with a fertile imagination inspired by the Pythagoreans, Kepler thought the universe was constructed on geometric principles. Thus he devised an ingenious geometric model of Copernicus's system, in which each planet's orb occupied a sphere circumscribed on one regular polyhedron and inscribed within another. In this construction there appears for the first time the fertile idea of a relationship between the respective distances of the planets from the Sun. In fact, Kepler was convinced that the number of planets, their distances from the Sun and the speeds of their revolutions were not the results of chance. He made it his goal to find the laws of their motion as well as those governing the distribution and dimension of their orbits.

It was by engaging in a systematic study of the motion of the planet Mars (whose trajectory remained misinterpreted by Ptolemy and Copernicus), and after laborious calculations that he was able to check against the accurate observations of

Moon, planets and stars. For the first time, the measurements of angles attained an accuracy on the order of 0.3 arc minute and were corrected for atmospheric refraction.

Among his discoveries was the variation of the obliquity of the ecliptic, the irregularities of lunar motion known as *variation* and *annual equation*, the variation in the angle of the lunar orbit with respect to the ecliptic, etc. We are also indebted to him for a catalog of 777 stars. In 1577, while studying the path of a brilliant comet, he showed that it could not be part of the Earth's atmosphere, contrary to accepted opinion ever since antiquity, but followed an oval orbit around the Sun.

A less inspired theoretician, Brahe rejected the Earth's motion for reasons at once physical and metaphysical. Thus he rejected Copernicus's system without going back to that of Ptolemy, whose inadequacies he knew of. He proposed one of his own making, halfway between the two previous ones, in which the Earth remained at the center

***Uranieborg**, residence and observatory of Tycho Brahe on the island of Hveen. Handpainted line engraving, reprinted, like the one following, from **Tychonis Brahe astronomiae instauratae mechanica,** Wandesburgi, 1598, Germany (Bibliothèque Nationale, Paris). Bibliothèque Nationale Photo.*

***The great quarter-circle mural of Uranieborg** Observatory, used by Tycho Brahe and his assistants. The figure on the right (representing Tycho Brahe) is taking a reading through the skylight visible in the upper left-hand corner. In the foreground, another figure (right) records the time, while a third (left) enters the observation in a log. B. N. Photo.*

his master Tycho Brahe, that Kepler discovered the first two laws that would immortalize him. These laws, published in 1609 in his *Astronomia nova*, state, first, that the planets' motion follows an elliptical path with the Sun occupying one of its foci, and second, that the radius vector of a planet—that is, the line joining the planet with the Sun—sweeps out equal areas of the ellipse in equal times (see "The Solar System," page 68).

***Johannes Kepler** (1571–1630). Oil painting by Matthias Bernecerrus (?) 1627 (Collegium Wilhelmitanum, Strasbourg). Photo: Fondation Saint-Thomas, Strasbourg.*

These laws thus marked a definite break with the Greek tradition of circularity of motion.

Still haunted by Pythagorean ideas, Kepler next tried to prove the existence of a harmonic relation (in the musical sense) between the fastest and the slowest speeds of the planets. Hence he studied the numerical relations in the solar system. After painstaking calculations, he thus discovered the third fundamental law of motion of the planets, which established a relationship between the dimensions of the planetary orbits and the time it took to travel them: The squares of the periods of revolution are proportional to the cubes of the major axes of the orbits. This central finding was published in 1619 in *Harmonices mundi*, wherein Kepler described his somewhat mystical vision of the universe, illustrated by his researches in geometry, music and astronomy.

Thereafter, the solar system was described from a mechanical point of view as it is conceived of in our time. What remained to be studied was its dynamics, to discover the physical cause of the planets' motion. This would be the contribution of Isaac Newton in the next century. Kepler already wondered if the planets' motion is controlled by a central force, and he attributed its origin to a "motive soul" emanating from the Sun. But while he grasped the existence of a controlling principle, he was not able to discern its nature and exact properties.

In the last years of his life, Kepler devoted himself to making up tables of the positions of the planets of the solar system based on the laws he had demonstrated and on the observations of Tycho Brahe—the Rudolphian tables, which Kepler published in 1627. Much more accurate than those that preceded them, these tables sealed the final victory, at least among astronomers, of Copernicus's system.

Galileo and the Invention of the Telescope

After the upheaval Copernicus wrought in the realm of ideas, a revolution in instruments awaited astronomy with the appearance, at the dawn of the 17th century, of the telescope, enabling mankind to take a fresh look at the universe and to perceive, in fact, its grandeur.

With the rudimentary telescopes he built starting in 1609 in order to focus on the heavens, the Italian astronomer Galileo Galilei (known as Galileo) (1564–1642) made the first observations of the surface of the Moon and of sunspots; discovered the phases of Venus and the four major satellites of Jupiter; showed that the Milky Way was composed of a multitude of stars; and spotted in the constellations a large number of stars that had previously been unsuspected.

These discoveries, published in 1610 in *Sidereus nuncius* (The Sidereal Messenger), clearly demonstrated that the universe did not have the characteristics attributed to it by Aristotle. They also reinforced Copernicus's hypothesis in various ways. Hence, the presence of mountains (whose height Galileo found by measuring their extended shadows) and valleys on the Moon proved that it was fundamentally no different from Earth, and since the Moon moves in the sky, it made sense to think of the Earth as being in motion, too. The appearance of Venus's phases—which could not be explained in Ptolemy's system—provided unquestionable empirical evidence of the heliocentric system, the validity of which was further shown by the existence of satellites around Jupiter, which together with the planet formed a virtual solar system in miniature. Finally, the distinction permitted between stars and planets—when observed through the telescope, stars remain points of light, while planets have a substantial apparent diameter—proves that the stars are considerably more distant than the planets and reveals the immensity of the stellar universe.

Concomitant with his astronomical discoveries, Galileo made an outstanding contribution as a physicist. He replaced the old Aristotelian approach, which consisted of relating phenomena to hidden properties, with objective research into the laws of nature based on observation and experimentation. In this spirit, he took on the problem of the motion of bodies, not seeking causes but rather the mathematical relations between physical facts, and he subjected the laws he discovered to verification. He was thus able to define uniform motion and uniformly accelerated motion. He also formulated the principle of inertia, opening new

***Galileo presents his first telescope to the doge of Venice;** it revolutionized astronomy. Painting by Luigi Sabatelli (Museum of Physics and Natural History, Florence). Photo: Scala.*

paths to the understanding of the universe: *A body not being acted upon by some outside force is either at rest or in a state of uniform motion in a straight line.*

The first modern scientist, Galileo is also remembered for his skirmishes with the Inquisition, which made him the hero and symbol of a fight for the freedom of scientific thought against the authority of all types of conservatism.

Newton, Last of the Giants

It was the English scientist Isaac Newton (1642–1727), in the second half of the 17th century, who completed the foundations of celestial motion by extrapolating the discoveries of Kepler and Galileo.

Kepler, by showing that the planets do not revolve around the Sun with a uniform motion and that their trajectories are not circular but elliptical, had formulated the laws of that motion. But he had failed to provide a physical explanation of the forces behind it. This was the effort to which Newton devoted himself.

In 1666, according to legend, his observation of a falling apple led Newton to ponder gravity, which causes bodies to fall to Earth, and to wonder whether the gravitational pull extended as far as the Moon, holding it in its orbit, and likewise, whether the Sun's gravitational pull might control the motion of the planets. But the numerical data on the Earth's dimensions available to him at that time did not allow him to test his ideas. He was unable to do so until 16 years later, after the French astronomer Picard, as a result of painstaking geodesic surveys carried out north of Paris in 1669–70, had gotten a new measurement of the Earth's diameter that was much more accurate than previous ones. Knowing the result, Newton could then correctly compare the fall of bodies to Earth with that of the Moon toward the Earth, as a function of their relative distances from the Earth's center. By making the link between Kepler's third law and the law of centrifugal force, he was finally able to establish the law of universal gravity, which he formulated in 1687 in his famous work *Mathematical Principles of Natural Philosophy:* Any two bodies attract each other in direct proportion to their mass and in inverse proportion to the square of the distance between their centers of gravity. With this law he could then derive Kepler's laws, calculate the motions of the Moon and planets more accurately by taking their mutual perturbations into account, calculate the orbits of comets, explain the tides and calculate them accurately, etc.

Newtonian mechanics, with its most remarkable feature, the law of universal gravity, dealt the final blow to the ancient conceptions of the universe and gave classical cosmology its theoretical underpinnings. But astronomy is also indebted to Newton for various discoveries in the field of optics, such as the breakdown of light with a prism, interference of light rays and finally, in 1671, the invention of the modern telescope.

During the 17th and 18th centuries, *Europe witnessed the emergence of large observatories, which powerfully aided astronomical research. Here we see the observatory of Paris, completed in 1671, as it looked in the beginning. Painted wood panel from the Hall of Emblems at the château of Bussy-Rabutin (Côte-d'Or). Photo: Lauros-Giraudon.*

The first reflecting telescope, invented by Newton in 1671. *Its mirror was only 5 centimeters (2 inches) in diameter. Photo: Erich Lessing-Magnum.*

Founding of the Observatories

Thus the 17th century not only provided astronomy with an essential theoretical tool it lacked but also gave it the instruments on which its later development was based, the refracting and reflecting telescopes. Astronomy after Newton was a mature science. Discoveries multiplied with the construction of more and more sophisticated equipment housed in observatories, some of which are still standing.

In France, Louis XIV and his minister Colbert became convinced by the members of the newly created Academy of Sciences of the advantage of setting up an institution in Paris devoted to observation of the planets, in view of the applications of astronomy to navigation, geodesy, time measurement, etc. Beginning in 1667, the observatory was promptly built under the direction of the architect Claude Perrault, and upon completion in 1671, was placed under the administration of the Italian astronomer Jean Dominique Cassini (1625–1712), whom Louis XIV had brought to France two years earlier. In England, a similar case made to King Charles II led to the construction in 1675 of the Royal Greenwich Observatory, of which the astronomer John Flamsteed (1646–1719) became the first director.

The Morphological Study of the Planets

The rapid growth in popularity of the telescope enabled 17th-century astronomers after Galileo to take up the morphological study of the moon and planets.

In about 1650, the Belgian Langrenus, the Pole Hevelius and the Italians Riccioli and Grimaldi tried to draw the first maps of the Moon and to lay the foundations of selenography.

The strange appearance of Saturn, mentioned by different observers from Galileo onward, was explained in 1659 by the Dutch astronomer Huyghens: The planet has a ring around it. Later, in 1675, Cassini discovered a division in the center of this ring, suggesting its multiple structure. In 1659, Huyghens drew the first sketch of Mars, based on telescopic observations, on which a spot—quite characteristic of the planet's surface—can clearly be distinguished. In 1666, Cassini in turn observed the planet, demonstrating its rotation in slightly more than 24 hours. He also discovered white spots at both poles of Mars, which were studied by William Herschel a century later. In addition, he did the first series of consecutive observations of Venus but was unable to determine its rotation period because of the lack of visible details on its surface.

This close study of the planets led to the discovery of several minor bodies of the solar system. In the second half of the 17th century, five satellites of Saturn were discovered: first, Titan, in 1655, by Huyghens; then Iapetus in 1671, Rhea in 1672 and Tethys and Dione in 1684, by Cassini.

Nevertheless, the optical instruments used in the 17th century retained their primitive character. Since refractors had single lenses, it was necessary to give them enormous focal lengths to reduce chromatic aberration; hence the appearance of "telescopes long enough to frighten people," as Molière wrote in *Les Femmes Savantes*. Hevelius's telescope had a record length of 47 meters (155 feet), and the tube had to be set in place with a hoist. At the Paris Observatory, the instrument with which Cassini discovered two of Saturn's satellites, in 1671–72, was equally impressive: It was a refractor without a tube whose lens was placed on the balcony of the building, pointed more or less in the intended direction, while the observer stood in the courtyard, 27 meters (89 feet) below and chased the image with a magnifying glass. One can understand the difficulty of a detailed study of planetary surfaces under these conditions. Thus what mainly characterized the end of the 17th and the first half of the 18th century was the flourishing of positional astronomy and celestial mechanics.

The Discovery of the Speed of Light

Among the bodies whose motion was closely followed in this period were the moons of Jupiter. Indeed, based on an earlier proposition by Galileo, their rapid motion—particularly of the one closest to the planet—was seen to constitute a kind of celestial clock that was visible everywhere on Earth, making it possible to solve—at least approximately—the difficult problem of figuring longitudes at sea. Hence the necessity arose of establishing the most accurate possible ephemerides of these satellites. Cassini tackled the job and published the first ephemerides in 1668. Following up on his observations at the Paris observatory, beginning in 1671, one of his assistants, the Danish astronomer Römer, discovered that eclipses of the satellites, particularly of the one closest to Jupiter, sometimes occurred ahead of the ephemerides and sometimes behind them, depending on the relative position of Jupiter and the Earth. A careful study of the matter led him, in 1676, to the conclusion that these periodic discrepancies were due to the time it took light to reach Earth from the moons of Jupiter. This was a seminal discovery, because it proved that light does not travel instantaneously—as the ancients thought—but at a *finite* speed. Estimating the time required for light to travel the radius of the terrestrial orbit at about 11 minutes, Römer came up with a speed of slightly more than 210,000 kilometers per second (130,516 miles per second) (it is actually about 300,000 kilometers per second [186,451 miles per second], with light taking only about 8.25 minutes to cross the radius of the terrestrial orbit).

The Flourishing of Positional Astronomy

The Greenwich Observatory was founded with the special goal of determining the position of the planets with greater accuracy in order to permit a better determination of position at sea—a crucial problem for a seagoing nation such as Great Britain. Its first director, John Flamsteed, met this task head on. For nearly 30 years, from 1676 to 1705, he compiled observations, deliberately combining telescopes with the traditional measuring instruments in order to increase sighting accuracy, and in 1712 he published a catalog listing the positions of nearly 3,000 stars. In 1729—10 years after his death—a celestial atlas based on this catalog was published as well.

Edmund Halley (1656–1742), who succeeded Flamsteed as head of the Greenwich Observatory, published independently, on his return from a trip to St. Helena in 1678, the first catalog of southern stars, containing 381 stellar positions. Then, in 1718, he made a fundamental discovery: the *proper motion* of stars. Ever since antiquity, the stars had been thought of as completely motionless in the celestial sphere, and it was common to speak of the "sphere of fixed stars." But by comparing the old positions of a few bright stars to the ones he observed, Halley discovered greater differences than might be attributed to errors of observation. Therefore, it had to be concluded that the stars slowly change position in the sky in the course of centuries and that the shapes of the constellations change little by little. Thus the dogma of the fixed position of the stars collapsed.

In 1726, James Bradley (1693–1762), who became the third director of the Greenwich Observatory, discovered the *aberration of light*, which causes an apparent annual shift in the positions of stars around an average position. The discovery of this phenomenon, a result of the combination ofthe speed of light and the speed of the Earth's travel around the Sun, brought physical confirmation to Copernicus's system and to Römer's determination of the speed of light.

In 1748, Bradley again called attention to himself by discovering and explaining *nutation*, the oscillation of the Earth's rotational axis and therefore of the celestial poles under the gravitational influence of the Moon (see "The Earth," p. 131). This was a splendid confirmation of Newton's law of universal gravity. Bradley's two corrections made it possible to avoid an error of more than 1 arc minute on the average in the position of a star; they ushered in the birth of precision astronomy.

In the following years, new stellar catalogs appeared. From 1742 to 1762, Bradley found the positions of about 60,000 stars. From 1750 to 1760, the priest Nicolas-Louis de la Caille traveled to the Cape of Good Hope and there found the positions of more than 10,000 stars, naming 14 new constellations and drawing the first map of the sky in the Southern Hemisphere.

The development of positional astronomy was also demonstrated by the publication of accurate astronomical ephemerides. *Connaissance des temps,* an annual collection of ephemerides chiefly for the use of sailors, was published in France in 1679 on the initiative of an educated amateur, Joachim d'Alencé. In Great Britain in 1766, the royal astronomer Maskelyne first published the *Nautical Almanac,* up to the end of the 19th century, it listed the distances of the Moon from the Sun and the major stars every three hours to within 1 second, thus permitting an accurate determination of longitudes at sea.

The Measurement of the Earth

Among the problems with which astronomers were preoccupied at the end of the 17th century and during the 18th century was the measurement of the Earth's dimensions and shape.

In 1669–70, as noted earlier, the French astronomer Picard measured a meridional arc north of Paris, and Newton then used that result to test his ideas of universal gravity.

From 1679–1682, Picard and La Hire found the geographic coordinates of several French towns. Then, in 1683, Cassini and La Hire undertook to measure the arc of the meridian crossing France from north to south, from Dunkirk to Perpignan. This large-scale operation, intended to serve as the basis for an accurate map of the realm, was not finished until 1718, after many setbacks; Brittany was reduced by one-third in comparison to previous maps! These measurements were verified in 1739 by La Caille and Jacques Cassini (1667–1756), son of Jean Dominique. This laid the groundwork for a national geodesic survey.

In 1747, Louis XV asked César François Cassini de Thury (1714–1784), son of Jacques Cassini, to draw up a general topographical map based on this survey. Begun in 1750, this map was finished at the time of the French Revolution.

In addition, at the request of the Academy of Sciences, two expeditions were organized to settle the question of the flattening of the globe suggested by Newton: The geodesists Bouguer, Godin and La Condamine went to Peru to measure the length of one degree of the meridian (1735–44), while Maupertuis, Clairaut and Le Monnier did the same thing in Lapland (1736–39). The comparison of the measurements taken by the two parties proved that the Earth was flattened at the poles and bulging at the equator, in conformity with Newton's ideas. These geodesic operations were repeated later, with improved methods. The measurement of the meridian between Dunkirk and Barcelona, done by Delambre and Méchain from 1792 to 1799, was the basis for determining the meter, which was established as the new unit of length by France's National Constituent Assembly in 1791.

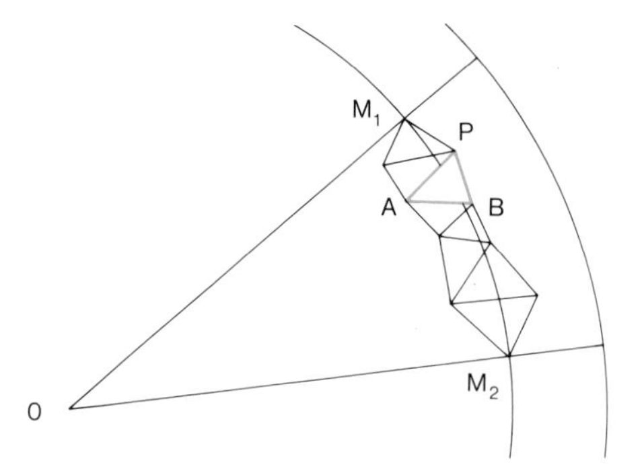

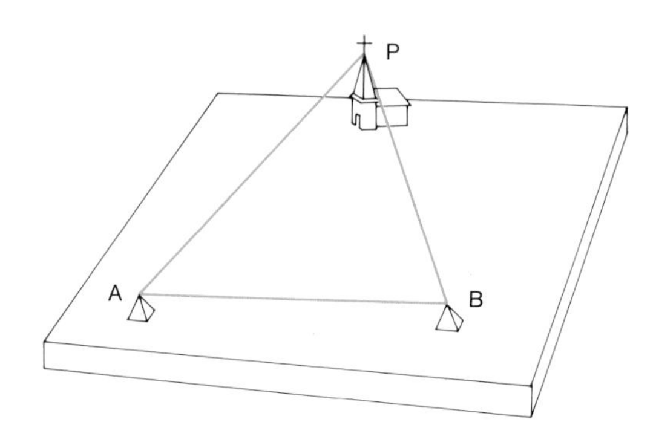

Principle of the triangulation method of measuring distances on Earth. *Given a point P on the Earth's surface, measure the base AB (A and B being suitably chosen points) and the angles BAP and ABP. The distances AP and BP are found by calculation. To measure an arc along a meridian, draw a chain of triangles between its extremities M_1 and M_2 and measure all the angles. Also measure the angle OM_1M_2 by focusing on the same stars from M_1 and M_2. In this way the length of the arc can be found relative to the length of the base AB of one of the triangles.*

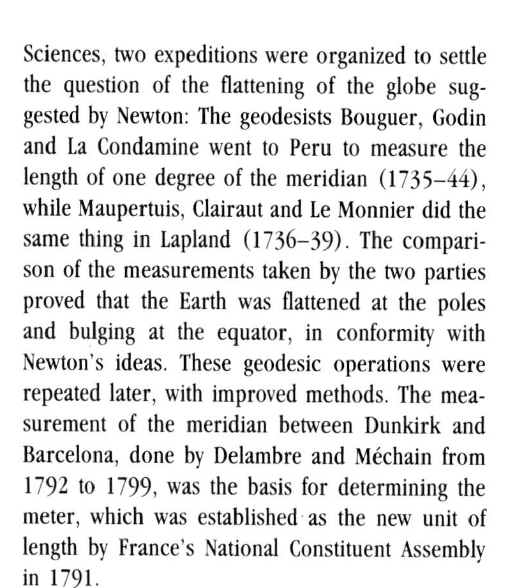

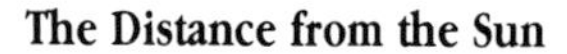

The Distance from the Sun

Other problems that were solved in the 18th century were the distance of the Earth from the Sun, and the dimensions of the solar system. The study of the angular motions of the planets, and the application of Kepler's laws, yielded only relative distances, the conventional unit being the average radius of the Earth's orbit. To find the actual scale of the system, in real size, it was necessary to know the value in kilometers of this astronomical unit of distance. The method used was based strictly on the same principle as the measurement of distances on Earth by means of triangulation. It consisted of pinpointing a planet close to the Earth, either from two observatories sufficiently distant from one another (one in the Northern Hemisphere, for example, and one in the Southern Hemisphere), or from the same observatory, first when the planet had just risen, and then when it was about to set. The planet's distance can easily be deduced from the small angular displacement relative to the stars that results. Since the calculations of orbits make it possible to know the planet's distance from the Earth at any given moment based on the average radius of the terrestrial orbit, the observation amounts to determining the Earth's distance from the Sun. In fact, astronomers express the results of these measurements not in terms of the distance from the Sun itself in kilometers, but in terms of the angle formed by the Earth's equatorial radius at the distance of the Sun, which is called the Sun's *parallax.*

Applied in 1672 by observations of the planet Mars carried out simultaneously in Cayenne by Richer and in Paris by Jean Dominique Cassini and PIcard, this method showed that the solar system was 20 times larger than had been imagined. Maraldi in 1704, Bradley in 1719 and La Caille in 1751 went on to new determinations of the solar parallax. A more accurate value was obtained in 1761, then in 1769, during Venus's transit in front of the Sun, through the use of an ingenious method proposed by Halley. Since that time, the average distance of the Earth from the Sun has been known to within about 1 percent.

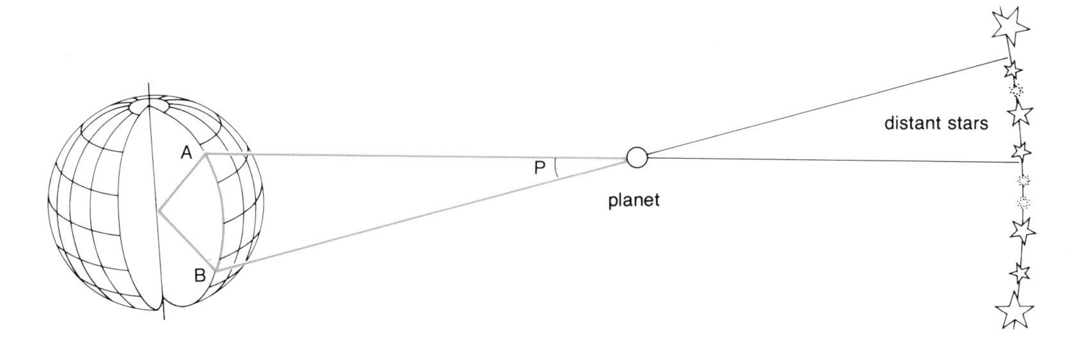

Principle of measurement of the distance of a planet. *The method is similar to that used to measure distances on Earth by triangulation. The base of the triangle is constituted by two points, A and B, very far apart on the Earth's surface (preferably located on the same meridian, since their distance is then better known), from which the planet is sighted simultaneously. In this way the angle P (parallax) can be calculated and the planet's distance determined.*

The Development of Celestial Mechanics

The announcement by Newton in 1687 of the law of universal gravity also encouraged the development of celestial mechanics. Halley devoted himself to the study of the motion of comets and in 1705 published *Synopsis of Cometary Astronomy,* in which, among other things, he predicted the return, toward the end of 1758, of the comet that had appeared in 1682 (see "The Comets," p. 212). The success of this prediction aroused great excitement and helped to establish firmly the success of Newton's theory.

In the second half of the 18th century, Clairault (1713–65) gave a precise description of the Earth's shape, Euler (1707–83) amplified the theory of the Moon and d'Alembert (1717–1813) formulated the theory of precession and nutation.

Lagrange (1736–1813), in his *Mécanique analytique,* made a study of the solar system based on Newton's laws and made calculations that demonstrated all of the perturbations of the motion of the planets relative to the simple orbits that Kepler's laws predicted by introducing the action of the other celestial bodies.

Finally, Laplace (1747–1827) in 1773 demonstrated the invariability of the major axes of the planetary orbits, and in 1784 he established the stability of the solar system. In his *Mécanique céleste* (1799), he also studied the motion of the Moon and the satellites of Jupiter, the theory of tides, etc. In his *Exposition du système du monde* (1796), he advanced a theory of the formation of the solar system that is still the starting point for most modern theories.

The success of celestial mechanics was shown particularly in 1846 with the discovery by the German Galle (1812–1910) of a new planet, Neptune, in the position predicted by the calculations of Le Verrier (1811–77) based on the study of the perturbations of the motion of Uranus (see "Neptune," p. 200).

The Development of Better Instruments

Parallel to this progress in theory, the technique of optical instruments underwent significant improvements in the mid-18th century. In 1747, Euler showed the theoretical possibility of compensating for the chromatic defects of single lenses by combining two lenses with different dispersion capacities. In 1757, the English optician Dollond presented the first achromatic telescope to the Royal Society; its objective combined a "crown" glass with a "flint" glass, and its performance was greatly superior to that of the single lenses used formerly because its weight was much reduced. After this invention, only minor improvements were added to the telescope in the 19th century. Between 1810 and 1820, the Swiss glassmaker Guinand and the Bavarian optician Fraunhofer together developed the technology of variable index lenses, which led, in 1824, to the development of the telescope at the Dorpat Observatory (in Estonia), the first good large-diameter refractor, of 24 centimeters (9.6 inches). Later, the technique of manufacturing large telescopes was pushed to its farthest extremes with the production of instruments as much as 1 meter (3.3 feet) in diameter. The great telescope of Meudon, France, completed in 1891, was the largest in Europe, with its lens having an aperture of 83 centimeters (3.3 inches) and a focal distance of 16 meters (52.8 feet), and was exceeded only by those of the Lick observatory, 91 centimeters (36 inches), and the Yerkes observatory in the United States, 1.02 meters (40 inches).

On the other hand, reflecting telescopes were largely unused for a long time because of the difficulty of making good mirrors and because of the rapid oxidation of the metal mirrors they originally included. In about 1750, small reflectors created by the English optician Short became widely used in observatories but did not arouse enthusiasm from the professionals. It was a musician with a passion for astronomy, William Herschel (1738–1822), who built the first big reflectors after 1780: In 1787, he finished one with a 48-centimeter (19-inch) aperture, and then, in 1789, another with a 1.22-meter (4-foot) aperture, magnifying up to 6,000 times, which was surpassed only in 1845, by a telescope of 185 centimeters (74 inches) invented by a rich Irish amateur, Lord Rosse (1800–67).

The Rise of Stellar Astronomy

Until the 18th century, astronomers had paid little attention to the stars as such except to measure their coordinates and draw up more and more lengthy catalogs.

But around 1750, it became clear that these bodies were other, extremely distant suns. A few bold intellects then began to become interested in their distribution in space. The Englishman Thomas Wright (1711–86) put forward the notion that the Milky Way might be an enormous disk of finite dimensions, encompassing the Sun and the visible stars. The German philosopher Emmanuel Kant adopted this idea, adding to it the hypothesis that space contained other, similar island universes. The French mathematician Jean Henri Lambert (1728–77) espoused similar ideas.

Nevertheless, it was William Herschel who emerged as the founder of stellar astronomy. An organist and professional musician but with a strong attraction to astronomy, he built his first telescope in 1774. Four years later, he possessed an instrument with an aperture of 15 centimeters (6 inches) and a focal distance of 2 meters (6.6

feet), superior to the better telescopes then in use. With this instrument, he discovered the planet Uranus in 1781 and gained celebrity status. But this was only the first success of a fruitful career in astronomy. In 1783, while studying the proper motion of several bright stars, he discovered that the Sun's motion in space is toward a point (the apex) in the constellation Hercules. Anxious to establish the spatial distribution of stars in the Milky Way and to ascertain the Sun's place in the stellar system, he proceeded to do numerous countings of stars in different directions over the entire sky. In 1785, he was able to give the first description of the spatial structure of the galaxy. Repeated observation of stars that appear very close to one another in the sky led him in addition to the discovery of double stars: physically linked stars rotating around a common center of gravity. Thus, for the first time, Newton's law of gravity was verified outside the solar system. His patient observations of the sky also led him to draw up the first big catalog of nebulae, of which only a hundred specimens were known prior to his time. After resolving several of these formations into stars, while others retained their nebulous character under the highest magnifications, he suspected that some nebulae, in keeping with the hypotheses of Wright and Kant, were galaxies external to our own, while others were clouds of gas within our galaxy. Moreover, he put forward the idea that clusters of stars are formed from diffuse gas clouds, thus introducing for the first time the notion of evolution in the study of the universe.

After Herschel's death, his son John (1792–1871) continued his work. He discovered nearly 3,500 new double stars, and in 1833 he published a catalog of more than 2,300 nebulae and stellar clusters, containing 525 new objects. He then traveled, from 1834 to 1837, near the Cape of Good Hope to study the sky in the Southern Hemisphere. There he observed more than 2,000 double stars and more than 1,700 nebulae, 300 of them new, and proceeded to a detailed study of the Magellanic clouds.

Among the pioneers of stellar astonomy one may also mention the German Friedrich Bessel (1784–1846), who in 1838 made the first determination of a stellar parallax—i.e., the first measurement of the distance of a star—and in 1840 predicted the existence of a hidden companion surrounding some of the bright stars (Sirius, Procyon) and perturbing their motion; the Russian Friedrich Struve (1793–1864), who distinguished himself in the discovery and accurate measurement of thousands of double stars; and finally the German Friedrich Argelander (1799–1875), author of a monumental catalog and celestial atlas giving the positions and brightnesses, which he determined in the course of 25 years of observation, of more than 324,000 stars.

The Birth of Astrophysics

While thousands of new stars were thus being counted, the development of physics led to the creation of techniques making it finally possible to decipher the message conveyed by their light.

Earlier, in the 17th century, Newton had observed that the breakdown of white light by a prism produced a *spectrum* in the colors of the rainbow. In 1814, the Bavarian optician Fraunhofer (1784–1820) discovered that the spectrum of sunlight contains a multitude of dark lines. Analyzing next the light from the Moon, Venus and Mars, he found the same rays in the same position. Later, while studying the spectra of different stars, he observed that they differed from that of the Sun but also revealed lines. The Germans Robert Bunsen (1811–99) and Gustav Kirchhoff (1824–87) explained these phenomena in 1859, showing that the line spectra appearing both in absorption (dark lines) and in emission (bright lines) constituted in light the *signature* of each chemical substance that had been capable of producing its characteristic emission or partial absorption between the source and the observer. It thus became possible to identify in the stars the presence of known substances and to discover new ones. Thus the foundations of spectral astronomy were laid.

In 1842, the Austrian Christian Doppler (1803–63) showed that the pitch of a sound varied when the source of the sound and the observer are in motion relative to one another. In 1848, the Frenchman Armand Fizeau (1819–96) showed that this phenomenon also applied to light, a light ray appearing to shift toward the red or toward the blue according to whether the emitting source was moving away from or toward the observer. By using this phenomenon, the American Huggins achieved the first determination in 1868 of the radial velocity of a star by examination of the shifts of the lines in its spectrum.

In 1879, the Austrian Josef Stefan (1835–93) formulated a law of radiation with which it became possible to determine the temperature of a star by virtue of the energy it radiates at different wavelengths.

Finally, in 1894, the Dutchman Pieter Zeeman (1865–1943) discovered that a spectral line, emitted or absorbed in the presence of a magnetic field, divides into several components whose wavelength separation is proportional to the strength of the field: hence the possibility of measuring the size of the latter, and perhaps also its direction, if it is not too weak.

Spectral analysis also proved to be a powerful method in the investigation of stars, yielding the secret of their temperature and physical conditions prevailing in their outer layers or atmospheres, their chemical composition, radial speed, magnetic field, etc. At the same time, beginning in 1840, the application of photography to astronomy opened up new prospects by making it possible to compare data obtained on different dates or in different locations and making it possible to record, with sufficiently long time exposures, the light of many stars invisible to the naked eye. In 1840, the American Draper obtained the first photograph of the Moon; in 1845 the Frenchmen Fizeau and Foucault succeeded in making the first negative of the Sun, at the Paris Observatory; in 1850, the Americans Bond and Whipple succeeded in the first photographs of stars—those of Castor and Vega.

But it was only after the appearance in 1851 of the wet-plate method—10 to 100 times more sensitive than Daguerre's original method—that photography began to yield the first results really contributing to the progress of astronomy. In 1857–59, the amateur English astronomer W. de la Rue succeeded in taking the first somewhat detailed photographs of the Moon, while at Kew Observatory, near London, the first daily photographic observations of the Sun began in 1858.

At the same time, it became possible to measure the brightness of the stars with accuracy; the Englishman Pogson established in 1857 the fundamental relationship defining the scale of magni-

Modern Astronomy: A Broad Spectrum of Disciplines

A broad spectrum of disciplines makes up the study of the universe today.

The oldest branch of astronomy is *positional astronomy* or *astrometry*, the purpose of which is to determine the positions and motions of celestial bodies. It is notable in large part for the establishment of stellar catalogues, in particular the "fundamental catalogue" giving the positions of 1,535 bright stars evenly distributed in the sky and from which the coordinates of all the distant stars are calculated. The fact of the stars' proper motions, and the need for astronomers to avail themselves of more and more accurate measurements, make the compilation of stellar catalogues a permanent task. But astrometry also is concerned with the study of the relative motion of double stars—whence their mass is deduced—and the measurement of parallaxes, which determine the distances of nearby stars. More broadly, it includes research into the mechanics and dynamics of our galaxy and other galaxies. It also has established an astronomical time scale. For practical purposes it may be said that all knowledge having to do with the shape and motion of the Earth, the motions of the solar system and of our galaxy, and the scale and evolution of the universe is narrowly dependent on astrometric measurements.

Closely related to astrometry is *celestial mechanics*, which deals with the laws governing the motions of planets. Its purview includes the calculation of orbits and the establishment of astronomical yearbooks and ephemerides (tables showing numerical data—on a daily basis or otherwise—regarding the position of the Sun, Moon, planets, etc.).

Since the development of astronautics, a new use has been found for celestial mechanics in calculating the trajectories of artificial satellites and interplanetary probes.

Astrometry and celestial mechanics to-

tudes, and in 1859, the German Zöllner invented the first modern photometer.

The combined applications of spectroscopy, photography and photometry to astronomy in the second half of the 19th century marked the rise of a new branch of astronomy, astrophysics, which up to our own time has led to an explosive development of knowledge of the universe.

The General Use of Large Telescopes

Despite the progress made in the manufacture of lenses, the residual chromatism of double lens objectives made refractors impractical for photography and even more so for spectroscopy. Furthermore, it appeared that the manufacture of refractors more than 1 meter (3 feet) in diameter posed extremely tricky problems, both in terms of mechanics (bending of the tubes) and of optics (the impossibility of producing glass bulks homogeneous enough for light to pass through them without introducing defects that interfere with the image). Thus, with the development of astrophysics, the replacement of the refractor by the reflector became an imperative of astronomical research. In building the first reflecting telescope with a silvered glass mirror in 1858—very advantageously replacing the bronze mirrors used up to then—the Frenchman Foucault opened the way to the invention of modern telescopes. From then on, telescopes continuously increased in quality and size (introduction in 1918 of the 2.54-meter (100-inch) telescope of the Mount Wilson Observatory in the United States; in 1949, of the 5.08-meter (200-inch) telescope at Mount Palomar, also in the United States; and in 1976, of the 6-meter (236-inch) telescope of Zelenchukskaya in the Soviet Caucasus), and the combination with higher and higher performance detectors has made it possible to use them near their maximal capacities (see "Instruments and Techniques of Exploring the Universe," p. 25). With the introduction of the large telescopes, astronomy was armed for new conquests.

The 20th Century: A New Universe

The first years of the 20th century brought a flurry of upheavals in the image that human beings had formed of the universe.

The Andromeda galaxy. *Located 2.2 million light-years away, it is a stellar system similar to the one to which our solar system belongs. It was the first spiral nebula identified as a galaxy. Photo copyright 1959 by California Institute of Technology and Carnegie Institution of Washington.*

Einstein (1879–1955) advanced his theory of relativity, offering a new conception of the relationships of space, time and matter. For the absolute concepts of space and time implied by Newtonian mechanics, he substituted the concept of a space-time continuum. Furthermore, he reduced gravity to a geometric property of space-time, curved by the masses it contains. Finally, he established the existence of a limit to the speeds possible in the physical universe, that limit being the speed of light. The problem of the origin and evolution of the universe, and of its geometry, were raised anew thereafter.

The Dane Hertzsprung, in 1905, and then the American Russell, in 1913, conceived a diagram showing the distribution of stars according to their luminosity and color, making it possible to study the sequence of their evolution. In 1912, the American Henrietta Leavitt discovered a proportionality between the luminosity and the period of the brightness variation cycle of a category of variable stars, the *cepheids.* In 1916 another American, Harlow Shapley, deduced a law from this, broader in scope, which makes it possible to use these stars as indicators of distances in the universe.

Finally, with the aid of the 2.54-meter telescope at the Mount Wilson Observatory—then the most powerful one in the world—the American Edwin Hubble, in 1923–24, succeeded in resolving individual stars within the Andromeda nebula. What was once thought to be only a cloud of gas and dust was revealed in fact to be an immense, autonomous stellar system. Soon other nebulae in turn were identified as systems more or less analogous to our galaxy. They were therefore referred to as galaxies. The universe—within the limits of the range of the instruments—appeared to be populated by those "island universes" envisioned by Emmanuel Kant in the 18th century. A new field of research opened up.

It soon appeared that all the galaxies observed, with the exception of a few of those closest to our own, show a spectrum shifted toward the red. This phenomenon, interpreted as a Doppler-Fizeau effect, indicates that they are carried along by a general movement of recession. Hubble, in 1929, showed that the speed at which galaxies are receding is greater in proportion to their distance, implying that they are constantly moving away from one another. Thus the hypothesis of the expansion of the universe, formulated in 1927 by the priest Georges Lemaître, was confirmed.

The Development of Radio Astronomy

When the study of galaxies began, it seemed that the only possible way to reach new celestial objects that were farther and farther way was to build more and more powerful telescopes.

gether make up what is known as *basic astronomy.*

Since the second half of the 19th century, the systematic application of photography, spectroscopy and photometry to astronomy has led to the development of *astrophysics,* or the physical study of celestial bodies. Limited at first to the study of visible radiation, astrophysics has seen its field of inquiry gradually widen to include all forms of electromagnetic radiation, especially since the development of *radioastronomy*—the study of celestial sources of radio-frequency waves—and *space astronomy,* which makes it possible to study cosmic radiation stopped by the Earth's atmosphere, before it reaches the ground (gamma and X rays, ultraviolet and infrared radiation). The use of more and more high-performance telescopes, and the direct exploration of the Moon, planets and interplanetary medium with the help of spacecraft also have contributed to the vigorous growth of astrophysics over recent decades. The new look at the cosmos that modern techniques have given us has prominently benefited theoretical astrophysics, which combines empirical data with the principles of theoretical physics to develop "models" capable of explaining the structure and evolution of the cosmos. To this field may be added *cosmogony,* the study of the formation and evolution of the solar system and similar planetary systems, and *cosmology,* which seeks to account for the structure and evolution of the universe as a whole. Allied in certain respects to astrophysics, *astrochemistry* is concerned with the chemistry of outer space. The discovery of many molecules within the interstellar medium was one of the major factors in astrochemistry's development. *Astrobiology,* or *exobiology,* strives to answer questions regarding the possibilities of life existing elsewhere in the cosmos.

In fact, the universe is the most prodigious laboratory a scientist might dream of. From the highest to the lowest temperatures, from the densest to the most rarefied atmospheres, from the most massive to the smallest systems, matter exists under the most varied conditions, giving scientists the opportunity to examine an extraordinarily varied range of phenomena, the analysis of which requires an interdisciplinary approach. Thus astronomy now appears as a crossroads where the majority of scientific disciplines converge and are mutually enriched.

However, in 1931, the American Karl Jansky picked up a radio signal of extraterrestrial origin that was later shown to emanate from the Milky Way. In 1936, his compatriot Grote Reber developed the first radio telescope and thereby produced the first radio-frequency map of the sky. During World War II, the intensive use of wireless communications and radar led to the discovery of radio-frequency radiation from the Sun. These early results showed that light waves, on which astronomy had been traditionally based because human eyes are sensitive to them, are not the only form of radiation emitted by the stars. In 1944, the Dutchman Hendrik Van de Hulst (born in 1918) predicted the existence of a hydrogen emission line at a wavelength of 21 centimeters. This was actually observed in 1951, thus providing a valuable method of studying the interstellar medium. In 1946, the Briton James S. Hey discovered the first extragalactic radio emission source, in the constellation Cygnus. From then onward, the methods used in the study of Hertzian waves emitted by celestial objects developed in a remarkable fashion. Radio astronomy has continued to expand up to our time, prodigiously enriching our knowledge of the universe by making possible the discovery of new objects (quasars in 1960, pulsars in 1967) or phenomena formerly unsuspected (the thermal radiation from the edge of the observable universe in 1965) and by probing remote regions of space inaccessible by optical means.

Space Astronomy

Since the beginning of the space age and the launching of spacecraft outside the Earth's atmosphere, other forms of invisible cosmic radiation have been studied. Observations of these types of radiation were originally very difficult, if not in some cases impossible, because they are absorbed by the atmosphere itself—gamma rays, X rays and ultraviolet and infrared radiation.

In 1946, a V2 rocket recovered in Germany was equipped and launched by an American space laboratory to measure solar X-ray emission, the existence of which was considered very probable. In 1962, the first satellites especially designed for the study of invisible solar radiation were placed in orbit. The systematic study of the sky in the ultraviolet range began in 1968 with the American orbital observatory *OAO-2* ; the study of X rays, in 1970, with the American satellite *SAS-1*, also called *Uhuru*; finally, the study of gamma rays in 1972 with the *SAS-2* spacecraft. By comparison of the results obtained in these different wavelength regimes, astronomers finally had a comprehensive view of the extraordinarily varied phenomena occurring in the universe.

Furthermore, exploration of the Moon, planets and the interplanetary medium by means of spacecraft has led to considerable progress in our knowledge of the solar system. The sending of automatic probes to the Moon began in 1959, with the Soviet Luna series: *Luna 3*, in particular, provided the first photographs of the rear surface of the Moon, but it was only with *Luna 9*, in 1966, that the first soft landing on the Moon's surface was achieved. Mankind's direct exploration of the Earth's natural satellite began on July 20, 1969, with the landing of two astronauts (Armstrong and Aldrin) in the Sea of Tranquillity aboard the lunar module *Apollo 11*. The Apollo program ended in 1972 after six human landings on the Moon, which made it possible to carry out a vast scientific program (see the chapter "The Moon"). In 1962, the flyby of Venus by the American *Mariner 2* probe marked the beginning of planetary exploration by spacecraft. Mars was successfully approached for the first time in 1965 by *Mariner 4* (U.S.), Jupiter in 1973 by *Pioneer 10* (U.S.), Mercury in 1974 by *Mariner 10* (U.S.) and Saturn in 1979 by *Pioneer 11* (U.S.); the placement of satellites into orbit around Mars first occurred in 1971 with *Mariner 9* (U.S.) and around Venus in 1975 with *Venera 9* and *Venera 10* (USSR). But in 1967, *Venera 4* (USSR) succeeded in making a soft landing on Venus, and in 1971, *Mars 3* (USSR) accomplished the same on Mars. Nonetheless, it was only with the American spacecraft *Viking 1* and *Viking 2*, in 1976, that the on-site study of the Martian surface began. Finally, the American Voyager probes transmitted photographs of an unparalleled excellence of the moons of Jupiter and of the rings and moons of Saturn, in 1979 and 1980, respectively.

Thus, the discovery of the sky—that huge adventure begun thousands of years ago—is continuing today with the most sophisticated technical means. Mankind still awaits answers to those questions that have been asked for centuries: When and how did the cosmos arise? What will its destiny be? Is it eternal? Is it infinite? And if not, what was there before its beginning? What will there be after its end?

Exploration of the Moon *by spacecraft and by man marked the birth of a new era in the discovery of the solar system. Here, in the foreground, is a partial view of the lunar "jeep" during the* ***Apollo 17*** *mission. Photo: NASA.*

3. Instruments for and Techniques of Exploring the Universe

Observation is a fundamental method of astronomy. Not only does it reveal the phenomena occurring in the universe, giving rise to theories capable of explaining them, but it is also, in the last analysis, what confirms or disproves the hypotheses put forward by theoreticians. Before taking a look at the universe in the remainder of this book, it is appropriate to mention the principal techniques and instruments that are used to explore it.

The information we have gathered about the universe comes from three sources:

- directly accessible solid matter (meteorites, and samples taken from other planets in the solar system by spacecraft)
- high-energy elementary particles (cosmic rays, solar wind, etc.)
- electromagnetic radiation (gamma and X rays; ultraviolet, visible and infrared light; and radio waves)

The meteorites and samples taken on the surface of other planets are major sources of information regarding the processes of formation and evolution of the solar system. Cosmic rays, whose origin is unfortunately difficult to determine, provide valuable information about certain basic phenomena related to the origin and evolution of the universe, while the study of stellar winds—especially the solar wind—is now a major element in our improved knowledge of the stars. Nonetheless, electromagnetic radiation still remains by far the richest and most important source of information.

PROPERTIES OF ELECTROMAGNETIC RADIATION

All electromagnetic radiation can be considered in two ways: first, as a vibration traveling through space so that it is possible to define its wavelength and frequency (the wave aspect); second, as a collection of particles in motion, or photons, each carrying a quantity of energy that is proportional to the frequency of the vibration (the particle aspect). The wavelengths of electromagnetic waves can be very diverse, extending over a range of from 10^{-15} meter to greater than several kilometers. The manifestations and properties of a wave vary a great deal according to its characteristic wavelength; hence the great variety of specialized instruments with which to study them. High-energy radiation with a very short wavelength, such as gamma and X rays, usually appears in the particulate form, while at the other end of the electromagnetic spectrum, Hertzian waves are characterized mainly by their wavelike properties.

The problem of astronomical observation consists essentially of locating the spatial origin of radiation emanating from somewhere in the universe, analyzing the distribution of energy it carries according to wavelength, measuring the characteristics of its polarization and studying the possible variations of these parameters over the course of time. This will then yield valuable information regarding the source of emission of the radiation (chemical composition, prevailing physical conditions, motion in space, magnetic field, etc.).

Whatever type of electromagnetic radiation is being studied, the astronomical instrumentation used aims to collect it; analyze it; and concentrate it in a receptor, where it is recorded. Through appropriate treatment, it is then possible to reduce the defects inherent in the measuring technique used and finally to obtain a result, which must then be interpreted.

It may be said that the eye is a remarkable observation device, in which the pupil and lens serve as collector and the retina as a detector/analyzer combination that is very difficult to imitate.

Aerial view of the European Southern Observatory at La Silla, Chile. *The twin towers that house a large 3.6-meter (142-inch) telescope and an auxiliary 1.5-meter (59-inch) telescope can be seen on the mountain peak, at an altitude of 2,450 meters (8,038 feet). This facility allows European astronomers to study the sky in the Southern Hemisphere, which is less frequently observed than is the sky in the Northern Hemisphere. ESO Photo.*

Limit of Resolution and Enlargement

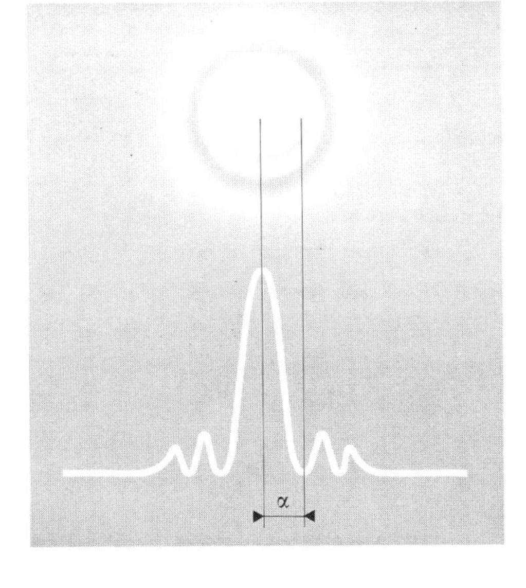

As a result of diffraction owing to the wavelike character of light, the image provided by a lens of a point of light is not a point. Since lenses have a circular contour, this image is, in fact, a small, round spot of light, with maximum brightness in the center, surrounded by alternately dark and bright concentric rings, the latter rapidly diminishing in brightness. This structure is called the Airy figure after the British astronomer who explained its theory in 1834.

Angle α. formed by the intersection of a line from the first dark ring surrounding the bright spot with the sky, is given by this formula:

$$\alpha = 1.22 \frac{\lambda}{D},$$

in which α is expressed in radians, while the λ wavelength of incident radiation, and the diameter, D, of the objective, are expressed in meters.

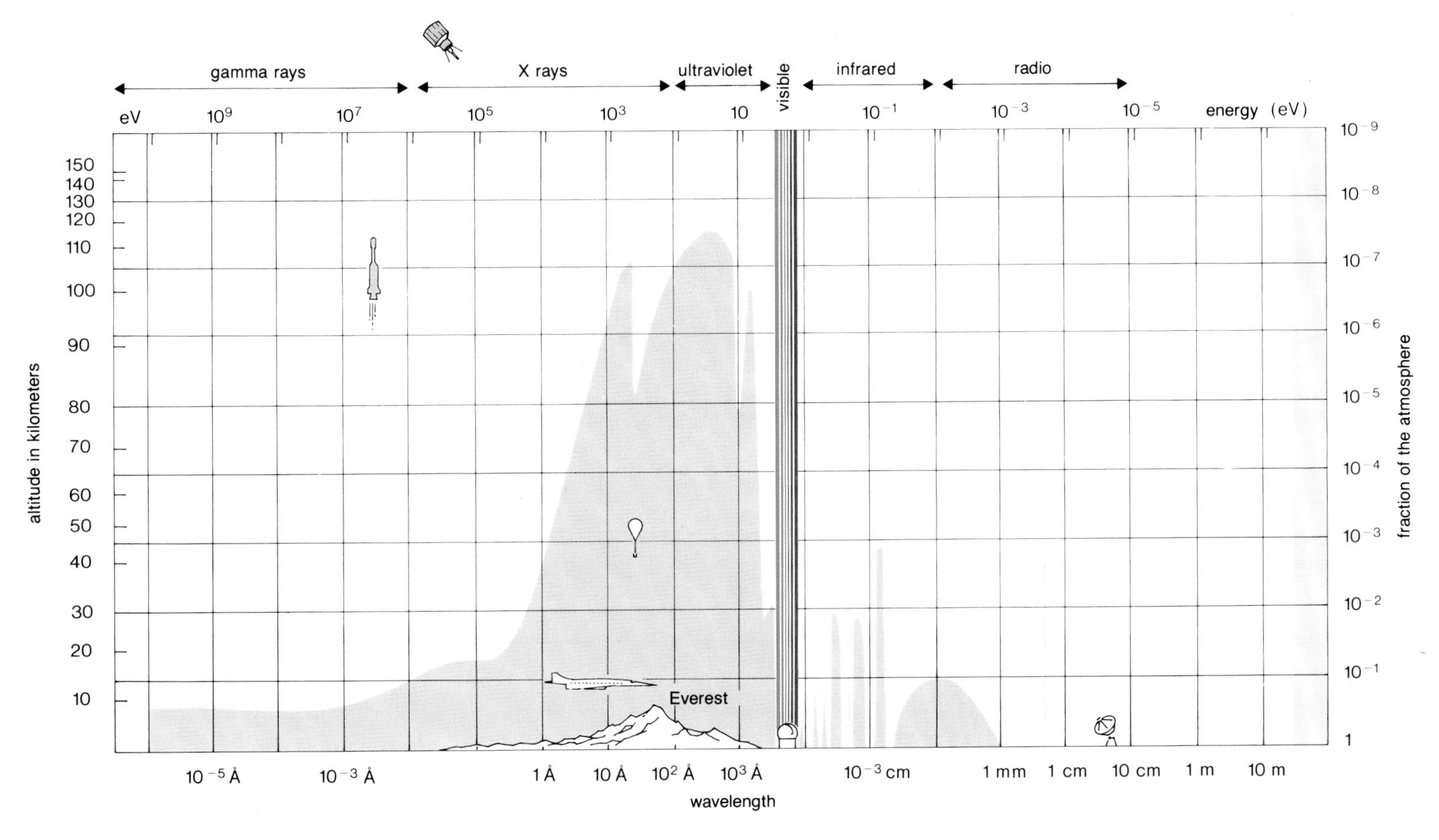

The Earth's atmosphere absorbs or reflects, either totally or partially, most of the electromagnetic radiation emanating from the cosmos. Since the invention of artificial satellites, access to the full range of this radiation has become possible. But on the ground, we possess only two, rather narrow observation "windows": One covers the wavelengths between 0.3 micrometer and a few tens of micrometers; this is the "optical window," which includes visible light (from 0.4 to 0.8 micrometer), as well as some ultraviolet and infrared radiation, which are the neighboring wavelength regimes. The other covers the wavelengths between 1 millimeter (0.04 inch) and about 20 meters (65.62 feet): This is the "radio window," which includes the millimetric, centimetric, (microwave), decimetric, metric and decametric Hertzian waves.

Optical Astronomy

Of the two realms observable from the ground, the optical realm, and more precisely that of visible light, was the only one used for a long time, and it retains a special role, since it is directly accessible to the eye.

The Collectors: Reflecting and Refracting Telescopes

Optical astronomy uses two kinds of collectors: the reflector and the refractor. These instruments essentially consist of a tube containing an objective at one end (the end pointed at the sky) and an eyepiece at the other end (against the eye). The objective collects the light rays from the stars being observed and, theoretically, concentrates them at a point (in practice, a small spot) whose enlarged image is then projected by the eyepiece. The nature of the objective distinguishes the refractor from the reflector. In a refractor, the ob-

The angle α may also be expressed in seconds of arc; D in centimeters; and the formula adapted to the optical regime by taking the wavelength of yellow light (G = 0.550 μm), to which the eye has maximum sensitivity. The formula then becomes:

$$\alpha = \frac{14}{D}$$

The angle α represents the angular size of the smallest details that can be observed through the instrument, or the minimal angular distance of two stars that can be seen separately through it; this is the theoretical limit of resolution of the instrument. The instrument's resolving power is said to improve as this limit gets smaller. Thus the greater the diameter D of an instrument's objective, the better its resolving power. Moreover, for a given diameter, the resolving power diminishes as the wavelength increases.

The theoretical resolving power of the 5-meter (200-inch) telescope at Mount Palomar is about 0.024 arc second. In fact, taking atmospheric turbulence into account, it is unusual to go below 1 arc second.

The image supplied by the objective of a refracting or reflecting telescope is seen through an eyepiece, a kind of magnifying glass composed of two or more lenses but similar to a single converging lens. The ratio between the focal length F of the objective and the focal length F of the eyepiece is called the telescopic magnification: G = F/f. Thus the shorter the focal length of the eyepiece used, the greater the resulting magnification. But the greater the magnification, the darker the image. For observations not requiring contrast, making use of the resolving power alone (such as for the study of double stars), high magnifications may be used. Conversely, to examine the surface of a planet, where detailed visibility requires sufficient contrast, it is necessary to restrict oneself to smaller magnifications so as not to dilute the image. The minimal magnification (taking the eye's resolving power into account) necessary to make the Airy figure visible and to benefit fully from the resolving power is expressed by a number equal to the diameter of the objective measured in millimeters; it is called the resolving enlargement.

The electromagnetic spectrum and the windows where the atmosphere is transparent to incoming radiation.

jective is a lens—or rather, a combination of lenses—that refracts light, while in a reflector, it is a mirror from which light is reflected.

In spyglasses, or terrestrial telescopes, an intermediary mechanism between the objective and the eyepiece makes it possible to get an upright image of the observed object. In astronomical observations, this mechanism—which has the drawback of blocking out part of the light—is useless, and refractors and reflectors alike produce inverted images.

The objective is the chief component of any optical instrument. Its purpose is first to provide the best possible image of the object under study, which is then visually inspected, measured, photographed, analyzed, etc., and second to collect as much energy as possible from that object. In practice, the objective of a small instrument is expected to provide the diffraction image (see box) of the stars being observed, while that of a large instrument is supposed to concentrate the rays making up the image of a point source so that 90 percent of the energy is contained within a very small angle—0.5 second, for example. The dimension of the lens determines the capabilities of the instrument; the amount of energy collected depends on its collecting area; while its ability to separate closely neighboring sources (the resolving power) depends on its linear dimension (see inset). Optical surfaces must be ground with a high degree of precision in order for the objective to perform at a level that matches its theoretical capacity.

The optical precision achieved for an objective must not be diminished by varying flexures of the objective when it is mounted in a mechanical frame so it can be pointed at different parts of the sky. This represents another source of problems. Since the defect tolerance limit remains the same for a given range of wavelengths, the relative precision of the technical solutions must increase with the diameter of the objective. This explains why, for instance, in the visible range, the construction of an objective 5 to 6 meters (200 to 236 inches) in diameter represents the limit of current capabilities.

The objectives of refractors, usually consisting of two adjoining lenses of different dispersions—one of crown glass and converging, the other of flint glass and diverging—which make it virtually achromatic (see box), are now built only with small diameters (up to about 20 centimeters [7.87 inches]) and are used in training or guiding instruments. Some big refractors built in the late 19th or early 20th century (see table) are still operational, however, and are usually used for observations of double stars.

All big modern optical collectors have a mirror for an objective because of its many advantages over lenses—the absence of chromatic aberrations, a usable spectral range extending to the ultraviolet and infrared, and the possibility of obtaining large optical surfaces (whereas the need for uniformity in the glass limits the size of lenses). These mirrors are usually made of ceramic glass with a very small thermal expansion coefficient, which makes them practically immune to distortion from changes in temperature. A high reflecting ability is achieved by vacuum deposit of a very thin layer of aluminum on the glass. After it is applied, the aluminum coating reflects about 90 percent of the light it receives, but after a year or two, a reapplication is necessary because of oxidation.

The World's Largest Refracting Telescopes

Observatory Name and Location	Diameter of Objective Lens (m)	Focal Length (m)	Year Placed in Service	Remarks
Yerkes (Wisconsin, U.S.)	1.02	19.4	1897	Owned by University of Chicago
Lick (California, U.S.)	0.91	17.6	1888	
Meudon (France)	0.83	16.2	1891	Combined with a photographic objective of 0.62 m
Potsdam (G.D.R)	0.80	12.0	1905	Photographic objective
Nice (France)	0.76	17.9	1887	
Allegheny (Pennsylvania, U.S.)	0.76	14.1	1914	Photographic objective
Greenwich (U.K.)	0.71	8.5	1894	
Berlin (G.D.R.)	0.68	21	1896	
Vienna (Austria)	0.68	10.5	1878	
Johannesburg (South Africa)	0.67	10.9	1925	
McCormick (Mt. Jefferson, Virginia, U.S.)	0.67	9.90	1883	
Greenwich (U.K.)	0.66	6.8	1899	
U.S. Naval Observatory (Washington, D.C., U.S.)	0.66	9.9	1873	
Mount Stromlo (Australia)	0.66	10.8	1956	Orginally installed in South Africa

THE DIFFERENT KINDS OF TELESCOPES

Telescope models are distinguished from one another by the kind of objective they use.

Parabolic mirror telescopes. The most common type of objective, which is used to outfit the largest telescopes currently in operation, is the parabolic mirror. Among the advantages it offers is the possibility of obtaining several focal lengths, hence several focal ratios (focal length/diameter), by combining different secondary mirrors with the primary mirror.

In the Newtonian combination (1672), a secondary plane mirror positioned at a 45-degree angle projects the image laterally.

In the *Cassegrain* combination (1672), the primary mirror has a central aperture; the secondary mirror—hyperbolic and convex—projects the image behind the primary mirror. By replacing the secondary mirror, the instrument's focal length can be varied without changing either the structure of the telescope or the location of the terminal focus.

In the *Coudé* arrangement, the optical elements are the same as in the previous combination, with the addition of a set of plane mirrors, one of which is at the intersection of the optical axis and the declination axis, and another that directs the light

Defects of Objectives

All objectives produce images affected by distortions that are called aberrations.

Thus, a single converging lens supplies images marked by chromatic aberration: The image it gives of a point source, such as a star, is not a point but a small, iridescent spot. This phenomenon stems from the fact that the refraction index of glass varies with the wavelength of the various rays that make up light; green rays converge farther from the lens than blue rays, and red rays still farther, with each element of the lens acting as a prism.

In addition, if the image is observed through a filter that allows only one color to pass through, so that the iridescence becomes invisible, the image nonetheless is not perfectly clear. The focal length of the central areas of the lens is greater than that of the perimeter. The image is said to be suffering from spherical aberration, the defect being caused by the spherical nature of the lens surfaces.

It has been shown that an objective can be made achromatic (devoid of chromatic aberration) and at the same time stigmatic (devoid of spherical aberration on the optical axis) if it is composed of two lenses—one, a converging lens of crown glass (a light, low-dispersion glass similar to window glass); the other, a diverging lens of flint glass (a heavy, high-dispersion crystal containing lead)—whose surfaces have the proper radii of curvature. The objectives of large astronomical refractors, like those of spotting scopes and image-inverting teleparallel to the polar axis. Thus the terminal focus is a fixed point on the polar axis that can be placed in a separate part of the dome by adjusting the focal length. This arrangement permits the use of heavy auxiliary instruments (such as spectrographs) in a single laboratory. For instruments with an alt-azimuth type of mount, the Coudé focus is referred to as the "Nasmyth focus" (after a combination invented by James Nasmyth in the 19th century).

The 2-meter (79-inch) telescope of the Pic du Midi Observatory, *in operation since 1979. Located at the top of a 28-meter (92-foot) tower reserved for observations that require a high resolving power. OPMT Photo.*

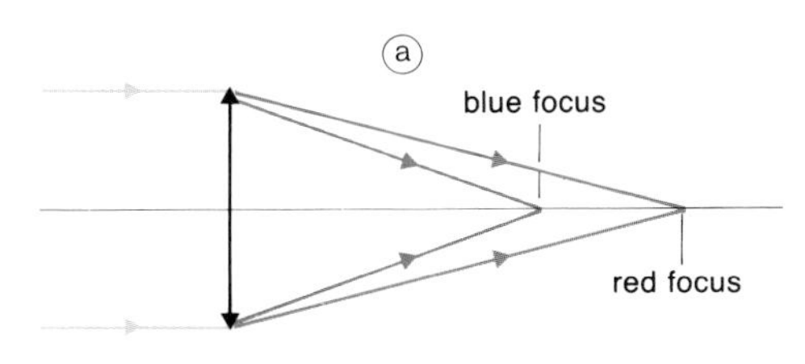

Examples of optical aberrations:
a. chromatic aberration
b. pincushion distortion
c. barrel distortion

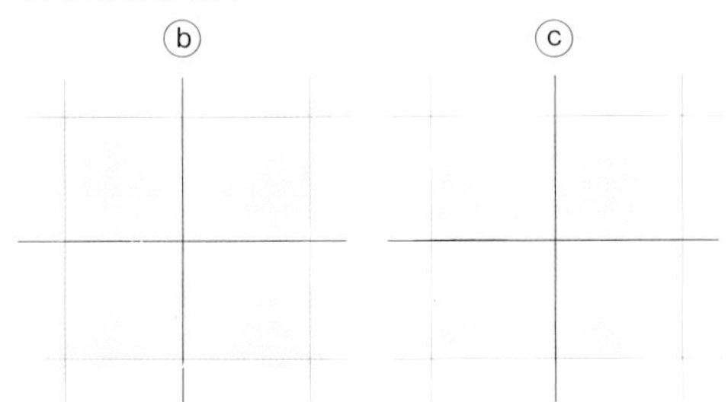

Finally, on large instruments, the primary mirror may be used alone, at the primary focus, where there is an observer's cage. This is a possibility that is employed, for instance, in the study of very distant objects; their brightness is greatest at this focus, since the image has been reflected only once.

The four previous combinations may be used interchangeably with any telescope.

Aplanatic telescopes. Richtey in the United States and Chrétien in France have investigated two-mirror combinations that are not only stigmatic on the optical axis but also correct the coma aberration that can be seen off axis with all parabolic objectives, whether or not they are combined with secondary mirrors. These kinds of objectives, which are now called Richtey-Chrétien objectives, have the advantage of producing good images over a field of about one degree. Their disadvantage is that they make it harder to change the secondary mirror or to use the primary mirror alone. What is more, the primary mirror is hyperbolic, and there is only one aplanatic combination that can be used with it (without spherical or coma aberration); all others produce aberrations, even on the axis, if no additional correcting optical element is used. For the past 15 years, large telescopes have been of the Richtey-Chrétien variety, with a set of correctors to improve the

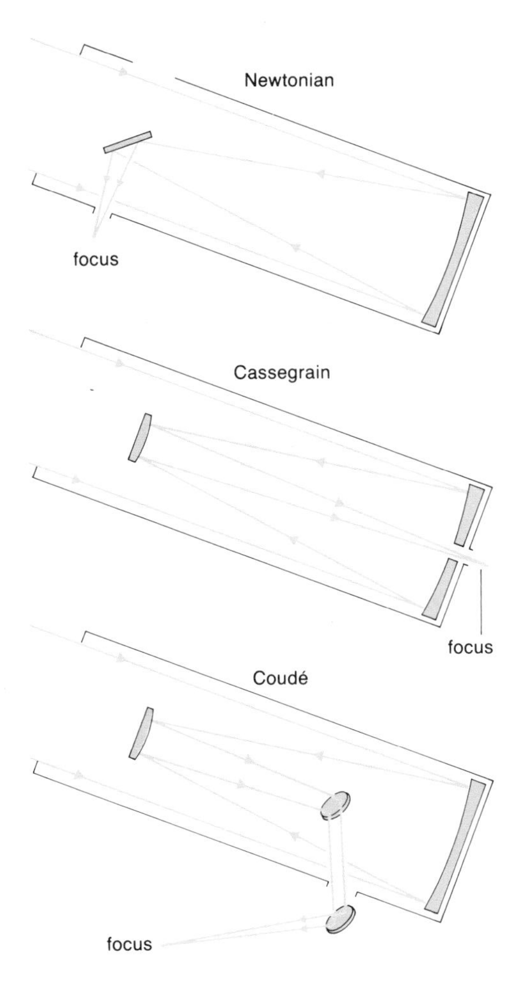

Telescope focus arrangements. *Top to bottom: Newtonian focus, Cassegrain focus, coudé focus.*

scopes, are constructed according to this principle.

In fact, the achromatism created by the use of two lenses is not perfect. Green, yellow and orange light rays—to which the eye is most sensitive—can be made to converge in a single focus, but in that case, the violet and red light rays are focused at a substantially different place. Therefore, the image of a star or planet supplied by an astronomical refractor appears surrounded by a purplish halo caused by an effect known as *secondary spectrum,* the importance of which increases with the diameter of the objective. This defect makes it impossible, for instance, to take good planetary photographs in ultraviolet light.

Unlike lenses, mirrors yield images devoid of chromatic aberration, since reflection is independent of wavelength. In this, reflectors have a decided advantage over refractors. However, a spherical mirror supplies images marked by spherical aberration. This defect can be avoided by giving a parabolic shape to the mirror. The resulting image then is perfect on the optical axis, but upon moving away from that axis, a new aberration manifests itself—coma, so called because it gives the image the appearance of a tiny comet. In aplanatic telescopes, this defect is eliminated by the use of a specially shaped secondary mirror.

Other aberrations emerge with large-field instruments: distortion, which causes the scale of the image to change with increasing distance from the field center; astigmatism, which prevents clear images of a horizontal and a vertical line from being seen simultaneously; and field curvature, in which the surface where the image has the best definition is not the focal plane but a portion of a sphere tangent to it.

It is possible to calculate the expected performance of an objective of a given shape. The task then is to execute it, giving it a shape as close as possible to the ideal. According to the rule set forth by the British physicist Rayleigh, the amplitude of the defects of the wave surface upon exiting from the objective should not exceed $\lambda/4$ (where λ is the wavelength of the captured radiation). Based on this rule, the surface of the mirror of a reflecting telescope should be executed with approximately four times more precision than that of a refractor's objective (with the advantage, however, that there is only one surface to polish). In practice, the required precision is better than 0.1 μm. It is achieved after lengthy, painstaking polishing with a disk tool, with which the mirror is rubbed after having been covered with fine-grained abrasives. The condition of the optical surface is checked by different methods, in particular the one devised by Léon Foucault, in which the surface observed along a knife edge in light is reflected at a small angle with respect to the optical axis. The defects then become readily visible because of the shadows they create.

stigmatism of the various possible combinations (which are the same as those mentioned earlier for instruments with parabolic objectives).

Spherical mirror telescopes. While the previous instruments are well suited to the detailed study of individual objects, their field of view is too narrow (only a few minutes for the traditional combinations with a parabolic primary mirror) for exploration of broad areas of space or for statistical data gathering.

In 1930, the Estonian optician Bernhard Schmidt developed a new kind of telescope that is now the ideal instrument for wide-field photography. The primary mirror is spherical, with sometimes a very small f/D focal ratio (somewhere around 3, in general, but it may be as low as 0.5 in some cases). The spherical aberration afflicting images produced by this mirror is corrected by a specially shaped thin lens, converging at the center and diverging at the edge and that is near the mirror's center of curvature. A curvature-of-field aberration remains, which is compensated for by using a convex support for the photosensitive layer. The field of view, which is generally on the order of 5 to 8 degrees, may reach 20 degrees.

A similarly conceived arrangement is the telescope invented around 1940 by Dmitri Maksutov. It also has a spherical primary mirror and a very short tube, but the correcting lens is a bispherical meniscus. Simpler to produce than the Schmidt telescope, it is less effective in correcting aberrations. Initially developed for photography, it may also be employed for visual observation, with an aperture ratio of f/7 or f/8.

MOUNTS

Whatever kind of instrument is used, the optical system must be held rigidly in place and must be kept trained on the point in the sky that is being studied; that is the purpose of the mount, which allows the instrument to be moved around two perpendicular axes.

The most frequently used arrangement is the equatorial mount, in which one of the axes (polar) is parallel to the axis of the poles, while the other (axis of declination) is parallel to the equator. This type of mount easily compensates for the objects' diurnal motion by rotating the instrument on its polar axis at the same rate as the Earth's rotation, but in the opposite direction (i.e., one rotation every 23 hours and 56 minutes in a retrograde direction). To train the instrument on the object to be studied requires merely moving it

The World's Largest Reflecting Telescopes

Observatory Name and Site Location	Altitude (m)	Diameter of the Primary Mirror (m)	Ownership	Year Placed in Service	Telescope Name
Zelenchukskaya; Mount Pastukhov, Caucasus, USSR (special astrophysical observatory of the USSR Academy of Sciences)	2,070	6.00	USSR Academy of Sciences	1976	Bol'shoi Teleskop Azimutal'nyi (BTA)
Mount Palomar; California, U.S.	1,706	5.08	U.S.	1948	Hale
Mount Hopkins; Arizona, U.S. (Fred Lawrence Whipple Observatory)	2,600	4.50 (6 × 1.8)	Smithsonian Institution	1979	Multiple Mirror Telescope (MMT)
La Palma; Canaries, Spain (obs. Roque de los Muchachos)	2,300	4.20	U.K.	under constr.	William Herschell
Cerro Tololo; Chile (Cerro Tololo Interamerican Observatory, *CTIO*)	2,400	4.00	U.S.	1976	
Siding Spring; New South Wales, Australia (Anglo-Australian Observatory)	1,164	3.89	U.K. Australia	1975	Anglo-Australian Telescope
Kitt Peak; Arizona U.S. (Kitt Peak National Observatory, *KPNO*)	2,064	3.81	U.S.	1973	Mayall
Mauna Kea; Hawaii, U.S.	4,194	3.80	U.K.	1979	UK Infrared Telescope (UKIRT)
Mauna Kea; Hawaii, U.S.	4,200	3.60	Canada-France	1979	C.F.H. (Canada-France-Hawaii)
La Silla; Chile	2,400	3.57	ESO*	1976	
Calar Alto; Sierra Nevada, Spain	2,160	3.50	W. Germany	under constr.	
Mount Hamilton; California, U.S. (Lick Observatory)	1,277	3.05	U.S.	1959	Shane
Mauna Kea; Hawaii, U.S.	4,208	3.00	U.S. (NASA)	1979	IRTF (Infra Red Telescope Facility)
Mount Locke; Texas, U.S. (McDonald Observatory)	2,070	2.72	University of Texas (U.S.)		
Crimea, USSR (Crimea Observatory)		2.60	USSR Academy of Sciences	1961	Shajn
Mount Aragatz; Armenia, USSR (Byurakan Observatory)	1,500	2.60	USSR Academy of Sciences	1971	
La Palma; Canaries, Spain (obs. Roque de los Muchachos)	2,300	2.59	U.K.	**	Isaac Newton
Las Campanas; Chile (Carnegie Institution Southern Observatory, *CARSO*)	2,300	2.57	Carnegie Institution	1976	Du Pont
Mount Wilson; California, U.S.	1,750	2.54	U.S.	1918	Hooker
Kitt Peak; Arizona, U.S. (Kitt Peak National Observatory, *KPNO*)	2,100	2.29	U.S.	1964	
Mauna Kea; Hawaii, U.S.	4,200	2.20	University of Hawaii (U.S.)		
Calar Alto; Sierra Nevada, Spain	2,160	2.20	W. Germany	1979	
Kitt Peak; Arizona, U.S. (Kitt Peak National Observatory, *KPNO*)	2,100	2.15	U.S.	1964	
San Juan; Argentina		2.15	Argentina		
Baja California, Mexico (obs. San Pedro Mártir)	2,800	2.12	Mexico	1979	
Mount Locke; Texas, U.S. (McDonald Observatory)	2.070	2.08	University of Texas (U.S.)	1939	Struve
Pirkuli Mts., Azerbaidjan, USSR (Chermakha Observatory)	1,450	2.00	USSR		
Ondrejov; Czechoslovakia	550	2.00	Czechoslovakia		
Pic du Midi; Pyrenees, France	2,850	2.00	France	1979	
St-Michel-de-Provence, France (Haute-Provence Observatory)	650	1.93	France (C.N.R.S.)	1958	

*European Southern Observatory
**Placed in service in 1967 at Herstmonceux, Sussex, Great Britain; then transferred to the Canary Islands.

Different telescope mounts:
a. German
b. English
c. forked
d. cradle
e. horseshoe

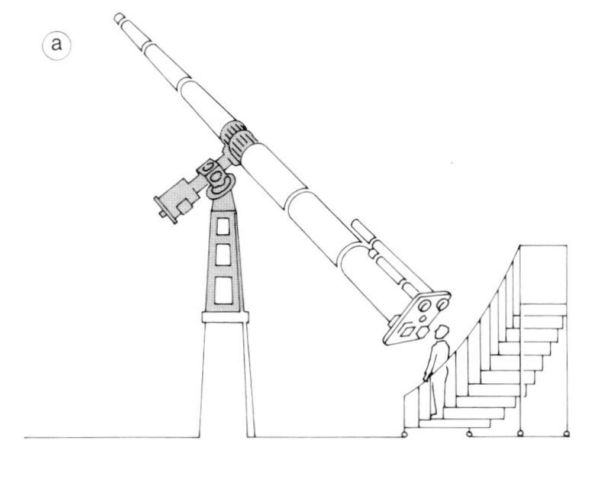

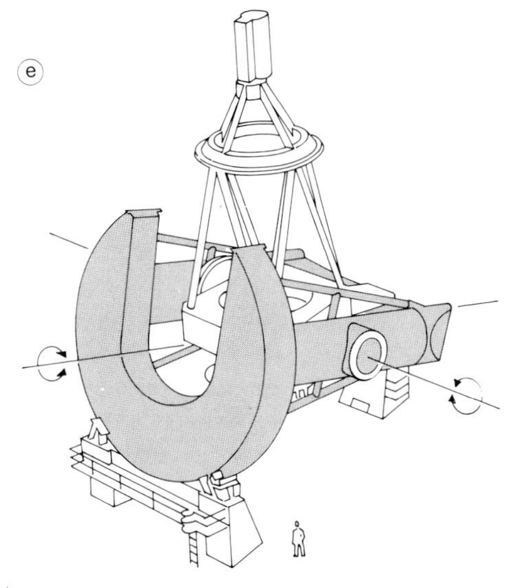

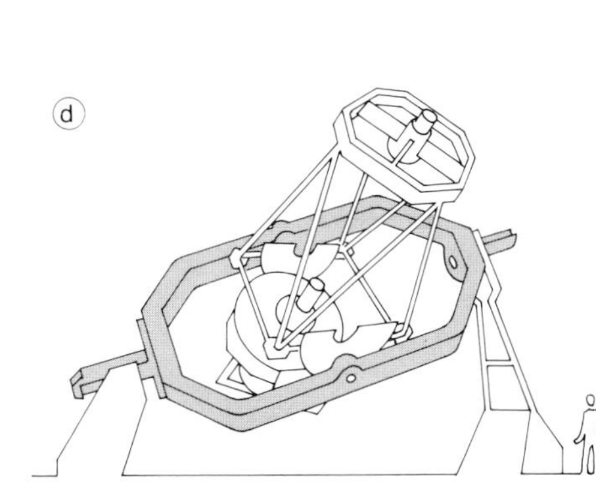

around its axis of declination. Once the pointing is accomplished, the motion around the polar axis tracks the object automatically.

There are two families of equatorial mounts: German mounts, where the axis of declination is exterior to the interval between the bearings supporting the polar axis; and English mounts, where the axis of declination is fixed on the polar axis and is inside of the polar axis bearings.

These two families have variants. The fork mounting is a symmetrical German mounting in which the polar axis has a forked tip; the instrument moves between its prongs in declination. The cradle mount is a symmetrical English mount in which the axis of declination is supported by a rectangular frame, which is also movable around the horary axis; the horseshoe mount is a variant of the cradle mount, in which the cradle's north crossbeam is replaced by a large, horseshoe-shaped notched disk, making it possible to sight the pole. The choice of mount depends on the overall mass of the instrument and on the type of observations it is primarily designed for. In general, the axis of declination should go through the instrument's center of gravity, which for refractors is almost in the center of the tube, and for reflectors is just in front of the mirror. When the polar axis is not in the same plane as the optical axis, the instrument must be balanced by a counterweight, which substantially increases the mass of the mobile fitting. The best mount from a technical standpoint is the English cradle mount; however, it blocks out an area of several degrees around the pole, which led to the development, for large instruments, of the horseshoe mount. It is used, for instance, in the 5-meter (200-inch) Mount Palomar telescope and the 3.6-meter (142-inch) Mauna Kea telescope (Canada-France-Hawaii) (see box).

In addition to the equatorial mount, the alt-azimuth mount is also used. It allows the instru-

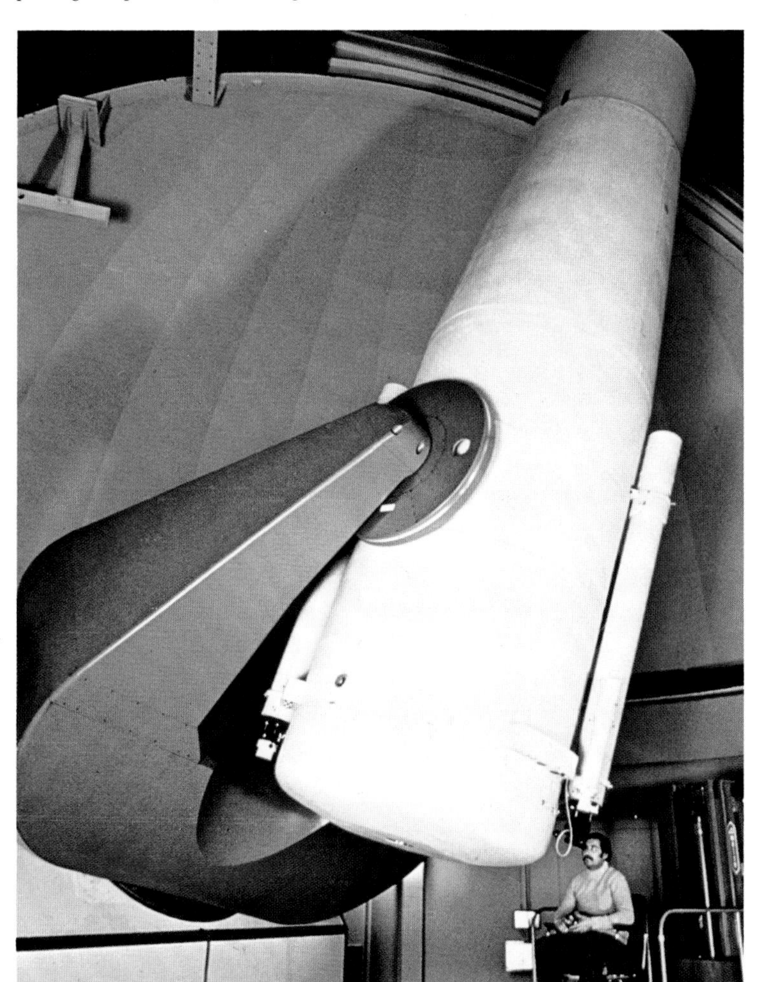

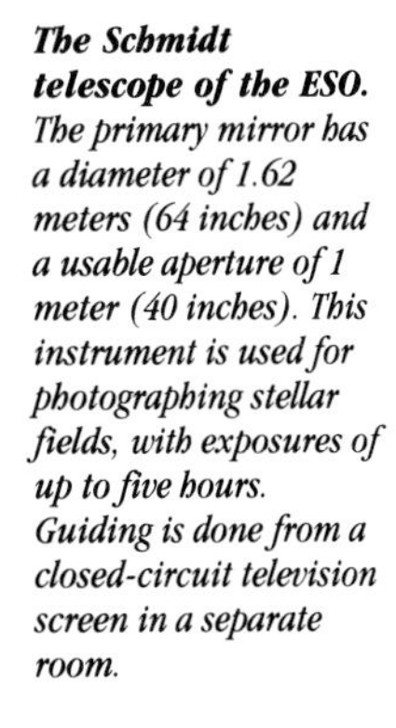

The Schmidt telescope of the ESO. *The primary mirror has a diameter of 1.62 meters (64 inches) and a usable aperture of 1 meter (40 inches). This instrument is used for photographing stellar fields, with exposures of up to five hours. Guiding is done from a closed-circuit television screen in a separate room.*

ment to be moved around a horizontal and a vertical axis. This simple arrangement is used for the theodolite and for small amateur telescopes. It offers solutions to the problem of the bending of mechanical structures, helps make instruments more compact and has been shown to be much less expensive than the equatorial mount for large instruments. However, it has a major drawback in that it requires both axes to be moving in order to track an object, and at different speeds, depending on the direction in which it is pointed. This disadvantage, which made it unsuitable for big instruments for a long time, can now be overcome through the use of computers to coordinate the vertical and horizontal movements. At least two big telescopes that have been placed in operation in recent years are outfitted with this type of mount: the 6-meter (236-inch) Soviet telescope, and the multimirror telescope at the Mount Hopkins Observatory in the United States. Most radio telescopes also have an alt-azimuth mount.

Analyzers: Filters and Spectrographs

During an astronomical observation, the focal plane of the collecting instrument yields an image of the light source. Before being recorded by a detecting device, such as a photographic plate, this image often undergoes a spectral or polarization analysis. Spectral analysis is done with the aid of filters or spectrographs.

A filter is a substance that absorbs a portion of the light rays entering it, making it possible to isolate a particular region of the spectrum, in order, for example, to do photometry in one or more specific colors.

A simple colored screen of glass or gelatin, containing mineral or organic dyes that absorb the unwanted light rays, can isolate a range of wavelengths of about 1,000 angstroms (Å), or 1/10,000 millimeter in width. In order to isolate much narrower spectral ranges, of a few dozen angstroms or thereabouts, interference filters are used. These consist essentially of two plates with parallel surfaces, whose facing surfaces are covered with a semireflecting film and are separated by a transparent layer of rigorously controlled thickness. A light ray entering this device travels back and forth many times between the plates; each time, a portion of the light is reflected, and the rest transmitted. The light rays coming from a given source travel different distances depending on the number of times they have been reflected back and forth in the transparent layer, and consequently the transmitted beams interfere with each other. If the source emits a steady stream of rays, only those rays whose wavelength is a multiple of twice the distance between the blades will be transmitted, other wavelengths being subject to destructive interference. The width of the isolated wavelength interval (pass band) depends on the reflectivity of the film: the higher it is, the more selective the filter is, but the smaller the light yield of the filter. The yield is improved by laying down several thin layers in succession, alternately transparent or semireflecting.

Another type of filter consists of a group of birefringent crystalline plates of increasing thickness, sectioned parallel to the optical axis of the crystals and separated by polarizers. This polarizing filter, invented by the French astrophysicist Bernard Lyot in 1927, may be placed directly into

The 3.6-meter (142-inch) Canada-France-Hawaii telescope. CNRS Photo.

The Canada-France-Hawaii Telescope

One of the best telescopes in the world devoted to research in astronomy is the Canada-France-Hawaii (CFH) telescope, built by France and Canada near the top of Mauna Kea, an extinct volcano in Hawaii, at 4,200 meters (13,780 feet) in altitude.

Inaugurated on September 28, 1979, this telescope—one of the biggest in the world by the diameter of its primary mirror—has no truly revolutionary features but has all the improvements allowed by current state-of-the-art techniques.

Built by the New Design and Construction Company of La Rochelle-La Pallice, the horseshoe mount, of the "Palomar" type, and the framework constituting the tube, which is capable of parallel displacements along the optical axis (according to the principle invented by Max Serrurier), are built on the model almost universally adopted for large modern telescopes. One original feature of the telescope, however, is the fact that it is driven in an hour angle by a gear wheel 10 meters (33 feet) in diameter joined to the horseshoe. The front part of the telescope tube, which bears the secondary optical (head ring), may be removed and replaced by a different one, enabling the use of different optical configurations (see following material).

The primary parabolic mirror has a diameter of 3.6 meters (14.2 inches) and weighs 14 metric tons. It is made of Cer-Vit, a ceramic glass with a very low thermal expansion coefficient, making it virtually immune to distortion from changes in temperature. Created in a single casting, it was then polished to within 0.0001 millimeter; this process, carried out at the Federal Astrophysical Observatory in Victoria, Canada, lasted three years and required about 31,000 hours of labor. The primary focus, located 13.5 meters (44 feet) above the primary mirror, gives a focal ratio of f/3.8 and a useful field of view of 3 arc minutes, which may be extended to 1 degree with a corrector (the focal ratio then becomes f/4.2). It is essentially meant for direct photography, or—by wide-field electronography—for photometry and low-resolution spectrographic observations of dim objects. The Cassegrain focus, which has a focal ratio of f/8, has a field of view of 14 arc minutes, which can be extended to 25 arc minutes with a corrector. It is used for high-precision photometry and polarimetry, nebular and stellar spectrometry, and Fabry-Pérot interferometry. A special Cassegrain focus is provided for infrared observations, with a focal ratio of f/35; the secondary mirror is designed to wobble, making it possible to switch rapidly from the object under study to a nearby point in the sky and thus to subtract the sky background, which is not as "dark" in the infrared as in the visible range; this focus is used for photometry and for Fourier transform spectrometry.

Finally, beneath the telescope base are two Coudé foci with focal ratios of f/100 but capable of reduction to f/20 by insertion of a special optical element. The Coudé foci are reserved for high-dispersion spectrometry.

The telescope is guided by a minicomputer that insures that the instrument is accurately pointed, making all necessary corrections for atmospheric refraction, mechanical bending, etc.

An Exceptional Location

Selected after an exhaustive search, the location of the telescope appears to be exceptionally good for astronomical observation. Because of its altitude, it generally emerges from a cloud layer located at about 2,000 meters (6,562 feet), thus enjoying a twofold advantage: While the telescope itself is in a zone of pure atmosphere, the cloud cover below forms a natural protection from the lights and smoke of the inhabited areas. The island of Hawaii, although the largest in the archipelago, is sparsely populated and does not seem threatened in the near future by large-scale urbanization or industrialization.

The site is equally remarkable for the evenness of the air temperature; the daily thermal amplitude is somewhere around 1° or 2°C only, which helps to ensure the stability of images; the annual amplitude also is very low.

Moreover, it is one of the parts of the globe where the atmospheric water vapor content is lowest, which makes it particularly well adapted to infrared observations. These qualities explain why the top of Mauna Kea is now one of the high points of world astronomy. As a matter of fact, in addition to the 3.6-meter CFH telescope, a 2.2-meter (87-inch) telescope has been built there by the University of Hawaii, as well as two infrared telescopes—one of 3.8-meters (150 inches) in diameter by Great Britain, and one of 3 meters (118 inches) by the United States (NASA). The altitude of the site does, of course, cause some problems of adaptation for atronomers. At 4,200 meters, partial oxygen pressure is only 60 percent of its value at sea level, which can cause "altitude sickness" and various physiological disorders. In order to help the scientists working there to acclimate faster, a rest station was installed at 2,800 meters (9,187 feet).

The CFH telescope with its dome and auxilliary equipment cost $30 million. It was financed equally by France and Canada, with the University of Hawaii supplying the site, the access route and various facilities. Operating costs and utilization times are shared by Canada, France and the University of Hawaii at the rate of 42.5 percent, 42.5 percent and 15 percent, respectively.

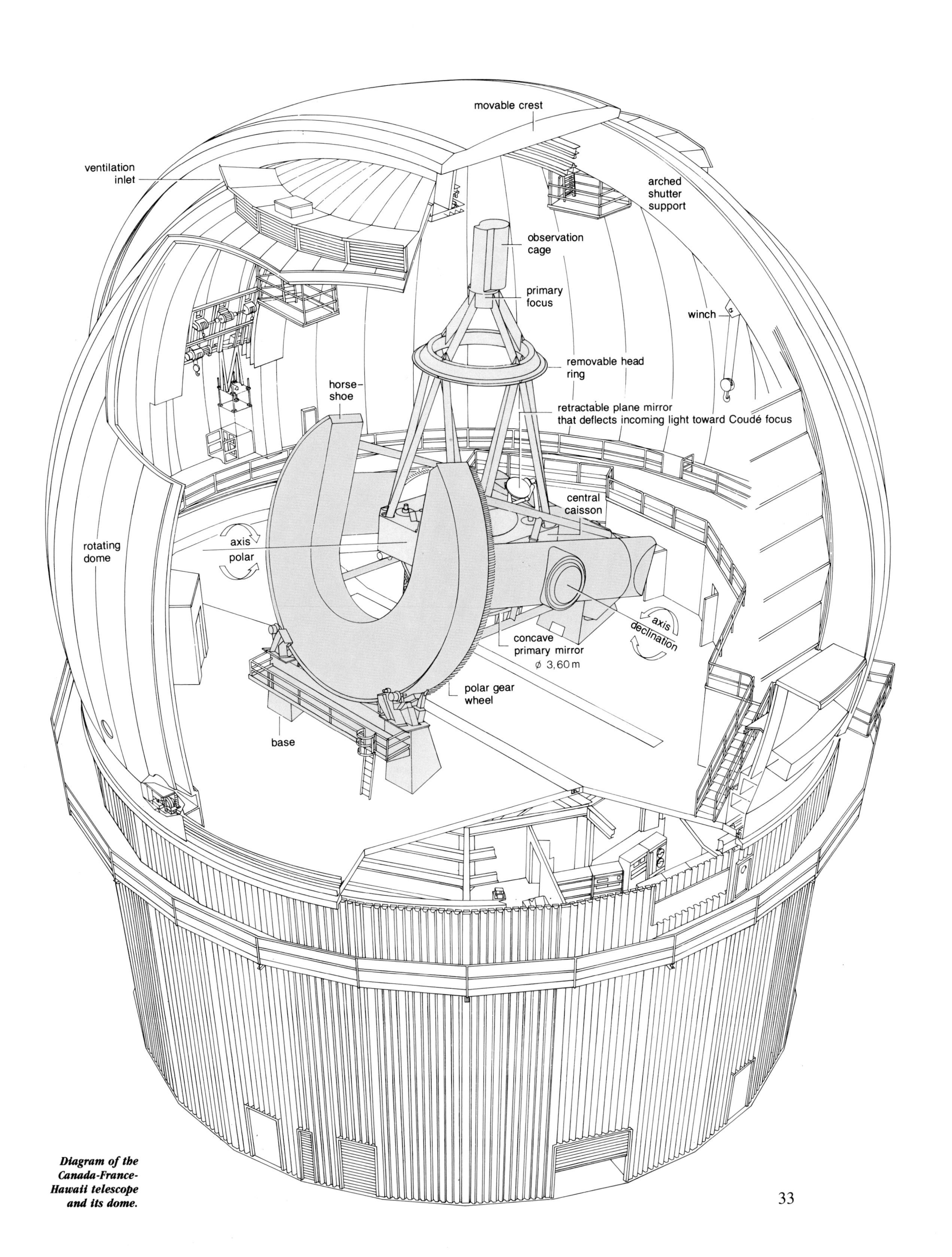

Diagram of the Canada-France-Hawaii telescope and its dome.

***The world's largest telescope:** built by the USSR Academy of Sciences at 2,070 meters (6,792 feet) in altitude on Mount Pastukhov in the Caucasus near Zelenchukskaya, this instrument has a primary mirror with a usable diameter of 6 meters, weighing 42 tons, and an alt-azimuth mount, the moving parts of which weigh nearly 650 tons, guided by computer. The primary focus, equipped with an observation cage, is 24 meters (79 feet) from the primary mirror. The dome housing the instrument measures 50 meters (164 feet) in diameter. Photo by G. Courtès.*

a highly collimated beam of light, thus providing a monochromatic image (see following material) of the object under study. Such an arrangement can isolate spectral bands of 1 angstrom or less.

To get a continuous spectrum, a spectrograph is used. This device for dispersing light consists essentially of a collimator, which produces a parallel cluster of rays, followed by a dispersive mechanism and a photographic chamber where the image is recorded. The simplest arrangement of this kind is the prismatic spectrograph. A glass prism is known to bend light rays (Newton's ex-

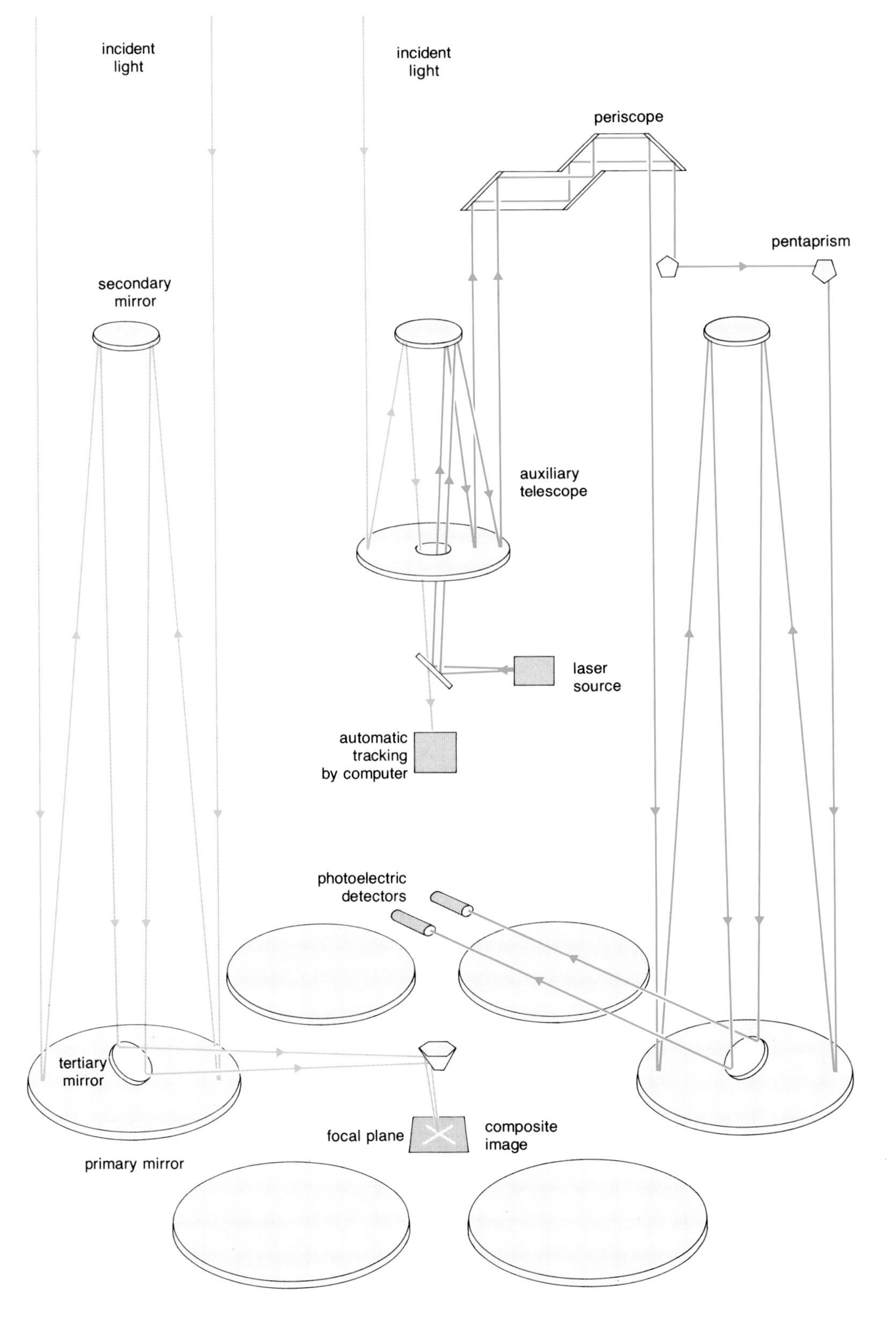

***The multimirror telescope placed in service in 1979 at the Mount Hopkins Observatory near Tucson, Arizona,** is innovative in design. Extremely compact and placed on an altazimuth mounting, it has six Cassegrain mirrors 1.8 meters (71 inches) in diameter at the points of a hexagon, whose images are concentrated on the central axis and superimposed to produce a single image equivalent to one that would be supplied by a single 4.5-meter (177-inch) mirror. If it fulfills the hopes vested in it, this type of telescope will represent a valuable alternative, making it possible to have powerful telescopes without the difficulties inherent in the manufacture of large mirrors. Diagram of operation: To ensure the alignment and convergence of the images supplied by the six elementary telescopes (of which only two are shown for greater clarity), a laser is used. Its beam is aimed at the focus of an auxiliary telescope with a diameter of 76 centimeters and directed simultaneously toward two opposite points on the periphery of each of the primary mirrors by periscopes and retroreflectors. After being reflected by the secondary and tertiary mirrors, the two beams are received by photoelectric detectors. The least fault in alignment or focusing of a telescope thus can be detected and corrected (by adjusting the position of the secondary mirrors). The auxiliary telescope, in addition, is equipped with an automatic tracking system that can follow any star of 14th magnitude or brighter within a field of 1 degree centered on the object to be observed.*

periment, 1666). The deflection angle depends on the wavelength of the rays—that is, on the color of the light; it increases from red to violet. Thus, a beam of white light is separated by the prism into a band—a spectrum—containing the colors of the rainbow, each of which occupies a certain spectral interval. The transition from one color to the next is continuous, for there in fact exists an infinite number of primary colors, or, as the experts say, of monochromatic rays. Each monochromatic ray is characterized by a unique wavelength. In practice, however, any ray with a very narrow spectral band is termed monochromatic.

Prismatic spectrographs have several disadvantages. First, they yield spectra whose dispersion varies according to wavelength; prisms spread out radiation at long wavelengths to a greater extent than at short ones. Second, glass absorbs wavelengths shorter than 0.36 μm, which has led to the use of quartz or fused silica prisms. Above all, however, the total absorption inside the prism becomes significant when its dimensions are large. Therefore, it is preferable to use diffraction gratings, which yield the same dispersion but which produce spectrographs of modest size.

The gratings used in astronomy are of the reflection type. They consist of a glass support (mirror) on which are etched very fine parallel grooves, equidistant from one another (several hundred per millimeter). A beam of white light striking such a grating is diffracted in a series of spectra. The greater the angular distance between one of these spectra and the nondispersed image reflected directly by the mirror, the more broadly it is spread. In this way, "first order," "second order" spectra and so on can be obtained. The higher the order, more dispersed spectra overlap. Therefore, the first- or second-order spectra are usually chosen for measurement.

Until recently gratings suffered from a low efficiency. But today we know how to make reflection gratings very bright within a given order, by etching the grooves in the shape of facets, which are slanted so that they preferentially reflect light in the direction of a given diffraction spectrum. Gratings are now the most commonly used filters in astrophysics. The largest of them contain several interchangeable chambers and weigh as much as 1 metric ton.

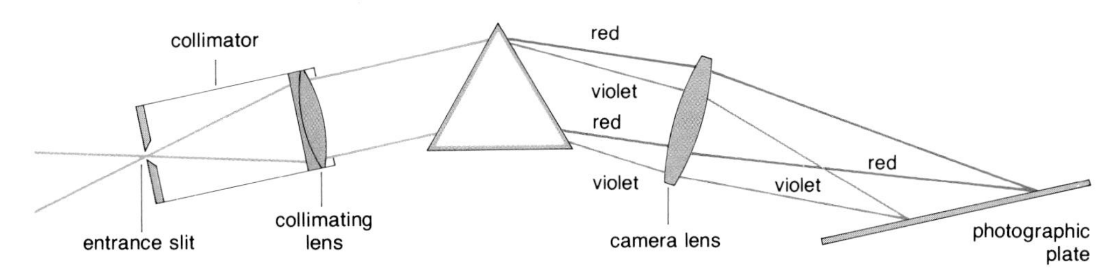

Diagram of a prismatic spectrograph.

If a low-angled prism (from 1 to 5 degrees) is placed in front of a refractor lens (with a small diameter), it forms an objective prism, which produces the spectra of many stars simultaneously within its focal plane. The main drawback of such a system is that it makes it necessary to point the telescope in a direction that is rather distant from the field under study, owing to the deflection of the light rays by the prism. Moreover, the dispersion thus obtained depends on what point in the field is being examined. With Fehrenbach's normal field objective prism—which combines three prisms made of two different types of glass, which together disperse but do not deflect—these two problems are removed. Furthermore, a 180-degree rotation mechanism makes it possible to obtain, for each star in the field, two symmetrical spectra for a given wavelength, and to measure directly the radial velocities of the stars in the entire field by measuring the separations between the images of the same spectral lines.

Detectors

Among the essential auxiliary instruments of a telescope are detectors, which record the image, in some cases after enlarging it, correcting it and putting it into a form that can be measured by an instrument.

THE EYE

For a long time, the eye was the sole detector. It was the chief auxiliary of astronomers up to the beginning of the 20th century and played a decisive role in the growth of positional astronomy. Adapted to the optical window, its sensitivity

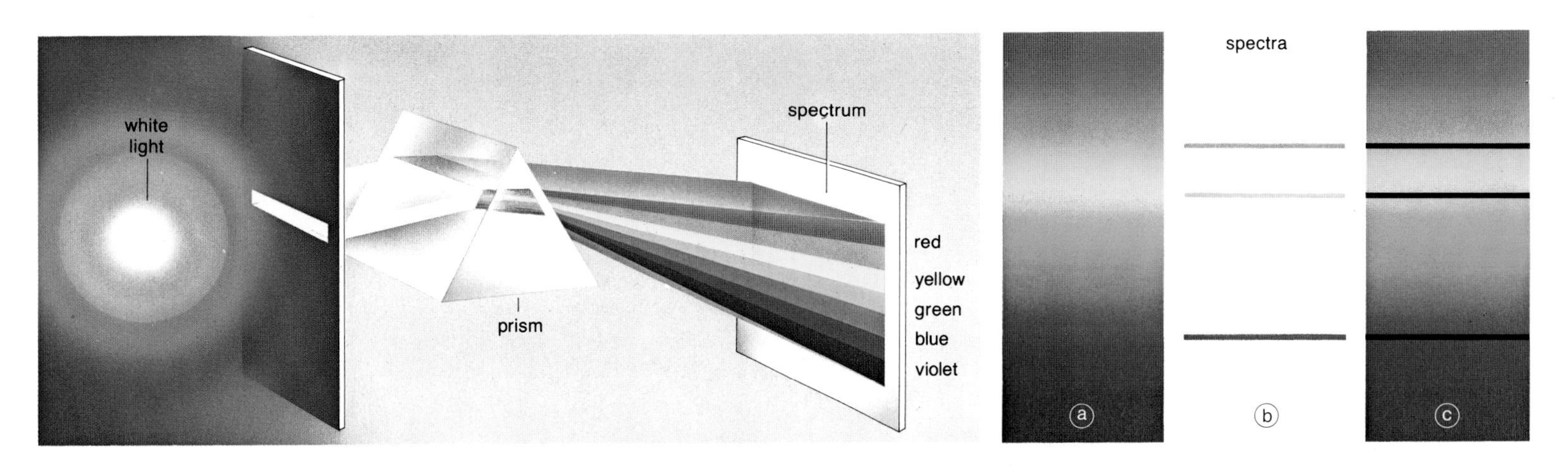

Principle of spectral analysis: *(a) White light is broken down by the prism into the colors of the rainbow; this is a continuous spectrum. (b) A light gas made incandescent produces bright lines with well-defined wavelengths; this is an emission spectrum. The arrangement of the lines is characteristic of the substance. (c) If the gas is placed in front of a bright source of white light, it absorbs some of the radiation emitted by the source. The result is an absorption spectrum in which the bright lines are replaced by dark lines located in the exact same places in the spectrum as the previous ones. A star is a ball of incandescent gases producing a continuous spectrum. The colder, denser atmosphere surrounding it usually produces absorption lines characteristic of the chemical elements that compose it. Some very diluted envelopes, surrounding stars whose surface temperatures are particularly high, may nonetheless be subjected to intense excitation and emit their own radiation; they then produce an emission spectrum. The identification of spectral lines, the precise determination of their positions in the spectrum, the study of their profiles and the photometric analysis of the spectrum provide a great deal of information about the physical conditions, chemical composition and motion of the observed object.*

ranges from violet to red, with its maximum in the yellow. It allows good perception of colors and contrasts. It is very sensitive (all that is needed for the eye to react are a few photons, provided they reach it simultaneously or nearly so), and its essential quality is its ability to judge whether two bright areas are equal, provided their images form on the same portion of the retina. However, the eye does not have the power to integrate light signals for a long period, nor is the brain able to retain images in an objective, quantifiable form. Therefore, the eye is no longer the perfect detector for astrophysicists, who are interested in quantitative measurement of low-luminosity objects. Nonetheless, the eye is still the final link in the chain of observations, for all images must eventually be shown in a form that can be read by the eye. Its role in our understanding of the universe thus remains primary.

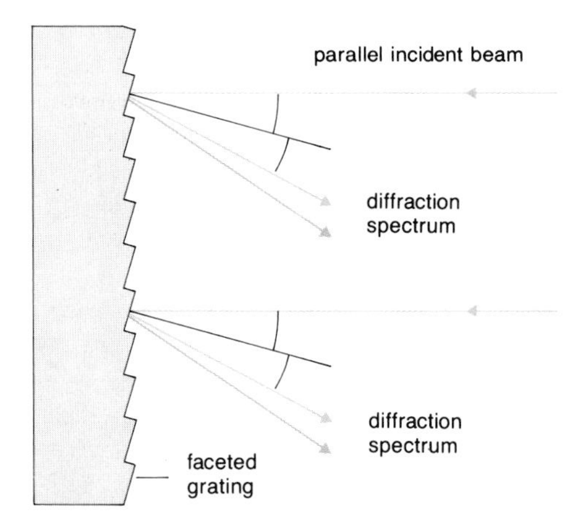

Diagram of a faceted diffraction grating: *Engraved on the surface of the grating are facet-shaped grooves whose slope is such that they reflect light in the same direction as the chosen diffraction spectrum is formed, thus greatly increasing its photon efficiency.*

THE PHOTOGRAPHIC PLATE

The invention of photography sparked a real revolution in astronomy. Although the photographic plate, or emulsion, is far less sensitive than the eye, the photographic plate has a substantial advantage over it in that it can accumulate photons received at separate instants. Objects invisible to the naked eye owing to their low luminosity can be photographed through the use of exposure times that may last overnight. The photographic plate has other essential qualities: It is easy to use, inexpensive, and above all, has a large information-storage capacity. Without it, an astronomer would have to rely on his visual memory or drawing talent. It is therefore unrivaled for systematic observation programs. However, photography has several defects that limit its range of uses. Its sensitivity is very slight; out of a thousand photons it receives, only a few, at most, go into the image it supplies. Furthermore, below a certain flux level, nothing is recorded; thus, for a

Diagram of a grating spectrograph.

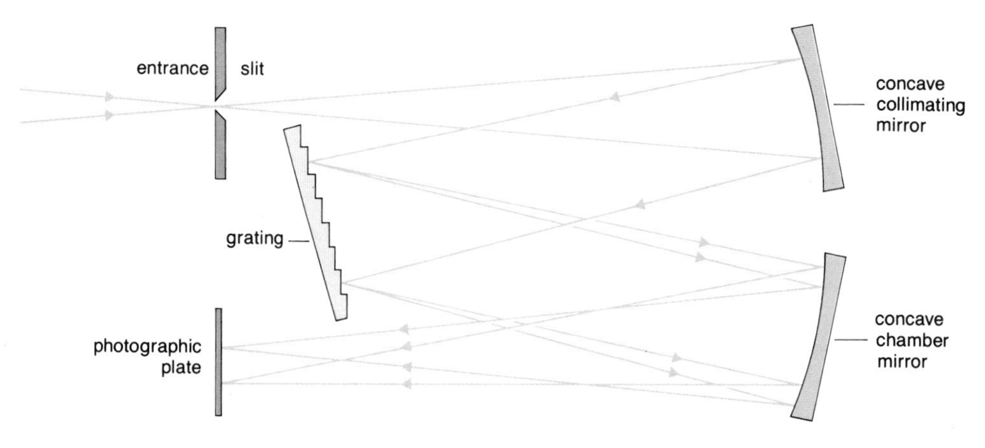

The large objective prism telescope of the Observatory of Haute-Provence. *One of the tubes contains a visual lens with a diameter of 26 centimeters, the other a photographic lens with a normal-field, direct-vision prism in front of it. This instrument can simultaneously obtain spectra of all the stars in a field and determine their radial velocities. Photo: Observatory of Haute-Provence—CNRS.*

grain of silver salt to be reduced and appear in development, it must receive several photons. On the other hand, if the flux is too high, the extra photons are not recorded and can even reduce the sharpness of the image. Owing to these threshold and saturation effects, the photographic plate is emphatically nonlinear; the recorded signal (the photographic image) is not directly proportional to the received signal (the telescope image). Photographic recordings are often converted to electrical impulses for further processing. The instruments that perform this operation are called microdensitometers; they scan the plate with a tiny light pencil.

PHOTOELECTRIC DETECTORS

Light detectors superior to photographic plates because of greater sensitivity have been gradually invented and developed since 1930. The basic principle of such modern receptors is the photoelectric effect, discovered by the German physicist Hertz in 1887 and theoretically explained by Einstein in 1905. Light can ionize the atoms of certain substances—that is, strip electrons from them. Such substances are called photoemissive. Hence, an illuminated metallic surface (a photocathode) emits electrons, which will be captured by an anode if a suitable potential is established between it and the cathode. By measuring the electric current produced in this way, the flux of light incident upon the cathode can be measured; this is the principle on which the photoelectric cell is based. The photoelectric effect, unlike the photographic plate, is linear, for the number of electrons emitted is strictly proportional to the number of received photons. Moreover, it is capable of a high yield, on the order of 20 percent, for visible light and for the better photocathodes.

This principle is applied in astronomy in photoelectric photometers that measure the light flux that has been isolated by a diaphragm from objects such as stars, extended sources, or points within a source. It is also employed in photomultipliers, which have a set of secondary electrodes, or dynodes, between the cathode and the anode, each of which is coated with a substance that can emit several electrons when it is struck by a single one. The dynode voltages are made to increase from one to the next in order to conduct the flux of electrons as far as the anode. As many as 20 stages are sometimes used, so that several million electrons are collected at the anode for every electron emitted by the cathode. Upon issuing from the anode, the current is proportional to the light flux received by the cathode. Moreover, there is no longer a threshold, and the quantum yield is close to the theoretical maximum—up to 20 percent of the incident photons—which represents a yield 10,000 times greater than that of the photographic plate. Thus the photomultiplier is well suited to photometry, particularly for point sources. Its disadvantage is that it merely records the flux entering a diaphragm, while the photographic plate can record images—that is, light distributions.

This disadvantage disappears with the electronographic camera, developed in 1936 by the French scientist André Lallemand. In this instrument, the electrons released by the photocathode are accelerated in an electric field (enabling them to acquire far greater energy, so that the available energy greatly exceeds that of the incident light) and guided by electric or magnetic fields (electromagnetic lenses), so that all the electrons emitted from a given point of the photocathode converge at the same point on the anode. In this way, an electronic image identical to the original photonic image can be reconstituted. This electronic image may then be directly recorded on a special emulsion that is sensitive to electrons; this is called electronography. This technique has many advantages in comparison with ordinary photography. Exposure times can be reduced by a factor of 50 to 100; spatial resolution is excellent; the density of the negative is directly proportional to the illumination (however small); and finally, the quantum yield is excellent (20 to 30 percent, depending on the nature of the photo cathode). However, the electronographic camera is complicated and its successful use requires a thorough knowledge of high vacuum techniques, cryogenics and photographic photometry. Therefore, astronomers have investigated techniques of detection that are easier to work with.

This led them to the use of image intensifiers (also called image tubes) with the photographic plate. The functioning principle of these instruments is very similar to that of the electronographic camera. The basic difference is in the receiving layer, a phosphor screen that can transfer an image from emitted photoelectrons onto a traditional photographic emulsion. The brightness of the image is increased by stacking units so that the fluorescent screen of one unit supplies the image for the photocathode of the next. The primary photoelectron emitted by the input photocathode is thereby focused onto the first fluorescent screen-photocathode sandwich, which reemits 100 photoelectrons at the point of impact. With four successive stages, a multiple of about 10^8 is achieved. The initial photon makes a very bright spot on the final phosphor screen; the spot is readily recorded by a photographic plate or the eye.

However, this detection technique also suffers from the defects of photographic plates, so the next idea that astronomers developed was to replace the photographic plate with a television tube. This technique employs an array of photoemissive elements, each of which is connected to a microcondenser. By sweeping a narrow beam of electrons over the array and probing the elements one by one, an electric current is created; its magnitude depends on the quantity of charge built up, and thus upon the degree of illumination. The resulting signal (a video signal) may, when amplified, be used to regulate the electron firing mechanism of a cathode tube; a reproduction of the original image will then appear on its screen. It may also be tape-recorded for later restoration. The advantage of this technique, which generates an electronic output signal, is that the information contained in the image can be fed directly into a computer.

Another powerful detector was developed in 1970 at the Lick Observatory by Wampler and Robinson. The image of a spectrum produced by the telescope is first amplified by a series of three image tubes, the photocathode of the first replacing the photographic plate. The greatly enhanced image is then projected onto the photocathode of another device called an *image dissector*. This device is similar to a photomultiplier, except that at any given time it amplifies electrons arising from a single spot on the photocathode. By varying a magnetic field, the image dissector scans along the spectrum. The time variation of the output from the amplifier thus corresponds to the variation in intensity at different wavelengths along the spectrum. By rescanning repeatedly, one eventu-

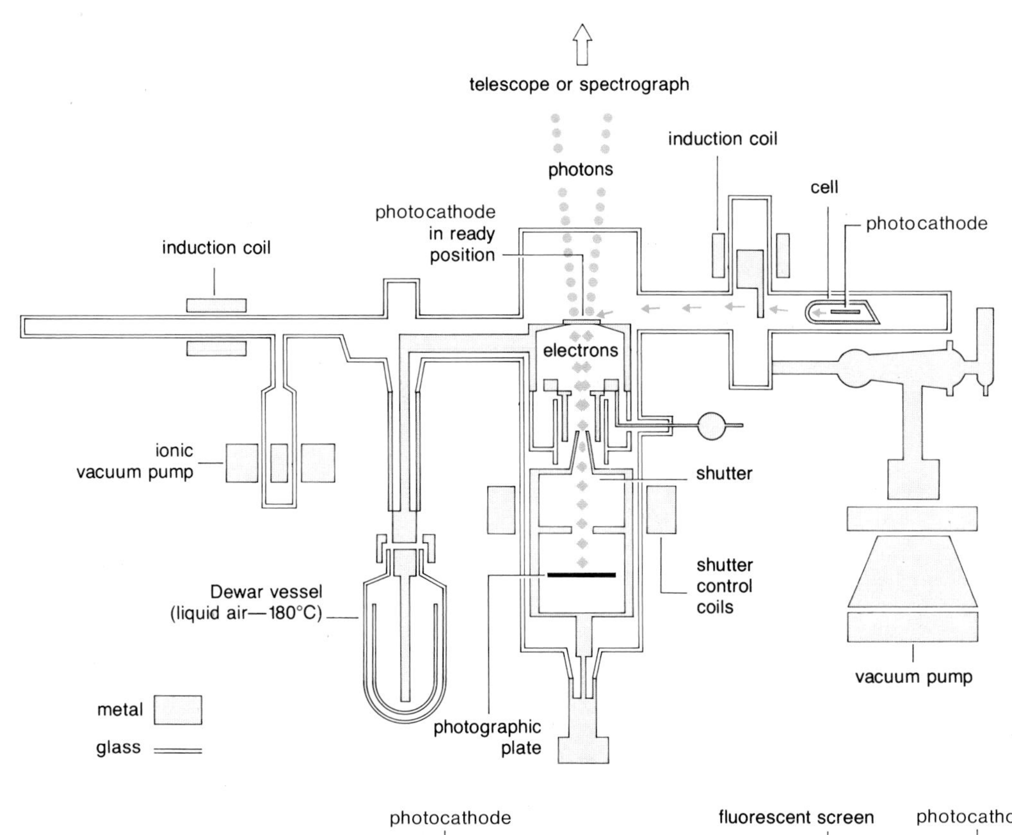

tube, which is both fragile and sensitive to the disruptive influence of external fields. The development of integrated circuits led to the creation of a new type of detector that could easily be miniaturized—the charge coupled device, or CCD. Developed in 1969 by Boyle and Smith at Bell Telephone Laboratories, the CCD is based on the principle of the photoconduction effect (the increase in the electric conductivity of semiconducting substances when they are exposed to light). It consists of a black plastic rectangle the size of a finger, equipped with metal feet on the sides, recalling in appearance a traditional integrated circuit. In the center, a small transparent window reveals a silicon chip; its surface (several square millimeters) is divided into thousands of independent microscopic rectangles, or pixels, which constitute the photo array. When the array is exposed to light, the photons falling on it excite the electrons in the silicon, and every pixel in the chip takes on a specific electric potential. The charge on every pixel is then individually measured by a suitable method, to reconstruct the recorded image. The performance of this type of

Lallemand's electronographic camera: *Accelerated and focused, the electrons emitted by the photocathode reconstitute on the photosensitive plate an electronic image similar to the luminous image formed on the photocathode.*

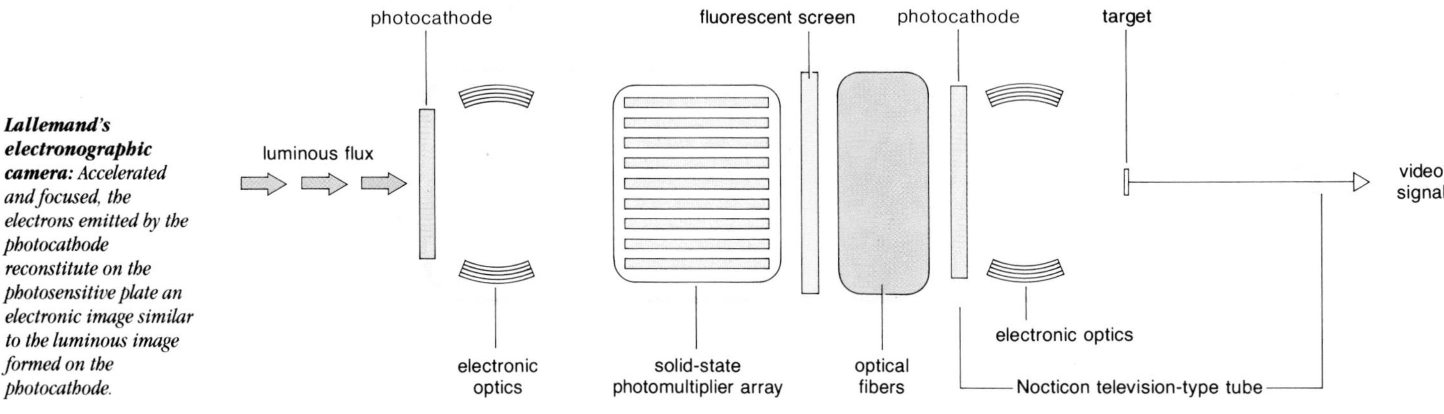

Diagram of a photon-counting camera: *A photon strikes the photocathode, releasing an electron. The electron is guided by the electronic optics as far as the array of solid-state photomultipliers. This results in emission of a large number of electrons, which strike the fluorescent screen, forming a bright dot on it that is transmitted by optical fibers to the television camera placed in back. The resulting signal lasts until a counter in the computer corresponding to the initial impact point of the photon is incremented by one unit. The number of photons entering during the exposure at each point on the photocathode is recorded in the memory, producing a numerical representation of the image. The image may then be processed in various ways or be reconstructed visually on a screen.*

ally builds up an adequately strong record of the observed spectrum in a computer memory, and the final spectrum can therefore be recorded in a variety of ways.

The ideal instrument should count each photon received. In fact, a fundamental limitation stems from the quantum yield of photocathodes, which is currently no greater than 20 percent. Taking into account various losses, the actual quantum yield is presently about 8 percent. Moreover, the photocathode releases electrons even when it is not illuminated, creating a noise that limits detection. In spite of this, the photon counting device is ideal for detecting low-luminosity objects. The digitalization of each picture dot—necessary for computer processing of the information—takes place during the observation; in addition, this technique, unlike traditional photography, permits knowledge of the precise times at which the photons making up the image arrived, through the use of real-time recording.

The technique of scanning photoelectric surfaces with an electronic probe presents the major drawback, however, of requiring a bulky vacuum

detector is essentially comparable to that of the electronographic camera. Their reliability and suitability for miniaturization (the first charge coupled devices contained only 10,000 elements, but soon devices will be manufactured with 640,000, chiefly for the space telescope to be launched in 1986) make the charge coupled devices the leading detectors of space astronomy, and their use is growing rapidly today.

Instruments of Positional Astronomy

LIKE SOLAR ASTRONOMY, positional astronomy requires special instruments.

The traditional instrument—first developed in 1690 by Römer in Copenhagen—is the meridian or transit instrument. It consists essentially of a

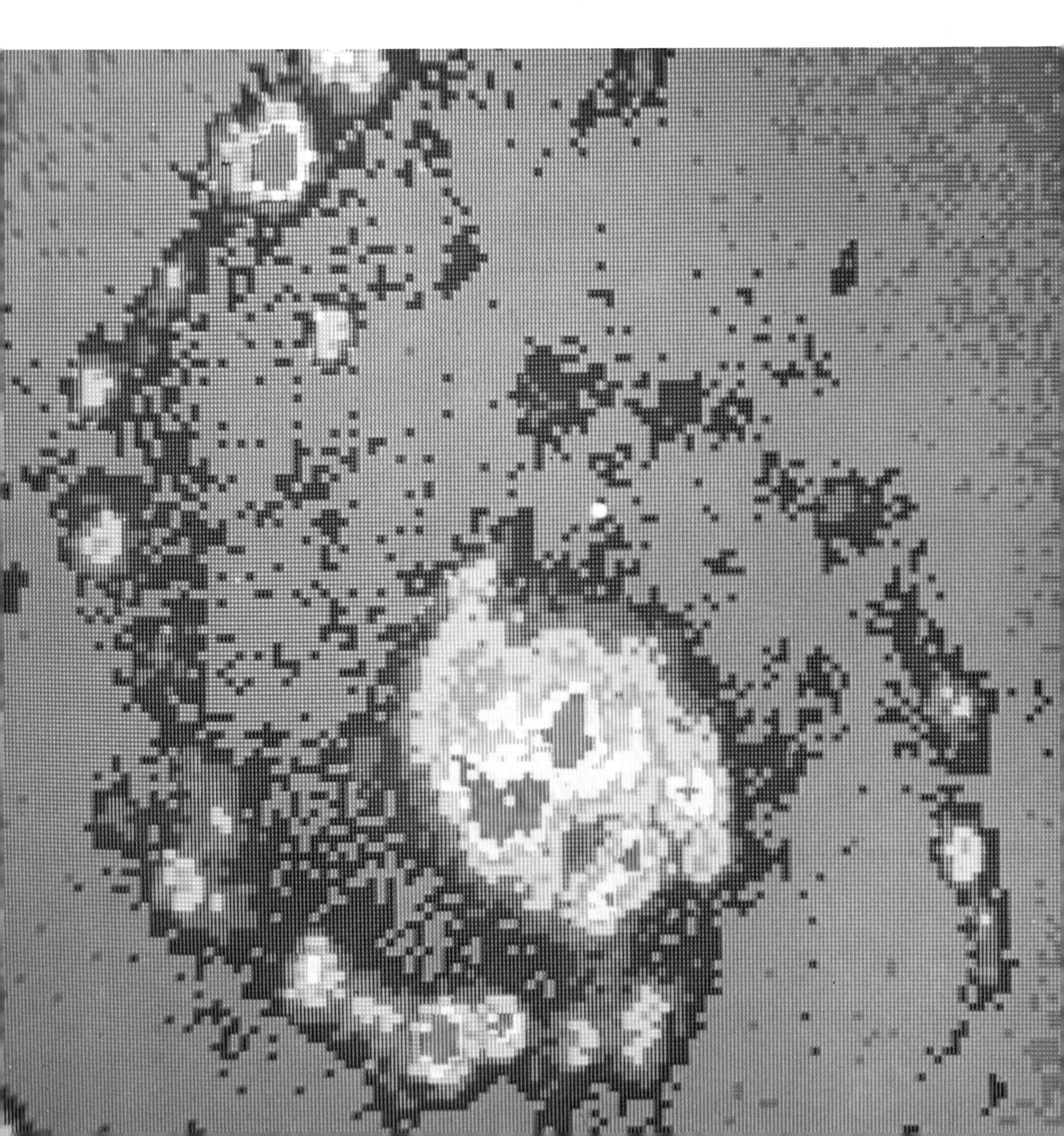

Monochromatic image of the galaxy M51, *obtained in a 10 Å pass band around the Hα line of hydrogen (wavelength = 6,563 Å), with a photon-counting device at the Cassegrain focus of the 193-centimeter (76-inch) telescope of the Observatory of Haute-Provence. The different colors represent different intensities of Hα radiation. Photo by Michel Marcelin, Observatory of Haute-Provence—CNRS.*

small refracting telescope (meridian telescope) that is made to pivot in the meridian plane (the north-south plane containing the vertical) of the location where it is installed, with a graduated circle (meridian circle) indicating the height of the observed object, while a timing device makes it possible to pinpoint the instant when the object transits the meridian plane.

Although simple in concept, this instrument is very difficult to make and install, because of the precision required by the measurements for which it is used. The mechanical construction, painstaking though it may be, is never free of defects: The optical axis of the telescope is not strictly perpendicular to the axis of the shaft on which it pivots; the moving parts wear out; the telescope distorts as a result of temperature changes; the divided circle is always somewhat off-center; the gradation is flawed, etc. Moreover, the instant of the observed object's transit across the meridian is hard to determine without some error, owing to the operator's reflexes. These causes of error can be overcome, as much as possible, by the use of appropriate devices. In particular, the telescope is made rigid by its massive, symmetrical structure, and attached to it are not one but two graduated circles of large diameter (1 meter), each with reading microscopes, nearly eliminating reading and engraving errors. With such an instrument, one cannot hope to determine an angular position with an accuracy greater than 0.5 arc second for a single measurement; the accuracy can be increased to 0.1 arc second by repeating the measurement several times.

The transit instrument played an essential role in compiling the catalogs of fundamental stars—bright stars to which the positions of other objects are compared. But today astrometry also uses more precise instruments—the prismatic astrolabe and the photographic zenith tube, which are used, among other things, to study the slight variations in the geographic coordinates of observatories owing to the shifting of the pole, as well as to determine the sidereal and universal time with high accuracy.

Conceived and invented between 1900 and 1902 by the French astronomers Claude and Driancourt as a portable device for determining astronomical positions with average precision (± 2 seconds), the prismatic astrolabe functions on the following principle: An equilateral prism is placed before the objective of a telescope with a horizontal axis. The light from a star is divided into two beams: One enters the prism on the upper surface, is reflected on the lower surface and is deflected onto the telescope objective; the other part is reflected onto a liquid mercury bath (with a perfectly horizontal surface); enters the prism through the lower surface; and is reflected on the upper surface, which deflects it onto the objective. In most cases, the two beams yield separate images. But if the angle between the two rays and the vertical is 30 degrees, the two beams are parallel upon entering the objective and yield a combined image. In this way the instant at which they coincide may be noted and the latitude of the place of observation inferred. Moreover, the time at which the star transits the meridian (whence the longitude of the place is inferred) can be obtained by averaging the moments of transit at a height of 60 degrees (since a star crosses the horizon at a given height twice a night, once when rising, once when setting).

The chief defect of this instrument is that the presumed moment when the images coincide depends on the positioning of the telescope eyepiece, or, in other words, of the observer himself. Through the use of a birefracting prism, André Danjon developed the instrument into an impersonal astrolabe, immune to errors made by the observer. The impersonal astrolabe is a fixed observatory instrument (weighing about 200 kilograms [441 pounds]). It can attain a precision of about 0.05 arc second in a latitude reading, corresponding to 1.5 meters (5 feet) on the Earth's surface.

As for the photographic zenith tube (PZT), developed in the United States around 1950, it is a vertical axis tube making it possible to photograph stars a few moments before and after their transit at the zenith. This device, which lends itself to a high degree of automation, is slightly more accurate than Danjon's astrolabe; after a night in operation, assuming that the coordinates of the observed stars are fully known, the latitude of the site may be calculated to within 0.03 arc second, and the longitude, or sidereal time, to within 0.003 second. But its use remains strictly limited to these determinations, since the only stars that can be observed belong to a very narrow strip of the sky and generally do not include any fundamental stars.

Amateur Astronomical Instruments

To observe the sky as an amateur, one may begin by making use of the most wonderful optical instrument of all: the eye. Others rely on binoculars, or on a telescope tube slapped together in a few hours. The most devoted amateurs go so far as to purchase or even to build (wholly or in part) what may be a very sophisticated instrument.

Binoculars

Although many amateurs trained in astronomical observation often use binoculars, beginners rarely think to use them. Nevertheless, their use is a logical next step after studying the sky with the naked eye, and they remain a choice instrument for watching lunar eclipses, earthshine, comets, stellar clusters and the most remarkable nebulae. Usually engraved on a pair of binoculars are two numbers separated by an ×. For example: 7 × 50. The first number refers to the factor by which objects are enlarged (7, in this example); the second indicates the diameter of the objectives in millimeters (50) (2 inches).

For observing the sky with binoculars, strong enlargements are not recommended. The chief merit of binoculars in astronomy is not to enlarge objects but to make it possible to "see better" low-intensity extended objects such as nebulae, or other events such as those mentioned earlier.

Typical characteristics of binoculars of different diameters are as follows (where an enlargement is chosen to give an adequate image brightness for each objective size): 6 × 40, 7 × 50, 11 × 80, 14 × 100.

Prior to any observation, the user of a pair of binoculars must be sure to make three adjustments:

1. Close the right eye, or, better still, block the right-hand objective with the hand (being careful not to leave a print on the glass), and bring the object into focus for the left eye, using the middle knob.
2. Close the left eye, or block the left-hand objective with the hand, and bring the image into focus for the right eye, using the eyepiece ring.

Principle of a grazing incidence X-ray telescope: *The rays are focused by reflection at a grazing angle on specially shaped surfaces—a paraboloid and then a hyperboloid. The* ***HEAO 2*** *telescope (focal length: 3.4 meters [11.2 feet]), has four overlapping surfaces of each type, each one having been cut and polished from a uniblock cylinder of a glass having a low thermal expansion coefficient and coated with a thin layer of nickel.*

Launched November 13, 1978, and placed initially in a circular orbit at 537 kilometers (334 miles) in altitude, the HEAO 2 satellite— *also called Einstein Observatory—essentially consists of a grazing incidence telescope with a 58-centimeter (23-inch) diameter capable of producing high-resolution images (2 arc seconds) of the sky in X rays, in the 0.4 to 4 keV energy band. It also has the capability for low- and medium-resolution spectroscopy. Its length is 6.7 meters (22 feet), its diameter 2.4 meters (7.9 feet) and its mass 3,175 kilograms (7,000 pounds). The telescope may be pointed with a precision of 1 arc minute, and its resolving power is about 4 arc seconds, corresponding to that of optical telescopes on the ground under average observing conditions. This device has shown that normal stars emit much more X radiation than had been thought. It has identified new binary systems in our galaxy, in which matter is being transferred from a large star to an ultradense companion; in neighboring galaxies, it has found a large number of previously unresolved X-ray sources; and it has recorded X-ray emission coming from the most distant known quasars. Finally, its observations suggest that the greatest part of the X-ray background originates not in diffuse hot gases, as was believed, but in galaxies and quasars that are too far away to be detected individually. Here, diagram of the device and view of the satellite prior to launching. After approximately two years in space, it is no longer operational.*

the energy of the incident X-ray photon. By utilizing different types of collimators, the field of view of the detector can be restricted, and the positions of X-ray sources can be pinpointed with high angular precision (an arc minute or better in the best cases).

At energies less than about 2 keV, it is possible to focus the X rays through the use of real telescopes, enabling images to be formed, but the telescopes must be equipped with mirrors that receive the rays at a "grazing" angle and not at a nearly perpendicular angle (like the mirrors of optical and radio telescopes); otherwise the X rays would penetrate the surface instead of being reflected by it. The overall collecting area is small in comparison with the diameter of the aperture but may be increased by overlapping several reflecting surfaces (see figure).

Grazing incidence X-ray telescopes were used aboard the space laboratory *Skylab* to observe the Sun, and the American satellite *HEAO 2* is equipped with a telescope of this type having a 58-centimeter (22.83-inches) aperture, and constructed in such a way that the focus can be brought to four different instruments.

GAMMA-RAY ASTRONOMY

The energy range of gamma-ray astronomy extends that of X-ray astronomy. Up to about 10 million electron volts (MeV), photon energies correspond to the transitions between the energy levels of the nuclei of various elements: this is the low-energy gamma range. Beyond that, up to several billion electron volts (GeV), is the high-energy gamma range, in which the most numerous and most interesting results have so far been obtained. High-energy gamma rays stem from the interaction of cosmic rays with interstellar matter, with magnetic fields or with the low-energy photons that pervade the universe.

incident gamma photon
10 cm
anticoincidence detector
spark chamber
photomultipliers
e+
e−
charged particle detectors
calorimeter

Cross-section of the X-ray telescope placed on the COS-B satellite. *Installed inside a cylinder 1.4 meters in diameter, which is covered with solar photocells designed to supply electric power to the satellite, this telescope is a spark chamber containing a series of thin parallel plates where gamma photons are converted to an electron-positron pair. The product of collaboration among six European laboratories, the COS-B satellite—designed to study celestial sources of gamma radiation—was launched on August 8, 1975, and placed in a very eccentric orbit having a 101,570 kilometers apogee and a 434 kilometers perigee. The data transmitted by it helped to establish the first gamma radiation map of the galaxy, to detect the first extragalactic gamma source (the 3C273 quasar) and to identify certain pulsars.*

Placement of the Space Telescope into Earth orbit, circa 1986, *will mark the dawn of a new era for astronomy.*

The instrument will be placed in orbit around the Earth at an altitude of approximately 500 kilometers (311 miles) by the U.S. Space Shuttle. Although it is less than two times smaller in diameter than the Mount Palomar telescope, it should perform substantially better even in the visible light range, simply because the residual brightness of the night sky and the blurring of images caused by air turbulence will be eliminated. It is predicted that it will be possible to detect bodies with a brightness not exceeding 29th magnitude—i.e., objects that are 100 times darker than the dimmest objects that can be photographed with the Mount Palomar telescope. It is estimated that the Space Telescope will attain spatial resolution of 0.08 arc second and a pointing accuracy of 0.01 arc second.

Normally remote-controlled from Earth, the Space Telescope can be visited by astronauts for maintenance and repairs or to replace certain auxilliary devices. Designed to serve for 15 years, it also can be retrieved by the Space Shuttle and brought back to Earth for a general overhaul, then replaced in orbit.

Astronomers are counting on this remarkable instrument to help them solve some of the major puzzles confronting us today. But the crop of new information about the universe that will be gathered surely will contain a whole new set of mysteries.

II

The Empire of the Sun

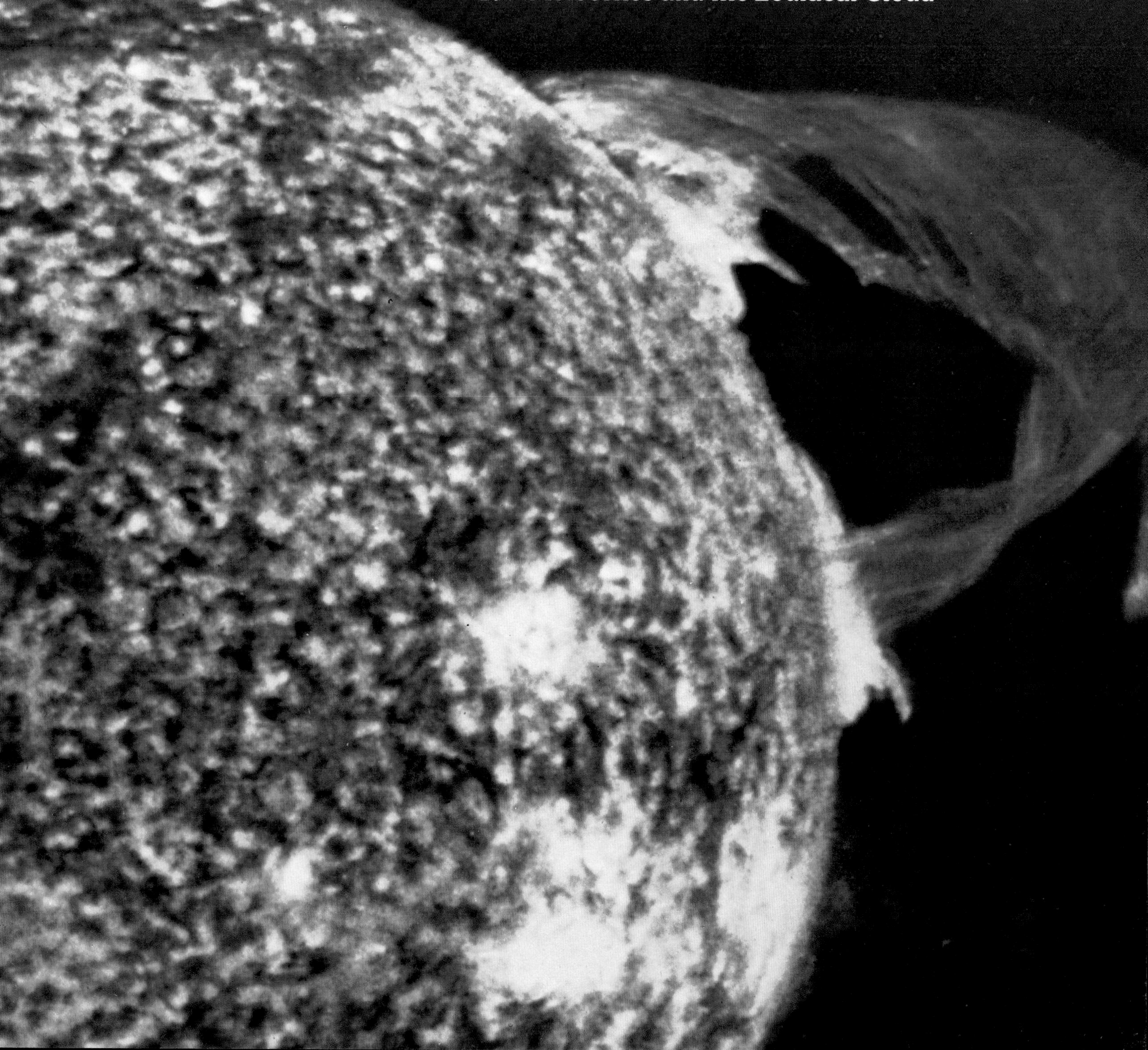

4. The Solar System

The region of space in which the Sun's gravitational force is dominant is called the solar system. Its boundaries are difficult to define. The system probably extends as far as two or three light-years (the closest star, Alpha Centauri, is 4.27 light-years away). The only part we know, in fact, is the central portion closest to the Sun, in which are found revolving nine major planets (including the Earth), their satellites, thousands of asteroids (minor planets), comets, as well as countless meteorites and dust particles populating the interplanetary medium.

In ancient times, bright objects resembling stars attracted attention because they seemed to wander among the constellations; the Greeks called them *planets,* meaning *wandering stars*. We know now that they are bodies orbiting the Sun and are not luminous by themselves but merely reflect some of the light they get from the Sun. In addition to the five planets known to the ancients—Mercury, Venus, Mars, Jupiter and Saturn, in increasing order of their distance from the Sun—we may add Uranus, Neptune and Pluto, which have been discovered since the 18th century; the Earth itself, located between Venus and Mars; and the thousands of asteroids that have been identified since the early 19th century, mostly concentrated between the orbits of Mars and Jupiter. From the Sun to the orbit of Pluto, the farthest known planet, the solar system covers a distance of about 6 billion kilometers (3.73 billion miles), which equals the distance traveled by light in just under six hours.

Mechanics of the Solar System

In studying the mechanics of the solar system, the bodies revolving about the Sun are viewed as geometric points possessing mass. Through the universal law of gravity, it is possible to calculate the positions of these bodies at any given moment. Since the Sun has by far the dominant mass (if all the remaining matter in the solar system were put together, its mass would be less than 0.0015 times the mass of the Sun), a rough estimate can be made of a single planet's trajectory, independent of the others—this is the two-body problem. In this case, the trajectory is an ellipse, located in a fixed plane and conforming to Kepler's laws (see box, page 68).

The orbits are described with respect to a reference plane. For the planets, this is the plane of the Earth's orbit, or ecliptic; for the satellites, it is the plane of the equator of the respective planet. The two points where an orbit crosses its reference plane are called nodes; the south-to-north crossing occurs at the ascending node. The point on the orbit that is closest to the principal body is called the periaster; the point that is farthest is called the apoaster. When referring to orbits around the Sun (planetary orbits, for example), these terms are replaced by perihelion and aphelion; for orbits around the Earth (by the Moon or artificial satellites), they become perigee and apogee. The motion is termed direct if it is counterclockwise when viewed from the ecliptic north and retrograde if it is clockwise.

A planet's orbit is uniquely determined by six elements:

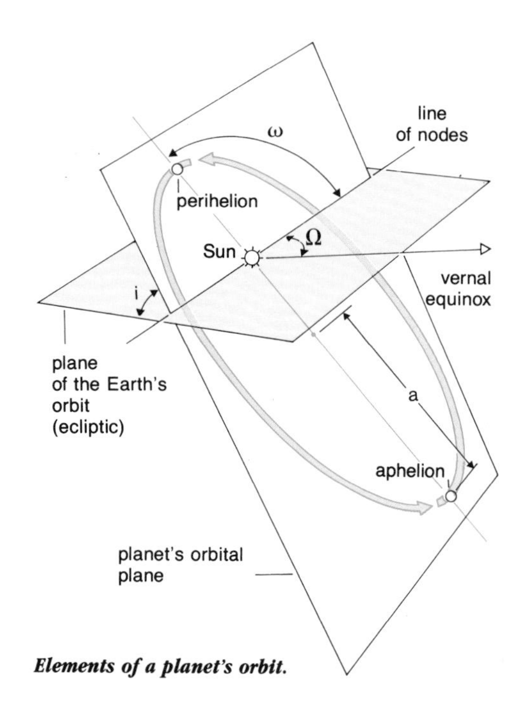

Elements of a planet's orbit.

The Astronomical Unit

Astronomical distances are so great that it is hardly practical to express them in kilometers. Within the solar system, astronomers relate lengths to a unit that is very nearly equal to the semimajor axis of the Earth's orbit and to the average distance of the Earth from the Sun: the astronomical unit (au).

Strictly speaking, an astronomical unit is defined as the radius of the circular orbit around the sun that would be followed by a planet of negligible mass that was immune to all perturbations and that completed its sidereal revolution in 365.2568983263 average days. Recognized in 1964 by the International Astronomical Union as a primary constant equal to 149 million kilometers (93 million miles), since 1976 it has been considered only a derived constant, defined by the speed of light (299,792,458 kilometers [186,606,570 miles] per second) and by the time it takes light to travel the distance it represents (499.004782 seconds). This change was prompted by the fact that the astronomical unit is based on radar measurements of planetary distances, which are mainly expressed in time units. The current value of the astronomical unit, obtaining by finding the product of the two primary constants from which it is derived, is 149,597,870 kilometers (92,975,679 miles).

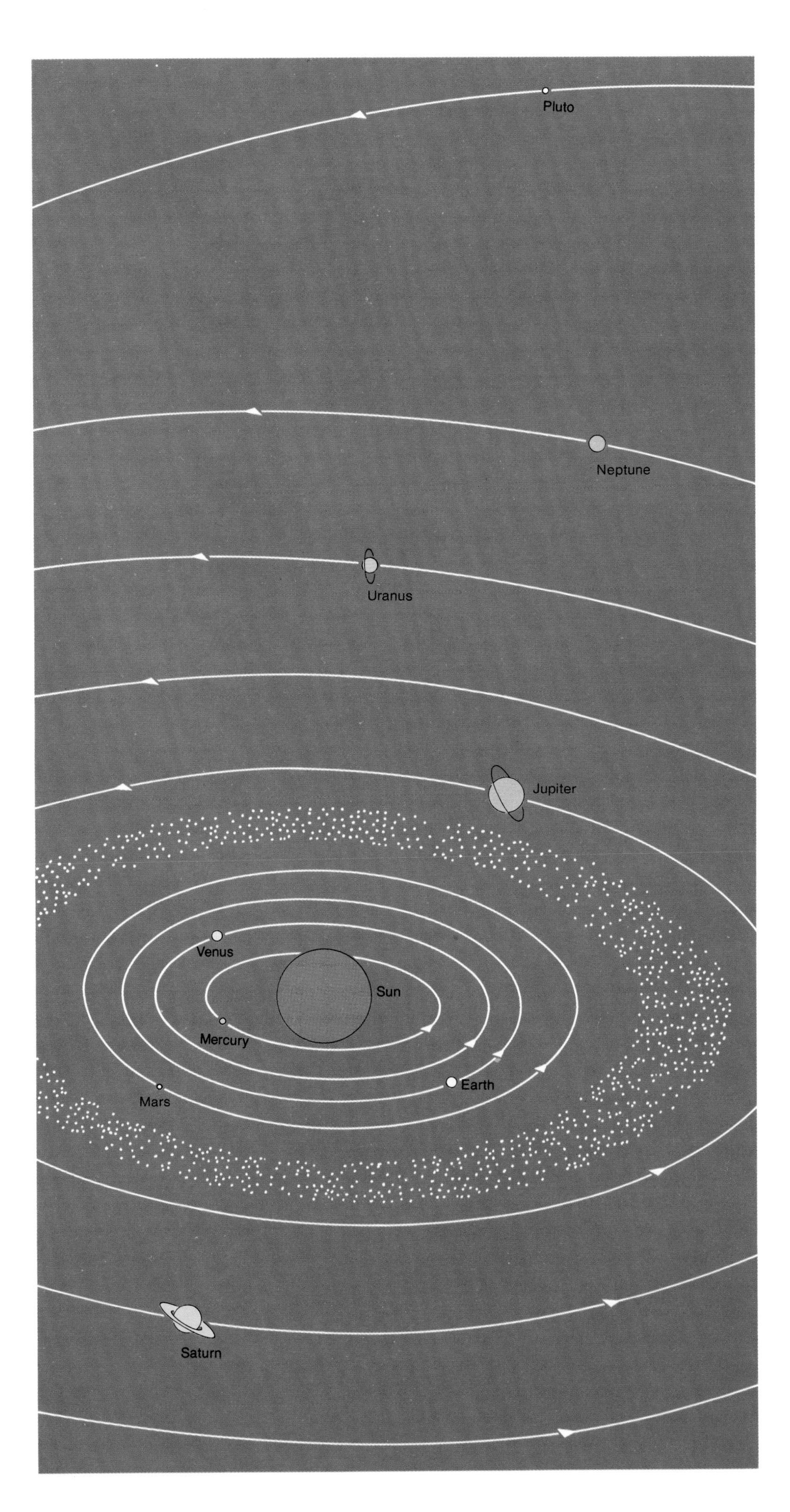

1. the semimajor axis a, usually expressed in astronomical units (i.e., calculated according to the semimajor axis of the Earth's orbit)
2. the eccentricity e (the ratio of the distance between the foci of the ellipse to its semimajor axis)
3. the inclination i of the plane of the orbit relative to the plane of the ecliptic
4. the longitude Ω of the ascending node, measured from the vernal equinox (see figure, page 62)
5. the argument ω of the perihelion angle (the angle made by the direction of the perihelion with the line of the nodes, measured from the ascending node)
6. the instant T of the crossing to the perihelion

The semimajor axis determines the length of the orbit, while the eccentricity determines its shape; the inclination and longitude of the ascending node specify the orientation of the orbital plane with respect to the ecliptic; the perihelion angle ω fixes the position of the orbit within its own plane; finally, the instant of the passage through perihelion defines the position of the planet in its orbit at any given time.

The sidereal period of revolution, P, the interval between two passages by a planet through the same point on its orbit, is not an independent parameter; it is tied to the semimajor axis a by Kepler's third law: $P^2 = a^3$, where P is expressed in years and a in astronomical units.

Similar elements are employed to specify the orbits of the planetary satellites. However, the inclination is then referred to the plane of the equator of the planet about which the satellite revolves.

For orbits of periodic comets, the perihelion distance q, or the aphelion distance Q, are usually expressed in astronomical units: $q = a\ (1 - e)$ and $Q = a\ (1 + e)$.

If the motion were truly Keplerian, the quantities a, e, i, Ω, ω and T, known as the elliptic elements, would have constant values. Owing to the presence of the other planets, however, these quantities are in fact considered to be equal to constants to which must be added certain functions of time; these functions are determined by the method of successive approximations based on equations of motion. These functions take into account the masses of the perturbing planets, which are small. The variations in the elliptical elements are either very small in amplitude or extremely slow. The semimajor axis, eccentricity and inclination range periodically about a mean value; the

Diagram of the solar system.

line of nodes and the perihelion make a full circle in the plane of the orbit, but over a very long period (generally, several tens of thousands of years). In order to calculate the perturbation upon the motion of a planet as a result of the gravitational forces exerted on it by all the other planets, complex mathematical formulas are required, but the work of scholars such as Clairaut, Euler, Lagrange, Laplace, Le Verrier and Newcomb have contributed to the development of precise methods used today to calculate the ephemerides of the planets many years in advance.

It was while analyzing the perturbations of unknown origin affecting the motion of Uranus that Le Verrier discovered the planet Neptune in 1846.

The major planets of the solar system all revolve about the Sun in a direct sense. They describe nearly circular orbits (with the exception of Mercury and Pluto) in closely neighboring planes that deviate little from that of the ecliptic and solar equator. Finally, their distances from the Sun increase in an approximately geometrical progression, as shown by the Titius-Bode law (see box page 69). Such coincidences cannot be the result of chance. They testify to the common origin of the different planets.

Apparent Motion and Conditions of Observation of the Planets

The ancients did not fail to notice that the planets visible to the naked eye never stray very far from the ecliptic—that is, from the Sun's apparent annual path among the stars. They always remain within a zone 17 degrees in width (8 degrees, 30 minutes on either side of the ecliptic), the zodiac, a term that derives from the Greek word *zoon*, "animal," because the constellations it comprises are mostly represented by animal figures. The ancients divided this region into 12 equal parts, or signs, each of which marks the Sun's sojourn during one month of the year. The planets Uranus and Neptune, which were discovered in modern times, also move within the confines of the zodiac; but Pluto and several of the small planets circulating between Mars and Jupiter leave it (their orbits being rather sharply inclined relative to the ecliptic), and the comets sometimes stray so far from it that they reach the celestial poles.

Since the planets emit no light of their own but merely reflect a portion of the light from the Sun, the ease with which they can be seen varies with their position relative to the Earth and Sun. The observed phenomena are slightly different depending on whether the planet is nearer to or farther from the Sun than the Earth.

The inferior planets (Mercury and Venus), which are closer to the Sun than the Earth, never stray very far from the Sun in the sky. Therefore they can be observed only in the early evening or early morning. Furthermore, they may pass between the Sun and the Earth, thus turning their dark hemisphere toward us, before and after which they move through a series of phases similar to those of the Moon.

The superior planets (Mars, Jupiter, etc.), which are farther from the Sun than the Earth, may occupy any angular position relative to the Sun and may thus be observed in the middle of the night. On the other hand, they can obviously never pass between the Sun and the Earth; they always show us a major portion of their lighted hemisphere. As a result, their phases are not much different from one another. We never see them in crescent form, but only in gibbous form (a half circle joined to a half ellipse).

Special terms are used to designate the particular positions of the planets relative to the Sun and the Earth.

INFERIOR PLANET

When an inferior planet comes between the Earth and the Sun, the planet is said to be at inferior conjunction (with the Sun). The planet is then at its smallest distance from the Earth but remains invisible, like the new Moon, because its dark side is facing us. Owing to the relative motion of the Earth and the planet, it then moves west, and its direction moves away from that of the Sun. When seen through a telescope, the planet then appears as a thin crescent. As it moves farther and farther from the Sun, the crescent becomes thicker and thicker, but the planet's apparent diameter decreases. However, its brightness increases, because the visible surface area increases. These phenomena continue until the angle between the Sun's direction and that of the Earth seen from the planet is 90 degrees. At that point, the planet reaches its greatest angular distance from the Sun: It is said to reach its greatest western elongation. It then appears as a perfect half, like the Moon in the last quarter. It then seems to draw closer to the Sun, while its brightness diminishes. When it passes directly behind the Sun, it is said to be at superior conjunction; its angular diameter is smallest. It reappears several days later as an eve-

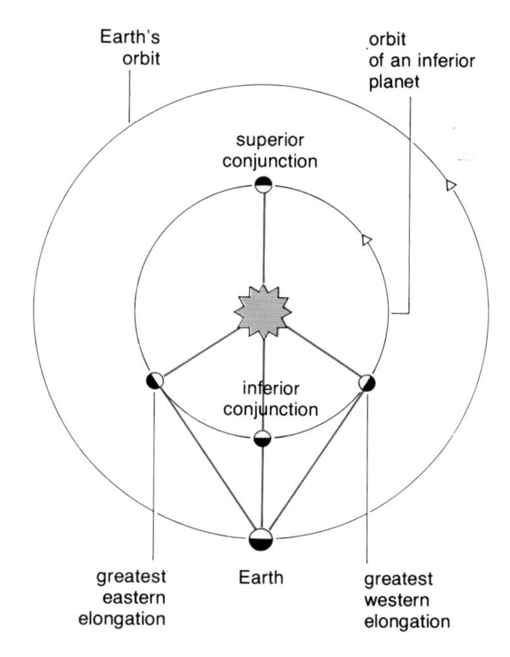

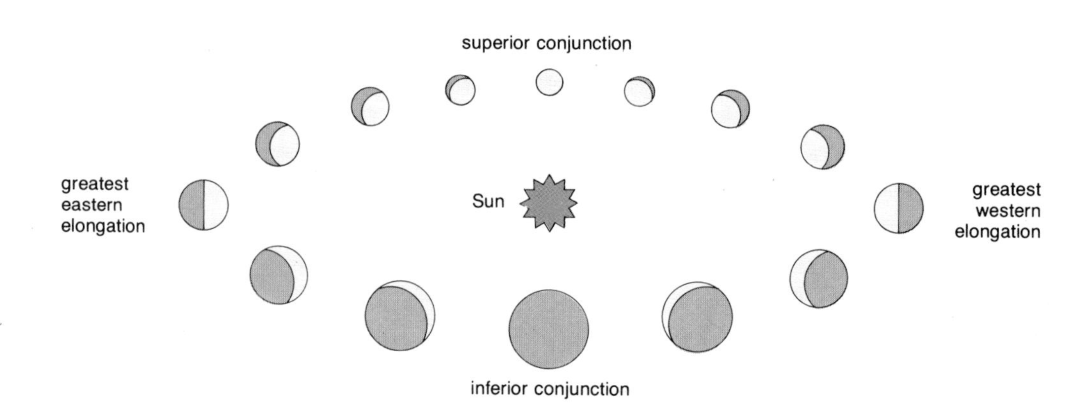

Special configurations (above) and phases (below) of an inferior planet.

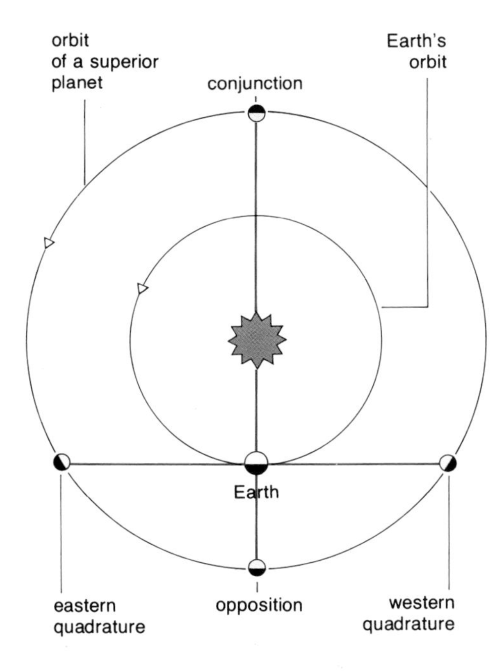

Special configurations of a superior planet.

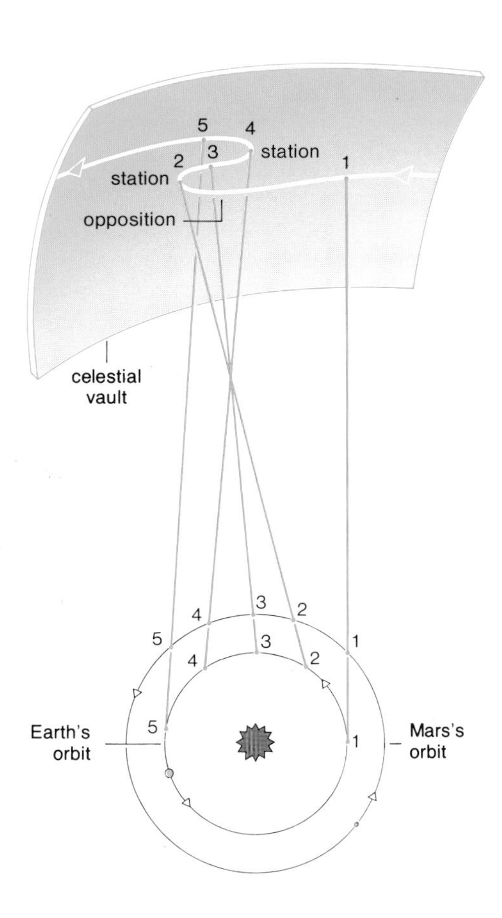

Apparent motion of a superior planet on the celestial sphere. *If a planet is observed regularly for several months, it is seen to describe huge loops in the sky among the constellations; usually it moves in the same sense as the Sun, but at other times it stops and reverses direction so that it moves in a retrograde sense. This phenomenon is explained simply by the fact that the Earth and the planet in question are moving around the Sun at different speeds.*

ning star and is then seen to move away from the Sun again until it reaches its greatest eastern elongation, when the angle between the Sun's direction and that of the Earth seen from the planet once again reaches 90 degrees. Finally, the planet again arrives at inferior conjunction and the cycle begins anew.

SUPERIOR PLANET

When a superior planet is aligned with the Sun and Earth and is on the same side of the Sun as the Earth, the planet is said to be in opposition (with the Sun). Its distance from the Earth is smallest, its apparent diameter greatest. It faces us on its bright side and crosses the meridian at midnight; this is when observational conditions are most favorable. Since the planet orbits more slowly than the Earth, relative motion shifts it westward. This motion, which is retrograde relative to the stars, continues for a while; then the planet comes to rest relative to the stars—it becomes stationary. The motion then resumes in an eastward direction. Several days later, the planet is 90 degrees to the east of the Sun and is then said to be at its eastern quadrature. It then passes through a short phase in which it is gibbous. After some more time, it disappears behind the Sun, in conjunction with it, before reappearing as the morning star, while becoming more and more visible.

From then on, it moves in a prograde sense until it reaches another stationary point; its motion then becomes retrograde again. It gradually moves away from the Sun until it appears 90 degrees to the west of it, at its western quadrature, before moving toward the Sun again and returning to opposition. Then the cycle begins again.

As the months pass, the planets are seen to follow loops or figure eights among the constellations. These apparent motions result simply from the combination of their orbital motions and that of the Earth, which occur at different speeds.

A planet's sidereal period of revolution is the interval of time between two consecutive arrivals at the same point on its orbit. Likewise, the synodic period of revolution is defined as the interval of time between two consecutive conjunctions (superior in the case of an inferior planet) of the planet in question.

Physics of the Solar System

FROM THE PHYSICAL STANDPOINT, the major planets of the solar system fall into two families:

The terrestrial planets (Mercury, Venus, Earth, Mars) are of modest dimensions but have relatively high density. They have evolved extensively since their formation from the primitive nebula. They have lost their light elements, and their present atmosphere is a secondary one, probably formed through outgassing, chemical evolution and evaporation. Finally, they have a solid crust separating their interiors from their atmospheres.

The giant planets (Jupiter, Saturn, Uranus, Neptune), also called gas giants, or the Jovian planets, are substantially more massive and voluminous, but their density is low. Their composition, based predominantly on hydrogen and helium, has remained very close to that of the primitive solar nebula, at least in the region where they were formed. Gaseous in their outer regions, they may be liquid or solid in the interior owing to the very strong pressures occurring there. Their apparent surface, in the visible range, consists of clouds stemming from the condensation of certain gaseous components.

Pluto, the most remote planet, is still poorly known and difficult to categorize. It seems to resemble the terrestrial planets in its dimensions, but its density, and hence its composition, is intermediate between that of the terrestrial and the giant planets.

Physical Characteristics of the Planets

Name	Symbol	Ratio of the Equatorial Diameter to That of the Earth	Equatorial Diameter (in km.)	Maximum Apparent Diameter	Minimum Apparent Diameter	Flattening	Ratio of Volume to That of the Earth	Ratio of Mass to That of the Earth	Average Density	Ratio of Weight at the Equator to That on the Earth	Escape Velocity (in km./sec.)	Sidereal Rotation Period (in days or in hr., min., sec.)	Tilt of Equator to the Orbit	Albedo	Observed Surface Temperature	Principal Constituents of the Atmosphere	Magnetic Field (gauss)	Known Satellites
Mercury	☿	0.382	4,878	12.9″	4.7″	0	0.056	0.055	5.44	0.37	4.25	58.646 days	0°	0.055	−170 to +390	Ne, H (solar wind)	2.10^{-3} 1.10^{-12}	0
Venus	♀	0.949	12,102	66.0″	9.9″	0	0.858	0.815	5.25	0.91	10.36	243 days (r)*	2°07′	0.64	+450 to +480	CO_2 (97 p. 100)	$< 3 \times 10^4$	0
Earth	♁	1	12,756	—	—	0.003353	1	1	5.52	1	11.18	23 hr., 56 min., 04 sec.	23°26′	0.39	− 88 to + 48	N_2 (78 p. 100), O_2 (21 p. 100)	0.3 to 0.6	1
Mars	♂	0.533	6,794	26.0″	3.5″	0.005	0.152	0.107	3.91	0.38	5.02	24 hr., 37 min., 23 sec.	23°59′	0.154	−128 to + 24	CO_2 (95 p. 100)	6×10^{-4}	2
Jupiter	♃	11.194	142,796	49.6″	30.5″	0.062	1,338	317.95	1.31	2.54	59.64	9 hr., 50 min. to 9 hr., 56 min.	3°04′	0.42	−140	H, He, CH_4, NH_3	4 to 14	16
Saturn	♄	9.407	120,660	20.5″	14.7″	0.0912	766	95.2	0.69	1.08	35.41	10 hr., 14 min. to 10 hr., 39 min.	26°44′	0.45	−160	H, He, CH_4, NH_3	0.2 to 0.6	16
Uranus	♅	3.98	50,800	4.3″	3.4″	0.06	60.4	14.6	1.21	0.88	21.41	10 hr., 42 min.	98°	0.46	−180	H, He, CH_4, NH_3	1.8	5
Neptune	♆	3.81	48,600	2.9″	2.0″	0.02	56.9	17.2	1.67	1.15	23.52	15 hr., 48 min. (?)	29° (?)	0.53	−200	H, He, CH_4, NH_3	1.6	2
Pluto	♇	~0.2	3,000	0.2″	?	?	?	0.002	1 ?	?	?	6 days, 9 hr., 18 min.	?	?	−238	—	—	1

*(r) indicates a retrograde rotation.

Comparative dimensions of the planets.

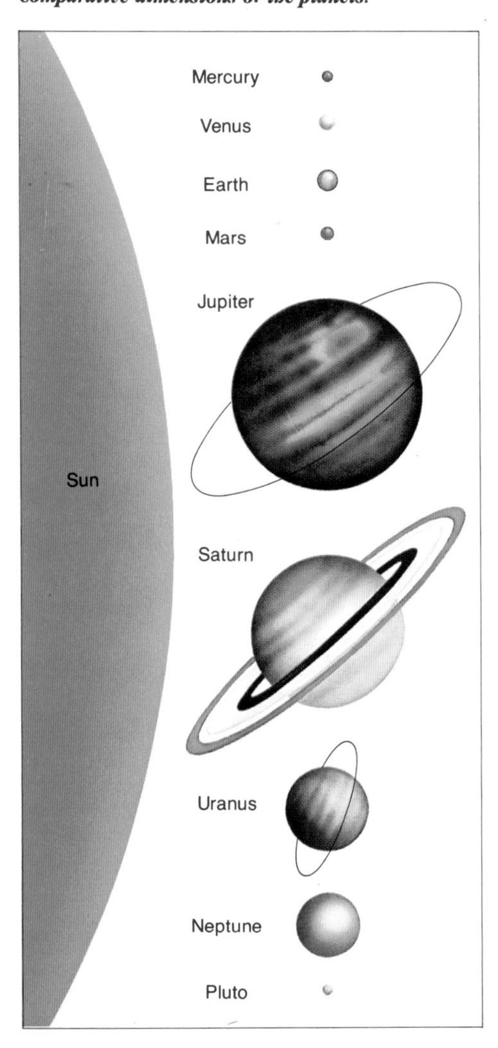

Determination of the Planets' Principal Physical Features

A planet's mass is determined in different ways depending on whether or not it has satellites.

For a planet possessing at least one satellite, the satellite may be considered to be in dynamical equilibrium between the gravitational force of the planet and the centrifugal force resulting from the satellite's rotation about the planet. In that case, the value of the planet's mass can be directly determined from the parameters of the satellite's orbit.

If the planet has no satellite, its mass is much more difficult to determine. The perturbations it inflicts on the orbits of neighboring planets must be taken into account, but these perturbations are slight and difficult to measure, and the results obtained are much less accurate. The passage of space probes near two of the planets lacking a satellite (Mercury and Venus) has overcome this drawback, since the probe can itself be treated as a satellite.

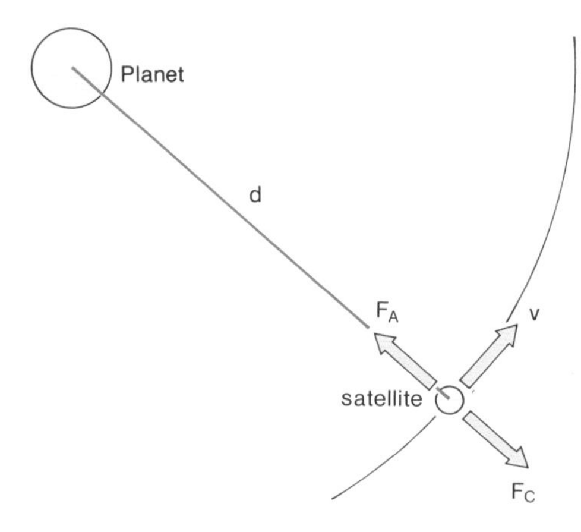

Finding the mass of a planet that has a satellite. *The satellite is subject to the gravitational force* $F_A = \frac{kmM}{d^2}$ *and to the centrifugal force* $F_c = m\frac{v^2}{d}$ *since it is in equilibrium,* $F_A = F_c$ *from which one deduces:* $M = \frac{dv^2}{k}$ *where m equals the mass of the satellite; M equals the mass of the planet; d, distance of the satellite from the planet; v, velocity of the satellite; k, gravitational constant.*

Orbital Characteristics of the Planets

Name	Semimajor Axis (au)	Semimajor Axis 10⁶ km.)	Eccentricity	Inclination Relative to the Ecliptic	Sidereal Revolution Period	Orbital Velocity (km./sec.)	Synodic Revolution Period	Maximum Distance from the Earth (10^6 km.)	Minimum Distance from the Earth (10^6 km.)
Mercury	0.3871	57.91	0.206	7°00′	87.969 d.	47.89	115.9 d.	220	80
Venus	0.7233	108.21	0.007	3°24′	224.701 d.	35.04	1 yr., 218.7 d.	258	41
Earth	1	149.60	0.017	0°	365.256 d.	29.80	—	—	—
Mars	1.5237	227.94	0.093	1°51′	1 yr., 321.73 d.	24.14	2 yr., 49.5 d.	400	56
Jupiter	5.2026	778.31	0.048	1°19′	11 yr., 314.84 d.	13.06	1 yr., 3 d.	960	590
Saturn	9.5547	1,429.4	0.056	2°30′	29 yr., 167.0 d.	9.64	1 yr., 4.4 d.	1,650	1,200
Uranus	19.2181	2,875.0	0.046	0°46′	84 yr., 7.4 d.	6.80	1 yr., 4.4 d.	3,100	2,700
Neptune	30.1096	4,504.4	0.009	1°47′	164 yr., 280.3 d.	5.43	1 yr., 2.2 d.	4,650	4,350
Pluto	39.4387	5,900	0.246	17°10′	247 yr., 249.0 d.	4.74	1 yr., 1.5 d.	7,500	4,300

The Natural Satellites of the Planets

Name	No.	Year of Discovery	Discoverer	Sidereal Revolution Period[1] (days)	Semimajor Axis of the Orbit (average distance to the planet) (10^3 km.)	(in planetary radii)	Eccentricity of the Orbit	Inclination of the Orbit[2]	Diameter[3]	Density (water = 1)	Apparent magnitude
EARTH											
Moon				27.322	384.40	60.268	0.055	5°8.7′	3 476	3.33	− 12.7
MARS											
Phobos	I	1877	A. Hall (U.S)	0.319	9.38	2.76	0.017	[1.1°][2]	27 × 21 × 19[4]	~ 2	11.5
Deimos	II	1877	A. Hall (U.S.)	1.262	23.48	6.91	0.003	[0.9°] to [2.7°]	15 × 12 × 11[4]	~ 2	12.0
JUPITER											
1979 J 3	(XVI)	1979	S. P. Synnott (U.S.)	0.294	127.6	1.79	?	?	(40)	?	?
1979 J 1	(XIV)	1979	D. Jewitt, E. Danielson (U.S.)	0.297	128.4	1.80	?	?	(30)	?	?
Amalthea	V	1892	E. Barnard (U.S.)	0.498	181.0	2.52	0.003	[0.4°]	265 × 140	?	13.0
1979 J 2	(XV)	1979	S. P. Synnott (U.S.)	0.678	222.4	3.11	?	?	(70)	?	?
Io	I	1610	Galilei (Ital.)	1.769	421.6	5.90	0.000 6	[0.11′]	3,632	3.53	5.5
Europa	II	1610	Galilei (Ital.)	3.551	670.9	9.40	0.000 1	[1.03′]	3,126	3.03	5.7
Ganymede	III	1610	Galilei (Ital.)	7.155	1 070.0	14.99	0.002	[5.22′]	5,276	1.93	5.0
Callisto	IV	1610	Galilei (Ital.)	16.689	1 880.0	26.33	0.007	[25.75′]	4,820	1.79	6.3
Leda	XIII	1974	C. Kowall (U.S.)	240.0	11,134	156.0	0.146	26.7°	(10)	?	20.5
Himalia	VI	1904	C. Perrine (U.S.)	250.6	11,478	159.8	0.158	28.4°	(170)	?	13.7
Lysithea	X	1938	S. Nicholson (U.S.)	260.0	11,720	163.2	0.107	28.8°	(20)	?	19.0
Elara	VII	1905	C. Perrine (U.S.)	260.1	11,737	163.4	0.207	27.8°	(80)	?	17.0
Aranke	XII	1951	S. Nicholson (U.S.)	631.0 (r)	21,209	295.3	0.169	146.7°	(20)	?	19.0
Carme	XI	1938	S. Nicholson (U.S.)	692.5 (r)	22,564	314.2	0.207	163.4°	(20)	?	19.0
Pasiphae	VIII	1908	P. Melotte (U.K.)	743.7 (r)	23,457	326.6	0.410	148.2°	(30)	?	18.0
Sinope	IX	1914	S. Nicholson (U.S.)	746.6 (r)	23,725	330.3	0.317	153.0°	(20)	?	19.0
SATURN											
Atlas		1980		0.602	137.7	2.28	0.002	[0.3°]	20 × 40		
1980 S27		1980		0.613	139.4	2.31	0.003	[0]	140 × 100 × 80		
1980 S26		1980		0.628	141.7	2.35	0.004	[00.5°]	110 × 90 × 70		
Janus	X	1966	A. Dollfus (Fr.)	0.694	151.4	2.51	0.009	[0.3°]	140 × 120 × 100	?	14
Epimetheus		1980		0.695	151.5	2.51	0.007	[0.1°]	220 × 200 × 160	?	?
Mimas	I	1789	W. Herschel (U.K.)	0.942	188.2	3.08	0.020	[1.5°]	390	1.2	12.1
Enceladus	II	1789	W. Herschel (U.K.)	1.370	240.2	3.95	0.004	[0]	500	1.2	11.6
Calypso		1980	B. A. Smith (U.S.)	1.888	294.6	4.88			34 × 22 × 22		
Tethys	III	1684	J. D. Cassini (Fr.)	1.888	294.7	4.88	0	[1.1°]	1,050	1.2	12.5
Telesto		1980		1.888	294.7	4.88			34 × 28 × 26		
Dione	IV	1684	J. D. Cassini (Fr.)	2.737	377.4	6.26	0.002	[0]	1,120	1.4	10.7
1980 S6		1980	P. Laques, J. Lecacheux (Fr.)	2.739	378.1	6.27	0.005	[0]	36 × 32 × 30	?	
Rhea	V	1672	J. D. Cassini (Fr.)	4.517	527.1	8.74	0.001	[0.4°]	1,530	1.2	10.0
Titan	VI	1655	C. Huyghens (Holl.)	15.945	1,221.9	20.25	0.029	[0.3°]	5,150	1.9	8.3
Hyperion	VII	1848	W. Bond (U.S.)	21.276	1,481.0	24.55	0.104	[0.4°]	410 × 260 × 220	?	13.0
Iapetus	VIII	1671	J. D. Cassini (Fr.)	79.33	3,560.8	59.02	0.028	18.4°	1,460	1.2	10–12
Phoebe	IX	1898	W. Pickering (U.S.)	550.45 (r)	12,954	214.7	0.166	175.1°	220	?	14.5
URANUS											
Miranda	V	1948	G. Kuiper (U.S)	1.41 (r)	130	5.49	0	0	(300)	?	19.0
Ariel	I	1851	W. Lassel (U.K.)	2.520 (r)	191.8	8.14	0.003	0	(1,330)	?	15.2
Umbriel	II	1851	W. Lassel (U.K.)	4.144 (r)	267.2	11.35	0.004	0	(1,110)	?	15.8
Titania	III	1787	W. Herschel (U.K.)	8.706 (r)	438.4	18.61	0.002	0	(1,600)	?	14.0
Oberon	IV	1787	W. Herschel (U.K.)	13.46 (r)	586.2	24.89	0.001	0	(1,630)	?	14.3
NEPTUNE											
Triton	I	1846	W. Lassel (U.K.)	5.877 (r)	355.3	15.9	0	160°	(3,200)		13.6
Nereid	II	1949	G. Kuiper (U.S.)	359.4	5,560	249	0.749	27.5°	(300)		19.5
PLUTO											
(Charon)	(I)	1978	J. W. Christy (U.S)	6.386 7	20	13	0	115°	⩾ 1,200		16.3

1. (r) indicates that the revolution occurs in a retrograde direction.
2. Brackets indicate the inclination of the orbit to the planet's equator.
3. Parentheses indicate an estimate.
4. Triaxial ellipsoid of revolution.

The measurement of the planets' apparent diameter—i.e., the angular size they appear to have from Earth—can yield their real diameter if their distance is known. This involves a difficult measurement, however, since the apparent diameter is always very small. Before spacecraft came into use, the most accurate results with regard to the nearby planets (Mercury, Venus and Mars) were obtained through radar measurements. The advantage of such measurements was that they supplied the diameter of the solid planetary globe, whereas in the case of Venus, visual measurements refer to a cloudy layer of unknown thickness.

The planets' period of rotation is determined either through visual or photographic observation of surface features or through radar measurements. For Pluto and for several asteroids, the periodic variations in brightness revealed by photometric observation determines the period of rotation.

Indications regarding the physical nature of the planets are supplied by the measurement of their albedo (the ratio of the total light intensity reflected by the planet in all directions to the intensity of solar light it intercepts). A high albedo, such as that of Venus (0.64), is the sign of great reflecting capacity and reveals the presence of thick clouds or ice; a very low albedo, such as that of Mercury (0.05), is interpreted in terms of a surface that is covered with fine dark dust or, in the case of asteroids, with dark, carbonaceous material. These results are supported by polarimetric measurements (measurements of the polarization of light). The albedo and polarization measurements are interpreted by comparison with similar curves obtained in a laboratory from terrestrial samples. The partial polarization of the reflected radiation is due to the planet's surface or atmosphere; its variation with wavelength is characteristic of the dimension and nature of the scattering particles.

The infrared radiation of the planets makes it possible to estimate their temperature by assuming them to be similar to a black body. The measurement of the energy given off in the radio range leads to a "brightness temperature," the temperature of the black body emitting the same quantity of energy at the wavelength of measurement.

Finally, a spectroscopic analysis of the light rays from a planet reveals absorption lines and bands caused by the atmospheric constituents, which can thereby be identified.

Determination of Distances Within the Solar System

The method of trigonometric parallax involves relatively simple procedures for determining the distance between the Earth and celestial bodies that are not too distant.

The parallax of a body is the angle subtended by a given reference length for an observer on that body. In the solar system, the Earth's radius serves as the reference length. Since the Earth's dimensions are known as a result of geodesic studies, measuring the parallax of a body in equivalent to indirectly determining its distance from the Earth.

The parallax of the Moon and nearby planets may be measured by means of triangulation, by

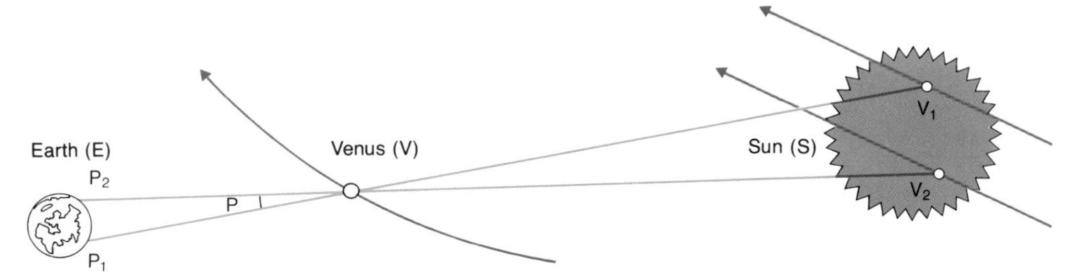

***Finding the Sun's parallax** during a transit of Venus in front of the solar disk. The planet Venus (V), observed from Earth (E) at two stations P_1 and P_2 whose separation is known, is seen, at a given moment, at points V_1 and V_2 on the solar disk. These points cross the solar disk on two parallel lines. Observation yields the angle $\widehat{P_1VP_2}$, from which the Sun's parallax is deduced by comparing the distance P_1P_2 to the Earth's radius and by using the known relative distances of the Earth and Venus from the Sun.*

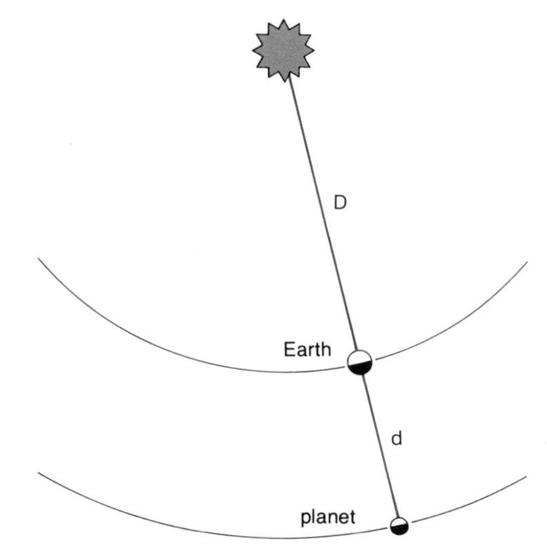

***Principle of measurement of the distance between the Earth and the Sun.** The sidereal periods T_E of Earth and T_P of another planet are known. Having found the distance d of the planet in opposition by measuring its parallax, one directly obtains the Earth–Sun distance d by applying Kepler's third law:* $\frac{D^3}{(D + d)^3} = \frac{T_E^2}{T_P^2}$

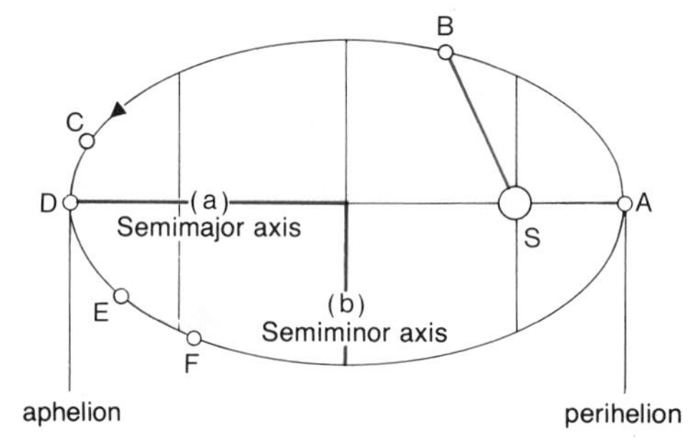

Kepler's Laws

Discovered empirically by Kepler at the beginning of the 17th century, these laws concerning the motion of the planets were confirmed in the following century by Newton's theory of universal gravitation.

- first law (1609): The planets follow elliptical paths around the Sun, with the Sun occupying one of the foci.
- second law (1618): The line joining the Sun to a planet (radius vector) sweeps out equal areas in equal intervals of time.
- third law (1618): The squares of the planets' periods of revolution are proportional to the cubes of the major axes of their elliptical orbits.

The second law, known as the "law of areas," governs the velocity of the planets' motion around the sun. It shows that the closer a planet is to the Sun, the faster it moves, while the farther it is, the slower its motion. Thus, for example, the velocity of the Earth at its perihelion (January 2) reaches 30.27 kilometers (18.81 miles) per second, while it is no more than 29.28 kilometers (18.2 miles) per second at its aphelion (July 2). Arcs such as AB, CD, and EF (see figure) covered in equal time intervals therefore are shorter the farther the planet is from the Sun; but the areas contained between the lines joining the Sun with the two extremities of these arcs are all equal, because the vector radii, such as SA, SF, etc., are longer when the orbital velocity is smaller. Thus the planet takes as long to go from A to B as from E to F, although the first arc is much longer than the second. These variations in velocity according to the distance from the Sun have various consequences, one of which is the length of the seasons.

The third law governs the period of revolution around the Sun—i.e., the length of the year—of any planet according to its distance from the Sun. It states that the greater the size of the planet's orbit, the longer the period of revolution, although the periods of revolution increase more abruptly than the distances from the Sun. This law has fundamental importance because it relates all the planets to one another. For any given planet for which one considers the orbital dimensions, the square of its period of revolution divided by the cube of the major axis of its orbit is a constant number.

Kepler's laws apply equally to the motion of satellites around their planets and to double stars.

viewing the body whose distance is to be found from two separate points on the Earth's surface. But this method cannot be used to determine the solar parallax; because of the Sun's distance, observational errors can be a considerable fraction of the parallax one wants to measure. It is difficult to measure the position of the Sun accurately because of its apparent dimensions and brightness; however, there is considerable value in determining the solar parallax, hence the Earth's distance from the Sun—it provides a scale by which all other distances in the solar system can be measured. By virtue of Kepler's third law, once the distance of any planet from the Sun is known, all others may be deduced from it using this formula:

$$a^3 = \frac{T^2}{T_0^2} a_0^3$$

where a is the unknown semi-major axis, a_0 the known semimajor axis and T and T_0 the sidereal periods of revolution (the values of which are supplied by observation).

In practice, it is a two-step process. In the first step, the parallax of a planet is determined at a time when it is close to the Earth. Then, in the second step, the value of the semimajor axis of the terrestrial orbit is calculated by application of Kepler's third law.

This method was used for the first time in 1672 by Cassini and Richer, using a favorable opposition of the planet Mars. The solar parallax was then found to be equal to 9 seconds, which made the Sun 20 times more distant than the highest estimates of the time.

More recent findings, using minor planets that pass very close to the Earth, such as Eros, have yielded a solar parallax of 8.8 seconds, which corresponds to approximately 149,600,000 kilometers for the semimajor axis of the terrestrial orbit.

Another method of determining the solar parallax utilizes the transits of Venus in front of the Sun; unfortunately, these are very rare (see "Venus," page 120). The most accurate determinations of solar system distances that have been obtained in recent years are based on radar echoes from the Moon, from nearby planets or from space probes. For bodies very close to the Earth, such as the Moon or artificial satellites, radar may be conveniently replaced by a laser beam; since the frequency is higher and the duration of the radiative pulse can be made shorter, greater accuracy is achieved.

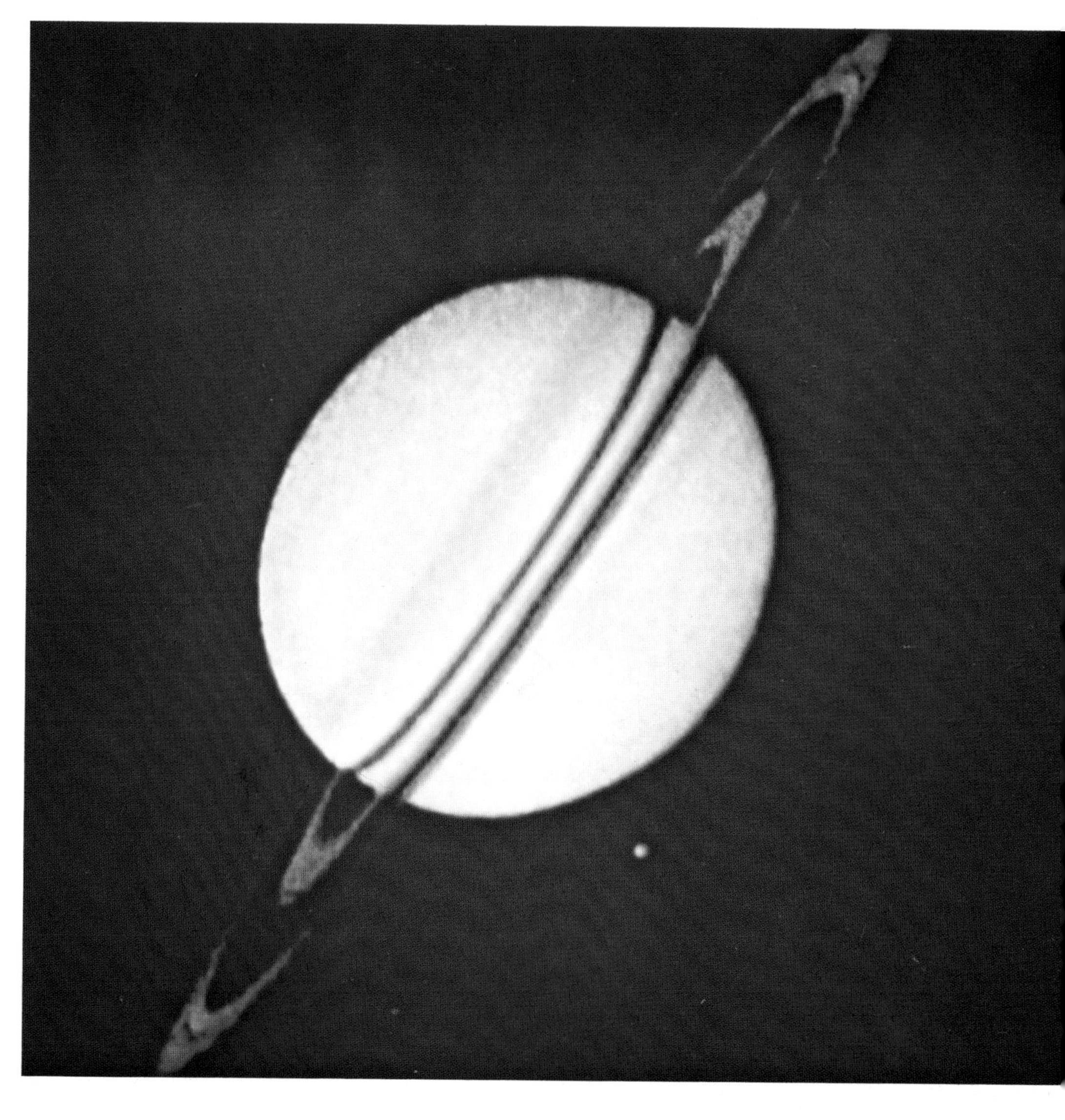

Saturn and its rings, seen on August 29, 1979, from a distance of 2 million kilometers (1.6 million miles), by Pioneer 11. *The rings, whose shadow is clearly visible on the planet's disk, parallel to the equator, have an unusual appearance since the observation is made from their dark side; the brighter they appear, the better they allow light to penetrate. Just below the planet, the satellite Rhea can be seen. Photo: IPS = P.P.P.*

The Titius-Bode Law

The term Bode's law, or Titius-Bode law, is used to designate an empirical relationship that approximately states the average relative distances of the planets from the Sun. Discovered by 1741 by the German astronomer Wolf and rediscovered in 1772 by his compatriot Johann Titius, it is known mainly through the efforts of Johann Bode, who gave it a precise mathematical formulation in 1778. According to this law, the planets' average distances from the Sun (using the Earth's distance as the basic unit) are found by taking the series 0 3 6 12 24 48 96, etc., in which each number, beginning with the third, is the double of the previous one; then by adding 4 to each of these terms, producing a new series: 4 7 10 16 28 52 100 196; and finally, by dividing each of the resulting numbers by 10, which yields 0.4 0.7 1 1.6 2.8 5.2 10.0 19.6, etc.

Mathematically, the relationship can be expressed by a geometrical progression: $D = 0.4 + 0.3 \cdot 2^n$ with $n = -\text{infinity}$ for Mercury, $n = 0$ for Venus, $n = 1$ for Earth, $n = 2$ for Mars, etc. and D being the distance from the Sun of the planet in question in astronomical units.

Bode's law provides a good approximation as far out as Uranus, since the difference between the distance it gives for a planet and the average real distance from the Sun never exceeds 5 percent. The true values are: 0.39 (Mercury); 0.72 (Venus); 1 (Earth); 1.52 (Mars); 5.20 (Jupiter); 9.55 (Saturn); 19.22 (Uranus). For Neptune, however, the discrepancy exceeds 22 percent, and it reaches 49 percent for Pluto.

5. Origin and Formation of the Solar System

We do not know with certainty if there exist other solar systems in the galaxy or in the external galaxies. It is suspected that there are a few planets around one or two nearby stars—Barnard's star and ε Eridani—but their orbits cannot be determined. The luminosity of a planet is so low compared to that of the star around which it revolves that we do not yet possess the technical means to observe one directly. We expect to be able to do so within the next 10 years, mainly because of the space telescope.

If solar systems are very numerous—if a large fraction of stars is accompanied by a string of planets like our own—then the evolution of the solar system can be studied by the usual method of the natural sciences: systematic observation of more or less similar objects. Classification, the search for common characteristics and the development of a chronology are the familiar aspects of this approach.

If, on the other hand, planetary systems are infrequent in the galaxy, if it turns out that the formation of such systems demands completely exceptional circumstances, then studying them will be considerably more difficult. A biologist in possession of only a single tree of a given species would be hard put to study its evolution.

Here we will assume, as do almost all authors, that the birth of solar systems is a frequent occurrence. This hypothesis is not unfounded, but it has not been proved. It can be confirmed or invalidated only through direct observation of other planetary systems. This hypothesis will influence the discussion that follows. If it is false, many statements will have to be revised.

The Story Told by the Embryos of Stars

IT IS GENERALLY AGREED that the Sun and planets were born at approximately the same time, or, more precisely, that they all issued from the same interstellar cloud. This assertion is based on several indirect arguments, all more or less valid, to which we will return in due course. It may be considered plausible, and indeed even probable. However, given the great complexity of the subject, it is also possible that the arguments may still have been misinterpreted.

The appropriate method consists of searching for and observing places where stars are being formed today. To do this, we rely in part on the fact that certain very massive blue stars (20 to 50 times more massive than the Sun) have an extremely short life-span, in astronomical terms. These stars (known as O and B stars) are extremely bright and burn with such vigor that they exhaust their store of nuclear energy in less than 4 million years. Furthermore, we notice that these stars are always clustered in certain well-defined regions of the sky. These groupings, called OB associations, do not contain only massive stars, but also contain stars of lesser mass (as little as one solar mass or less) having a life-span of up to tens of billions of years.

Spiral galaxies are, in fact, double spirals. *The outer part of the arm (the blue spiral) is composed of a string of supergiant blue stars grouped together in associations (stars of the O and B type) and nebulae of ionized hydrogen (H II regions) produced by these stars. The inner part of the arm (the black spiral) is made of a string of dark interstellar clouds, inside of which bright stars are being born. The galactic matter, including both clouds and stars, moves in relation to the design of the galaxy: they go from one arm to another, following a path similar to the one shown here.*

In external galaxies, the OB associations are arranged like beads on a string that is wound up to form the spiral arms. These associations are visible not only because of the extreme brilliance of the blue giants (up to 100,000 times the luminosity of the Sun), but also because these very hot stars (more than 20,000°K) excite and ionize hydrogen atoms in the interstellar medium. These atoms glow red and form what are called ionized hydrogen regions (or H II regions) that are visible around the blue giants along the arms of the spiral.

In our galaxy, the OB associations are always found in proximity to the Milky Way, closely intermingled with the large, dark clouds that mark the central plane of the galaxy. A well-studied OB association can be seen in the constellation Orion. We are going to use it as the model for our study. In fact, there is nothing special about it. Many groups have the same characteristics. We are taking it simply because, being relatively nearby in the galaxy, it is the one about which we have the most information.

First, a large portion of Orion is obscured by a huge, opaque cloud of interstellar matter. This cloud is visible in millimeter-wavelength radiation. Its luminosity stems from the carbon monoxide (CO) molecules within it. It is actually two clouds, called "Orion A" and "Orion B," the dimensions of which are measured in several dozen light-years, and the densities in thousands of atoms per cubic centimeter (whereas outside the cloud, the density is closer to one atom per cubic centimeter). The total mass of the cloud is several hundred thousand solar masses.

How was this cloud formed? We do not really know. It is believed that the generation of such clouds is tied to the existence of spiral galaxies.

The M-51 spiral galaxy and its companion, in the constellation of Canes Venatici. *Located about 15 million light-years away, this system appears face-on, so that its structure is particularly visible. The spiral arms can be seen clearly, highlighted by stellar complexes and by hydrogen clouds associated with these young stars.*

It has been noted that such galaxies actually form double spirals. Each arm has two components—a bright one, made up of OB stars and H II regions; and, inside it, a dark component, made up of a series of dark clouds like the Orion cloud. We now know that a given mass of galactic matter—whether stars or clouds—does not remain fixed in any part of the pattern formed by the galaxies; it circulates, gradually passing from one arm to the next while rotating around the center of the galaxy. The arms are like stationary waves on the surface of a raging storm. The masses of matter move in such a fashion that they always approach an arm from its concave side. Owing to the forces of gravity, the matter slows down abruptly as it enters the arm. This slackening of speed causes a kind of avalanche that compresses the gas, makes it more opaque and thus generates a series of dark clouds all along the concave part of the arm. We will call them protocluster clouds, for reasons that will shortly become clear.

Several years ago, the presence of a more condensed concentration of matter was discovered inside the Orion A cloud. This type of subcloud, located only a few light-years from the northwestern extremity of the Orion A cloud, contains about 1,000 solar masses of gas, at a density about 100 times greater that that of the parent cloud itself. This cloud is visible because of the microwave and millimeter-wave emission of a large number of molecules, including formaldehyde (H_2CO). It is called the Kleinman-Low nebula, or KL, after its discoverers.

In the center of the KL nebula is a small cluster of about a dozen infrared sources. These sources are probably embryonic stars (or protostellar nebulae), the age of which is measured in tens of thousands of years. They interest us particularly because they are believed to represent the very first stage in the life of stars.

We have so far witnessed three stages of fragmentation and condensation: that which brought forth the protocluster nebula from interstellar matter; that which leads from that nebula to the KL nebula; and that which leads from the KL nebula to the protostellar nebulae. At each stage the gas becomes more and more opaque. In a protostellar nebula, the heat given off by condensation progressively warms the matter, which consequently becomes a very powerful infrared source. Analysis of the infrared radiation shows the presence in some cases of an absorption feature at wavelengths close to 10 micrometers. This absorption indicates the presence of silicates, probably in the form of dust grains concentrated around the sources. Does this represent the first stages of formation of a planetary system? The study of certain very young stars has sometimes suggested the existence of thick, gaseous, dusty disks surrounding a central star. These facts spur our imagination, encouraging us to believe that it was around this time that the "solar systems" began to form. The clusters of protostellar objects exhibit a curious behavior that is not yet generally understood. They contain a number of intense sources of radiation called masers. These are microwaves emitted by OH and H_2O molecules, the characteristics of which—intensity and intensity ratios at different wavelengths—reveal rather peculiar behavior of the matter inside the source. The existence of this phenomenon contains potentially valuable information about the mechanisms by which stellar systems form; however, we do not yet know how to decode this message. The multiplicity of maser emission sources of various wavelengths within the infrared cluster is such that we are not very sure which sources correspond to distinct stellar objects and which to different parts of a single large object. However, it already seems clear that the average distance between future stars is at least 100 times smaller than the typical distance between the stars of our galaxy. For example, there are about four light-years be-

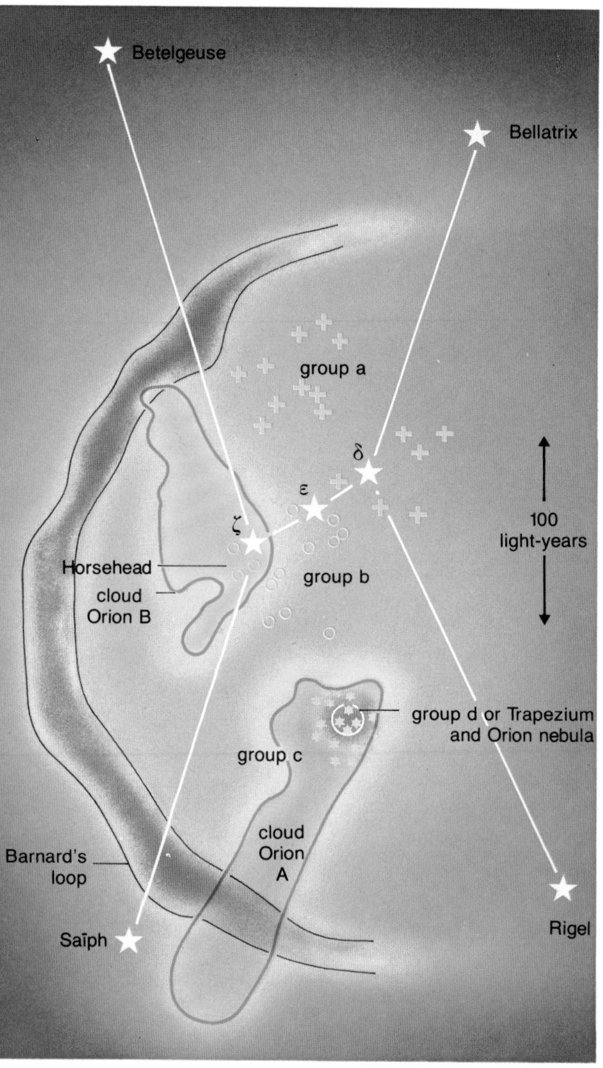

The constellation Orion contains a virtual breeding ground of stars. *Orion A and Orion B are two massive interstellar clouds. They are presumably the residues of a much more massive cloud that formed about 10 million years ago. In the vicinity of these clouds, five groups of stars of different ages can be identified. Groups a, b, c and d contain many blue supergiant OB stars (these are OB associations). The symbols (+), (o) and (*) denote the stars of groups a, b and c, which are 8 million, 5 million and 3 million years old, respectively. The fourth group (d), or the Trapezium, which is much smaller, is less than 1 million years old. It is in the center of the Orion nebula (a region of ionized hydrogen). A fifth group, called an infrared cluster, is in the center of a subcondensation called the KL (Kleinman-Low) nebula, very close to the Orion nebula. In the Orion B cloud there is another region of ionized hydrogen, one part of which is called the Horsehead Nebula. A large, luminous semicircle that surrounds the clouds and constellation is called Barnard's loop. The Orion clouds and associations are 1,600 light-years from us. The vertical line shows the linear scale (100 light-years) at that distance.*

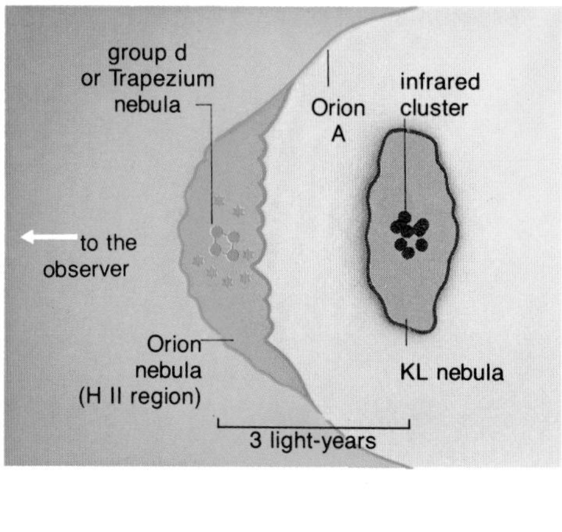

Enlargement of the area adjoining the northeast extremity of the Orion A cloud. *The Orion nebula contains the Trapezium cluster (or group d in the preceding figure), while the KL nebula contains the infrared mass cluster. The distance between protostellar objects in the infrared cluster is a few hundredths of a light-year. After Zuckerman and Evans.*

tween the Sun today and the nearest stars, while there is less than 0.03 light-year between the distinct infrared objects in a protostellar cluster. This "crowding" of the protostellar nebulae suggests that major interactions may take place between neighboring systems and that a hypothesis generally espoused by the vast majority of investigators—that the solar system and its processes of formation can be studied in isolation—must be seriously reexamined.

A few light-years from the infrared cluster is a second condensation of the protocluster cloud, the dimensions and mass of which are very similar to those of the KL nebula. But its physical appearance is completely different: It shines with a magnificent green glow, edged in red. It is called the Orion nebula and is a favorite spectacle of amateur astronomers. In the center of the nebula, four blue stars form the Trapezium. These are O and B stars of the kind mentioned earlier. The high-energy photons emitted at their surfaces penetrate deep into the surrounding gas and excite or ionize the atoms there. The green is produced by the discharge of oxygen atoms, while the red comes partly from hydrogen and partly from nitrogen. The Orion nebula is an example of an ionized hydrogen region (or H II region), which was discussed earlier.

The four stars of the Trapezium are the hottest members of a cluster containing several hundred stars concentrated in a volume of about one cubic light-year. The age of this cluster is about a half-million years. The infrared cluster and the Trapezium cluster may be considered as representing a kind of time sequence. In the same way that specimens of animal embryos aged one day, one week, one month, etc., can be seen in a natural history museum, Orion gives us successive views of the evolution of young stars.

The first stage in the evolution of a star consists of a gradual contraction of its volume, accompanied by an increase in surface temperature and an overall decrease in luminosity. This stage, called the Hayashi phase at the outset and later the T Tauri phase, ends when the central temperature of the star becomes high enough for nuclear reactions to take place. The star has then joined the group of stars known as "the main sequence." For a star of more than 15 solar masses, this initial phase lasts several hundred thousand years; for a star of one solar mass, it will last 15 million years. The Trapezium cluster will therefore include several stars already on the "main sequence" (see

The central part of the Orion Nebula, *photographed through a 3.05-meter (120-inch) telescope at the Lick Observatory. This is a region in which stars are being formed. The brightest stars that can be seen there are extremely young and massive. Destined to evolve rapidly, they will explode in several million years, enriching with heavy elements the interstellar medium out of which they were born.*

page 247), such as the Trapezium O and B stars themselves, but the majority of stars are still in the gravitational contraction phase.

The sudden "ignition" of the blue giants obviously has an important effect on the small cluster in formation. Not only do the blue giants light up the residual gas that gave rise to the Trapezium cluster, in a manner analogous to that of the KL nebula; apparently they are also responsible for heating it very quickly so that it dissipates into space. The geometric relationships are very revealing in this regard. The Orion nebula is, in fact, at the extreme end of the Orion A cloud, like the glowing end of a burning cigar, while the infrared cluster is just inside, like the fire kindling the glow.

Solar systems in formation cannot help but be affected by the abrupt lighting up of these giant stars. Of course, the total luminous intensity received by a protostellar nebula represents only about one thousandth of the intensity it receives from its central star. The important thing is that the ultraviolet flux issuing from the O star be several thousand times higher than that of the central star of the protostellar nebula. In fact, we can expect that a large portion of the nebula will be ionized. The heating of the nebular matter could then evaporate part of the residual gas of the nebula itself.

The chronology of events that transform a gaseous nebula into a planetary system is poorly understood. The phenomenon is estimated to last between 10^5 and 10^7 years. Assuming that the beginning occurs during the infrared "cluster" phase, it is not possible to say whether it finishes before what might be called the "Trapezium" phase, or whether, on the contrary, it continues well beyond that, during the later phases that will now be described.

In addition to the two stellar subgroups previously described, the Orion group includes three other subgroups located in well-defined areas, northwest of the Orion A cigar and west or southwest of Orion B. These groups are aged 3 million, 5 million and 8 million years, respectively. The volumes occupied by these groups are in increasing proportion to their ages. Moreover, a detailed analysis of the motions of the individual stars inside each group suggests that they are expanding away, to some extent, from a central location where they were concentrated at the time of their birth.

These various characteristics (age sequence, sequence of increasing volume and the expanding

motion of each of the subgroups) suggest that these subgroups represent the successive chronological stages of our stellar embryology museum. Resuming our journey, we now come to subgroup c, or the "sword" outside the Orion A cloud. Its age is 3 million years, its linear dimension is 15 light-years. This subgroup is in the immediate vicinity of the two clusters that have already been described. To continue describing it, it is necessary to present a few concepts of stellar evolution. (These will be developed more fully in a later chapter, "Structure and Evolution of the Stars.")

As mentioned earlier, the massive stars have very short life-spans. A star of 15 solar masses will use up all its nuclear energy reserves in 10 million years, while a star of 50 solar masses will use them up in 3 million years. The life-spans of stars with intermediate masses are appropriately scaled between these values. It is generally agreed that at the end of their lives, these stars rapidly go through a red-giant phase, then explode as supernovae (see p. 275). In this explosion, the products of stellar nucleosynthesis generated during the various phases of nuclear fusion that occurred during the life of the star are ejected into the surrounding medium at speeds of several thousand kilometers per second. The violent encounter of this torrent of matter with the surrounding interstellar medium creates a shock wave (supernova remnant). The matter ejected from the star mixes with the galactic gas. The mixture expands progressively into the surrounding space until, after about 100,000 years, it reaches the proportions of more than 100 light-years. A supernova remnant issuing from subgroup c would include in large part the initial parent cloud as well as all of the subgroups already formed. The matter recently ejected by the supernova is finally completely decelerated and integrated into the mass of the original cloud. A calculation shows that the dilution will be very great and that the totality of the new matter will constitute only a few thousandths of the mass of the initial cloud.

The age of subgroup c indicates that any star of more than 50 solar masses that had been there would already have exploded. Statistically, we can expect that there may have been one or two such stars. The impact of such an event on the younger clusters, particularly on the hypothetical protosolar nebulae of the infrared cluster, should not be understated: The average distance between the stars of group c and these clusters is less than 10 light-years; the luminosity attained by a supernova is 10 billion times higher than that of the Sun.

The Veil nebula in the constellation Cygnus. *It is being dispersed in space at a speed of 100 kilometers (62 miles) per second and is presumed to be part of the remnant from a supernova explosion occurring about 25,000 years ago. It is 2,500 light-years away.*

Several researchers have studied the effects of an encounter between a supernova shock wave and an interstellar nebula. Depending on the case and upon the density and temperature of the nebula, the collision may lead either to an expansion or to a greater contraction of the nebula. The flux of ultraviolet, X-I and gamma rays received by a protostellar nebula at the time of the explosion would be several million times greater than what it normally receives from its central star. This flux would be followed several hundred years later by the arrival of a shock wave preceding the remnant, then by the blast wave of the remnant itself.

It is possible that the enhanced density thus produced by compression of the cloud is responsible for the formation of new stars in a given region. The characteristics described earlier of the Orion subgroups do indeed suggest that the phenomenon of star formation travels through space something like a "forest fire." It is seen to progress from region a (northwest) to region b (the belt of Orion), then to region c (the outer sword of Orion) and finally to the youngest region, where the Trapezium and infrared clusters are found. It is notable that the age difference between these subgroups is about 3 million years, which is equivalent to the life-span of a very massive star. This kind of quantification of subgroup ages, and the numerical difference between those ages, clearly suggest that the propagation of star formation is indeed induced by the shock waves of massive supernovae.

Observation of the two older subgroups, a and b, shows that they contain no stars of more than 20 solar masses; this is not surprising, for their age is greater than the life-span of such stars. The implication is that these stars have already exploded and that their remnants have already contaminated the cloud with material enriched with new elements. Stellar statistics make it possible to estimate that about 20 massive stars explode within the various subgroups of an OB association. The new matter contributed by all these supernovae is a few hundredths of the mass of the initial cloud.

What happens next? The stellar expansion—which explains why subgroups a, b and c have increasing volumes—continues, making the identification of older groups, if they exist, very difficult. The stars are effectively dispersed in the gravitational field of the galaxy. Stars of the same generation are irreversibly separated among the vast galactic population. It would be very difficult now, for example, to recognize the stars born at the same time as our Sun.

In summary, the Orion protocluster cloud probably included, some 10 million years ago, not only the still-nebulous regions of Orion A and Orion B, but also the volumes now corresponding to the different stellar subgroups. For an as yet undetermined reason, a portion of the gaseous mass condenses and produces a stellar cluster in region a. Embryonic stars form there. The most massive ones, after about 10^5 years, become blue supergiants that drive out the residual matter with their stellar winds and radiation pressure. Then, after about 3 million years, these stars explode. Their expanding remnants enter the cloud, and set off the birth of new stars in region b. The same phenomena continue, and the conflagration extends to region c, while, in region a, less and less massive stars continue to explode. Finally, the region of the Trapezium and the infrared cluster is set aflame. The previous comparison with a cigar may be embellished if the subgroups are seen as successive puffs of smoke.

Around the Orion A and B clouds and the stellar subgroups, one sees a shining halo called Barnard's loop. We do not know much about its origin. Its location and orientation suggest that its origin is linked to the Orion association. It may be understood as having been generated and maintained by the succession of supernova remnants stemming from the massive stars in the association.

The Story Told by Atomic Nuclei

In the preceding pages, we tried to reconstruct the scenario of the birth of the Sun from observations of celestial regions where stars are being born today. Now, using observations of the material of the solar system, we are going to try to extract the elements that might permit us to reconstruct the course of events. We will see if these elements corroborate the story outlined so far, and, in addition, if they make a new contribution to the development of our scenario.

The study of the composition of the solar system from the chemical and isotopic standpoints is one of the most fruitful chapters of our archaeology. There exist in nature 92 separate chemical elements, ranging from hydrogen to uranium.

Each of these elements has a fixed number of orbiting electrons. This number is equal to the number of protons in its nucleus. A few are shown in the accompanying table.

Moreover, each of these chemical elements can exist as one of several isotopes, each corresponding to a different number of neutrons in the nucleus. Some of these nuclei are stable; others are unstable (or radioactive)—that is, they decay after a certain time into other nuclei. Carbon, for example, exists either in the form of carbon-12 (or C_{12})—stable, with six electrons, six protons and six neutrons; as carbon-13—stable, with six electrons, six protons and seven neutrons; or as carbon-14—unstable, made up of six electrons, six protons and eight neutrons, with a half-life of 5,400 years (the time taken for half of the ^{14}C nuclei to decay). Carbon-14 decays into nitrogen-14 (seven electrons, seven protons and seven neutrons). Thus the atomic number (here, 12, 13 and 14) are the sum of the numbers of protons and neutrons in the respective nuclei.

The chemical composition of the Sun, Earth, Moon and certain meteorites is now fairly well known. We also have some information about other planets. From the standpoint of isotopic composition, we have some data about the Sun (especially because of the solar wind), and terrestrial rocks, lunar rocks and meteorites have frequently been analyzed. We can state that all of these data are compatible with the following hypothesis: The Sun and planets issued from a single concentration of matter having at least a roughly homogeneous chemical and isotopic composition. We will see further that a more detailed analysis brings out minor but highly informative inhomogeneities. The major differences in chemical composition between the Sun and planets are directly attributable to differences in temperature. One of the key elements in these differences in chemical composition is the volatility of gases. Hydrogen and helium, which together form more than 99

Abundance of Some Elements Compared to Silicon

	Solar	Interstellar	Earth
H	25×10^9	25×10^9	trace
He	2.5×10^9	2.5×10^9	trace
Li	0.25	25	difficult to ascertain
C	12×10^6		difficult to ascertain
N	2×10^6		difficult to ascertain
O	21×10^6		3×10^6
Ne	1.7×10^6		
Na	50×10^3		120×10^3
Mg	10^6		90×10^3
Al	84×10^3	10^6	300×10^3
Si	10^6		10^6
K	3×10^3		70×10^3
Ca	65×10^3		90×10^3
Ti	2.2×10^3		110×10^3
Mn	8×10^3		1.8×10^3
Fe	10^6		90×10^3
Ni	50×10^3		130
Au	0.2		0.002
Hg	0.4		0.25
Pb	2.5		7

percent of the Sun's mass, are virtually absent from the inner planets. These gases have long since evaporated from those planets, which were not massive enough to hold them in their gravitational field.

While it is relatively easy to alter the chemical composition of a body through the effects of temperature, it is much more difficult to alter its isotopic composition (or, to be more precise, the relative amounts of each isotope of a given element). The next few paragraphs will deal successively with several elements and their isotopes. The observed differences between certain objects will provide information about the subject that concerns us.

Hydrogen exists as two stable isotopes: light hydrogen and deuterium (or heavy hydrogen). Observation of interstellar clouds shows us about 50,000 atoms of light hydrogen for each atom of heavy hydrogen. The same relative numbers are found on the planet Jupiter.

However, the relative number of deuterium atoms in the solar atmosphere is at least 100 times smaller. There is nothing surprising in this: The nucleus of deuterium is very fragile and can be destroyed by nuclear reactions at temperatures of less than 1 million degrees. Furthermore, the atoms of the solar atmosphere are continually transported by powerful convection currents to depths where the temperature is far greater than 1 million degrees. Consequently, the deuterium atoms of the original gas have long since been destroyed. The fact that deuterium atoms still exist on Jupiter, as on the Earth and in meteorites, shows that the matter of the solar system has never in the course of its history been heated to such temperatures. This statement makes it possible to eliminate several possible scenarios for the birth of the solar system. For example, it has been considered possible that the Sun originally formed a binary system with another star. Its companion might have exploded, and its debris might have given rise to the solar system. Another possibility that has been considered is that a star moving at random in the galaxy might have passed so close to the Sun that it caused powerful tides that tore off some surface layers. These layers were then left in orbit about the Sun and gave rise to the planetary system. These two hypotheses see planetary matter as originating in a superheated stellar interior and are therefore incompatible with the presence of deuterium atoms. On the other hand, scenarios based on the OB association with its infrared clusters and protosolar nebulae encounter no such difficulties here.

The element lithium tells virtually the same story in its own way. The amount of lithium relative to silicon (generally used as the norm of abundance for nonvolatile elements) is almost the same in meteoritic matter as in interstellar gas. However, on the surface of the Sun, this quantity is 100 times smaller. After deuterium, lithium is the element most susceptible to thermonuclear destruction. At a few million degrees, lithium rapidly disappears. There again, we conclude that the matter of the solar system was never very hot.

The C_{12}/C_{13} Abundance Ratio

The study of the isotopic composition of carbon is also very instructive. On Earth, on the Moon and in the meteorites, the abundance ratio of carbon-12 to carbon-13 is precisely 89. This value is the same in the solar wind, to within a few percent. Less accurate measurements taken on other planets are equally compatible with this finding. With an uncertainity of about 50 percent, comets also have the same ratio of carbon isotopes.

However, measurements made of the molecules of the interstellar clouds in our galaxy seem to give a value two-thirds as large, about 60. This difference may evidently be tied to the fact that the solar system represents a sample of interstellar matter as it was 5 billion years ago, at the birth of the Sun, while the clouds are a current sample. What could have altered the abundance ratio of these carbon isotopes in the meantime?

Stellar nucleosynthesis regards carbon-12 as a "primary" product and carbon-13 as a "secondary" product. A primary product may be generated in a "first generation" star—that is, one that contains no chemical element other than hydrogen and helium at the outset. A secondary element can be produced only from elements already generated in a star of the previous generation, then expelled into the interstellar gas, then reincorporated into a new star. Carbon-13, nitrogen-14 and nitrogen-15, for example, are produced from carbon-12 in the CNO or Bethe cycle. It can easily be seen that production of carbon in the galaxy originally favored carbon-12. Gradually, however, the situation changed, and the carbon-12/carbon-13 ratio declined, dropping, for example, from 89 at the birth of the Sun (nearly 10 billion years after the birth of the galaxy) to about 60 today (15 billion years after the birth of the galaxy).

This interpretation of the facts—very probable although not firmly established—interests us for the following reason: The homogeneity of the C_{12}/C_{13} ratio inside the solar system, compared to the different value observed outside it, suggests that the Sun and the solar system had a common place and time of origin in the galaxy.

Before going farther, it should be noted that there are physical and chemical reactions that can change the isotopic ratios of a given element, such as molecular exchange reactions that, depending on the temperature, may favor one isotope over another. These phenomena are used as a geophysical thermometer. For example, the measurements of the relative abundances of oxygen-18 and oxygen-16 are used to reconstruct the temperature of the ancient oceans. However, fortunately for our study, the variations of isotopic abundances created in this way are slight; they do not exceed a few percent. Moreover, their origin can be recognized so that they are not confused with authentic variations (that is, denoting a different isotopic composition at the outset) if the element in question possesses more than two stable isotopes. This is the case with oxygen, which is represented by three isotopes: oxygen-16, oxygen-17 and oxygen-18.

Isotopic Anomalies of Oxygen

One of the most important discoveries of recent years in the subject of interest here is that of the "isotopic anomalies" of oxygen. It has been shown that there are differences in the O_{17}/O_{16} and O_{18}/O_{16} ratios between certain families of meteorites, differences that cannot be explained by physical or chemical phenomena and that must be ascribed to different original compositions (to different "reservoirs" from which the matter was originally drawn). These differences are no more than a few percent at most and seem to involve only oxygen-16. It appears, for example,. that the Earth-Moon system contained a few tenths of a percent more O_{16} than the family of meteorites called ordinary chondrites, but up to a few percent less than the carbonaceous chondrites.

The origin of these isotopic differences of oxygen has been widely discussed. Within the astronomical framework presented at the beginning of this article, it is tempting to see them as the effect of successive additions throughout the existence of the protosolar OB association of ejecta stemming from supernova remnants. Furthermore, the nucleosynthetic evolution of massive stars predicts a large production of oxygen-16 (either through the fusion of helium during the red-giant phase, or through the later fusion of carbon-12) but little or no production of the oxygen-17 and oxygen-18 isotopes. It is to be expected, therefore, that the ejected matter would be rich in oxygen-16, foretelling a gradual increase in the $O_{16}/.O_{17}$ and O_{16}/O_{18} ratios in the gas surrounding the association. This interpretation, while tempting, is far from generally accepted.

Other chemical elements have also show anomalies in various meteorites. In addition to the rare gases (which will be examined later), these include magnesium, calcium, strontium, samarium, barium and neodymium. In each case, the physical and chemical effects can be separated from real abundance variations. These exist, but at rather low levels. The differences in isotopic ratios range from 1 to 0.1 percent. Curiously, these anomalies—except the magnesium anomaly—are nearly all concentrated in a single meteorite called Allende (see box, page 78), and, in that meteorite, only in a few inclusions. They correspond to the addition of very small quantities of certain heavy isotopes of the elements mentioned earlier. Like the atoms of oxygen-16, these atoms probably arise from the ejecta of O-star explosions in the protosolar OB associations.

Magnesium is a particularly interesting case, since it can be shown that the isotope that is generally overabundant, magnesium-26, comes from the decay of the aluminum-26 isotope, which has a half-life of 740,000 years. Analysis of certain meteoritic samples proves beyond the slightest doubt that at the time of their formation they contained atoms of aluminum-26. The quantities, although small (one Al_{26} nucleus to ten thousand of Al_{27}, aluminum's only stable isotope), mean that no more than a few million years can have passed between the moment of formation of these nuclei, very probably in a supernova, and their incorporation into the crystalline rock where they now appear in the form of a magnesium-26 atom. This observation is probably the strongest argument in favor of the presence of supernovae in temporal and spatial proximity to the young solar system as it was being formed. This proximity between dying and emerging stars fits naturally into the framework of an OB association as the environment in which the solar system formed.

It is important to distinguish clearly two distinct processes of nucleosynthesis. The traditional process is of very long duration and is therefore called the slow process. It implies, first, the production of heavy nuclei by all the known agents of nucleosynthesis—red giants, planetary nebulae, novae and supernovae of all masses; second, the expulsion of this matter into interstellar space, either in a continuous fashion by stellar winds, or in a more sudden fashion by stellar explosions; third, the mixture of this matter into the interstellar gas by the relative movements of interstellar matter, particularly by the differential rotation of the galaxy (in a period of about 100 million years [see page 284]); and finally, the incorporation of this matter into new stars formed in the vicinity of the arms of the spiral. The average duration of this cyclical process is several billion years.

The rapid process, described at the beginning of this section, implies only the very massive stars as agents of nucleosynthesis. The transfer of matter takes place without being diffused throughout the galaxy, inside the stellar association and over a period of only several million years. In essence, the existence of the rapid process was revealed by the isotopic anomalies of magnesium.

As for the noble gases, they pose special problems. Identification of the effects of abundance is much more difficult. These gases are subject to several processes of isotopic separation that have important effects. The most impressive example is that of neon, which exists in three stable isotopes: neon-20, neon-21 and neon-22. In the solar wind, the Ne_{20}/Ne_{22} ratio is 13; in the Earth's atmosphere, it is 7; in certain carbonaceous chondrites, it is less than 2. How could such huge differences have been produced? Here the idea of volatility must be introduced.

We mentioned earlier the interest in the carbonaceous chondrites, which are a good sample of the nonvolatile elements in the chemical composition of the Sun. These stones were rapidly degasified of the volatile elements they may have contained at the outset. This nearly total degasification may have caused a substantial isotopic separation: The lighter isotopes, such as neon-20, escaped more easily than the heavier ones.

Other phenomena may also have altered the isotopic ratios of the noble gases. First, there is the partial evaporation of a planetary atmosphere toward interplanetary space: The lighter atoms escape more quickly. There are also the nuclear reactions induced by the cosmic radiation from the Sun or the galaxy; they destroy nuclei already present, generating new but fewer nuclei, and their effect can be assessed only on isotopic species that are already very rare.

Significant isotopic variations among the elements helium, neon, argon, krypton and xenon are observed among the meteorites, the Earth's atmosphere and the solar wind. In certain cases, the effect of cosmic radiation can be recognized. In general, however, the importance of the physical and chemical effects discussed earlier is too poorly understood to identify the effects that can truly be ascribed to initial differences in abundance. For the time being, little information can be gleaned from these elements.

Radioactive Atoms

The radioactive species aluminum-26 has already been discussed. We will now consider elements whose average half-life extends over several billion years, such as uranium-235 (1 billion years), uranium-238 (6.5 billion years), and thorium-232 (20 billion years).

Around 1910, the British physicist Rutherford noted that unstable atoms are found on Earth with a half-life exceeding 1 billion years, such as uranium and thorium, but there are no atoms with a half-life clearly less than 1 billion years (aluminum-26 is not directly observed but is found as magnesium-26, into which it was transformed at the beginning of the solar system). He judiciously concluded that the Earth must be more than 1 billion years old but less than 10 billion years old (or there would be no more uranium-235). The decay of uranium and thorium produces lead. By comparing the amounts of remaining uranium and the amounts of lead accumulated by this decay, a time scale is imposed that makes it possible to estimate when the Earth, Moon and meteorites were solidified. For all three, an age of 4.6 billion years is found. The synchronization of these ages confirms that these bodies were formed at the same time (to within a few dozen million years). We have no data with which to determine directly the age of the Sun, but in the framework of its formation within a stellar OB association, it would be about the same age, give or take a few million years. The galaxy, which was born about 15 billion years ago, had already lived two-thirds of its current life-span at the time the Sun was born. The heavy atoms of the solar system were formed by generations of innumerable stars that followed one another during the first 10 billion years of the galaxy's existence. They enriched the interstellar cloud that gave rise to the Sun. A final addition of atoms, responsible for the isotopic anomalies mentioned earlier, was accomplished by the rapid process of nucleosynthesis, taking place within the protosolar association itself. This addition represents, at most, no more than a few percent of the atoms in the solar system.

Two other radioactive atoms must still be dealt with: plutonium-244 (120 million years) and iodine-129 (25 million years). Like aluminum-26, these atoms have not existed in the solar system for a long time; traces of their presence are found in the form of "radioactive fossils." Magnesium-26 was the radioactive fossil of aluminum-26; iodine-129 now appears in the form of xenon-129, into which it decayed; and plutonium-244 appears as fission products, including xenon isotopes.

The presence and abundance of these atoms in the beginning of the solar system provides information about the events that occurred 100 to 200 million years before the birth of the Sun. It appears that the next-to-last addition of matter to the clouds of the protosolar cluster (before the birth of the OB association and before its contamination by supernova remnants occurring in that group) took place 100 million years earlier.

We mentioned earlier that stars are formed only at the moment when the galactic gaseous matter, carried in its motion around the center of the galaxy, encounters an arm of the spiral. The matter is decelerated, compressed and becomes a dense cloud, like the large protocluster in Orion. Stars are formed, live, explode and contaminate their surroundings. The gas that was later to form the solar system thus received its last major addition from these contaminating atoms at the time of its next-to-last journey into an arm of the spiral. During its final passage into a spiral arm, this gas received the shock that transformed it into a protosolar nebula within a stellar association. This event was accompanied by a new contamination by O_{16} atoms and Mg_{26}, but apparently by negligible quantities of heavier atoms, such as iodine-129 and plutonium-244.

Thus, the comparative study of stable and radioactive isotopes makes it possible to retrace the chronology and physiognomy of events that gave rise to the solar system. This information, while fragmentary, is in accordance with the astronomical information described earlier; it confirms the image of the solar system arising from a stellar cluster similar to those that can be observed today.

The Story Told by Atoms and Molecules

WHAT WERE THE PHYSICAL PROCESSES and chronological stages that transformed the protosolar cloud into a sun surrounded by planets, satellites, asteroids and comets? Study of the chemical and molecular composition of objects in the solar system contains much valuable information on this subject, which must be unraveled and read attentively. For example, it has been observed that meteorites of the "ordinary chondrite" type possess an iron sulfate (FeS) component but do not possess ferrous oxide (Fe_3O_4). In the framework of the condensation of these solids in a gaseous nebula, this observation, according to the thermodynamics of these substances, implies a condensation temperature of 400°–600°K (125°–325°C).

Similar information may be obtained from the global properties of planets, in particular from their density. The farther one moves from the Sun, the less dense the planets are.

Furthermore, it so happens that as a rather general rule the densest chemical compounds are those that have the highest melting point. It may therefore be assumed that the farther a planet is from the Sun, the lower the temperature was in the place where it formed, in order to account for the correlation between planetary density and distance from the Sun. For example, Mercury is thought to have formed at such a high temperature (about 1,500°K) that only iron and a few particularly refractory silicates could have solidified. Venus, which formed at 1,000°K, probably contains, in addition, alkalines and aluminum silicates of smaller density. The Earth, condensed at about 550°K, would contain iron sulfide as well. At a lower temperature, substances of lower and lower density are condensed (hydrous tremolite, magnesium silicates, etc.); finally, in the region of the giant planets, ices such as water, ammonia and methane condense.

That the temperature at which planets are formed should decrease with their distance from the Sun is not in the least surprising. It is still the case today. But temperatures required are much higher than present values (fortunately, it is no longer 275°C at the Earth's orbit!). The initial heat might have come either from a much brighter sun, or from a greenhouse effect produced by the opacity of a nebula lighted from inside by a newborn sun. The rates of chemical reactions implied in the observations described here depend, first, on temperature, but also, to a lesser degree, on pressure, hence on density. Thus we can hope to obtain an estimate of the total mass of the protosolar

The Allende Meteorite

On February 8, 1969, at about 1:00 A.M., a large meteorite crashed near the town of Pueblito de Allende in the state of Chihuahua, Mexico. The meteorite exploded shortly before reaching the ground, but hundreds of fragments—the largest weighing 110 kilograms (243 pounds)—scattered over 150 square kilometers (93 square miles) and representing a total mass of more than 2 tons, were collected. Known since then as the Allende meteorite, this meteorite belongs to the class of carbonaceous chondrites; inside a blackish matrix of carbon and carbon compounds, many chondrules (small silicate spheres a few hundredths of a millimeter to several millimeters in diameter) can be observed, as well as inclusions rich in refractory elements such as aluminum, calcium, titanium, etc. Isotopic dating techniques have shown that carbonaceous chondrites are the most primitive meteorites in the solar system. Formerly, however, specialists had possessed no more than a dozen specimens of small dimensions of this type of meteorite. Thus the study of the Allende meteorite has contributed substantially to progress in research into the birth of the solar system.

The discovery within some inclusions of the meteorite of a series of elements with isotopic anomalies—i.e., differences in isotopic composition compared with their isotopic composition on Earth that could not be explained by processes at work within the solar system—rekindled the debate about the conditions that unleashed the formation of the solar system. In 1973, Robert Clayton, Toshiko Mayeda and Lawrence Grossman, of the University of Chicago, discovered that the oxygen contained in the inclusions of the meteorite was too rich in oxygen-16 and too poor in oxygen-17 and oxygen-18 in comparison with terrestrial standards. Then, in

nebula if we are able to determine even roughly the range of pressures that prevailed at the time of the condensation of solids. According to recent studies, particularly those done by Professor Anders' group at the University of Chicago, the necessary pressures would be from 10^{-3} to 10^{-6} atmosphere. This assigns to the nebula (minus the Sun) a mass of between a few hundredths of a solar mass and a few solar masses.

It is interesting to compare this physical and chemical estimate to another estimate of astronomical origin. The present mass of the planetary system is about one-thousandth of the solar mass. But this is only a lower limit to the nebular mass, for that nebula has certainly lost a large part of its initial volatile component. For example, the terrestrial planets have retained only the refractory component of the interstellar matter of which the nebula was composed. However, this refractory component is estimated at a thousandth of the initial nebular mass. The remaining portion, composed almost entirely of hydrogen and helium, was dissipated in space, probably expelled by a powerful solar wind emanating from the young Sun. We thus arrive at a new lower limit of about one one-hundredth of the solar mass. Other arguments, which are more or less convincing, suggest an even higher lower limit, of about one-twentieth of the solar mass.

Is it possible to set an upper limit on the mass of the nebula? Strictly speaking, no, but if we assume a very large mass at the outset, it will then be necessary to come up with a way to get rid of it. Astronomical observations of young stars (of the T Tauri type) show that these stars reject a large part of their mass in the form of a stellar wind of great intensity. This is obviously the means by which the undesirable surplus will be eliminated. But this mechanism has its limitations: It is very unlikely that it can efficiently disperse a nebula containing more than a few solar masses.

A Probable Scenario

FROM THESE TWO COMPLETELY different estimates, an image emerges of an original object of about one or two solar masses, very roughly compatible with the objects of an infrared cluster. (As an aside, we note that in this model, we recover something closely akin to Laplace's nebular picture presented at the end of the 18th century.) In this oblate, or disk-shaped nebula, flattened by its rotation and crisscrossed by the lines of force of the galactic magnetic field, a kind of dynamic quasi-equilibrium has been established, governed by the force of gravity, centrifugal force and magnetic force. The system slowly contracts; its density and temperature gradually increase, particularly in the vicinity of the axis of rotation, where the Sun will be.

Atomic collisions become more and more frequent. Molecules form. Interstellar dust grains, already present in the original cloud, serve as a center of nucleation. The nonvolatile elements rapidly condense. Under the influence of gravity, the dust grains fall rapidly toward the equatorial plane of the nebular disk, forming in this plane a new dust disk that is much thinner than the nebular disk itself.

In the region of the giant planets, the temperature is relatively low. The frozen water, methane and ammonia condense on the surface of the dust grains. The condensed fraction of the nebular matter is several times greater than in the vicinity of the terrestrial planets, where only the silicates and irons can take a solid state.

What makes the dust disk divide into rings, and what transforms each ring into a planet? This is the point that remains most unclear. We may, however, imagine that the orbital motion of the dust is relatively irregular, erratic and turbulent. Statistically, these motions generate fluctuations in the density: "clouds" of dust, in the same way that we saw clouds of interstellar matter form.

We could pursue the analogy and see these clouds collapse under their own weight when they have reached a sufficient mass, giving rise to a multitude of planetary nuclei. The planetary nuclei of the outer regions, already heavy with their icy components, will take further advantage of the low surrounding temperature, retaining a near totality of the hydrogen and helium in their immediate vicinity, whence comes the disparity in mass between the giant planets and the terrestrial planets, which are completely devoid of hydrogen and nearly devoid of ices. If we mentally add to the composition of the internal planets the complement needed to obtain a solar composition, we nearly arrive at the masses of the giant planets.

While the dust grains accumulate and generate the planetary nuclei, the diameter of the protosolar disk shrinks: The nebula contracts toward its center, leaving behind it a copious quantity of various-size celestial bodies, in the way that a low

__Fragments of the Allende meteorite,__ polished and then irradiated by electrons, reveal many mineral inclusions. This carbonaceous chondrite—unusual in its mass and in the abundance of such inclusions—enabled many isotopic anomalies to be discovered, clarifying hypotheses of the origin of the solar system. Photo: Department of Geophysical Science, University of Chicago.

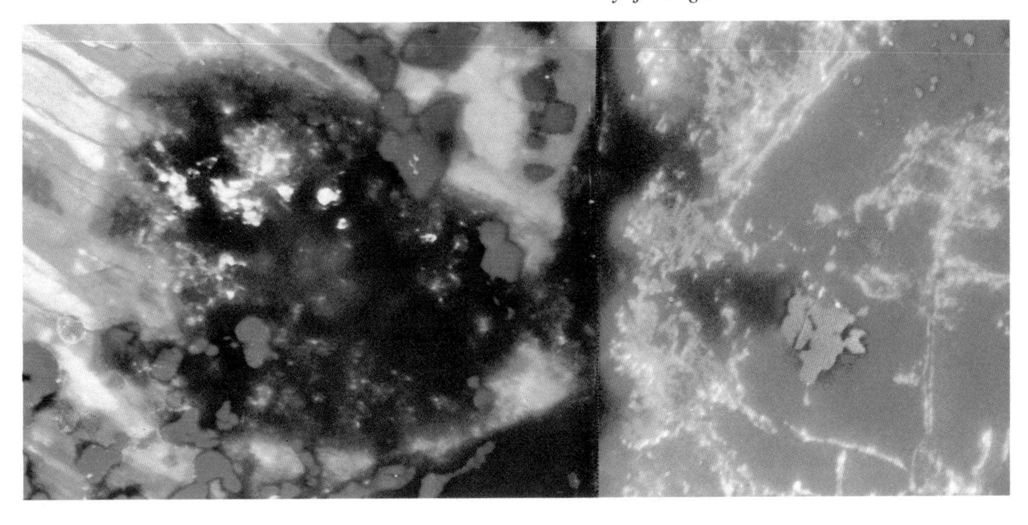

1976, Typhoon Lee, D. A. Papanastassiou and G. J. Wasserburg, of the California Institute of Technology, found an 8 percent to 10 percent excess of magnesium-26. In 1977, the same team discovered other isotopic anomalies in calcium, barium, neodymium, etc. These anomalies, which concern a dozen elements in all, have a varying degree of importance and are not correlated among themselves; thus, in some parts of the meteorite a normal calcium or magnesium can be found, accompanied by an abnormal oxygen, whereas in other parts, the reverse is true.

According to specialists, the elements characterized by isotopic anomalies were created by nucleosynthesis—i.e., by the series of thermonuclear reactions taking place within the stars and leading to the formation of heavier and heavier elements. They are thought to have been injected into the original solar nebula by the explosion of neighboring stars.

Since they were discovered, these isotopic anomalies have considerably changed the accepted notions of the formation and early stages of evolution of the solar system. Contrary to what was previously thought, they suggest that the solar nebula did not form an isotopically homogeneous cloud buffeted by internal turbulence but one that contained isotopic inhomogeneties both in the composition of the gas making it up and in the distribution of the particles of solid matter already present.

tide uncovers the ocean depths. The volatile atoms are pulled toward the axis of rotation, where the future Sun will be located. The contraction of the nebula is accompanied by a gradual warming. The heat is transferred to the surface by convection, as if the nebula had begun to "boil." At about this time, there occurs the birth of a solar wind, which intensifies and sweeps across the planets like a torrent. All atoms that are loosely attached to their parent bodies are swept away by this gale and expelled from the solar system. Can we go further and suppose that this heavily ionized wind substantially heated up the small bodies of the solar system? Certain physical and chemical phenomena observed in several meteorites imply that a warming up may have been caused in this way. But nothing really proves it.

Large fluxes of high-velocity particles, whose energy is measured in millions of electron volts, probably accelerated by solar flares of high intensity, accompanied this wind. Primordial dust grains now accumulated in meteorites reveal the existence of this radiation; their surfaces—saturated by the solar wind—were literally torn apart by the fast atoms. Up to 100 billion impacts per square centimeter can be observed.

The nebula continues to contract, gradually assuming the shape of the Sun as we know it. The increased thermal pressure begins to predominate over centrifugal force to counterbalance the force of gravity. The gaseous configuration is practically spherical. In the terminology of astronomers, it is entering the "main sequence" (see "Structure and Evolution of the Stars," page 245). The Sun's rotational deceleration will continue for a long time afterward. In the course of the billion years that follow its birth, the Sun's rotational speed goes from a few hundred kilometers to a few kilometers per second. We should note that this final deceleration phase took place long after the birth of the planets and has nothing to do with it. Contrary to what was long believed, the Sun did not transfer its angular momentum to the giant planets but lost it at the boundaries of the system, where the solar wind simply carried it away.

Interplanetary space at that time resembled a demolition derby. Movement within it was extremely erratic. A large number of small bodies, moving in extremely varied orbits, elliptical and inclined, crossed each other's paths, collided and shattered each other. Debris flew off in all directions. Because of their large surfaces and strong gravity, the most massive planets gradually cleared the great majority of small bodies from their paths. At that time the Moon felt the impact of the projectiles that gave it the "facial" features—resembling eyes, a nose and a mouth—we know today. In even larger numbers, they struck the surface of the Earth, but intense volcanic activity and atmospheric erosion quickly erased their traces. There was also great turbulence on the surface of the Sun, large and sudden variations in luminosity and large-scale flares.

Gradually, calm was restored. All that was left of the Sun's initial activity was its 22-year cycle (see page 93) and its low-intensity flares. The planets and asteroids that survived the great periods of collision silently pursued their course in peaceful orbits, perturbed only slightly by each other's gravitational fields. Only a few meteorites—probably stemming from the asteroid belt—or a few comets from the boundaries of the solar system recall the existence of this turbulent period, as do the photographs of the surfaces of the Moon, Mercury and Mars—and a number of planetary satellites pockmarked by meteoritic craters of all sizes.

The principal stages in the formation of the solar system, after G. Kuiper.

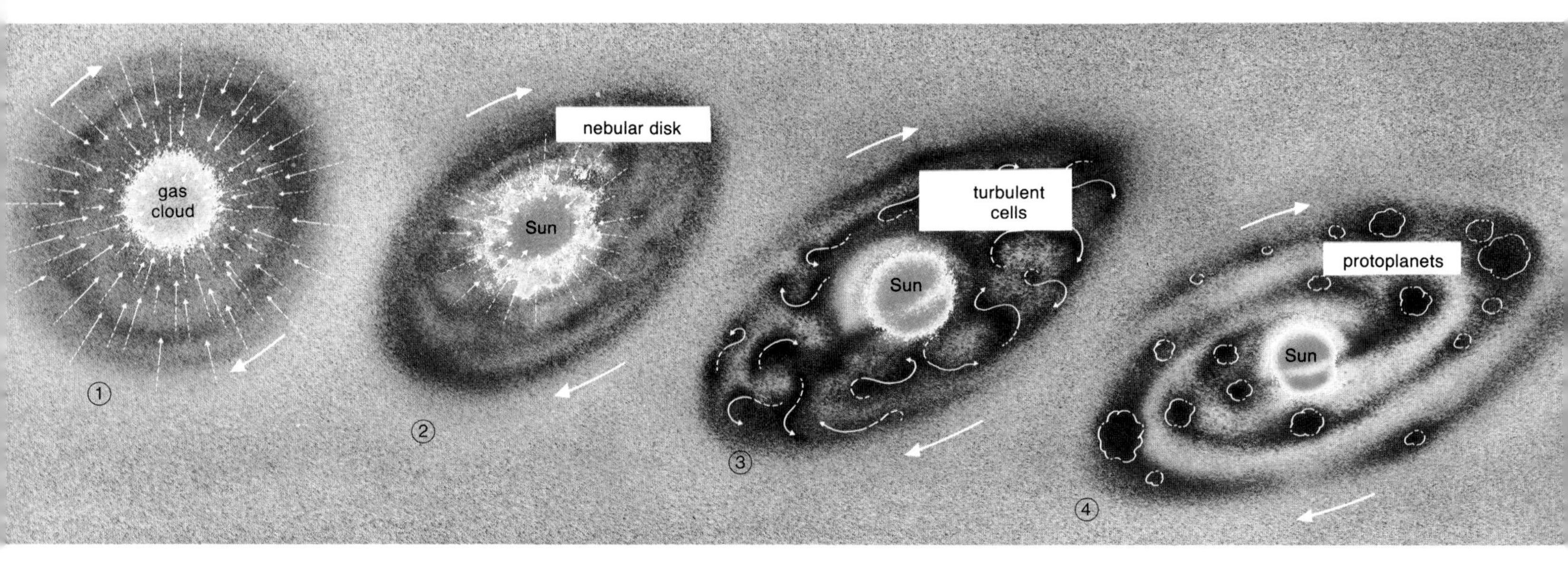

6. The Sun

The Sun is our star, an object long revered by men. Yet it was not until Copernicus came along in the 16th century, followed by Kepler, Galileo and Newton in the 17th century, that the Sun's role as the center of the system that bears its name was finally recognized. Small wonder that the Sun should be the focus of our attention, for our very existence depends on the energy it provides naturally and in abundance, as do most aspects of our daily activities. It is hardly surprising, therefore, that this celestial body—the largest and brightest of the visible objects that surround us—should have received special attention from observers throughout the ages.

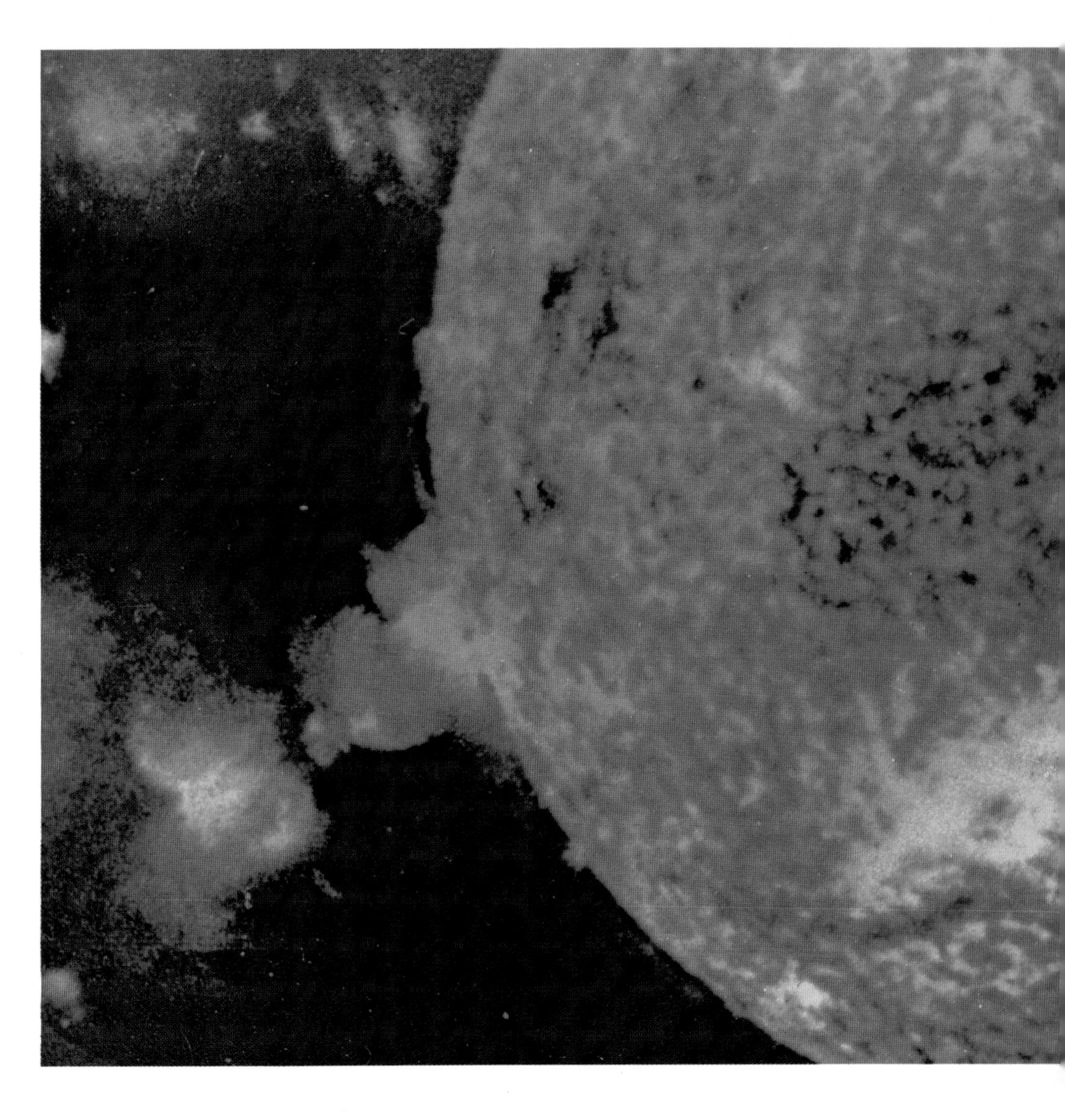

***Careful study of the solar prominences** (here photographed aboard* Skylab*) is a valuable source of information for astronomers. USIS Photo.*

Surprising, indeed, is this Sun—perfectly round in appearance, but surrounded, as revealed by eclipses, by a thin veil or corona. Its surface is astonishing—it appears to be uniform, but when examined more carefully, it is found to be studded with dark spots and is darkened around the edge. Astonishing, too, are the powerfully intense eruptions that appear with increasing frequency every 11 years. No less surprising are the Sun's steady pulsations. And what about its lack of neutrinos, which seems to indicate that what goes on inside the Sun is not what everyone thought? Is it really a star, in all senses of the term?

It is because we can today study the Sun in great detail that it is possible to develop theories regarding the evolution, structure and composition of the millions upon millions of other stars that make up our universe. When breakthrough occurs in our understanding of the Sun, the consequences for stellar astronomy are felt immediately. The Sun, therefore, is a kind of stellar laboratory—a primitive laboratory, one might say, at the risk of diminishing its grandeur and understating the complexity that remains, even now, one of its main characteristics. In the following pages, the reader will explore the mysteries of this body. It is, of course, within everyone's range of vision, but observing it in detail requires ever more sophisticated equipment that is beyond the means of the amateur astronomer. Solar exploration, like other kinds, has not yet produced absolute certainties, and it will often be necessary here to set forth theories and offer interpretations that may perhaps be contradicted a few months or weeks from now by new observations. The reader should not take offense at this; he or she is being kept abreast of the development of a vigorous branch of science. Solar researchers are undertaking ambitious projects with the means available to them. Clearly, the advent of space research and the use of ever-higher-performance telescopes have led to significant progress in solar research over the past 15 years. At the same time, tomorrow's progress will depend on the future methods available to researchers. So without further ado, let's look at the Sun.

A Ball of Gas

THE SUN IS ONE OF SEVERAL hundred billion stars that make up our galaxy. Located on one of the outer arms, at three-fifths the distance between the galactic center and the periphery, it manifests itself to us by the light it sends us, an analysis of which makes it possible to infer most of its outstanding properties: its division into different colors (spectral analysis), its distribution on the Sun's apparent surface, its degree of polarization, etc. Thus, for example, we can easily get an idea of the Sun's "temperature."

The Sun Is Hot

We have all observed that a surface exposed to the Sun's light—a sheet of metal, the soil, or simply our skin—is hotter than the same surface placed in the shade. Hence, the Sun's rays carry energy that heats up the obstacles barring their path: Mercury, Venus, the Earth, the Moon, Mars, Jupiter, Saturn, etc. The Sun is like a huge stove whose temperature can theoretically be measured by comparing its light intensity to that of a body of known temperature. This comparison shows rather easily that the Sun is much hotter than most of the hottest temperature standards commonly used in photometry laboratories. It is necessary to use plasma machines* (which need not be described here in detail) to simulate in the laboratory surfaces as bright and as hot. One may also compare the Sun's light to that of a body in thermodynamic equilibrium, or black body—that is, an ideal body that has no energy loss and whose temperature is solely characteristic of its physical state. The light of a black body conforms to Planck's law, named after a German physicist who, at the beginning of the 20th century, formulated it on the basis of the laws of thermodynamics, electromagnetism and the quantum theory. The validity of this comparison is questionable, since the Sun is obviously not in thermodynamic equilibrium; if we perceive its radiation, that means it loses energy. For the time being, we will use this hypothesis, which can, of course, give only an approximate idea of temperature. We thus find that the actual temperature (what it would have if it were a black body) is about 6,000 Kelvins** (more precisely, 5,770°K). At that temperature, in the laboratory, all bodies are in a gaseous state at normal pressure. But, of course, we cannot reproduce in the laboratory the pressure that exists on the Sun! The question of whether or not the Sun is in a gaseous state cannot be solved by this simple measurement. Other very simple arguments and observations enable us to reach that conclusion. First, an observation.

If we look more carefully at the solar disk, we see that it is not uniformly bright. Specifically, it is less bright on the rim (the limb) than in the center. This is easily explained if the Sun is a gaseous body; it also shows that its temperature is not equal everywhere to 5,770°K but that it varies, not so much over the surface, but rather internally, with the temperature decreasing toward the exterior. We might, however, hypothesize that since the Sun is a solid body, its edge is less bright than its center. But the Sun rotates on its axis, and in that hypothesis, we should then see this difference in intensity shift with the rotation. But this is not the case. The darkening of the limb is a property of the limb, and the only rational explanation is that the Sun is in a gaseous state, at least on the surface.

The Sun's Vital Statistics

Age	About 5 billion years
Average Location	At 3/5 of the distance from the center of the galaxy to its periphery; 149,600,000 km. from Earth
Radius	696,000 km. (about 10 times the radius of Jupiter; 100 times that of Earth)
Mass	1,989 million billion billion tons, or 332,946 times the mass of the Earth, or 700 times the mass of all the planets in the solar system
Average density	1.4 gm./cm.^{3}, or 0.25 times the average density of the Earth
Spectral type	G2 V
Effective temperature	5.770° K
Absolute visual magnitude	+4.83
Speed relative to other neighboring stars	19.7 km./sec.

*A plasma is a gaseous mixture containing variable proportions of neutral atoms, electrons and ions—that is, atoms that have lost one or more electrons.

**The temperature scale in degrees kelvin (K) is found by adding 273° to the scale of degrees Celsius (°C).

Moreover, it is also gaseous underneath. A rather simple calculation can demonstrate this, once we know its mass and radius, which are deduced easily from the laws of celestial mechanics. Under the hypothesis that it is composed mainly of hydrogen (which is the case, as we shall see) the laws of thermodynamics then make it possible to deduce easily both the pressure at the center and the temperature, which we thus find is greater than 10 million Kelvins. Our Sun, therefore, is only a very hot ball of gas.

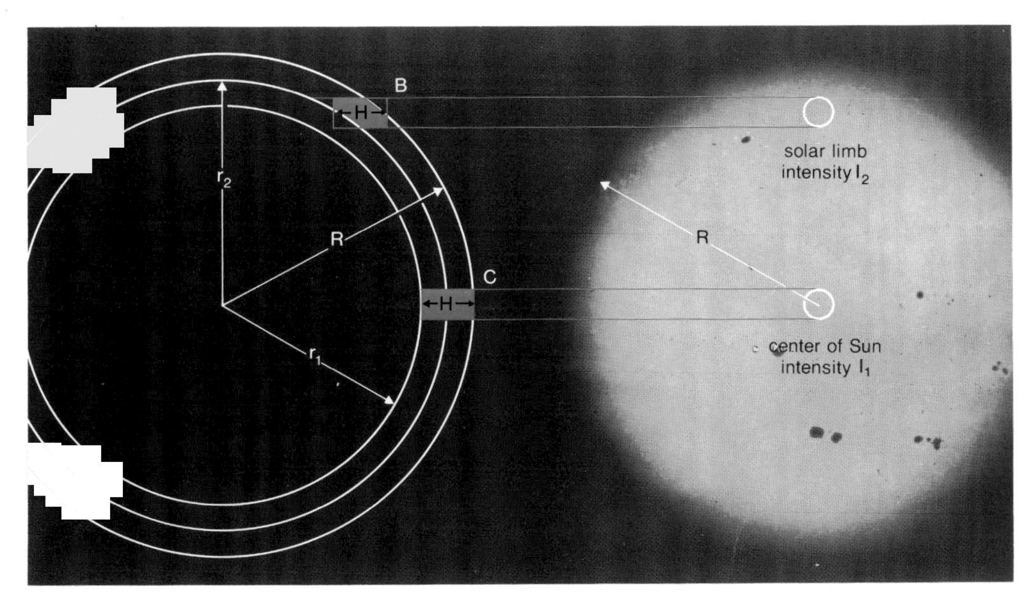

The phenomenon of limb darkening. *At left, a cross-sectional diagram of the Sun. Each circle represents a spherical volume element of solar gas. The light emitted per unit surface area at C, at the center of the apparent circular surface, results from the contribution of all the atoms contained in a column of height H contained between the spheres of radii r_1 and R. At the limb, the light emitted at B results from the contribution of all the atoms contained in a column of approximately the same height H between the spheres of radii r_2 and R. At right, a photograph of the Sun in visible white light, which thus represents the projection on a plane perpendicular to the luminous rays reaching us from the sphere whose radius is R. The light intensity observed at C being greater than at B, we conclude that the atoms between the spheres with radii r_1 and R are on average hotter than those between the spheres with radii r_2 and R, and hence that the temperature increases with decreasing radius.*

Solar Spectroscopy

Analysis of the spectral distribution of light shows us some very interesting phenomena. This analysis has been given the name "spectroscopy." It is the principal source of our knowledge of the Sun. It also provides additional proof that the temperature is not the same throughout.

Solar light striking a prism is broken into its component colors, which are exactly the same as those of the rainbow. All of the colors thus obtained constitute the Sun's spectrum. This reveals several dark bands: the absorption lines (see p. 96). These lines appear whenever an atom is subjected to ambient radiation. The different orbits of the electrons around the nucleus of the atom correspond to different energy levels; they are characteristic of a given atom. A particle of light, or "photon," with sufficient energy can, through the influence of its electric field, "excite" an atom from a lower energy level to a higher energy level. The result of this excitation is that the photon responsible is absorbed and thereby disappears. Once excited, the atom does not remain in that state very long. It rapidly returns to its initial state, reemitting a photon of the same wavelength. However, the probability that the second photon should be emitted in the same direction as the first is low, and when we observe in that direction, we get the impression that it has indeed disappeared. This is manifested in the spectrum by the presence of a dark band. Atoms have several different states of excitation and can therefore be responsible for the appearance of several lines having different wavelengths. An atom's state of excitation depends directly on the temperature of its immediate environment: The higher the temperature, the more it is excited. Temperature is really a measurement of the motion of gas atoms, or, if one prefers, of their kinetic energy: The higher the atom's speed, the greater the temperature.

The excitation of atoms in thermodynamic equilibrium results from the collisions they receive from all other particles (electrons, atomic nuclei, etc.). In a gas in thermodynamic equilibrium, each microscopic process is exactly counterbalanced by the opposite process: For each atom that gains energy during a collision, another atom loses the same amount of energy in an identical collision. Thus, at thermodynamic equilibrium, all atoms are at the same temperature; all photons have an energy that, according to Planck's law, depends on that temperature; and no absorption line appears in the spectrum. That such lines exist thus proves that, on the one hand, the Sun is not in thermodynamic equilibrium, and that, on the other hand, its temperature is not uniform, which we already knew. We can even confirm in this way the observation of the darkened limb, indicating that the temperature of the atoms in the Sun's outer layers is lower than that of deeper layers.

Chemical Analysis of the Sun

Spectroscopy informs us primarily about the chemical composition of the Sun. Since an atom that is subjected to the effect of radiation has a set of characteristic absorption lines, we can easily infer from their presence the existence of the responsible atom in the solar gas. This analysis is only qualitative, but we can thus observe that most of the elements are present.

Quantitative analysis is even more interesting, because it gives the proportions of the elements in relation to one another, or their abundances. In this type of analysis, one measures the quantity of light energy that has been subtracted to produce an absorption line. The more absorbing atoms there are, the stronger the absorption line.

However, this analysis is not simple, for the number of absorbing atoms depends not only on the total number of similar atoms but also on the proportion that are in an energy state where they can absorb the radiation, a proportion that depends on the temperature. The quantity of light energy absorbed also depends on the quantity of incident light. Furthermore, the temperature is a function of chemical composition. So we must proceed by a series of iterations. The use of more and more powerful computers now makes it possible to refine the values of the abundances deduced from absorption-line measurements.

However, some of these values are still very uncertain. This is particularly true of helium, the second most abundant element in the Sun after hydrogen; the determination of its abundance may be marred by an uncertainty that may reach a factor of 2. The abundances commonly accepted for the Sun are (by number): 92 percent hydrogen atoms, 7.8 percent helium atoms and 0.2 percent atoms heavier than helium.

We can therefore, as a first approximation, consider the Sun as composed principally of hydrogen, helium and traces of heavier elements.

The Solar Model

Knowing the Sun's mass, radius and chemical composition, we still need one more piece of information to calculate uniquely the temperature, pressure and density of the gas at each point between the surface to the center. That is the total luminosity, or the total amount of energy given off by the Sun per second. This energy is mainly emitted in the form of light (energy losses through particle emissions—the "solar wind"—and neutrinos are negligible). To find this value, we measure the solar constant S—that is, the illumination of a unit area of the Earth's surface. To prevent any modification (absorption) by the Earth's atmosphere, the measurement must be made by rockets or artificial satellites. The most precise measurement that has been obtained is $S = 1.367$ kilowatts/m^2, which is a fairly large quantity. The Sun's total luminous energy output is found by multiplying this value by the surface area of a sphere whose radius is equal to the distance from the Earth to the Sun, or 149 million kilometers (93 million miles) we get 383 billion billion megawatts, a gigantic number.

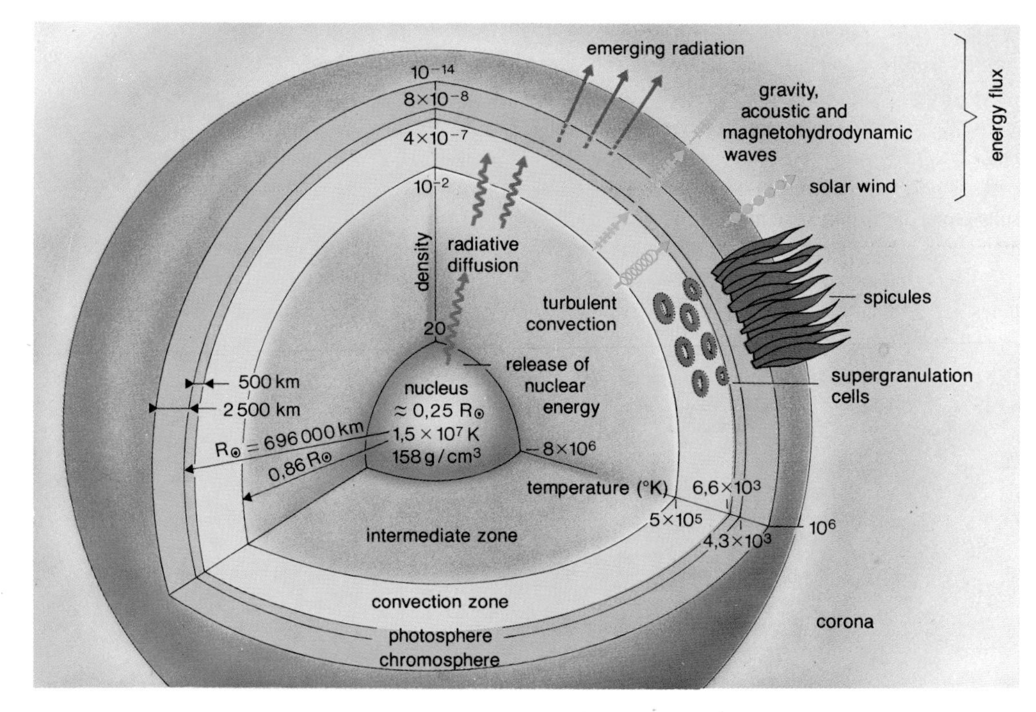

Schematic model of the major solar characteristics. *Temperature, density, structures and modes of energy transport from the center to the exterior. The structures are not to scale and can be used only as a qualitative description. The symbol R_0 represents the Sun's radius. From Gibson,* ***The Quiet Sun.***

Having established these factors, all that remains is to solve a series of equations embodying the following hypotheses:

1. The gas pressure at a point on the Sun is equal to the weight of the layers lying above that point (law of hydrostatic equilibrium).
2. The Sun's mass increases regularly from the center to the surface, in proportion to the density (law of mass conservation).
3. Solar energy produced at the center (we will see how later on) is transported to the surface either by light (radiative transport) or by convection movements (convective transport). Moreover, we assume that there are no other energy sources apart from the central source.

The equations implied by these hypotheses result in precise values of the different parameters that describe the physical conditions of the Sun's interior and make it possible to establish what is conventionally called the "solar model." Thus we find that at the center, the temperature is 15 million degrees, the pressure is equal to 340 billion times the Earth's atmospheric pressure and the density is equal to 114 times the Sun's average density, or 158 gm/cm^3. These values diminish regularly with distance from the center.

Solar Energy

THE SOURCE OF THE SUN'S ENERGY remained a great mystery for a long time. It was not until 1920 that the British physicist Eddington suggested for the first time that this energy might result from thermonuclear reactions. It was difficult to be more precise at that time, as atomic physics was still in its infancy, and it was thus not until 1938 that two American physicists, Hans A. Bethe and

Charles Critchfield, proposed an entire set of reactions that are probably also the origin of the energy of most other stars.

In the Sun's present state (we will see at the end of this chapter that it is evolving, and that over time, other reactions may occur), energy results from the fusion reactions of four hydrogen atoms into a helium atom. During these reactions mass is lost, since an atom of helium is only 3.97 times heavier than an atom of hydrogen. This mass loss of 0.0075 gram per gram of fused hydrogen corresponds to an energy creation of which sunlight is the most directly visible manifestation: Each second, 500 million tons of hydrogen are "burned" to maintain the Sun's present brightness. Given the total quantity of hydrogen contained throughout the Sun, we see that it would require about 11 billion years to exhaust this precious fuel completely.

Two types of reactions are in competition in the Sun's interior: the proton-proton chain, and the carbon cycle. In the first kind, explained by Critchfield, two protons unite to form a deuteron (a combination of a proton and a neutron); a positron (equivalent to an electron, but with a positive charge) and a neutrino (an uncharged particle—hence its name—which, like the photon, has no mass) are also created in this reaction. The positron rapidly reacts with an electron, and the pair disappears, releasing a pair of gamma rays (very-short-wavelength electromagnetic radiation). The neutrino, which does not interact with anything, quickly leaves the Sun. The deuteron combines with another proton to form a helium isotope, helium-3. This new reaction is accompanied by the emission of gamma rays. The formation of helium-4 in a third stage may take place in several different ways. Most frequently, two helium-3 atoms are transformed into an atom of helium-4 and two protons. The accompanying figure diagrams these various reactions and details the possible branches, all of which lead to the transformation of four protons and two electrons into a helium atom, two neutrinos and gamma rays.

Another type of reaction is the carbon cycle, explained simultaneously and independently by Hans Bethe and a German physicist, Carl von Weizsäcker. A carbon atom reacts with a proton to form radioactive nitrogen, which immediately decays into a carbon-13 atom, a positron and a neutrino. Carbon-13 reacts with a new proton to form nitrogen-14, which then reacts with another pro-

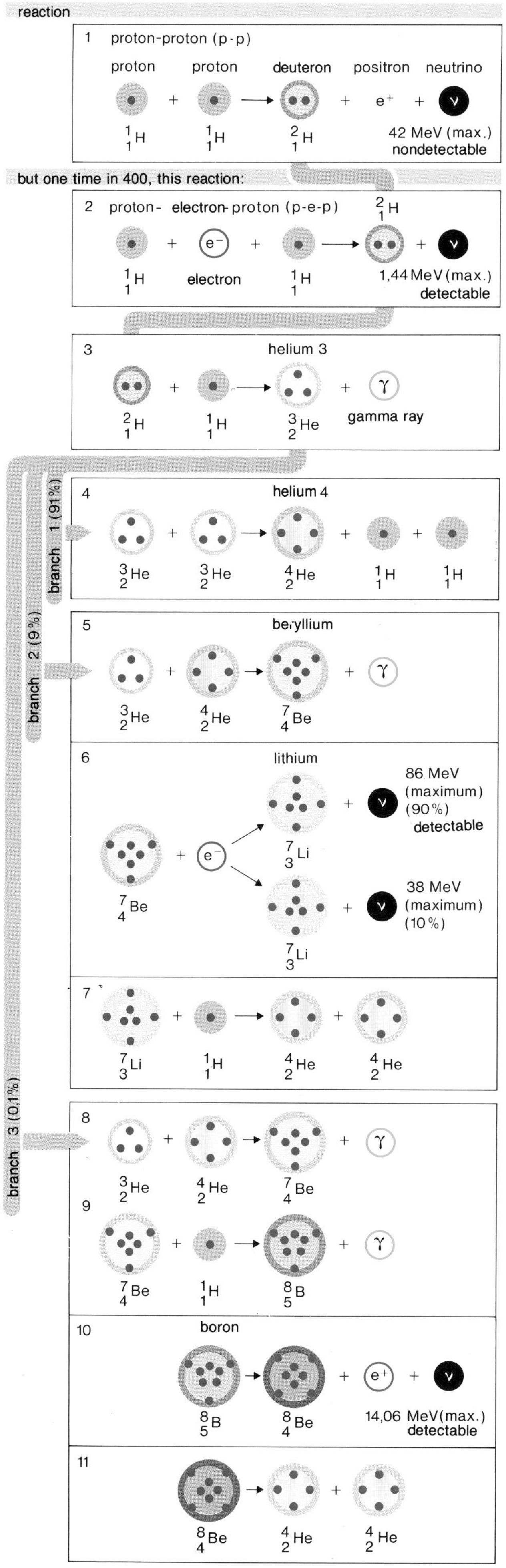

The "proton-proton" thermonuclear chain reaction probably is the main source of energy in the Sun. *The first reaction (1) completely governs the way in which the others take place. The neutrinos produced in this reaction cannot be detected by the Brookhaven detector. Reaction (2), which involves a proton, an electron and a proton, hence its name, occurs only one-four hundredth of the time. The neutrinos emitted during this reaction can be detected by the Brookhaven detector. After the fusion of the deuterons and protons that form helium-3, the cycle divides into three branches, where only reactions 2, 6 and 10 emit detectable neutrinos. From* ***Scientific American.***

reaction
1 proton-proton (p-p)
proton + proton → deuteron + positron + neutrino
^1_1H + ^1_1H → ^2_1H + e^+ + ν
42 MeV (max.) nondetectable
but one time in 400, this reaction:
2 proton- electron- proton (p-e-p)
^1_1H + e^- (electron) + ^1_1H → ^2_1H + ν
1,44 MeV (max.) detectable
3 helium 3
^2_1H + ^1_1H → ^3_2He + γ (gamma ray)
branch 1 (91%)
4 helium 4
^3_2He + ^3_2He → ^4_2He + ^1_1H + ^1_1H
branch 2 (9%)
5 beryllium
^3_2He + ^4_2He → ^7_4Be + γ
6 lithium
^7_4Be + e^- → ^7_3Li + ν 86 MeV (maximum) (90%) detectable
^7_4Be + e^- → ^7_3Li + ν 38 MeV (maximum) (10%)
7
^7_3Li + ^1_1H → ^4_2He + ^4_2He
branch 3 (0,1%)
8
^3_2He + ^4_2He → ^7_4Be + γ
9
^7_4Be + ^1_1H → ^8_5B + γ
10 boron
^8_5B → ^8_4Be + e^+ + ν
14,06 MeV(max.) detectable
11
^8_4Be → ^4_2He + ^4_2He

ton to form a radioactive isotope of oxygen, oxygen-15. This isotope decays into nitrogen-15, again emitting a positron and a neutrino. During the final reaction in this cycle, nitrogen-15, reacting with a fourth proton, reproduces a carbon-12 atom and a helium atom. Thus, the carbon-12 used at the beginning is regenerated at the end of the cycle; it serves as a catalyst. Each one of the reactions in the cycle is accompanied by the emission of gamma rays. The end result of this cycle is identical to the first one, since there, too, we have two neutrinos being given off and two electrons lost in the annihilation of the two positrons created.

In both cases, energy is created in the form of gamma rays, which the Sun converts to light, and neutrinos, which leave the Sun and do not contribute to the total amount of emitted light energy. The total quantity of energy generated is the sum of the energy produced in each of the reactions, quantities that are strongly dependent on temperature. At the Sun's center, at 15 million degrees, it is the proton-proton cycle that is most effective and contributes about 90 percent of the energy emitted. In other stars with higher temperatures, the carbon cycle may be predominant.

From the Center to the Surface

The energy generated in the central parts is not transported to the surface just as it is. That is fortunate, for, owing to its high concentration of gamma rays, it would not allow the development, much less the continuation, of life on Earth. As the gamma photons are transported to the surface, they undergo a transformation that results in their energy being spread over a much broader spectrum of longer-wavelength radiation. This transformation takes place by means of successive absorptions and reemissions along their path as they interact with solar gas atoms. The process of diffusion of this energy toward the surface is known as "radiative transfer." As the temperature varies regularly from the center to the surface, so does the spectral distribution of light. Richer in short-wavelength radiation (gamma, X and ultraviolet rays) in the deepest regions, it shifts into light closer and closer to the visible spectrum, finally acquiring, on the surface, that golden yellow color that is so familiar to us.

Hence, the photons' path is indirect, to say the least, and strewn with obstacles. We can calculate the time it takes a photon of a given wavelength to reach the Sun's surface and escape once it is emitted at the center. We arrive at the incredible figure of about 2 million years. If, like the neutrino, it encountered no obstacles, it would take only 2 seconds.

There exists another process that can, under certain circumstances, efficiently transport energy toward the exterior. That is convection. What would happen if, for example, a giant opaque screen prevented light from getting through? With the gigantic weight of the overlying gas layers maintaining an enormous pressure at the center of the Sun, the temperature would remain at values close to or higher than 15 million degrees, and the energy release would continue just the same. The Sun, unable to cool off by emitting light, would expand rapidly, releasing energy in a kinetic form, with gas escaping in all directions. This idealized example has only a remote connection with reality, but within the Sun there do exist "obstacles" that are not totally absorbing, as in the example, but only slightly, so that they do not lead to an overall expansion but rather to convective motions.

Within the Sun, there is a continual competition between the two modes of energy transport: radiation and convection. If the gas contained in a unit of volume is given a rising motion, two situations may occur: It may cool off faster or slower than its surrounding environment. In the first case, it becomes denser than the surrounding gas and tends to fall back down; then convection cannot be sustained. In the second case, however, it is lighter and continues to rise, according to Archimedes' principle, somewhat like a hot-air balloon. Convection is then efficient and sustained; it ceases only when the gas cools, either because it begins to radiate rapidly or because it breaks up into several smaller elements, and its temperature becomes equal to that of the environment.

The theory of convection within the Sun is certainly one of the areas in which much progress remains to be made. Nevertheless, we know that at a distance from the center equal to about 86 percent of the Sun's radius, the temperature has diminished by about fifteen times its value at the center, and that there the electrons that moved freely in the deeper layers recombine with the heavier atomic nuclei. They then absorb light immediately, forming the absorbing shield described above. Still closer to the Sun's surface, another complementary phenomenon amplifies the convection: The temperature becomes sufficiently low for helium and hydrogen, the Sun's major constituents, also to recombine with free electrons. This phenomenon, which resulted in sharply in-

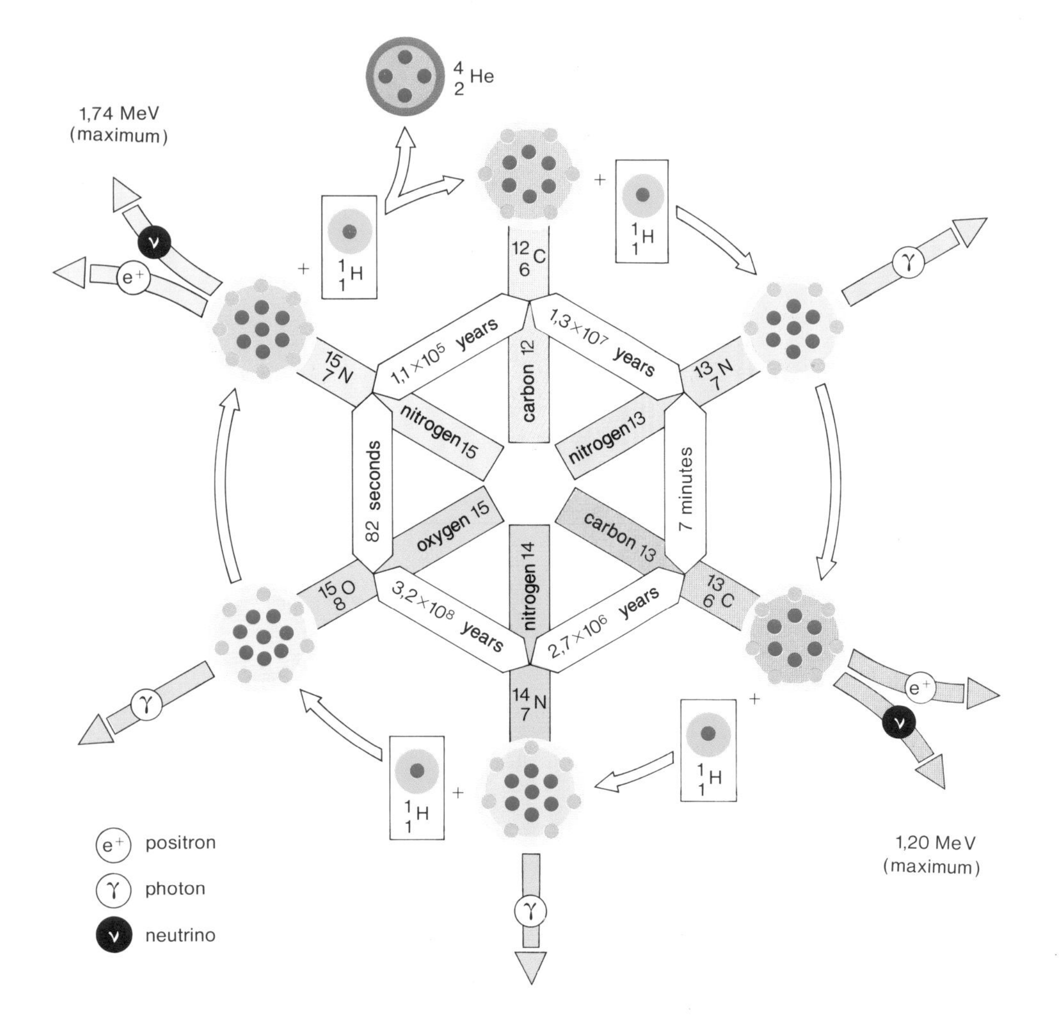

The carbon cycle of reactions** that also transform by fusion four hydrogen nuclei (protons) into one helium nucleus, uses normal carbon as a catalyzer; it is regenerated at the end of the cycle. Neutrinos are emitted during the second and fifth stages of the cycle. Because they share their energy with positrons, neutrinos have energies distributed over a wide range, the maximum values of which are expressed in millions of electron volts. Many of them cannot be detected by the Brookhaven detector. From **Scientific American.

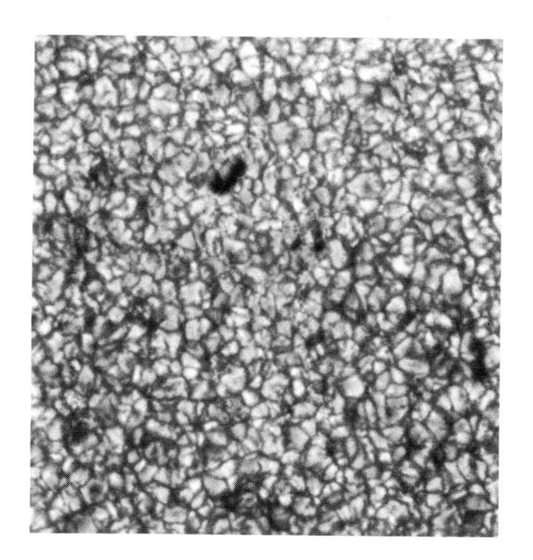

Solar granulation. *This photograph, taken with a telescope 50 centimeters (20 inches) in diameter in visible light (5,750 Å), shows the detail of the solar surface covered with a continuous network of cells, each of which represents the top of a convection tube; 1 millimeter in the photograph represents a distance of 285 kilometers (177 miles) on the Sun. O.P.M.T. Photo.*

creasing light absorption for the heavier elements in the deeper layers, has the effect here, owing to the large quantity of hydrogen and helium, of "braking" the cooling process, for recombination traps the energy without changing the temperature. When it occurs in a rising-volume element, the element cools more slowly, and the convection increases.

Although the phenomenon ceases before reaching the Sun's surface, where the density becomes too low and the number of atoms capable of absorbing light becomes insufficient, the convective motions continue as far as the photosphere and can be observed on photographs taken in visible light: They manifest themselves as "granulation," each "cell" of which probably represents the top of a convection tube. The average width of these cells is 1,000 kilometers, and there are more than 4 million of them at each point on the Sun's surface. The difference in temperature between a cell's center and its outer edge is 300°K on average.

Is the Sun Transparent?

We have already answered no to this question, since we know that light encounters several convection-producing obstacles on its route to the surface. The Sun's "transparency" depends upon the absorbing properties (that is, upon the absorption coefficient) and on the density of the gas of which it is composed. It is linked to what is called the "mean free path of a photon," which represents the average distance traveled by a photon between being emitted and reabsorbed. At the Sun's center, the mean free path is only a few tenths of a centimeter. On the surface, it may vary between a few dozen and a few thousand kilometers, becoming essentially infinite, of course, when the photon escapes into the interplanetary medium.

Hence, the layer from which the greatest part of the light we observe escapes is barely 200 kilometers (124 miles) thick, or a fraction equal to 0.0003 of the Sun's radius. It is useful to realize that our knowledge of the Sun has been inferred up to now from analysis of the light that is in fact emitted from this infinitesimal layer, which has been given the name of "photosphere" in the special language of solar physicists. (Incidentally, it is because this layer is very thin in comparison with the Sun's radius that the Sun's edge seems so sharp to us, which is surprising, after all, for a ball of gas.) This leads us to question the validity of our conclusions, particularly in regard to the Sun's inner structure.

The Neutrino Enigma

ONLY A FEW YEARS AGO, photons were our only source of information. But we saw that the generation of solar energy is accompanied by the creation of neutrinos, which can traverse the Sun's volume without interacting or changing in any way (the proportion of neutrinos absorbed while traversing the Sun is estimated at 1 in 100 billion).

If it were in our power to gather neutrinos and measure their flux, we would gain direct information on the physical condition of the Sun's core.

The experiment has been tried. Measurement is difficult, for, if neutrinos do not interact with the Sun's atoms, neither do they interact with other atoms. We therefore compensate for their low rate of interaction by using a huge quantity of the same

kind of atoms. This is the principle behind the detector developed at Brookhaven National Laboratory in the United States. The measurement uses the capture of a neutrino by an atom of chlorine-37. In this reaction, an atom of argon-37 and an electron are formed:

$$\text{neutrinos} + \text{chlorine-37} \underset{\text{radioactive decay}}{\overset{\text{capture}}{\rightleftarrows}} \text{argon-37} + \text{electron}$$

The reverse reaction is the decay of argon-37 by capture of an electron. Chlorine is practical from a chemical standpoint, for it is a very abundant element, relatively inexpensive and may be stored in very large quantities (several hundred tons). The most appealing chlorine compound is tetrachloroethylene C_2C1_4, currently in widespread use as a cleaning fluid. The Brookhaven detector thus incorporates a reservoir of more than 400,000 liters (105.675 gallons) of the liquid.

The reaction product, argon-37, is a radioactive noble gas, which separates from the tetrachloro-

The neutrino detector consists of a tank containing 400,000 liters (105,680 gallons) of tetrachloroethylene. *It is placed in a mine shaft 1,500 meters (4,922 feet) below the ground surface, near Lead, South Dakota. Photo supplied by R. Davis, Brookhaven National Laboratory, Associated Universities, Inc.*

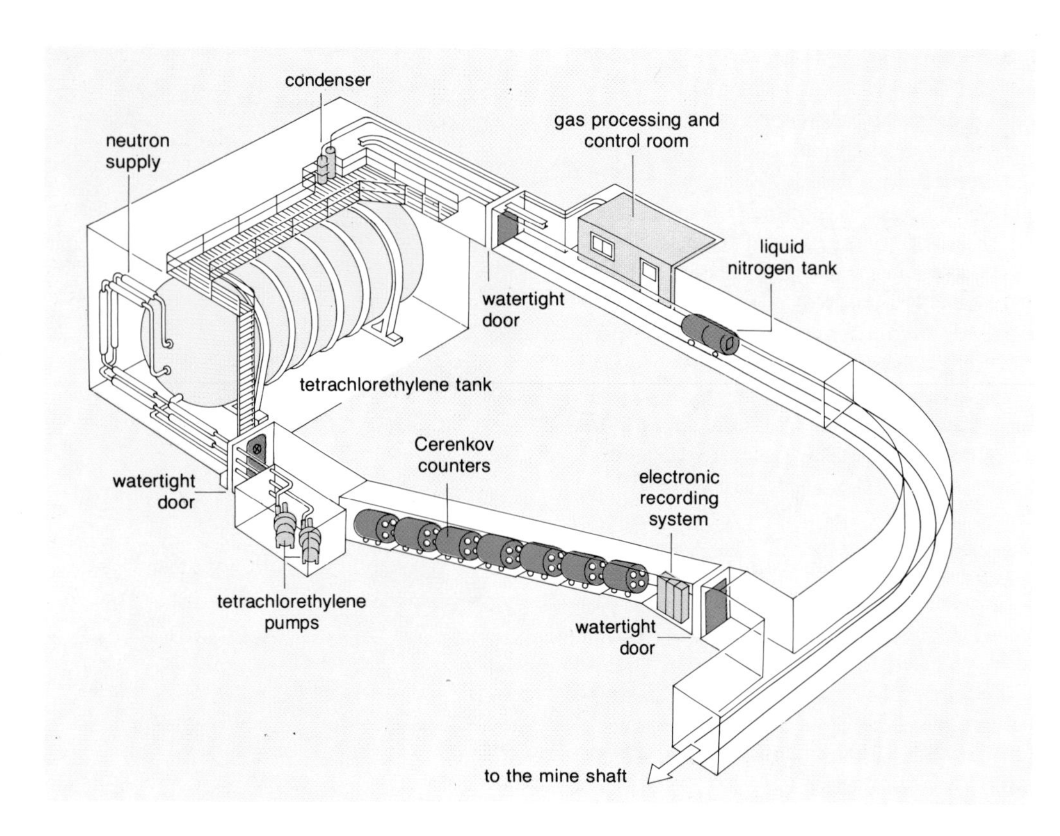

Diagram of the Brookhaven detector. *On average, each molecule of tetrachloroethylene, C_2Cl_4, consists of one chlorine-37 atom (containing 17 protons and 20 neutrons) and three atoms of chlorine-35 (containing 17 protons and 18 neutrons). When a neutrino of sufficient energy reacts with an atom of chlorine-37, it produces an atom of argon-37 and an electron. Argon-37 accumulates over several months and then is purged with gaseous helium. The argon then is absorbed in a cold trap, then measured. From R. Davis, Brookhaven National Laboratory, Associated Universities, Inc.*

ethylene molecule and reaches equilibrium with the rest of the liquid. The number of argon atoms thus created is measured by purging the liquid with gaseous helium.

The detector is located 1,500 meters (4,921 feet) underground, since it is necessary to protect it from cosmic rays, which can also create argon-37 and bias the measurements. The neutrinos are able to penetrate the overlying rocks with negligible alteration or absorption, while the cosmic rays are quickly stopped. As an additional precaution, the entire detector is placed in a water-filled basin, which serves as a shield against neutrons. The site chosen for the detector is an abandoned mine near Lead, South Dakota. The detector is exposed to solar neutrinos for many months, to enable a measurable amount of argon-37 to accumulate. After the helium purge, the argon is absorbed in a charcoal trap kept at the temperature of liquid nitrogen (77°K). The system is calibrated by adding to the tetrachloroethylene bath a known quantity of argon-36, a nonradioactive isotope of argon that is measured in the same way as argon-37. The method is effective, because 95 percent of the argon added is recovered. The collected argon is placed in a proportional counter with a volume less than 1 cm^3 (compare this with the 400,000 liters of C_2Cl_4), in which the few atoms of argon-37 produced by the capture of solar neutrinos may be counted, each one individually giving off a characteristic pulse as it decays. All of this "chemistry," as well as the experimental procedure, is being examined meticulously by the American experimenters, for the results of the measurements are surprising, to say the least.

Surprising Results

The flux of neutrinos is measured in SNU (standard neutrino units). One SNU is equal to 10^{-36} capture per second per chlorine atom in the detector. In other words, a chlorine atom subject to a flux of one SNU would have to wait about 10 billion times the age of our universe before being able to capture a neutrino. However, since there is an enormous number of chlorine atoms in the reservoir, it is necessary to wait only six days before detecting a SNU. The measurements reported during a symposium held in Bonas, France, in 1980 indicate that the Sun emits no more than 1.6 SNU ± 0.1. This figure should be compared to what we should expect from the different fusion reactions.

If the reactions characteristic of the carbon cycle are dominant, we should be able to measure 35 SNU. We are far from that figure, which, to a first approximation, confirms that this cycle is not the main generator of solar energy. To calculate the flux of neutrinos emitted during reactions of the proton-proton type, it is necessary to have a very precise model of the Sun, since in order to obtain a relatively accurate prediction, the central temperature must be known to better than 0.1 percent. In fact, the neutrinos emitted in the different reactions of the chain (see p. 86) do not all have the same energy, and chlorine-37 can capture only the neutrinos produced in reactions 2, 6 and 10 of the cycle. Those produced in reaction 1 are the most numerous, but their energy is too low to enable them to be detected. Nevertheless, the calculations have been refined to the utmost over the past 15 years, and they lead to an estimate of 7 SNU, in comparison with the 1.6 SNU detected. The factor of 3.5 that exists between these two figures is significant; it is substantially higher than the cumulative errors of calculation and measurement. We thus have a real problem on our hands. Where did the solar neutrinos go?

Despite numerous attempts to answer this question, it remains one of the most exciting enigmas in modern astrophysics. It may challenge commonly held ideas concerning the generation of solar energy, and of stellar energy in general. Such a result demands a critical analysis involving a close scrutiny of both the experimental method and the theoretical calculation.

The former has now reached such a degree of refinement, and has been subjected to so many verifications, that it may be considered truly reliable. If we wish to improve the measurement, we must first change the type of experiment and try to detect the neutrinos produced in reaction 1, where they are the most numerous. This can be accomplished by using as the target atom not chlorine, but gallium. However, to obtain significant results, the detector must contain 20 tons of that element, a quantity equal to eight times the world supply, and it seems unrealistic to imagine that production could be increased by such amounts merely for the needs of modern astrophysics.

We are left with the theoretical estimates. Here, two possibilities must be considered: Either the equations governing the thermonuclear reactions have been incorrectly calculated, or the solar model itself is deficient. The first possibility was tested in reactors at institutes of nuclear physics, particularly the Kellogg Laboratory of the California Institute of Technology, and the Los Alamos Laboratory; the reaction rates used in the equations were fully verified. Recently, experiments conducted in the United States and the USSR have led to the surprising hypothesis that neutrinos do in fact have a mass, and that different types exist, depending on whether they are attached to electrons ("e" neutrinos), muons ("μ" neutrinos) or tau particles ("τ" neutrinos). Neutrinos would thus oscillate from one type to another. Solar neutrinos are of the first type, and we can imagine that only a fraction of them are detected by the Brookhaven device, the rest being distributed between the two other nondetectable types. This discovery, if confirmed, would be enough to solve the enigma we are discussing. If not, we would then have to suspect our model of the Sun itself.

Thus the hypothesis has been put forward that the Sun's energy is not produced entirely by thermonuclear reactions but in large part by the attraction that the Sun's center exerts on the upper layers (energy due to the "gravitational collapse"). This assumes that the Sun's nucleus has a much greater mass than previously thought, but since the total mass is known, it implies that its internal distribution is not homogeneous.

Is the Solar Model Deficient?

The more likely hypotheses are those that challenge the model without abandoning the idea of a thermonuclear energy source. The most important parameter in this regard is temperature, for that is what determines the quantity of energy produced: The higher it is, the more effective the fusion reactions are. The temperature itself is determined by the abundance of elements heavier than helium, which, by being strong absorbers of radiation, act as a thermal shield. If they are more abundant, they raise the temperature, and vice versa. In order to reduce the neutrino flux, the temperature at the Sun's center must be reduced, hence the abundance of heavy elements. But how can this be accomplished without at the same time reducing luminosity?

The answer may stem from the difference between the time it takes the neutrinos and the photons to leave the sun: 2 seconds for the former, as against 2 million years for the latter. In other words, the neutrino flux corresponds to the Sun's present state, but its luminosity corresponds to its physical state as it was some 2 million years ago! The Sun, then, is in a transitory phase of its evolution, and since the neutrino flux is now too low (see earlier in our discussion), we should expect to see the luminosity diminish in a few million years. We can accommodate a cyclical process in the Sun if we assume that the intermediate elements produced in the different phases of the thermonuclear reactions are not immediately mixed and reused in the succeeding phases. This is true of helium-3, of which a large quantity can be produced if the temperature is not too high, before it takes part in the final reaction of the proton-proton cycle. The volume of unmixed helium-3 may cause cataclysmic pulsations of the Sun's interior. When it is finally mixed, yielding the final product of the reaction, helium-4, there is a very large release of energy. This causes a sudden expansion of the Sun, and through a relaxation effect, results in such a cooling that thermonuclear reactions are stopped. Such episodes are thought to recur every million years.

This theory, however, is contradicted by the study of our planet's past climate (paleoclimatology), for the variations in the Earth's temperature that would follow would be four times greater (about 20°C) than what paleoclimatology seems to indicate (5°C). This explanation, therefore, is probably incorrect.

The most direct and most logical conclusion from the neutrino deficiency is, after all, that the Sun's interior is colder than expected. This, as we saw, may result from a lesser abundance of heavy elements than is commonly accepted. The abundances used in the calculations result from spectroscopic measurements of the Sun's surface—and are supposed to represent the composition that existed at the time the Sun was formed, hence at the beginning of the thermonuclear reactions, before those reactions altered the composition. If a greater quantity of hydrogen were concentrated at the center, there would be more nuclear fuel available, and a lower temperature would be necessary to produce the same luminous energy, with a lower flow of neutrinos. The central temperature thus depends directly on the exchange of matter among the different parts of the Sun, hence on dynamic processes. The most effective process here is convection, assuming that we know how to calculate its effects well, which is far from the case. Moreover, convection occurs rather close to the surface, hence far from the region of energy production. Yet we are inclined to revise this model, since it leads to a disagreement between luminosity and the neutrino flux. If we assume that the Sun's interior is entirely convective, we can lower the central temperature sufficiently. But we know that the bulk of the energy is transported by radiation to the surface. In 1980, Evry Schatzman, a Frenchman, and André Maeder, a Swiss, developed a new theory of turbulence at the Sun's center that would lead to an increase in the concentration of hydrogen.

Another hypothesis is that the Sun was formed in two stages: An initial nucleus, with a mass about equal to half the present mass, would have condensed, with a chemical composition different from the current composition. The exterior, with the composition observed today, would then have attached itself to this primitive nucleus. Calculations show that such a system is compatible with measurements of the flux of light and of neutrinos but possesses a strongly convective central nucleus.

The diversity of these hypotheses shows clearly the extent to which the solar neutrino deficiency challenges what were believed to be the most solid notions regarding the internal structure of our star. But perhaps the answer lies elsewhere. On the basis of laboratory experiments, we saw that people have been led to hypothesize that neutrinos have a mass. If this were true, we might easily be able to solve the neutrino enigma without relying on other hypotheses that, we must admit, are uncertain, to say the least.

Solar Seismology

WE ARE NOW TRYING, however, to probe the Sun's internal structure. This is a new branch of solar research and astrophysics generally, and is receiving a lot of attention. It has been given the evocative name of "solar seismology," for its object is to observe the Sun's modes of vibration. In the few years since it was discovered that these modes actually exist, this science has made substantial progress. The simplest mode is that in which the Sun alternately dilates and contracts, a bit like a balloon that is periodically inflated and deflated. In another, more complicated mode, only a part of the Sun dilates, while another part contracts (the "football" mode). There are also "higher degree" modes.

It is not surprising that the Sun oscillates or vibrates, since similar phenomena have been observed in many stars other than our own, but what is surprising in this case is the relatively low magnitude of the phenomenon, which perhaps explains why it was not observed sooner.

There are two types of vibrations: acoustical vibrations, in which the gas vibrates like air in an organ pipe; and gravitational vibrations, in which the sun's surface oscillates like the sea, some parts rising above an average level while others drop beneath it. The study of the Sun's oscillations informs us of its internal structure, somewhat as the analysis of the sound rendered by a vase or a champagne glass tells us whether it is made of crystal or glass. The period of the oscillations is closely related to temperature and density distribution within the Sun. If it can be measured accurately, we then have a way of knowing these two crucial parameters of the solar model.

The methods used until now to detect the oscillations may be divided into three main categories according to whether they are based on the velocity at which the Sun's surface approaches or recedes from the observer; the variations of light intensity; or variations in the Sun's diameter.

The velocity measurements make use of the Doppler-Fizeau effect, which leads to a change in the period of an acoustic or electromagnetic wave emitted by a moving body. We have all noticed this phenomenon by hearing the change in tone of a car horn as it first approaches, then moves rapidly away from us. By an identical mechanism, the light given off by the moving parts of the Sun undergoes a translation in wavelength. If these movements are oscillatory and periodic, the resulting translational movements will be likewise. The detection technique thus consists of measuring the position of a spectral line that follows the Sun's periodic movements.

The Sun's Diameter Varies!

Velocity variations with a period of 160 minutes have been regularly observed for several years at the astrophysics observatory in Crimea, in the USSR, as well as by a group of English scientists working at the Pic du Midi Observatory in France. The identification of these oscillations with phenomena of solar origin has, however, been frequently questioned. Nevertheless, it is remarkable that they have continued uninterrupted (without a phase change), with clockwork regularity, since the first observations began. This compelling argument is what the Soviet astronomers rely on in stating that they are indeed of solar origin. These oscillations lead to velocity variations of only ± 2 meters (6.56 feet) per second, which correspond to shifts in the Sun's diameter of 38 kilometers (23.62 miles). They are unlikely to be "acoustical" in nature, for the periods of acoustic vibrations of a gaseous body of the Sun's mass and size are much closer to 40 or 60 minutes. Therefore they are probably gravity waves. However, in that case, we should also observe other periods close to 160 minutes, which have not yet been detected. We do not know, therefore, what makes the 160-minute period special in comparison with others.

In the United States, periodic variations in the Sun's diameter have been observed at the University of Arizona; the periods range from 5 to about 60 minutes. The shorter periods may correspond to acoustic radial pulsations, while the longer ones may correspond to oscillations of a gravitational type.

Other velocity variations affect the Sun's upper layers; their period is 5 minutes. They are obviously not radial, and the smaller the area chosen in which to observe them, the greater their amplitude. They likewise affect the intensity of light, as was shown by the Solar Maximum Mission, launched in 1980.

Comparison of the periods calculated for these oscillations using a solar model with observed periods shows systematic differences; the former are always higher. The two findings may be reconciled by changing the solar model and, more specifically, by increasing the thickness of the convective layer by about half. Moreover, the model is incompatible with a reduced abundance of heavy elements, which brings us back to the enigma posed by the apparent neutrino deficiency.

No less interesting is the use that can be made of the global oscillations to determine whether the Sun has an internal rotational motion. It turns on its own axis, of course, but the optical methods used for such observation allow us to study only the surface rotation, and it is much more difficult to know what is going on underneath.

Rotation and Flattening

Several years ago, an American physicist, Dicke, hypothesized that the Sun's interior must be rotating very rapidly, in order to explain the advance of the perihelion of Mercury by 43 seconds of arc per century. Newton's law of universal gravitation does not succeed in explaining this effect. On the other hand, a rapid internal rotation, if it exists, would result in a flattening at the solar equator, so that the Sun's radius would be greater in the ecliptic plane. Mercury's orbit, then, does not obey Newton's law of gravity at the assumed distance from the Sun, but at a slightly shorter distance. Indeed, the theory of general relativity also explains the advance of Mercury's perihelion with-

Oscillations of the Su[n's] chromosphere, *observed by me[ans] of the variations in intensity of [the] emission lines of singly ioni[zed] calcium and magnesium (wh[ich] have lost one electron). The [...] curves a and b correspond [to] observations made simultaneo[usly] of their emission lines using [the] American satellite* ***OSO 8*** *[...] represent the oscillations of zo[nes] about 7,000 kilometers (4,[...] miles) in width. The period of [the] oscillations is close to 3.5 minu[tes].*

out recourse to the notion of "solar flattening," for it modifies Newton's law sufficiently to account for the recorded effect. In Dicke's hypothesis, the flattening, measured by the ratio of the difference to the sum of the Sun's equatorial and polar diameters, should be 0.00005, corresponding to a difference of only 30 kilometers (18.65 miles) between the diameters, and to an internal rotation 30 times faster than at the surface. In addition, rapid rotation lowers the central pressure and thus also the temperature; it may account for the neutrino deficiency. Needless to say, therefore, the flattening observation is very important; but also needless to say, such a measurement is very difficult, given the minuteness of the quantities involved. In fact, recent results indicate a flattening no greater than 0.00001 on average, which corresponds to the flattening expected from the surface rotation alone. Internal rotation must therefore be very minor, contradicting Dicke's hypothesis. This result indirectly confirms the validity of the theory of general relativity. In taking this measurement, it was discovered that the flattening oscillates with a period of 12.6 days and an amplitude of 0.00002. An explanation for this phenomenon must await further observations and an attempt at theoretical interpretation; it is possible that the periodicity may result from the beating of two oscillatory modes with very similar periods, which would be identical if the Sun did not rotate, but which are separated in period by the internal rotation. Knowledge of the period of the beating brings us back to the value of the internal rotation, which we then find ranges from six to 12 days—that is, only four to two times faster than the surface rotation. If this explanation is not subsequently invalidated, it obviously has considerable importance in the study of the Sun's internal structure. It proves, in any case, that in the near future, solar seismology may lead to results of prime significance and perhaps make it possible to solve the enigma of the neutrino deficiency.

Differential Rotation and the Sun's Magnetic Field

It is thus confirmed that, contrary to what occurs in a solid body, all parts of the Sun do not rotate at the same rate—an additional argument, if one were needed, in favor of a "gaseous" Sun.

In fact, from detailed, continuous observation of the solar surface and from measurements of the shifts of spectral lines due to the Doppler-Fizeau effect, it was already known that the rotation speed varies with latitude, ranging from 26 days at the equator to 37 days at the poles. This is the phenomenon known as differential rotation.

Since the discovery in 1908 by the American astronomer George Ellery Hale, we have also known that the Sun has a magnetic field. There again, spectroscopy made this discovery possible. If an atom is placed in a magnetic field, its energy levels are separated into two or more closely spaced components. This separation into sublevels is called the "Zeeman effect," named after the Dutch physicist who discovered it. The electron orbiting around an atom's nucleus constitutes an elementary electric current. Now, we know that a wire through which a current is flowing, if placed in a magnetic field, is subject to a force that is perpendicular to both the direction of the current and that of the field, and the magnitude of the force is proportional to both that of the current and that of the field. The electron in an atom can orbit in either direction, but when placed in a magnetic field, it is attracted or repelled according to the direction in which it orbits. This force then causes the electron to orbit either closer to or farther from the nucleus. In a gas, there exist statistically as many atoms with electrons orbiting in one direction as in the other, and the Zeeman effect is expressed by the appearance of two symmetrical lines called "σ components" located on either side of the atomic line undisturbed by the magnetic field (or π component). We also observe this central line—for there are, of course, electrons in orbits whose plane is parallel rather than perpendicular to the direction of the field and that for that reason are not displaced. The distance separating these two "sublines" from the central line is proportional to the intensity of the field, which can thereby be measured. This is a very effective method and has been applied with good results for the Sun and many other stars.

The Sun's magnetic field, viewed as a whole, is a bit like that of the Earth. For that matter, the values are amazingly similar: 1 gauss for the Sun; 0.6 gauss for the Earth. The sign of the solar field (its polarity) is reversed in each hemisphere. However, contrary to the Earth (whose field can also be roughly compared to that of a bar magnet: a dipolar field), the overall solar field results from the addition of many magnetic fields spread over nearly the entire surface.

The origin of the field is directly linked to the existence of differential rotation. Under the surface, in the convective zone, are many free electrons that give the gas high electric conductivity, which may approach that of mercury. A conductor through which a current is flowing generates a magnetic field; the same is true for the Sun's deep layers in their rotational motions relative to one another. In reality, the phenomenon is complicated, for to correctly calculate the resulting field,

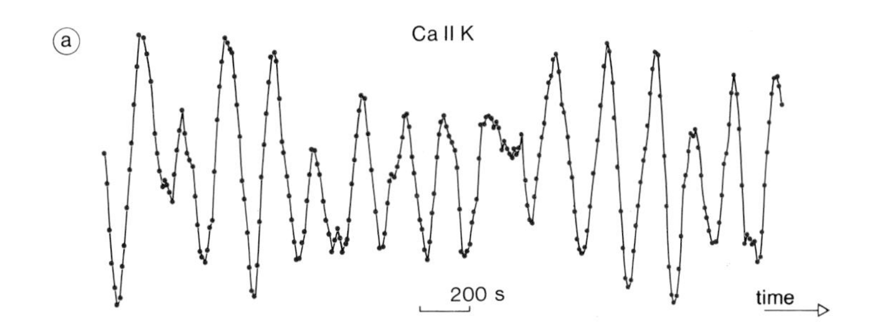

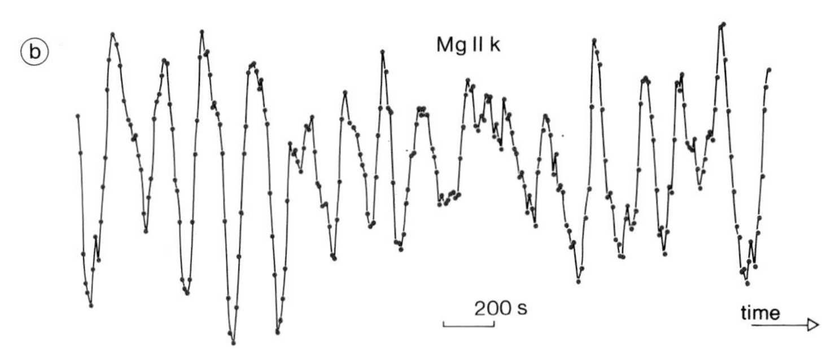

it is necessary to take into account the coupling between the rotation, which varies with depth and latitude, and the convection movements. These calculations were refined in the 1970s by many teams using large computers, and we now have a fairly good, if incomplete, notion of the complex mechanisms involved in generating the solar field.

But how does the field, which is created in the inner regions, find its way to the surface? To understand this, it is necesarry to introduce the notion of magnetic pressure. Like a conductor with a current flowing through it that is placed in a magnetic field, solar electrons are subject to a force from the local field. This force, when expressed per unit surface area, supplies the magnetic pressure. It is directly proportional to the square of the intensity. The total pressure at any point is the sum of the gaseous and the magnetic pressures. To maintain its pressure, a region with a strong magnetic field will thus need fewer atoms than a neighboring region where the field is not active. In other words, the density of the region with a strong magnetic field will be less than that of the surrounding regions, so that the region of strong field will be subject to a buoyant, or Archimedean force that will push it toward the surface. The surface magnetic field, then, is the residue of the field created farther below, within the convective zone.

Sunspots

It was, in fact, on the image of a sunspot that George Ellery Hale placed the aperture of his spectrograph in 1908, and discovered for the first time the presence of a solar magnetic field. Since then, we have been able to measure precisely the intensity of the field within the spots: It is equal to several thousand gauss. If we compare the sunspot to an electromagnet, we find that in order to produce the observed field, a current of several thousand billion amperes is needed.

Seen in visible light, the spots appear to be made up of a very dark center, the umbra, and a lighter part having a filamentary appearance, the penumbra. Their size varies from one spot to another, but the largest ones may attain dimensions greater than the diameter of our planet. Around the spot we observe the granulation of the photosphere, which, in contrast, seems to be completely absent in the umbra and penumbra. This disappearance has commonly been attributed to the magnetic field's inhibition of the convective motions, of which granulation, we recall, is the outward manifestation.

The spots are found in pairs slightly inclined relative to the solar equator, and having opposite magnetic polarities. Also, the polarity of the pairs of the solar Northern Hemisphere is opposite to that of the pairs of the Southern Hemisphere. If the "western" spot of the pair—or leading spot—has a north polarity in a hemisphere, the "eastern" spot will have a south polarity. The opposite occurs in the other hemisphere. In general, the leading spot has the same polarity as the pole of the hemisphere to which it belongs, and its latitude is lower than that of the trailing spot. The field tends to be vertical at the center of the spots.

Of all solar structures, sunspots are the easiest to observe; the ancients observed them with the naked eye. But they are also among the most difficult of all solar phenomena to understand. Thus, we still cannot explain today either why they exist, or why they are dark. They appear dark, of course, because they are less hot; their effective temperature is 1700° K lower, on average, than that of the surrounding regions. But why are they "cold"? To be sure, it is almost certainly the magnetic field that causes the cooling, rather than the reverse, for, with strengths of 3,000 to 4,000 gauss, the energy stored in the field dominates the thermal energy of the gas by a factor of about 10. But this does not fully explain the phenomenon.

They hypothesis has also been advanced that the spots are the source of intense magnetic wave emissions, and the large energy loss that would result from this would cause them to cool down. But no one has yet been able to demonstrate the existence of these waves. Moreover, we still do not understand what forces cause the concentration of magnetic energy in the spots, whereas the magnetic tubes of which they are composed would naturally tend to move away from one another. Furthermore, they turn more quickly with the solar rotation (their period is 25 days) than the neighboring photosphere, as though they were more rigidly tied to the Sun's interior. It has been sug-

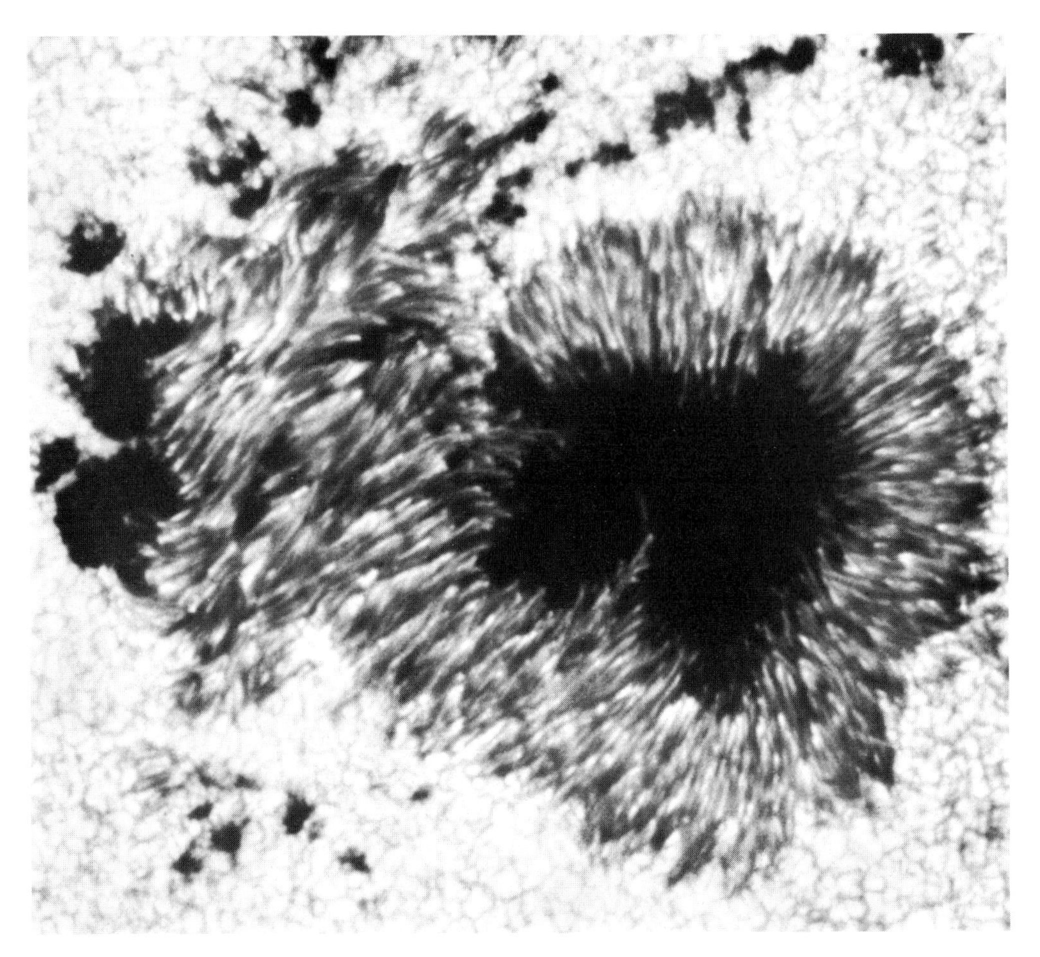

Photograph of a pair of sunspots, *taken in visible light, on July 5, 1970, at the Pic du Midi Observatory, with a refracting telescope 38 centimeters (15 inches) in diameter. We can clearly see the umbra and penumbra, where the granulation seems to have totally disappeared. The filament like part of the penumbra seems to be plunged in shadow. Each filament represents a magnetic flux tube. O.P.M.T. Photo.*

gested that the magnetic flux tubes are responding to underlying currents, like an iceberg that moves in the opposite direction from the wind pushing it, because it is deeply submerged in water driven by invisible countercurrents. There again, the explanation lies in areas that we cannot observe directly, of which we therefore know little or nothing, and that are subject to forces we do not yet understand.

Moreover, recent observations done with high-resolution telescopes seem to indicate that the average magnetic field of the Sun is in fact concentrated in many nodules from 1,000 to 1,500 kilometers (622 to 932 miles) in diameter, where it can reach values greater than 1,000 gauss. This seems to indicate that the problem of the existence and stability of regions with a high concentration of magnetic energy is characteristic not only of the sunspots, but also is perhaps a much more widespread characteristic. Unlike the spots, however, the nodules do not appear dark; it is therefore difficult to interpret the question of their pressure equilibrium. It is not impossible that they may be related, in one way or another, to other structures recently discovered on pictures taken under exceptional observing conditions at the Sacramento Peak Observatory in New Mexico: the "filigrees." These structures seem to be directly associated with regions of high magnetic field, and that association probably alters the structure of granulation in their vicinity.

The solar tower of the Sacramento Peak Observatory in New Mexico is 40 meters (131 feet) high. *With its 76-centimeter (30-inch) aperture and its optical system entirely enclosed in a vacuum to eliminate instrumental turbulence, it has made possible the observation of the smallest structures ever detected on the Sun's surface. Photo: Sacramento Peak Observatory, Association of Universities for Research in Astronomy, Inc.*

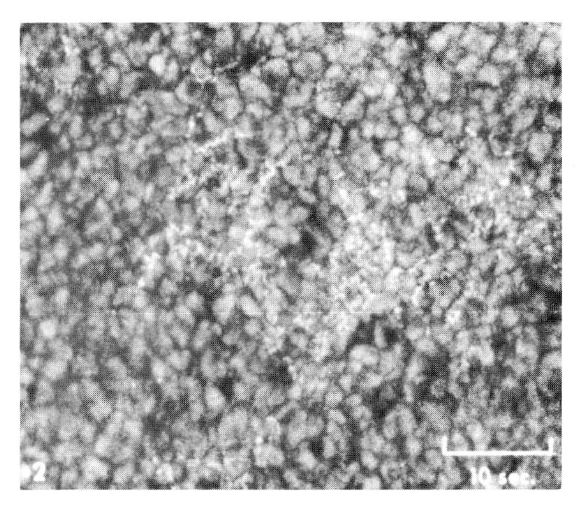

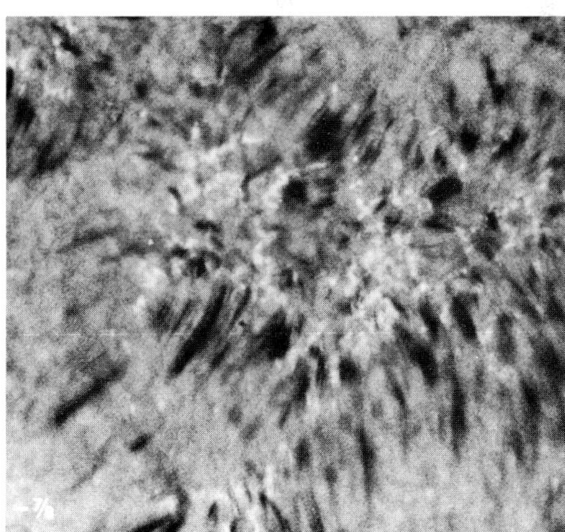

Photospheric granulation photographed with the solar tower of the Sacramento Peak Observatory, at wavelengths that are far from (left) and within (right) the Hα line. *These photographs, the first that have enabled us to separate two structures only 300 kilometers (186 miles) apart on the Sun's surface, reveal a very fine, threadlike network: the filigrees. These structures are associated with very intense magnetic fields. On the photograph at left, we notice that the granulation is diminished at places corresponding to the filigrees. Photo: Sacramento Peak Observatory, Association of Universities for Research in Astronomy, Inc.*

The 22-Year Magnetic Cycle

One of the more remarkable phenomena—of which sunspots, among others, are one of the visible manifestations—is the 22-year magnetic cycle. One of the more interesting facts of nature is indeed the way in which the solar magnetic field reverses itself every 11 years, with each pole returning to its original polarity every 22 years. At present, the North Pole actually corresponds to a north magnetic pole. One might be surprised that the magnetic field should undergo variations over such a short time period, for the field, if left to itself, would take about 10 billion years (or about the age of the Sun) to filter from the interior to the exterior of the Sun. In reality, the field is not left to itself; it is constantly regenerated by the interaction of the convective and rotational motions, and its reversal every 11 years sets a lower limit on the depth of the region in which it is generated. A field of about 100 gauss is produced at about 200,000 kilometers (124, 300 miles) below the surface—that is, in the deepest part of the convective zone. It can hardly exceed 100 gauss, for then the solar cycle would have to be shorter, and it can hardly be weaker, for the average field would also be weaker than the one that is measured. The number of spots present annually on the Sun's surface is a characteristic index of the solar activity cycle. This number has been measured regularly since 1843, when the amateur German astronomer Heinrich Schwabe first noticed the existence of an 11-year cycle in the number of spots. The observations systematically carried out after that by the professional Swiss astronomer Rudolf Wolf confirmed the existence of the cycle, which is marked by alternating and quite distinct maxima and minima of that number, referred to since then as the "Wolf num-

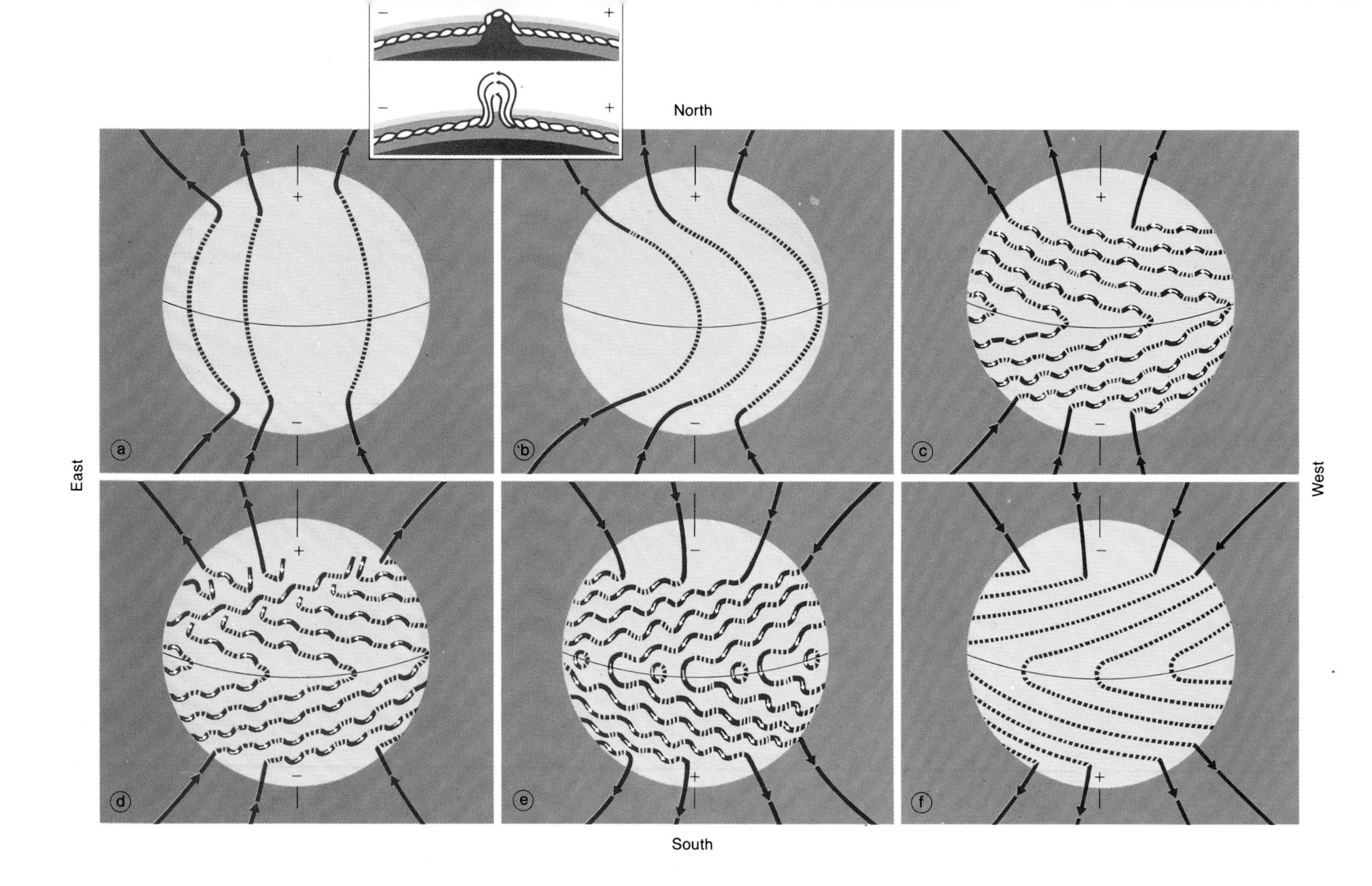

ber." Once remarkable property of the spots, which perhaps provides a clue to the physical phenomena responsible for the 11-year cycle, is that, at the beginning of a cycle, they appear preferentially in a band at 40 degrees latitude on either side of the equator. They then migrate slowly toward the equator, and appear in a 5-degree strip at the end of the cycle. At each new cycle, there is a change in the polarity of the spots, and when the spots of the new cycle appear at high latitudes, their polarity is the reverse of that of the previous cycle's spots, which are still close to the equator.

Even though the main features of the 22-year cycle can be explained today, all the details are not yet well understood, and much work is still being done on them. The explanation that follows is the one most commonly acknowledged, although it leaves a certain number of points unexplained. As a result of the differential rotation, the field's lines of force (which are an imaginary representation of the direction of the magnetic field, along which a compass needle would point) quickly become wound around the Sun's axis of rotation. In doing so, they approach one another. The strength of the field at a given point, which is proportional to the number of lines of force passing through a unit area at that point, is thus increased. In reality, the solar plasma and its magnetic field are so closely coupled that, in nearly all the cases we will encounter, the field is "frozen" inside the gas and consequently follows all its movements. The swirling convection movements then produce a twisting of the lines of force, so that they resemble a cord or a cable. This twisting further increases the field strength. When it is sufficiently increased, the magnetic pressure overcomes the gas pressure, and the region where the field is concentrated is subject to Archimedes' principle. It emerges from the photosphere in the shape of a loop, the "feet" of which correspond to a pair of spots.

This phenomenon preferentially takes place first in the zone where the torsion is maximal. We calculate that it is located on either side of the equator, at 40 degrees latitude. The faster internal rotation has the effect of increasing the field strength at latitudes less than 40 degrees, and of reducing it at higher latitudes. Thus, the zone in which the spots emerge advances toward the equator. This model, therefore, represents several characteristics of the cycle, in particular the appearance of spots at particular latitudes, and the inclination relative to the equator of all pairs in a chosen hemisphere.

Once emerged, the magnetic regions are subjected, at the photospheric level, to shallow currents of matter associated with supergranulation, a structure that will be described later (see page 96). The intense magnetic fields of the spots are then gradually eroded by this action, while at the same time they are stretched out by the differential rotation. The field of the trailing spot, which is closer to the pole, and has a polarity opposite to it, then preferentially migrates toward it. After almost five or six years—that is, a quarter of the cycle—a sufficient quantity of the field has mi-

The 22-year magnetic cycle *illustrated in six figures representing the winding of the lines of force of the magnetic field (imaginary lines, representing at each point the direction that the needle of a compass would take):*
a. Originally pointing from south to north, the lines of force are distorted and approach one another as a result of the differential rotation that stretches them along the equator.
b. The strength of the magnetic field consequently increases and, when magnetic pressure prevails over gaseous pressure, the magnetic tubes emerge (inset at upper left) from the photosphere in the form of a pair of spots.
c. The circulation associated with supergranulation erodes the field of the trailing spots, which diffuse toward the pole (of opposite polarity).
d. The leading spots then recombine with the trailing spots of pairs situated farther to the west; the lines of force thus gradually reverse.
e. The lines of force spread out as a result of differential rotation.
f. After 11 years, the magnetic field is completely reversed.
The dotted lines represent the lines of force under the photosphere.
(From W. M. Adams, Big Bear Solar Observatory, Big Bear, California.)

grated in this way so that the pole's field is annulled and gradually reverses itself.

The lines of force that linked the trailing spot to the leading spot then recombine with the trailing spot of a pair of spots at lower latitude, located farther to the west, while remaining linked to the head spot of the first pair. With the strength of the polar field now increasing, this change of direction advances gradually toward the equator. Wherever it takes place, differential rotation tends to unwind the lines of force and to reduce the strength of the field. After 11 years, the magnetic field is completely reversed. Thus described, the cycle appears as a relaxation phenomenon that repeats itself continually in an identical manner, and the characteristics of one cycle are practically independent from those of the previous cycles, of which it has gradually lost "memory."

An interesting observation, which sheds a new light on the processes of the magnetic cycle and the generation of the magnetic field, was recently made by the American astronomer J. Eddy, who undertook to study the successive cycles by going back in time as far as possible. Among different indices of activity, Eddy, of course, used the sunspots, which Wolf had been able to count back to 1700, using carefully verified historical observations. He also used the observations of the English astronomer Maunder, who tells us that between 1645 and 1715, very few spots were visible on the disk. By reanalyzing Wolf's and Maunder's data, which Eddy also related to other indices, Eddy confirmed the disappearance of the spots noted by Maunder. Analysis of the abundance of carbon-14 in the Earth's atmosphere, which is tied to the level of solar activity and can be measured in the growth rings of trees (see page 110), shows that the cycle also disappeared about 1,500 years before the birth of Christ and during the fourth, seventh and 14th centuries.

Accompanying the disappearance of the solar cycle was another anomaly revealed by analysis of spot motions during the period of the "Maunder minimum": The rotation speed at the solar equator was about 3 percent higher than it is today, and the magnitude of the differential rotation was greater by about a factor of three. We are thus led to conclude that, during these periods, circulation and convection in the Sun's interior were certainly different from what they are today. Since the heat radiated by the Sun is deposited on the surface by the underlying convection and rotation motions, we should expect to see the solar luminosity vary likewise. We thus note with interest that the periods coinciding with the disappearance of the solar activity cycle correspond to periods of the Earth's cooling. Variations in the Earth's average temperature over scales of several thousand years suggest that these larger variations in the Sun's luminosity, hence of rotation and convection, may last for long periods of time.

Variations in rotation and convection may have an effect only on the distribution of intensity over the Sun's surface without affecting the overall luminosity. But since the Earth is warmed mainly by the equatorial regions (half of the visible solar disk is located in a strip of 48 degrees latitude centered on the equator), a difference in intensity between the pole and the equator of only 2 percent can produce a change of 1 percent in the luminosity received on Earth.

We can therefore understand how important the theoretical study of the Sun's rotation and convection movements may be to climatologists. There again, solutions to the problem lie in a better knowledge of the motions in the Sun's interior.

The "Maunder minimum" *can clearly be seen on this diagram, which shows, over time, the annual number of spots appearing on the Sun's surface since the 17th century. The oldest observations were collected by Maunder, manager of the Greenwich Observatory in Britain, at the end of the past century, from descriptions, reports or drawings in the archives of several European observatories, particularly the Paris Observatory. We note that between 1645 and 1715, the magnetic cycle completely disappeared. From J. Eddy,* ***Science****.*

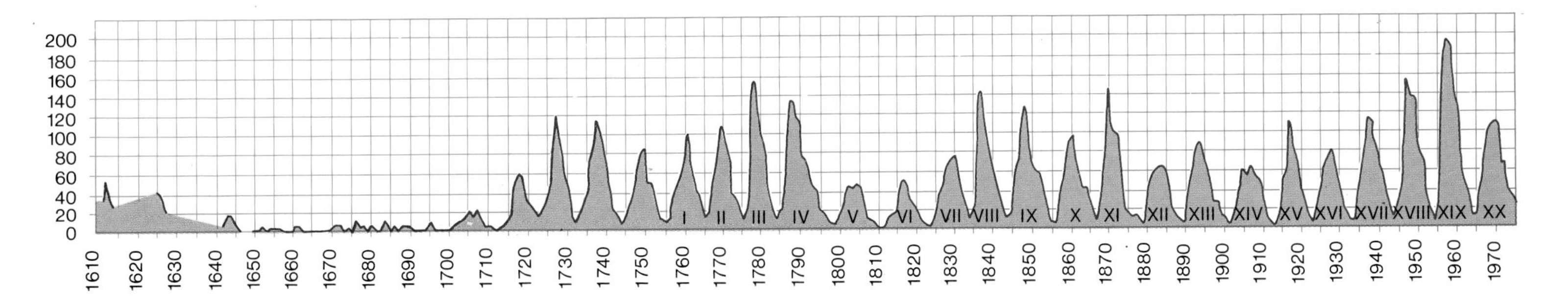

The Sun's Atmosphere

Another Flaw in the Model

The model described at the beginning shows that near the surface, the density decreases rapidly by a factor of 2.7 (the base of natural logarithms) for every 100-kilometer (62-mile) increase in height. This interval of 100 kilometers is called the "scale height." It is this rapid decrease that explains the relative thinness of the photosphere and the marked sharpness of the visible solar rim. The model also predicts that the temperature declines below 4,000° K at some height. The electrons then recombine with the atomic nuclei, and the abundance of neutral atoms increases greatly relative to that of the ionized atoms—the reason for the appearance of many absorption lines in the spectrum. The notion of thermodynamic equilibrium then gradually loses its meaning and must be replaced by the more restricted notion of local thermodynamic equilibrium, according to which thermodynamic equilibrium is considered to hold only within a small volume element around the point of interest.

A notable interest was shown by astrophysicists when the first spectra in ultraviolet light were obtained with instruments sent above the Earth's atmosphere aboard rockets or artificial satellites. The Sun's ultraviolet light is absorbed by the molecules and atoms that make up our atmosphere so cannot be observed with telescopes on Earth, and one must observe at altitudes of several dozen to several hundred kilometers.

The study of the Sun's ultraviolet light, begun in the United States just after World War II with the aid of German V2 rockets, developed greatly in the 1960s and 1970s in the United States, Europe and the Soviet Union and is still continuing today at a very fast pace. In the near ultraviolet, the solar spectrum resembles the absorption spectrum that we see in visible light, but at even shorter wavelengths, we see lines appear in increasingly large numbers, this time in emission. In the far ultraviolet, the spectrum is made up only of emission lines. Photographs of the solar disk in far ultraviolet light show that the edge is no longer darkened, as in visible light, but appears surrounded by a bright ring.

It is impossible to interpret these observations if we assume that the temperature continuously decreases toward the exterior. Indeed, they would then imply that atoms radiate more energy than they themselves receive from the thermal agitation of the particles. Yet this is contrary to the laws of thermodynamics. The model, then, seems to have yet another flaw.

Indeed, we saw that the solar gas is no longer transparent beyond a certain distance, which depends on its density and its absorptive properties. The outer layers, which have a very low density, are highly transparent, since the absorption is quite small. But the atoms' capacity to absorb light increases considerably when we go into the ultraviolet. Thereore, in this part of the spectrum, de-

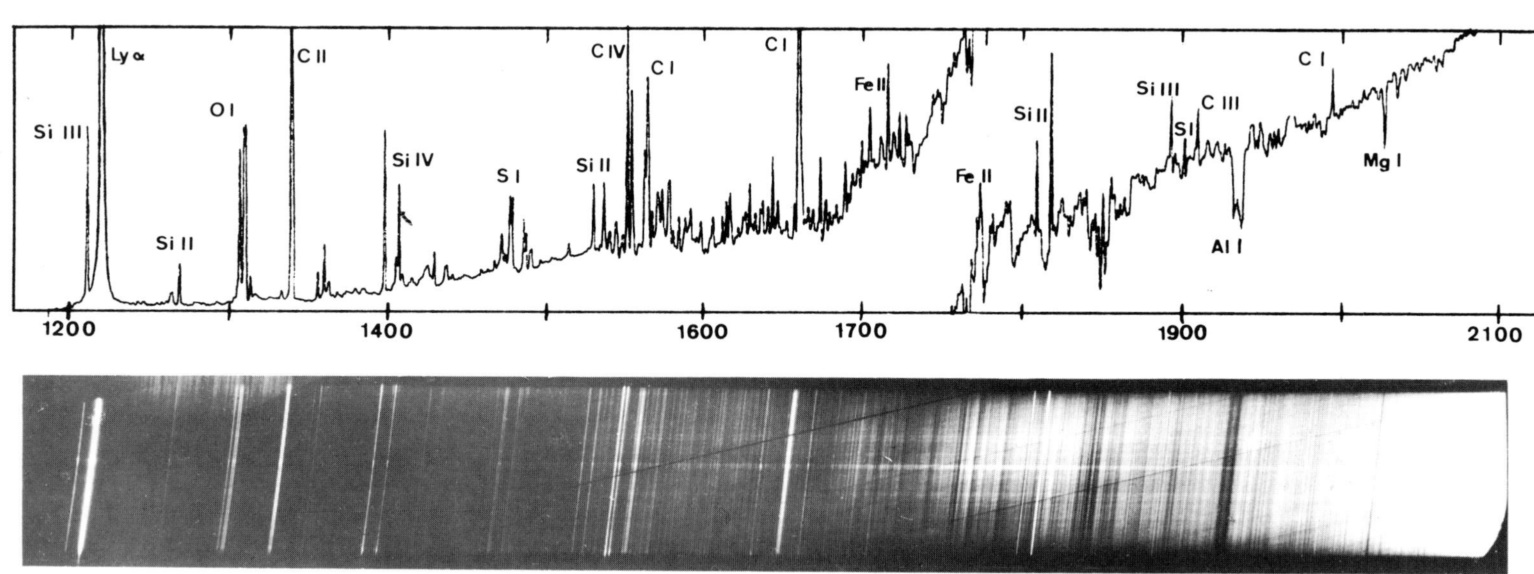

The Sun's spectrum photographed in far ultraviolet light. *a. From instruments aboard a Véronique rocket launched from the Kourou base in Guyana on April 17, 1973. The right-hand portion of the spectrum represents the longest wavelengths and is characteristic of an absorption spectrum such as may be observed when visible sunlight is broken down into its different colors. The left-hand portion includes many emission lines, revealing the existence of hotter gas layers above the Sun's visible limb. Each line is an image of the slit of the spectrograph used to record the spectra. The slit was placed on a diameter of the Sun so that the center of the spectrum represents the spectral composition of light at the center of the disk; the two ends represent this same composition for the Sun's limb. The spectrum is also represented in the form of a recording, the principal lines being identified by the atom or ion responsible.*

spite their low density, the outer layers can become opaque, and increasingly so as we look farther into the ultraviolet. Thus the zone that contributes to the emission of ultraviolet light occurs at higher altitudes. Observation of the ultraviolet spectrum thus permits an exploration of the high-altitude layers of the Sun. We thus interpret the transition from the visible-light absorption spectrum to the ultraviolet-light emission spectrum, as well as the appearance of a bright ring at the solar limb in the ultraviolet, as an increase and not a decrease in the Sun's temperature toward the exterior. This is, in fact, not really a surprise, for it had already been known for a long time that the Sun is surrounded by a tenuous whitish envelope—the corona.

b. From an instrument placed in a hot-air balloon. The part of the spectrum that contains the two H and K emission lines of singly ionized magnesium (which has lost an electron) has been isolated. Here too, the spectrometer slit intercepts the solar diameter. However, the spatial resolution of the instrument used to take this photograph is clearly superior to the previous one, and we clearly note here the change in the shape of the lines where the slit intercepts different areas on the disk. In particular, we note the increased intensity corresponding to a small active region near the center of the spectrum. Each line appears to be doubled, which results from the transfer of light from the center of the line, where the atmosphere is very opaque, to the near wings, in which light can escape. This example clearly illustrates the difference between an absorption spectrum, where the bands appear dark on a bright background, and an emission spectrum, where the opposite occurs.
Photos C.N.R.S. Laboratoire de physique stellaire et planétaire.

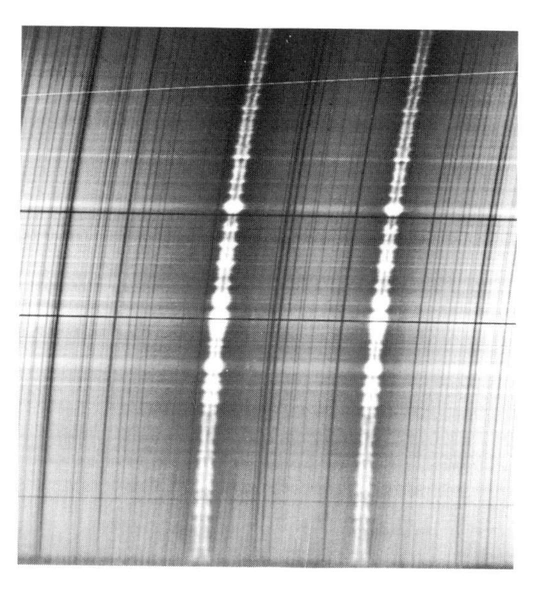

There had already been theoretical speculations about spectra of the corona, photographed in visible light during eclipses, for, in the first analysis, no known atom had lines identifiable with those observed. In this way, a new element was invented, "coronium," which at that time was unknown on Earth. References to this new element in the scientific literature quickly disappeared after the Swedish physicist Edlen showed that the bands of "coronium" could in fact be attributed to very heavily ionized atoms (that is, those that had lost many electrons), such as calcium-XV (14 electrons) or iron-XIV (13 electrons). Now, these ions could exist only at temperatures of about 1 or 2 million degrees. At such high temperatures, practically all the hydrogen is dissociated, and the coronal plasma may be regarded as consisting predominantly of protons and electrons in nearly equal amounts. The whitish color that characterizes the corona is explained by the scattering of photospheric light by the free electrons, with the luminous intensity being directly proportional to their density: The more numerous they are, the more light is scattered.

Moreover, the corona's extension up to several hundred thousand kilometers beyond the edge of the solar disk is incompatible with a 100-kilometer scale height. To maintain such an envelope, it is necessary to counteract gravity with the effects of atmospheric pressure. This is possible if a very high temperature is maintained in the corona; the particles, being hotter, have greater speeds, and can move greater distances from the center of the Sun. The Sun's outer layers have a higher temperature, then, than the one calculated by our oversimplified model.

An Empirical Model of the Outer Layers

Unlike the interior, where pressure, density and temperature are inferred from macroscopic quantities (mass, luminosity, radius) and from spectroscopy of the photosphere (chemical composition), the physical characteristics of the outer layers are inferred mainly from direct observation. We now have several detailed empirical models, developed through observation of eclipses and ultraviolet light.

During a total eclipse, when the Sun is entirely hidden by the Moon, we can judiciously place the slit of a spectrograph perpendicular to the solar limb and study the variation in the intensity and width of the spectral lines according to altitude above the solar surface. We notice, first of all, that the spectral lines that were in absorption on the disk have all gone into emission, for the most part, in a very thin layer about 1,500 kilometers (932 miles) thick. This results from the rise in temperature toward the exterior. Moreover, the lines have widths whose measurement yields a semi-direct determination of temperature, which, as we know, depends on the individual atoms' speed of thermal motion; the faster they move, the more the wavelength of the photons they emit is shifted by the Doppler-Fizeau effect. Their speeds are distributed around an average speed, and the spectral line for which they are responsible appears widened overall. Moreover, the intensity of the lines emitted in the Sun's outer layers is directly proportional to the number of emitting atoms, for, when a photon is emitted, it can escape into space without being absorbed. The measurements of intensity therefore yield the density at different altitudes. With temperature and density, we can thus construct an empirical model.

Analysis of the ultraviolet spectrum makes it possible to refine these results and to extend the model to higher altitudes. Indeed, the relationship between the intensities of two ultraviolet emission lines of the same atom depends either on temperature or on density, according to the pairs of lines chosen. We thus obtain the model shown in the diagram. There we see the temperature go first through a minimum at around 4,500° K, then slowly increase to 10,000° K and then increase very rapidly over a short distance until it reaches levels of over 1 million degrees.

The layers in which the temperature ranges between 4,500° K and 10,000° K have been given the name of "chromosphere." It is in the chromosphere that the principal lines of hydrogen, the most abundant element, are emitted, particularly the Hα line. It is this line that gives the chromosphere its deep red color during eclipses (hence its name). Located in the far ultraviolet, the Lyman α line of hydrogen, which can be observed only through space techniques, is also emitted in the chromosphere. Other lines such as the H and K lines of singly ionized calcium and magnesium, which appear in the violet and ultraviolet, respectively, are also characteristic of the chromosphere. Generally speaking, the chromosphere radiates all the more intense lines of the

neutral or less-ionized elements. However, if the temperature increases by even a small amount, this emission ceases, for hydrogen loses its sole electron, and the calcium and magnesium ions (the most abundant) slowly disappear while becoming more highly ionized. Unable to cool off by emitting light, the upper layers of the chromosphere thus heat up to the temperatures at which the very heavily ionized atoms can again radiate. The quantity of energy an atom can radiate increases with the fourth power of the electric charge of its nucleus; an atom that has lost 10 electrons increases its charge by a factor of 10 and is capable in that state of radiating 10,000 times more luminous energy. This explains how and why the corona, which possesses only a few atoms (its density varies from 100 million to 1 billion protons per cubic centimeter, or about 1/100th the density of the air we breathe), can nevertheless radiate fairly intensively. This is what explains, in particular, the sudden sharp rise in temperature at about 1,000 kilometers (622 miles) (the transition zone between the chromosphere and the corona) and the corona's temperature "threshold." Far from the Sun, in the interplanetary medium, the temperature goes down to very low levels, approaching 100° K.

Model of the Sun's atmosphere. *After smoothly decreasing from the center outward, the temperature falls to a minimum of about 4,500°K and then climbs sharply to levels of several million degrees. This abnormal variation reveals that somewhere in the Sun is an energy source capable of heating the outer atmosphere. The sharp temperature rise reflects the inability of the atmosphere to dissipate this excess energy except at high temperatures.*

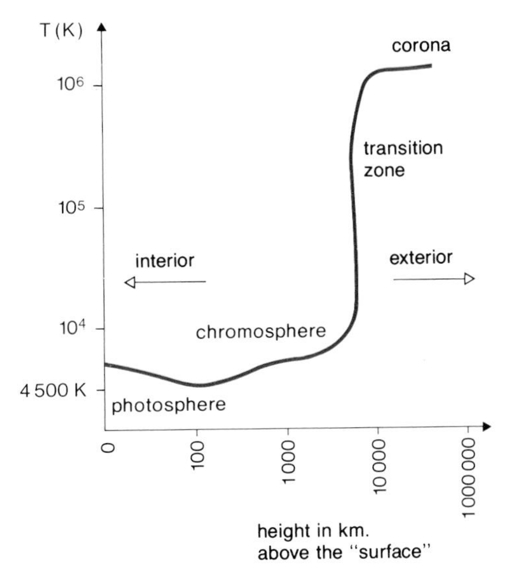

We will hereafter call all of the layers located above the minimum temperature of 4,500° K the "solar atmosphere."

Why Is the Atmosphere So Hot?

The atmosphere is thus "wedged" between the photosphere, whose average real temperature is 5,770° K, and the much colder interstellar medium. If it were warmed only by photospheric radiation, its temperature should be somewhere between 5,770° K and 100° K. The only explanation for its high temperature is that somewhere in the Sun there exists one—or perhaps several—energy sources capable of warming it beyond the photospheric temperatures. It was an American, Martin Schwartzchild, of Princeton University, who suggested 30 years ago that this energy might have a mechanical origin.

This source was sought for a long time, until the discovery in 1960 of periodic 5-minute waves, which we have already mentioned (see page 90) in connection with solar seismology. Because of the rapid decrease in atmospheric density with altitude, the 5-minute oscillations are soon transformed into shock waves, which dissipate, warming the surrounding atmosphere through the mechanical friction of the atoms and ions against one another.

Nevertheless, the 5-minute waves are not enough in themselves to explain the thermal structure of the atmosphere in detail, for we calculate that only waves with the shortest periods (30 to 90 seconds) are capable of dissipating their energy into the lower chromosphere and of beginning the temperature rise observed at the surface. The existence of such waves has not yet been proved.

On the other hand, the Doppler shifts and intensity variations of the spectral lines emitted in the chromosphere reveal very distinct waves with periods close to 200 seconds (see page 90). These waves probably originate in the subphotospheric regions and propagate without difficulty up to the chromospheric levels. But shortly before 1980, results obtained jointly by a team at the Meudon Observatory and by the *OSO-8* satellite showed that the quantity of mechanical energy carried by these waves is also insufficient! The problem thus remains unsolved, and we have to look for other energy sources.

The chromosphere also receives a large amount of heat energy from the much hotter corona. The coronal gas consists mainly of protons and electrons, which, moving very quickly because of their low mass, carry heat from the hotter to the colder regions, hence toward both the photosphere and the interstellar medium. This is energy transfer by conduction, and this is the only part of the Sun where it is effective. To give an idea of this effectiveness, we can state that the thermal conductivity of the coronal plasma at 1 million degrees is about 1 million billion times that of silver.

We thus begin to understand that the equilibrium between the energy received and the energy lost by each layer of the solar atmosphere is not in itself a simple problem, with each layer receiving mechanical energy and warming up after dissipation, then eliminating that energy, either in the form of light, or by thermal conduction to the neighboring layers. In reality, the situation is still more complex, because of the presence of the magnetic field.

The Swedish physicist Alfvén actually showed that the field itself was capable of generating magnetic or magnetohydrodynamic (MHD) waves and of modifying the acoustical waves[1] mode of propagation. Alfvén waves can also dissipate in the upper atmosphere and warm it, this time by "electric viscosity," a bit like the way a resistor heats up when a current runs through it. This is now, "by default," the most probable explanation of the problem of the temperature structure of the solar atmosphere.

In the deeper regions, and as far up as the photosphere, gas pressure in general prevails over magnetic pressure where the field is less than 1,500 gauss—that is, everywhere except in the spots and nodules. In the atmosphere, on the other hand, with the density sharply reduced, magnetic pressure becomes stronger than gaseous pressure. We can then no longer study the atmosphere without taking into account the influence of the field. The field governs the distribution of the gas and shapes each of its structures, giving it its most spectacular shapes. This predominance of the magnetic effects has become obvious ever since we have been able to obtain detailed "cross sections" of the atmosphere at higher and higher

altitudes, through observations in ultraviolet light and in X rays.

Supergranulation and Spicules

Photographs of the Sun taken by the light of lines formed in the chromosphere or in ultraviolet light show that at the level of the chromosphere, photospheric granulation—quite visible in white light—has practically disappeared, giving way to another structure of similarly inhomogeneous appearance: supergranulation. In the light of H and K lines, supergranulation takes the form of a network of bright polygon-shaped cells (hence its other name, "chromospheric network") 25,000 to 50,000 kilometers (15,538 to 31,075 miles) in diameter. Although the cells slowly lose their shape over time (their life-span is about 20 hours), the network is a permanent structure of the Sun. We see it also in the light of the Hα line, but the latter, being emitted from a slightly higher zone than the H and K lines, has a different appearance, showing many details whose orientation and structure are obviously governed by the magnetic field. This is specifically the case with the "fibrilles," which resemble a field of oats waving in the wind and which point in the direction of the field like iron filings in the vicinity of a bar magnet.

Maps of the magnetic field obtained simultaneously show that the supergranule boundaries correspond to magnetic fields 10 to 20 times more intense than the average, and which are oriented vertically, on the whole. The brightest points correspond to fields of several hundred gauss. The chromospheric network is thus strongly influenced by the magnetic field.

The photographs in ultraviolet light reveal that its structure is maintained from the base of the atmosphere to the corona. However, its contours lose their sharpness as one goes higher, probably a result of the divergence—growing with the altitude—of the magnetic field, which, no longer confined by the gas pressure, follows its natural inclination. Higher up in the corona, the network has totally disappeared.

The study of the frequency shifts of spectral lines in the links of the network shows that these links coincide with major descending currents, which can reach several kilometers per second. These currents are seen at all altitudes up to the base of the corona, where their speed has considerably increased, attaining several dozen kilometers per second. They seem nevertheless to be rather localized in the brightest regions of the network, which correspond to the more intense fields. At the same time, we see at the base of the chromosphere rising currents that emerge from the supergranulation cells, but have a much lower amplitude than the descending currents. They did not appear higher in the atmosphere. This creates a problem, for the law of mass conservation requires that we observe as much descending as ascending matter.

Supergranulation. *This permanent structure of the solar disk can be seen easily on photographs taken in the light of the K line of ionized calcium. Each image corresponds to a different date. The upper part represents a map of the magnetic field measured on the Sun's surface by application of the Zeeman effect. We note that the bright regions of the bottom images correspond to regions of a more intense magnetic field. The curves of the upper images are lines of equal magnetic field, and their density per unit surface area on the Sun's disk depends on the intensity of the magnetic field; the closer together they are, the stronger the field. Photo by R. Howard, Hale Observatory, Caltech, California.*

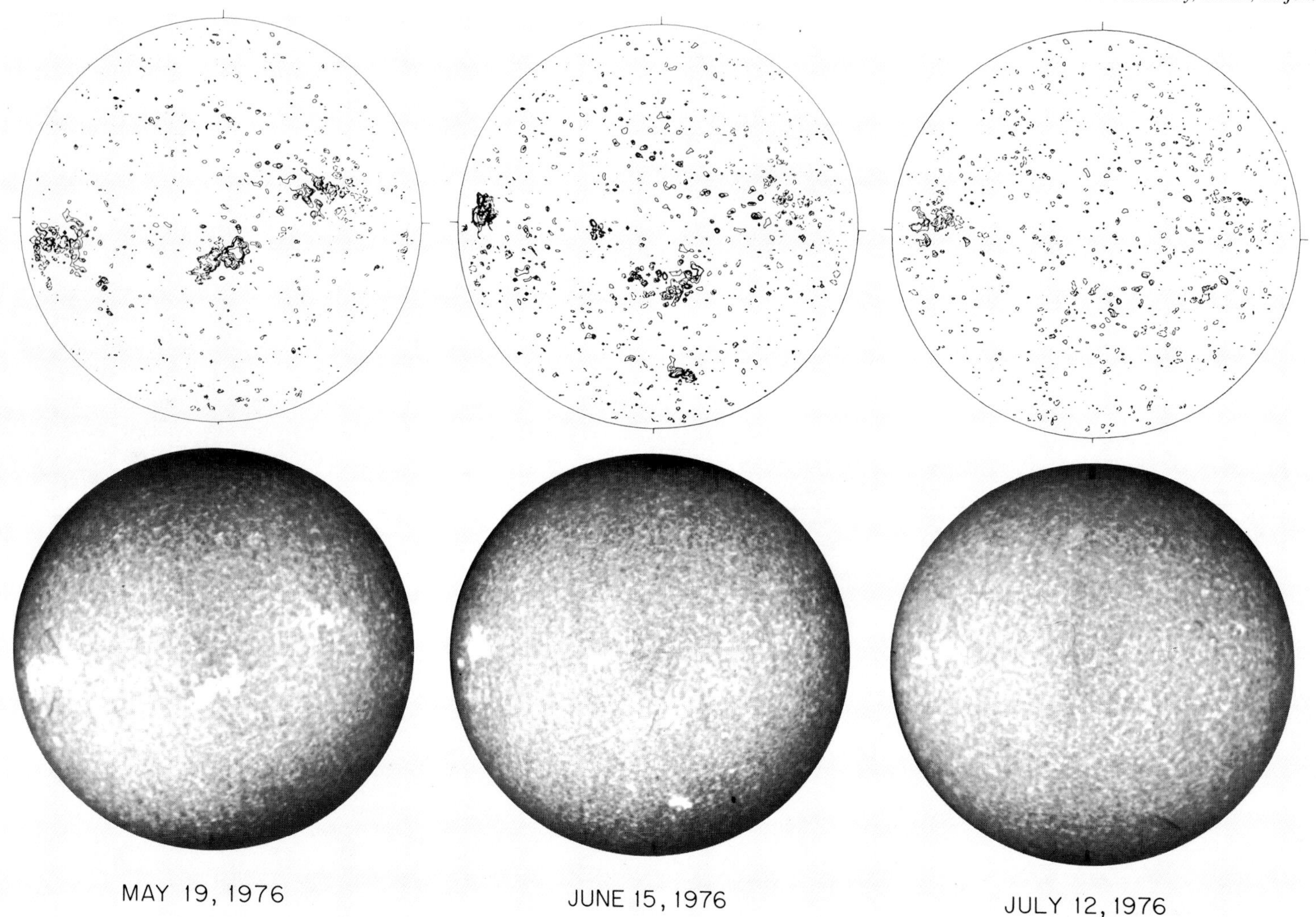

Photograph of the Sun taken at the center of the Hα line, *which shows the "fibrils" of the chromosphere, a gaseous formation oriented by the solar magnetic field. We note that the fibrils surrounding the spot at lower left sink into it, so to speak, or escape from it. We also note, at the top of the photograph, the appearance of the spicules seen in a side view at the solar limb, as well as a small developing eruption. Photo: Sacramento Peak Observatory, Association of Universities for Research in Astronomy, Inc.*

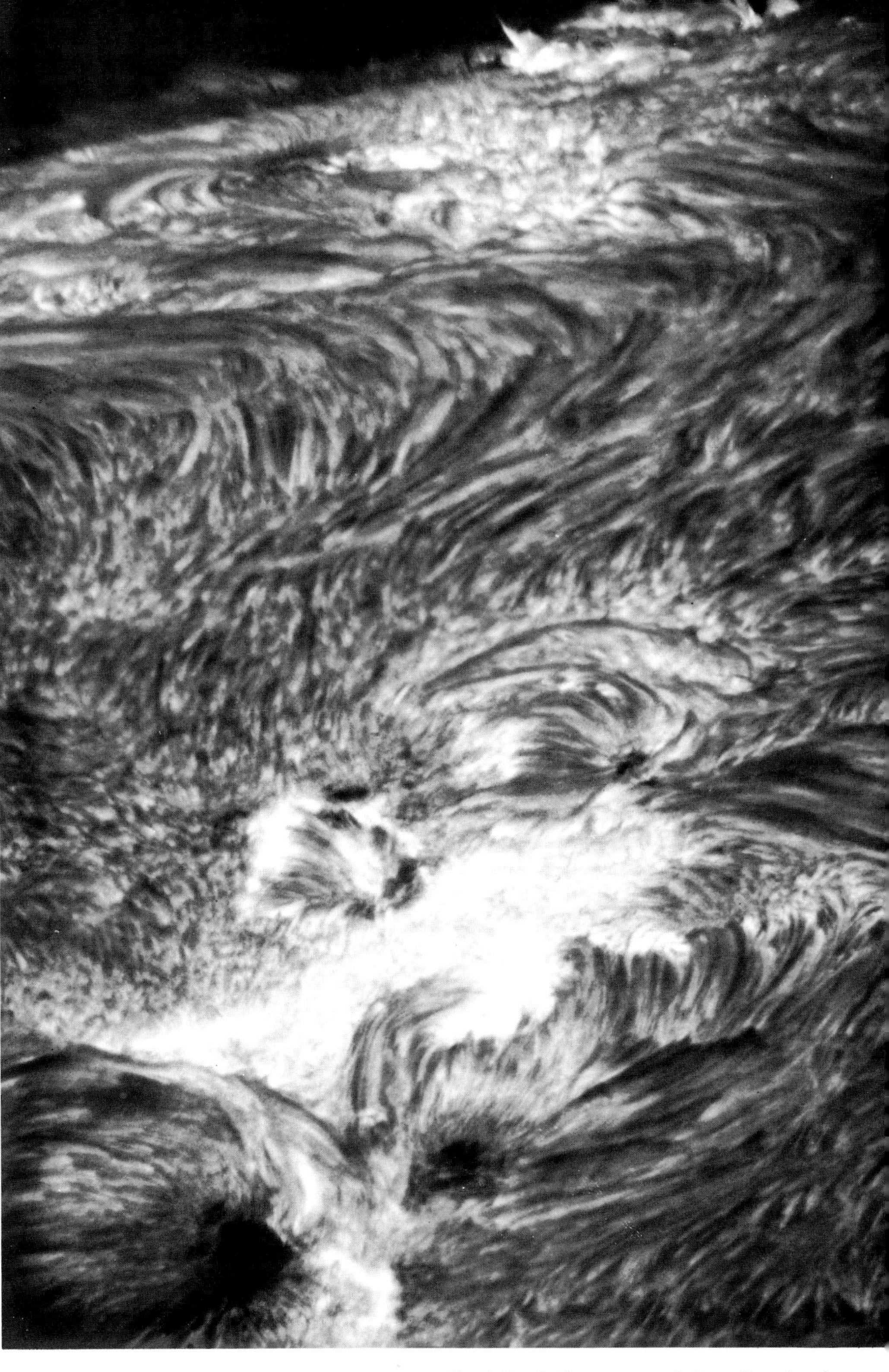

Still more astonishing, if it can be confirmed, is the close coincidence between the convergence points of the descending currents and the spicules. Very clearly observed in the light emitted close to the center of the Hα line, the spicules are also quite visible at the Sun's edge, where they resemble a burning prairie in perpetual movement. Their inclination rarely exceeds 15 to 20 degrees from the vertical, and their diameter typically equals 400 kilometers (249 miles). They extend up to 7,000 to 9,000 kilometers (4,351 to 5,594 miles) above the limb, and seem to correspond to vertical jets of matter whose average speed is about 20 kilometers (12 miles) per second. They are of interest here only because they transport a substantial share of kinetic energy and contribute significantly to the atmosphere's energy equilibrium.

Many theories have been put forward to explain their origin, but none of them is satisfactory. For example, they have been identified with the Alfvén waves that would be propagated along the magnetic lines of force, marking the boundaries of the supergranules, and would contribute to a concentration of matter stemming from the corona; they would usually be produced along those same lines of force. Another theory ties their origin to the energy flux transmitted by conduction from the corona to the atmosphere. Because the temperature varies very rapidly between the corona and the chromosphere, thermal conduction is very efficient there. With the magnetic field preferentially directing this energy flux along the direction of the field lines, the chromosphere, at the points where the lines of force emerge, receives a large

Circulation in the supergranulation cells. *a. Seen from above: The descending currents are concentrated at the boundaries of the cells. The direction of the current is marked by (+) for the rising currents, by (−) for the descending currents and by an arrow for the horizontal currents. b. The current of ionized gas that circulates from the center to the periphery of the cells carries the magnetic field to the edges of the cells, where it is concentrated and appears as brighter areas. The spicules seem to be formed in such places. Higher, in the zone separating the corona from the chromosphere (transition zone), the regions corresponding to the site of the spicules are the center of intense descending currents of unknown origin. From Gibson,* ***The Quiet Sun****.*

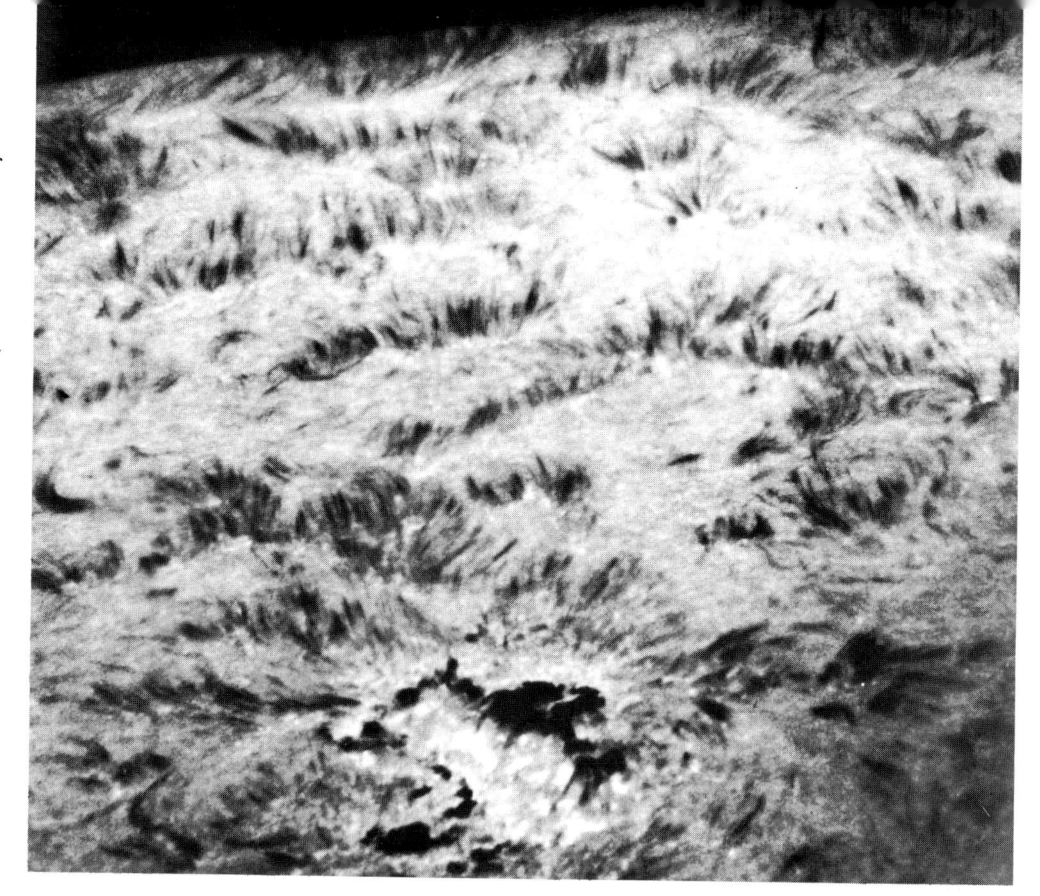

Photograph of the Sun taken in light very close (0.8 Å) to the Hα line of neutral hydrogen, *showing the cells of the chromospheric network seen in perspective toward the Sun's edge, with the spicules (small dark strips) outlining the contours. Brighter dots, corresponding to more intense magnetic fields, mark the foot of the spicules. They may be the sources of nearly vertical gas jets carrying a large quantity of kinetic energy. Photo: Sacramento Peak Observatory, Association of Universities for Research in Astronomy, Inc.*

energy surplus that it cannot eliminate radiatively because it is already at the limit of what it is able to radiate. This energy could then be transformed into kinetic energy, which would force the gas to escape upward, preferably in the direction of the magnetic field. The nature of the process that transforms the energy is, unfortunately, unknown. The close association of the spicules with the descending currents described above is a phenomenon that is now under intensive investigation. It will probably be necessary to wait until we have instruments of much higher resolving power, perhaps to separate two phenomena that now appear combined with current instruments.

The Corona

Photographs of eclipses taken in visible light clearly show that the corona is made up of very large structures whose shapes seem to depend closely on the configuration of the magnetic field. Nevertheless, observations of eclipses have many drawbacks. On the one hand, they are relatively infrequent (one or two per year, on average). Futhermore, they are limited to the visible spectrum, whereas the corona radiates most of its energy in X rays and ultraviolet light. They often require long and costly, even dangerous, expeditions, and their success depends entirely on the meteorological conditions at the site—conditions that cannot, of course, be controlled. Finally, they rarely last longer than 5 minutes. In 1973, French scientists partially overcame these drawbacks by using the Concorde supersonic airplane as a flying observatory. The airplane, flying above the clouds, eliminated the meteorological drawback. Moreover, its speed of 1,500 kilometers (932 miles) per hour allowed it to remain for a long time in the Moon's shadow cone, thus increasing the useful observation time from a few minutes to more than an hour. However, such an expedition is expensive and does not make it possible to get rid of atmospheric absorption.

Another technique for observing the corona consists of creating an artificial solar eclipse in a special refracting telescope called a "coronagraph." With the coronagraph, invented by the French astronomer Bernard Lyot, the corona is observed regularly, regardless of eclipses, in all observatories equipped with instruments of this kind. It is essential for the use of these devices that the sky conditions be excellent, with high transparency and purity. These conditions are obtained when the brightness of the sky around the Sun is about a millionth that of the Sun. Several high mountain observatories possess these characteristics from time to time: This is particularly true of the Pic du Midi Observatory in the Pyrenees, (France).

Still, eclipses and coronagraphs allow us to see the corona only beyond the solar disk and not

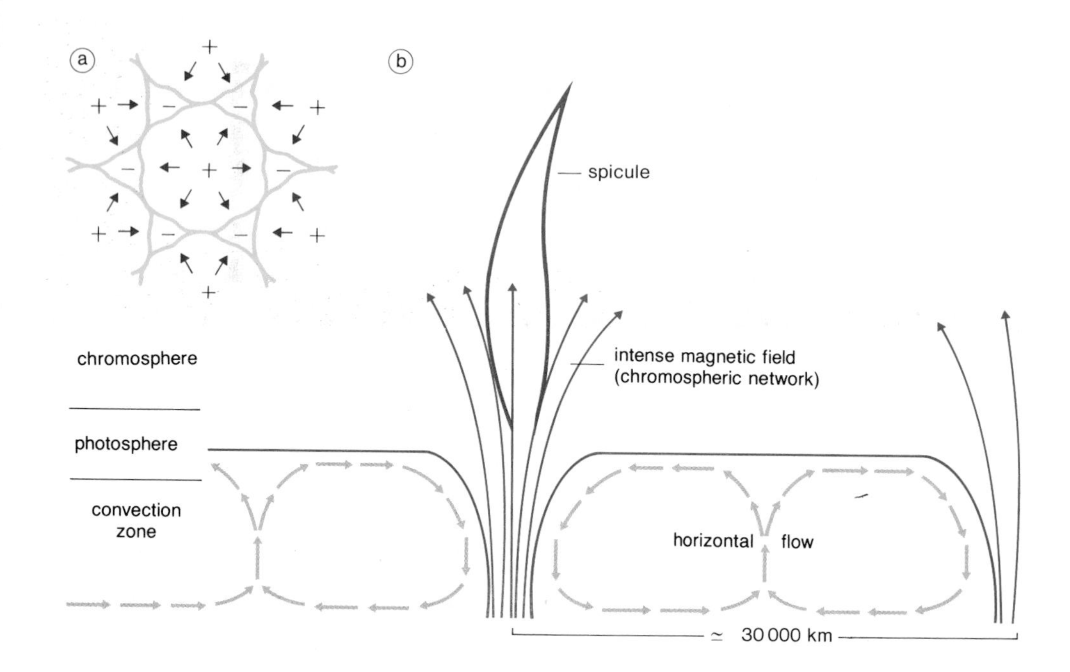

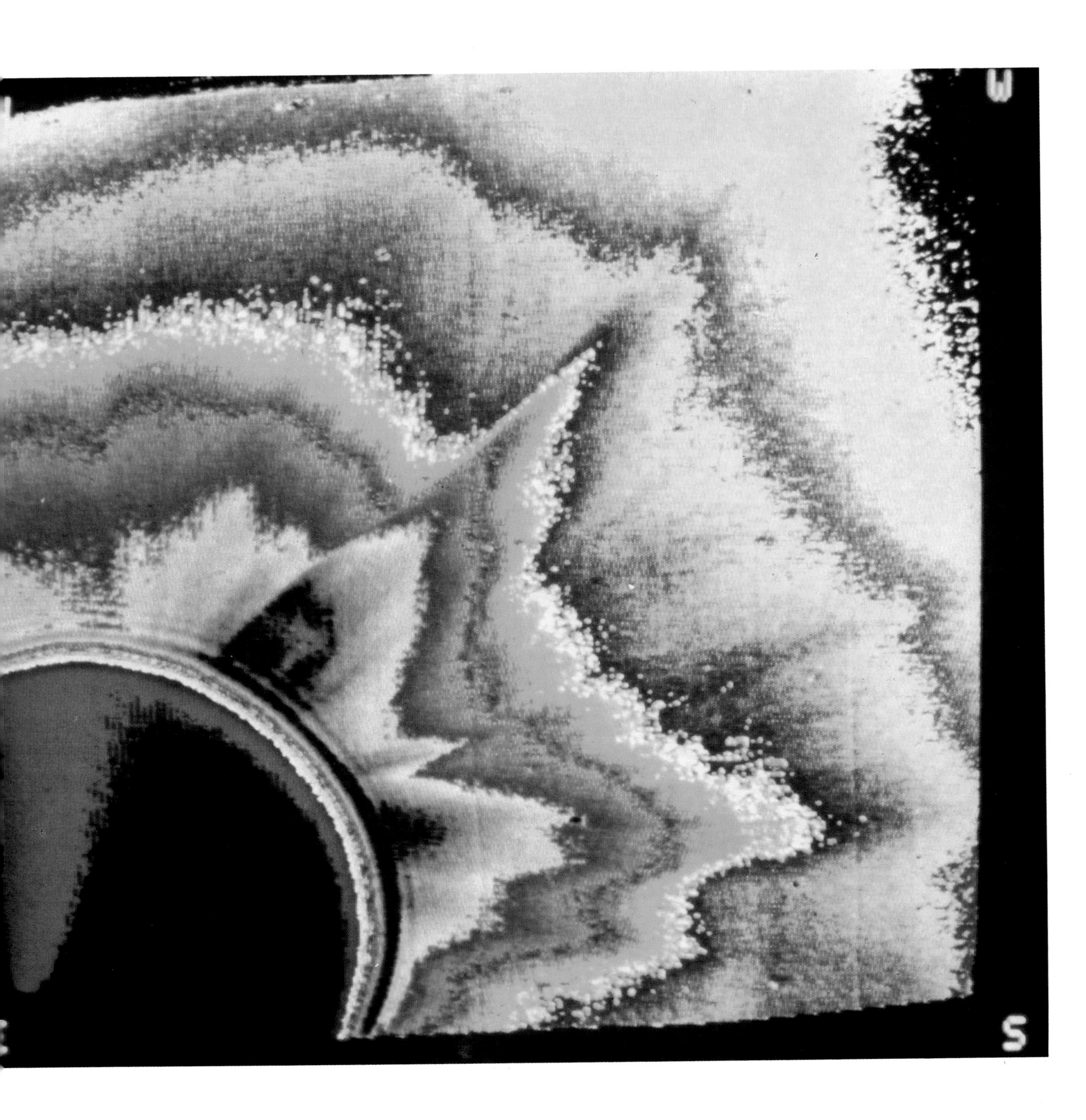
W
S

he solar corona observed from space. *false-color image made from a picture ken with a coronograph-polarimeter laced aboard the* ***Solar Maximum lission*** *satellite launched on February 14, 980, to study the Sun during its period of aximum activity. The various colors reflect e differences in the electron density of the orona. At upper right we note a particularly tense plasma jet. Pictures like this one ake it possible to observe phenomena king place in the corona and to follow their evelopment out to more than 2 million ilometers (1.2 million miles) from the hotosphere. Photo: NASA, Goddard Space light Center.*

projected onto it. The latter would be impossible to observe in visible light, for, on the one hand, the photosphere is much too bright, and, on the other hand, the corona is transparent. To see the corona "from above," it is necessary to observe it at a wavelength where those two defects do not exist, as is the case in two extreme-wavelength regimes: radio waves and X rays.

As the corona is opaque to electromagnetic radiation with wavelengths of more than 2 meters (6.56 feet), the photosphere's radiation cannot penetrate it. On the other hand, the corona's electrons do emit significant amounts of radio waves. However, there are a few instruments in the world that operate at such long wavelengths, in Australia; Japan; the United States; and in France, at Nançay. The poor angular resolution of these instruments (it is inversely proportional to the wavelength of the light) is compensated for when several of them are combined in an array to form an interferometer. We then get much better quality images, for the resolution is the same as that of a telescope whose diameter is equal to the largest interval between two antennas of the array. It can be brought down to a value close to 1 arc second.

A similar resolution is also possible with X-ray telescopes. In this part of the spectrum, the photosphere, being relatively cold, radiates negligibly in comparison with the corona, which has a temperature of more than 1 million degrees and radiates the main part of its energy there (see page 98). Its intensity in X-ray radiation is very sensitive to temperature and varies proportionally with the square of the density of electrons, while the X ray intensity of the visible corona depends linearly upon this quantity. Thus hot, dense concentrations of the coronal gas will appear with a much greater contrast in X rays than in visible light.

The Sun has been photographed in X rays by an instrument placed aboard Skylab. The beauty of our Sun seen in this unfamiliar way leaves us amazed. Its surface, thus revealed, shows several very characteristic structures.

First, we find very bright regions—hence hotter and denser than average. A comparison with simultaneously made maps of the magnetic field shows a close correlation between the bright regions and very intense bipolar fields. We also notice many loop-shaped structures that invariably link regions of opposite magnetic polarity. The latitudinal distribution of these bright regions seems to be the same as those of the sunspots at the photospheric level. We will see that this is not accidental. We also note dimmer structures of larger dimensions, which seem to link the bright regions and the loop-shaped structures.

Furthermore, we see many "hot spots" that, like the previously noted bright regions, are connected to very dense bipolar magnetic fields. These "hot spots" appear and disappear continually in

One of the 16 elements of the radio heliograph of the French station at Nançay. *Each antenna may be thought of as part of a much bigger mirror equal in diameter to the total dimension of the array. The radio-frequency signals, after being synthesized and analyzed by a computer, make it possible to get a picture of the solar corona in the meter-wave range. Since the diffraction of light is proportional to the telescope diameter, it is reduced in the radio-wave range by using an array of large dimension. The antennas are distributed along a baseline of 3.2 kilometers (2 miles). Photo: Paris Observatory by the kind permission of Mme. M. Pick.*

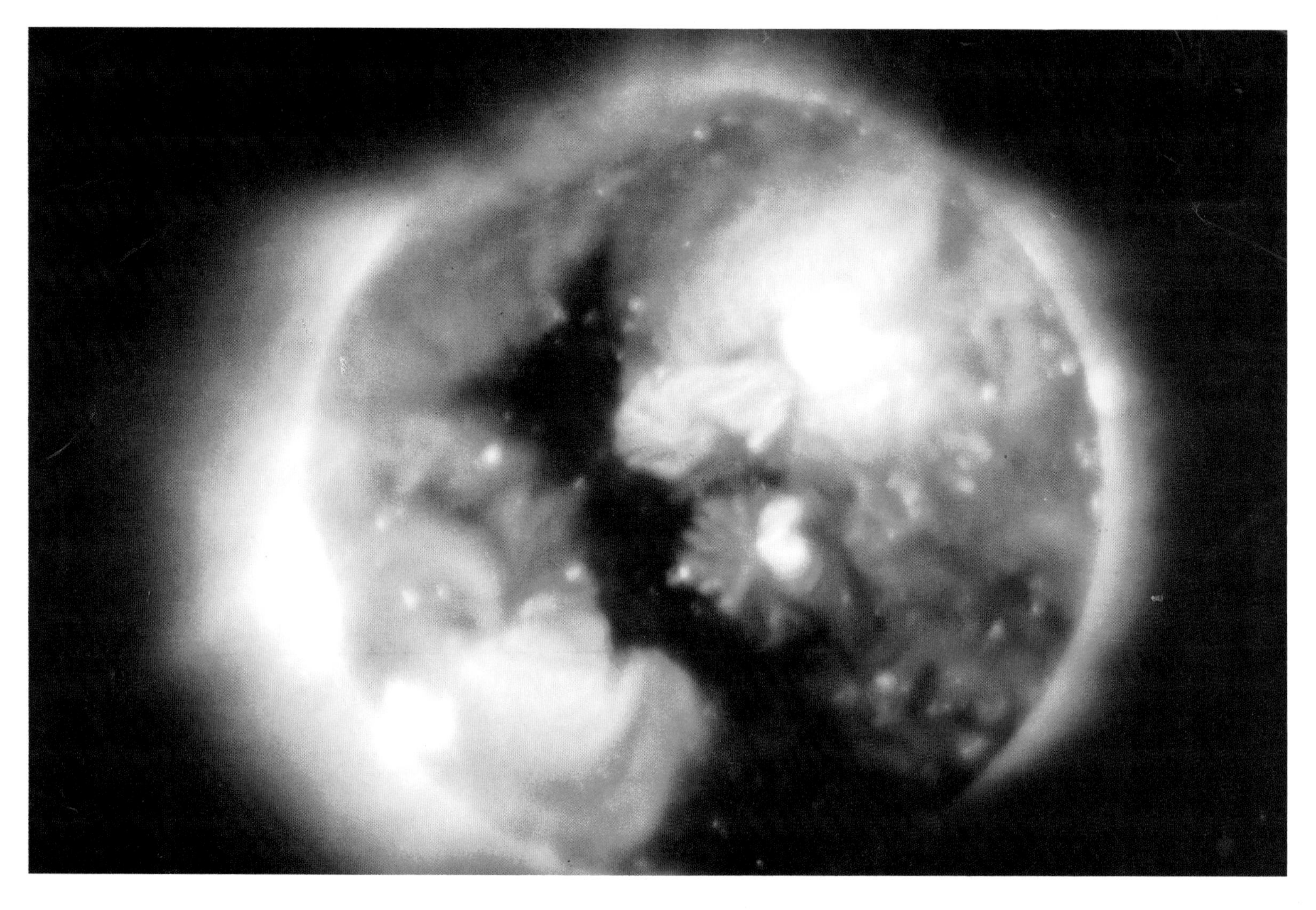

The Sun's appearance when photographed in X-ray light. *Above, a photograph taken in 1973 with the telescope aboard the* Skylab *orbiting laboratory. It shows the morphology of the corona seen projected against the solar disk, the presence of many bright spots (which represent the bottom of magnetic loops joining two points of opposite polarity) and, at the positions of the poles, two huge darker zones called coronal holes. Right, a series of photographs taken with the same instrument, showing the evolution of the coronal holes during solar rotation. We note the semipermanent presence of these holes in the vicinity of the poles and the existence of many bright spots. Photo: Harvard College Observatory, NASA.*

the corona, their average life-span being only a few hours. Their latitudinal distribution seems to be uniform, which implies that their origin is different from that of the magnetic regions associated with the sunspots and the bright coronal emission. However, they carry a very large quantity of magnetic energy into the corona, at least equal to that of the brightest of the previously mentioned X-ray-emitting regions, which may mean that they play an important role in magnetic equilibrium and in heating the corona.

The discovery of the "hot spots" is relatively recent, which accounts for the fact that their origin has not yet been explained.

Finally and above all we note dark cavities bordered by luminous structures that seem to be escaping from them. These cavities represent regions of open magnetic fields, where the lines of force are practically parallel: Wherever you place yourself inside these regions, a compass needle will always point in the same direction. They have been given the evocative name of "coronal holes." It has been determined that the density of the corona is about three times less there than in nearby regions, and the temperature about twice as low. These holes are seen only in the corona; farther down, they disappear. It has also been noted that they do not conform to the system of differential rotation but seem to revolve like solid structures—which is probably due to their essentially magnetic origin. Most of those that have been observed up to now have survived for several rotations. Each of the poles is permanently occupied by a coronal hole. That is hardly surprising, given the "open" magnetic configuration of the field at the poles.

We will see later that the coronal holes can play a dominant role in generating the solar wind.

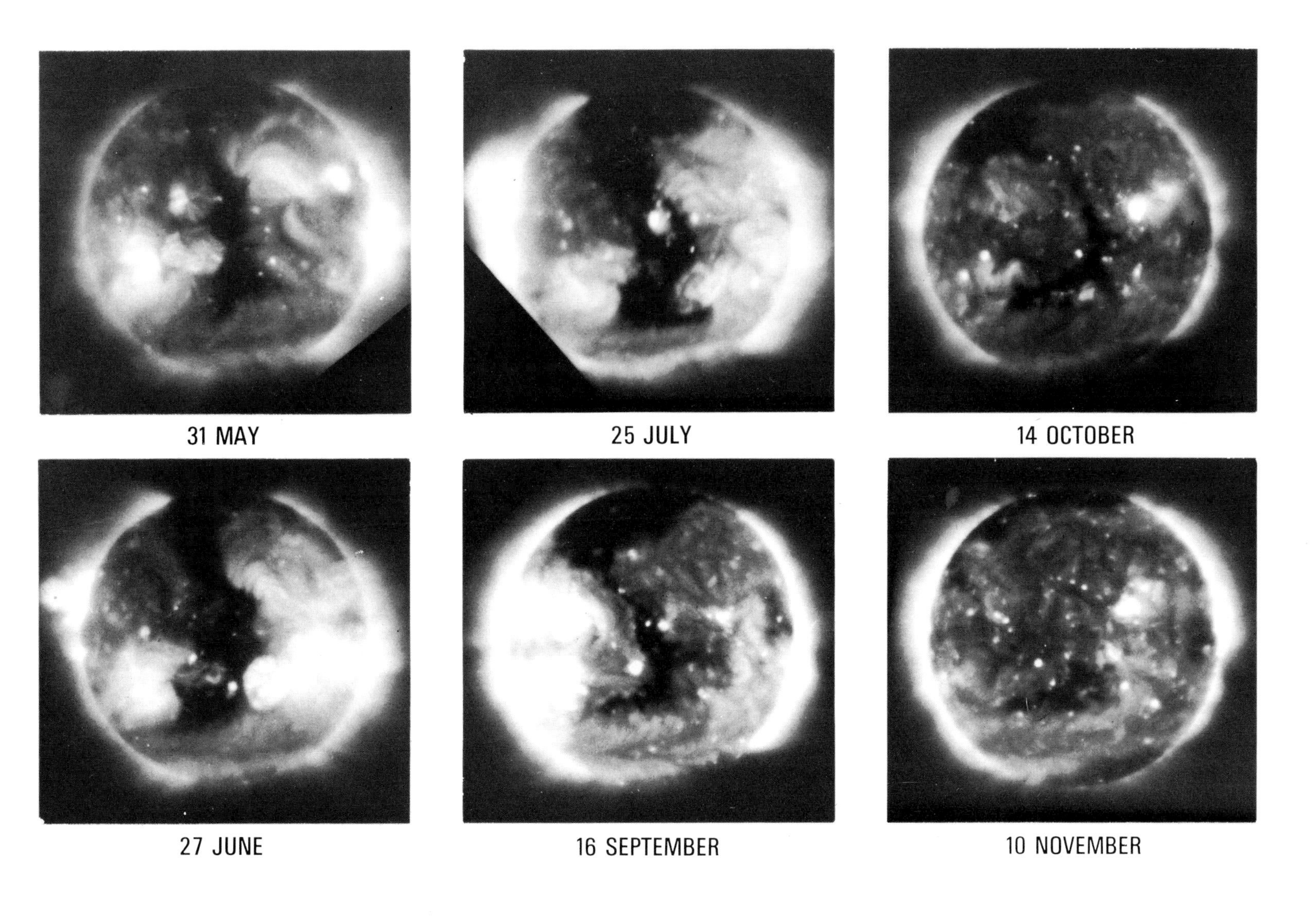

The American space laboratory Skylab, launched in 1973 and fell on July 11, 1979. *It contained a total of six instruments, including a telescope for studying the Sun, and a white-light coronograph, located in the circular portion, at the intersection of the four solar photobattery panels. All the instruments were operated by the astronauts from inside the station. We note: at left, nondeployment of one of the solar panels; the makeshift thermal shield installed by the astronauts during a special foray into space to repair the damaged shield in orbit and protect the station from solar radiation; the cloud cover near the ground. Photo: NASA, Marshall Space Flight Center.*

Manifestations of Solar Activity in the Chromosphere and Corona

Active Regions

Under this label are commonly grouped all regions and structures in which the field lies close inside the Sun and in which the intensity is greater than average. This is true of the sunspots and regions of bright coronal X-ray emission and perhaps also of the X-ray hot spots. At the chromospheric level, the active regions are generally marked by an excess emission of light and are often labeled "plages." The plages usually appear shortly before the sunspots emerge and often surround them completely. The plages are more difficult to perceive farther down, in the photosphere.

We might wonder why sunspots are very dark and the active regions very bright if the phenomena that cause them are the same. It is because, as we have seen, the magnetic field—much more intense in the spots than in the plages—inhibits the convective flux in the former and not in the latter.

The active regions appear at very special latitudes and may remain on the Sun's surface for several days, if not several months. Except for the X-ray hot spots, the magnetic field is carried into the solar atmosphere mainly by the active regions and is then dispersed by the eroding action of supergranulation.

The energy balance of the active regions is certainly much more complex than that of the average normal Sun precisely because of the influence of the magnetic field. Indeed, not only does it shape their structure, but also it is probably responsible for their surplus heating. How? That is not as clear! Is there more energy coming from the convection zone, carried along the direction of the field and deposited in the atmosphere, or does the presence of intense fields facilitate this transport? Doesn't the heat actually come from the dispersion of the Alfvén magnetic waves? This question is still unanswered, although many theories have been offered.

Extension of the Spots into the Atmosphere

Observations of sunspots in ultraviolet light make it possible to study how their appearance changes with altitude in the vicinity of the active regions. In addition, their changes of shape may give us information about the variation in the orientation of the field.

Ultraviolet photographs taken aboard the Skylab space station show that the region overlying the umbra of a sunspot appears, at the base of the corona, as a kind of very bright plume. These "plumes" form loops that link the spot's shadow to a point on the Sun of opposite polarity. Moreover, ultraviolet spectroscopy reveals phenomena directly linked to the atmospheric dynamics above the spots.

Study of the Doppler shifts of spectral lines reveals very rapid vertical currents that observations made on the ground cannot show. In particular, the region that represents the bright "plume" is the site of a very strong descending current having a speed that can reach 100 kilometers (62 miles) per second, or 12 times greater than the speed of sound in the solar atmosphere. Curiously, this current is observed only in the region between the chromosphere and the corona; farther down, in the chromosphere, it disappears entirely. No explanation has yet been given of this phenomenon, but it is clear that because of the amplitude of the measured velocities, it must play a decisive role in energy balance of the spots. It is probably a key indicator of the complex phenomena that are the origin of these still poorly understood structures.

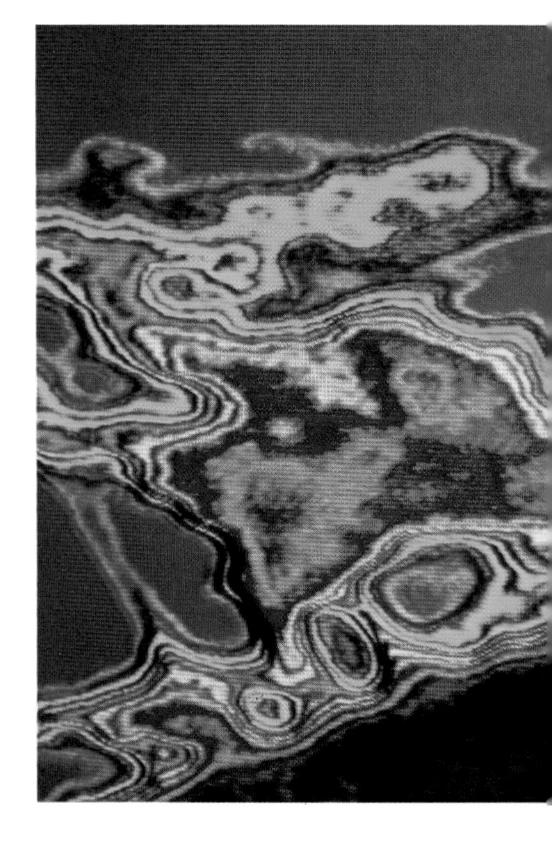

A prominence observed in the light of the **Hα** *line, reaching up to several tens of thousands of kilometers above the solar limb. In this photograph, we clearly see that a prominence does not have a uniform structure but is made up of many filaments, the smallest of which are equal in dimension to about 1,000 kilometers (622 miles). The chromosphere also is visible, as are the spicules that form a hem around the solar limb. Photo: Big Bear Solar Observatory, Big Bear, California.*

Prominences and Filaments

Because the signs of coronal activity are most often very luminous phenomena, the active regions were the first structures to be observed with the coronagraph. This is true of the splendid prominences whose simplicity or complexity compel our admiration.

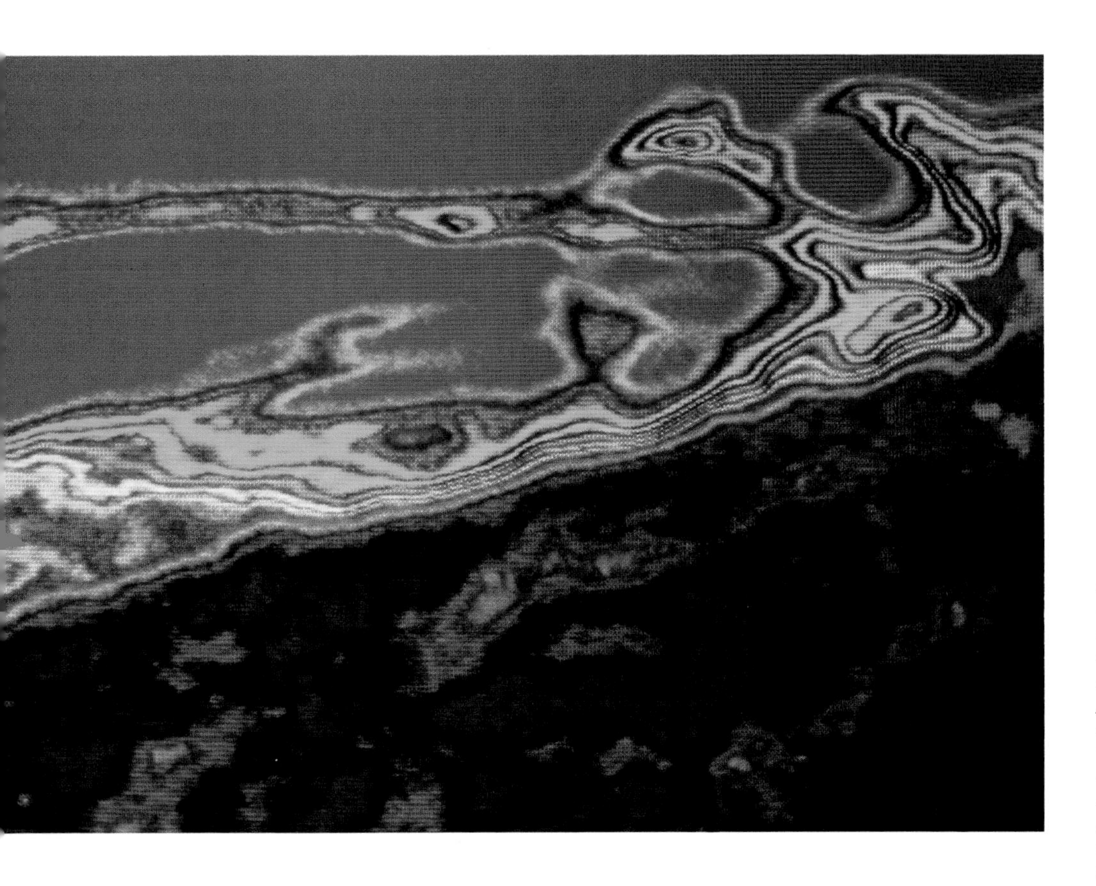

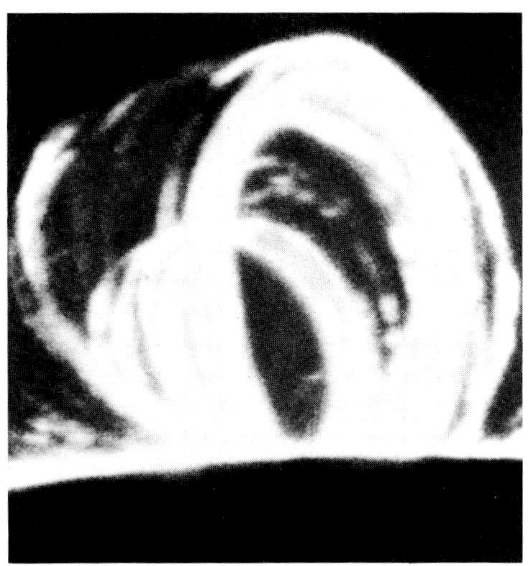

Left, a prominence observed in the light of the Lyα resonance line of the hydrogen atom. *The plasma traces the structure of the magnetic field. This photograph was taken during a rocket flight on July 3, 1979. Photo: Laboratoire de physique stellaire et planétaire, color-coded at the Paris Institut d'Astrophysique.*
Above, a loop-shaped prominence. The plasma flowing from the corona to the chromosphere traces the structure of the magnetic field. Certain prominences, called eruptive, are characterized by a violent, short spurt lasting only a few minutes; others, called quiescent, develop much more slowly, and the arches they produce may last for several days. Photo: Sacramento Peak Observatory, Association of Universities for Research in Astronomy, Inc.

The prominences are commonly thought to be condensations of coronal plasma. Indeed, although totally immersed in the corona, their temperature is about 100 times lower and their density 100 times greater than that of the corona. They are made up of highly (but not totally) ionized plasma, which in some cases assumes the shapes of the magnetic field. This is true of the loop-shaped prominences. It is also the field which produces the force necessary to maintain them in the corona, thus preventing them from collapsing under their own weight.

Seen from above, and thus projected onto the disk, they have the appearance of dark filaments. In the light of the Hα line of hydrogen this phenomenon is more striking, for, being denser and colder than the surrounding atmosphere, prominences contain relatively more hydrogen atoms, which heavily absorb the light of the underlying photosphere and chromosphere. The coronal gas condenses preferentially in places where the magnetic field is horizontal—that is, where its direction and polarity reverse themselves. This geometry is peculiar to linear prominences rather than to the "loop-shaped" structures. The filaments corresponding to the first type thus constitute the separation zone of two regions of opposite polarity. This zone is called the "neutral line," for it has no definite polarity.

Flares

The most catastrophic atmosphere manifestations of the Sun's magnetic activity are the flares.

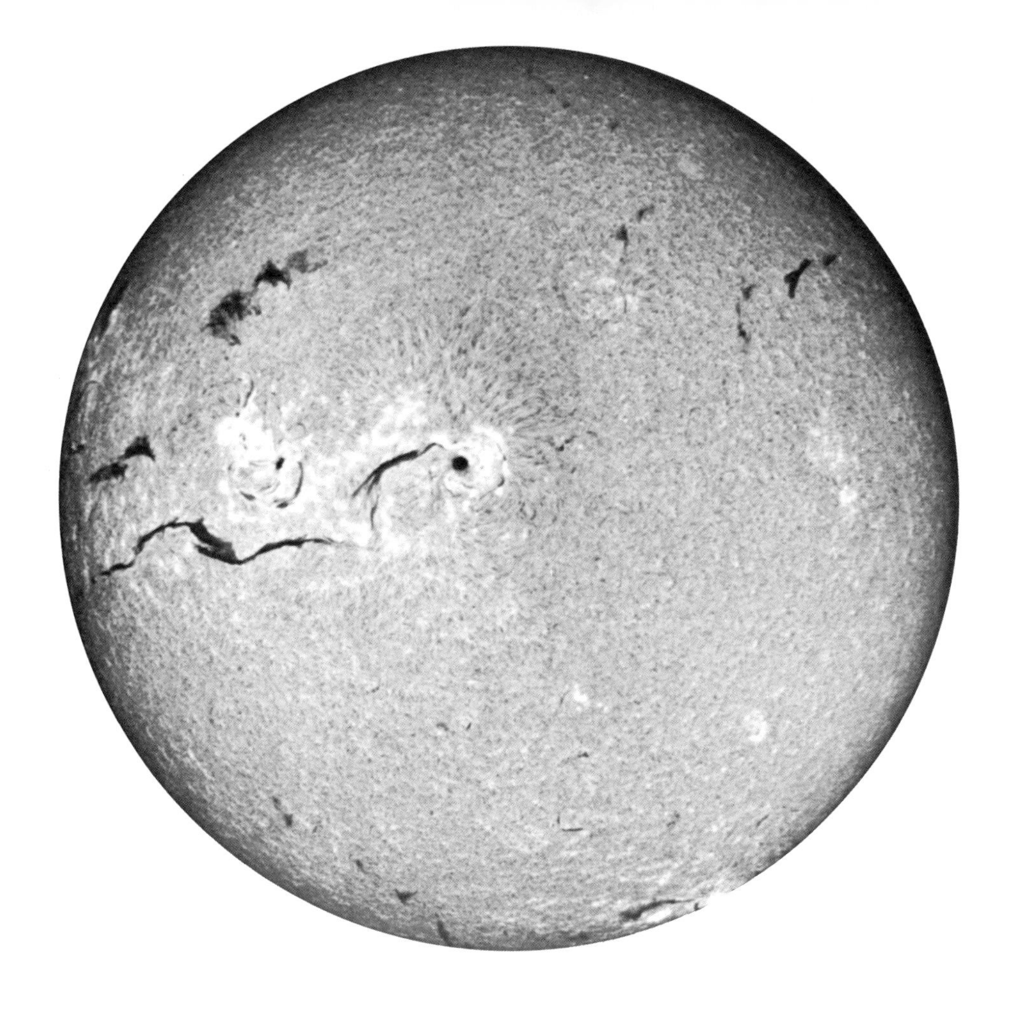

The Sun, photographed in the light of the Hα line. *We note the sunspot practically at the center, surrounded by bright areas delineating the outer boundaries of an active region. The slender black structures are filaments—i.e., prominences seen in projection onto the disk. The neutral hydrogen within the prominences absorbs the underlying light from the Sun, giving it a darker appearance. Photo: Sacramento Peak Observatory, Association of Universities for Research in Astronomy, Inc.*

They probably constitute the most complex phenomenon that can be observed in the Sun, encompassing an incredible combination of the most diverse physical phenomena. Essentially, a flare consists of a violent explosion of variable intensity. The strongest of these eruptions release by themselves as much energy in the space of a few minutes as the entire Sun in a tenth of a second, or the equivalent of a bomb of 2 billion megatons of TNT.

Observed in the red light of the Hα line, the flare first appears as a small bright spot in an active region, followed by a rapid spreading in certain directions, and, in the space of a few minutes, a surface as large as one-tenth of the Earth is totally ablaze. It takes several hours afterward for the Sun to return to its previous state.

The origin of such enormous energy is still unknown. The hypothesis of a nuclear process is discounted, for, in contrast to the central parts of the Sun, the density is much too low here to permit the unleashing of fusion reactions. Rather, it is once again in the magnetic field that the explanation is to be found. Indeed, flares systematically occur in regions with strong fields having abnormal structures and configurations. By this is meant, for example, bipolar fields having polarities reversed relative to the normal orientation; or very tortured and twisted neutral lines; or abnormally bright, active regions. A flare is frequently preceded by the expansion and then the rupture of a filament, indicating a change in the magnetic configuration.

But how does this energy slowly accumulate from one day to the next? That question has not yet been answered. By way of explanation, we often use the analogy of the rubber band employed to activate the propeller of a model airplane. After being twisted several times and suddenly released, it quickly releases all of the energy slowly stored up beforehand. In the Sun's case, the rubber band is the magnetic field. But the "hand" that winds the field around itself has never been caught in action! Likewise, the mechanism that triggers the sudden liberation of energy has not yet been clearly identified. It probably involves an instability whose releasing action becomes more effective as more energy is accumulated.

A flare is seen at nearly all the spectral wavelengths, but especially at the very short ones (X rays) and very long ones (radio waves), which indicates an origin located rather high up in the atmosphere. The first appearance of a flare has been observed almost simultaneously in these two wavelength regions. This is the result of the sudden acceleration of the ions and electrons just after the energy release. As soon as they are accelerated, the electrons lose their energy, giving off X rays and spiraling around in the direction of the magnetic field. In doing so, they also emit radio-frequency radiation.

The more rapid electrons propagate preferentially toward regions of low density and tend to escape from the Sun rather than fall back to it. In their flight they encounter the successive layers of the atmosphere and transmit part of their energy to the surrounding gas. In particular, they are responsible for the "type III bursts," which correspond to a kind of "vibration" of the coronal plasma under the impact of the electron bundle. The period of this vibration (corresponding to radio frequencies) declines as the square root of

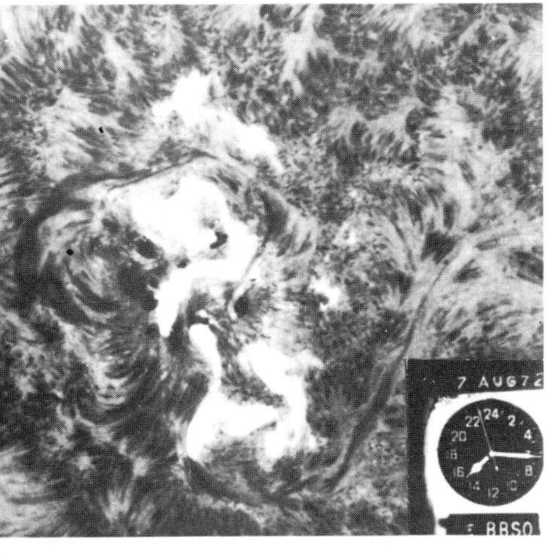

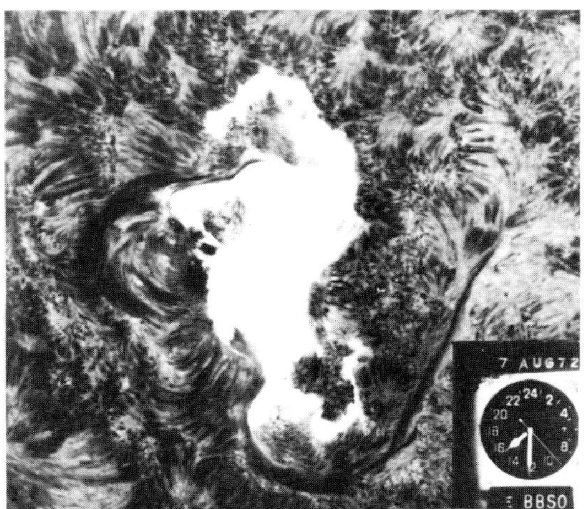

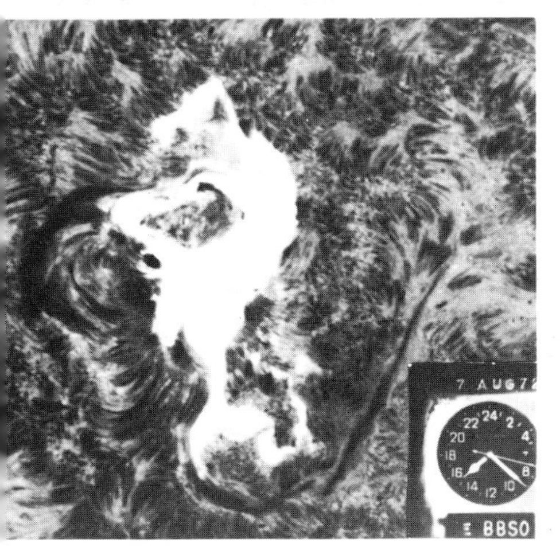

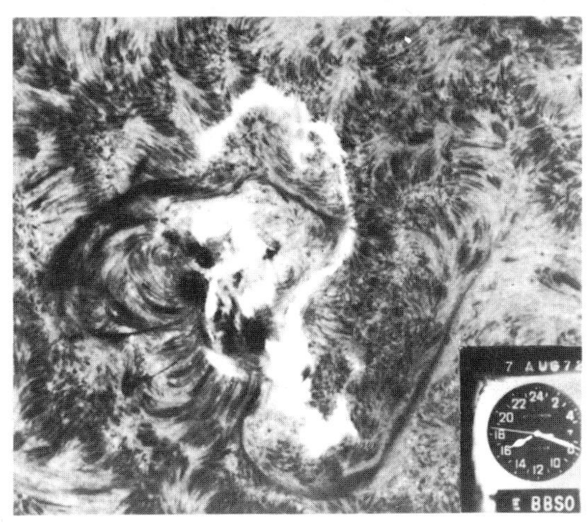

the gas density. In its movement toward the exterior, the bundle encounters less and less dense layers, which vibrate with longer and longer periods. By measuring the waves' arrival time with a radiotelescope and knowing the density, one can therefore determine the velocity at which the bundle moves. We thus find velocities that can reach the amazing value of one-half the speed of light!

The energy thus lost by the electrons represents in reality only a tiny portion of the total energy of the flare. What happens to the remainder? We think that it serves to warm the atmosphere in the vicinity of the initial explosion. The heating is stronger in the more highly conductive corona: In the chromosphere, the temperature rises by only a few thousand degrees, leading to an excess ultraviolet-light emission, whereas in the corona, it may reach 10 million degrees, causing the intense X-ray emission. This phase during which the flare cools while transmitting its energy to the surrounding plasma may last several minutes or several hours, according to the quantity of energy involved. In fact, in some cases, the X-ray emission lasts so long that we think it is continuously being replenished by an energy source, (that, however, has never been discovered).

Within the chromosphere, the heating is often the cause of an explosive ejection of matter, which may attain a mass of more than 10 billion tons and a speed of several hundred kilometers per second. This formidable shock wave blows away everything in its path, particularly the magnetic field. This, in turn, through a process that has not yet been identified, sets off a secondary acceleration of the plasma, which can reach the Earth's

The great solar eruption of August 8, 1972, observed in the light of the Hα line. *During an eruption, very great changes in the luminosity of a particular region of the Sun appear in the space of several minutes, due to a sudden release of energy stored up for several days by the twisting of the magnetic lines of force. The very hot gas travels in preferred directions corresponding to the neutral lines of the magnetic field (where the polarity is neither positive nor negative). In this example, it took almost an hour for the Sun to recover its normal appearance. Photo: Big Bear Solar Observatory, Big Bear, California.*

orbit, where it causes geomagnetic disturbances: northern lights, magnetic storms, etc. The most intense eruptions may even be accompanied by nuclear reactions (which are not the cause but rather the consequence of the enormous quantity of energy released), such as the creation of deuterium through the collision of a proton and a neutron, and the annihilation of electron-positron pairs.

These monstrous but sporadic ejections of matter are superimposed upon that which continually but gradually evaporates the substance of our star into the interplanetary medium: the "solar wind."

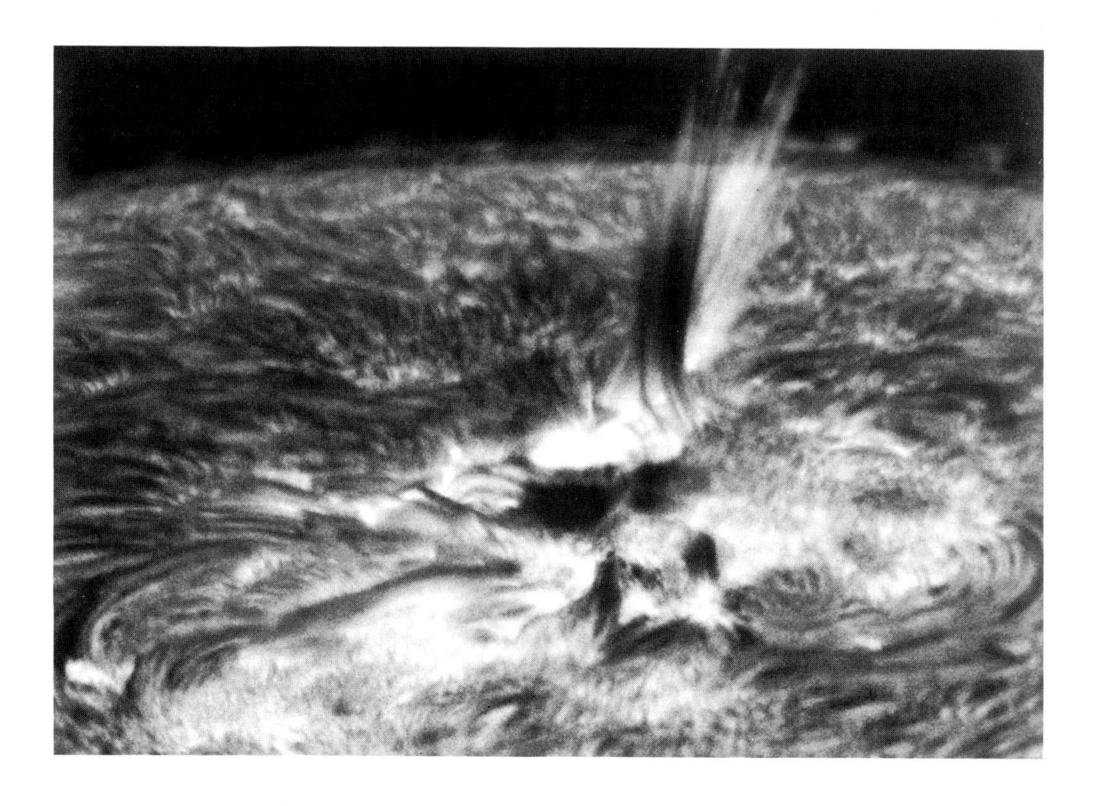

Solar Wind

THE NORWEGIAN PHYSICIST Olag K. Birkeland was the first to suggest, in 1896, that the Sun may emit something other than light, when he explained the northern lights as the reaction of the Earth's atmosphere to the influx of electrically charged particles of solar origin that are guided toward the poles by the Earth's magnetic field. The English geophysicist Sydney Chapman was the first to interpret, in the 1930's, the magnetic storm that disturbed radio communications as being due to the arrival in the Earth's atmosphere of charged particles emitted during the flares.

At the end of the 1940s, the American scientist Forbush noticed that the intensity of cosmic radiation on Earth is systematically weaker at the maxima of the solar magnetic cycle, and diminishes sharply during magnetic storms. The explanation of this phenomenon was given by Alfvén: A stream of charged particles carries within it a magnetic field, which tends to pull the cosmic rays out of their normal trajectory. The observed correlation with the maxima of the solar cycle shows that the phenomenon is connected with the Sun, which must, therefore, emit charged particles capable of deflecting the cosmic rays from their trajectory and preventing them from reaching the ground. We note in passing that it is during the solar minima that the flux of cosmic rays is greatest in the Earth's atmosphere, where they react with carbon atoms to form a radioactive isotope: carbon-14. Absorbed like the normal isotope by plants during photosynthesis, carbon-14 shows up in the growth rings of plants, where it can be detected and measured. Carbon-14 dating makes it possible to retrace the evolution of solar activity over several hundred cycles going back to 4000 B.C.

Decisive proof of the existence of a steady emission of solar particles was given a little later by the German scientist Ludwig F. Biermann, who interpreted the systematic curving of the tails of comets in the direction away from the Sun in terms of the pressure exerted by the particles emitted by the Sun. The curvature of the comet tails being a lasting phenomenon, the flux of particles responsible must therefore be continuous. Comet tails are, so to speak, a "wind sock" for the solar wind, the existence of which was thus directly verified. Nevertheless, it was not until the end of the 1950s, with detailed though relatively simplified calculations at that time, that the American Eugene N. Parker explained the origin of this "wind" as being directly related to the existence of the corona and its high thermal conductivity. It is not unimportant to realize that the solar wind results from the existence of a convective zone that gives rise to phenomena that cause the heating of the corona and thus the wind. If other stars have convective zones, they are very likely to have a solar wind as well. Through calculations making use of thermodynamic equations and that of hydrastatic equilibrium, it was shown that, on the one

The Sun and the Amateur

The Sun is an easy target for the amateur. But it is a dangerous body to look at: *It should never be observed through binoculars nor with a telescope without special protection,* for its extremely bright image may cause irreversible damage to the retina.

The most impressive things that can be observed with simple instruments are the sunspots that dot the photosphere as well as the bright areas surrounding them (faculae). The simplest and safest method of observation consists of projecting the Sun's image onto a smooth white screen—i.e., a sheet of posterboard—placed a suitable distance behind the eyepiece of the telescope and held in place by an appropriate device. With a small reflecting telescope, the best results can be achieved by using an eyepiece having a magnification of approximately 60. In order for the projected image to be fully visible, the screen should not be in direct sunlight; the instrument should be fitted with a large sheet of stiff cardboard that casts a shadow on the screen. Observation by projection brings out the exact position of the spots on an adequate scale and makes it possible to measure their size and determine the Sun's period of rotation by following the shift of the spots across its surface from day to day.

The Sun also may be observed directly through the eyepiece, *under the condition that the telescope is equipped with a protective device:* a special filter and a smoked glass lens, or, better yet, a Herschel helioscope. With a 50-magnification telescope, we can already make out the spots and their penumbrae as well as the largest faculae. A telescope that enlarges about 100 times will show their contorted structure very clearly. With more powerful instruments allowing magnifications up to 200 with a 75-mm [3-inch] aperture and more), we can study the features of the umbra and penumbra zones within the spots in detail. With telescopes of 150 to 200 millimeters (6 to 8 inches), we can also note the sometimes very rapid motions and transformations accompanying the appearance, development and disappearance of the spots. We cannot fail to note also the granular, bubbling structure of the photosphere.

Observation of prominences is possible only for users of coronagraphs.

Ejection of solar gas following an eruption, observed at the level of the chromosphere in the light of the Hα line on May 22, 1970. *Such ejections may travel at very high speeds and reach the Earth's orbit, causing various phenomena in the upper atmosphere: northern lights, magnetic storms, disruption of radio-frequency links, etc. Photo: Big Bear Solar Observatory, Big Bear, California.*

hand, the temperature and pressure of the corona at a great distance from the Sun are very high, but especially that the pressure is greater than that of the interstellar medium at the frontiers of the solar system. In the absence of a counterpressure from the outside, the corona must therefore continually escape. That is the "solar wind." Near the Sun the rate of expansion is slow, but farther out, the pressure of the upper layers diminishes, and dynamic pressure accelerates the flow. At 10 million kilometers, (6.22 million miles), the speed has reached several hundred kilometers per second—much greater than the speed of sound. From there on, the corona is more a supersonic wind than it is part of the Sun's atmosphere. The gas seen at the base of the corona on Sunday will be in the Earth's vicinity by Tuesday. Two weeks later, it will be within reach of Jupiter. Thus the Earth and most of the planets move in the corona of their own star.

The Soviet satellites *Luna 1* and *Luna 2,* which were the first to voyage sufficiently far from the Earth to go beyond the region shielded magnetically from the solar wind (followed by many others, in particular by the American space probes *Mariner 2, Explorer 10* and *Pioneer 11)* could measure the characteristics of the wind and supply irrefutable proof of its existence after it had been predicted by theory. More recently, the German Helios probes came close enough to our star to be able to measure the characteristics of the wind halfway between the Earth and the Sun.

Near the Earth, the wind's speed is 400 to 700 kilometers (249 to 435 miles) per second; its temperature is very high (50,000 to 500,000° K), and its density is extremely low (barely a dozen hydrogen atoms per cubic centimeter). These experimental findings make it possible to evaluate the mass lost and the energy dissipated by the wind. We thus find that each second, 1 million tons of hydrogen escape from our star forever; after existing for 10 billion years, a billionth of its mass will be evaporated in this way. This is an enormous quantity, but negligible, after all, compared with the Sun's total mass. Likewise, the energy expended in accelerating the wind is only a millionth of the total energy emitted by the Sun, and thus, once again, negligible.

Also, the wind carries with it the solar magnetic field, which has a strength of no more than 1/10,000 to 1/100,000 gauss near the Earth. However, the expanding corona cannot stretch out the very concentrated fields of the bipolar regions, for they are too intense. It is rather the open-field regions—that is, those associated with the presence of the coronal holes—that are best suited to the dissipation of the wind. Indeed, there is now hardly any doubt that the solar wind is much more intense above these regions, since a close correlation was recently observed between the wind's maximum speeds, as measured by space probes in eccentric orbits, and the coronal holes moving into the Earth's visibility sector. We can then offer an explanation for the very low intensity and density of the coronal holes. Since their temperature is lower than that of the surrounding corona, they lose less energy by conduction. In addition, they are darker, and, they lose less energy in the form of light. If, therefore, the energy source supporting the corona is the same everywhere, then, on the average, we should find the energy missing from the coronal holes somewhere. It is found precisely in the kinetic energy of the solar wind particles—whose speed is twice as large above the coronal holes as elsewhere.

If the Sun did not rotate on its axis, the field's lines of force would be straight and would be

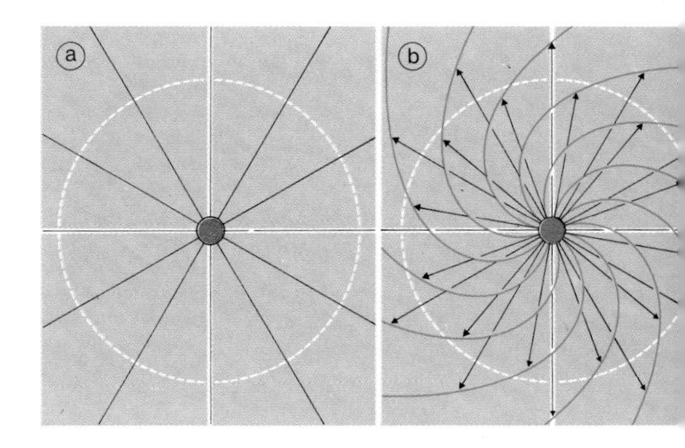

The lines of force of the solar magnetic field in the interplanetary medium: *a. As they would appear if the Sun did not rotate on its own axis. The lines are in the solar equatorial plane (the ecliptic). The dotted circle shows the Earth's orbit. b. As they really are. The spiral results from the solar rotation. The lines show how the needle of a compass would be oriented at each point. The arrows show the course of the solar wind particles. From* ***Scientific American****.*

oriented radially in the plane of the ecliptic; a compass needle placed between the Earth and the Sun would always point toward the latter. Solar rotation distorts this simple figure into a spiral centered on the Sun. But what happens outside the ecliptic? No one has ever been able to find out, for no satellite has yet been able to acquire sufficient energy to pass outside that plane. That is why there is a great deal of interest in sending space probes outside the plane of the ecliptic (see box).

So ends the journey that has taken us from the Sun's center to the outer reaches of the system bearing its name.

In about 5 billion years, when it has "burned up" nearly all its hydrogen, the Sun's diameter will increase, and it will be transformed into a red giant, a gigantic sphere of gas in which the Earth will be "swallowed up."

The International Solar Polar Mission

Developed by the European Space Agency, the International Solar Polar Mission (ISPM) will make it possible for the first time to observe the Sun from above its poles. It will include a probe that will be launched, theoretically in 1986, by the American Space Shuttle, equipped with an inertial upper stage and placed on a trajectory that will carry it along the plane of the ecliptic toward Jupiter. Once in the vicinity of the giant planet, its path will be strongly deviated by Jupiter's powerful gravitational pull, and it will be thrown out of the plane of the ecliptic and its orbit back toward the sun will cause it to fly over the poles.

The essential purpose of this project is to deepen scientific knowledge of the Sun and its environment through a study of its structure and the emission from the equator to the pole. The characteristics of the corona, magnetic field and solar winds will be directly measured at different latitudes for the first time. The measurements taken above the poles, characterized by the permanent presence of coronal holes through which solar wind escapes, will be of particular interest.

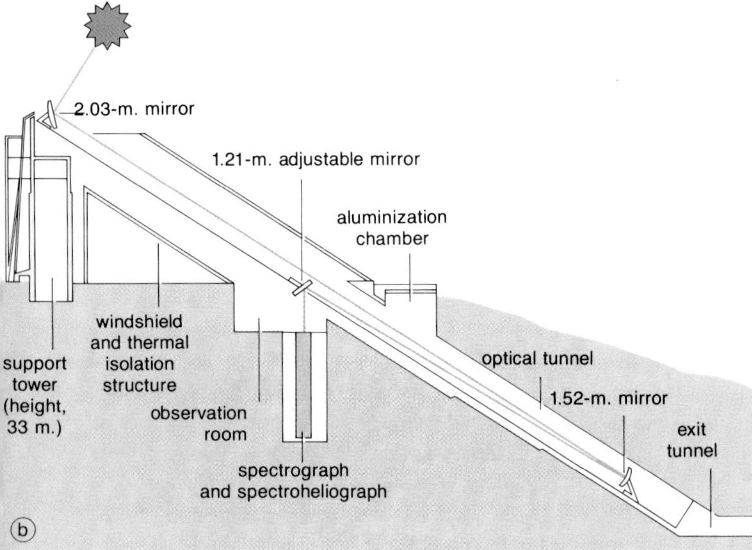

The solar tower at the Meudon Observatory (France): *Above left, the coelostat. Photo: Paris-Meudon Observatory. Above right, full view: Sunlight striking the coelestat located 35 meters (115 feet) above the ground on the upper platform is reflected to a concave mirror at the base of the tower; the beam then is sent back by two plane mirrors and gives an image of the Sun about 42 centimeters (17 inches) in diameter, which is analyzed by a high-dispersion spectrograph in an adjacent building. Photo: M. J. Martres.*

The 1.52-meter (60-inch) solar telescope at the Kitt Peak Observatory in Arizona. *Cross section shows the path taken by light rays.*

We can assume that, before then, solutions will be found to the enigmas we have discussed on the previous pages:

- Where do the solar neutrinos go? Is our star in a stable or a transitional phase of its evolution? Is it completely convective?
- How does the rotation vary as we penetrate to the interior? Does it increase? Fast? Slowly?
- What phenomena create the magnetic field, and how do the internal rotation and convection, which are assumed to give rise to it, interact?
- Why do all the bipolar magnetic regions arise in a specific strip of latitude, while the hot spots observed in X rays, which are also linked to bipolar regions, appear nearly everywhere?
- What processes are responsible for the magnetic cycle, and why has this cycle totally disappeared on several occasions in the recent past?
- What is the energy source that keeps the atmospheric temperature so high? Mechanical waves? The magnetic field? Both?
- What physical phenomena are the source of the gigantic solar flares? How is the energy stored up beforehand, and what releases it suddenly?
- Is the solar wind confined to the plane of the ecliptic, or does it fill all of space? Is it directed above the poles?

Instruments of Solar Observation

In the optical domain, the Sun is studied with the aid of special instruments.

The Solar Telescope

The primary piece of equipment for optical study of the Sun is the solar telescope. It is an instrument of a very long focal length (several dozen meters) and moderate aperture; at its focus we get an image of the Sun several dozen centimeters in diameter, which is then analyzed with a spectrograph.

Frequently, the telescope axis is vertical, forming what is called a solar tower. The light then is captured at the top of the tube by a system of plane mirrors (coelostat), one of which is movable to compensate for diurnal motion. Thus the solar rays are collected at several dozen meters in altitude before they travel through the turbulent air layers near the ground.

There are a dozen solar towers in the world. The oldest and tallest (50 meters [164 feet]) is at Mount Wilson Observatory. The one at Meudon Observatory in France, that was placed in operation in 1966, measures 35.4 meters (116 feet) in height, and gives an image of the Sun that is 42 centimeters in diameter. At the foot of the building, the light is directed toward a spectrograph 13 meters (43 feet) long, where gratings permit one to resolve narrow lines that may be as little as 0.01 Å (or $10-^{2}$ meter) apart. The largest solar telescope in the world is the one at the Kitt Peak Observatory in Arizona. It has a focal length of 90 meters (295 feet), and its objective has a useful diameter of nearly 1 meter. It is arranged in an inclined tunnel, because the system used to illuminate the telescope mirror is a polar siderostat, a single-mirror device that sends the beam in a direction parallel to the Earth's polar axis, hence inclined from the horizontal at an angle equal to the site's latitude.

The Spectroheliograph

This instrument, invented independently by the American George Hale and the Frenchman Henri Deslandres in 1891, gives an image of the entire solar disk in the light of a particular spectral line. It is a special spectrograph, containing an entrance slit and a selecting slit. By simultaneously scanning the Sun's image across the entrance slit and the photographic plate behind the selecting slit with a uniform motion, the Sun's entire image at the chosen wavelength usually that of the Hα line of hydrogen at 6,564.6 Å or the K line of calcium at 3,968 Å) is reconstituted on the plate. Thus the evolution of the sunspots, faculae, prominences, etc., can be followed over the entire solar disk.

The same result can be obtained without mechanical scanning, through the use of polarizing monochromatic filters designed by Bernard Lyot, or modern interferential filters with a narrow band pass.

The Coronagraph

This device, invented by the French astrophysicist Bernard Lyot in 1930, enables us to observe and study the solar corona at times other than total eclipses: It is a special refractor consisting essentially of a single lens perfectly polished and free of optical defects (to prevent any parasitic diffusion of light), followed by a metallic disk placed at the focus, which precisely intercepts the image of the solar disk; the corona, usually invisible due to the intense light of the photosphere (see page 101), then becomes observable. Thus it can be photographed, and a spectrophotometric study can be made. To take full advantage of the instrument, it is necessary to place it at a mountain station where the atmosphere is very pure (devoid of dust that can cause light scattering, making observation of the corona impossible). Lyot installed his first coronagraph at the Pic du Midi Observatory at an altitude of 2,860 meters (9,384 feet).

7. Mercury

The first planet a traveler would encounter upon leaving the Sun and heading toward the periphery of the solar system is Mercury. Revolving at only 58 million kilometers (36 million miles) from the Sun, on average, it completes its orbit in less than three months (88 days). Because of this, its movement in the sky appears rapid to an observer on Earth; its swiftness is the reason it was once regarded as the "messenger of the gods."

Because of its brightness, which vies with that of the brightest stars, Mercury was discovered in antiquity. But the Egyptians long believed that there were two separate planets (Set and Horus), for Mercury appears sometimes in the west, after sunset, and sometimes in the east, before sunrise.

The planet is never more than 28 degrees away from the Sun; at most, it rises only two hours, 15 minutes before the Sun, or sets only two hours, 15 minutes after it. It therefore only appears close to the horizon, in a region of the sky often shrouded in mists, and never stands out against a totally dark background, being always bathed to a greater or lesser extent by the glow of dawn or twilight. Thus, observing it with the naked eye is generally difficult. It is said that Copernicus was never able to see it.

Mercury is distinguished from the other terrestrial planets by its very elliptical orbit. Thus, at the point on its orbit closest to the Sun—perihelion—it is 45.9 million kilometers (28.5 million miles) from the center of the star, while at the farthest point—aphelion—it is 69.7 million kilometers (43.3 million miles) away. In addition, its orbital plane is distinctly more inclined with

Mercury's surface bears a striking similarity to that of the Moon. In this photograph of a region near the planet's South Pole, taken September 21, 1974, by Mariner 10 *from a distance of 55,000 kilometers (34,183 miles), we see in the center a very old crater. Cervantes, surrounded by a double ring of mountains, the side of which was partially destroyed during a more recent impact. Photo:* Cosmos Encyclopédie, *with the kind permission of A. Ducrocq.*

respect to the ecliptic plane than that of the other planets, except Pluto.

Owing to the eccentricity of its orbit, Mercury's distance from the Earth varies considerably in the course of its synodic revolution, which lasts 116 days. When Mercury moves into position between the Sun and the Earth, its distance may drop to 80 million kilometers (50 million miles), but the planet then shows us its dark hemisphere, so that it is totally invisible. On the other hand, when it is in opposition to the Sun relative to the Earth, its distance may be as much as 220 million kilometers (137 million miles). Its disk then appears entirely bright, but its apparent diameter is no more than 4.7 seconds, and only powerful instruments enable us to make out a few details on its surface.

In the telescope, Mercury's surface, which has a yellowish color, appears pockmarked with gray spots; they are of a permanent nature, indicating that they are not clouds but formations that are part of the planet's surface.

The Italian astronomer Giovanni Schiaparelli, of the Milan Observatory, was the first person, at the end of the 19th century, to discover these configurations and attempt to draw a map of Mercury. Rather than study the planet at night, when it was visible to the naked eye, he preferred to make his observations in daylight, when Mercury was high above the horizon. Nevertheless, his map was very sketchy. Later, between 1924 and 1933, Eugene Antoniadi, who made daytime observations using the large, 83-centimeter (33-inch) refractor of the Meudon Observatory, also did a lengthy study of Mercury and drew a map of it, which, though of purely historical interest today, remained the best until the American space probe *Mariner 10* flew past the planet in 1974.

Dimension and Mass

Simply measuring Mercury's diameter has required a great deal of work. It is difficult to determine with precision the apparent diameter of a small planetary disk in a more or less pronounced crescent phase.

To avoid making measurements during the gibbous phase, astronomers hit on the idea of using the planet's transits in front of the Sun (see box). At these times, Mercury appears as a black disk on the Sun's bright background, and that disk is seen in its greatest apparent dimension, since the planet's distance from the Earth is then minimal. Mercury's diameter can be measured with a micrometer. But precision remains poor, on the one hand, because of systematic errors due to diffraction and to atmospheric turbulence, a large but unquantifiable error since it varies according to the Sun's height above the horizon and the transparency of the atmosphere.

A photoelectric method invented by the Danish astronomer Hertzsprung has also been used. An opaque screen is set up in a telescope's focal plane; the screen is pierced at the center by a small circular hole whose diameter is much larger than that of the image of Mercury's disk provided by the objective. The light flux through this opening is measured with a photoelectric cell, first when the planet is in the field, then when it is not. As these fluxes are proportional to the illuminated surface area, a simple calculation gives the diameter of the image of Mercury's disk. The advantage of this over a micrometer measurement is that if atmospheric turbulence is not too great, it does not affect the measured fluxes.

These methods were applied simultaneously during Mercury's passage in front of the Sun on November 7, 1960, and yielded fairly similar results; the average value obtained for Mercury's diameter was 4,840 kilometers (3,008 miles).

Since then, analysis of the radar echoes reflected by Mercury at different points in its orbit, and, in 1974, the blocking of *Mariner 10*'s radio signals as it went behind the planet have made it possible to make new and more accurate determinations of its diameter, which is now estimated at 4,878 kilometers (3,032 miles).

Little was known in the past about Mercury's mass; its measurement was made difficult by the fact that the planet has no satellite. One could do little more than place it in a range between 0.05 and 0.06 times that of the Earth; nevertheless, the astronomer Rabe arrived at a more precise estimate—0.054 times that of the Earth—by studying the perturbations it inflicts on the minor planet Eros. Very substantial progress was made possible by *Mariner 10*; the study of the influence of Mercury's gravitational field on the device's trajectory made it possible to attribute to the planet a mass equal to 0.055354 times that of the Earth (or about 33.10^{19} times).

Rotation

In about 1890, Schiaparelli thought he could establish that the planet rotates on an axis nearly

Mosaic of 18 photographs of Mercury taken at 42-second intervals, from a distance of about 200,000 kilometers (124,301 miles), by Mariner 10 *as it approached the planet on March 29, 1974. The largest visible craters have a diameter of about 200 kilometers (124 miles). IPS Photo.*

Mercury's Transits in Front of the Sun

If Mercury and the Earth revolved around the Sun in the same plane, we would see the planet cross the solar disk at each of its inferior conjunctions—i.e., three times a year on average. In reality, Mercury orbits in a plane inclined about 7 degrees from that of the Earth's orbit, and for the phenomenon to occur, the conjunction must take place on the line of intersection of the two planes. The Earth itself crosses that line every year on May 8 or 9, then again on November 10 or 11. Thus a transit occurs when an inferior conjunction falls within a few days of one of these dates.

Mercury's transits in front of the Sun are much less rare than those of Venus. They number about 13 per century, on average. The first known observation of this phenomenon was made by the scholar Gassendi, on November 7, 1631, of a crossing that had been predicted by Kepler. Louis XV personally followed the transit on May 6, 1753, from the terrace of the Château de Meudon (now an annex of the observatory). Since then, most of Mercury's transits have been observed by astronomers, which made it possible to determine the planet's orbit and diameter with good accuracy before the space age. The next transits will take place on November 13, 1986; November 6, 1993;

Panorama of Mercury's northeast region, photographed from a distance of 78,000 kilometers (48,477 miles) by Mariner 10. *The three main craters are Sor Juana (top right), Gluck (lower right, with a clearly visible central peak) and Holbein (far left, below), which has a diameter of about 80 kilometers (50 miles). We also note two of the major escarpments discovered on Mercury: Victoria Rupes (top center) and Endeavour Rupes (far left, near the Holbein crater). IPS Photo.*

perpendicular to the plane of its orbit, by studying the dark spots he had discovered on Mercury's surface. He surmised that the rotational period coincides with the time in which it completes a revolution—88 days—and that Mercury always faces the Sun on the same side (synchronous rotation). Later observers (particularly Lowell in the United States, Antoniadi, then Lyot and Dollfus in France) reached the same conclusion, and this result was long accepted.

However, in 1962, the study of the planet's thermal radiation in the radio wave region revealed too high a temperature for the hemisphere supposedly in permanent darkness. Then, in 1965, according to an analysis of the frequency of radar signals emitted by the Arecibo radio telescope and reflected off of Mercury, the Americans Rolf Dyce and Gordon Pettengill found that the planet rotates in a direct sense, not in 88 days but in about 59 days, give or take three days, a period representing nearly two thirds of the sidereal revolution period. Such a result was not the result of chance. Studying the question from a purely theoretical angle, the astronomers Colombo and Shapiro showed that Mercury's rotation period could in fact be equal, for reasons of mechanical resonance tied to the effect of solar tides on the planet, to two thirds of its revolution period. The study of the photographs gathered by *Mariner 10* in 1974 confirmed that this indeed was the case. Today, the convergence between the considerations of resonance and ever more precise measurements leads us to a current value for the rotation period of 58.646 days. Under these conditions, the duration of the solar day on the planet (the time interval between two passages of the Sun through the meridian of a given point on the surface) is equal to two Mercurian years, or 176 Earth days; Mercury has the peculiarity—unique in the solar system—of having days longer than the year!

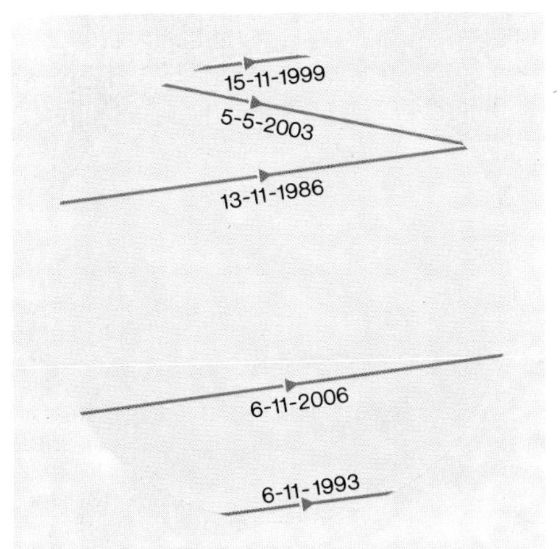

November 15, 1999; and May 5, 2003.

The accompanying figure shows the path the planet will trace on the solar disk in each case. The crossings in May are longer than those in November, for Mercury then is at the aphelion of its orbit and moves more slowly. They can last nearly nine hours if the sun is crossed diametrically.

A careful reexamination of older drawings of the planet showed that they are perfectly compatible with a rotation period of nearly 59 days. The illusion to which the astronomers fell victim is due to the fact that the most favorable periods for observation recur at the end of three cycles of phases—that is, at intervals of 348 days, hardly less than double the length of the solar day on Mercury. In other words, an astronomer observing Mercury during the most favorable elongations always sees the same side of the planet in the same phase—hence the same surface configurations, illuminated at the same angle—as though the rotation were synchronous. In fact, all of the visual and photographic observations collected at the Pic du Midi Observatory since 1942 show a sidereal rotation period equal to 58.67 days, with an accuracy of 0.03 day.

Surface

Before the space age, the study of the light reflected by Mercury had already made it possible to deduce the nature and structure of the planet's surface material. The albedo, the light curve (variation in brightness with phase) and polarimetric observations lead us to expect a rugged, dust-covered surface, similar to that of the Moon. Hence, the existence of craters appeared very probable. Radar studies of the relief were undertaken in 1972 at Goldstone, California. They revealed hills, valleys and craters approximately 50 kilometers (31 miles) in diameter in an area near the equator.

The photographs taken by *Mariner 10* fully confirmed these results while yielding a crop of new information.

Launched from Cape Canaveral on November 3, 1973, the American probe, after passing Venus, flew past Mercury on three occasions: on March 29, 1974, at an altitude of 700 kilometers (435 miles); on September 21, 1974, at a distance of 48,000 kilometers (29,832 miles); and finally on March 16, 1975, at an altitude of only 300 kilometers (186 miles). Overall, the spacecraft took some 4,000 photographs of the planet, covering nearly its entire surface; their resolution ranged between 5 kilometers (3.1 miles) and 100 meters (328 feet), some pictures even showing details on a scale of only 50 meters (164 feet), which represents a sharpness about 5,000 times greater than that of the best observations made from Earth.

Mercury's surface shows a striking resemblance of that of the Moon. We find mountainous regions, and huge basins riddled with meteoritic craters. However, the areas between the craters and basins are much more uneven than on the Moon, and the huge basins are more rare, since we find only one that is more than 500 kilometers (311 miles) in diameter, compared with five on the Moon, whose total surface is nonetheless twice as small. Moreover, we find very few craters of dimensions greater than about 20 to 50 kilometers (12 to 31 miles) on Mercury.

The craters themselves show the morphological types found on the Moon: craters with "rays," with a central peak, patterns of concentric, terraced rings, small secondary craters arranged in a necklace around a larger central crater, etc. However, compared to the Moon, the Mercurian craters have comparable diameters but generally are not as deep as those on the Moon; in addition, on Mercury the secondary craters are concentrated closer to the main crater, and the "rays" (which are long trails of debris) characteristically surrounding certain young craters are, on the average, six times less extensive than on the Moon. These differences probably stem from the fact that gravity is more intense on the surface of Mercury.

As on the Moon, we note linear faults, often quite long, that cut across most of the other topographical formations without deviating. The longest one is 550 kilometers (342 miles) long and has a width of up to 6 kilometers (3.7 miles) in places.

But in the plains between the craters, we also note scarps that have no equivalent either on the Moon or on Mars. These are veritable cliffs, which may reach an altitude of 2,000 to 3,000 meters (6,562 to 9,843 feet) and a length of several hundred kilometers. These formations seem to have been caused by violent tectonic shifts that probably occurred at the time Mercury's crust cooled.

The most spectacular structure is unquestionably a vast basin that bears a resemblance to the Mare Orientale on the rear surface of the Moon. Astronomers have given it the name *Planitia Caloris* (Basin of Heat) because it faces the Sun directly once out of every two times that Mercury passes through perihelion and therefore has a very high temperature. Its diameter, comparable to that of the Sea of Rains—the largest of the lunar "seas"—reaches 1,350 kilometers (839 miles), and it is surrounded by a triple mountainous ring of about 2,000 meters (6,562 feet) in altitude. Its surface appears streaked by a large number of faults, which delineate a network of polygonal figures. To all appearances, this basin was gouged by the impact of an asteroid, whose diameter is estimated to have been about 100 kilometers (62 miles). This impact was a major event in Mercury's history, for it changed the face of an entire hemisphere of the planet. Even a vast region at the antipode was affected, probably as a result of the focusing of the systemic waves generated by the impact. This region now has a chaotic appearance, with small hills more or less aligned; dispersed craters; and huge, scattered boulders. Moreover, we find an analogous configuration on the Moon, bordering the Mare Humorum, located precisely opposite the Mare Orientale.

Internal Structure

By making possible a simultaneous determination of Mercury's mass and diameter, *Mariner 10* made it possible to know the average density of the planet, about which there had been a degree of uncertainty up to then: At 5.44 gm/cm^3, it proves to be much closer to that of the Earth (5.52) than to that of the Moon (3.33). Despite a crust similar to that of the Moon, Mercury must therefore have a very different internal structure; the planet has the face of the Moon but not its body. We assume that under a thick mantle of silicates, Mercury contains a voluminous metallic core (based on iron) occupying 50 percent of its volume and representing nearly 80 percent of its mass (compared to 30 percent and 15 percent, respectively, for Earth). The crust and mantle are probably undifferentiated, forming a single thick shell about 600 kilometers (373 miles) thick. With a diameter estimated at 3,600 kilometers (2,237 miles), the core itself appears more voluminous than the Moon.

Atmosphere

Despite Antoniadi's observations of the sporadic darkening of certain regions of Mercury's surface,

Close-up view taken at 19,000 kilometers (11,809 miles) altitude by Mariner 10 *of a fractured terrain extending east of Planitia Caloris (basin of heat), the largest structure on Mercury's surface. P.P.P. Photo.*

The Kuiper crater is the brightest of those shown in this photograph, taken by Mariner 10 *on March 29, 1974, from a distance of 88,500 kilometers (55,003 miles). With a diameter of about 40 kilometers (25 miles), it was carved out of the wall of a larger, older crater, Murasaki; both show a central peak. At the bottom of the picture, in the center, we see part of the Goldstone valley, which extends over 120 kilometers (75 miles). Photri Photo.*

Mercury's Cartography

Analysis of the photographs sent back by *Mariner 10* led the International Astronomical Union to distinguish six major kinds of topographical formations on Mercury: craters, mountains (*montes*), plains or basins (*planitia*), escarpments (*rupes*), faults (*dorsa*) and valleys (*valles*).

It was decided to give the craters names of artists, composers or writers. A single astronomer was honored, the American Gerard Kuiper (1905–73), because of his very important contribution to planetary studies. Another exception is the crater Hun-Kal.

The basins, with the exception of the largest one, known as the Basin of Heat because of the high temperature often found there (see preceding text), and the one closest to the North Pole, known as the Boreal Basin, bear the names given to Mercury by different peoples (Hermes, Budh, Odin, Suisei). The escarpments are named after famous explorer ships (*Endeavour, Santa Maria, Pourquoi-Pas?*) to emphasize that Mercury, in Greco-Roman mythology, was associated with commerce and travel; the faults honor astronomers who were particularly interested in Mercury (Schiaparelli, Antoniadi). Finally, the valleys were given names of radio-astronomical observatories whose instruments were employed to study Mercury using radar

as though clouds had veiled them, specialists agreed before the space age that Mercury does not have an appreciable atmosphere, in view of the planet's low mass. Also, the sudden disappearance of the stars it occults constitutes an additional argument in favor of that hypothesis.

Mariner 10 confirmed that Mercury has only an extremely rarefied atmosphere. The measurements taken by the spacecraft revealed that atmospheric pressure on the ground is less than 2 billionths of a millibar (whereas, on the Earth, at sea level, it is about 1,000 millibars). This atmosphere amounts to a modest flux of light gases (hydrogen, helium) constantly supplied by solar wind, and of inert gases (argon, neon) resulting probably from the outgassing of the planetary interior under the action of radioactive minerals. Likewise, the ionosphere is almost nonexistent, the electron density being less than 4,000 electrons per cubic centimeter, or 1,000 times less than in the Earth's ionosphere.

Such a thin atmosphere naturally involves considerable differences between day and night temperatures on the planet's surface. The *Mariner 10* infrared radiometer measured up to 300°C, but it is believed that at noon, local time, at the equator, when Mercury is at its least distance from the Sun, the temperature may exceed 400°C. On the other hand, at night, when the planet is at its greatest distance from the Sun, it falls to -170°C, which gives a thermal contrast of about 600°C, about two times higher than that of the Moon, which is already considerable. In fact, the two celestial bodies have similar thermal properties; because of the lack of atmosphere, their surfaces heat up and cool very rapidly. The difference stems from the fact that Mercury gets more energy from the Sun; at perihelion, it gets about twice as much as the Moon.

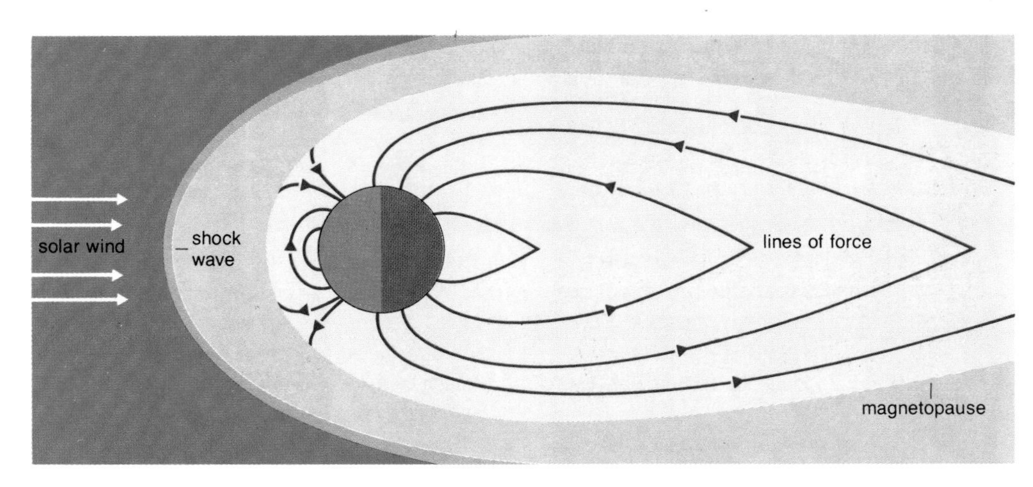

Mercury's magnetosphere (seen in cross section). In the direction of the Sun, the magnetic lines of force are compressed by the solar wind, similar to what we observe on a larger scale around the Earth, Jupiter or Saturn.

Magnetic Field

One of the great revelations of the space exploration of Mercury was the discovery of a dipolar magnetic field specific to the planet, inclined 12 degrees on its rotation axis, whose surface intensity appears close to 350 γ. Though relatively weak in comparison to that of the Earth's field

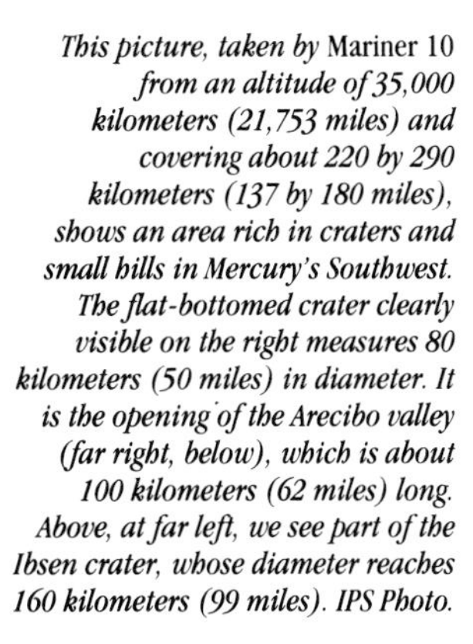

This picture, taken by Mariner 10 *from an altitude of 35,000 kilometers (21,753 miles) and covering about 220 by 290 kilometers (137 by 180 miles), shows an area rich in craters and small hills in Mercury's Southwest. The flat-bottomed crater clearly visible on the right measures 80 kilometers (50 miles) in diameter. It is the opening of the Arecibo valley (far right, below), which is about 100 kilometers (62 miles) long. Above, at far left, we see part of the Ibsen crater, whose diameter reaches 160 kilometers (99 miles). IPS Photo.*

(Arecibo, Goldstone).

The locations of the topographical formations are specified, as on Earth, through a network of parallels and meridians.

The meridian that crossed the planet's subsolar point during Mercury's first passage through perihelion in 1950—that is, the point on the equator where the Sun is at the zenith—was adopted as the reference meridian (longitude 0°).

No topographical formation was seen along this meridian, since it was on Mercury's dark side during *Mariner 10*'s fly-by. Therefore, astronomers chose as a reference point a small crater 1,500 meters (4,922 feet) in diameter at 0.6° southern latitude and assigned it a longitude of 20° west while conferring on it the name Hun-Kal, which represents the number 20 in the Mayan language. This small crater can be regarded as Mercury's "Greenwich."

The topographical formations photographed by *Mariner 10* thus could be positioned on a computer-generated control grid containing 2,378 points localized to within 10 kilometers (6.2 miles) in the Southern Hemisphere and approximately 25 kilometers (16 miles) at the North Pole.

(less than 1 percent), this intensity is not negligible and appears clearly greater than that of the interplanetary magnetic field.

Because Mercury rotates very slowly on its own axis, it is unlikely that this field results, like the geomagnetic field, from a dynamo effect. Rather, it is believed to result from the planet's rapid motion (we should recall that its internal structure is regarded as essentially metallic and is therefore a conductor of electricity) across the lines of force of the solar magnetic field. But this is only a hypothesis, which still should be viewed with caution.

The existence of this magnetic field creates a magnetosphere around Mercury (as around Jupiter and the Earth, but on a much smaller scale) that deflects the energetic particles of the solar wind. Its intensity is not sufficient, however, for charged particles to be trapped and to collect around the planet to form a radiation belt.

Another Planet Between Mercury and the Sun?

According to Newton's theory of gravity, if Mercury and the Sun were alone in space, Mercury's path would be rigidly elliptical; the planet would reappear at the perihelion after having made exactly one revolution, or 360 degrees, around the Sun.

In fact, the planet's motion is disturbed by attraction from other bodies in the solar system, which slightly modify its trajectory; from one crossing of the perihelion to the next, it travels slightly more than 360 degrees. The excess is very small, but it accumulates from one revolution to the next and is reflected in a slow rotation of the perihelion around the Sun.

In the past century, Le Verrier attempted to draw up tables of motion for the major planets, taking into account their mutual perturbations. But he realized it was impossible to describe Mercury's motion in a way that fully conformed to observations solely on the basis of the perturbation theory. To match the calculations to observations, it was necessary to ascribe to Mercury's perihelion an advance of about 40 seconds of arc per century, in addition to what the theory predicted.

Among the plausible hypotheses that Le Verrier formulated to solve this enigma without undermining Newton's law, he was particularly attached to that of an unknown planet orbiting between the Sun and Mercury's orbit and causing noticeable perturbations in the latter. Earlier, in 1845, he had explained the discrepancies found in the motion of Uranus as the result of perturbations caused by the attraction of an outer planet, whose position in the sky he was able to determine; and that planet, Neptune, was in fact discovered in 1846.

On March 28, 1859, a country doctor, Dr. Lescarbault, also a great amateur astronomer, observed a small spot on the Sun, very dark and round, which he assumed to be the disk of a planet. Le Verrier then decided to analyze all the similar observations he could find; he found 50 of them from 1802 to 1862 but retained only six, which seemed to him particularly precise and trustworthy. He deduced that the unknown planet, to which he gave the name Vulcan, a bit prematurely, completed its orbit in 33 days, and announced that it would pass in front of the Sun on March 22, 1877. But no planet appeared before the Sun on that day. An intensive visual or photographic exploration of the Sun's immediate vicinity was next attempted during seven total eclipses between 1878 and 1900, in vain. People then fell back on the hypothesis of a ring of cosmic dust surrounding the Sun, an explanation defended for a time by the mathematician Henri Poincaré. But this ring, which would scatter sunlight, should have been visible at twilight or at dawn, and its presences should have perceptibly perturbed the motions of comets at perihelion.

In fact, the motion of Mercury's perihelion was explained very satisfactorily in 1917 by Einstein's general theory of relativity. This theory predicted an advance of the perihelion by 42.91 seconds of arc per century, in addition to the advance provided by Newton's theory of perturbations, while the faithful account of observations led to an advance of 42.84 seconds of arc. Hence, the agreement was excellent.

Some people, however, have not ceased believing in the existence of a planet closer to the Sun than Mercury. However, in 1923, the American astronomers Campbell and Trumpler, after systematic research, concluded that it was impossible for any unknown planet more than 20 kilometers (12 miles) in diameter to exist in proximity to the Sun. Much more recently, in 1973–74, the search for a potential unknown planet in the neighborhood of the Sun was again attempted as part of the Skylab program. Despite the excellent observing conditions that prevailed in space (the sky always very dark, no atmospheric absorption or turbulence), this search again ended in failure. It seems, therefore, that we can finally consider the existence of an intermercurian planet as a myth.

The observations on which Le Verrier based his calculations of the orbit of Vulcan, the planet he assumed to exist, remain to be interpreted, however. They may perhaps correspond to asteroids passing in front of the Sun, but they more likely result from some artifact of the observing procedure.

Mercury and the Amateur

Mercury shines only when it is near the horizon, which means that it is frequently hidden by mist, clouds or dust that make it difficult to observe. Even on a clear day, it generally is invisible to the untrained observer, for it remains submerged in the light of the Sun, from which it is never very distant. The best times for observing it are during the periods when it is near its greatest elongations—morning in October and November, or evening in April, May and June, in the Northern Hemisphere (the opposite is true in the Southern Hemisphere). In the morning, observations may begin about five days before the date of the greatest elongation and continue for 10 days after; in the evening, 10 days before and five days after. Some elongations are more favorable than others, depending on the planet's height above the horizon.

In the morning, the horizon must be very clear toward the east, and in the evening, toward the west. A search with binoculars should begin 30 to 45 minutes after sunset; the observation may be continued with a telescope once the planet has been spotted. Mercury may also be seen in daylight with a telescope having a stationary equatorial mount, which can find the planet by its coordinates (right ascension and declination). If it can be located with binoculars before dawn, it can also be followed after sunrise. With an amateur telescope, one can barely observe the phases. To see them with a reflecting telescope, an aperture of at least 60 millimeters (2.5 inches) and a magnification of 100 times are essential. When Mercury is visible in the morning, one can see that its illuminated fraction increases while its diameter decreases, whereas when it is visible in the evening, its illuminated fraction decreases while its diameter increases. The periods of greatest elongation correspond to the quarters.

With a 75-millimeter (3-inch) telescope, sometimes it is possible to glimpse a few irregularities in the sunlit portion of the planet's disk. A telescope of 100 to 150 millimeters (4 to 6 inches) is needed to view light or dark spots on the surface, but they always appear with very little contrast and are difficult to make out.

Mercury is not a fascinating planet for amateur observation.

8. Venus

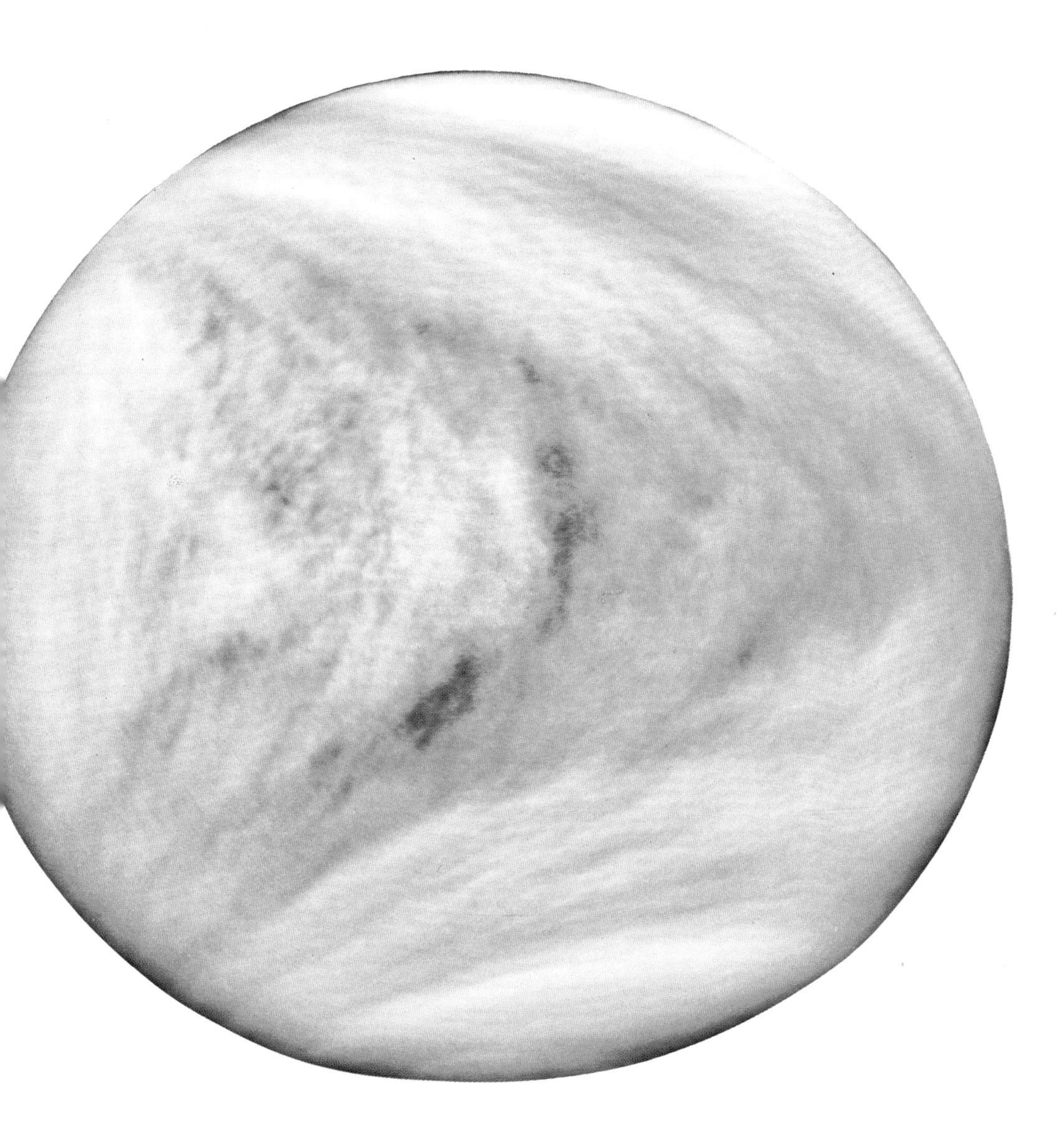

The second planet we encounter as we move away from the Sun is Venus, noted for centuries for its brilliance; it is the brightest object in the sky after the Sun and Moon.

Venus, photographed in the ultraviolet by Pioneer Venus 1, *February 19, 1979. One sees only a pattern of cloud formations carried in a rapid rotation around the planet. IPS Photo.*

Seen from Earth, Venus, like Mercury, appears to shift from one side of the Sun to the other, sometimes rising before it (morning star), sometimes setting after it (evening star); but it may deviate as much as 48 degrees from it, thus remaining visible for a little more than three hours before sunrise or after sunset. This, added to its brightness, makes Venus much more easily observed than Mercury.

In observing Venus both at dawn and at sunset, the peoples of antiquity believed that, as with Mercury, there existed two separate celestial bodies. These were Tioumoutiri and Ouaiti to the Egyptians, Phosphoros and Hesperus to the Greeks, Lucifer and Vesper to the Romans. The latter understood that what was involved was in fact a single planet, and they dedicated it to the goddess of love and beauty. However, Venus is also very well known under the name of the *shepherd star.* Its brightness—as much as 12 times brighter than Sirius, the brightest star in the sky—is indeed what makes it the first "star" to appear in the evening, or the last to disappear in the morning; whenever it moves far enough away from the Sun, it may even be seen with the naked eye in daylight, if the sky is very clear.

Venus's orbit, in contrast to that of Mercury, is almost precisely circular, so that the planet's distance from the Sun varies very little. During a sidereal revolution, which lasts 225 days, it always remains between 107 and 109 million kilometers (66.5 and 67.7 million miles) from the Sun.

To an observer on Earth, the apparent movements and configurations of the planet are similar to those of Mercury. Still, the cycle of phases extends over 584 days (the synodic period of revolution) rather than 116. Furthermore, unlike Mercury, Venus does not disappear when approaching the inferior conjunction, but always remains visible through a telescope. At the superior conjunction, the planet's distance from the Earth reaches 258 million kilometers (160 million miles); at the inferior conjunction, it drops to 41 million kilometers (25 million miles) (Venus then becomes the closest planet). Thus the planet's successive phases are accompanied by very great variations in its apparent diameter. In the first case, Venus appears as a small, bright disk with a diameter no greater than 10″, or 10 arc seconds. In the second case, it turns its dark side toward the Earth, and only a thin, bright sliver defines its outline, but its diameter then reaches 66″, or 66 arc seconds. Finally, at maximum elongation, Venus is seen at quarter phase, and its apparent diameter measures 25 arc seconds. The phases of Venus are not visible to the naked eye, but they can easily be observed with a small telescope.

DEL SIG. GALILEI. 217

ducete fino à dannar con lunghi diſcorſi chi prende il termine vſitatiſſimo d'infinito per grandiſſimo. Quando noi abbiamo detto, che il Teleſcopio ſpoglia le Stelle di quello irraggiamento, abbiamo voluto dire, ch'egli opera intorno à loro in modo, che ci fà vedere i lor corpi terminati, e figurati, come ſe fuſſero nudi, e ſenza quello oſtacolo, che all'occhio ſemplice aſconde la lor figura. E egli vero Sig. Sarſi, che Saturno, Gioue, Venere, e Marte all'occhio libero non moſtrano trà di loro vna minima differenza di figura, e non molto di grandezza ſeco medeſimi in diuerſi tempi? e che coll' occhiale ſi veggono Saturno, come appare nella preſente figura, e Gioue, e Marte, in quel modo ſempre; e Venere in tutte queſte forme diuerſe? e quel, ch'è più merauiglioſo con ſimile diuerſità di grandezza? ſi che cornicolata moſtra il ſuo diſco 40. volte maggiore, che rotonda, e Marte 60.

volte, quando è perigeo, che quando è apogeo, ancorche all'occhio libero non ſi moſtri più che 4. ò 5.? Biſogna, che riſpondiate di ſi, perche queſte ſon coſe ſenſate, ed eterne, ſi che non ſi può ſperare di poter per via di ſillogiſmi dare ad

E e inten-

The phases of Venus as drawn by Galileo. It was in September 1610, with a telescope magnifying 30 times, that Galileo first perceived the phases of Venus. To give himself time to verify his discovery without the risk of his claim to priority being disputed, he made an enigmatic announcement: Haec immatura a me jam frustra leguntur, o.y. *("In vain these things are gathered by me today, prematurely"), the 35 letters of which, if arranged differently, form another Latin sentence much easier to understand:* Cynthiae figurae aemulatur mater amorum *("The mother of Love imitates the phases of Diana").*

Size and Mass

THE SILHOUETTE OF VENUS has long fascinated astronomers, for it closely resembles the Earth. The two planets do, in fact, have a comparable size. By 1807, the German astronomer Johann Wurm had estimated the diameter of Venus at 12,292 kilometers (7,640 miles). Subsequently, more accurate calculations done prior to the space age pointed to a figure approaching 12,200 kilometers (7,582 miles). But all optical measurements have one drawback: They are based on the visible disk and thus include a portion of the atmosphere. It is only since the 1960s, thanks to radar probes of the planet made from Earth, and the "landing" of automatic devices on its soil, that we have had measurements of the solid sphere of Venus. They show that the planet is slightly smaller than was thought; its diameter, now estimated at 12,102 kilometers (7,521 miles), represents 0.949 of the Earth's and gives it a volume equal to 0.858 times that of our planet.

The two planets also have a very similar mass. Whereas in former times it was determined by the perturbations that it caused on the Earth's motion or on that of certain asteroids (such as Eros) that have a very elliptical orbit, the mass of Venus is now deduced from a careful study of the trajectory of the radio waves that pass over it. It weighs in at about 81.5 percent of the mass of our planet.

Thus the average density of Venus (5.26 g.m./cm³) appears to be slightly less than that of the Earth (5.52), and one may thereby deduce that the two planets probably have very similar internal structures. According to recent radar studies of the planet's surface relief (see page 000), it would nevertheless appear that the Venusian crust is thicker than the Earth's, although it is substantially thinner than that of Mars. The Venusian crust appears to have been formed from a single basaltic tectonic plate that supports both the great plain and the mountainous regions.

Atmosphere

We now possess a fairly accurate knowledge of the characteristics of the Venusian atmosphere as a result of the measurements and analyses made by the various spacecraft that have passed over or through it since 1962—a series of Soviet Venera probes, and the American probes *Mariner 5, Mariner 10, Pioneer Venus I* and *Pioneer Venus II.*

COMPOSITION

Venus's atmosphere is composed essentially of carbon dioxide gas (95 to 97 percent CO_2 by mass, according to altitude), as had been assumed since 1932, when the Americans Adams and Dunham discovered very intense absorption bands of this gas in the infrared spectrum of the light reflected by Venus. But a dozen other constituents have also been discovered, including nitrogen (2 to 5 percent), water vapor (0.1 to 0.4 percent), oxygen, carbon monoxide, rare gases (helium, argon, neon) and sulphur compounds (SO_2, OCS, H_2S).

The clouds that extend between altitudes of 48 and 68 kilometers (30 and 42 miles) (see later) contain droplets of sulphuric acid. The planet, then, appears surrounded by an acidic, toxic atmosphere. Many substances that are relatively rare above the clouds have a substantial abundance beneath them. This includes oxygen and sulfur dioxide, whose relative abundances above and below the clouds differ by factors of 60 and 240, respectively. This phenomenon results from the fact that, in the lower atmosphere, the droplets of sulphuric acid vaporize and even dissociate from the effects of heat, chemically attacking the gases it encounters.

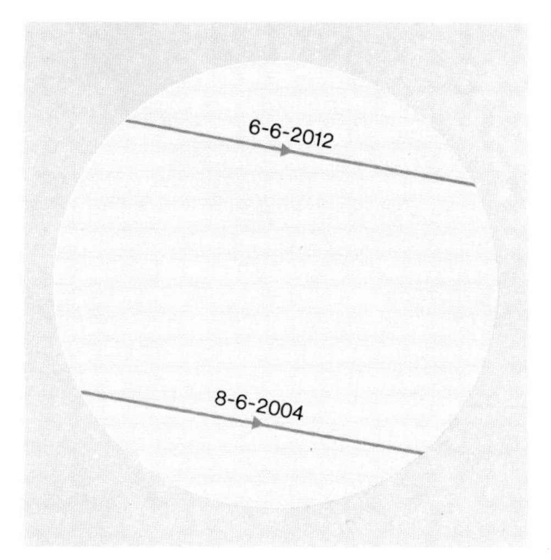

Venus's next passages in front of the Sun.

Venus's Transits in Front of the Sun

The orbits of Venus and the Earth are not exactly in the same plane but lie at an angle of 3°24′ to each other. Small as it is, this is sufficient so that we on Earth cannot see the body of Venus pass in front of the solar disk at each inferior conjunction of the planet—i.e., every 584 days. As with Mercury, the transits take place only when the inferior conjunction occurs in the vicinity of the line intersecting the plane of the Earth's orbit and the plane of Venus's orbit. However, this circumstance occurs far less frequently than it does with Mercury, since the time interval between two inferior conjunctions is about five times greater for Venus. The appropriate orbital configuration reoccurs cyclicly, with the following intervals; 8 years, 121½ years, 8 years, 105½ years, 8 years, 121½ years, etc. The transits always occur at the beginning of June or December. The last transit occurred on December 6, 1882; the next ones will take place on June 8, 2004, and June 6, 2012.

Apparently the Briton J. Horrock was the first to observe a transit of Venus in front of the Sun, in 1639. But the phenomenon did not really attract the interest of astronomers until another Briton, Edmund Halley—whose name still is linked to the most famous periodic comet—proposed an ingenious

Below 20 kilometers' (12 miles) altitude, the total sulphur concentration in the form of SO_2, OCS, H_2S and S reaches 1,000 parts per million and is therefore comparable to the concentration of water vapor.

Despite vast differences, the Venusian atmosphere has some features in common with the Earth's atmosphere, such as, for example, the abundance of nitrogen and carbon dioxide per gram of planetary weight (on Earth, however, the major portion of the carbon dioxide is trapped in rock as carbonates). Still, the Pioneer Venus probes revealed that certain isotopes of volatile elements have an unexpected abundance. In particular, the relative quantity of argon-36, a gas whose origin is thought to coincide with the formation of the solar system, appears much higher than on Earth or Mars. This anomaly, not yet fully explained, imposes new constraints on the existing models of the formation of the solar system.

STRUCTURE

The atmosphere of Venus has a stratified structure. The permanent cloudy mantle, whose upper portion can be seen from Earth, is 20 kilometers (12 miles) thick and is composed of three distinct layers. The first layer is found at altitudes between 68 and 56 kilometers (42 and 35 miles), the second between 56 and 52 kilometers (35 and 32 miles) and the third between 52 and 48 kilometers (32 and 30 miles). The upper layer is made up of droplets of sulphuric acid (from 1 to 2 micrometers in diameter), as had been suggested in 1972 by the Americans G.T. Sill and A.T. Young, working independently of one another. The middle layer also contains droplets of sulphuric acid, as well as large solid particles of 10 to 20 micrometers in diameter and a population of particles of intermediate dimensions whose composition has not yet been determined. The lower layer contains almost nothing but large particles, which might be sulphur in the free state. The concentrations vary according to the layer and type of particles; with an average concentration of 400 particles per cubic centimeter, the lowest layer appears densest—the visibility there is less than 1 kilometer. The upper layer has about 300 particles per cubic centimeter, and the middle layer has 100 per cubic centimeter.

In the upper layers, the sulphuric acid is found in aqueous solutions, which are increasingly diluted as the altitude decreases. The maximum concentration in the uppermost clouds is between 75 and 85 percent acid by mass. Since sulphuric acid is a powerful desiccator, its presence explains, among other things, the low water vapor content of the atmosphere.

Generated by computer from ultraviolet spectrometer measurements made by Pioneer Venus 1, *this colored composition shows the distribution of atomic hydrogen and oxygen in the Venusian atmosphere. The oxygen remains trapped in the upper atmosphere (crescent), while the lighter hydrogen escapes slowly into space (horizontal bars). The variation of color reflects the varying intensity of the gaseous flow, which gets weaker as one goes from yellow to blue. IPS Photo.*

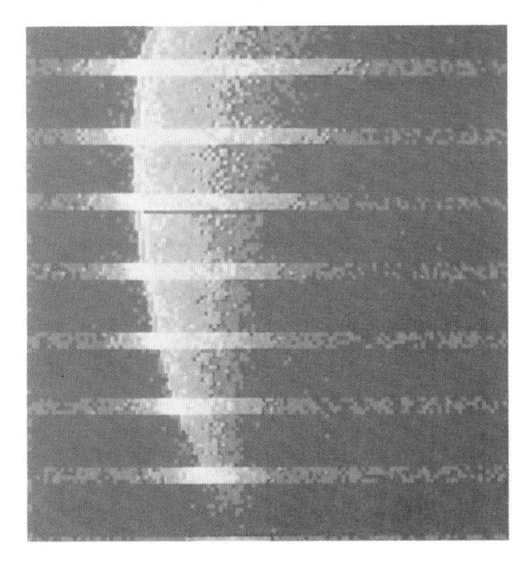

In ultraviolet photographs, the Venusian clouds are characterized by the presence of dark features, which are especially visible on the *Mariner 10* photographs. These features correspond to places where the cloud tops are at lower altitudes, where the radiation is strongly absorbed. This absorption is attributed to sulphur particles and, to a lesser extent, to sulfur dioxide (SO_2). The presence of sulphur in the free state would also explain Venus's pale yellow color.

Beneath the clouds, up to an altitude of 32 kilometers (20 miles), lies a misty region populated by particles about 1 micrometer in diameter. Their concentration ranges from 1 to 20 particles per cubic centimeter, with the maximum density found in the upper portion, at the lower boundary of the clouds, which is an area of convective activity. After that, the atmosphere remains clear all

altitude in km.
upper atmosphere
clouds
lower atmosphere
180 160 140 120 100 80 60 40 20 10 0
−90°C −80°C −50°C 0°C +100°C +200°C +300°C +400°C +470°C
weakly ionized layer
turbopause maximum ionization
weakly ionized layers
mists
clouds
aerosols
(sulphuric acid droplets)
clear atmosphere
75%
2,5%
145 km
128 km
124 km
90 km
68 km
48 km
32 km
25 50 75 100 m./sec.
wind speed
proportion of solar radiation

Model of the atmosphere of Venus.

method in 1716 of determining the distance of the Earth from the Sun based on the time it took Venus to cross the solar disk. In the 18th century, the transits in 1761 and 1769 were eagerly awaited, and many astronomers made long journeys to see them, so they were observed from all parts of the world. This was the case, for example, for the French astronomer Le Gentil de la Galaisière, whose misfortunes still are legendary.

De la Galaisière embarked in 1760 for India to observe the transit there on June 6, 1761. But because of the naval war between France and England, he was unable to land in time. He then decided to stay in Pondichéry for eight years to await the crossing of June 3, 1769, and set up a local observatory. When the long-awaited date finally arrived, the sky—which had been perfectly clear on the previous days—clouded over, contrary to all expectations, and the Sun remained stubbornly hidden during the entire crossing of Venus. Le Gentil then fell ill, and no more was heard of him. When finally he returned to France in 1771, he found out that, being presumed dead, he had been replaced at the Academy of Sciences, and his possessions had been given away!

the way to the ground. Nonetheless, the visibility is limited because the atmosphere is so dense that the scattering of sunlight by the gas molecules is sufficient to mask the surface.

Other heavily misted layers have been detected above the clouds at altitudes from 80 to 90 kilometers (50 to 56 miles); these layers appear related to a temperature inversion zone.

The *turbopause*, or the boundary between an upper region, where the gases separate into layers according to density, and a lower region, where they blend uniformly, lies 145 kilometers (90 miles) above the ground.

The exosphere, in which the atoms and molecules can acquire enough thermal energy to escape into interplanetary space, begins at 160 kilometers (99 miles) (exobase). However, the atmosphere remains discernible up to an altitude of about 250 kilometers (155 miles). The uppermost atmosphere, like that of the Earth, is ionized by solar radiation. However, the ionosphere is thinner and closer to the ground than that of Earth. Furthermore, since Venus has no magnetic field (see later) it also has no magnetosphere, and the solar wind can therefore interact directly with the ionosphere. The ionization is greatest at about 145 kilometers, the most abundant ions at this level being O_2^+ and CO_2^+. Layers having a lower degree of ionization have been detected at 124 kilometers (77 miles), 128 kilometers (80 miles) and about 180 kilometers (112 miles). The upper boundary of the ionosphere lies at about 400 kilometers (249 miles). In contrast with the Earth's outer atmosphere, the flux of particles of solar origin confines the Venusian ionosphere below a well-defined boundary. Moreover, contrary to what was previously believed, this ionosphere does not disappear above the dark hemisphere of Venus despite the long duration (8 weeks) of the Venusian night, but remains permanently in place all around the planet.

TEMPERATURES AND PRESSURES

The temperature at 250 kilometers of altitude approaches +30°C. It drops to −90°C at about 100 kilometers (62 miles), then rises again to about −40°C at the cloud peaks and +100°C at about 48 kilometers, at the base of the clouds (at this level, the sulphur and sulphuric acid vaporize). On the ground, all spacecraft that have transmitted information to date have recorded temperatures ranging from +450°C to +480°C, and pressures on the order of 90 bars (90 times the average atmospheric pressure at ground level on Earth).

The high temperature at ground level can be explained by the *greenhouse effect,* described in 1942 by the American astronomer Rupert Wildt. The carbon dioxide in the atmosphere lets part of the incident solar radiation pass through it, but in return absorbs the infrared radiation emitted by the planet. More precisely, the Venusian atmosphere reflects into space about 75 percent of the solar radiation it receives. Fifteen percent is absorbed by the clouds, 7.5 percent above the clouds or in the lower atmosphere and only 2.5 percent reaches the ground.

Near the cloud peaks, in the upper atmosphere, the Pioneer Venus probes have detected small diurnal thermal variations that increase with altitude. The temperature contrasts between the equator and the poles are much greater. Finally, in going from the planet's sunlit hemisphere to its dark one, *Pioneer Venus 1* discovered a sudden temperature drop of more than 100°C. This sudden variation is in contradiction to the theoretical models that project a gradual decrease from noon to midnight. Another surprise was the discovery in the lower atmosphere of a great many lightning flashes.

Why is Venus a genuine inferno? Why are conditions there so different from those on Earth, even though the two planets are formed together out of the same "mold," in the same region of the solar system? Today we are beginning to understand why. We assume that the planets were formed through the accretion of matter within the solar nebula. Initially, they were probably all surrounded by similar envelopes composed essentially of hydrogen and helium. Incapable of retaining this envelope because of their insufficient mass, the terrestrial planets allowed it to escape fairly rapidly into space, unlike the giant planets. But because they were warmed from inside by their radioactive substances, the terrestrial planets displayed volcanic outgassing, which generated a secondary atmosphere based on nitrogen; methane; carbon dioxide; and above all, water vapor. At that point, their evolution took different paths, depending on their distance from the Sun. On Earth, the temperature and pressure that prevailed at the time when the outgassing began allowed the water to exist in a liquid state and to collect in basins, giving rise to the oceans. The carbon dioxide was soon mostly dissolved within the oceans before gradually fixing itself to rock in the form of carbonates. At the same time, other processes led to the formation of the present atmosphere based on nitrogen and oxygen. But on Venus, which is warmer because it is nearer to the Sun (it gets about twice as much energy from the Sun as the Earth does), it is assumed that the water vapor could not condense into liquid form. This vapor and carbon dioxide accumulated in the atmosphere and gave rise to the greenhouse effect. As this effect increased in importance, it caused a substantial increase in the surface temperature, until the planet acquired the characteristics it has today. A sufficient rise in temperature allowed the water vapor to escape.

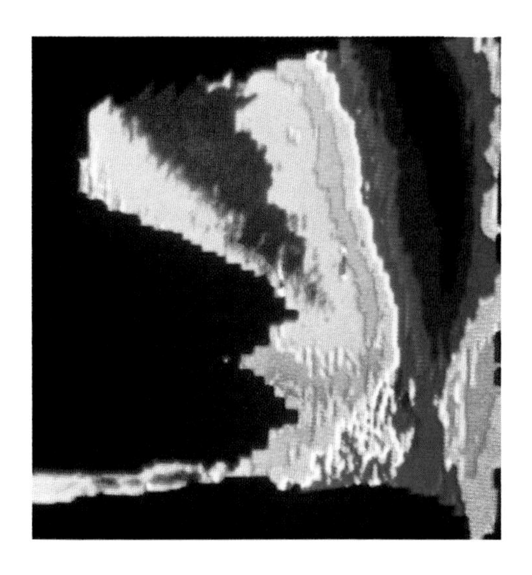

This colored composition generated from infrared data gathered by Pioneer Venus 1 *in December 1978 is a map of the cloud cover of Venus in a zone extending from the planet's North Pole (top) down to 30 degrees southern latitude (bottom right). The various colors represent different altitudes and cloud temperatures. IPS Photo.*

The Enigma of Rotation

Of all the enigmas that Venus formerly presented to astronomers, one of the most intractable had to do with its rotation. As we saw with Mercury, the traditional method of determining a planet's rotation period consists of obtaining a series of drawings or photographs of the planet and of measuring the displacement over time of the spots appearing on the disk. Unfortunately, the visually perceived spots on Venus generally have a very weak contrast and, in small instruments, are usually imperceptible because of subjective contrast effects stemming from the sharp difference in brightness between the limb and the terminator. Thus ancient observations often led to conflicting results, and Venus's rotation period long gave rise to heated controversies.

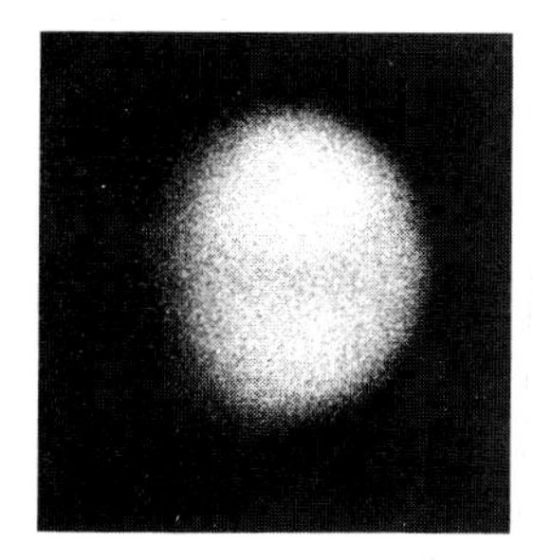
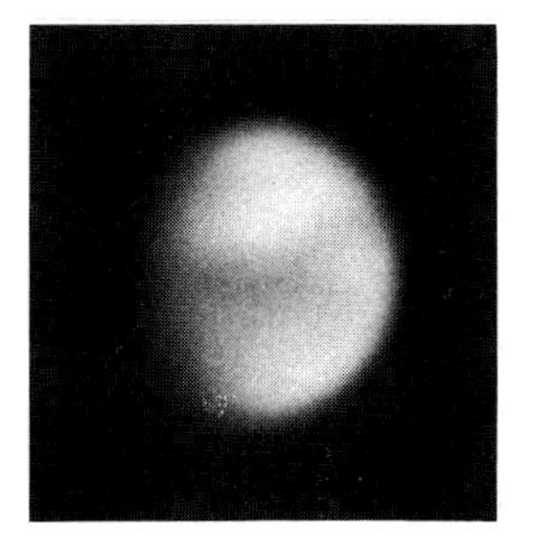
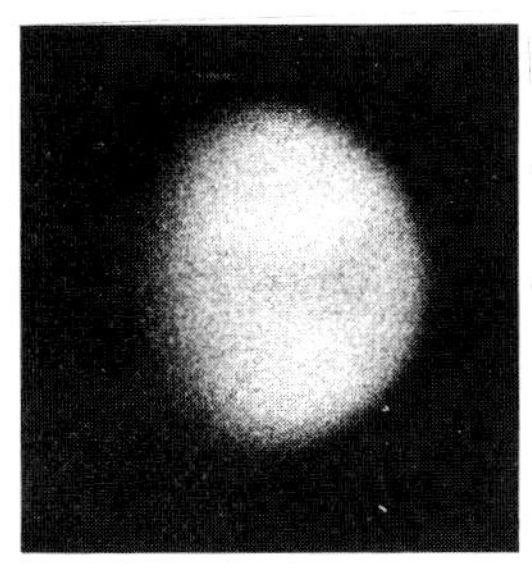
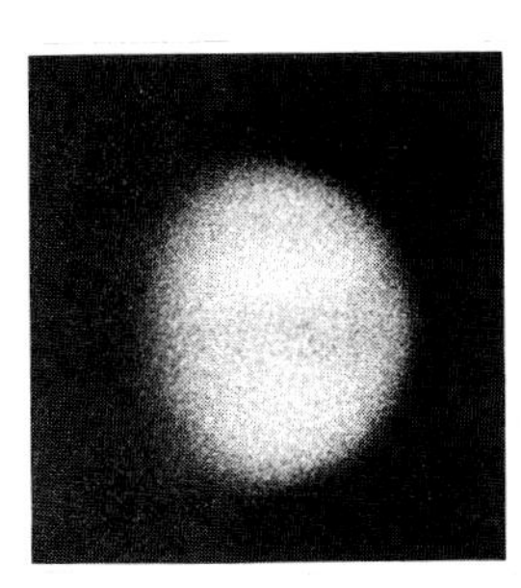

These four ultraviolet photographs of Venus taken by Charles Boyer at an observatory in the South of France on July 24, 1966, at 5:23 A.M., 7:29 A.M., 9:41 A.M. and 12:20 P.M., respectively, clearly show the retrograde rotation of a vast cloud formation shaped like a reclining Y. Photograph by Charles Boyer, Observatoire du Pic du Midi.

Jean Dominique Cassini carried out the first systematic series of observations of Venus, at the Paris Observatory in 1666–67. He concluded that it was impossible to obtain valid results concerning the planet's rotation. However, 60 years later, in Rome, Bianchini made his own drawings of Venus and believed they indicated a rotation in 24 hours. In 1732, Jacques Cassini, using the same documents, came to an entirely different conclusion and announced a rotation in 23 hours, 20 minutes. The German astronomer Schröter supported this result following observations he carried out from 1779 to 1795. But at about the same time, William Herschel expressed the same opinion as had Jean Dominique Cassini more than one century earlier. In the 19th century, new attempts were made in Rome by Father de Vico, who did not hesitate to establish a Venusian rotation period of 23 hours, 20 minutes, 15 seconds, although he admitted to the blurriness and transient nature of the spots on which he based his estimates. Finally, in 1890, Schiaparelli did a critical study of all previous calculations and reduced them to shreds. On the basis of his own observations, he proposed a new rotation period equal to the sidereal period of revolution, or 225 days, implying that Venus always faces the Sun on the same side. Most astronomers gradually accepted this finding. However, the controversy once again erupted in 1957. In that year, a French amateur, Charles Boyer, observing the planet in Brazzaville through a telescope with a 26-centimeter (10.24-inch) aperture, took ultraviolet exposures of Venus that revealed a dark structure shaped like a reclining Y that reappeared every four days. Alerted to this, another French astronomer, Henri Camichel, found the same periodicity on photographs taken in 1953 in the South of France. This result was later confirmed by a great many observations carried out by Boyer and Camichel in collaboration, as well as by high-resolution spectrographic measurements made in 1964 at the Observatory of Haute-Provence by Bernard Guinot.

Nevertheless, by 1962, radio astronomers were raising challenges to everything. Using radar echo measurements, they found a longer rotation period, somewhere around 240 days. In 1965, with the help of the great radio telescope at Arecibo, a more accurate measurement was taken, suggesting 243 days. Such a value leads to an estimate of the length of the Venusian day of 117 terrestrial days, so that exactly five Venusian days go by between one of the planet's conjunctions and the next (in other words, at a given phase, Venus always faces us on the same side); this probably involves a resonance phenomenon in which gravitational perturbations by the Earth have played a role in determining Venus's rotation rate. The mystery of this double rotation has now been clarified: Visual or ultraviolet observations refer to the top of the cloud layer, while radar measurements involve the solid sphere of Venus. Thus the upper atmosphere rotates in a period of about four days (the winds at this level attain a speed of about 400 kilometers (249 miles) per hour, while the planet itself definitely rotates in 243 days. These two rotations are retrograde; they occur in the opposite sense from both the rotation and the revolution of other planets of the solar system.

ATMOSPHERIC CIRCULATION

The photographs taken by *Mariner 10* in 1974 and by the *Pioneer Venus 1* orbiter since the end

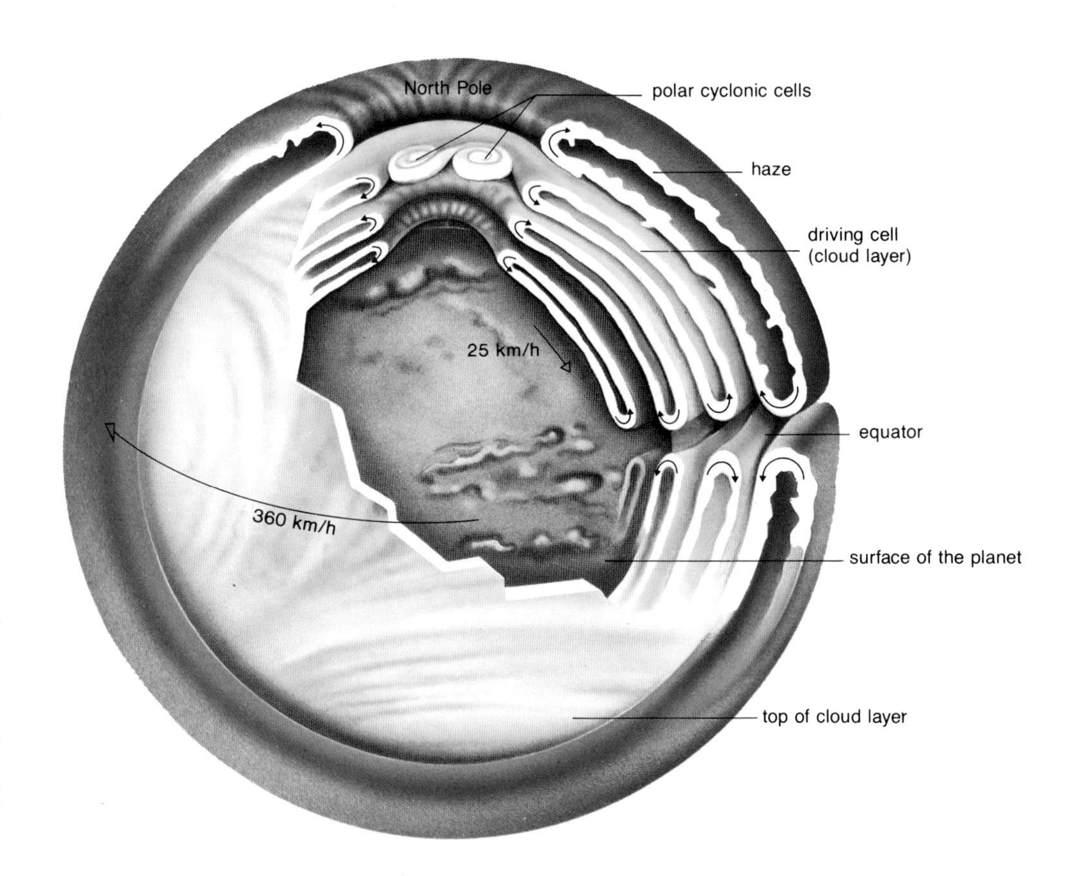

The atmospheric circulation of Venus. The convection cells (Hadley cells) may be split into three levels. Heat received from the Sun and stored in the cloud layer is transferred from the equator to the poles by these cells. But this north–south circulation, which does not exceed 25 kilometers (16 miles) per hour, is quite small compared to the powerful east–west currents, which cause the cloud cover to rotate at speeds up to 360 kilometers (224 miles) per hour at the equator.

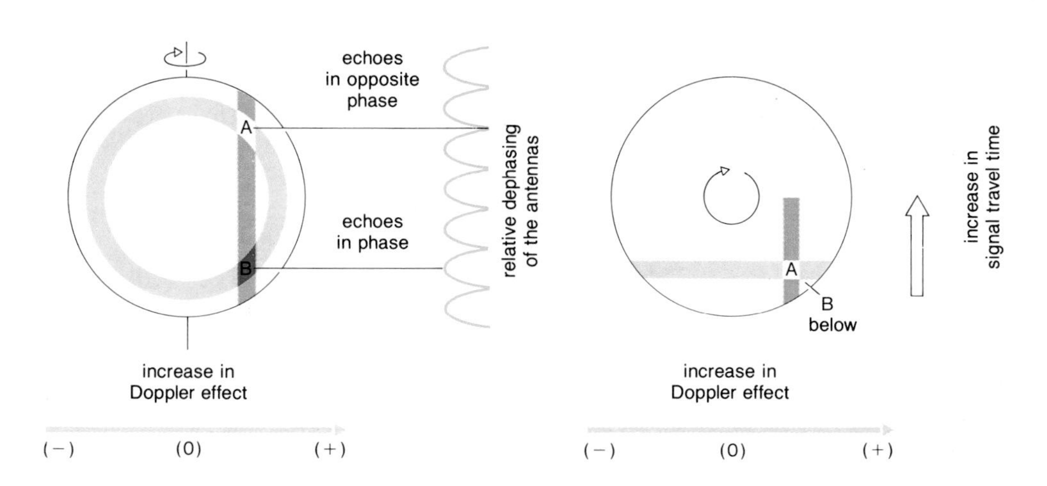

Venus's rotation creates a Doppler effect in the radar echo. The signals reflected by the zones that approach the receiving stations get shifted to higher frequencies, while those reflected by receding zones are shifted to lower frequencies. The two sites, A and B, produce echoes that take the same length of time to travel and have the same Doppler shift. When the A echoes are eliminated by interferometry, only B remains visible. From Scientific American.

of 1978 reveal the patterns and dynamics of Venusian atmospheric circulation. Solar energy deposited in the atmosphere heats the equatorial regions more than the polar regions, and convective motions lead to a transfer of heat between the equator and the poles. This mechanism follows a model devised for the terrestrial atmosphere in the 18th century by the English physicist Hadley, in which air rises near the equator, where the solar energy input is greatest, and then circulates toward the poles. Having reached the polar regions, it cools and falls toward the surface, where it returns at low altitudes to the equator to begin the cycle anew. The convection cells thus formed, called Hadley cells, split into three levels in Venus's atmosphere. The intermediate cell, which dominates the motion of the two others, corresponds to the thick cloud layer where most of the incident solar radiation is absorbed. However, the

Placed in a very elliptical orbit (periapsis: 150 kilometers [93 miles]; apoapsis: 66,000 kilometers [41,019 miles]) around Venus on December 4, 1978, the Pioneer Venus 1 *probe is equipped with a radar altimeter that measures the altitude of the spacecraft whenever it flies over the planet at a distance of less than 4,700 kilometers (2,921 miles). Simultaneously, the satellite's position is calculated on Earth and its trajectory determined relative to Venus's center of gravity. By subtracting the spacecraft's altitude from its distance from the center of gravity, we obtain the planet's radius as a plumb line from the measuring point. This relief reconstruction was made by computer from the altimeter data. The depressions appear in dark blue, the mountain ranges in green, yellow or red, depending on their altitude and the regions having an altitude at about the reference level (the average radius of Venus: 6,051.2 kilometers [3,760.8 miles]) are shown in blue-green. Photo: NASA, kindly communicated by the Jet Propulsion Laboratory.*

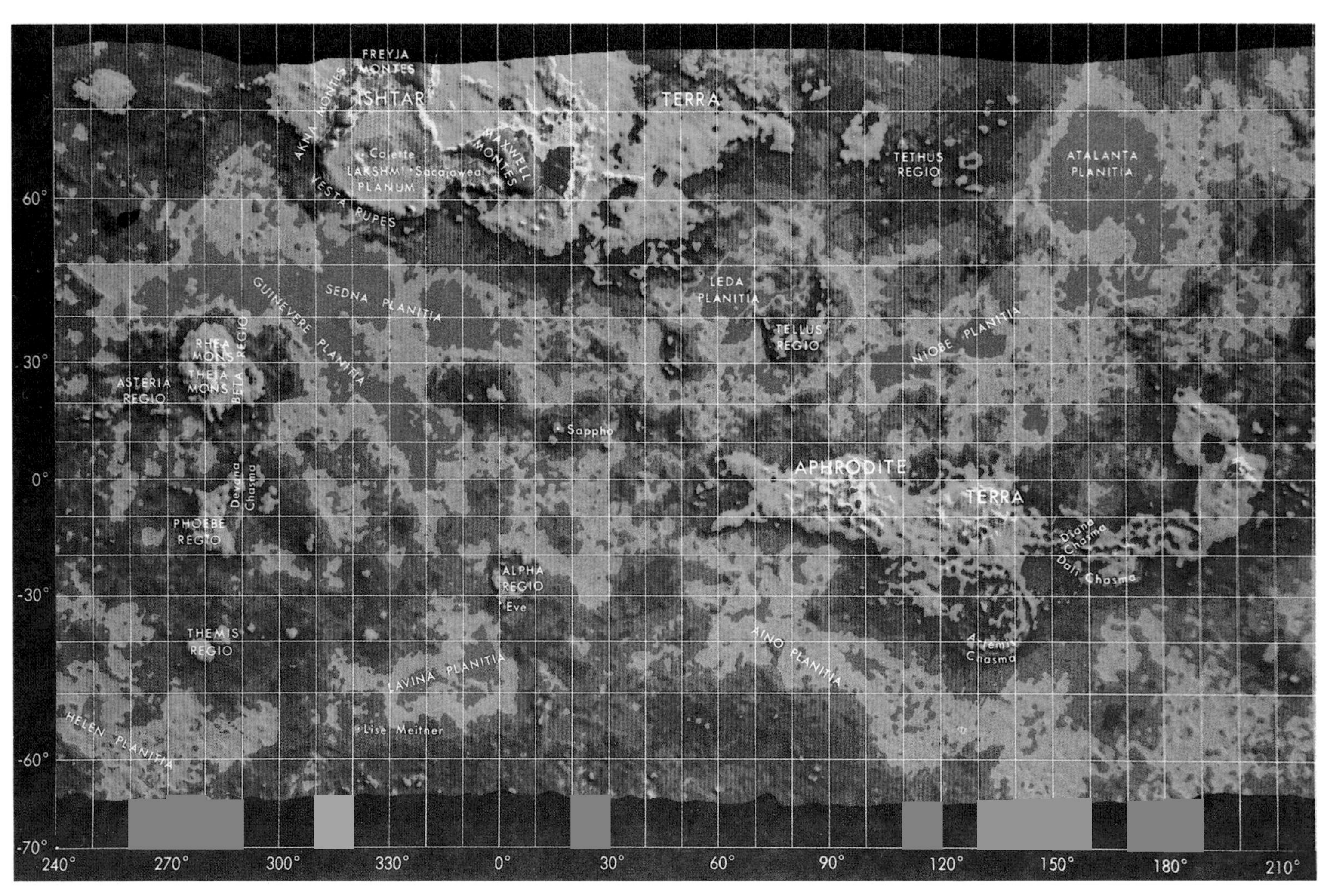

north–south circulation in this layer, which has a maximum speed of 25 kilometers (16 miles) per hour, is eclipsed by the "superrotation" of the upper atmosphere mentioned above. Winds blowing at near 400 kilometers per hour at the equator carry the cloud layer with them in a rotation from east to west 60 times faster than the planet itself. The origin of this phenomenon is still poorly understood.

The Discovery of Venus' Topography

Ultraviolet spectrometric observations made by Pioneer Venus I have made it possible to detect light emissions which are the indicators of recombinations between atoms of oxygen and atomic nitrogen occurring at about 110 kilometers of altitude. These atoms could be created on the light side of Venus (through the dissociation of carbon dioxide and molecular nitrogen) and transported within a few days to the night side. This kind of circulation would be generated by the difference between the temperature prevailing in the upper diurnal atmosphere and the nocturnal. The glimmers observed would, in this case, constitute a remarkable "tracer" of the movements affecting the Venusian upper atmosphere. Still not fully understood are all the complex phenomena occurring in this atmosphere and an explanation of why it rotates 60 times faster than the planet itself. Perhaps its motion results from the impact of huge convection cells resulting from the cloud absorption of the infrared rays given off by the ground. Some of the workings of this thermal machine still elude us.

The Discovery of Surface Relief

The topography of Venus formerly constituted a huge unknown because of the clouds that permanently veil the planet's surface. We are discovering it today with the help of radar measurements made from Earth and from the *Pioneer Venus 1* orbiter.

The first radar echo measurements of Venus date from 1961, but only after 1964 did the measurements become sophisticated enough to reveal the most prominent features of the surface morphology. Because the strength of the echo decreases strongly with distance, radar probes can best be done around the time of inferior conjunction, when the planet is at its shortest distance from Earth, the disadvantage being that then the planet always faces us on the same side. Three American radio telescopes are now equipped to function as planetary radars: the 36-meter (118-foot) Haystack antenna (Massachusetts), the 64-meter (210-foot) Goldstone antenna (California) and, above all, the spherical antenna 305 meters (1,001 feet) in diameter stationed in a natural hollow in the ground near Arecibo, Puerto Rico. This instrument, coupled with an auxiliary 30-meter (98-foot) antenna 10 kilometers (6 miles) away, permits interferometric measurements that enable us to pick out details on Venus having sizes less than 20 kilometers (12 miles). Still, the pictures having the highest resolution (on the order of 10 kilometers) and the topographical data having the greatest accuracy (an uncertainty of about 400 meters [1,312 feet] in the altitude determination) presently involve only a very small portion (about 2 percent) of the planet's total surface.

These first attempts at cartography from the Earth remain modest compared with what has been accomplished by the spacecraft *Pioneer Venus 1* which has been orbiting Venus since December 1978. With the help of its radar altimeter, it carried out a topographical survey covering 93 percent of the planet's surface between latitudes of 75 degrees north and 63 degrees south. The resulting maps are sensitive to all surface features larger than 25 kilometers (16 miles), and their altitudes are determined to within 200 meters (656 feet).

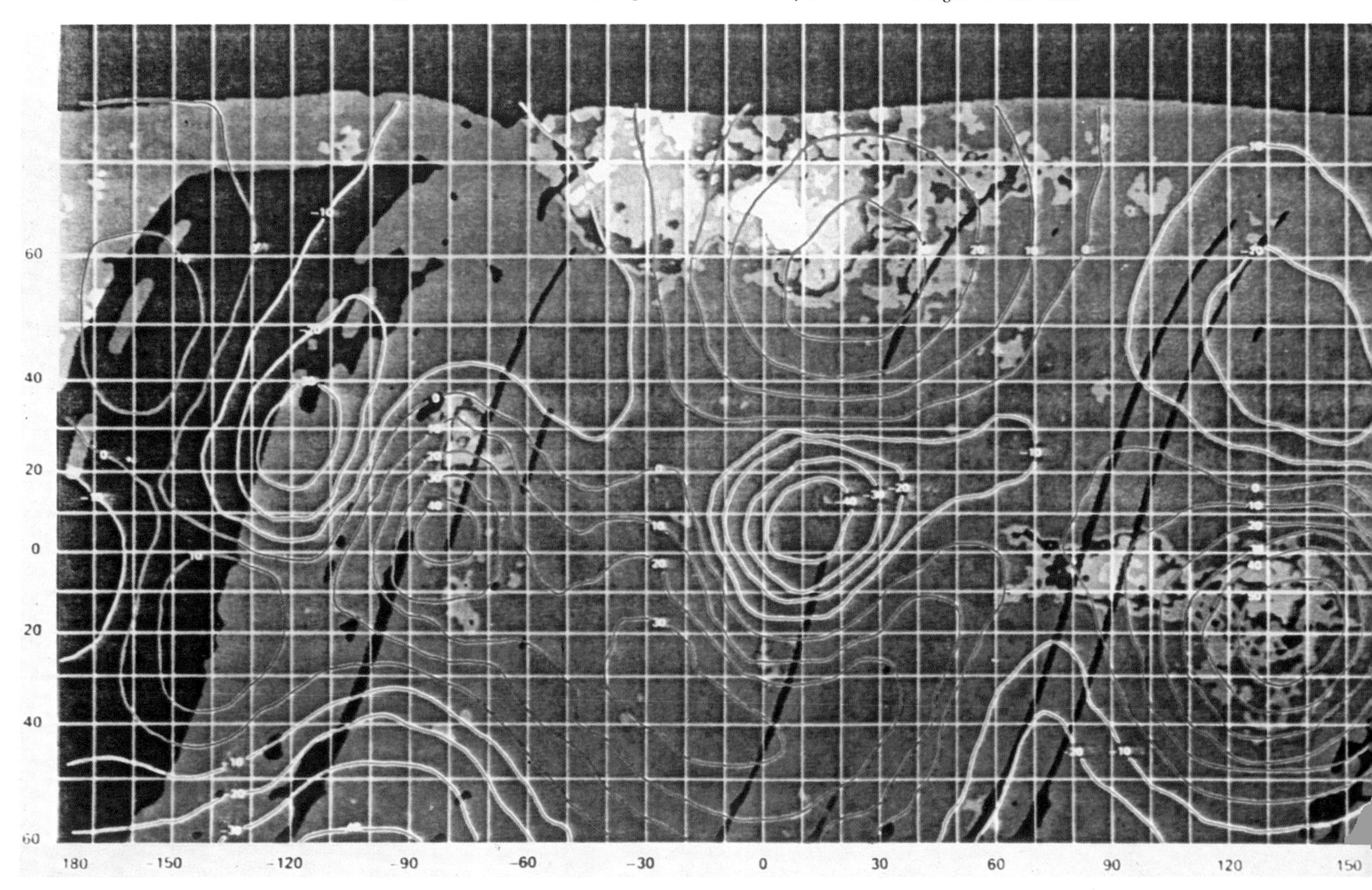

Map of Venus's gravitational field at an altitude of 100 kilometers (62 miles) above the average surface of the planet, based on data gathered by Pioneer Venus 1. *Positive or negative anomalies relative to the average field are shown in milligals. P.P.P.-IPS Photo.*

Sixty percent of the Venusian surface is taken up by a vast plain with altitude variations not exceeding 1,000 meters (3,281 feet) in amplitude. It appears to be strewn with large but shallow craters (400 to 600 kilometers [249 to 373 miles] in diameter but only 200 to 700 meters [656 to 2,297 feet] deep), which are probably very ancient impact craters.

Twenty-four percent of the surface consists of regions above the average level of Venus. They are typically no more than hills, with an altitude no greater than a few hundred meters. Nevertheless, the immense Venusian plain appears to be dominated by two big mountainous regions: *Terra Ishtar* and *Terra Aphrodite* (after the Assryian and Greek goddesses of love.) The more extensive highland region, *Terra Aphrodite*, is the size of the northern half of Africa. It includes a series of mountain ranges culminating at 9,000 meters (5.6 miles) to the west and 4,300 meters (2.67 miles) to the east. Its eastern section is bordered by a giant valley 280 kilometers (174 miles) wide and 2,250 kilometers (1,398 miles) long, where the lowest point of Venus's surface is found, 2,900 meters (1.8 miles) below the reference level. This huge canyon, with walls as high as 6,000 meters (3.7 miles), is reminiscent of terrestrial *rifts*, where tectonic plates encounter each other. The second large highland region, *Terra Ishtar*, is the size of the United States. Its central portion consists of a large plateau (*Lakshmi Planum*) with an altitude 3,000 meters (1.9 miles) above the average level of Venus. It is bordered on the west and north by Mount Akma (6,000 meters) and Mount Freija (7,000 meters [4.3 miles]). To the east lies the range of Mount Maxwell, the highest point on Venus (11,800 meters [7.3 miles]).

Finally, only 16 percent of the Venusian surface consists of regions below the average level of the planet (whereas on Earth, the oceans occupy about two-thirds of the planet's surface). The single large Venusian basin, comparable in size to the North Atlantic Basin, lies to the east of *Terra Aphrodite*, and its maximum depth is about 3,000 meters (1.9 miles).

In contrast to the Earth's crust, that of Venus, which is about twice as thick, is not divided up into a number of tectonic plates. Rather, it seems to consist of a unified planetary shell. As a result, the internal heat of the planet escapes almost entirely through volcanoes, which are much less common than on Earth and are localized only at a few "hot spots" of the planet.

Direct images of the surface were obtained in 1975 as a result of the Soviet *Venera 9* and *Venera 10* probes and then, with improved resolution and in color, in 1982 by *Venera 13* and *Venera 14*. They show rocky debris that has clearly been subject to erosion, the extent of which depends on the site. The ground and the sky have an orange hue as a result of the preferential absorption and scattering of the blue component of sunlight by the thick Venusian atmosphere.

In June 1985, the first part of the Soviet Veha (Venus-Halley) mission was a success, as the spacecraft sent atmospheric and surface probes to Venus while swinging around the planet on its way to Halley's Comet. The descent module landed in the Rusalka plain (6°27′ S. latitude, 181°5′ longitude) and transmitted data for 22 minutes, 17 seconds before being destroyed by the high surface temperature. It revealed the presence of soft, sandy soil at that site. The two atmospheric probes, which were held aloft by helium balloons,

Terra Ishtar, one of the large mountainous regions of Venus, consists of a vast plateau ringed by high mountain ranges. In this artist's view, based on altimeter measurements by Pioneer Venus 1, *the altitude of the relief was exaggerated for greater clarity, and the outline of the United States is shown to indicate the scale. To the right, we see Mount Maxwell, whose summit is the highest point on the surface of Venus. P.P.P.-IPS Photo.*

The stony and inhospitable soil of Venus photographed by the Soviet Venera 9 *probe. (This photograph has not been retouched; the vertical bands contain information provided by the spacecrafts's scientific instruments.)*

Venus and the Amateur

Despite its impressive brightness, which makes it easy to identify at first glance, Venus is not the most interesting planet for the amateur. Indeed, it turns out to be disappointing when observed through an instrument, since the thick clouds permanently surrounding it make it impossible to distinguish even the most prominent surface details. The most spectacular thing about Venus for the amateur are its phases, which can be observed with even the smallest hand-held telescope. Observation should begin in the evening two or three months prior to inferior conjunction. When the planet is 39 degrees from the Sun, 69 days before inferior conjunction, it is at its brightest. Its apparent diameter at that time is 40 arc seconds, but one sees only a crescent whose median width is 10 arc seconds. The nearer the planet draws to the Sun in the sky, the more marked its crescent phase becomes. The crescent gets increasingly hollowed out, its points nearly meeting at times because of the scattering of sunlight by the Venusian atmosphere. When it is fully dark, the brightness of Venus is so intense that all one gets in a telescope is a blinding image. It is better to observe the planet at dawn or dusk, when the brightness of the sky gives a better contrast that makes it appear less blinding. Very good observations also confirmed the presence of large quantities of sulfuric acid in the Venusian clouds and showed the powerful circulation of the atmosphere, including strong up– and downdrafts.

Future Exploration Programs

In the years ahead, new spacecraft will be needed to sharpen and expand the knowledge thus far accumulated regarding Venus and its environment.

The next scheduled American mission to the planet is Mission VMM (Venus Mapping Mission), which could be launched in 1988–89. This probe would be an adaptation of a Voyager platform designed to carry out a semicomplete (95 percent) and detailed (from 600 to 150 meters [1,969 to 492 feet] of resolution) survey of the planet through its thick cloud layer, using an aperture synthesis radar system that would scan the planet as the spacecraft followed an elliptical orbit about Venus.

Venus unveiled. Created by a graphic visualization method and based on data provided by the Pioneer Venus 1 *radar altimeter, this globe shows what the Venusian topography looks like from a point directly over the equator, at 90 degrees east longitude. The principal feature is Terra Aphrodite. The colors reflect the altitudes of the terrain relative to the average radius of the planet. The low-altitude regions are blue, the medium ones green and the high-altitude regions yellow. Still, the altitudes were exaggerated and artificial lighting was used to enhance the impression of relief. P.P.P.-IPS Photo.*

This artist's representation, based on altimeter measurements by Pioneer Venus 1, *shows a fragment of the big valley, 2,250 kilometers (1,398 miles) long and 280 kilometers (174 miles) wide, appearing on the eastern flank of Terra Aphrodite, its bottom 2,900 meters (9,515 feet) below the average level of Venus. P.P.P.-IPS Photo.*

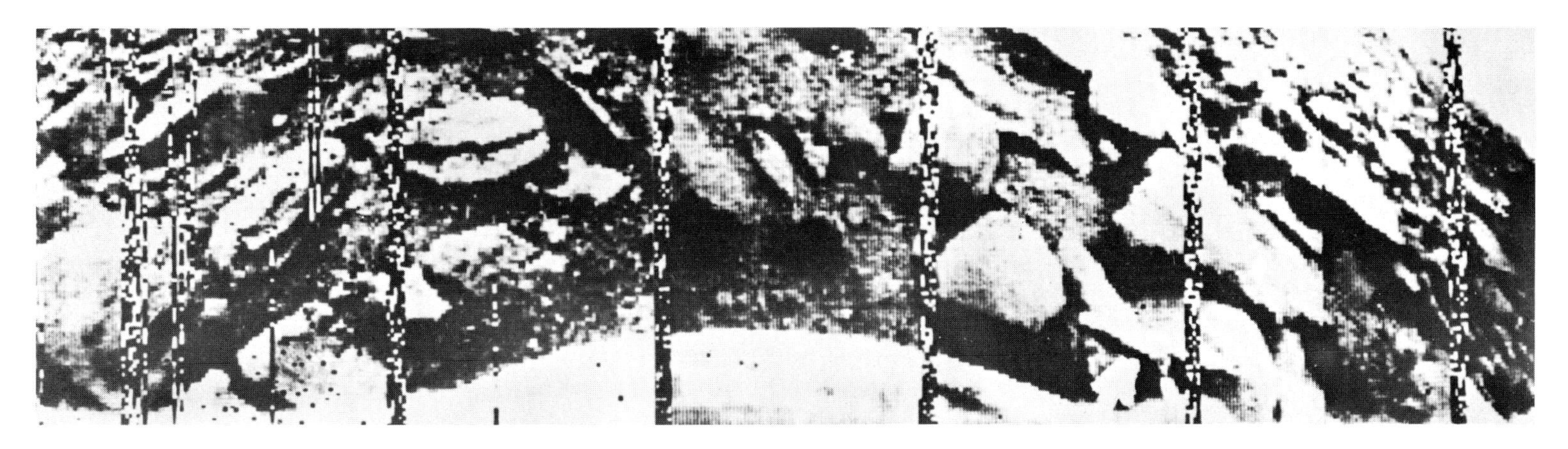

can be had during daylight on a day when there is little atmospheric turbulence. If one does not have an instrument with a prealigned equatorial mount, making it possible to point to the planet using its coordinates, it is essential to know its approximate position relative to the Sun, either by following it starting at dawn or by picking a day when it can be located in proximity to the Moon (in conjunction with the Moon). Then, by carefully taking note of its position, one can easily find it again the next day at the same time.

Using an instrument with an aperture of at least 80 millimeters (3 inches) and magnifications of 150 to 200, one sometimes can make out the irregularities of its terminator and the uneven tone of the cloudy mantle.

At night, using a camera with a regular or telescopic lens, one can show Venus's change of position relative to neighboring stars by taking photographs every two or three days of the region of the sky it is in, using exposures of a few seconds. During its periods of intense brightness, Venus has a shadow. In the absence of moonlight and internally scattered light in the camera, this shadow can easily be shown by, for example, observing the silhouette of an object on a white wall. It also can be photographed, using rapid film and exposing for less than a minute.

9. The Earth

Beyond Venus lies the Earth, the world on which we live. Its study goes well beyond the framework of astronomy; it includes a broad range of disciplines that together make up the Earth sciences.

It would be difficult to try to sum up in a single chapter all of the knowledge acquired in such varied fields as geodesy, geology, oceanography, geophysics, meteorology and so on.

Nevertheless, since the invention of spacecraft, it has become possible to make a genuine comparative study of the planets, to bring out both their common characteristics and their specific features and to reconstruct the evolution each planet has undergone since the time of its formation, in order to attempt to retrace the history of the solar system. This growth of planetology has now rendered the distinction between astronomy, regarded as the science of all cosmic bodies except the Earth, and the earth sciences, somewhat arbitrary. In any case, in a modern work on popular astronomy, it justifies devoting a chapter to the Earth as a planet, as the inhabitants of other planets would see it.

Orbital Characteristics

The Earth's orbit, like that of the other planets, is an ellipse. Earth's is not very different from a circle, since its eccentricity is only 0.0167 (or about 1/60), but is different enough so that there is an appreciable variation in the distance separating us from the Sun at opposite points of the orbit. On average, the Earth lies at a distance of 149.6 million kilometers, (93 million miles) from the Sun (this distance—the Astronomical Unit—is the standard on which other distances inside the solar system usually are based), but it may come as close as 147.1 million kilometers (91.4 million miles) (perihelion), or as far as 152.1 million kilometers (94.5 million miles) (aphelion). As a result of these variations in distance, the solar disk appears larger or smaller to us according to the time of year. But such differences are barely perceptible to the naked eye, and are revealed only by instrumental measurements: They show that the solar disk's apparent diameter oscillates between 32′36 and 31′38 arc seconds. The largest of these measurements naturally corresponds to the period when the Earth is at perihelion, around January 2 or 3, and the smallest to the time when it is at aphelion, six months earlier or later, i.e., around July 1 or 2. Thus the diameter of the solar disk gradually decreases during the first half of the year, and increases during the second; hence, the sun gives the entire globe more light and heat during the Northern Hemisphere's winter than during its summer—the difference in radiation flux can reach 7 percent, which is not negligible.

The Earth's orbit is called the *ecliptic,* because eclipses of the Sun and Moon occur in its plane. The plane of the ecliptic intersects the heavenly sphere in a great circle. An observer on the Sun would see the Earth complete this circle around the sky in one year, while an observer on the Earth would see the Sun move completely around this same circle relative to the stars in one year.

The ecliptic also serves as a reference plane to define the orientation of the orbits of the comets and other bodies of the solar system.

The orbit thus defined is merely theoretical. If the Earth were isolated, the curve describing the motion of its center around the Sun would indeed be like this. But the Earth is coupled with the Moon, and the laws of gravity dictate that both the Earth and the Moon must orbit about their mutual center of gravity, which is not located at the Earth's center, but is inside the Earth, about 1,000 kilometers (622 miles) from its surface. As a result of this double motion, to which are added the perturbations caused by the other planets, the Earth finally describes a slightly sinuous curve. Moreover, its orbit is not invariable. On the one hand, its major axis makes a complete revolution of the Sun in 21,000 years. On the other hand, its eccentricity undergoes a very slow periodic variation, which tends alternately to lengthen the trajectory or to bring it closer to a circle in the space of about 100,000 years; currently it is diminishing, and it will reach its smallest value in 24,000 years.

The Earth as seen from space. In this photograph, taken with the American spacecraft Gemini 12 *at nearly 600 kilometers (373 miles) in altitude, we see: above, Arabia; below, Africa; at left, the Red Sea; in the center, the Gulf of Aden. P.P.P.-IPS Photo.*

Other phenomena also contribute to modifying this orbit: The Earth is, in fact, subject to a number of diverse and complex motions. These motions have, however, for the most part only infinitesimal consequences during a human lifetime. It is therefore sufficient to consider, as we understand them, the conditions resulting from the two principal motions of our planet: its rotation on its axis, and its translation about the Sun.

The Earth's Translational Motion; the Aberration of Light

According to Kepler's second law, or the law of areas, the Earth travels in its orbit at a speed that does not remain constant but rather varies at each point; it is slightly faster when it is closest to the Sun than when it is most distant. This orbital speed averages 29.76 kilometers (18.5 miles) per second, or 107,000 kilometers (66,501 miles) per hour. We are thus pulled around the Sun with extreme rapidity. And if we do not perceive it, this is due to a lack of nearby reference points to which we could compare the movement of the globe moving past them, as we see telephone poles or trees file past a moving vehicle. But although we do not directly feel the sensation of this motion, it is still possible to demonstrate it, thanks to the phenomenon of the aberration of light. This phenomenon was discovered by the English astronomer James Bradley in 1727 as he was studying the parallactic motion of nearby stars caused by the Earth's orbital motion. The aberration of light appears as a cyclical annual variation in the apparent positions of the stars, with the maximum angular difference between the observed direction and the true direction of the star being equal to 20.48 arc seconds—a difference that occurs when the direction of the Earth's motion is perpendicular to the direction of the star.

Orbit Around the Sun

The time the Earth takes to complete an orbit around the Sun represents what we call a year. Relative to the stars, the length of this revolution, or *sidereal year,* is 365 days, 6 hours, 9 minutes, and 9.5 seconds—or, in decimal figures, 365.256361 days. We can also measure the revolution not in relation to the stars but in relation to a given point on the orbit, such as the point the Earth crosses at the vernal equinox; the period between two crossings of the vernal equinox, or *tropical year,* represents 365 days, 5 hours, 48 minutes, 45.975 seconds, or 365.24219879 days and is therefore about 20 minutes shorter than the sidereal year. The difference between these two periods is due to the phenomenon of precession.

Rotation

The Earth rotates from west to east around an axis (the polar axis) tilted 66°34′ from the plane of its orbit (the plane of the ecliptic). This rotation determines the length of the day. The *sidereal day,* the rotation period measured relative to the stars, equals 23 hours, 56 minutes, 4.091 seconds. The *solar day,* the rotation period measured relative to the Sun—that is, the interval of time between two consecutive crossings of the Sun through the meridian of a given place—varies in the course of the year, due to the ellipticity of the Earth's orbit and the tilt of the polar axis relative to the ecliptic; but, in a year, there is exactly 1 sidereal day less than solar days; on average, the solar day is longer than the sidereal day by about 4 minutes. For everyday purposes, we define a *mean solar day* whose length is exactly 24 hours.

Proof of the Earth's rotation is provided by Foucault's pendulum (an experiment carried out for the first time in 1851 at the Panthéon in Paris by the French physicist León Foucault). Let us assume there is a freely swinging pendulum set up at one of the poles. Since the pendulum's plane of oscillation is invariable, if the Earth rotates, then relative to the ground, the pendulum will appear to turn in a direction opposite to that of the Earth's rotation. On the other hand, at the equator, the plane of the pendulum's oscillation will remain constant relative to the ground, and no rotation will be observed. In a place of latitude ϕ, the hourly rotation observed will be equal to 15° sin ϕ. This experiment was carried out in various places with the expected result.

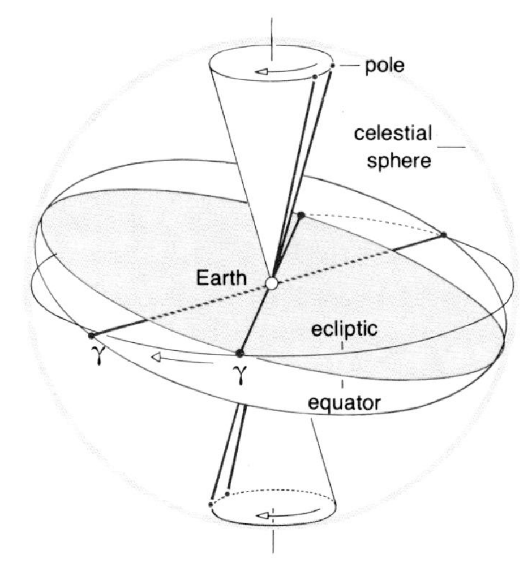

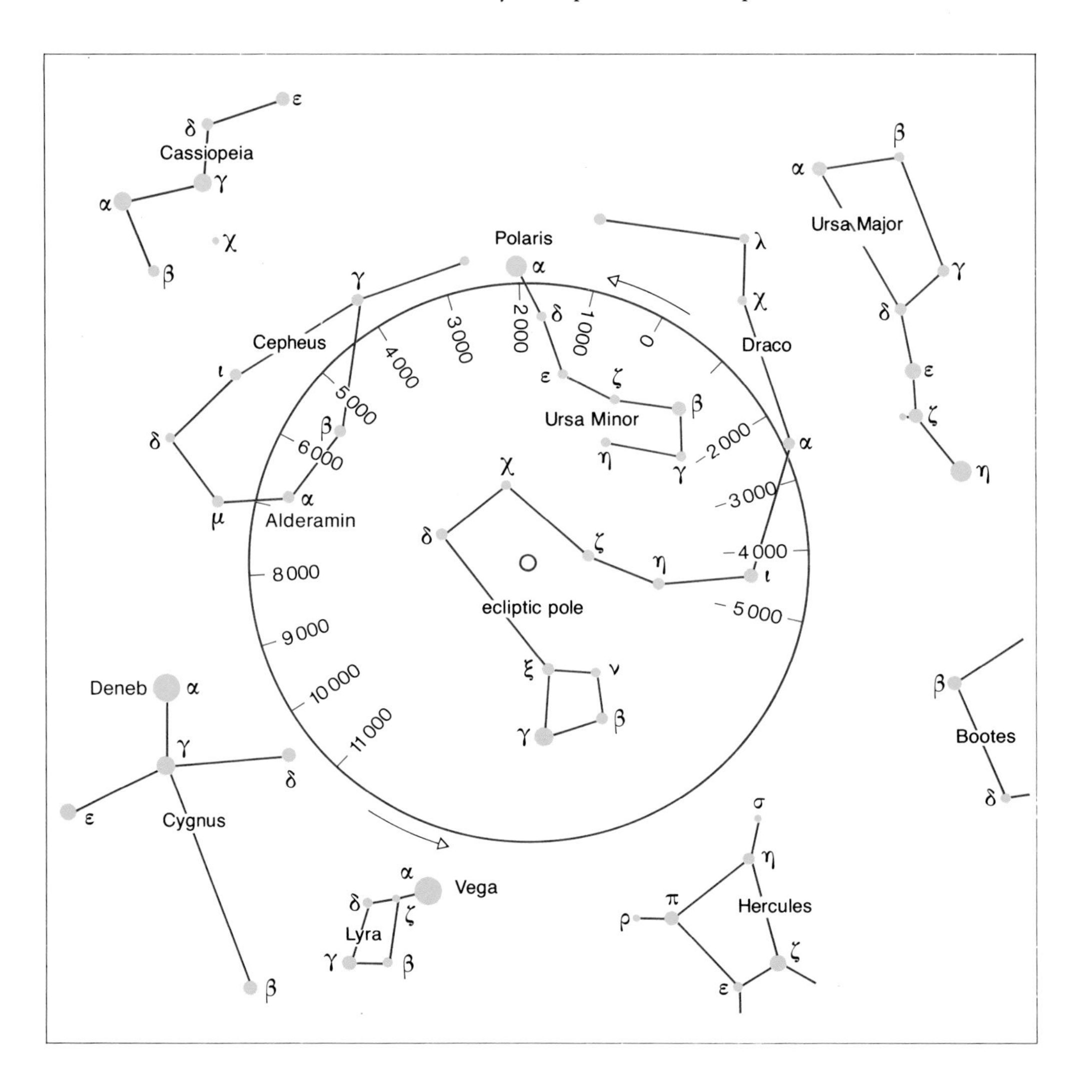

Tilt of the Rotation Axis; Precession; Nutation

The Earth's rotation takes place around an axis tilted 23°26′ away from the perpendicular to its orbital plane (in other words, 66°34′ away from this plane itself). Because the orientation of the axis remains roughly constant during the Earth's movement around the Sun, the Earth assumes various positions relative to the direction of the Sun's light; this gives rise to the seasons, as well as to the inequalities in the length of the days and nights at a given place.

In actual fact, the axis of rotation does not rigorously maintain a constant orientation; it changes in a constant manner as a result of the combination of different motions of unequal amplitude and period. The most important of these, precession, makes the earth's axis describe a cone with the planet's center at its apex, which draws a circle of 23°26′ radius in the sky around the pole of the ecliptic—in other words, around the point on the celestial sphere that is perpendicular to the plane of the Earth's orbit. This motion usually is compared to that of a top, which, while spinning in the plane where it has been launched, balances in such a way that its axis draws a cone with the point of the top at its apex. Still, the pivot of the motion is not the same in the two cases; for the top, it is a point on its base; for the Earth, a point on its center. While the North Pole is describing its precession cone, the South Pole is describing its cone in the opposite direction.

The Earth's precessional motion stems from the fact that the Earth is not perfectly round but is flattened at the poles and swollen at the equator, creating inequalities in the attraction exerted by the Sun and Moon on the planet. The combined actions of the two bodies, which are exerted more strongly on the part of the Earth that is closest to them than on that which is farthest away, amount to a resulting force passing through the center of the globe, and a torque, which acts in a direction that would pull the equatorial bulge toward the plane of the ecliptic; this causes the precession of the polar axis around the perpendicular to this plane.

Nutation

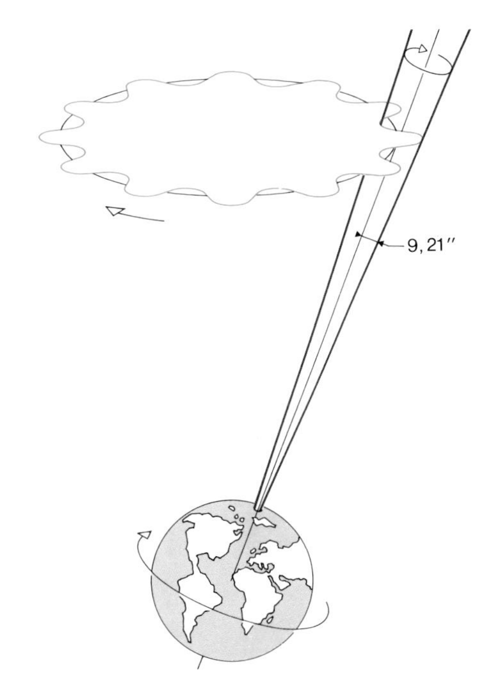

This movement is extremely slow; it takes about 25,800 years. During this long cycle, the celestial poles occupy different points on the precessionary circles on the celestial sphere, north and south, and all the stars on their path become "polar" in turn. The fact that a star is described as "polar" does not mean that it is exactly at one of the celestial poles but merely that it is close enough to appear practically immobile relative to other stars. Polaris currently is situated about 1° from the northern celestial pole, or almost twice the apparent diameter of the full Moon. Around the year 7600, the new pole star will be Alderamin, the brightest star in the constellation of Cepheus, and around the year 14,800, it will be Vega.

As a result of precession, the line of intersection of the celestial equator and the ecliptic is not fixed; the point γ *(or vernal equinox)* corresponds to the position held by the sun at the spring equinox, which moves retrograde each year by 50.3″ along the ecliptic.

In addition to precession, there is another phenomenon, discovered in 1748 by the Englishman James Bradley: nutation. Since the Moon's motion around the Earth is strongly perturbed by the Sun, the nodes of the lunar orbit move uniformly around the ecliptic in the retrograde direction, completing a rotation in 18.7 years; this motion causes an important variation in the tilt of the lunar orbit relative to the Earth's equator, hence of the forces exerted by the Moon on the Earth's equatorial bulge, and entails an oscillation of the polar axis; every 18.7 years, each pole describes a small ellipse on the celestial sphere in the retrograde direction, the semi-axes of the ellipse measuring 91.2″ and 6.86″ corresponding to the maximum amplitude in obliquity of the ecliptic and in longitude, respectively. Thus each pole describes not only a small circle parallel to the ecliptic, as it would if it were affected only by precession, but a sinusoidal path that oscillates about this circle.

In the foregoing we have as yet described only the principal effect of nutation. A complete description involves other effects of smaller amplitude, and the secular variation of the obliquity of the polar axis (6″ in 1,000 years) must be taken into account. Finally, each pole describes a sinusoidal trajectory in space that is not closed.

The Earth's rotation axis does not shift only in relation to the stars. Its direction also varies in relation to the planet itself. Measurements show that the North Pole follows on the surface of the globe, a complex trajectory called *polhodia,* which has three components: a kind of annual ellipse, whose major axis equals 0.18″ and minor axis 0.15″ (or 5.4 meters [17.7 feet] and 4.5 meters [14.8 feet] respectively, on the ground), the former being specified with respect to the Greenwich meridian; an oscillation of about 0.2″ (or 6 meters [19.7 feet]) resulting from the Earth's elasticity, with a period of 427 days (the Chandler period); and finally, a slow drift in the direction of Canada.

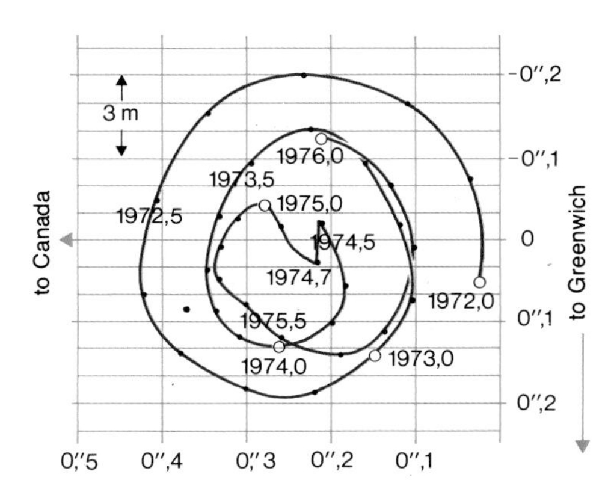

Polbody. The North Pole's trajectory over the globe from 1972 to 1976, as determined by the International Time Bureau. Point 0 is called International Conventional Origin (I.C.O.).

Shifting of the celestial North Pole around the pole of the ecliptic as a result of precession. The Christian era is designated as 0; earlier dates are preceded by a minus sign.

According to some, this drift is real and consists of a linear motion of 0.003″ (or 10 centimeters [3.9 inches]) per year, and a nutation with a period of 24 years; according to others, it is merely a reflection of changes in the average latitude of the observation posts. The problem, like that of the continental drift, will be solved only when we have more precise data. In any case, studies of the flora, fauna, climates, glaciers and magnetism do not reveal large polar shifts in the Earth's history.

Moreover, the speed of the Earth's rotation is not uniform, which is reflected in variations in the length of the day. To begin with, we observe a seasonal variation: The length of the day decreases in summer and increases in the spring, with the amplitude reaching about 0.60 second. Recent studies have shown that this variation and its irregularities can be explained by atmospheric motions. We next note that the Earth's rotation slows every 100 years, so that the length of the day increases by 0.000164 second per century. This phenomenon is ascribed to the dissipation of the Earth's rotational energy by tidal friction, particularly in the shallow seas. It is accompanied by a transfer of angular momentum from the Earth's rotation to the orbit of the Moon: The Moon moves away from the Earth. We also see periodic variations, predicted by theory, caused by a change in the Earth's moment of inertia as a result of the declination motions of the Sun and the Moon. Finally, there is an irregular variation whose general behavior has been known since 1770 but that has been followed carefully only since 1955. Its greatest amplitude in relation to the length of the supposedly uniform day is 0.004 second. Its origin remains obscure; it might be related to the existence of hydromagnetic coupling between the Earth's core and mantle, or to long-term fluctuations in atmospheric circulation.

In view of these various irregularities, the time scale used by astronomers for their studies *(ephemeris time)* is based not on the Earth's rotation but on the motions of the Sun, the Moon and the planets.

Equinoxes and Solstices; the Seasons

Because the Earth's rotation axis remains parallel to itself during the planet's travel around its orbit, its orientation relative to the Sun changes from time to time. During one complete revolution, the Earth goes through four special points in opposite pairs, the *solstices* and *equinoxes,* which divide the year into four *seasons:* spring, from the vernal equinox (March 21, give or take a day, in the Northern Hemisphere) to the summer solstice; summer, from the summer solstice (June 21 in the Northern Hemisphere) to the autumnal equinox; autumn, from the autumnal equinox (September 23 in the Northern Hemisphere) to the winter solstice; winter, from the winter solstice (December 21 in the Northern Hemisphere) to the vernal equinox.

These four periods are uneven in length owing to the ellipticity of the Earth's orbit and the 10° angle between the line of the solstices and the orbit's major axis. Thus, in the Northern Hemisphere, spring lasts 92 days, 19 hours; summer, 93 days, 15 hours; autumn, 89 days, 20 hours; and winter, 89 days. Spring and summer thus are a little longer than autumn and winter (the situation is reversed in the Southern Hemisphere).

Shape and Dimensions

The earth's shape is close to being spherical, with a radius of 6,370 kilometers (3,959 miles), slightly flattened along the polar axes. More precisely, its shape is similar to that of an ellipsoid of revolution, with a flattening (the difference in the equatorial and polar radii divided by the equatorial radius) of slightly more than 1/300. The flattening of the terrestrial globe is tied to its rotation, as with the other planets. The first measurements of the Earth's flattening were made in

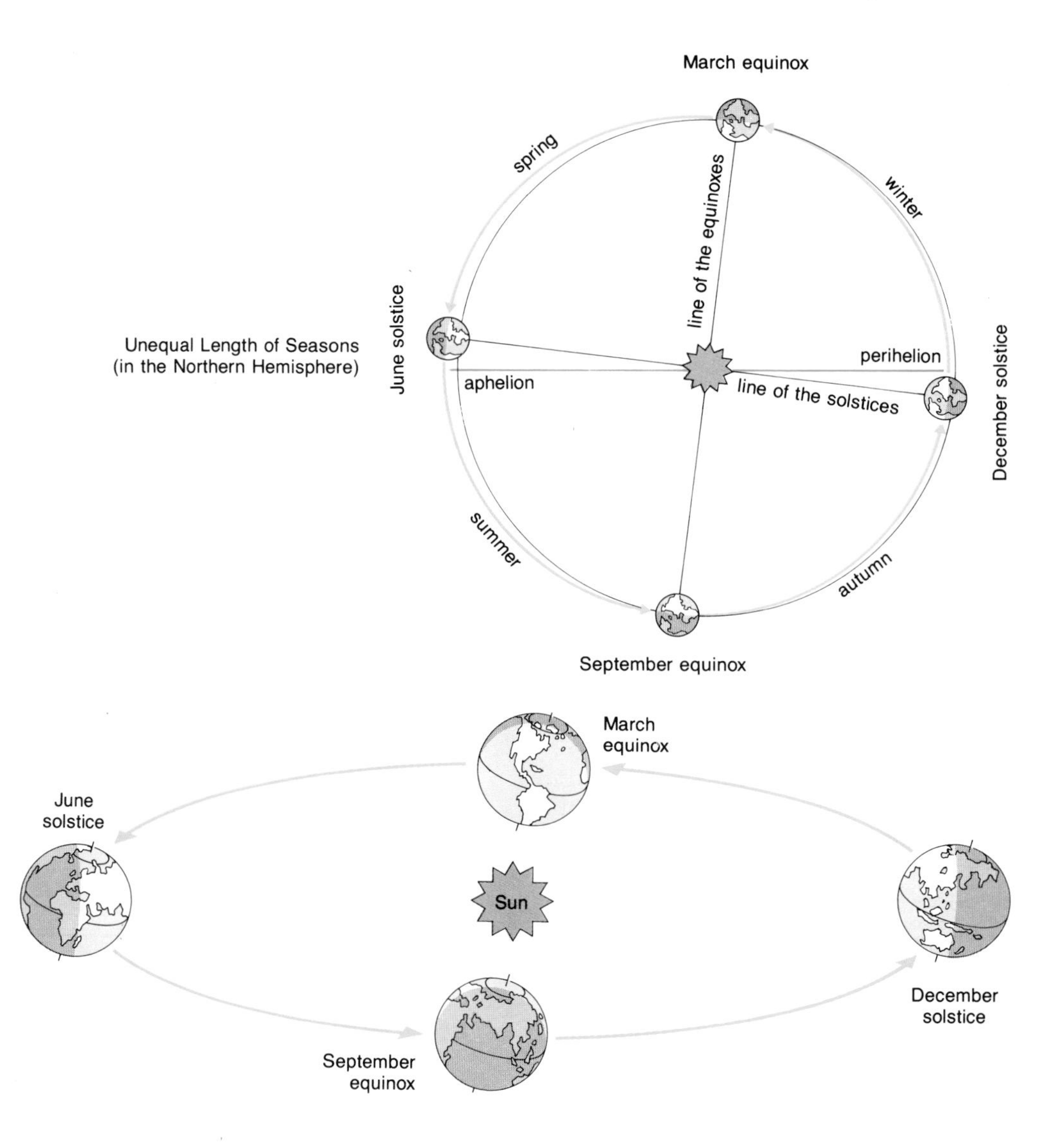

The seasons. The division of the year into seasons results from the tilt of the Earth's axis of rotation relative to the plane of its revolution around the Sun. Since the polar axis maintains a fixed direction in space throughout the year, the Sun shines alternately on the North and South poles, and the length of the day varies at different points on the globe. At the June solstice, the Sun reaches the zenith on the Tropic of Cancer, and the Northern Hemisphere has the longest days; at the December solstice, it reaches the zenith on the Tropic of Capricorn, and the Southern Hemisphere has the longest days. At the equinoxes (March and September), the Sun lies exactly in the plane of the equator, so that day and night are of equal length at all places on the globe.

the 18th century, through the geodesic measurements of meridian arcs in Peru and Lapland (see the chapter "Man's Exploration of the Sky," p. 6). Today they are also based on measurements of the strength of gravity at different latitudes, as well as on the study of variations in the orbits of artificial satellites. The currently accepted measurement of the Earth's flattening is 1/298.257. Moreover, the reference ellipsoid has an equatorial radius of 6,378.140 kilometers (3,964.04 miles). The corresponding polar radius equals 6,356.755 kilometers (3,950.749 miles), which is less by 21.385 kilometers, (13.291 miles), an extremely small amount on a planetary scale.

The earth's real shape is represented by a geoid, an equipotential surface at zero altitude, hence, by definition, perpendicular at each point to the direction of the force of gravity—i.e., to the vertical—and tangential in the horizontal plane. Practically speaking, the geoid coincides with the average sea level, corrected for fluctuations due to the tides, swells and currents and to the irregularities of density caused by differences in temperature or salinity.

The geoid is represented with a close approximation by an ellipsoid of revolution. One of the main objectives of geodesy is precisely to determine the differences between the geoid and the revolving ellipsoid here taken as a reference.

Through the use of geodesic satellites—either by studying the perturbations in their motions, or by using laser triangulation operations—we now are able to determine geoid irregularities to within less than 1 meter (3.281 feet). For example, we have been able to establish that the South Pole is 25 meters (82 feet) below the average reference surface, while the North Pole is 15 meters (49 feet) above it, which led to the figurative—but exaggerated—statement that the Earth is shaped like a pear. Near New Guinea, south of Iceland and south of Madagascar, the geoid has "humps" from 50 to 80 meters (164 to 262 feet) above the reference ellipsoid; south of the Indian subcontinent, on the other hand, it has "depressions" of slightly more than 100 meters (328 feet), and other smaller depressions exist in Mongolia, south of New Zealand, and off the California coast.

Mass and Gravity

The Earth's mass can be precisely determined, either from the rotation period of the Moon or of artificial satellites (by applying Kepler's third law), or from the strength of gravity on Earth's surface. The calculation shows that it is very close to 6.10^{24} kilograms (6,000 billion billion tons).

Since the Earth is not perfectly round, the acceleration of gravity varies with latitude. It is equal to 978.049 cm./sec.^{3} (32.088 ft./sec.^{3}) at the equator but 983.221 cm./sec.^{2} (32.258 ft./sec.^{2}) at the poles. The Earth's average gravitational field is close to 980 gals (1 gal corresponds to an acceleration of 1 m./sec.^{2}). It follows that a body released above the Earth's surface, in the absence of atmosphere, assumes a speed of 9.8 m./sec. (32.2 ft./sec.) after 1 second, 19.6 m./sec. (64.31 ft./sec.) after 2 seconds, etc.

Remote-sensing provides a new means for observing the Earth. It permits accurate cartography and enhances the perception of certain large-scale geological features (folded structures, faults, etc.); it makes it easier to establish and regularly update inventories of the planet's resources (water, vegetation, mineral resources), and to survey various natural and human-induced phenomena (erosion, floods, drought, various kinds of pollution). It provides benefits to certain economic activities (agricultural and forestry operations, oil and mineral prospecting and mining, navigation and fishing). This remarkable view of Sicily was taken in infrared light on January 2, 1976, by the American satellite Landsat 1 *from an altitude of 900 kilometers (559 miles). The colors are the result of coding applied to three spectral bands: the 0.5 to 0.6 μm band was coded in blue; the 0.6 to 0.7 μm band in green; and the 0.8 to 1.1 μm band in red. NASA Photo; processing by Institut Français du Pétrole.*

The measurements of gravity at the Earth's surface now have achieved an accuracy of almost a milligal. The gravimetric maps of the planet that such measurements have made possible reveal negative anomalies over Ceylon (−95 mGal), south of New Zealand (−65 mGal), in the Antilles and in the eastern Pacific off the California coast (−45 mGal); and positive anomalies over New Guinea (+75 mGal), between Madagascar and the Antarctic (+45 mGal) and over western Europe, Iceland and part of the North Atlantic (+55 mGal).

Atmosphere

The Earth is surrounded by a gaseous layer that gives it a lovely bluish color when viewed from outer space. This atmosphere has no well-defined upper limit; as one goes higher above the ground, it becomes thinner, and finally it merges with the interplanetary medium. However, while its presence still can be detected at several thousand kilometers' altitude, its dense portion is confined to a thin skin; it is estimated that 50 percent of its total mass—estimated at 5 million billion tons, or about one-millionth that of the Earth—is concentrated below 5 kilometers (3.1 miles) in altitude, and 99 percent below 30 kilometers (18.6 miles). If the atmosphere had the same density at all altitudes, it would have a thickness of only 8 kilometers (5 miles).

Composition

It was long believed that the air comprising the Earth's atmosphere was an element (the ancients considered it one of the four fundamental elements). In 1777, Lavoisier first proved that it was in fact a mixture of gases.

In the lower atmosphere, the composition of air remains homogeneous as a result of a continuous vertical mixing. We find "permanent" gases (those whose proportions remain constant) and others for which the concentrations vary with altitude. The gases for which the proportions remain essentially constant form what is called dry air; it is made up of 99 percent nitrogen and oxygen, in the approximate proportion of 1 part oxygen to 4 parts nitrogen (see the following table). Water vapor, carbon dioxide and ozone are the main components with variable concentrations. Although they are present only in small quantities, they play a major role in absorbing solar radiation.

The atmosphere also contains many solid particles in suspension or liquids of highly diverse origins: salt crystals, volcanic dust, combustion residues, grains of sand, etc. These particles, which are blown about by the wind, are very important, for they represent the condensation nuclei around which raindrops form.

Constituents of Air in the Lower Atmosphere

Name	Symbol	% by Volume
Permanent Constituents		
Nitrogen	N_2	78.110
Oxygen	O_2	20.953
Argon	Ar	0.934
Neon	Ne	0.00182
Helium	He	0.00052
Krypton	Kr	0.00011
Xenon	Xe	0.00087
Hydrogen	H_2	0.0005
Methane	CH_4	0.0002
Nitrous oxide	N_2O	0.0005
Variable Constituents		
Water	H_2O	0 to 7
Carbon dioxide	CO_2	0.01 to 0.1
Sulfur dioxide	SO_2	0 to 0.0001
Ozone	O_3	0 to 0.00001
Nitrogen peroxide	NO_2	Traces

STRUCTURE

The air comprising the atmosphere is characterized with the aid of three parameters: pressure, temperature and humidity. Each parameter evolves in time and space independently of the other two.

- *Pressure.* At sea level, atmospheric pressure varies in general between 950 and 1,050 millibars. At this level, so-called *normal* pressure (standard pressure corresponding to an air temperature of 15.5°C) is 1,013.25 millibars. At a given point on the surface of the globe, the pressure is subject to a very small diurnal variation (often masked in the temperate zones by irregular variations due to the passage of atmospheric perturbations). It is more obvious in the tropical and equatorial latitudes, where the pressure reaches a high point between 8:00 A.M. and 10:00 A.M. (depending on the season) and between 8:00 P.M. and 10:00 P.M.

Pressure also varies in space, both in the horizontal and in the vertical plane. Its horizontal variation is reflected in the existence of low-pressure zones on the surface of the globe (depressions or cyclones) and of high-pressure zones (anticyclones); it is of crucial importance in meteorology, for it generates the circulation of atmospheric air, hence the winds. Nevertheless, the pressure varies much more rapidly with altitude; practically speaking, in the first kilometers above the ground, we observe a 1-millibar decrease in pressure for an 8-meter (26.2 foot) elevation. But the rate of this decrease does not remain constant with altitude, for it is linked to air temperature.

- *Temperature.* Air temperature also is subject to variation in time and space. The thermal distribution fluctuates more than that of pressure, for local influences have more effect on temperature than on pressure. Apart from atmospheric perturbations, we note a diurnal variation in temperature depending on the place and the season. The vertical thermal profile of the atmosphere reveals several layers, in which the temperature alternately decreases or increases. The *troposphere* extends from ground level to an altitude that varies from 8 kilometers (5 miles) (in polar regions) to 17 kilometers (10.6 miles) (in equatorial regions). The temperature there decreases, on average, by 6.5°C per kilometer. It contains about 85 percent of the atmospheric mass, and practically all the atmospheric perturbations develop there; from the meteorological standpoint, it is the most important layer. It ends at the *tropopause,* at which level the temperature stops decreasing and stabilizes at about −57°C, on average.

Next comes the *stratosphere*, which is characterized by an increase in temperature with altitude. In this region, at about 20 to 45 kilometers (12.4 to 30 miles) in altitude, the molecules of oxygen, O_2, are transformed into molecules of ozone, O_3, as a result of very-short-wavelength ultraviolet radiation, thus causing the formation of an ozone layer. This chemical transformation releases heat, which explains the rise in temperature. Toward 50 kilometers (31.7 miles) in altitude, the temperature reaches about 0°C and stops rising; this is the *stratopause.* Beyond lies the *mesosphere*, which is characterized by another decrease in temperature with altitude and which ends at about 85 kilometers (52.8 miles) with the *mesopause,* where the temperature approaches −100°C; then comes the *thermosphere,* where the temperature again increases with altitude (as a result of the absorption of

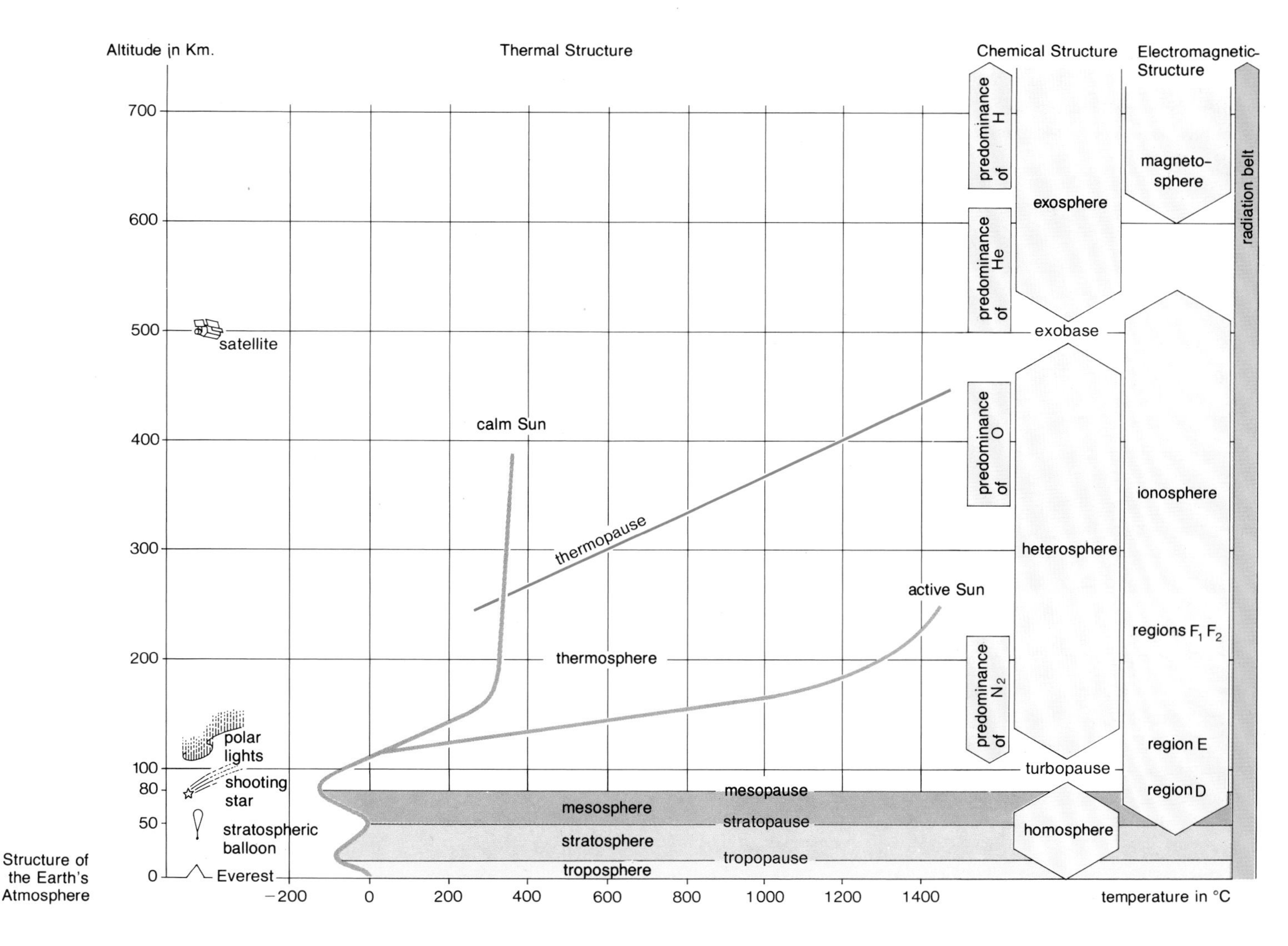

very-short-wavelength ultraviolet radiation), extending up to the ***thermopause,*** for which the altitude (400 to 800 kilometers) (249 to 497 miles) and temperature (400° to 1,800°C) depend on solar activity. Beyond the thermosphere there remains only a medium where the atomic density is so low that we can no longer speak of temperature: This is the *exosphere*, where the lightest molecules escape from the Earth's gravity and disappear into interplanetary space.

● *Humidity.* Humidity characterizes the water vapor content of air. The supply of water vapor in the atmosphere is provided by the continuous evaporation of seawater and, more generally, of all bodies of water (rivers, lakes, etc.). The variation in humidity is related in the horizontal plane to the distance from sources of evaporation, and in the vertical plane to the decrease in temperature with altitude. The presence of water vapor in the atmosphere is the source of cloud formation.

The atmosphere can also be stratified from an electromagnetic standpoint. Up to 60 kilometers (37.3 miles) in altitude, the particles it contains are mostly electrically neutral; thus this region sometimes is called the *neutrosphere.* Between 60 and approximately 500 kilometers (311 miles), the absorption of solar ultraviolet radiation leads to the formation of three strongly ionized layers designated from the lowest to the highest respectively, by the letters D, E and F), which altogether make up the *ionosphere.* These layers have the property of reflecting radio waves; this phenomenon is put to use in establishing long-distance short-wave communications on the Earth's surface. At higher altitudes, the Earth's magnetic field exerts the dominant force on the very rarefied atmospheric plasma, and particles are trapped along the field lines; this is the ***magnetosphere*** (see p. 138).

ATMOSPHERIC CIRCULATION

Solar radiation is the main source of energy in the atmosphere. The energy radiated by the Sun may be regarded as constant. At the upper limit of the atmosphere, an area exposed perpendicularly to solar radiation receives an energy of about 1.4 $kW/m^2/min$; this value is known as the *solar constant.* Nevertheless, the atmosphere is far from being totally transparent, and only 40 percent of the incident solar radiation reaches the ground. The clouds reflect 35 to 40 percent of it into space, and the atmosphere itself absorbs 20 to 25 percent. The Earth's surface absorbs a part of the solar radiation it receives, and reflects the rest into the atmosphere; the fraction of reflected radiation depends on the reflectivity of the surface, which itself varies depending on the nature of the soil (stone, forest, sand, snow or ice) from 5 to 90 percent; the average albedo of the Earth's surface is close to 40 percent.

The effect of heat is expansion of the air, which then tends to rise; cold, on the other hand, makes it contract and sink. The differences in temperature between various regions of the atmosphere thus create currents, which are the source of winds and the formation of high- and low-pressure zones. Atmospheric circulation is basically governed by the fact that the atmosphere receives more heat at the equator than at the poles, and that the air masses above the equator tend to rise toward the polar regions while those above the poles tend to sink toward the equator. But this simple model is profoundly altered by the presence of the oceans (which act as thermal reservoirs and maintain an intense evaporation, which generates the clouds) and by the Earth's rotation.

Eruption of Mount Etna in May 1971. Volcanic eruptions are a spectacular demonstration of activity taking place in the depths of the Earth. Photo: Haroun Tazieff.

CLIMATIC VARIATIONS

The Earth's climates are constantly changing, but their variations occur according to very different time scales and magnitudes. We know that in the course of geologic time, the planet's global climate was substantially hotter for tens of millions of years (from 100 to 50 million B.C.), or colder (from 50 to 25 million B.C.) than it is now. We also know that during the past 2 million years, glacial periods have alternated with warmer periods, each cooling and subsequent rewarming lasting, on average, 100,000 years. We also know that less marked variations have affected parts of the Earth for several centuries, even several decades: Around the ninth century, a slight warming enabled the Vikings to colonize Iceland and the coast of Greenland; on the other hand, from 1550 to about 1850, a "miniature ice age" made Europe somewhat colder.

These climatic variations are the result of complex processes that have many causes. For the variations occurring on a long time scale, continental drift, which changes the distribution of oceans and exposed land areas, must certainly play an important role. Astronomical phenomena may also be mentioned. Around 1935, the Yugoslav astronomer Milutin Milankovich forwarded a long-derided theory that later served as the basis for many studies. According to this theory, important climatic variations can be explained by the combination of three periodic astronomical phenomena, all of which modify the seasonal effect by which solar radiation reaches different latitudes of the planet with different fluxes: the variation every 100,000 years in the eccentricity of the Earth's orbit; the variation every 41,000 years in the obliquity of the polar axis relative to the ecliptic; the precession of the equinoxes, which has a cycle of 25,800 years. More recently, it has been argued that the Earth's climate might also be related to the Sun's motion within the galaxy. The Sun and its accompanying planets are rotating around the galactic center, with a period of about 200 million years. When the solar system passes through a region with an enhanced content of interstellar dust, the dust may have a slight screening effect, causing a reduction in the amount of solar radiation received on Earth, hence a global cooling of our planet and consequently a glaciation.

This would last throughout the time the dust cloud is crossed—i.e., several million years.

Internal Structure

By direct observation, we can know the composition of only the most superficial part of the Earth—that is, down to a depth of about 10 kilometers (6.2 miles), which is reached by mining and drilling. The internal structure of the globe can be estimated only indirectly, in the form of models based on geophysical studies. The structural models thus developed make it possible to calculate the resulting effect on the surface. If the calculations do not yield results consistent with the observations, it becomes necessary to develop other models, which will be compared with new data. Thus we have only a simplified, provisional image of the Earth's interior, which can always be modified in accordance with the growth and improvement of the measuring methods.

Current notions of the internal structure of the planet derive from considerations of the gravitational field and from deep-lying phenomena such as earthquakes and volcanoes; seismology contributes information about the physical state (see box), while volcanism provides indications of the chemical composition of the Earth's inner layers.

While differences remain on points of detail, there is general agreement that the Earth consists of three large concentric units, respectively from the surface to the center: the crust, the mantle and the core.

Seismology and the Study of the Earth's Internal Structure

An earthquake is a violent shaking of the ground caused far below the surface by a sudden movement between two deep zones within the Earth's crust. It is accompanied by a violent release of slowly accumulated elastic energy. The point at which the rupture begins is called the origin or hypocenter (generally at a depth underground of less than 70 kilometers [44 miles]), but sometimes much greater, ranging up to 700 kilometers [435 miles], its vertical projection on the ground is the corresponding epicenter.

Seismic waves are mechanical waves that move out from their origin in the form of oscillations and may travel very far on the surface or in the interior of the globe. There are three successive groups (or trains) of waves: 1) longitudinal, or compression waves (8 waves), such as those that can be trans-

THE CRUST

This is a heterogeneous layer 5 to 10 kilometers (3.1 to 6.2 miles) deep, on average, under the oceans, and 30 to 40 kilometers (18.6 to 24.9 miles) within the continents. Its average density is relatively low (2.8 gm./cm.³) and it represents only 1 percent of the Earth's mass. The speeds of its longitudinal seismic waves generally range from 5 to 7 kilometers (3.1 to 4.4 miles) per second, increasing with depth. In places where it has not been too disturbed, it appears to consist of three superimposed layers:

- a superficial sedimentary layer formed from materials broken off from the preexisting rocks and driven into the depressions by wind and running water; it deposits, nonexistent in some places, reach a thickness of several kilometers elsewhere
- an intermediate granitic layer 15 to 20 kilometers (9.3 to 12.4 miles) thick (continental crust)
- a lower basaltic layer, which makes up the ocean bottoms (oceanic crust)

In fact, these sedimentary layers, primitively horizontal and superimposed in the order of their formation, were folded, turned over, broken up and mixed with the primitive materials of the crust, which, in the beginning, was merely a thin skin formed on the surface of the developing mantle. Hence the present crust is a heterogeneous mosaic of primitive rocks, rocks from the underlying mantle (endogenous rocks) and rocks resulting from the agglomeration and consolidation of sediments (exogenous rocks).

On its inner side, the Earth's crust is bordered by a region in which the speed of longitudinal seismic waves increases suddenly, reaching a level of about 8.1 kilometers (5 miles) per second; this is the Mohorovicic discontinuity, or, more simply, the Moho, named after the Croatian geologist who discovered it in 1909.

THE MANTLE

Below the crust lies the mantle; it is composed of two parts representing 65 percent of the Earth's mass and 83 percent of its volume:

- the upper mantle, made up of a rigid first layer (average density: 3.3), which goes from the Moho to a depth of 60 to 100 kilometers (37 to 62 miles), and a viscous layer, or asthenosphere, which reaches to 200 kilometers (124 miles), and finally a thick region between 200 and 700 kilometers (435 miles); the first layer, together with the overlying crust, form the rigid lithosphere, which separates laterally into plates. These plates drift on the asthenosphere, the viscosity of which is due to conditions of pressure and temperature approaching those of fusion
- the lower mantle, separated from the upper mantle (at about 700 kilometers in depth) by the Repetti discontinuity, characterized by an increase in density, which goes from 3.3 to 4.3; the composition of the lower mantle is roughly that of a peridotite enriched by very dense minerals appearing as the result of the strong pressure exerted by the mass of rocks lying above

A final discontinuity, at a depth of 2,900 kilometers (1,802 miles), separates the lower mantle from the nucleus; this is the Gutenberg discontinuity, named after the German geophysicist who discovered it in 1913. It is apparent in the form of a sudden large drop in the speed of the longitudinal seismic waves, which falls from 13.7 to 8.1 kilometers (8.5 to 5 miles) per second, and of that of the transverse waves, which disappear.

THE CORE

The core, which is thought to be made up of iron and a small amount of nickel, appears to be divided also, at about 5,100 kilometers (3,170 miles) in depth, into two regions:

- the external core, assumed to be liquid, since it arrests seismic waves (sheer, or transverse waves)

Internal Structure of the Globe

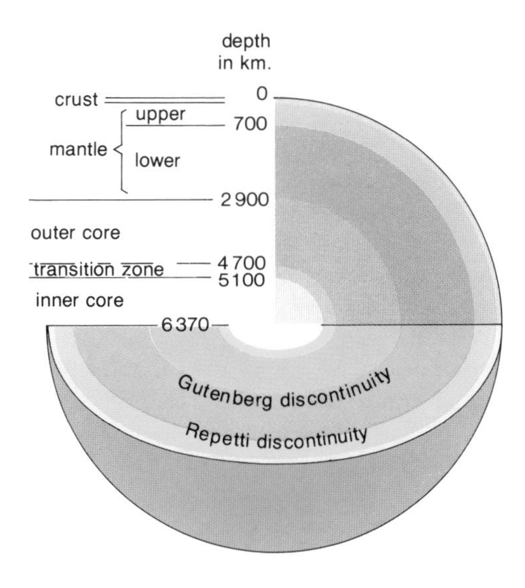

Illustration of the various types of trajectories that may be followed by the waves produced by an earthquake having its epicenter at E. The notations given are those generally employed to describe the paths followed when these waves reappear at the surface and are recorded on seismograms. P indicates a path within the mantle, K a path in the outer core, I a path in the inner core. These wave types are recorded one after the other, with repetitions if they occur (in case of reflections). The primary waves (in red) go through gases, liquids and solids. They travel fast; their speed increases rapidly in the mantle, falls abruptly when they enter the outer core and rises again in the inner core. The secondary waves (in blue) go through solids only; thus they do not penetrate the molten core. No waves arrive in the region between A and B; it is a shadow zone.

mitted in all fluids, are the first to be recorded; 2) transverse waves (S waves), which compress material transverse to their direction of motion and which can be transmitted only by solid bodies, are the second to arrive; and 3) a third group considered a mixture of two similar kinds of surface waves: Love waves (named after the British physicist A.E.H. Love) and Rayleigh waves (named after the British physicist J.W.S. Rayleigh).

These three kinds of waves provide information on the internal structure of the globe. The surface waves tell us about the crust, sometimes as far down as the mantle; the longitudinal and transverse waves reach the surface only after having followed paths at varying depths within the Earth, which also may include a number of reflections and refractions at the interfaces separating materials of different densities. By reconstituting these paths and studying the variation in the waves' speed of motion according to depth, based on recordings taken at many observatories, it is possible to determine the position of the concentric surfaces of discontinuity that mark off the various internal surfaces of the globe.

- the internal core, or nucleus, rather easily crossed by P waves (compressional waves) and considered to be solid

At the boundary of these two zones, the density goes from 12.3 to about 13.3, and then rises to about 13.6 at the center of the Earth.

Magnetism

The Earth has its own magnetic field. This field may, as a rough approximation, be compared to the field of a dipole the axis of which is inclined 11.6° from that of the geographical poles and passes about 400 kilometers (249 miles) from the Earth's center. The points where the dipole's axis pierces the surface of the globe are called geomagnetic poles. The north geomagnetic pole is at 69° west longitude and 78.5° north latitude. In fact, we note very large irregularities of the Earth's field, due partly to the varied type of rocks beneath the surface, which act as small dipoles and are distributed differently in different regions. This complication leads to the consideration, in addition to the dipolar field (which accounts for more than 90 percent of the Earth's current magnetic field), of a nondipolar field. In particular, the true magnetic poles (see box) do not coincide with the geomagnetic poles.

The average intensity of the Earth's field, at the surface, is close to 50,000 gammas (γ), equivalent to 0.5 gauss. As a result of the solar wind, this field is confined within a vast cavity, the magnetosphere, the general structure of which now is known thanks to space vehicles.

Up to about 20,000 kilometers (12,430 miles) in altitude, the geomagnetic field remains practically dipolar: Its lines of force, arranged symmetrically around the globe, link the southern and northern polar regions, forming loops in space that are larger as their field lines emanate from higher latitudes. At greater distances, the structure of the field becomes more complex: Under the pressure of solar wind, the lines of force are sharply compressed in the direction of the Sun, while they stretch out disproportionately in the opposite direction, forming a long magnetic tail.

In 1962, it was proved that the boundary of the magnetosphere, or magnetopause, lies only some 65,000 kilometers (40,398 miles) from the center of the Earth. However, because of its very high speed, the solar wind, on approaching the magnetosphere, generates a shock wave comparable to that arising on the leading edge of the wing of a supersonic jet. Between the magnetopause and the front of the shock wave there thus exists a turbulent zone, where a magnetic field prevails that is extremely variable in intensity and direction as a result of solar activity; this transition region is called the magnetogain.

In a period of low solar activity, when the speed of the solar wind reaches its lowest intensity (less than 300 kilometers [186 miles] per second), the front of the magnetopause is about 90,000 kilometers (55,935 miles) from the center of the Earth, in the direction of the Sun, and about 140,000 kilometers (87,011 miles) in the perpendicular direction (above the poles). In a period of high solar activity, the solar wind pressure is stronger, and the outer edge of the magnetosphere may be only 40,000 kilometers (24,860 miles) from Earth, in the direction of the Sun.

In the opposite direction, the magnetospheric tail has been observed up to more than 6 million kilometers (3.73 million miles). Within this structure, certain lines of force are so stretched out that the regions of opposite polarity are parallel and nearly contiguous; the corresponding zone is referred to as the neutral layer.

The charged particles that fall into the magnetosphere form a spiral around the lines of force and oscillate from one magnetic pole to another (in reality, they do not reach the poles themselves, but their movement occurs between two "mirror" points at about 70 to 75° latitude), with a period of 0.1 to 3 seconds. On this rapid motion there is superimposed a general drift—toward the west for the protons, toward the east for the electrons. Finally, these particles accumulate around the Earth in regions called radiation belts, which have a crescent-shaped cross section. Their discovery was the first great discovery ascribable to artificial satellites. The American physicist J. A. Van Allen, using measurements taken by the first Explorer satellites in 1958, gave a first glimpse of their shape and characteristics; hence they have come to be known as the Van Allen belts. Two belts, in fact, have been observed: one relatively narrow, centered at about 3,000 kilometers (1,865 miles) in altitude and made up basically of protons having an energy greater than 30 MeV; the other, longer, about 15,000 to 25,000 kilometers (9,323 to 15,538 miles) and filled with electrons and low-energy protons.

The charged particles emitted during particularly intense solar flares may escape being trapped by the lines of force in the magnetic field and may penetrate the upper atmosphere. There they are responsible for a number of phenomena: sudden variations in the intensity of the magnetic field (magnetic storms), perturbations of the ionosphere causing the fading of some radio waves emitted on Earth, auroral displays, etc.

The aurorae are among the most beautiful of natural phenomena. The name "polar dawn" was given to them by the French scholar Gassendi, who was fortunate to observe one on September 12, 1621. The famous English navigator James Cook was the first to call attention to the existence of northern lights, in the 18th century. In that era, it occurred to some scholars, such as the Englishman Graham in 1722 and the Swede Celsius

Cross section of the Earth's magnetosphe

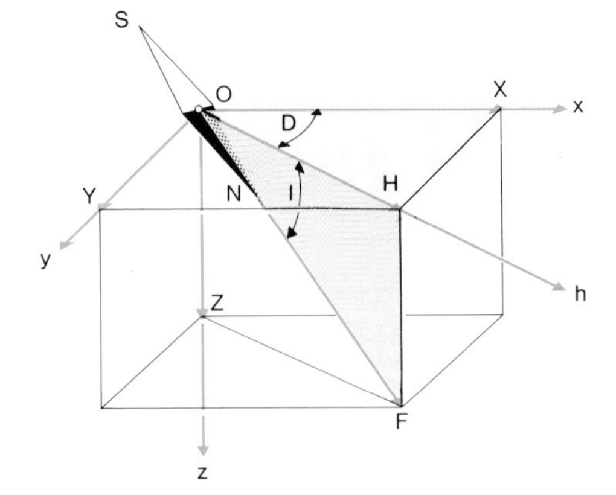

Components of the Earth's magnetic field.
Ox, horizontal axis, pointing north;
Oy, horizontal axis, pointing east;
Oz, vertical axis, pointing downward;
OF, field vector at point O;
OZ, vertical component of the field;
OH, horizontal component of the field;
D, declination;
I, inclination.

Components of the Magnetic Field

At any point on the Earth, the magnetic field may be defined by its intensity and its direction.

Often we use the vertical and horizontal components of the field rather than its total intensity—i.e., the field's projection on the vertical and on the horizontal plane, respectively.

The direction of the field may itself be broken down into two parameters: the declination, which is the horizontal angle between the field and the direction of geographical north; and the inclination, which is the angle between the field and the horizontal.

There are maps, updated periodically, that show the distribution on the surface of the globe of the field's intensity, declination and inclination.

The map of lines of equal value of the inclination, or isoclines, shows that the inclination, which is zero near the equator, increases by positive numbers toward the north and by negative numbers toward the south; the line along which the inclination is not coincident with the geographical equator; it defines the magnetic equator. The points where the inclination reaches ±90° are called the magnetic poles; the North Magnetic Pole is at Bathurst Island in the North of Canada; the South Magnetic Pole, near the French station at the Adélie Coast in Antarctica. These two poles are not antipodes. Their positions change slowly over time.

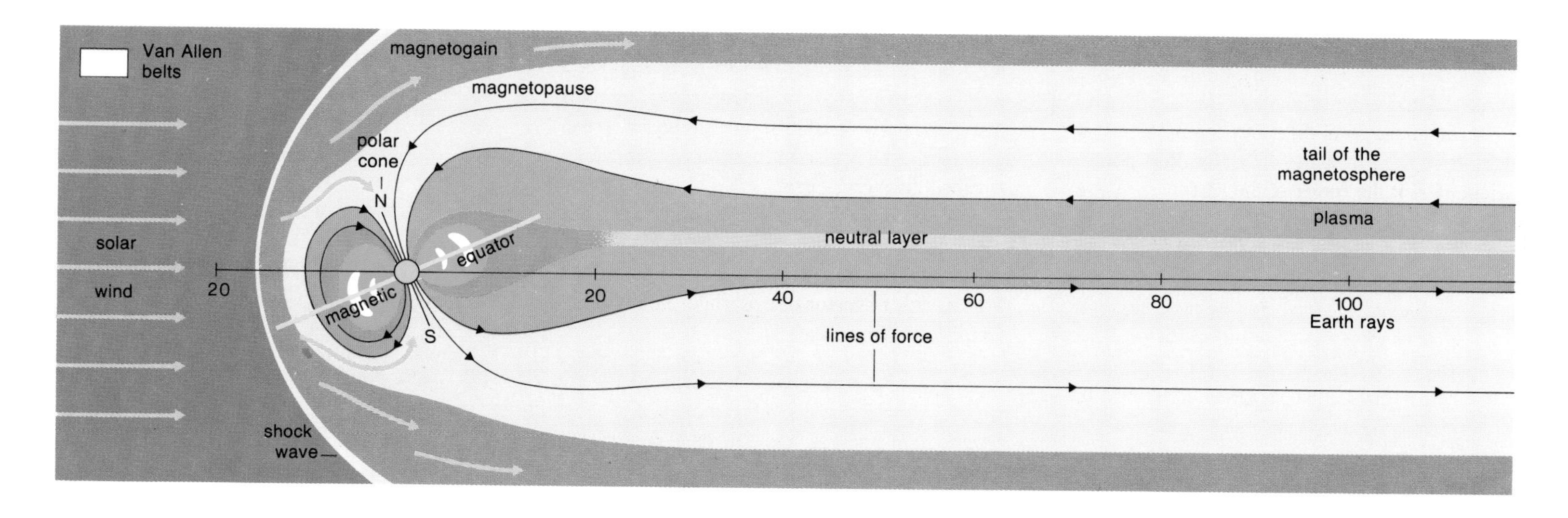

in 1741, to associate the polar lights with geomagnetic activity. However, it was not until 50 years later that such phenomena were understood: Their light is due to the excitation and ionization of atoms in the gases making up the upper atmosphere, at high altitudes, by the incoming flux of electrons and high-energy protons.

The lights may take various shapes; the most spectacular are those that unfold in rippling sheets, the folds of which follow the lines of force of the magnetic field along which the charged particles enter the atmosphere. Their lower edge is at an altitude of about 100 kilometers (62 miles), and they can extend over several hundred kilometers. In each hemisphere, they occur mainly within a zone known as the auroral oval, the position and extent of which vary with geomagnetic activity. In general, this region extends up to about 12° from the geomagnetic poles on the illuminated side of the globe and up to 22° on the dark side. But during times of high solar activity, it can descend below 50° latitude, and the northern lights may then be seen in the Northern United States and in Southern and Central Europe. In these areas, however, it is common to observe mere glimmers of these lights, which appear with a clarity analogous to the end of dusk, but in the direction of the magnetic pole. They may appear at any time of night and may last for several minutes or several hours. At times they are the prelude to a real dawn.

Although its origin has not yet been clearly established, the Earth's magnetism is very probably due to the existence of electric currents flowing in the planet's core (made up essentially of molten metallic matter) and maintained by a kind of dynamo effect (we do know that a current can generate a magnetic field, and vice versa). The main problem, as yet unsolved, is to determine how these currents are produced and maintained. According to the most commonly accepted hypothesis, the transfer of thermal energy in the Earth's internal layers helps to create convection currents in the core, which, combined with the planet's rotation, produce the presumed electric currents. According to a different theory, the molten masses making up the core do not follow the Earth's precession motion exactly, which would cause eddying motions within the core. These two hypotheses are not mutually exclusive, and both processes may be involved.

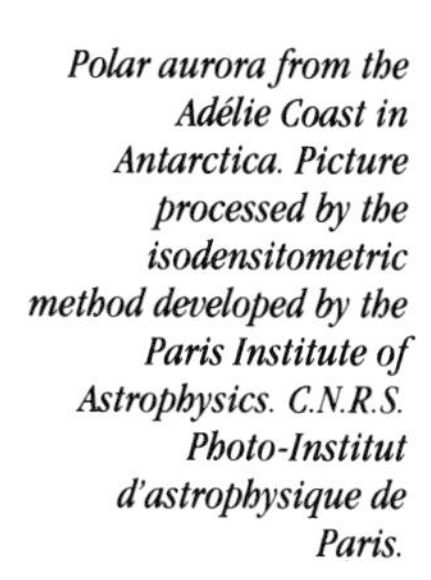

Polar aurora from the Adélie Coast in Antarctica. Picture processed by the isodensitometric method developed by the Paris Institute of Astrophysics. C.N.R.S. Photo-Institut d'astrophysique de Paris.

The lines of equal value of the field's intensity, or of its vertical or horizontal components, have the general appearance of isoclinic lines; the horizontal component, at its maximum near the magnetic equator, disappears at the magnetic poles; the vertical component, which is zero at the magnetic equator, is at its maximum near the magnetic poles; the total intensity, which is minimal near the magnetic equator (where the field is entirely horizontal), is maximal near the magnetic poles (where the field is entirely vertical).

The configuration of the lines of equal value of the declination, or isogones, is more complicated, especially in the polar regions; sometimes these lines cross both the magnetic and the geographic poles.

The intensity of the field usually is expressed in gauss or gammas (γ): 1 gauss = 100,000 gammas (and, moreover, 1 gauss = 0.0001 tesla).

The maximum value of the field, situated near the South Pole, is about 70,000 γ; near the equator, its intensity drops to about 33,000 γ.

*The inclination is considered positive when the north pole of the magnetic needle points downward.

10. The Moon

The Earth's only natural satellite, the Moon, is by far the natural body closest to our planet. Its spectacular appearance in the sky, the soft light it sheds on nocturnal landscapes, and the regular progression of its phases have been familiar to humanity for millennia. Long venerated as a goddess, the Moon today remains the symbolic ornament of night sung by the poets, while some of the innumerable beliefs and superstitions it has inspired throughout the ages remain alive.

From earliest antiquity, the Moon has been the subject of close scrutiny, its relative proximity making observation particularly easy. The Greeks before us were able to measure its dimensions and distance and to establish the laws of its apparent motion. The invention of the telescope early in the 17th century opened an extremely fruitful era for the study of its surface and physical characteristics. Finally, in the space age, man has undertaken to explore it directly, first by means of automatic remote-controlled devices, then by landing on the lunar surface.

Throughout the centuries, but more especially during recent decades, a considerable volume of information about the Moon has thus been accumulated, making it by far the best-known object after the Earth.

The Moon's Distance, Shape and Dimensions

The Moon's true nature was divined by the Greeks, beginning in the fifth century B.C. In about 450 B.C., Anaxagoras had the intuition that the planets and the Moon were of a rocky nature, like the Earth, and he therefore concluded—before anyone else, it seems—that the Moon shone with light borrowed from the Sun.

It is to the Greeks, also, that we are indebted for the first serious estimates of the Moon's distance and dimensions. Aristarchus of Samos, in the third century B.C., was able to measure the lunar diameter by measuring the time it took the Moon to cross the Earth's shadow during an eclipse; he got a figure that was too large, 4,600 kilometers, (2,859 miles) rather than 3,476 kilometers (2,160 miles). He also invented an ingenious trigonometric method for determining the Moon's distance from the observed value of its apparent diameter; this time he got a figure much lower than the actual one: 121,000 kilometers (75,202 miles) rather than 384,400 kilometers (238,658 miles) (on average). Hipparchus used these methods again a century and a half later with some improvements and obtained more satisfactory measurements: 42,000 kilometers (26,103 miles) and 425,000 kilometers (264,139 miles) respectively. But it was Ptolemy, in the second century A.D., who obtained the most accurate results: 3,700 kilometers (2,300 miles) for the diameter and 376,000 kilometers (233,686 miles) approximately for the distance, the errors then being only 5 percent and 2 percent, respectively.

After that, it was necessary to wait until 1751 for a more precise measurement of the Moon's average distance, through a triangulation operation carried out by the astronomers Lalande and La Caille, one in Berlin and the other at the Cape of Good Hope: The measurement yielded 384,700 kilometers (239,093 miles).

After 1946, with the help of radar, greater precision could be achieved; the best measurement of the Moon's average distance became 384,397 kilometers (238,904.3 miles), ± 1.2 kilometers, (.75 mile). Finally, since 1969, when reflectors were first placed on lunar soil during the Apollo missions, laser telemetry has made possible determinations of the distance between the Earth and the Moon, at a given instant, to within a mere 30 centimeters (11.8 inches).

We now know that the Moon's average distance lies between 384,400 and 384,401 kilometers (238,906.2 and 238,906.8 miles), with an uncertainty of a few hundred meters. This is, of course, an average figure, for, owing to the ellipticity of its orbit, the Moon's distance varies each month between 356,500 kilometers (221,566 miles) (at perigee) and 406,800 kilometers (252,828 miles) (at apogee). The Greek astronomer Hipparchus had already appreciated these discrepancies in distance more than 2,000 years ago, ϕ noting that the Moon's apparent diameter (equal to 31′26″, on average), varies by about 10 percent during the lunar cycle.

The lunar body itself was long considered as spherical, since observations made from Earth showed only a tiny flattening (0.00017). The measurements taken since the 1960's, with the help of artificial satellites orbiting the Moon, or by laser telemetry, have shown that relative to an average radius of 1,738 kilometers (1,080 miles) used as a reference, the lunar hemisphere that faces the Earth is rather flattened, while the hidden hemisphere is slightly raised, though the difference is no more than about 4 kilometers (2.5 miles). Thus there is a slight displacement of about 2 kilometers (1.2 miles) between the Moon's geometric center and its center of gravity. Despite its being raised, the Moon's rear hemisphere has a large depression almost in its center of about 1,400 kilometers (870 miles) in diameter and 5 kilometers (3.1 miles) in depth, while the hemisphere that faces us shows a local 3 kilometers (1.9 mile) superelevation at the Apennine Mountains chain.

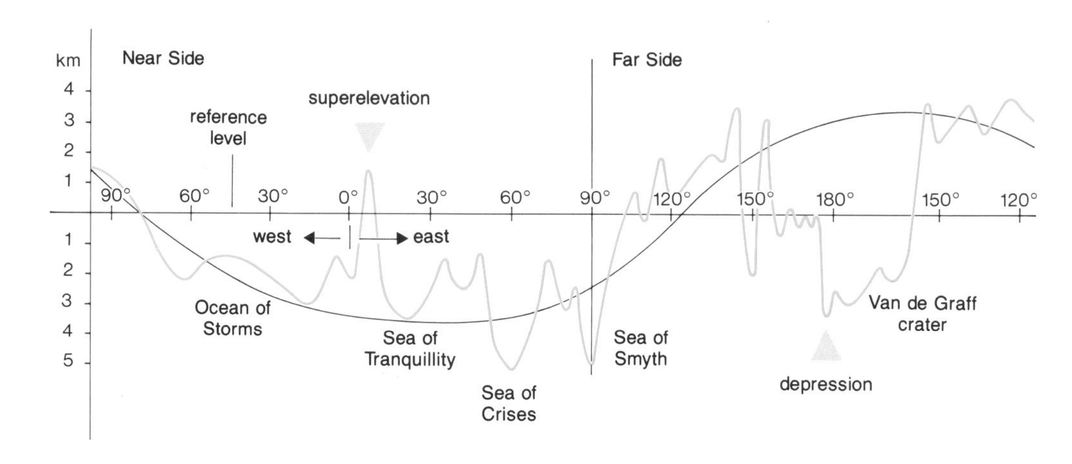

Profile of the lunar surface determined by laser telemetry. From P. Kohler, l'astronomie, *December 1973, page 425.*

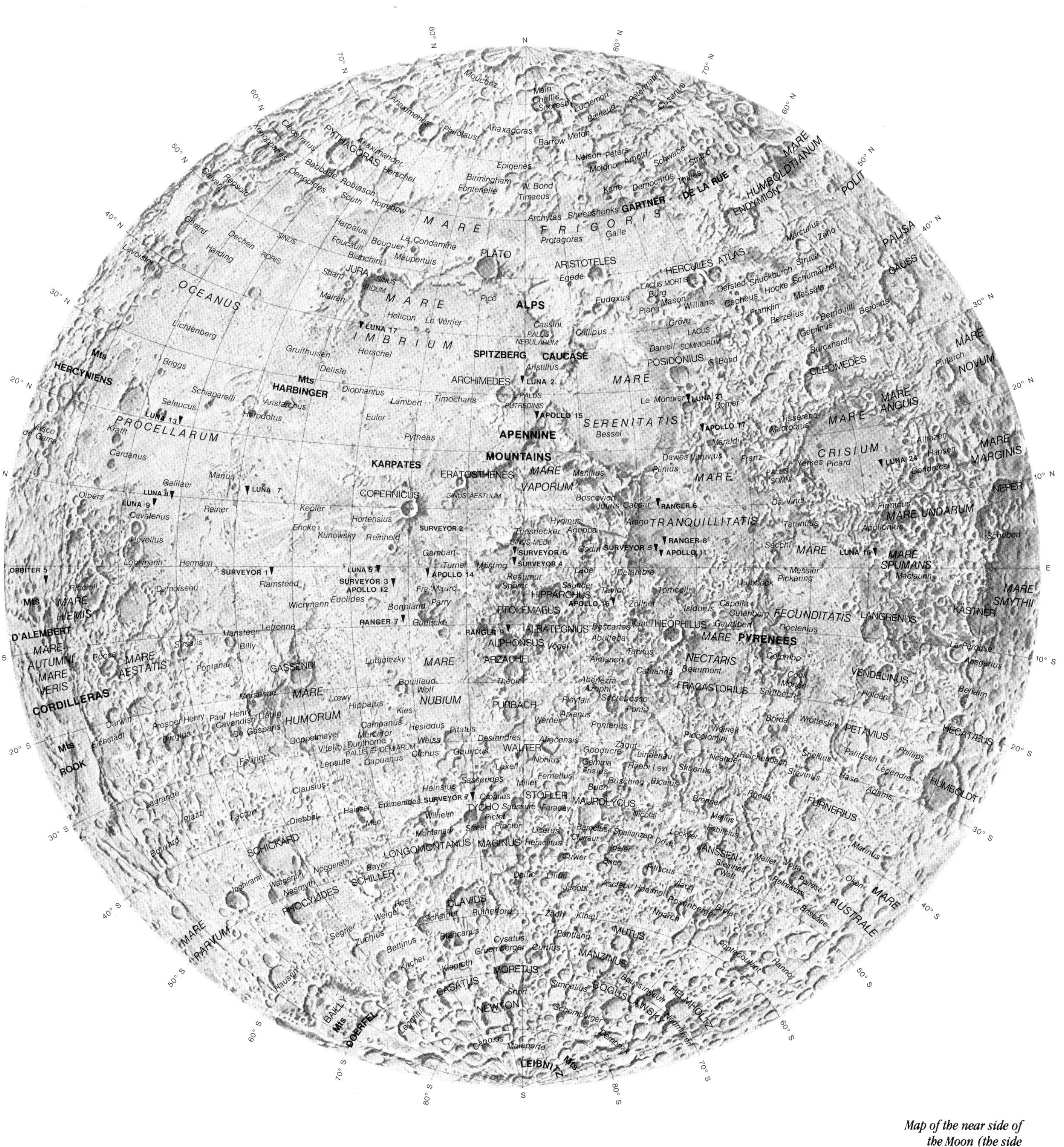

Map of the near side of the Moon (the side visible from Earth), with the official nomenclature of the principal features of relief. I.G.N. Photo.

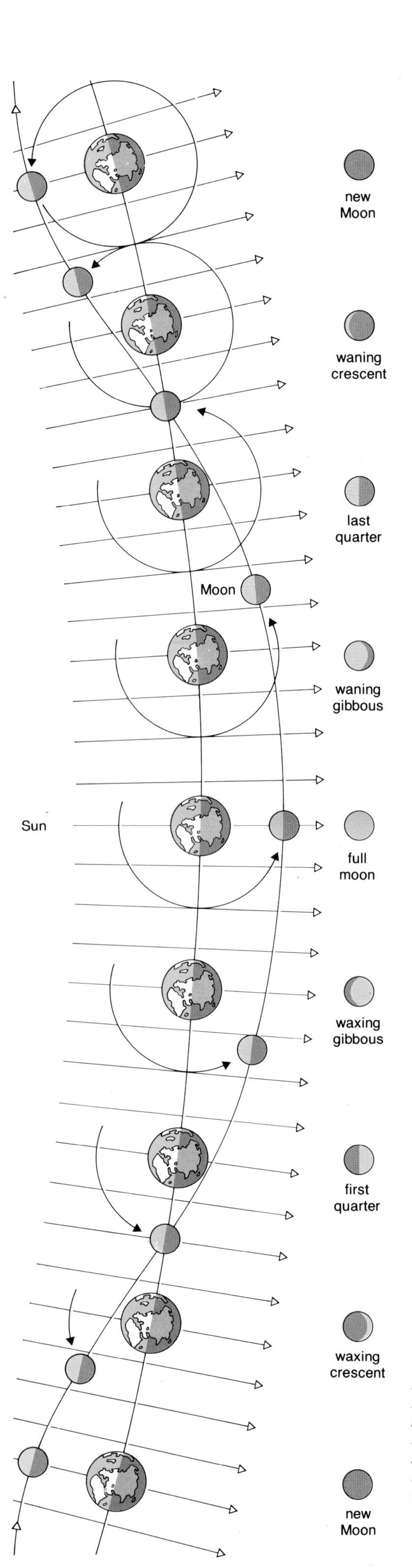

Phases of the Moon. Relative positions of the Moon, the Earth, and the Sun, and aspects of the Moon seen from the Earth during the lunar cycle.

The Moon's Motion Around the Earth

The Moon's motion around the Earth is governed by the attraction of our planet. However, it is subject to major perturbations by the Sun and planets (which act either directly on the Moon itself, or indirectly on its orbit, through their effect on the Earth's orbit).

Precise calculation of the Moon's position is therefore a thorny problem, which mathematicians and astronomers have striven to solve since the 18th century: Clairaut (1752), Euler (1753 and 1772), d'Alembert (1754), Laplace (1802), Damoiseau (1820), Pontécoulant (1846), etc.

The first truly accurate theory of the Moon's motion was published by Delaunay between 1860 and 1867. Current ephemerides are based on those of Brown published between 1897 and 1908. However, new theories are being developed, for modern technology—particularly laser telemetry—makes new progress possible.

As a first approximation, we may assume that the moon obeys Kepler's laws and that it revolves around the Earth, following an elliptical orbit at a speed of 1,023 kilometers (636 miles) per second, inclined by 5°9′ from the ecliptic, with a semimajor axis equal to 384,400 kilometers and an eccentricity of 0.0549. But all of these values are merely averages, considering the many variations that describe the Moon's real motion. We note in particular:

- a variation of the inclination of the plane of the lunar orbit from the ecliptic of between 5°0′ and 5°18′, with a period of 173 days (a

Phases of the Moon

The regular cycle of the Moon's phases is the most familiar astronomical phenomenon aside from the alternation of day and night.

The way it works is simple. The Moon does not emit light by itself; it merely reflects and scatters a fraction (about 7 percent) of what it receives from the Sun. Therefore, it always offers a dark hemisphere (opposite the Sun) and a bright hemisphere (facing the Sun). Since, in addition, it orbits the Earth, it takes on different appearances depending on its position relative to the Sun and to us.

When it is in conjunction with the Sun—i.e., between the Sun and the Earth—its dark hemisphere faces us. Then it is invisible—this is the new Moon. Then, little by little, as it moves around the Earth, its bright hemisphere becomes visible to us. At first, all we see of it is a bright crescent, visible at night. Day by day, this crescent thickens and remains visible later and later. About a week after the new Moon, the bright half of the lunar hemisphere can be seen: This is the first quarter, observable during the entire first half of the night. Continuing to change its position relative to the Sun and the Earth, the Moon next takes on an intermediate appearance (gibbous Moon), which evolves into the full Moon. At this stage, the Moon is in oppositon to the Sun—i.e., opposite the Sun from the Earth—and the Moon's entire bright hemisphere is turned toward our planet. The Moon then appears as a bright disk that shines all night. Then darkness gradually overtakes the areas that appeared first, after the new Moon. About seven days after the full Moon, we once again see only half the disk: This is the last quarter, observable during the second half of the night. Then, as the days go by, the visible portion is reduced to a crescent, which gradually narrows. When it finally disappears, the bright side of the Moon has returned to face the Sun. A new cycle begins.

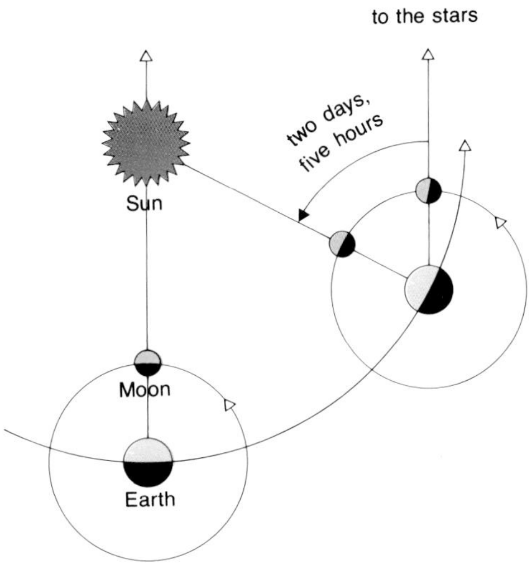

Difference between sidereal and synodic revolution. Because of the Earth's movement around the Sun, the time it takes the Moon, seen from the Earth, to return opposite the Sun (synodic revolution) is greater by two days, five hours than the time it takes the Moon to return opposite a distant star (sidereal revolution).

phenomenon similar to nutation, described on page 131)

- a rotation of the line of the nodes, in the plane of the ecliptic, in the retrograde direction, with a period of 6,793.5 days, or about 18.60 years (a phenomenon similar to precession, explained on p. 131)
- a rotation of the major axis of the orbit (or line of apsides), in the plane of the orbit, in the forward direction, with a period of 3,232.4681 days (about 8 years, 310 days) relative to the stars; this nonuniform motion is accompanied by a complex oscillatory motion with a period of 412 days and an amplitude of 12.33°.
- a variation in the eccentricity of the orbit of between 0.0666 and 0.0432, with a period of 412 days.

We can thus define several periods of lunar revolution having different lengths because they are based on different definitions.

The sidereal period is the interval between two consecutive conjunctions of the Moon with the same star—i.e., the period at the end of which the Moon returns opposite the same star in the sky. It equals, on average, 27 days, 7 hours, 43 minutes, 11.5 seconds.

The synodic period is the interval between two consecutive conjunctions of the Moon with the Sun—i.e., two consecutive new Moons. It equals, on average, 29 days, 12 hours, 44 minutes, 2.8 seconds. It is because of the Earth's motion around the Sun that it is longer (by about 2 days, 5 hours) than the sidereal revolution.

The tropical period is the interval between two consecutive conjunctions of the Moon with the vernal equinox (see page 237). It equals, on average, 27 days, 7 hours, 43 minutes, 4.7 seconds. It is because of the retrograde motion of the vernal equinox on the ecliptic that it is shorter by about 7 seconds than the sidereal period.

The anomalistic period is the interval between two consecutive passages of the Moon through its perigee. It equals, on average, 27 days, 13 hours, 18 minutes, 33.1 seconds. It is because of the slow shifting of the perigee, in the same direction as the Moon's orbital motion, that it has a duration longer (by about 5 hours, 35 minutes) than the sidereal period.

The draconic period is the interval between two consecutive passages of the Moon through the rising node of its orbit (its name comes from the fact that the nodes of the lunar orbit formerly were called the head and tail of the Dragon). It equals, on average, 27 days, 5 hours, 5 minutes, 35.8 seconds. It is because of the slow retrograde rotation of the line of nodes, in the plane of the ecliptic, that it is shorter (by about 2 hours, 38 minutes) than the sidereal period.

Added to the variations already mentioned in the elements of the Moon's orbit are a large number of nonuniformities (Brown studied nearly 1,500 of them, and the calculation of the Table of Lunar Positions includes about 500!) in the motion of the body itself, due to influences from the Sun or to variable configurations in the Sun-Earth-Moon system. The largest, already known to Ptolemy, is evection, which has a period of 31.812 days and produces a deviation of 1°19′ from the Moon's uniform motion, giving it a possible advance or delay of 2 hours, 20 minutes in its orbit.

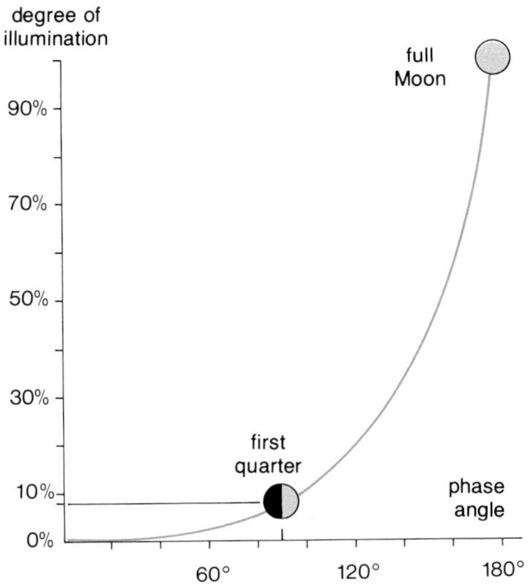

Variation in the Moon's illumination as a function of phase.

The cycle of lunar phases, or lunation, takes about 29.5 days. It provided the peoples of antiquity with a natural time scale between a day and a year: the month. But the ancients had many difficulties in establishing a calendar according to the seasons, since the year—defined by the Earth's orbit around the Sun—does not contain a whole number of lunations. The problem was solved by adopting months of 30 and 31 days.

The Greek astronomer Meton discovered in 432 B.C. the smallest common multiple between the solar year and the lunar month by noticing that after 235 lunations, or exactly 19 years, the Moon entered the same phase on a given date.

The phases around the new Moon are accompanied by the phenomenon known as earthshine, or ashen light. The dark part of the Moon remains visible for a few days with a grayish color, with a decreasing brightness as the crescent phase develops (in very small crescent phases, the deepest craters and highest relief on the dark side of the Moon can be clearly seen). The earthshine phenomenon is due to the part of the Earth lit up by the Sun. Indeed, just as the Moon is seen in different phases from the Earth, our planet, seen from the Moon, appears in various configurations, the phases of the two objects always being complementary. Thus during the new Moon, there is a "full Earth," and the earthshine seen at the Moon at such times is 45 times more intense than the light the full Moon sheds on us: ashen light merely is this light, partly reflected by the Moon.

How do we recognize the phases of the moon?

	New Moon (NM)	First Quarter (FQ)	Full Moon (FM)	Last Quarter (LQ)
aspect				
position in the sky	near the Sun	at 90° away from the Sun	opposite the Sun	at 90° away from the Sun
rising	at dawn	at noon	at sunset	at midnight
setting	at sunset	at midnight	at dawn	at noon
times of visibility	invisible	late afternoon and evening	all night	second half of the night and early morning

Another major nonuniformity, discovered by Tycho Brahe in 1582, is the variation whose period is one-half the synodic period, or 14.77 days, and that, at the midpoint between the full or new Moons and the quarters, leads to advances and delays of 72 minutes in the Moon's orbit.

There is also a very slow acceleration in the Moon's motion each century, first detected by Halley in 1693. This effect is partially due to the variations in the eccentricity of the Earth's orbit, but especially to a slowing of the Earth's rotation around its axis, which increases the length of the day. Because of this acceleration, the Moon now is moving away from the Earth, but the full theory shows that the phenomenon is merely transitory and will subsequently reverse itself. Moreover, the variation is very small: barely 2 meters (6.6 feet) per century.

The Moon's Rotation

Like all bodies of the solar system, the Moon rotates on its own axis, but its rotation has a notable peculiarity: It coincides exactly in period and duration with the sidereal revolution, which means that the Moon always shows the same face to the Earth. Such synchronism is not the product of chance. It results from the shape of the Moon, which, as mentioned earlier, is not strictly spherical, but ellipsoidal. At some point in the distant past, its long axis came to rest pointing in the direction of the Earth, following a well-known stabilization process used in the positioning of certain artificial satellites (stabilization "by gravity gradient"), and rotation takes place around the shorter axis.

If the Moon's orbit were perfectly circular, hence characterized by a uniform motion, and if its axis of rotation were fixed and perpendicular to the plane of that orbit, we on Earth would always see the same exact half of the lunar surface. In fact, none of these conditions applies, and as a result, we see the Moon undergo apparent oscillations known as librations.

Since the Moon's rotation speed is constant, while its speed of revolution varies between the perigee (where it is a maximum) and the apogee (where it is a minimum), the rotation is sometimes ahead of and sometimes behind the orbital motion. The result of this is a longitudinal libration of the Moon, which over a month makes it possible to view an additional 8° slice on both sides of the average visible surface (or 16° in total longitude, amounting to 4 percent of the total surface).

Moreover, the Moon's axis of rotation is not strictly perpendicular to the plane of its orbit but forms an angle of about 83°19′ with that plane, creating a latitudinal libration, which alternately reveals, at two-week intervals, a small region around each pole that amounts to 6°50′ in latitude.

Finally, there exists a diurnal libration due to the motion of the observer located on the surface of the rotating Earth: Because of a changing perspective on the nearby Moon, we do not see exactly the same lunar surface when the Moon is setting as when it is rising, but the difference is no more than 1°.

The combination of these various librations makes it possible to observe about 59 percent of the lunar surface from Earth.

The Moon's Mass

Kepler's laws governing the Earth's motion around the Sun actually apply to the motion of the center of gravity of the Earth-Moon system. It is around this point, in the interior of the Earth, that the Moon orbits. The orbit of the Earth itself is similar, but smaller by the ratio of the Moon's mass to that of the Earth-Moon system. This motion leads to an observable effect on the apparent motion of the Sun, which reaches the meridian with a gain or delay of about 0.43 second relative to its mean motion. By measuring this interval it is possible to calculate the Moon's mass, which is 1/81 that of the Earth. A better calculation was made using observations of the asteroids. The most accurate measurement thus obtained of the ratio of the Earth's mass to that of the Moon is 81.45. Finally, the study of the motion of artificial satellites orbiting the Moon has made possible a still more precise measurement: 81.301 ± 0.003.

The moon's average density (3.34 gm/cm³) appears clearly lower than that of the Earth (5.52). It shows that the Moon's average composition is different from that of our planet.

The force of gravity on the Moon's surface is relatively small: It is only about 16 percent of what

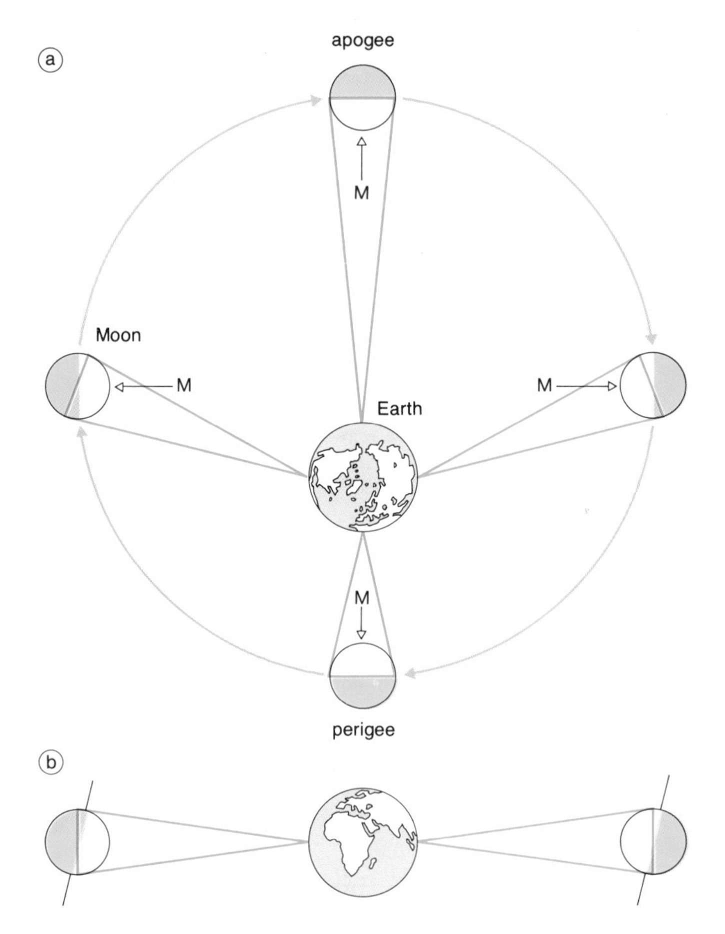

Lunar libration. Small apparent oscillations of the Moon enable us to observe slightly more than half its surface over time. Lick Observatory Photo.
a. longitudinal libration: This results from the fact that the Moon, whose orbit is elliptical, has a higher orbital speed when it is closer to the Earth, whereas its rotation around its axis is extremely regular. The Moon's center (M) thus appears to oscillate toward east or west.
b. latitudinal libration: This results from the tilt of the Moon's axis relative to its orbit and makes it possible to see the regions around the two poles alternately at two-week intervals.

we experience on the Earth's surface. Thus bodies are six times lighter on the Moon than on our planet. This property enabled the Apollo astronauts to move about on the lunar surface without too much difficulty, despite their bulky spacesuits and the large oxygen tanks they carried on their backs.

Anomalies in the Moon's gravitational field have been detected since the 1960s, through study of the trajectories of vehicles placed in orbit around the Moon (see page 147).

Selenography

The descriptive study of the lunar surface, or selenography, began in the 17th century, after the invention of the telescope. In 1609, Galileo was astonished to discover the existence of the lunar relief, with the help of an instrument of his devising. Many astronomers after him sought to draw maps of the Moon: the Frenchman Claude Mellan, in 1636; the Belgian Langrenus, in 1645; the Pole Hevelius, in 1647; the Italians Riccioli and Grimaldi, in about 1650. The drawings of the lunar surface made by Jean Dominique Cassini at the Paris Observatory between 1671 and 1679 were the basis for an engraved map, 54 centimeters (21 inches) in diameter, that remained unequalled for a century. The work of the Germans Schroeter (*Selenographiosche Fragmente,* 1791 and 1802), Beer and Maedler (*Der Mond,* 1837), and Schmidt (who in 1874 drew a map having a diameter of 1.8 meters [5.9 feet]) should also be mentioned.

At the end of the 19th century, photography replaced visual observations. The *Atlas photographique de la Lune,* assembled between 1896 and 1909 by Loewy and Puiseux at the Paris Observatory (using an equatorial coudé with a 60-centimeter [23.6-inch] aperture) and a similar work created in 1960 in the United States under the direction of the American Kuiper, based on photographs taken at four large observatories, are the best reference materials in this field.

However, we can photograph from Earth only the side of the Moon that is visible from our planet. One of the advantages of space exploration

Vital Statistics of the Moon

PHYSICAL CHARACTERISTICS	
Average diameter:	3,476 km. (0.2725 times the Earth's equatorial diameter)
Apparent diameter:	
Minimum:	29′22″
Maximum:	33′29″
Mass:	$73.4 \cdot 10^{21}$ kg. (0.0123 times that of the Earth)
Volume:	$22 \cdot 19^{9}$ km.^{3} (0.0203 times that of the Earth)
Average density:	3.34 (water = 1)
Average albedo:	0.073
Gravity at the equator:	1.627 m./sec.^{2} (0.166 times that of the Earth)
Escape velocity:	2.37 km./sec. (8,550 km./hr.)
Visual magnitude:	−12.7
Temperature at the surface:	about + 120° to − 180°C
Average atmospheric pressure at the surface:	10^{-11} torr
Intensity of the magnetic field at the surface:	6 to 313 γ
ORBITAL CHARACTERISTICS	
Semimajor orbital axis (average distance from the Earth):	384,400 km. (60.26659 terrestrial equatorial radii)
Average eccentricity of the orbit:	0.0549 (the eccentricity varies from 0.0666 to 0.0432 with a period of 12 days)
Minimum distance at the perigee:	356,375 km.
Maximum distance at the apogee:	406,720 km.
Average inclination of the orbit from the ecliptic:	5.1453° (the inclination varies by 0.30° around its average value with a period of 173 days)
Inclination of the lunar equator from the ecliptic:	1°32′
Inclination of the lunar equator from the orbit:	6°41′
Sidereal period of revolution (return to the same position in the sky, relative to the stars):	27.3216609 days, or 27 days, 7 hr., 43 min., 11.5 sec.
Synodic period of revolution (return to the same position relative to the Sun = lunation):	29.5305881 days, or 29 days, 12 hr., 44 min., 2.8 sec.
Tropical period (return to the same position relative to the vernal equinox):	27.3215816 days, or 27 days 7 hr., 43 min., 4.7 sec.
Anomalistic period (return to the perigee):	27.5545502 days, or 27 days, 13 hr., 18 min., 33.1 sec.
Draconic period (return to the ascending node):	27.2122178 days, or 27 days, 5 hr., 5 min., 35.8 sec.
Sidereal period of rotation:	strictly equal to the sidereal period of revolution
Libration	
latitudinal:	6.8°
longitudinal:	7.7°
diurnal:	1.0°

has been to make possible global lunar cartography. The pioneer in this endeavor was the Soviet spacecraft *Luna 3,* which, in 1959, transmitted the first photographs of the far side. There now exist photographic atlases of the Moon based on photographs taken from lunar orbit by the American Lunar Orbiter spacecraft or during the various Apollo missions, which show details as small as 1 meter (3.3 feet).

Selenographic Coordinates

The names of lunar features are very useful for pinpointing locations. However, to locate a point on the Moon's surface with precision, it is necessary to make use of selenographic coordinates, which are similar to geographic coordinates (latitude and longitude) used on Earth. The figure shows how these coordinates are determined. The reference circle is the lunar equator e, whose plane is perpendicular to the Moon's axis of rotation ω. This axis crosses the lunar surface at the North Pole N and at the South Pole S. According to a decision made in 1961 by the IAU, the east–west orientation is determined from the standpoint of an observer on the Moon for whom the Sun rises in the east and sets in the west. Thus, when we observe the Moon with the naked eye or with a telescope equipped with an image-correcting device, north (N) is on top, south (S) is at the bottom, east (E) is to the right and west (W) is to the left.

Any point on the lunar surface is determined by:
• its selenographic longitude (λ): the angle between the central meridian (m) and the plane containing the meridian that goes through the given point (it is between 0° and 180° and is considered positive in the east, negative in the west; in place of the signs + and −, we often use the symbols E [east] and W [west]; thus 60 W means 60° selenographic longitude west, or −60°)
• its selenographic latitude (β): the angular distance of the given point from the equator, measured as the parallel passing through that point (it is between 0° and 90° and is considered positive in the north, negative in the south; in place of the signs + and −, we often use the abbreviations N [north] and S [south]; thus 30 S means 30° selenographic latitude south, or −30°)

One parameter useful to observers and whose daily value can be found in ephemerides, is colongitude, or the selenographic longitude of the morning terminator (that is, the regions of the

The face of the Moon's far side has been revealed by spacecraft. The craters are as abundant there as on the near side (the side visible from Earth), but there are far fewer "seas": The main one is the Moscow Sea, visible (at left) in this photograph taken on August 15, 1967, by the American satellite Lunar 5 *orbiter from an altitude of about 1,250 kilometers (777 miles). Photri Photo.*

The Nomenclature of Lunar Relief

The current nomenclature of lunar relief is based on that proposed in the 17th century by an Italian Jesuit, Father Joannes Baptista Riccioli, who in 1651 published (in his *Almagestum novum*) one of the first maps of the Moon.

The craters bear the names of scholars, philosophers, writers or artists of the past: Ptolemy, Plato, Descartes, Laplace, Newton, Flammarion, Lyot, etc.

The maria were given names of universal nature. With a few exceptions, those on the half of the lunar disk visible in the first quarter were considered favorable areas (Sea of Tranquillity, Sea of Serenity, Sea of Fertility, Sea of Nectar), while those on the half visible in the last quarter were regarded as inhospitable (Sea of Storms, Sea of Rains, Sea of Cold, Sea of Clouds).

Finally, for the mountains, the names of terrestrial massifs were adopted, taking into account any possible resemblances: Alps, Apennines, Carpathians, Caucasus, Pyrenees, etc.

In 1935, the first international list of nomenclature, containing 672 names, was published. The International Astronomical Union (IAU) is the only authority competent to add to or change it.

The development of lunar spacecraft in the 1960s, and the resulting cartographical advances, led to the extension and refining of this nomenclature, which now includes nearly 1,500 names.

In 1961, the IAU accepted the first names of formations on the far side of the moon, photographed two years earlier by the Soviet probe *Luna 3*. In 1964, a total of 66 new names were added, and the IAU decided on a systematic latinization of the designations of topographical formations (*mons:* mountain; *montes:* chain or mounts; *rupes:* escarpment; *rima:* rille, fault; *vallis:* valley; *mare:* sea; *maria:* seas; *promontorium:* cape; *lacus:* lake; *palus:* swamp; *sinus:* gulf). In 1970, a total of 513 new designations, pertaining to formations on the far side, were adopted.

In 1973, the IAU accepted 53 new names and set about reforming the nomenclature, which was poorly adapted to detailed cartography. The Moon now is divided into 144 regions defined by a network of parallels and meridians corresponding to the data in the NASA Lunar Aeronautical Chart, at 1/1million. Each region is itself divided into 16 provinces, identified by the combination of a letter (A, B, C, or D) and a digit (1, 2, 3, or 4). In addition, each province bears the name of a particular crater it contains.

Moreover, new Latin appellations have been introduced: *dorsum* (plural, *dorsa*) for the marine ridges; *fossa* (plural, *fossae*) for the fissures in the shape of a ditch; *anguis* (plural, *angues*) for the meandering fissures; and *catena* (plural, *catenae*) for the chains of craters.

A very large and very old lunar cirque, Clavius (diameter: 230 kilometers [143 miles]), photographed with the 5-meter (200-inch) telescope at Mount Palomar Observatory. Note the very eroded ramparts, sprinkled, like the bottom of the crater floor, with more recent craters. Mount Palomar Observatory Photo.

lunar surface on which the Sun rises). It is measured from the central meridian, in a western direction, from 0° to 360°.

Major Features of Lunar Relief

Seen with the naked eye, the Moon's surface appears to be covered with dark spots. These actually represent vast, slightly indented plains, more or less hilly and ringed by mountains. Called seas (or *maria)* by the early observers, who thought they were bodies of water, they retain that inappropriate nomenclature to this day. Some of them have a well-defined circular or oval shape (Sea of Crises, Sea of Serenity, etc.); others, however, have an irregular contour (Ocean of Tempests, Sea of Rains, etc.) and branch off into "gulfs," "capes," "lakes" or "swamps." There are 22 of them on the Moon as a whole, but only a minority are located on the Moon's far side, the principal one being the Sea of Moscow, revealed for the first time in 1959 by photographs taken by the Soviet spaceship *Luna 3.* In fact, most of the seas on the far side are visible from Earth, some of them only partially (Southern Sea, Marginal Sea, Eastern Sea, Smyth Sea). This asymmetry in the distribution of seas probably results from the difference in the thickness of the crust between the two lunar hemispheres (see page 153).

Precise studies of the trajectories of satellites orbiting the Moon has revealed the existence of positive anomalies in the lunar gravitational field, directly over five circular seas (Sea of Rains, Sea of Serenity, Sea of Crises, Sea of Nectar and Sea of Humors). These anomalies reflect the presence of large concentrations of matter under these seas, to which the name mascons (for mass concentrations) has been given. The largest mascon is that of the Sea of Rains, which causes a gravitational disturbance of +23 milligals, representing about one-seventh of the intensity of lunar gravity. Next comes the mascon of the Sea of Serenity (+18 milligals); then, with similar anomalies, those of the Sea of Crises and the Sea of Nectar (+10 milligals).

In contrast to the seas, the open regions of the lunar surface are called continents (or *terrae).* These are extremely hilly areas saturated with craters (see following material). There are many mountainous formations, sometimes grouped in actual chains, particularly on the edge of the seas. Some peaks reach high altitudes, with the record held by the Leibnitz Mountains, near the South Pole, which rise 8,200 meters (26,904 feet) above the surrounding areas (on the Moon, there is no general reference level for measuring altitudes and depths comparable to the average sea level used on Earth). But the lunar mountains are similar to the old terrestrial massifs: without peaks or needles, they always have rounded tops and soft

Taking of lunar soil samples by the astronaut and geologist Harrison Schmitt during the Apollo 17 *mission in 1972. In the background the Taurus Mountains can be seen, with their softly rounded shapes. I.P.S. Photo.*

This picture of the Langrenus crater (diameter: 140 kilometers [87 miles]), east of the Sea of Fertility, was taken during the Apollo 8 *mission. In the center, the presence of a double mountain mass is characteristic. The walls, which rise above the crater floor by about 2,500 meters (8,203 feet), are layered in terraces; they are extremely uneven and pitted with small craters, which testifies to the extreme age of this formation. Photri Photo.*

A "young" crater, Copernicus (diameter: 90 kilometers [56 miles]), with its halo of bright, radiant streaks, and with the strings of small craters surrounding it. At right we note a smaller crater of the same type, Eratosthenes and, north of Copernicus, the Carpathian range. Mount Wilson Observatory Photo.

shapes, as revealed by space exploration.

The most characteristic formations of the lunar surface are circular or polygonal depressions of various dimensions: the craters or cirques (the latter term referring particularly to the largest craters, bordered by mountainous crater walls, or ramparts). The largest exceed 200 kilometers (124 miles) in diameter (Bailly reaches 270 kilometers [168 miles], Deslandres 260 kilometers [162 miles], Clavius 230 kilometers [143 miles]), but the smaller ones visible from Earth do not exceed 1 kilometer (62 miles), and space exploration has revealed a multitude of craters whose dimensions are expressed in meters and even in centimeters. In the 19th century, the astronomer Julius Schmidt counted more than 30,000 of them, but it is now believed that the visible side of the Moon alone contains about 10 times more with a diameter of more than 1 kilometer.

The largest cirques, whose diameter is greater than or equal to 100 kilometers (62 miles), always possess demolished, disarrayed and eroded walls that slightly overhang the surrounding areas. Their interior is often filled with solidified lava, giving them the appearance of a smooth plain, riddled, like the seas, with craterlets—hence the name "walled plain" by which they are frequently known. The height of their surrounding wall always is very low in comparison to their diameter (the ratio generally ranges from 1/50 to 1/100),

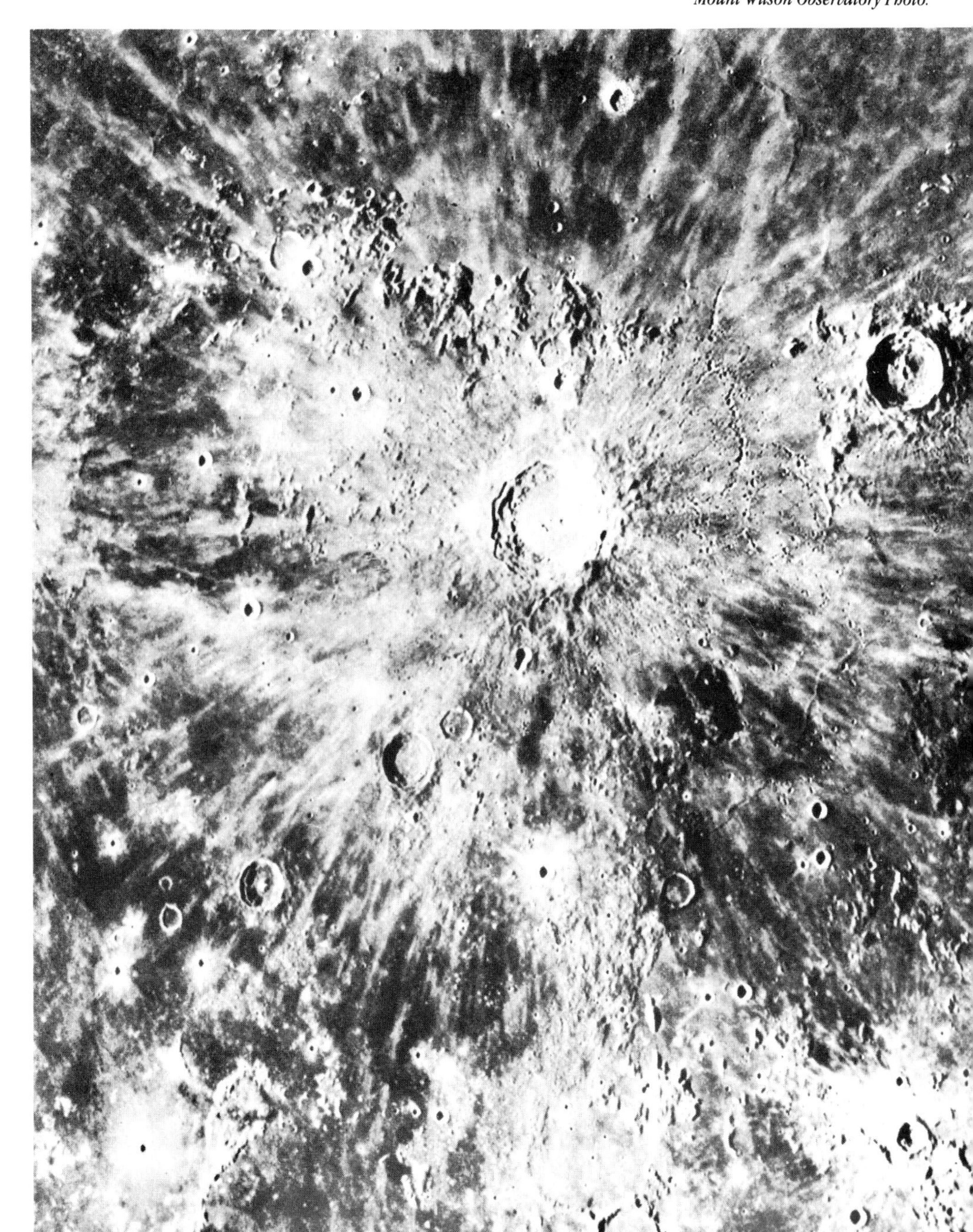

Several famous and characteristic lunar craters: In the center, from top to bottom: Ptolemy (diameter: 150 kilometers [93 miles]), Alphonse (diameter: 120 kilometers [75 miles]) and Arzachel (diameter: 100 kilometers [62 miles]); along the terminator, at right, partly in shadow: Hipparchus (diameter: 150 kilometers [94 miles]) and Albategnius (diameter: 140 kilometers [87 miles]). We note the mountainous pillars in the center of Alphonse, Arzachel and Albategnius. Below left we see (light line) the Straight Wall. Lick Observatory Photo.

so that an observer at the bottom, several dozen kilometers from the foot of the wall, would not realize that he is in a cirque; in many cases he would not even see the rim of the wall, owing to the curvature of the lunar globe. The cirques with a diameter ranging from about 20 to 100 kilometers (12.4 to 62 miles) often are called mountain rings. Their walls may show all possible signs of erosion. Some, like Catherine, have heavily eroded walls, like the very large cirques; most, however, like Tycho or Copernicus, have a massive, elevated rampart, made up of a series of terraces or concentric rings, relatively steep on the inner side, where the floor is at a lower level than the surrounding surface, the missing volume being precisely that of the matter constituting the rampart. In their center we often observe one or more mountainous peaks that are lower in altitude than the crater rim. Sometimes the interior of the crater is filled, at least partly, with solidified lava, and the central peak is then nonexistent (Archimedes, Plato) or not very noticeable; the lava may even entirely fill the interior of the crater (Wargentin). Finally, the rim is not always circular; often they have a rather pronounced hexagonal shape. Craters may have a relatively large depth, reaching up to 7,250 meters (23,787 feet) (Newton). The depth/diameter ratio generally is in the range of 1/15 to 1/20.

As for small craters—those with diameters less than or equal to 20 kilometers (12.4 miles)—they generally are perfectly circular, and their ramparts show neither concentric terraces nor any sign of decay. Their inner side is steep, while on the outside they have a rather gentle slope. Their interior floor is shaped like a bowl. The depth/diameter ratio, clearly larger than for the large craters, may be as much as 1/5. The tiniest craters (craterlets and craterlike depressions) randomly pockmark the surfaces of the "seas" and "continents." The relative ages of craters may be determined from the appearance of their rims (the relative sharpness of their ridges, the relative erosion of their walls, etc.) and by other criteria, such as whether the bottom is filled, whether they are covered by other craters, etc. In particular the great walled plains with demolished walls appear to be very old formations, while some terraced craters with a perfectly preserved surface are obviously much younger: The age of Copernicus is estimated at about 900 million years, that of Tycho at only 250 million years (which represents only 5 percent of the Moon's age).

The origin of these craters has long been a subject of debate. They were initially thought to be volcanoes, but that hypothesis had to be abandoned: A volcano is a cone widened by a crater only at the peak, while a lunar cirque is a huge depression surrounded by ramparts. Moreover, there are formations on the Moon that in no way resemble cirques but, on the other hand, have all the appearances of extinct volcanoes: These are domes, several kilometers in diameter and a few hundred meters in altitude, the top of which is pierced by an opening about 1 kilometer in diameter (we see them, in particular, north of the Hortensius crater, not far from Copernicus). Since

Transitory Lunar Phenomena

In 1958, the Soviet astronomer N. A. Kozyrev caused a sensation by reporting that he had observed a gaseous eruption emitted by the central peak of the Alphonse crater. In fact, a spectrum of the crater, obtained by Kozyrev on November 3, 1958, with the spectrograph of the Crimea Observatory telescope, which has a 125-centimeter (49-inch) aperture, does show unusual emission bands in the blue, located precisely at the position of the central peak. According to the Soviet spectroscopist Kaliniak, the bands are similar to those found in comet spectra and are due to the excitation of rarefied carbon-containing gases in the presence of short-wavelength solar radiation.

It seems that such phenomena are not as rare on the Moon's surface as was formerly thought. On October 29, 1963, two observers attached to the U. S. Air Force lunar cartography service, who were studying the Moon with the large 60-centimeter (24-inch) reflecting telescope of the Lowell Observatory in Flagstaff, Arizona, noted the appearance of two unusual bright spots, of a reddish color, in the region of the crater Aristarchus. In a few minutes, the brightness of the spots gradually increased, while a third, pink spot, about 20 kilometers (12.4 miles) long and encompassing precisely the southeast portion of the Aristarchus rim, appeared. Twenty minutes after the beginning of the observation, the first two bright spots disappeared, and the third one also ceased to be visible five minutes later.

Intrigued by this observation, the two astronomers paid careful attention to the Moon during the weeks that followed, without noticing anything unusual. But on November 27, 1963, while Aristarchus was again illuminated under the same conditions as on October 29 (i.e., slightly less than two days after the Sun rose on this region of the Moon), a new elongated spot, of a reddish color, appeared on the rim. The observation, which lasted 75 minutes, was confirmed by Dr. Hall, director of the Lowell Observatory, as well as by an astronomer of the Perkins Observatory who had been alerted by telephone.

Other analogous light phenomena, always very localized and generally rather brief, have been noticed on various occasions on the Moon. The sites where such transitory lunar phenomena appear are not randomly distributed but rather grouped in the vicinity of the circular seas and around the valleys.

It is generally accepted that the observed, light originates from gases released by rocks. In the presence of solar radiation, they become luminescent, but they also may be a sign of residual internal activity in the Moon.

the past century, two theories have competed to explain the origin of the lunar craters: the theory of internal origin, and the meteoritic theory. According to the first theory, developed by Loewy and Puiseux, among others, at the end of the 19th century and more recently revived by Haroun Tazieff, in 1949, the craters, though merely volcanoes, are nevertheless the result of internal forces that in the past acted on the lunar crust while it was still soft. According to the latter theory, introduced in 1840 by Gruithuisen, the craters are due to the impact of meteorites; the greater the force of the explosion, the greater the crater diameter. This theory now is accepted by the great majority of selenologists. Not only does it fit in well with what we know, since space exploration began, of planetary history (shortly after the birth of the solar system, all bodies underwent an intense meteoritic bombardment; some, such as Mercury and Mars, still bear traces of it); it also explains very diverse characteristics of the lunar relief. Thus the strings of small craters that are observed around some large cirques very probably result from the secondary impact of large rocks broken off from the lunar crust and ejected radially, at a relatively low speed, under the impact of the explosion that created the central cirque; the light rings and radiant streaks visible around several young cirques (Tycho, Copernicus, etc.) may be considered as having been formed by an accumulation of very fine matter lifted up by the original explosion and projected a greater or a shorter distance from the point of impact.

Apart from the craters, the surfaces of the lunar "seas" appear to be strewn with fissures or gorges, valleys, cliffs and single rocky pillars.

The gorges are roughly rectilinear, or shaped like arcs of a circle (crevices in the Sea of Humors). Narrow and shallow, usually with a flat bottom, they obviously represent fractures. The more unusual ones are those separating the Sea of Humors from the Sea of Clouds; the most spectacular is the Marius gorge, near the crater of the same name, whose twists and turns extend over 250 kilometers (155 miles). The most famous one, the Hyginus cleft, south of the Sea of Clouds, is not a true fault; through a telescope we see it as a clean, V-shaped fissure with a crater (Hyginus) 6 kilometers (3.7 miles) in diameter at one end, but in the photographs taken during space

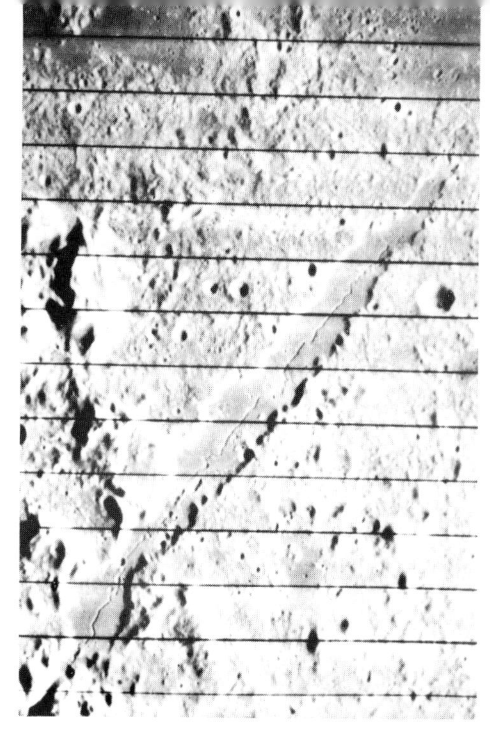

The Alpine valley. This picture, taken May 19, 1967, by the American satellite Lunar 4 *orbiter from an altitude of 2,900 kilometers (1,802 miles), revealed the existence of a fissure at the bottom of the valley. NASA Photo.*

The Hyginus cleft, photographed from an altitude of 110 kilometers (68 miles), during the Apollo 10 *mission. IPS Photo.*

An Easy Target for Amateurs

The Moon is the most accessible astronomical object for amateurs. You begin by observing it with the naked eye. You learn to recognize the phases and to follow their progress over the months. At twilight, less than 48 hours after the new Moon, you can try to find the slimmest lunar crescent possible: The chances of success will be greater in spring than in autumn, for the moon is higher in the sky in spring. Finally, you should not fail to observe ashen light a few days before the first quarter or after the last quarter.

Observation with an instrument is, of course, much more exciting. Choose a period midway between the new and the full Moon; you will be especially interested in the formations near the terminator (the boundary between the dark and bright sides of the disk); these, of course, are the regions over which the Sun is rising (if the observation takes place before the full Moon) or setting (if the observation takes place after the full Moon); the light, therefore, is low-angled, and the slightest irregularity in the terrain projects a disproportionately long shadow on the ground, intensifying the sensation of depth. On the other hand, avoid observing during the full Moon, for the lunar surface is perpendicularly illuminated by the Sun's rays at that time and appears without any relief at all, with one exception: The full Moon is when the brilliant rays radiating away from certain craters such as Tycho, Copernicus, Kepler and Aristarchus can best be seen (they probably are trails of debris ejected by the collision that caused these large craters). With the brilliance of the full Moon, you should take the precaution, if you want to observe for a long time without eyestrain, of placing a colored filter behind the telescope eyepiece to absorb part of the light, or use an eyepiece with a sufficiently high magnification so that only part of the Moon appears in the field.

missions, it appears to be made up essentially of a series of small attached craters. The valleys are clearly wider and more meandering, resembling riverbeds; their curves follow the line of greatest slope of the terrain. Often they begin as cavities of some kind, sometimes with a flat bottom, and sometimes funnel-shaped, such as the Cobra Head, out of which the Schroeter Valley develops, and extends over about 200 kilometers (124 miles) north of the Herodotus crater. Particularly famous is the Alpine valley, an indentation 130 kilometers (81 miles) long that juts directly into the Alps Mountains. As for the cliffs, the most spectacular is the Straight Wall, a sudden shift of 300 meters (984 feet) in altitude and 120 kilometers (76 miles) long, bordering the Sea of Clouds; despite its name, it is not a steep slope, since its wall is not vertical but slanted by about 40°.

The Lunar Surface

The lunar surface appears to be strewn with a multitude of stones—from big rocks to small pebbles—more or less embedded in a layer of dust from a few millimeters to 15 centimeters (5.9 inches) thick, depending on the location. Furthermore, it is riddled with microcraters dug by the impact of micrometeorites. Its color varies, depending on the lighting angle, from ashy gray, when it is lit up by earthshine, to chocolate brown, when the Sun is straight overhead. The dust is of variable texture depending on the region—fine at the *Apollo 15* landing site, in the Swamp of Putrefaction at the foot of the Apennines, and somewhat coarser, at the *Apollo 14* landing site in the Ocean of Storms, near the Fra Mauro crater. The dust is made up essentially of rocky fragments but also contains some meteoritic debris (in smaller quantities than had been expected) and glass spherules (whose presence was a total surprise).

Under the carpet of dust is a layer of broken rocks, the regolith (or regolite), the thickness of which varies from 2 to 20 meters (6.6 to 65.6 feet), depending on the region, and the density of which increases with depth, being very low at the surface.

Chemistry of the Rocks

The six Apollo missions that included a landing on the Moon (see table, page 152) made it possible to collect some 2,200 lunar rock samples, representing a total mass close to 400 kilograms (882 pounds) (of which more than 300 kilograms [661 pounds] still were awaiting analysis at the beginning of 1980). In addition, there are some samples picked up and brought back to Earth by Soviet automatic Luna devices.

As on Earth, oxygen is the most abundant element on the Moon's surface. The decreasing order of abundance of the elements is about the same on both bodies, but with significant differences. Compared to the Earth's crust, that of the Moon appears poorer in volatile elements, such as carbon and oxygen; in siderophiles (except for iron itself), such as cobalt and silver; and in potassium, sodium, silicon, rubidium, scandium and europium. It is, however, richer in refractory elements, such as calcium, titanium and magnesium, and in iron and rare earths (except europium). The oxygen deficiency is underscored by the presence of native metals in nearly all the rocks.

The great majority of elements present are original, stemming from condensation or accretion of the Moon 4.6 billion years ago. But a certain percentage of a few of them results from a later contribution ascribable to radioactivity, meteorites (native iron, in particular) and to solar wind. The study of lunar rocks has led to the discovery of 75 varieties of minerals, representing only 33 distinct species, as opposed to about 80 in meteorites and more than 2,000 on Earth. Three of these minerals had never been observed before: tranquillitite (named after the Sea of Tranquillity, where it was found); pyroxyferroite, a silicate of iron and calcium; and armalcolite (whose name is a tribute to the *Apollo 11* crew—Armstrong, Aldrin, Collins), a titanate of iron and magnesium that has since been discovered in diamond mines in South Africa. The most abundant minerals are silicates, as on Earth, but with differences stem-

The region of the Clavius crater, seen with an amateur telescope of 310 millimeters (12 inches). Compare with the illustration on page 147. Photo by P. Joannard, Assoc. astron. de l'Ain.

Microphotograph of a sample of lunar dust. The surface of the lunar soil has been reduced to a fine powder under the prolonged effect of micrometeorites. Liquefaction of material as a result of collisions produced tiny spherules of opaque glass. The other dark fragments are pyroxens; the light fragments, calcium-rich feldspars. Photo: Th. de Galiana archives.

Around the first or the last quarter the spectacle becomes most fascinating, for the terminator then moves to the center of the lunar disk, precisely where the relief is the sharpest. The first quarter has the advantage of being visible in early evening, whereas you must wait until midnight to be able to observe the last quarter.

A simple pair of binoculars will show all of the major features of the lunar topography: seas and large craters. With a simple telescope having an aperture of 40 to 60 millimeters (about 2 inches), the relief can be seen better: You can identify the principal features illuminated by earthshine and make out special formations like the Alpine valley and the Hyginus cleft. With a telescope that has an aperture of 60 to 75 millimeters (2.5 to 3 inches), the sight becomes gripping: Innumerable craters appear, and the undulations of the floors of the seas are visible. With an instrument (a reflecting or refracting telescope) at least 150 millimeters (6 inches) in diameter, you can begin a serious study of the lunar surface (a 200-millimeter [8-inch] instrument reveals details of less than 1 kilometer [.62 mile] on the Moon). Here are a few sights not to be missed:

- ashen light (the third, fourth, and fifth days of the lunation)
- Theophile, Cyrille, Catherine craters (after the fifth day)
- Alpine valley, Hyginus cleft, Apennines (after the first quarter)
- Straight Wall, Arzachel, Alphonse and Ptolemy craters; Archimedes, Autolycus and Aristille (after the eighth day)
- Clavius crater, Tycho and its rays (after the ninth day)
- Copernicus and its rays (after the 10th day)
- Aristarchus and Kepler craters and their rays (after the 12th day)
- Doerfel and Leibniz mountains at the Moon's South Pole (13th day)

Man on the Moon

Mission Name	Crew	Date of Arrival on the Moon	Landing Site on the Moon	Length of Stay on the Moon	Number and Duration of Sorties	Distance Traveled on the Moon	Mass of Lunar Samples Collected
Apollo 11	N. Armstrong E. Aldrin (M. Collins)	July 20, 1969	Sea of Tranquility (23°29′24″ E., 0°40′12″ N.)	21 hr., 36 min.	(1) 2 hr., 31 min.	400 m.	22 kg.
Apollo 12	C. Conrad A. Bean (R. Gordon)	Nov. 19, 1969	Ocean of Storms (23°20′23″ W., 2°27′00″ S.)	31 hr., 31 min.	(2) 7 hr., 45 min.	3 km.	34 kg.
Apollo 14	A. Shepard E. Mitchell (S. Roosa)	Feb. 5, 1971	Ocean of Storms, region of the Fra Mauro Crater (17°27′55″ W., 3°40′24″ S.)	33 hr., 31 min.	(2) 9 hr., 25 min.	4 km.	43kg.
Apollo 15	D. Scott J. Irwin (A. Worden)	July 30, 1971	Sea of Rain, at the foot of the Apennines (3°39′10″ E., 26°06′04″ N.)	66 hr., 55 min.	(3) 18 hr., 35 min.	28 km.	78 kg.
Apollo 16	J. Young C. Duke (T. Mattingly)	Apr. 21, 1972	near the Descartes Crater (15°30′47″ E., 8°59′34″ S.)	71 hr., 02 min.	(3) 20 hr., 14 min.	27 km.	96 kg.
Apollo 17	E. Cernan H. Schmitt (R. Evans)	Dec. 11, 1972	Taurus Mts. region, near the Littrow Crater (30°45′26″ E., 20°09′41″ N.)	75 hr., 01 min.	(3) 22 hr., 04 min.	36 km.	113 kg.

ming from the previously mentioned discrepancies in the relative abundances of chemical elements on the two bodies.

Some rock samples appear to be rich in potassium (symbol K), rare Earth elements, and phosphorus (symbol P)—hence the name "kreep" by which they are known.

While terrestrial rocks are characterized by their extreme diversity, those of the lunar surface are divided essentially into only two types: anorthositic rocks, very abundant in the mountainous areas, and basalts, which fill the "seas." During meteorite impacts, some of these rocks were broken; their debris, hurtled into space and then falling helter-skelter to the ground, reappear in the form of breches, relatively crumbly agglomerates of small rocky fragments held together by glassy matrices with a composition resembling dust.

Seismic Activity

Thanks to the network of seismometers set up during the *Apollo 12, 14, 15* and *16* missions, it has been possible to make a thorough study of the Moon's seismicity.

In general, the Moon appears to be tectonically very calm. The quakes of internal origin (some 3,000 per year) that affect it are of very low intensity: None of those that have been recorded has gone over magnitude 3 on the Richter scale, which is the limit of human perception.

Their total annual energy is only 10^8 J, hence at least 10 billion times less than the energy released in the same period of time by quakes on Earth.

One of the most remarkable characteristics of lunar quakes is the depth at which they occur. The majority of centers are located at a depth ranging from 700 to 1,100 kilometers (435 to 684 miles) (while the depth of the centers of terrestrial quakes usually is only between 3 and 70 kilometers [1.9 and 44 miles] and only rarely reaches 700 kilometers); there are, however, only a small number of known deep epicenters, and most of them appear to be repetitive. The most active one—responsible for a third of the recorded quakes—was localized at a depth of 800 kilometers (497 miles) (which represents nearly half a lunar radius!), under the small mountain range between the Sea of Clouds and the Sea of Humors. The various epicenters that have been localized are spread out along two lines about 2,000 kilometers (1,243 miles) long, one coinciding with the 30° west meridian, the other following a northeast—southwest axis. Another notable feature of

The Scientific Stations Installed on the Moon

Each of the Apollo missions involving a lunar landing included the placement of a group of scientific devices on the Moon.

Six stations were installed at different sites between July 1969 and December 1972. The first, installed in the Sea of Tranquillity by the *Apollo 11* astronauts, was in fact merely a prototype, EASEP (Early Apollo Scientific Experiments Package). It consisted of two devices: a laser reflector and a passive seismometer. The latter, fueled by solar cells, could be remote-controlled from Earth until August 10, 1969; the loss of telemetry occurred in January 1970. Beginning with *Apollo 12*, the installation of real stations, called ALSEP (Apollo Lunar Scientific Experiments Package), began. They were designed to measure seismic activity (passive and active), thermal, magnetic, gravitational and other characteristics (solar wind, atmosphere and ionosphere and to monitor meteorites). A typical station included half a dozen instruments connected, on the one hand, to a "SNAP 27" generator fueled by radioactive plutonium and providing about 70 W and, on the other hand, to a central control unit responsible for collecting information and relaying it to Earth via a 1 W radio transmitter, and for receiving remote-control orders for starting and stopping the various instruments.

The makeup of the stations varied slightly from one mission to the next, but in general, their instruments were chosen from the following categories:

1. a seismometer, to study the frequency of falling meteorites, to detect and localize possible moonquakes and to make it possible to establish a model of the Moon's internal structure from the behavior of the seismic waves and their speed of motion
2. corner reflector, consisting of an assembly of silicon cubes (100 for *Apollo 11* and 14,300 for *Apollo 15*), designed to reflect laser rays directly back toward their origin on Earth, in order to determine more accurately the distance between the Earth and the Moon and understand better the movement of our planet's natural satellite
3. a magnetometer, for measuring the intensity of the magnetic field on the Moon's

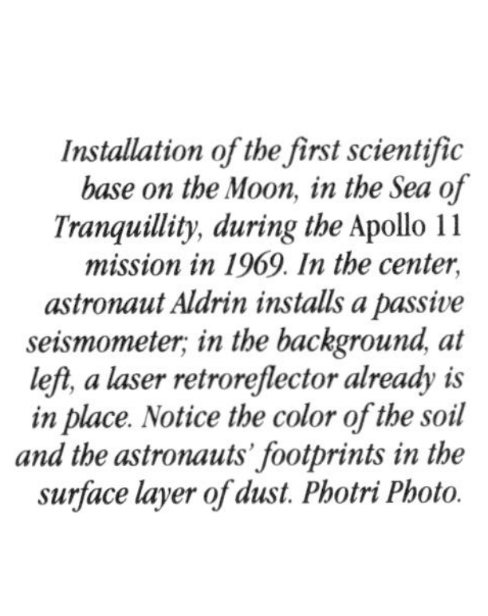

Installation of the first scientific base on the Moon, in the Sea of Tranquillity, during the Apollo 11 *mission in 1969. In the center, astronaut Aldrin installs a passive seismometer; in the background, at left, a laser retroreflector already is in place. Notice the color of the soil and the astronauts' footprints in the surface layer of dust. Photri Photo.*

lunar quakes of internal origin is their periodic nature. Most of them, representing 85 percent of the total energy released, occur during the week in which the Moon passes through perigee.

This synchronization probably results from a tidal stretching of the lunar crust (of an amplitude estimated at about 50 centimeters) (19.7 inches) under the influence of the Earth's gravity. Added to this 27-day periodicity due to the Earth's influence is a 206-day cycle (about 7 months) probably corresponding to a tide of solar origin.

In addition to the quakes of internal origin, there are those caused by meteoritic impacts. Such quakes, characterized by a sudden appearance of a wave train followed by a very gradual decrease, are easily detected on recordings. Thus, for instance, between January 1, 1973 and July 13, 1975, a total of 815 signals were interpreted as meteoritic impacts, the masses of which must have ranged from 50 grams (1.8 ounces) to 50 kilograms (110 pounds). During the period in which the automatic stations placed on the Moon by the Apollo missions were operating, only two unusual shocks due to natural celestial objects were recorded: one, on May 13, 1972, near the Gambart crater, in the Sea of Storms; the other, on July 17, 1972, on the Moon's dark side, east of the Sea of Moscow. But the Americans also caused artificial impacts by dumping five lunar module recovery stages and five S IV-B Saturn rocket stages on the lunar surface after they were used. In the first case, this represented meteorites of 2.4 t striking the ground at a speed of 1.7 kilometers (1.1 miles) per second at a very low angle, and in the second case, projectiles of 13.9 t arriving nearly head-on at a speed of 2.6 kilometers (1.6 miles) per second. Such impacts, natural or artificial, gave rise—against all expectations—to very long-lasting vibrations: nearly one hour for a lunar module stage, more than three hours for a Saturn rocket stage and nearly four hours for the two natural meteorites.

Study of the propagation of seismic waves during artificial impacts has made it possible to visualize the structure of the lunar interior down to a depth of 100 kilometers and, for the two large natural impacts, down to 400 kilometers.

Internal Structure

Based on the total data gathered by the Apollo seismometers, we have been able to construct the following model of the Moon's internal structure:

1. a multilayered crust, with a thickness varying from about 60 kilometers for the hemisphere visible from Earth to 100 kilometers for the hidden hemisphere
2. a mantle about 1,000 kilometers thick
3. a core with a radius of some 700 kilometers,

surface

4. a solar wind spectrometer, for counting the particles carried by solar wind and specifying their direction and energy
5. a charged-particle detector, comparable to the preceding one but designed only for protons and electrons emitted by the Sun
6. a fast-ion detector (Suprathermal Ion Detector Experiment) and a slow-ion detector (Cold Cathode Ion Gauge), revealing the nature, number and energy of the ions above the Moon
7. a cosmic dust detector, to measure the degradation of solar cells due to surface erosion by micrometeorites
8. thermal probe—i.e., an ultrasensitive thermometer placed in a hole dug by the astronauts with an electric drill and designed to give indications of the subsoil temperature.

In addition to these various instruments, we should add the "solar sail" placed aboard some of the missions at the urging of Dr. Johann Geiss, of the University of Bern, to trap particles of solar wind. It was an aluminum sheet measuring 30 by 140 centimeters and attached to a pole like a flag, which the astronauts set up during their stay on the Moon and then brought back to Earth. Three sails thus were used, during the *Apollo 11, 12,* and *15* missions, with an exposure time of 1 hour, 17 minutes; 18 hours, 42 minutes; and 41 hours, 8 minutes, respectively.

The ALSEP stations were designed for a cumulative service of six years; the first four were supposed to operate for a year, and the fifth for two years. These estimates were greatly surpassed. On October 1, 1977, when NASA, deluged by the flow of data that needed analysis, stopped monitoring the stations, most of the devices were still operating; apart from the corner reflectors, 16 instruments still were functioning perfectly and six intermittently.

Remote-controlled from Earth on more than 153,000 occasions, the stations transmitted more than 1,000 billion bits of information and provided, until they were taken out of service, a total of more than 29 years of operation.

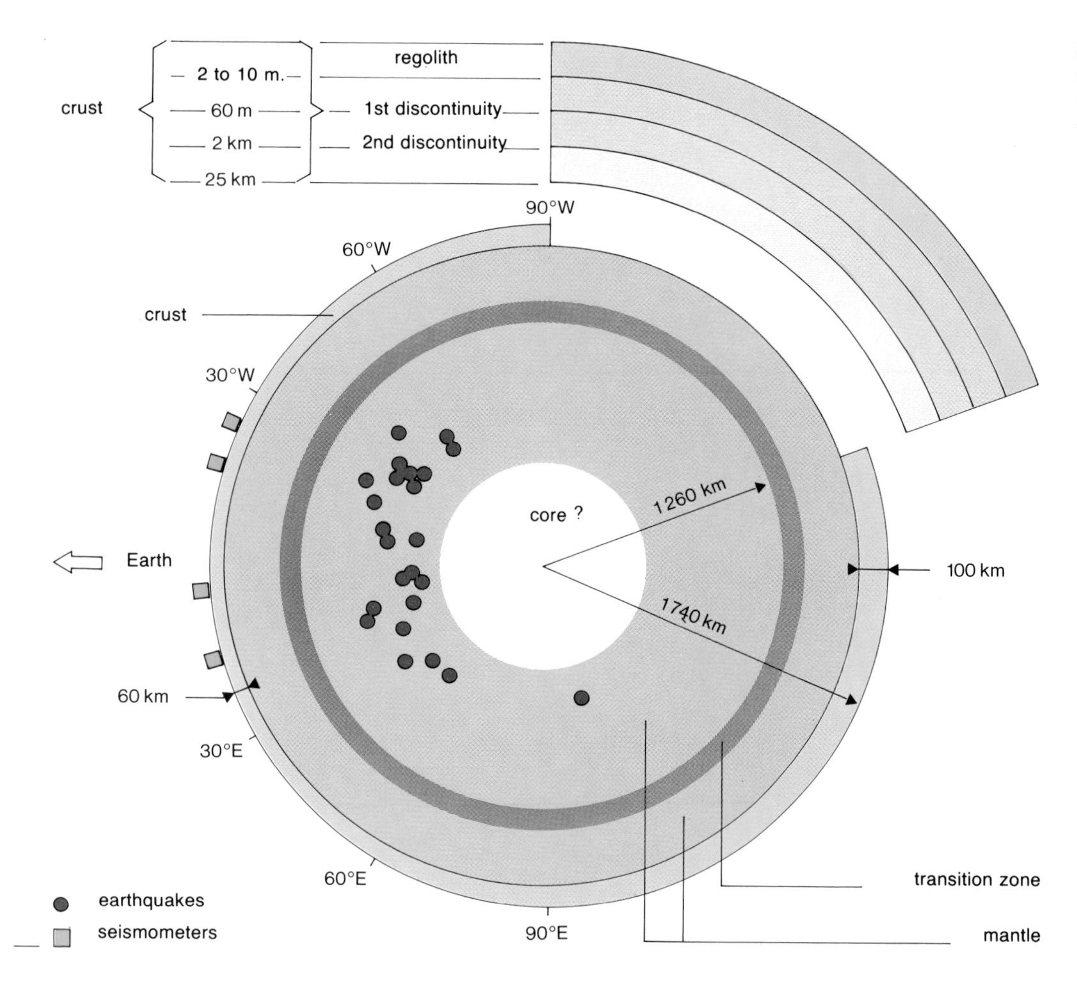

Model of the Moon's internal structure, and distribution of lunar earthquakes according to the data supplied by the Apollo missions.

Evolution of the Moon. Left to right: (a) formation of the Moon; (b) large-scale melting of the planet, resulting in intense volcanic activity, chemical differentiation and the formation of a crust; (c) formation of vast basins under the impact of an intense meteoritic bombardment; (d) second melting of the internal layers; filling of the basins by basaltic lava.

containing a rather large quantity of iron; this core is thought to be relatively viscous; the temperature at the Moon's center is estimated at close to 1,500°C (the hypothesis of the presence of a core is based mainly on the fact that transverse seismic waves do not propagate at depths beyond 1,000 to 1,100 kilometers, which implies that at this level, matter is at least partially molten.

Temperature

The near-absence of an atmosphere results in a considerable thermal amplitude (the difference between the extreme night and day temperatures) that may reach up to 300°C at a given point on the surface, as compared with only a few dozen degrees on Earth.

The Italian Melloni, in 1846, provided the first estimates of the temperature of the lunar surface. However, the first precise measurements were not obtained until 1898, by the American Langley, who derived temperatures ranging from +97°C to +184°C on the light side. in 1923, E. Pettit and S. Nicholson, at the Mount Wilson Observatory, measured (with a thermocouple) +107°C in the sunlight and −73°C during the lunar night. The most recent measurements yielded +117°C at the daytime maximum, and a nighttime minimum of −171°C. Moreover, thermal probes buried in the soil during the final Apollo missions indicated a temperature of −20°C at 65 centimeters (25.6 inches) depth *(Apollo 15)*, −22°C at 80 centimeters (31.5 inches) *(Apollo 17)*, and −17°C at 2.30 meters (7.5 feet) *(Apollo 17)*, while the thermal-flow measurements showed a temperature increase of 1.75°C per meter of depth, much higher than the estimates.

Magnetism

The Moon's magnetic field shows great differences from that of the Earth. The former's most outstanding feature is that it is extremely variable in intensity and direction from one place to another. While the Earth may be compared to a bar magnet pointing approximately north–south, the Moon appears rather as a collection of small magnets buried at random under its surface. The magnetometers of the ALSEP stations have recorded intensities ranging from 6 to 313 γ, about 1,000 times weaker than the geomagnetic field on the Earth's surface (30,000 to 60,000 γ). However, a large number of rocks brought back by the astronauts, all dating at least 3 billion years, show a much stronger magnetization, meaning that there

The Moon's Influences on the Earth

According to popular opinion, the Moon exerts various occult influences on our planet. In particular, it is said to have an influence on living beings and to be the cause of certain weather changes. Such statements—none of which has been scientifically confirmed—should be taken with the greatest caution.

The Tides

The only really proven influences of the Moon on the earth are of a gravitational nature: they are related to the Moon's gravitational pull on our planet. This pull, combined with the Sun's, determines the Earth's deformations (Earth tides) and produces periodic changes in the sea level (ocean tides).

The study of ocean tides is extremely complex. It can be simplified substantially by assuming that the Earth is entirely covered by water. Under the influence of lunar gravity, the body of water situated on the side of the Earth facing the Moon rises up and forms a swell because it is pulled more strongly than the rest of the terrestrial globe; the mass of water situated opposite the Moon also rises, but it, on the contrary, is less strongly attracted than the rest of the Earth, which then tends to fall away beneath it. The swells do not follow the motion of the Earth's rotation on its axis; rather, they remain aligned with the Moon; consequently, in a given place, two tides occur each day, or more precisely, in 24 hours, 50 minutes, taking into account the Moon's motion around the Earth.

In fact, it is also necessary to take into account the position of the Moon and the Sun relative to the Earth. The Sun's tidal effect on the Earth—though approximately 2.2 times weaker than that of the Moon—is not insignificant. At the time of the new or full Moon—i.e., when the Sun, the Earth and the Moon are essentially aligned—the Sun's influence is added to the Moon's, to produce the strongest tides, known as spring tides. On the other hand, during the first or last quarter, the Sun and the Moon are in quadrature, and their effects counteract each other, producing smaller tides, known as neap tides. It is around the time of the equinoxes that the spring tides reach their greatest magnitude, for the sun then is in the plane of the Earth's equator.

The behavior and amplitude of the ocean tides is related not only to the relative positions of the Earth, Moon and Sun, which

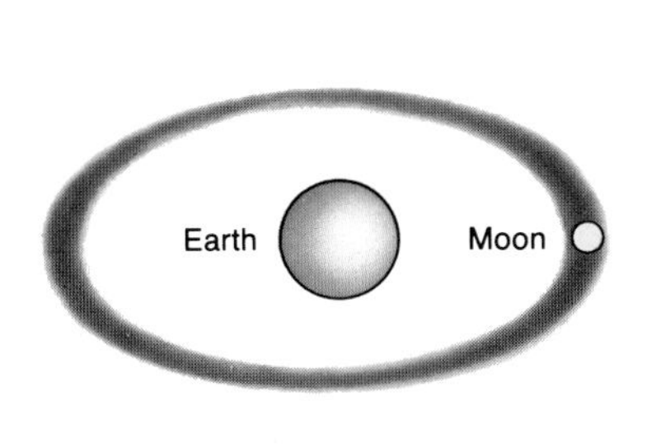

4.6 billion years

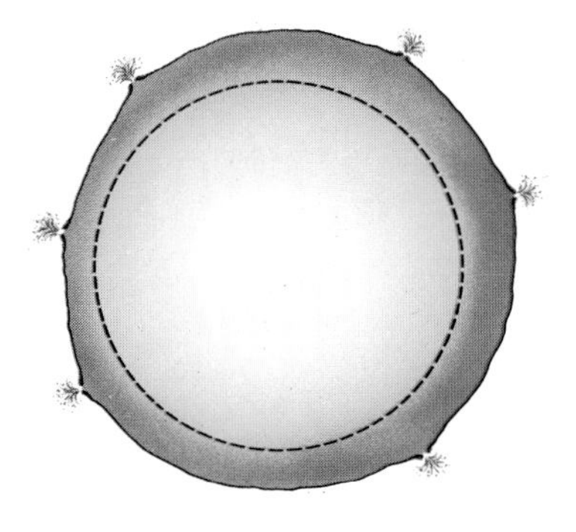

4.6-4.3 billion years

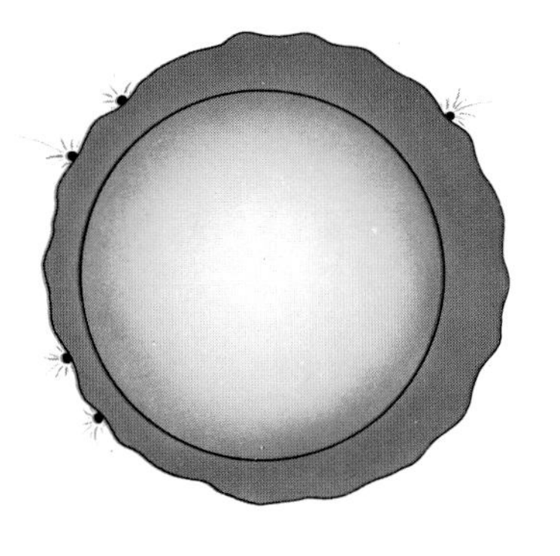

4.3-3.9 billion years

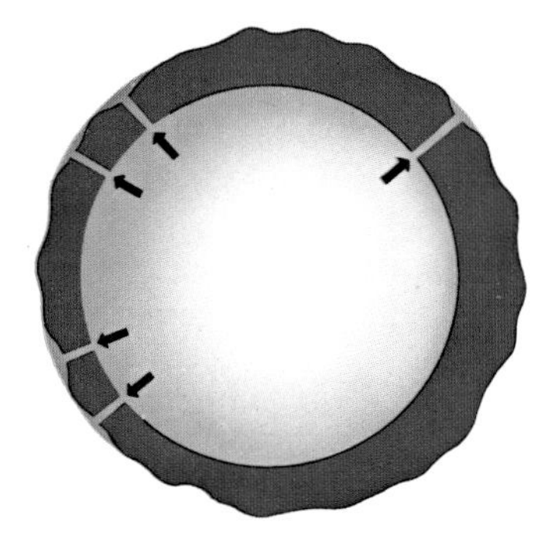

3.9-3.1 billion years

must have been a field of more than 3,000 γ on the Moon in the past epoch when they solidifed. It is possible that this field may have been generated during that period by a self-sustaining dynamo mechanism, since the Moon's interior was much more fluid then than now, and that the dynamo disappeared later, when the Moon cooled. According to this hypothesis, the present magnetism would be a fossil magnetism; and the detected field would be a vestige of the original field.

The Moon's History

The data gathered through space exploration now make it possible to retrace the principal stages in the Moon's evolution.

Born some 4.6 billion years ago, at the same time as the Earth and the other planets of the solar system, the Moon took form by the combination of the condensation of gases and the accretion of solid particles. Soon afterward, when the temperature of its outer layers rose to about 1,000°C, it liquefied to a depth of at least 200 kilometers, and its various component materials separated out from the center to the surface in increasing order of density. This chemical differentiation gave rise in particular, 4.5 to 4.3 billion years ago, to an outer crust composed predominantly of anorthosites. Just after solidification, this crust was intensively bombarded by huge meteorites, which were then abundant in interplanetary space; they dug out large basins and caused rock melting. This cataclysmic epoch ended 3.9 billion years ago; the last major impact seems to have been the one that created the Sea of Rains at that time.

Then, for the next 800 million years, the Moon underwent a great deal of internal activity. The heat released by the radioactive atoms in the rocks under the crust caused a second melting, this time of the deep interior. The basaltic lava that then formed rose and spread out over the surface, filling the basins, to make up the "sea floors" as we know them today. Furthermore, lava flows appear to have spilled out of the large basins, represented by the circular seas, onto the surrounding "lowlands," to cover the irregularly shaped seas; the Ocean of Storms, for instance, resulted from an overflow of the Sea of Rains. As for the near-total absence of seas on the dark side of the moon, it is explained by the fact that, in that hemisphere, the crust is substantially thicker, so that the magma issuing from the depths could reach the surface in only a few places.

In the last 3 billion years, the face of the Moon has barely changed. Its internal activity has quieted, meteoritic impacts on its surface have become rarer and the Moon has gradually cooled, becoming rigid down to a depth of at least 1,000 kilometers. The meteoritic bombardment that has continued through the present has, at most, caused the formation of a few large craters on the surface (Copernicus, Aristarchus, Tycho), the youth of which is attested to by the long radial trails [rays] of ejecta still surrounding them), of many smaller craters and of the superficial layer of broken rocks (regolith) as well as the carpet of dust covering it.

WHAT IS THE MOON'S ORIGIN?

While today we can retrace the broad outlines of the Moon's evolution, the puzzling question of its origin nonetheless remains unsolved.

Is the Moon a piece of the Earth that broke off from our planet as a result of solar tides while it was still fluid? (This hypothesis was put forward in 1880 by the Englishman George Darwin, son of the famous naturalist Charles Darwin.) Was it initially an independent body, born in another part of the solar system and captured by the Earth when its orbit brought it nearby? (This hypothesis was developed in 1950 by the American Harold Urey and repeated in a somewhat different form in 1971 by the American Fred Singer.) Or was it formed by accretion, starting from the same dust ring as the Earth, with the two bodies having constituted a double planet from the outset? (This hypothesis defended chiefly by the American Whipple, the

change each day, but also to the irregularities in the contour and depth of the ocean basins. In general, the tidal phenomenon, which has a fundamentally periodic character in a given place because of the Earth's rotation combined with the Moon's orbital motion, may be considered as the superposition of a large number of waves and has a character that is diurnal (a high tide and a low tide every 24 hours), semidiurnal (two high tides and two low tides per 24 hours, 50 minutes) or mixed (two high tides and two low tides of unequal amplitude and duration), depending on the place. The coasts are obstacles to the spread of the ocean wave, and its amplitude increases as it moves forward into a limited space. Thus, in some areas, particularly on the shores bordered by an extensive continental shelf, the tides may be very high: 19.6 meters (64.3 feet) in the Bay of Fundy (Canada); 16.1 meters (52.8 feet) in the Bay of Mont-Saint-Michel (France). On the other hand, in inland seas, the amplitudes usually are zero or almost zero.

The Red Moon

Growers often are known to fear the particular lunation that begins after Easter and generally continues through part of April and May. They call it "red moon," for, during this period, moonlight is supposed to be able to redden young plants. This is a typical example of an influence wrongly attributed to the Moon.

It is a fact that on certain clear nights in spring, young leaves and buds freeze, even though the thermometer stays several degrees above freezing. Since vegetation is in full growth then, the clear nights in the period corresponding to the red Moon are critical dates in agriculture. But the Moon is not to blame. Its very weak light could not substantially lower the temperature of the young plants receiving it without lowering that of the surrounding air. On the other hand, it is well known that under a very clear sky, free of clouds and mist, with or without moonlight, the ground and vegetation get colder during the night; their temperature can remain 6° or 7°C below that of the surrounding air. The Moon, therefore, is not involved. Its presence in the sky may, at most, be used as an indicator: If we see it shine, it means the weather is calm; then we can expect intense overnight cold, from which vegetation will suffer.

Englishman Fred Hoyle and the Frenchman Alexandre Dauvillier.) No decisive argument has yet been produced to permit us to select among these various hypotheses; on the contrary, their respective partisans have found additional supporting arguments in the results of the lunar rock analyses. However, the chemical differences discovered between the Earth and the Moon argue against the first hypothesis, which finds few supporters at present. The majority of specialists lean rather toward the last possibility, but the debate still is open.

In any event, beyond the problem of its origin, the Moon still contains many enigmas that justify pursuing its exploration in the future, using automatic or manned spacecraft on a much wider scale than in the past.

Solar Eclipses

By an extraordinary coincidence, the Moon is about 400 times smaller than the Sun but also 400 times closer, so that the two objects appear to us in the sky with practically the same apparent diameter. Thus when the Moon, in its motion around the Earth, passes through the direction of the Sun, it can place itself in front of the sun and hide it from us for a few moments. We then say that there is an eclipse of the Sun (in fact, it would be more accurate to speak of an occultation; in astronomy, the term "eclipse" generally refers to a planet's moving into the shadow of another). Depending on whether the solar disk is totally or partially masked, we say that the eclipse is total or partial.

The drawing explains the process of solar eclipses: the Moon, illuminated by the Sun, projects an umbral (shadow) cone and a penumbral cone behind itself. When these sweep across the surface of the Earth, we see a solar eclipse in the affected portions of the globe—total where the umbra passes, partial where the penumbra passes. The trajectories of the shadow and of the lunar penumbra on Earth are curves stretching over distances of several thousand kilometers. But while the umbra sweeps out only a narrow strip 270 kilometers (168 miles) wide at maximum (of which the median line, along which the eclipse has the longest duration, is called the centrality line), the penumbra affects a zone on either side of that strip whose width is close to 7,000 kilometers (4,351 miles). Thus, most of the eclipses that can be seen in a given place are partial eclipses, during which the Sun does not completely disappear but has merely had a portion scooped out by the Moon.

Sometimes a third type of solar eclipse occurs. Since the Moon's orbit around the Earth, and the Earth's orbit around the Sun are not perfectly circular, the distances between the Earth and the Moon and between the Sun and the Earth vary. Depending on the time of year, the Moon's apparent diameter may be slightly larger or smaller than that of the Sun. If an eclipse occurs at a time when the Moon's apparent diameter is less than that of the Sun (when the Moon is near the point on its orbit farthest from the Earth), the lunar disk does not mask the solar disk completely; at the center of the eclipse, there remains a thin ring of light around the dark lunar disk. The eclipse is said to be annular.

During a partial eclipse, the shadows cast by leaves on the ground or on walls take on an unusual appearance. Each gap between the leaves reveals not a bright disk but a small crescent—a true image of the eclipsed Sun.

Apart from this phenomenon, partial eclipses—unless they are nearly total—are nothing very special. On the other hand, a total eclipse is an impressive and unforgettable sight. Ten minutes before the eclipse is total, it begins to get dark; the sky and landscape take on strange colors. Animals become restless; birds return to their nests and stop singing. The temperature drops, and sometimes dew appears. A few minutes before darkness envelops the observer, while the Sun is reduced to a very thin bright crescent, moving shadows spread over all light surfaces; walls and ground seem to reflect the image of a body of water rippled by small waves; on them we see a series of wavy, parallel strips, moving relatively fast. These are probably due to air layers of unequal density blown by the wind. Finally, the Moon's umbral shadow rushes in at high speed from the horizon (this phenomenon, which occurs as the rapid spread of a large dark spot on the ground, is particularly spectacular when seen from a high place), while a last sliver of the Sun remains. The Moon's shadow often is broken by mountains, that nick the Moon's edge into a string of bright dots (called Baily grains, after the English astronomer who first pointed them out, in 1836). Then it is dark. During the few minutes that the phenomenon lasts, the planets, the brightest stars, and, around the Moon's black disk, the two outer layers of the Sun that usually are invisible appear: the chromosphere, a thin, pinkish ring from which prominences (huge jets of incandescent gas) emerge; and, beyond it, the corona, a vast, silvery halo with an irregular contour, giving off beams of light, some of which travel through space up to distances representing several solar radii. And

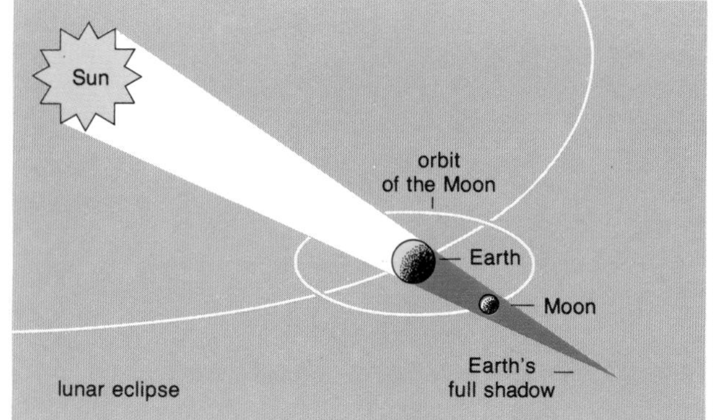

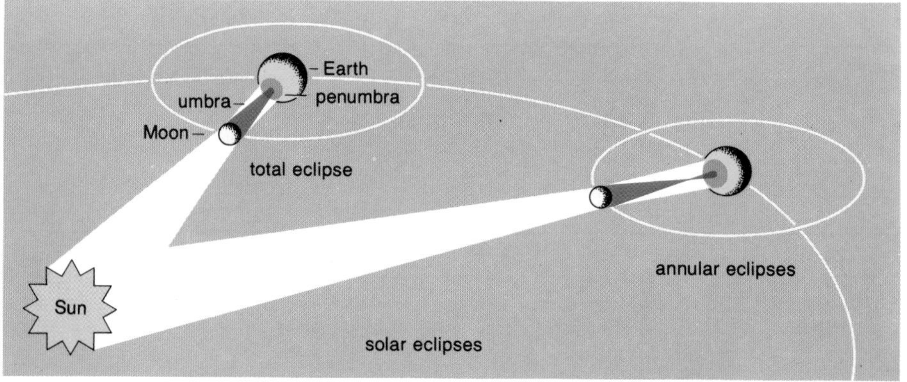

then suddenly, instantaneously, light reappears and nature revives. The total eclipse is finished. The Sun reappears as a rapidly growing crescent. Several dozen minutes later, the partial eclipse also ends; the Sun regains its usual appearance.

The duration of the total eclipse phase varies according to the respective distances of the Moon from the Earth and the Earth from the Sun, and according to the latitude of the observation site. In any event, it always is very short; its maximum length is 7 minutes, 30 seconds at the equator.

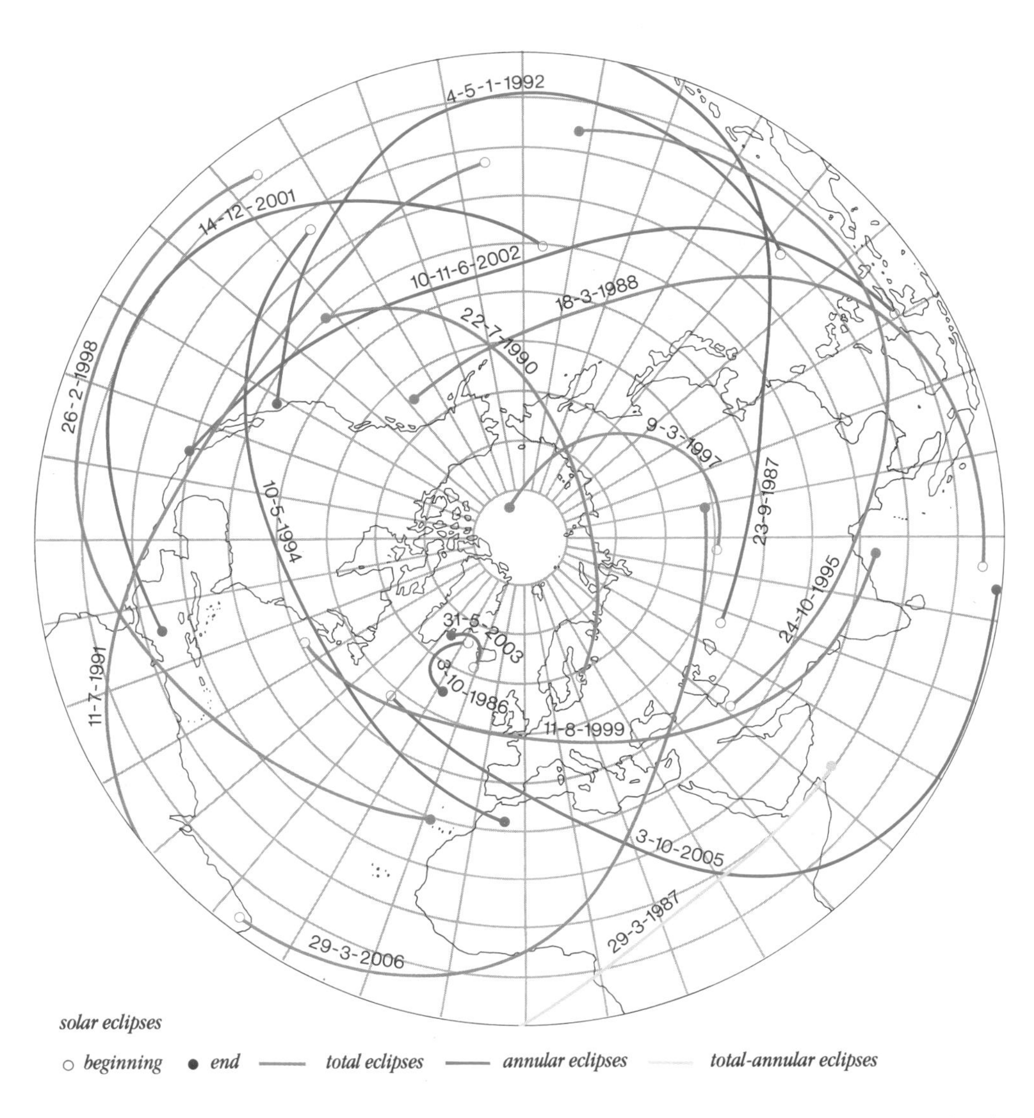

Visibility of total and annular solar eclipses in the Northern Hemisphere from 1982 to 2006. The line of centrality is shown for each eclipse. From Th. R. von Oppolzer.

Total solar eclipse of March 7, 1970, photographed in Miahuatlán, Mexico. Solar activity then was going through one of its maxima and the corona appears approximately spherical, crowned by many plasma jets emanating in all directions. Photri Photo.

Mechanism of the eclipses.

Baily grains during the total solar eclipse of June 30, 1973. Photo: S. Koutchmy.

Lunar Eclipses

At certain times, the Moon itself is eclipsed. As the Earth intercepts sunlight, it also casts an umbral shadow cone (with an average length of 1.4 million kilometers [870,106 miles]) and a penumbral cone behind it into space. Periodically, the Moon crosses through them; it is then no longer illuminated by the Sun, and we observe a lunar eclipse. It may be:

Different phases of the total lunar eclipse on September 16, 1978, photographed at Triel-sur-Seine, France: the Moon gradually emerging from the Earth's full shadow. Observers gave this eclipse a grade of 2.5 on the Danjon scale. Photo: G. Chiumenti, Groupe astronomique des Yvelines.

- total if the Moon enters the Earth's shadow completely
- partial if only part of the Moon enters the shadow
- penumbral if the Moon only crosses the Earth's penumbra

Lunar eclipses are no more numerous than those of the Sun, on average, but are visible each time from the entire nighttime terrestrial hemisphere; thus one can see more from a given spot.

During a total eclipse, the Moon begins entering the penumbra by its eastern edge and gradually takes on a grayish color. After about an hour, its entry into the shadow is signalled by the appearance of a dark indentation on that eastern edge. This indentation gradually increases, while maintaining a circular contour (this fact, noted in an-

Forthcoming Total and Annular Solar Eclipses

Date	Type	Duration of Totality	Best Observing Location
March 29, 1987	annular-total	42 sec.	South Atlantic, Africa
Sept. 23, 1987	annular	—	Asia, China, Pacific
March 18, 1988	total	3 min., 46 sec.	Indian Ocean, Indonesia
Sept. 11, 1988	annular	—	Indian Ocean
Jan. 26, 1990	annular	—	Antarctica, South Atlantic
July 22, 1990	total	2 min., 36 sec.	Finland, Northern Siberia
Jan. 15, 1991	annular	—	Australia, New Zealand
July 11, 1991	total	7 min., 6 sec.	Hawaii, Mexico, South America
Jan. 4, 1992	annular	—	Pacific
June 30, 1992	total	5 min., 24 sec.	South America, Africa
May 10, 1994	annular	—	Pacific, central U.S.
Nov. 3, 1994	total	4 min., 36 sec.	South America, Atlantic
April 29, 1995	annular	—	South Pacific, South America
Oct. 24, 1995	total	2 min., 24 sec.	Iran, India, Southeast Asia

Periodicity

Since the Earth's shadow is opposite the Sun, the Moon can cross it only at the time of full Moon. Likewise, the Moon's shadow cannot touch the Earth unless the Moon is between the Sun and the Earth—i.e., at new Moon.

If the Moon's orbit and that of the Earth were in the same plane, there would be a lunar eclipse at each full Moon and a solar eclipse at each new Moon (every month). In fact, the plane of the lunar orbit is tilted by about 5 degrees relative to the plane of the Earth's orbit. Thus, at new Moon, the Moon's shadow generally goes above or below the Earth, and at full Moon, the Moon usually passes above or below the Earth's shadow, which reduces the number of eclipses. Finally, calculations show that between four and seven eclipses occur each year: at least, two solar and two lunar eclipses (including eclipses by the penumbra); at most, five solar and two lunar eclipses, or vice versa.

Every 18 years, 11 days, the eclipses recur in the same order, with practically identical characteristics: this period, known since antiquity (it is supposed to have been noticed by the Chaldeans), is called *saros*. Actually, this interval of time, which contains, by def-

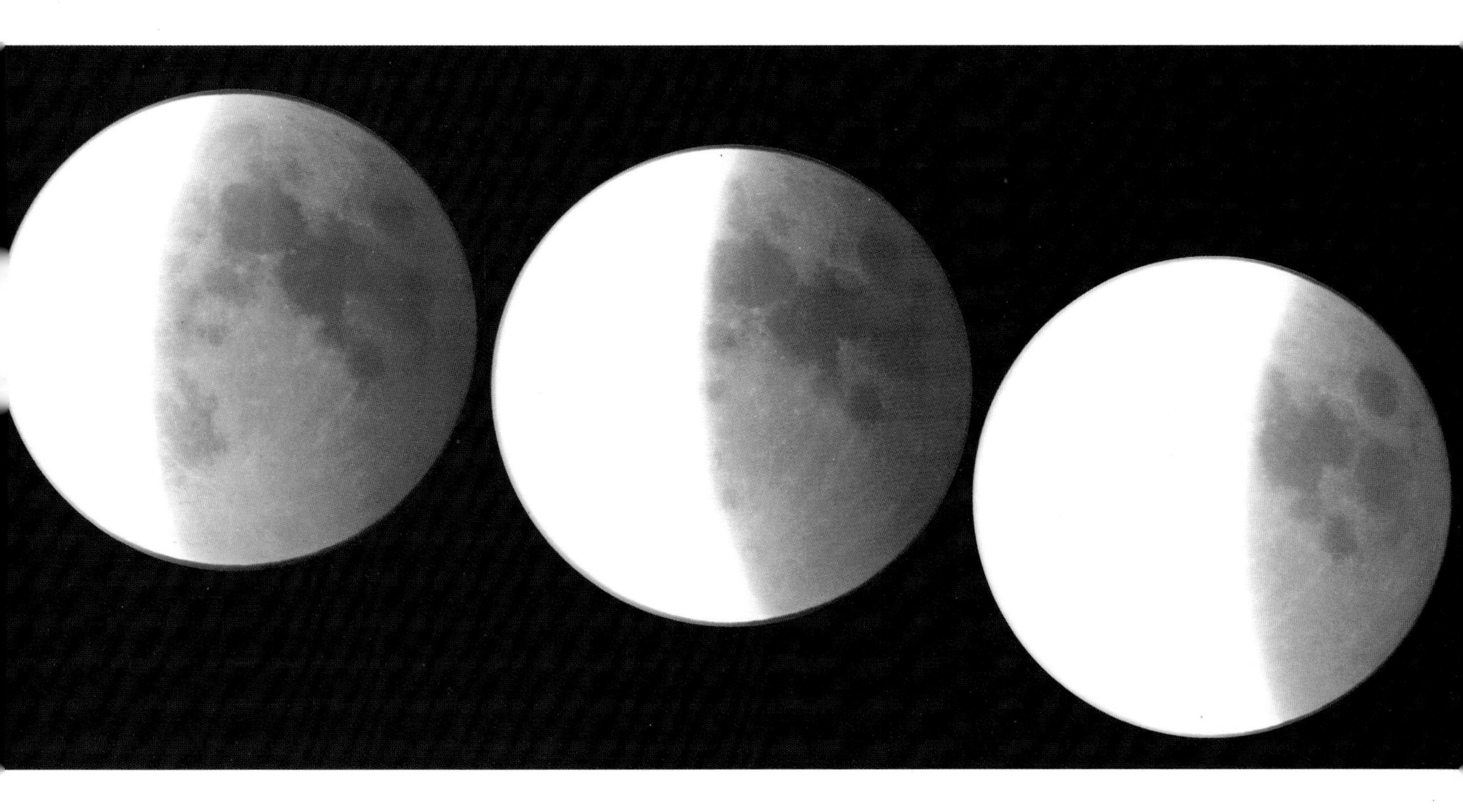

tiquity, was one of the first proofs of the Earth's roundness). In an hour, the entire lunar disk is covered by shadow, its color changing gradually to red.

Depending on the Moon's position relative to the Earth's shadow cone, the total eclipse phase lasts from a few minutes to a maximum of one hour, 45 minutes. It is unusual for the Moon to disappear completely while it is totally in shadow. In general, it takes on a magnificent coppery color. This is due to the fact that the Earth's atmosphere refracts and scatters a portion of the light it receives from the Sun back toward the Moon absorbing the blue rays in passing and allowing only the red rays to filter through. The residual light and coloring of the eclipsed Moon depend on solar activity and on the degree of transparency of the Earth's atmosphere. They are rated according to the Danjon scale:

- 0: very dark eclipse, moon almost invisible, especially in the middle of totality
- 1: dark eclipse, gray or brownish; lunar details difficult to discern
- 2: dark red or reddish-brown eclipse, usually with a very dark spot at the center of the shadow; outer zone fairly light
- 3: brick red eclipse; shadow often bordered by a gray or fairly light yellow area
- 4: copper red or very light orange eclipse; very bright bluish outer area.

Forthcoming Lunar Eclipses

Date	Type	Best Observing Location	Time (UT) of Mid-Eclipse
Oct. 7, 1987	Partial	South America	03:59
Aug. 27, 1988	Partial	South Pacific	11:04
Feb. 20, 1989	Total	Philippines	15:36
Aug. 17, 1989	Total	Brazil	03:09
Feb. 9, 1990	Total	India	19:12
Aug. 6, 1990	Partial	Australia	14:13
Dec. 21, 1991	Partial	Hawaii	10:33
June 15, 1992	Partial	Chile	04:58
Dec. 9, 1992	Total	North Africa	23:45

This table does not include penumbral eclipses.
The universal times given in this table correspond to the local mean solar time of the meridian running through Greenwich, England.

inition, 223 lunations, also contains, by an immense coincidence, approximately 242 draconitic and 239 anomalistic revolutions (see page 143). Thus the Moon, which, during an eclipse, is in conjunction or opposition relative to the Sun, near a node of its orbit, and at a certain distance from the Earth regains the same phase after a saros and returns to approximately the same node and the same distance from the Earth. Thus the conditions are created for the eclipse to recur, with very similar characteristics. The saros makes it possible to predict eclipses long in advance.

In this way the German astronomer Oppolzer, in 1887, published a catalogue containing the details of 13,200 eclipses from the period 1207 B.C. to A.D. 2161. The average number of eclipses per saros is 84, or 42 solar and 42 lunar eclipses.

Characteristics

A solar or lunar eclipse is characterized by:

- The timing of its different phases (particularly, the beginning, maximum and end). In a solar eclipse, the moment of beginning is called first contact and the moment of ending second contact.
- Its size—the portion of the solar or lunar diameter eclipsed, expressed in thousandths of that diameter. The eclipse is partial if the size is less than 1, total if the size is greater than or equal to 1. At all points on the Earth from which a lunar eclipse is visible, the size of the eclipse is the same; on the other hand, the size of a solar eclipse varies according to the position of the observer.

11. Mars

Beyond the Earth lies Mars, the last of the terrestrial planets. Observed since antiquity, like Mercury and Venus, because of its brightness, this planet was dedicated by the Greeks and later by the Romans to the god of war because of its reddish color, which resembles blood.

By revealing that Mars was capable of harboring life, astronomers made it very popular, especially since the end of the 19th century, when some went so far as to imagine the existence of an advanced Martian civilization. Space exploration dashed such hopes but made possible so much progress in our knowledge about the planet itself that the interest of specialists in this world close to ours—far from diminishing—has only grown in recent years.

Mars lies, on average, 228 million kilometers (142 million miles) from the Sun, but at perihelion, it comes as close as 207 million kilometers (129 million miles), while at aphelion it is as far as 249 million kilometers (155 million miles); the eccentricity of its orbit, 0.093, is less than that of Mercury's orbit but far greater than that of the Earth's orbit.

Mars completes its orbit in 687 days, 23 hours; that is the length of the Martian year—a bit less than twice the terrestrial year. The combination of the orbital motions of Mars and the Earth result in an alignment of the two planets relative to the Sun every 780 days, on average; that is the *average* interval between two consecutive oppositions of Mars. In fact, this interval may vary because of the eccentricity of the planet's orbit; it is only 764 days between two consecutive oppositions near Martian perihelion, but it rises to 810 days between two consecutive oppositions at aphelion. In the most favorable case (opposition at perihelion), Mars comes as close as 56 million kilometers (35 million miles) from the Earth. It can be as far away as 100 million kilometers (62 million miles) when the opposition is at aphelion. The oppositions recur in practically identical circumstances after an average period of 14 years and 530 days. The perihelic oppositions occur in August or September, the aphelic oppositions in February or March.

The next oppositions of Mars. Those in 1986 and 1988 will be particularly favorable for detailed observation of the planet from the Earth.

11-27-1990 18.1″
distances
in millions of km.
14.9″
Mars
1-7-1993
77
9-28-1988
94
23.8″
Earth
59
aphelion
perihelion
Sun
95
60
14.7″
23.2″
80
3-31-1982
17.6″
5-11-1984
7-10-1986

The Martian soil in the Utopia Planitia region, revealed by the American space station Viking 2 *after landing on the planet on September 3, 1976. Photo: Itek Corporation, Lexington, Massachusetts.*

Arandas, a Martian impact crater 25 kilometers (16 miles) in diameter, photographed by Viking 1 *orbiter on July 22, 1976, from an altitude of 1,857 kilometers (1,154 miles). It is surrounded by a layer of ejected matter, the appearance of which cannot be explained by impact on terrestrial-type soil; this structure is, however, compatible with a soil loaded with ice. The instantaneous vaporization of the ice by the explosion would have liquefied the medium; the vapor then would have helped to transport the matter in the form of a layer resembling the splattering made by a stone falling on muddy ground. Photri Photo.*

During its closest approaches, the planet has an apparent diameter of 26 arc seconds; its Southern Hemisphere is then turned toward Earth. During oppositions at aphelion, its apparent diameter is only 13 to 14 arc seconds, and it turns its Northern Hemisphere toward us. It is, therefore, the Southern Hemisphere of Mars that can be studied from Earth in the most detailed way.

Size and Mass; Rotation

Compared to the Earth, Mars is of relatively small size. Its equatorial diameter, precisely determined with the aid of spacecraft, is 6,794 kilometers (4,222 miles), or about half the diameter of our planet. Like the Earth, Mars has the shape of a rotating ellipsoid, slightly flattened at the poles; its polar diameter is only 6,760 kilometers (4,201 miles).

Mars's volume is 0.152 times that of the Earth. Its mass—determined by the motions of satellites—is 0.1069 that of the Earth. From this we conclude that its average density is 3.93 (as compared with 5.52 for the Earth). A body transported from Earth to Mars loses about two-thirds of its weight.

Through a telescope, the planet shows permanent dark spots, with well-defined shapes, on a reddish-brown background. Observation of these spots has made it possible to determine with great accuracy the Mars's rotational period, which is 24 hours, 37 minutes, 22.7 seconds. This is the Martian sidereal day, which is a little longer than the Earth's sidereal day (23 hours, 56 minutes, 4 seconds). From this we can easily infer the length of the Martian solar day: 24 hours, 39 minutes, 35 seconds. Thus the days and nights on Mars are only slightly longer than on Earth. The Martian calendar contains 668.6 solar days in a year.

There is another similarity between Mars and the Earth: Mars's equator is inclined 24° from the plane of the orbit (as compared with 23°26′ for Earth). Mars, therefore, is subject to seasons as pronounced as those on Earth, but nearly twice as long. Owing to the eccentricity of the orbit, the Martian spring and summer together last longer (371 days) than the autumn and winter (298 days) in the planet's Northern Hemisphere. The situation is reversed in the Southern Hemisphere. However, the Southern Hemisphere receives more light and heat during the hot seasons, for the planet is then near perihelion.

Surface Features

A large diversity of terrains on Mars was revealed by *Mariner 9* and the Viking spacecraft. The planet's terrain is characterized not only by craters and impact basins similar to those found on Mercury or on the Moon, but also by volcanic plains, numerous faults, winding valleys, dune, etc. Signs of an ancient meteoritic bombardment and evidence of major tectonic activity, volcanism, water and wind erosion, and large-scale sedimentation by the wind all can be found there.

On the planet as a whole, we see a marked asymmetry between the Northern and Southern hemispheres. While the Northern Hemisphere is largely covered by volcanic plains very similar to the lunar "seas," the Southern Hemisphere has a much rougher terrain, dominated by large craters with diameters greater than a dozen kilometers. This asymmetry probably results from the fact that, in the Northern Hemisphere, the lava flows that filtered through the crust to cover the plains buried the craters that had previously been formed there.

Most of the regions in which the innumerable craters of the Southern Hemisphere appear are old and rise about 3 kilometers (1.9 miles) above the average level. On the other hand, the great lava plains of the Northern Hemisphere are, in general, several kilometers below the average level of the Southern Hemisphere, and this situation certainly is not unrelated to the fact that the lava flows affected only the Northern Hemisphere. Probably the Martian crust, of uniform thickness in the beginning, became restructured after its formation as a result of convective motions within the mantle it covers.

The condition of the craters is highly variable. Those of the Northern Hemisphere generally are well preserved, which shows that the Martian terrain has not undergone significant erosion since the formation of the volcanic plains where they are found. On the other hand, such erosion was greater in an earlier period, the speed of its evolution having apparently been greatest in a period that coincided approximately with the end of the major meteoritic bombardments of the surface.

The great impact basins are similar to those seen on the Moon (circular "seas") or on Mercury. The largest one, with a diameter of 2,000 kilometers (1,243 miles) is Hellas, at 45° south latitude and 290° west longitude. It is a real basin, 3 to 4 kilometers (1.9 to 2.5 miles) deep, in which permanent dust storms take place that usually hide the bottom. At nearly the same latitude but at 45° west longitude, Argyre has a diameter of only 1,000 kilometers (622 miles) but its surrounding rim is much better preserved, forming a whole ring; except for the presence of an interior promontory, it greatly resembles the Moon's Eastern Sea.

The great volcanoes are concentrated in the equatorial region, on the Tharsis plateau, which is a 10-kilometer-(6.2 mile)-high bulge on the Martian crust, the center of which is approximately on the equator, at 115° west longitude. Three major volcanoes—Ascraeus Mons, Pavonis Mons and Arsia Mons (about 400 kilometers [249 miles] in

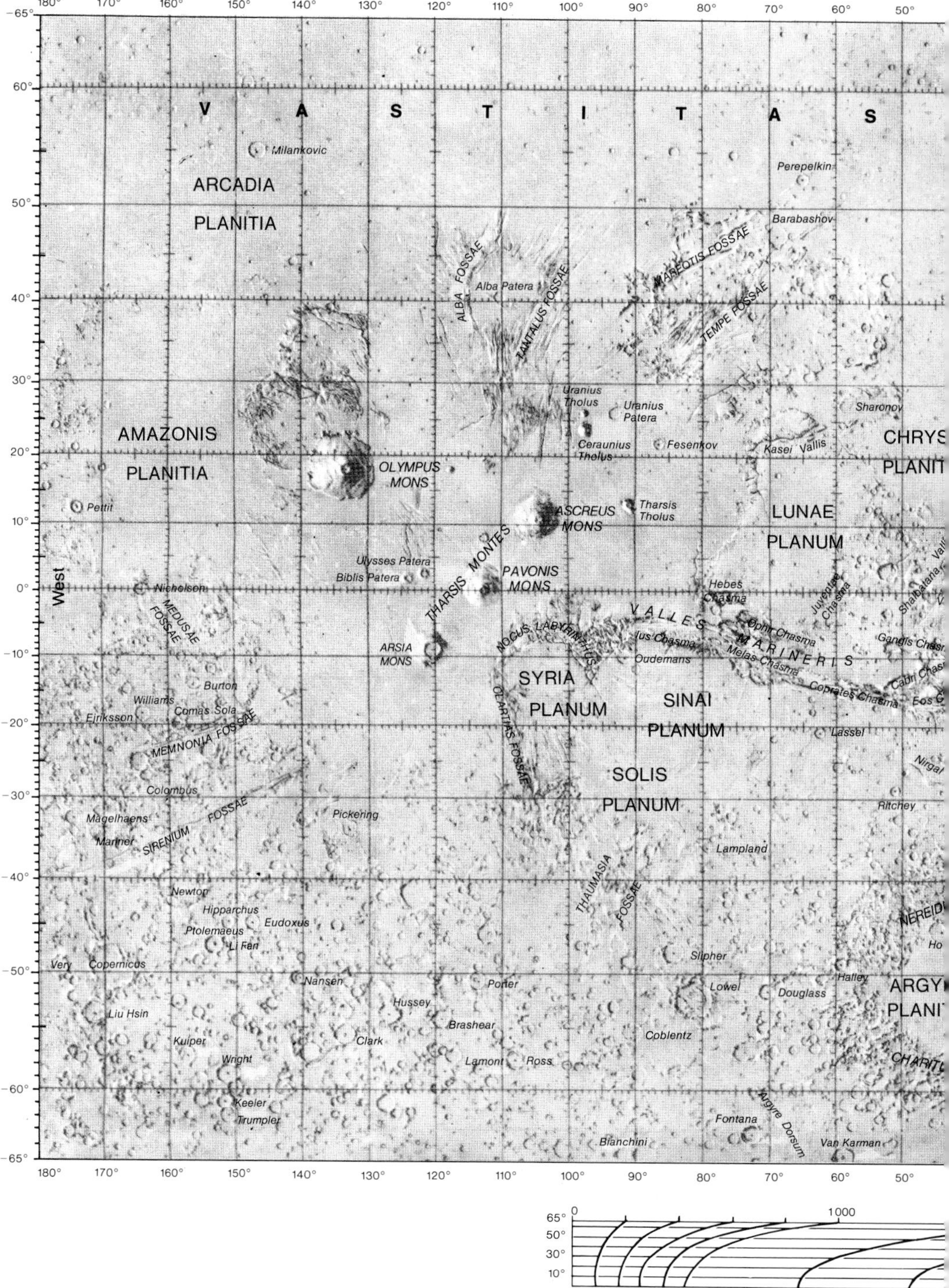

Mars, photographed from a distance of 560,000 kilometers (348,042 miles) by Viking 1 *orbiter. The picture has undergone special computer processing to emphasize the various types of clouds, the atmospheric haze and the ice-covered regions, through differences in color. Four of the largest Martian volcanoes are clearly visible at right (Olympus Mons at far right above). We also see, at the very bottom of the photograph, the Argyre basin in the form of a bright spot. The bluish-white strip running from the volcano region to the top part of the picture represents a layer of haze. Photri Photo.*

Mars's Cartography

The first drawing of Mars on which we can see a spot quite characteristic of the planet's surface was made in 1659 by Huyghens.

Cassini, in turn, observed the planet in 1666, discovering its rotation in slightly more than 24 hours, and its polar caps.

However, it was not until the 1830s that Beer and Mädler began a systematic study of the configurations of the Martian surface, using a small reflecting telescope with an aperture of only 10.8 centimeters (4.3 inches). These two astronomers drew up the first map of Mars, in 1840, on a grid of parallels and meridians similar to that of maps of the Earth; for the origin of the longitudes, they chose a small, very apparent dark spot essentially on the Martian equator, at the end of a dark linear feature; this spot has since been termed "Bay of the Meridian" (Sinus Meridiani). Later observations showed that actually it is double, resembling a fork: the central meridian of Mars, by convention, lies between the two prongs of this fork.

In the second half of the 19th century, many astronomers attempted to draw the Martian surface, particularly Dawes (1864–65) and Proctor (1867). The wealth of details revealed made it necessary to adopt a nomenclature. By analogy with Earth, it is accepted that the light ocher regions are continents (deserts), and the bluish-gray dark regions, seas and oceans; this is how the various Martian configurations are designated on the maps.

The cartography of Mars was completely overhauled as a result of the space missions. Based on the pictures taken by *Mariner 9,* the U.S. Geological Survey drew up an atlas

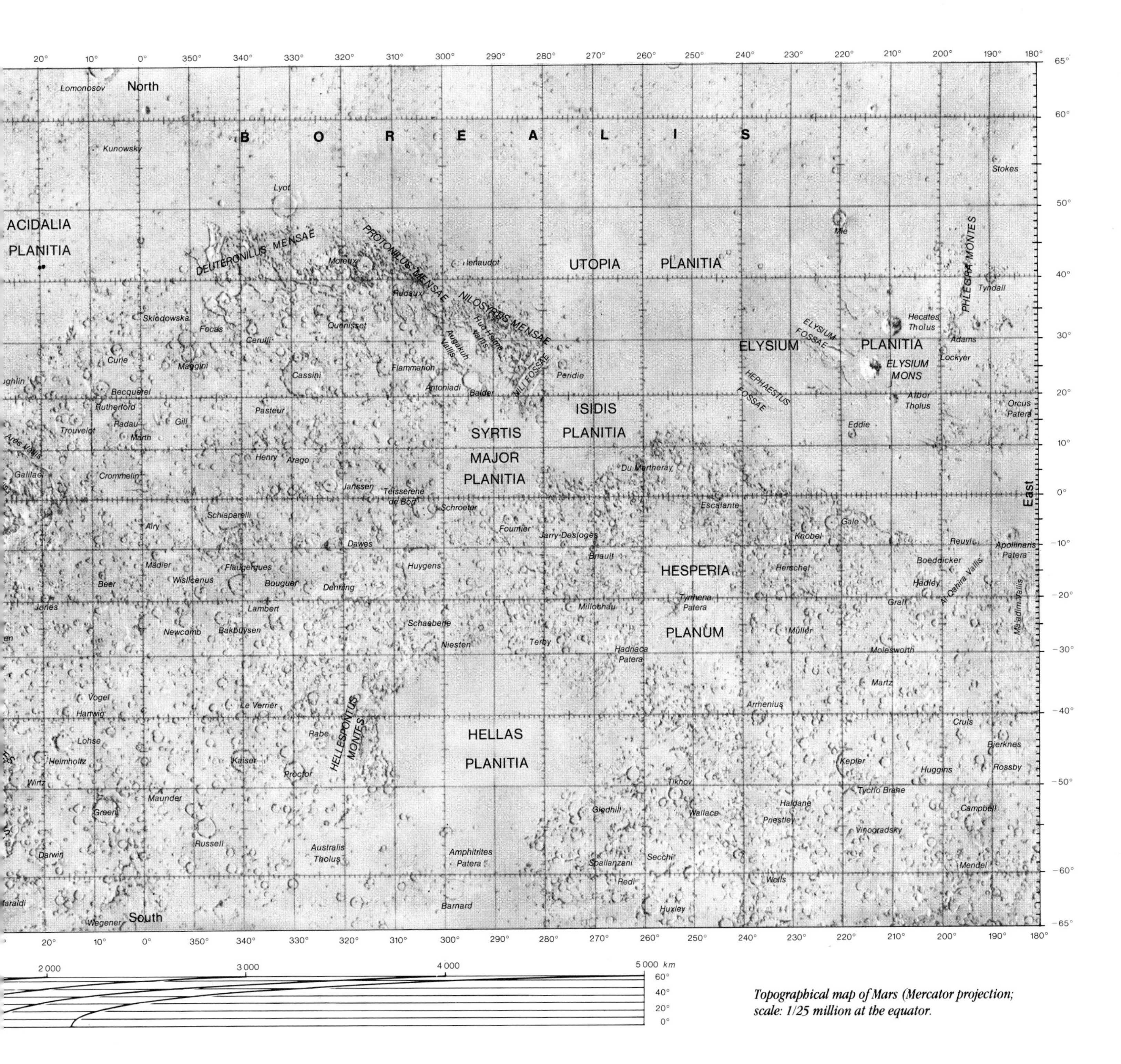

Topographical map of Mars (Mercator projection; scale: 1/25 million at the equator.

of the planet at a scale of 1/10 million, making it necessary to adopt a fuller and more appropriate nomenclature than previously used. An atlas at 1/2 million is being worked on, based on photographs taken since 1976 by the two Viking orbiters.

According to decisions made in 1973 at the 15th General Assembly of the International Astronomical Union, in Sydney, the planet Mars now is divided into 30 provinces, with 15 in each hemisphere; eight tropical provinces between the equator and the 30th parallel (30 degrees latitude), each extending over 45 degrees in longitude; six upper provinces (between 30 and 65 degrees latitude), each extending over 60 degrees in longitude; and one polar cap. Each province has been assigned a number (going from the North Pole to the South Pole) and a name, usually taken from the old nomenclature, which may be abbreviated by its first three letters.

Every large flat expanse on the surface of Mars now is designated by the general term *planitia:* Utopia Planitia, Amazonis Planitia, Hellas Planitia, etc. Moreover, in addition to the craters, astronomers distinguish 12 types of topographical formations: *catena* (chains of craters), *chasma* (canyon, gulf), *dorsum* (ridge), *fossa* (gorge—i.e., a long, narrow valley), *labyrinthus* (labyrinth—i.e., a complex network of valleys cutting through a flat region), *mensa* (a flat-topped protuberance ringed by cliffs), *mons* (mountain), *patera* (irregular crater with jagged sides), *planum* (plateau), *tholus* (hill), *vallis* (valley) and *vastitas* (vast, low-lying plain).

A total of 170 large craters (generally having a diameter greater than 100 kilometers [62 miles]) were named after a famous person who contributed to advancing our knowledge of Mars or who held an eminent position in astronomy: Flammarion, Galileo, Huyghens, Schiaparelli, etc. As for the small craters (about 6,000 inventoried from the *Mariner 9* pictures), they are designated by the name of their province, followed by two letters from Aa to Zz, the first assigned in increasing order of longitude (from east to west), the second in increasing order of latitude (from south to north). Such a system makes it possible to register 576 craters per province; if it proved necessary, a third letter would be used.

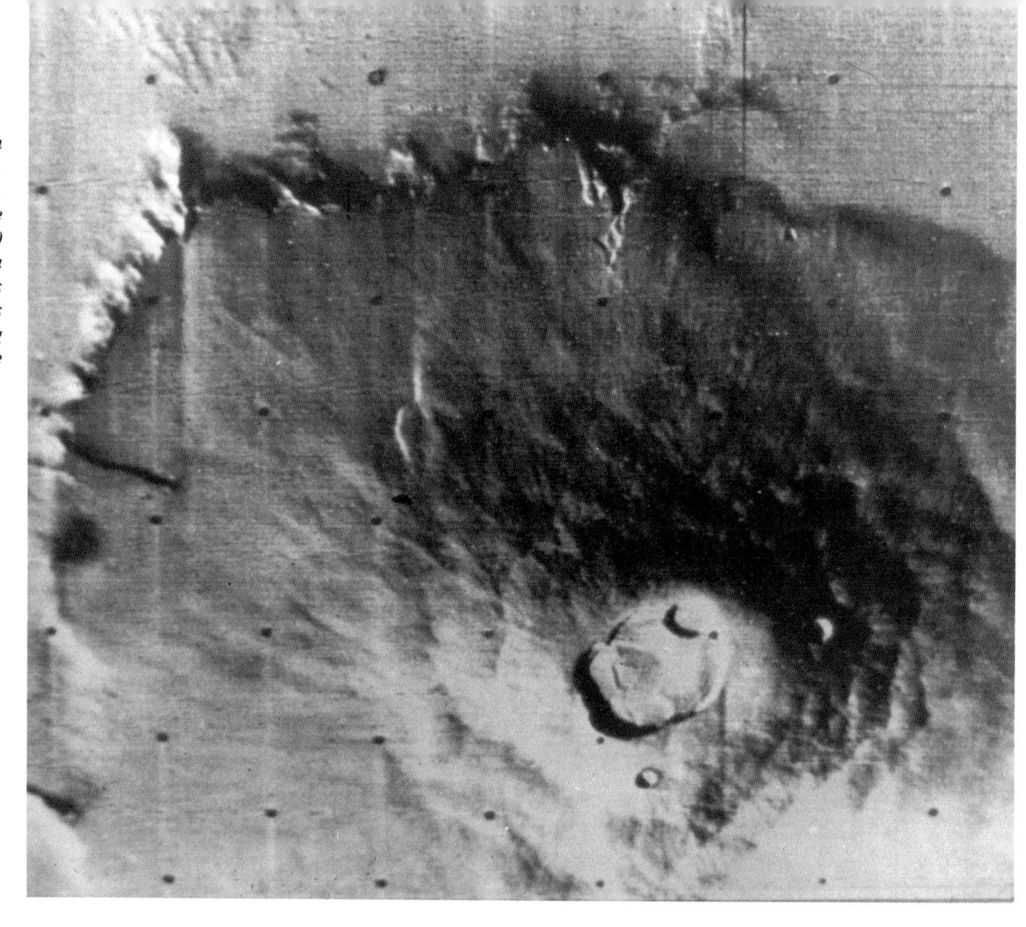

Olympus Mons, the largest known volcano in the solar system. In terms of its morphology, it resembles the Hawaiian-type volcanoes, but its dimensions are much more impressive. Its caldera—more than 60 kilometers (37 miles) in diameter—has a complex structure: It contains several craters formed by successive eruptions. Steep cliffs separate its slopes from the surrounding plain. Photo: Cosmos Encyclopédie, *with the kind permission of A. Ducrocq.*

diameter at the base and 20 kilometers [12.4 miles] in height)—are aligned in a northeast–southwest direction and overlook the central part of the plateau. But the most imposing of them all, Olympus Mons, is 2,000 kilometers from this chain, on the northwest fringe of the plateau: its diameter at the base is 600 kilometers (373 miles), and it rises 26 kilometers (16 miles) above the average level of Mars (such a height, which represents nearly three times that of Mount Everest, is considerable for a world with a diameter only half that of Earth!). It is the largest volcano currently known in the solar system. Another volcanic chain exists in the Elysium region, 90° farther to the west in longitude, also in the Northern Hemisphere, but these volcanoes are not as large.

The small number of impact craters on the side of these volcanoes, and the structure of the lava flows, indicate that these are geologically young formations. In any case, however, they are extinct volcanoes: No gaseous or thermal discharge has been detected rising above them. Olympus Mons, the youngest volcano, has probably been extinct for about 100 million years.

Sinuous valley north of the Argyre basin. Photri Photo.

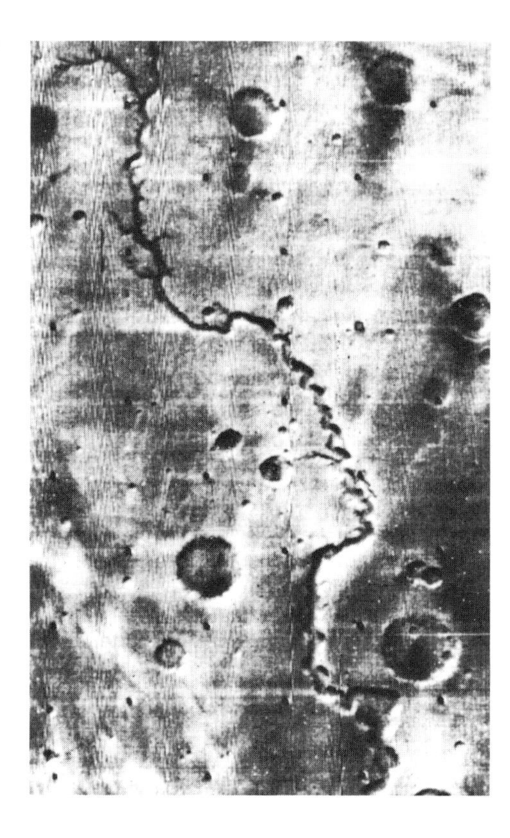

Flow structures. South of the Chryse Planitia basin, Viking 1 *orbiter discovered terrain characteristic of fluvial erosion. Torrential floods might have carved the islets we now see around some craters that stood in the way of the flow. Photo:* Cosmos Encyclopédie, *with the kind permission of A. Ducrocq.*

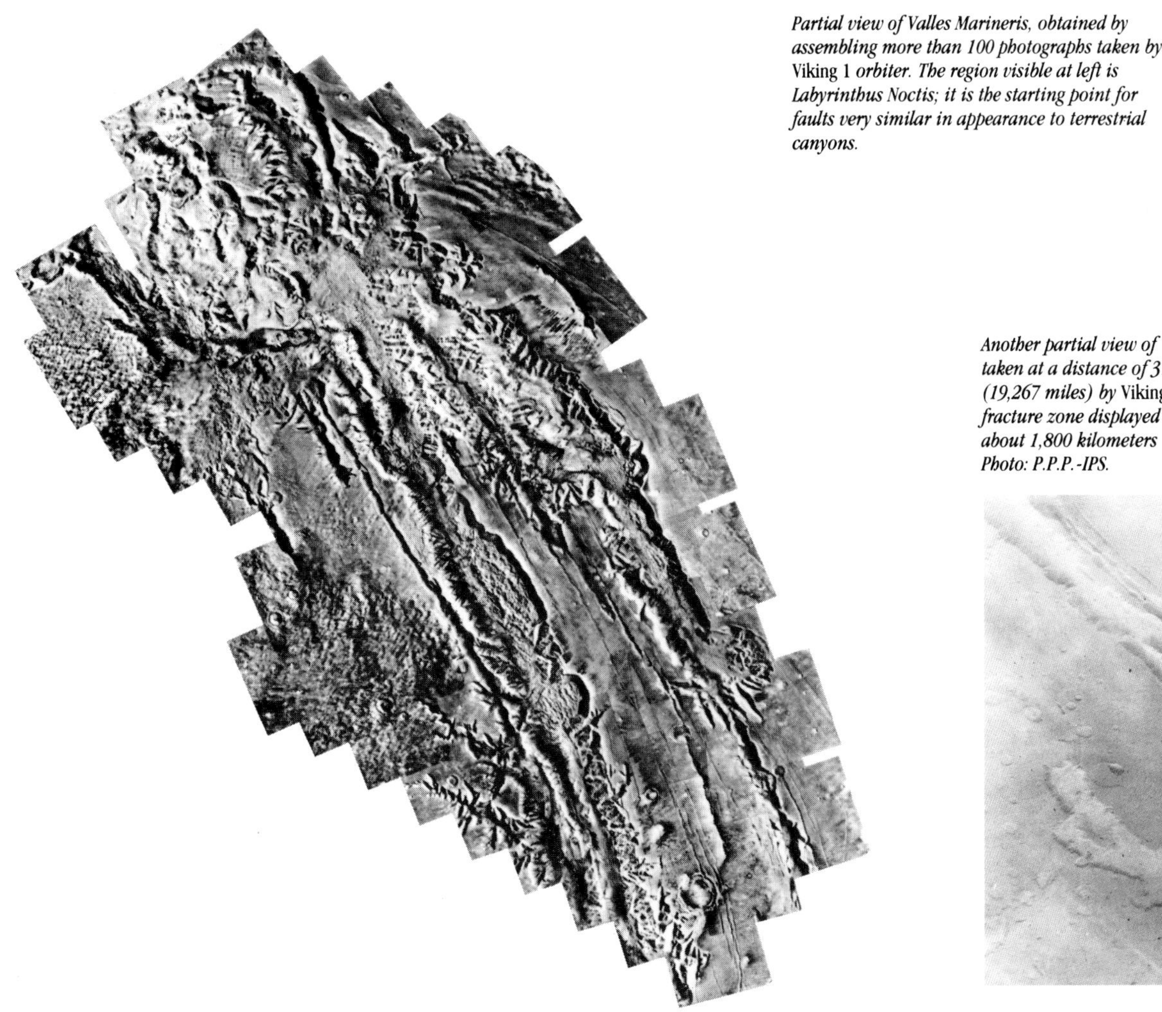

Partial view of Valles Marineris, obtained by assembling more than 100 photographs taken by Viking 1 *orbiter. The region visible at left is Labyrinthus Noctis; it is the starting point for faults very similar in appearance to terrestrial canyons.*

Another partial view of Valles Marineris, taken at a distance of 31,000 kilometers (19,267 miles) by Viking 1 *orbiter. The fracture zone displayed extends over about 1,800 kilometers (1,119 miles). Photo: P.P.P.-IPS.*

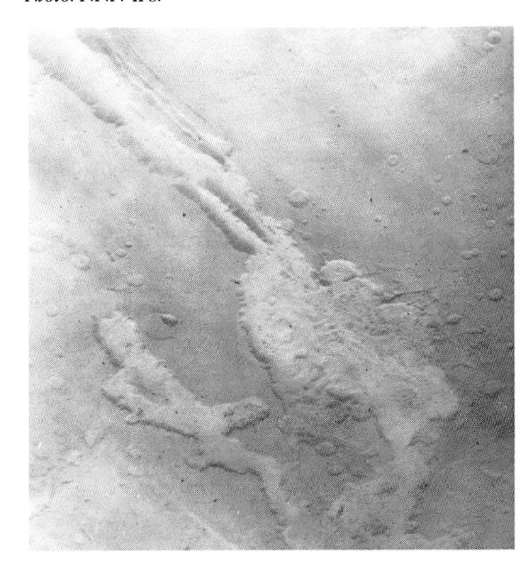

Aside from the presence of volcanic chains, one of the most spectacular discoveries of the space probes was the existence, near the equator in the Southern Hemisphere, of an immense chasm extending over a length of more than 2,500 kilometers (1,554 miles), a width of 120 kilometers (75 miles) and a depth of up to 6 kilometers (3.7 miles). Even though it appears startlingly similar, in places, to the canyons seen on Earth, this very unusual formation, Valles Marineris, is more closely related to the East African rift; it does not result originally from the effects of erosion but from a fracture in the Martian crust, and is, in fact, a complex system of more or less parallel and overlapping faults.

The photographs taken by spacecraft also revealed the surprising presence on Mars of sinuous valleys that nearly always follow the slope of the terrain, often possess tributaries and have all the characteristics of dry riverbeds including, in particular, the presence of alluvial deposits and tear-shaped islets pointing in the direction of the slope. One of the most remarkable of these valleys winds over 400 kilometers through Mare Erythraeum, its width being about 5 kilometers (3.1 miles). Other substantially wider formations (60 kilometers [37 miles] in places) resemble a particular type of terrestrial valley that probably originated in torrential floods. These two kinds of terrain—referred to as channels in Martian topography—evidently indicate that a liquid once flowed on the surface of Mars. This liquid is presumed to have been water, but that hypothesis implies that the planet's climate was warmer and more humid than it is today. Indeed, as a result of the very low atmospheric pressure (see following material), water cannot exist today in a liquid state on the surface of Mars. Given the prevailing temperatures on the planet, it would be transformed into ice and would then change directly to vapor through the phenomenon of sublimation. However, it is not out of the question that the Martian subsoil may contain large quantities of water in the form of permafrost. The "chaotic terrain" observed on certain portions of the Martian surface, especially in the Capri zone, may correspond to regions in which the ground was once churned up by the melting of the permafrost; thus it may have formerly been the site of water flows from which some of the identified channels may have sprung.

Finally, observations of spacecraft have improved our knowledge of the polar regions. These are covered with glacio-aeolian deposits, that form the famous "polar caps" clearly visible from

Earth, made up of a mixture of ice and dry ice. These polar caps alternately expand and contract according to the seasons. Their contraction leaves a dark fringe behind for some time, which was formerly thought to be terrain soaked by the melting ice but that actually represents dust deposits transported by the winds caused by the large difference in pressure between the polar zones and the neighboring regions. In winter, the polar caps may extend below 60° latitude, but in summer they rise just above the 80th parallel. Indeed, there remains a core of frozen water about 100 kilometers (62 miles) in diameter and probably several hundred meters thick. This permanent structure, which hugs the surface, is covered, in winter, by a thin deposit of dry ice, which descends to about 60° or 65° latitude. In summer, the temperature becomes sufficient (higher than − 125° C) for this deposit to sublimate. We now have proof that frost deposits also may form at certain times of the year in regions of lower latitude: In September 1977, during the winter in the Northern Hemisphere of Mars, photographs taken by the *Viking 2* lander at the Utopia Planitia site showed the soil covered by a thin layer of frost. It lasted about 100 days. It was then thought that the phenomenon had occurred subsequent to a violent dust storm that had swept over the planet several weeks earlier. But the appearance of a new ice deposit in the same place in May 1979, almost exactly one Martian year later, without any storm, led experts to offer another explanation: The phenomenon was due to the precipitation of dust suspended in the atmosphere, made heavier by their accumulation of small bits of ice, and the deposit of frozen carbon dioxide on their surface. After these particles fell to the ground, the carbon dioxide evaporated in the sunlight and returned to the atmosphere, releasing the dust and ice. The thickness of the film formed in this way would be no greater than 0.002 centimeter.

Around the southern polar cap are many meteoritic craters more or less covered with sedimentary material. A basin marks the site of the pole.

Structure and Composition of the Soil

From the depth of the imprint made by the Viking spacecraft that landed on Mars, it could be established that the soil of Mars, at the landing sites, had a consistency fairly similar to that of the lunar soil. In addition, the trenches dug by the lander's mechanical shovels, used for taking samples, showed that the surface matter is granular and has good cohesion (the edge of the trenches remained firm).

The panoramic photographs transmitted by the Viking landers after their touchdowns revealed vast rock-strewn expanses. As on the Moon, the rocks making up the soil are igneous rocks and breches, but it seems that the regolith—the granular, irregular layer, to which has been added the debris resulting from the impact of dust and meteorites—is thicker than the lunar regolith, more than 100 meters (328 feet). The granular layer is made up mainly of silicates combined with particles of iron oxide. The characteristic reddish color of the soil is ascribed to the presence of maghemite (Fe_2O_3), a mineral that is a ferromagnetic form of hematite.

The analysis of soil samples by X-ray spectrometers aboard each lander revealed proportions of about 50 percent oxygen, 20 percent silicon, 14 percent iron, 2 to 7 percent aluminum and smaller proportions of other elements. Relative to the average composition of terrestrial rocks, the main difference lies in the iron content, about three times greater (the third most abundant element on Earth, after oxygen and silicon, is aluminum).

Atmosphere; Meteorology

Before reaching the ground, the two Viking landers passed through the Martian atmosphere and determined its composition: It is composed of 95.3 percent carbon dioxide, 2.7 percent nitrogen, 1.6 percent argon and traces of oxygen, carbon monoxide and other gases. The abundance of rare gases (neon, krypton, xenon, argon) is comparable to what is found in the terrestrial atmosphere, which implies that the quantity of rare gases, per gram of planet, is much lower on Mars than on Earth. The water vapor content is 0.035 percent, on average, but varies with the seasons and the region. If all of the water in the atmosphere could precipitate on the ground, it would form a layer with a thickness varying between only 0.001 and 0.1 millimeters, which compares to 25 millimeters, on average, on Earth.

Atmospheric pressure on the plains at ground level varies from about 5 to 7 millibars, and 6.1 millibars has been chosen as the standard reference pressure defining zero altitude (the corresponding pressure on Earth is 1,013 millibars).

Temperatures are low, and thermal variations during the day are large. At the equator, the extreme temperatures recorded reach +22°C during the day and −73°C at night. At the Viking lander sites, temperatures ranging from −21°C to −83°C were recorded. The lowest temperature (−124°C) was measured by Viking *Orbiter 2* at

The extensible arm and excavator shovel of the Viking 1 *lander stand out against the reddish-orange soil of Chryse Planitia. Photo: P.P.P.-IPS.*

The Myth of the Canals

In 1864, the British astronomer Dawes noticed that many Martian "seas" ended in finely tapered extensions that led into long, very narrow dark branches extending across the "continental" regions. Father Angelo Secchi, an Italian Jesuit and astronomer who subsequently observed these strange formations, called them *canali* (arms of the sea), which was translated into English and French, quite improperly, as canals. This created a great stir beginning in 1877. In that year, thanks to an unusually close approach of Mars, many observers believed they could confirm the existence of the canals, particularly the Italian Schiaparelli, who tried to study them in detail. The astonishment aroused by these formations reached its peak, when, in 1882, Schiaparelli announced he had established the doubling (twinning) of many canals at certain times. From then on, astronomers divided into two camps: canalists and others.

In 1894, one of the most ardent supporters of the canals—the American Percival Lowell—decided to devote himself to studying the planet. To that end, he invested his wealth in constructing in Flagstaff, Arizona, an observatory that had a reflecting telescope with a 60-centimeter (24-inch) aperture (it is now, together with the Meudon Observatory in France, one of the two international centers for planetary documentation). After several years of diligently studying the Martian disk, which enabled him to "discover" a whole series of new interlocking canals, he became convinced that this geometric network could only be of artificial origin. He saw it as a vast irrigation network built by intelligent beings to irrigate the desertlike Martian plains with water from the melting polar caps.

These conclusions, which did not fail to strike the imaginations of others, led to the appearance of Martians in literature. In 1897, the German novelist Kurt Lasswitz imagined them landing on Earth, at the North Pole. The following year, the British writer H.G. Wells published his famous *War of the*

the South Pole during the southern winter.

Thermal variations seem to be the principal agents of atmospheric circulation: From one day to the next, the winds always blow in the same direction at the same times; and, at the same times of the year, they seem to regain the same intensity. Sweeping over a dry soil, they raise the iron-rich dust that causes an absorption and scattering of sunlight, giving the atmosphere a brownish pink color, pale near the horizon and yellowing near the zenith.

Generally, only a small breeze blows, its speed varying between 3 and 35 kilometers (1.9 and 22 miles) per hour, with the average at around 9 kilometers (5.6 miles) per hour. But at certain times, violent storms suddenly break out and can affect vast regions. The winds, then with speeds exceeding 200 kilometers (124 miles) per hour, carry reddish-brown dust up to altitudes as high as 50 kilometers (31 miles), and the planet's surface is then hidden. Particularly violent storms were observed in 1892, 1924, 1941, 1956 and 1971. The storm in 1971 was raging while *Mariner 9* was placed in orbit around Mars on November 14, 1971; it meant that the first photographs transmitted by the probe showed no details of the Martian surface. It is owing to such storms that certain regions undergo changes in shape and color, a phenomenon that once intrigued astronomers greatly. These changes, which some saw as an indication of the existence of vegetation on Mars, actually result from dust deposits or from the formation of trails behind certain craters. The appearance of dark spots is caused by a shifting of surface matter, revealing a darker substratum, while the formation of light spots results from deposits of a fine dust layer upon the soil. Thus wind has now proved to be the major agent of change on the Martian surface. Moreover, intense aeolian activity is evidenced by the existence of dune flats. The largest one, Hellespontus, west of the Hellas basin, occupies an area of 1,600 kilometers^2 (944 mi^2). Another one, near the Ganges chasm, covers 1,000 km^2. Both show parallel peaks spaced 1 to 2 kilometers (.62 to 1.2 miles) apart.

In addition to the yellowish dust clouds raised by windstorms and observed since 1879, the Martian atmosphere also contains real clouds. A large, very white spot of variable dimensions regularly appears on the Martian disk above the Tharsis plateau. In addition, the Viking Orbiter cameras recorded the appearance, directly above many regions of Mars's surface, of more localized small white spots. All of these spots are identified as condensation clouds resembling cirrus clouds on Earth, which on Mars owe their white color to ice crystals: dry ice above the poles, and water ice at intermediate latitudes.

Moreover, the *Viking 1* orbiter photographed two formations in the atmosphere of the Northern Hemisphere that resemble terrestrial cyclones and that also may be clouds of condensates composed of ice crystals.

To measure atmospheric pressure, temperature and velocity of the wind on Mars's surface several times a day, each Viking lander possessed a small meteorological station (2.5 kilograms [5.5 pounds]) set up on the end of an arm (visible at right) about 1.2 meters (4 feet) above the ground. In the photograph below, left, taken at the landing site of the Viking 1 *lander in the Chryse Planitia basin, the horizon is 1.5 kilometers (0.9 mile) away; the big dark rock that can be seen at left is about 8 meters (26 feet) from the lander and measures approximately 1 meter (3.3 feet), in height and 3 meters (9.8 feet) in width. Photo: P.P.P.-IPS.*

Raised by the wind, fine iron-rich dust remains suspended in the atmosphere and gives the Martian sky a pinkish color. Photo below, right, taken on September 24, 1976, by the Viking 2 *lander. NASA Photo.*

Worlds, describing octopuslike Martians who land on Earth and spread terror before being decimated by microbes. These precursors rapidly found their imitators, and today there are countless works of fiction inspired by the possible presence of intelligent beings on Mars.

However, while Lowell's conclusions may arouse the keenest interest in the general public, they were soon challenged by many astronomers. Using reflecting telescopes more powerful than Lowell's, the American Barnard and the Frenchman Millochau declared that they could see nothing at the supposed position of the canals but an irregular scattering of pale, diffuse spots. On the other hand, the first detailed photographs of Mars, taken in 1907 by the American Earl C. Slipher, clearly showed a certain number of dark streaks with ragged edges, at the sites of the principal canals. But these photographs were very grainy, and opponents of the canals made much of the fact that the streaks that could be seen possessed neither the narrowness nor the geometric regularity of the lines traced by Schiaparelli and Lowell.

In 1909, after a series of observations made with the 81-centimeter (32-inch) reflecting telescope of Meudon, the Frenchman Antoniadi concluded that the canals did not exist—according to him, they were merely the result of optical illusions and corresponded to perfectly natural alignments of irregular spots. Observations by the American Barnard, using the large, 1-meter (39-inch) Yerkes reflecting telescope, and those by the American Hale, with the 1.52-meter (60-inch) Mount Wilson telescope, independently confirmed this result. Nevertheless, the debate continued for several more decades after the death of the canalists' leader, Lowell, in 1916.

It was only after 1941, with Bernard Lyot's excellent observations at the Pic du Midi Observatory, that the hypothesis of artificial canals was put to rest for all practical purposes. The legend collapsed completely in 1965, during the flight over Mars by *Mariner 4:* At that time, the first close-up photographs of the planet were taken and did not show the least trace of a canal. With the later probes, the nonexistence of the canals was not only confirmed, it was even possible to set up a correlation between the lines traced by Schiaparelli and Lowell and faults or alignments of craters on the planet. Now it is no longer possible to question the totally illusory nature of the canals.

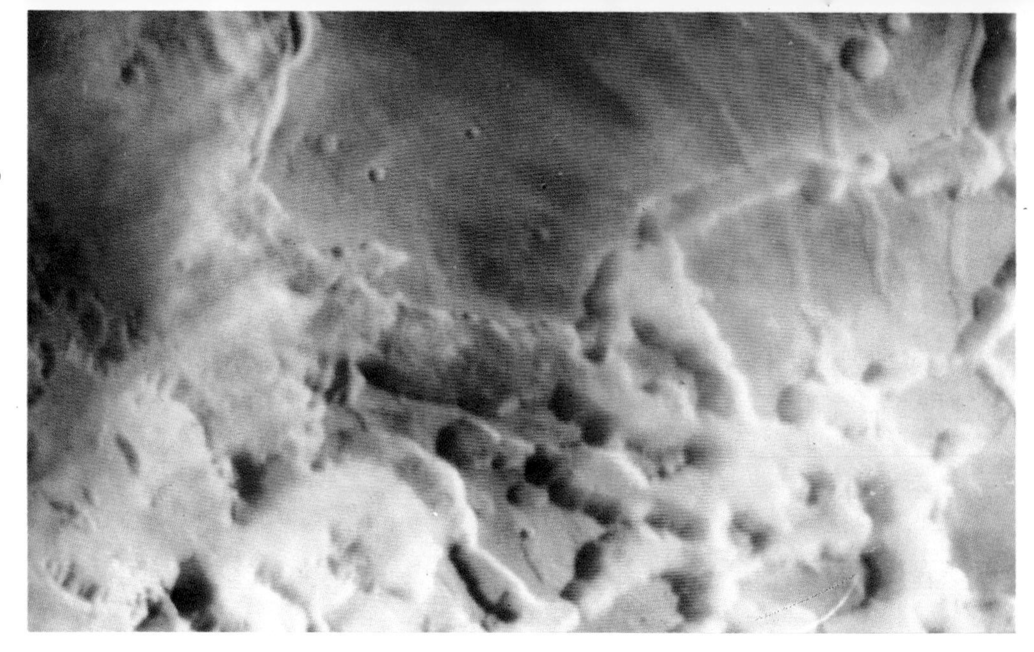

Ice clouds covering Labyrinthus Noctis at sunrise. The region depicted covers about 10,000 kilometers². Photri Photo.

Finally, observations have revealed the existence of a light mist above the ground at daybreak. It disappears during the day, as on Earth.

Gravimetry

The lowering of the *Viking 2* orbiter periastron to an altitude of 300 kilometers (186 miles) on October 25, 1977, made it possible to draw up a detailed gravimetric map of Mars.

The principal gravity anomaly, differing by 344 milligals from the mean, coincides with the Tharsis plateau. More specifically, it is centered on the great Olympus Mons volcano. The three other large volcanoes in this region also cause positive gravity anomalies, especially Arsia Mons. The measured gravity differentials make it possible to estimate the mass of the main volcano at $8.7.10^{18}$ kilograms. Alba Patera (40°N, 110°W), though not a remarkable topographic formation, nevertheless displays a strong gravity anomaly. Moreover, a mascon has been detected under Isidis Planitia, a depression at 12°N, 270°W; it is comparable in size to that of the Sea of Humors on the Moon (see "The Moon," page 147). This is the first mascon detected in the subsoil of Mars, and the only one, it seems, that is in the planet's Northern Hemisphere. Valles Marineris, the gigantic fault that runs along the Martian equator, shows a negative gravity anomaly, at its greatest between 30° and 75°W.

Seismic Activity

While the *Viking 1* lander seismometer could not be made to work, the seismometer on the *Viking 2* lander seismometer functioned well. It made it possible to establish that Mars is seismically a very quiet world. Only two quakes were recorded during an entire Martian year: the first one, measuring 7 on the Richter scale (hence, fairly sizable), on November 22, 1976; the second one, measuring 3, in January 1977. The epicenter was about 7,000 kilometers (4,351 miles) from the spacecraft in the first case, and 110 kilometers (68 miles) in the second. Mars, unlike the Earth or Moon, has no seismic "background noise."

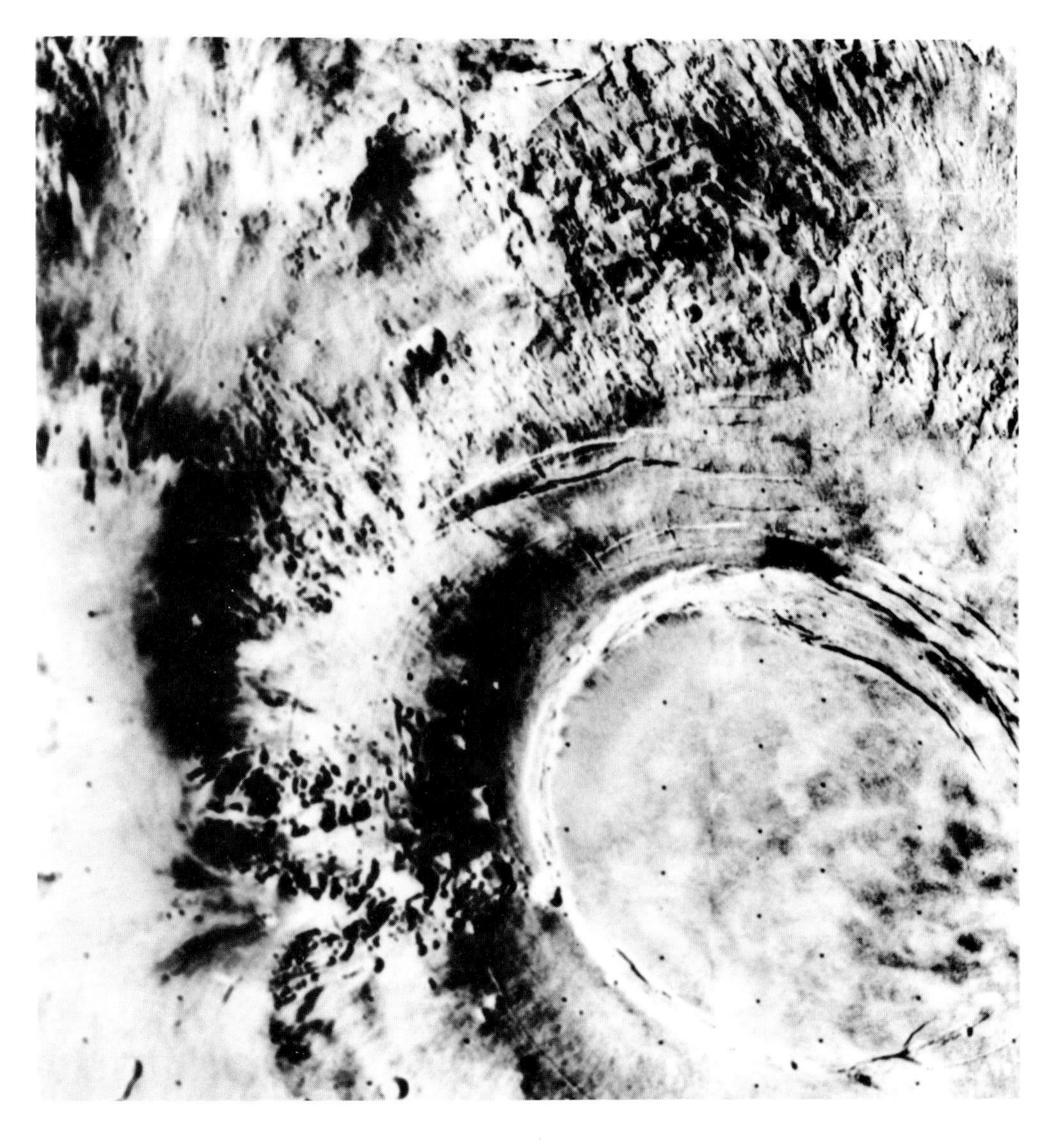

Arsia Mons. This is the southernmost of the three large aligned volcanoes that rise above the Tharsis plateau. Arsia Mons rises 27 kilometers (17 miles) above the surface level of Mars, and its caldera has a diameter of 100 kilometers (62 miles). In this picture, taken at a distance of 6,600 kilometers (4,102 miles) on July 29, 1976, by the Viking 1 orbiter, we can clearly see the lava flows that poured out of the central crater, and we also notice diffuse clouds similar in appearance to cirrus clouds. Photri Photo.

Internal Structure

The models of internal structure predict that Mars, like the other telluric planets, has a central core surrounded by a mantle and then an outer crust.

It seems that the crust has an average thickness of 40 to 50 kilometers (25 to 31 miles), reaching more than 80 kilometers (50 miles) under the high mountains and falling to only 8 kilometers (5 miles) under the great impact basins. It would therefore be much thicker than the terrestrial crust (which is only about 30 kilometers [19 miles] thick, for a planet whose diameter is approximately double that of Mars), which would explain its great stability. The core, however, must be relatively small: Its diameter would be about 2,500 kilometers (1,554 miles). Unlike the cores of the other telluric planets, it is not believed to contain much native iron but to consist essentially of iron sulphide, which has a density lower than that of the pure metal; this would explain Mars's lower overall density.

The Viking Probes and the Search for Life on Mars

Space exploration of Mars reached its zenith with the mission of the two American Viking spacecraft (3,450 kilograms [7,606 pounds]), launched on August 20 and September 9, 1975, respectively. Each craft contained two modules: an orbiter and a lander. The system was first placed into orbit around Mars; then the lander separated and made a soft landing on the planet's surface, while the orbiter remained in orbit. The two modules were equipped to do scientific research independently of one another. Each contained two television cameras (for photographing the ground), an infrared spectrometer (for measuring atmospheric water vapor) and an infrared radiometer (for remote measurement of the regions it passed over). The orbiters first made it possible to select the best landing sites for the landers, then acted as telecommunications relays between the landers and the Earth; this was done while carrying out a study of the atmosphere, and a mission to photograph the Martian surface from altitudes as low as 300 kilometers. The landers, for their part, were equipped with a meteorological station, cameras, a soil sampling device and a very sophisticated experimental package containing, among other things, three "life detectors" (see p. 315). The *Viking 1* Orbiter mission ended on August 7, 1980, and the *Viking 2* orbiter mission on July 24, 1978, when the gas reserves that made it possible to control the attitude and position of the spacecraft were used up.

The *Viking 1* lander touched down on July 20, 1976, in the Chryse Planitia region, at 22.5°N, and 47.8°W, and the *Viking 2* lander reached the surface on September 3, 1976, some 7,400 kilometers (4,599 miles) away, in the Utopia Planitia region at 48°N, 225.6°W, in the opposite Martian hemisphere. The *Viking 1* lander still is operating and should transmit meteorological data and photographs each week until the end of 1994. However, the *Viking 2* lander was ordered to cease operation in May 1979 after having transmitted a large quantity of information to Earth.

The most ambitious mission of the Viking spacecraft was to detect possible life forms on the Martian soil. Mars is, indeed, after Earth, the planet in the solar system where conditions are most favorable for the emergence of life as we know it. While water cannot now exist in a liquid form on the planet, its polar regions and probably its subsoil as well contain large quantities of ice; moreover, the winding valleys observed on its surface are most certainly dried-up riverbeds, and the presence of argon in the atmosphere is an indication that the atmosphere was once denser, which reinforces the hypothesis that water might have flowed on the planet in former times. Thus, while considering it highly unlikely that macroscopic life could exist on Mars, many scientists thought the existence of microorganisms perfectly plausible.

Although life may take other forms than those we know on other worlds, those in charge of the Viking program started from the hypothesis that any Martian organisms were very likely to manifest themselves by at least one of the two phenomena characteristic of life on Earth: the respiration of animals, which consume oxygen for their metabolic reactions and give off carbon dioxide; and the photosynthesis of plants, based, on the contrary, on the absorption of carbon dioxide and the discharge of oxygen.

To detect such manifestations of life, each Viking lander contained three analysis devices.

A fourth device, combining the technique of gas chromatography with that of mass spectrometry, was designed to analyze the chemical composition of the soil, and, in particular, to reveal the possible presence of organic molecules.

The various experiments conducted yielded results that initially left the specialists extremely perplexed. Indeed, the "life detectors" provided clearly positive responses, but the chemical analyzer—though extremely sensitive (it was capable of detecting a proportion of organic matter equal to one part in several dozen million parts of inorganic substance) did not find the slightest trace of organic molecules.

It seems, in fact, that the responses of the "life detectors" reflected a chemical activity peculiar to the Martian soil, which is continually subject to the Sun's ultraviolet rays. According to some experts, such activity is due to the presence of superoxides and hyperoxides on Mars's surface, but other chemical processes have also been suggested.

Mars and the Amateur

Provided you know what part of the sky Mars is in, it can easily be recognized with the naked eye because of its brightness and reddish color.

In a small reflecting telescope of 50 to 60 millimeters (2 to 2½ inches) in aperture, Mars appears as a small disk, uniformly ocher pink in color. This color is due to the presence of a fine layer of iron-rich dust covering the planet. Under the most favorable conditions, the most developed polar cap can be seen as a bright dot. Sometimes you can also see that the disk has a nonuniform brightness.

With a 75-millimeter (3-in.) telescope used at a magnification of 150, one can discern some spots on the surface, but with no detail.

To make interesting observations, one needs, at the very least, an 80-millimeter (3.1-inch) instrument with a 200 times magnification. The major Martian configuration then can be seen, particularly Syrtus Planitia, a dark, triangular spot in the equatorial region.

It is only with a telescope having an aperture of at least 150 mm (6 inches) that serious study can begin: variations in the extent of the polar caps, changes in color or appearance of the various configurations.

Because of the planet's rotation, a gradual change in its appearance can be noticed by observations spread over several consecutive evenings. However, since the Martian day is only slightly longer than the Earth day, it is only after four consecutive days of observation that the shifting of the Martian configuration becomes apparent. The planet returns to its initial appearance, at a given time, after a 40-day period.

The best time for observing Mars through a telescope is the period of several months around the opposition, when Mars and the Earth are on the same side of the Sun and the distance between them is minimal. The best conditions, which give the planet an apparent diameter midway between those of Jupiter and Saturn, occur only about every 15 years. Currently, Mars's apparent diameter is increasing: It is scheduled to reach 23.2 arc seconds during the opposition of July 10, 1986, and 23.8 arc seconds during the opposition of September 28, 1988. After that, it will decrease again.

For the beginning amateur, Mars remains a rather disappointing planet; its image in a small telescope never is overwhelming.

The Satellites of Mars

We know of two small satellites around Mars: Phobos and Deimos. There were people who imagined their presence long before they were discovered. It all began in 1610 when Galileo, having observed the strange appearance of Saturn through his telescope, wanted to insure priority for himself in this discovery without revealing it immediately. He had a message delivered to Kepler, consisting of a mysterious text. Kepler, naturally, strove to decipher this message, and he managed precisely to extract from it a sentence, apparently in Latin, that led him to believe that Galileo had discovered two satellites of Mars. Now, shortly before, Galileo already had discovered four satellites around Jupiter. Kepler, convinced of the harmony of the cosmos, happily saw in the presence of two satellites around Mars the confirmation of an order he had sensed in the solar system: The Earth had one Moon, Jupiter four, Mars—located between the Earth and Jupiter—two. . .

Alas! A few months later, Galileo revealed the precise content of his message, and Kepler's logic collapsed. No matter. From then on, Mars's two presumed satellites were spoken of periodically as though their reality were proved beyond a doubt. The Irish writer Jonathan Swift mentioned them in 1726 in *Gulliver's Travels.* Voltaire likewise alluded to them in 1752 in *Micromégas.* The astronomers mobilized: During Mars's opposition in 1783, William Herschel trained his large telescopes on the red planet in the hope of finally glimpsing the satellites, but he got no positive results. The searches made in 1830 by Mädler in Berlin using a 95-millimeter (4-inch) telescope, and in 1862 and 1864 by D'Arrest at the Copenhagen Observatory with a 25-centimeter (10-inch) telescope, proved to be equally fruitless. In 1877, around the time of a new opposition of Mars that looked to be very favorable (the planet was supposed to come within only 56.4 million kilometers [35 million miles] of the Earth), the American Asaph Hall took his turn and threw himself into the search for the famous satellites. To do so he used the Washington Naval Observatory's large telescope placed in service in 1873, which, with its 66-centimeter (26-inch) aperture, was the largest refracting telescope in the world at the time. Hall began his search at the beginning of August. He systematically scrutinized the neighborhood of Mars, taking care to mask the planet's bright disk or to take it out of the instrument's field so as not to be blinded by its brightness. Despite his efforts, Hall's first observations ended in failure. And, on August 11, the disappointed astronomer was about to give up; he yielded, however, to the entreaties of his wife, who urged him to continue his search for one more night. And on that night—August 12, at 2:40 A.M.—Asaph Hall perceived a small bright dot some 70 degrees to the east of Mars. Unfortunately, he had only a fleeting glimpse of it, for a thick fog rose from the Potomac River, rendering all observation impossible for the rest of the night. On the four following nights, the sky remained stubbornly cloudy. When it cleared, on the night of August 16, Hall began his observations again. The bright dot was indeed there. Was it not an asteroid accidentally present in that region of the sky? The astronomer hastened to consult his ephemerides. They showed that the minor planet 52 Europa should indeed be in the neighborhood of Mars. But on the following night all doubt was dispelled: Hall found his bright dot again, followed it for two hours and noted that it followed the motion of Mars, and that its motion could not coincide with that of 52 Europa. This time, doubt was no longer possible: The object seen was a satellite of Mars. Moreover, on that same night, Hall was surprised to see a second bright point, even closer to the planet than the first one: It, too, proved to be a satellite of Mars. Thus, as Kepler and others after him had imagined, two satellites orbited Mars: Reality coincided with fiction! Notified by Hall, several American astronomers confirmed his discovery. The news was then communicated to the Smithsonian Institution, which cabled it to the various observatories of the world. In France, Le Verrier was informed of it shortly before his death, and on August 27, the two satellites were observed for the first time at the Paris Observatory by the Henry brothers.

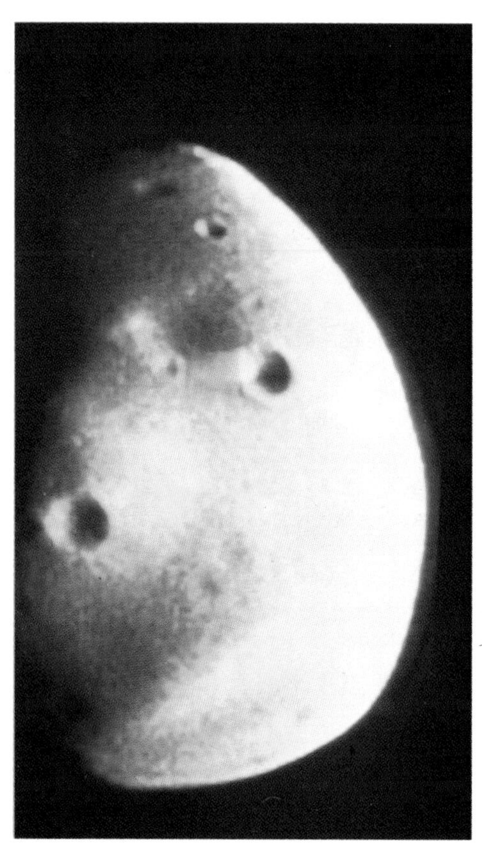

Deimos, the smaller of Mars's two satellites, photographed on March 3, 1977, by the Viking 1 *orbiter. This color picture was actually made by using a computer to combine two black-and-white photographs, one taken with a violet filter, the other with an orange filter. The two main craters are 1.3 kilometers (0.8 miles) and 1 kilometer (0.6 miles) in diameter; the smallest visible details measure about 200 meters (656 feet). P.P.P.-IPS Photo.*

On February 7, 1878, Hall suggested calling the inner Martian satellite Phobos ("fear," "bafflement") and the outer satellite Deimos ("terror"). These names are borrowed from Greek mythology, in which they stood for the horsemen of Aries, the god of war (equivalent to the Roman god Mars).

A Long Mysterious Appearance

Their apparent brightness and presumed albedo implied that they are miniscule, as could rapidly be ascertained from their distance; indeed the two satellites of Mars could be seen from Earth only as small, bright points. And while many observations, both visual and photographic, soon made it possible to determine their orbital characteristics with accuracy, their nature remained mysterious for a long time.

In 1945, the American astronomer S. P. Sharpless, after going over all the observations of Phobos since its discovery, revealed that the satellite's position was ahead of the predictions of the ephemerides. The implied slow acceleration in the motion of Phobos seemed inexplicable. However, in 1959, the Soviet astrophysicist I. S. Shklovsky announced that according to his calculations, the observed phenomenon could be ascribed to a slowing down of the satellite in the Martian upper atmosphere, provided Phobos was assumed to have an extremely low density, of only 0.00001 gm/cm³. From there it was only a short step (which some did not hesitate to take) to conclude that the satellite is hollow, hence artificial: Shklovsky's hypothesis caused a sensation!

But thanks to the American space probes Mariner and Viking, the faces of Mars's satellites have lost a good deal of their mystery. In August 1967, a picture taken by *Mariner 7* showed the shadow of Phobos on the Martian soil. Its elongated appearance suggested that the satellite—rather than being spherical—was shaped like a potato. It was also discovered that the satellite has a reflectivity of only 6.5 percent, much lower than what had been attributed to it; is larger than was supposed; and that its dimensions could be estimated at 18 by 23 kilometers (11 by 14 miles). With *Mariner 9*, in 1971–72, the detailed structures of Phobos and Deimos were finally revealed: The astronomers discovered two rocky masses of irregular shape. Their surfaces were found to be saturated with craters, the younger ones overlapping the older ones. Twenty-seven high-contrast pictures of Phobos (revealing 70 percent of its surface and making it possible to discern all details of the terrain exceeding 200 meters [656 feet] in size) and nine of Deimos (showing 40 percent of its surface) were transmitted by the spacecraft. Some 50 craters could thus be catalogued for Phobos, a map of which was drawn in 1973 by the American T. C. Duxbury, of the Jet Propulsion Laboratory. The largest of these craters, known as Stickney (the maiden name of Asaph Hall's wife), was no less than 8 kilometers (5 miles) in diameter, which represents nearly a third of the largest dimension of the satellite itself. Five other craters were given the names Roche (a 19th-century theoretical astronomer and expert in satellite dynamics), Wendell, Todd, d'Arrest and Sharpless. On Deimos, two craters were named Voltaire and Swift.

In addition, the dimensions of the two satellites were determined. Phobos and Deimos may be regarded in rough approximation as triaxial ellipsoids measuring, respectively, 27 by 21 by 19 kilometers (17 by 13 by 12 miles) (Phobos) and 15 by 12 by 11 kilometers (9 by 7.4 by 6.8 miles) (Deimos).

Finally, *Mariner 9* gathered data making it possible to know more about the orbits of the two objects. The position of the center of Phobos was pinpointed to within 2 kilometers and the eccentricity of its orbit—which was assumed to be 0.0184—proved to be only 0.015. But above all, it appeared that Phobos and Deimos, like the Moon around the Earth, were in synchronous rotation around Mars; they always showed the same side to the planet, toward which their long axes were constantly pointed, and their period of rotation is equal to their period of revolution, or 7 hours, 39 minutes for Phobos and 30 hours, 19 minutes for Deimos, with the two objects orbiting at distances of about 6,000 kilometers (3,729 miles) and 20,000 kilometers (12,430 miles), respectively, from the edge of Mars.

The exceptionally high velocity of Phobos means that it has a retrograde motion in the Mar-

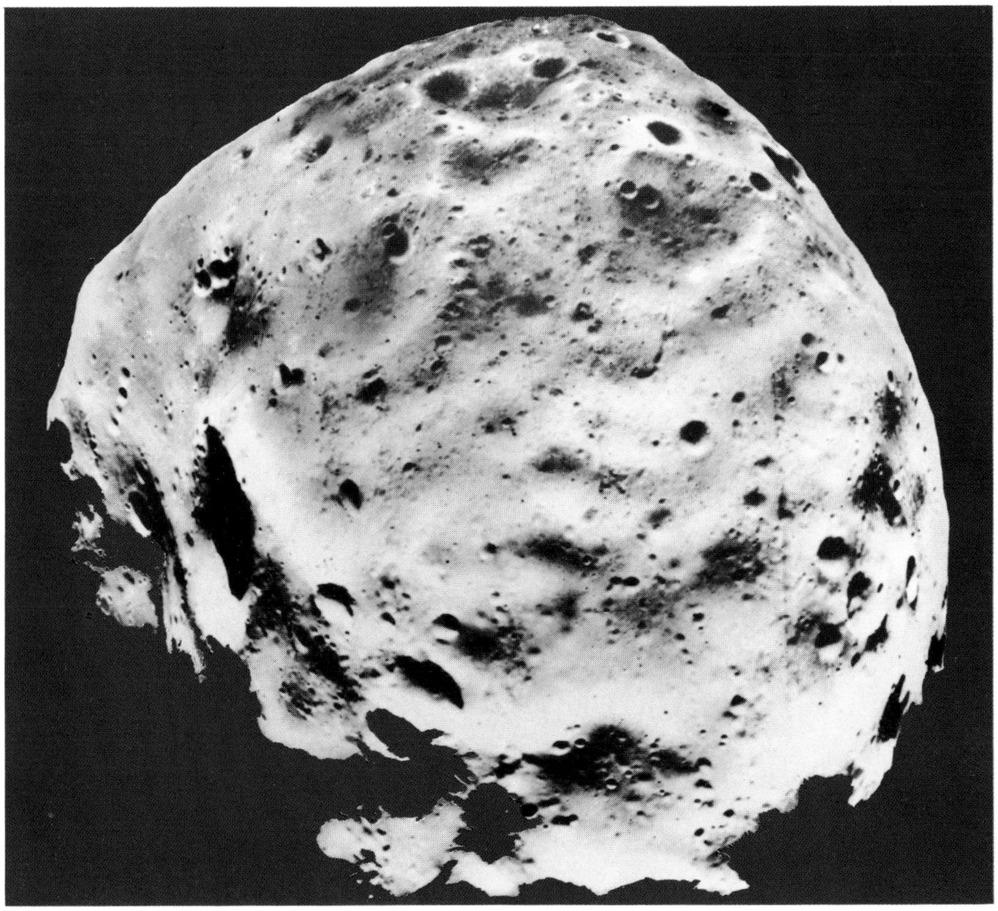

Phobos, photographed from a distance of 480 kilometers (298 miles) on February 18, 1977, by the Viking 1 *orbiter. North is at top. The smallest visible details measure about 20 meters (66 feet). Photri Photo.*

tian sky: An observer on the planet would see the satellite—contrary to the other planets—rise in the west and set in the east; at least in the places where it is visible (between 69° north and 69° south latitude), for it is too close to the planet to be observable from the higher latitudes. We can also predict that its phases will follow one another at such a fast pace that, if it appears as a thin crescent on the western horizon of the Martian sky, it will be "full" by the time it crosses the meridian and will already have resumed the appearance of a crescent when it disappears in the east.

As for the acceleration in the satellite's motion, estimated by Sharpless at 28.2 billionths of a degree per day, it has been recalculated: In 1972, A. T. Sinclair found that it had a value of only 14.4 billionths of a degree per day; then, in 1975, V. A. Shor, of the Leningrad Institute of Theoretical Astronomy, estimated it at 21.4 billionths of a degree per day. Though smaller than was formerly thought, this acceleration is no less real, and its cause can be traced to the anomalies of Mars's gravitational field. Assuming that the acceleration remains constant, Phobos will crash into Mars in about 100 million years, according to the American J. Veverka, of Cornell University. No acceleration in the motion of Deimos, however, has yet been demonstrated.

Two Captured Asteroids?

Since 1976, thanks to the two American Viking orbiters, our knowledge of the satellites of Mars has again been enriched with new information.

Phobos, photographed from a distance of 800 kilometers (497 miles) by the Viking 2 *orbiter. The smallest visible details measure about 40 meters (131 feet). Note the many parallel grooves running across the surface. P.P.P.-IPS Photo.*

In the most detailed photographs—taken by the *Viking 2* orbiter in 1977 at less than 100 kilometers altitude, which showed details as small as 10 meters (33 feet)—the surface of Phobos appears to be carved with long, parallel grooves: These may be fractures caused by underground mechanical tensions created by the tidal forces stemming from the planet Mars. Phobos does orbit very close to the Roche limit, inside of which the disintegration of a satellite is inevitable. However, the fact that these grooves predominate in the neighborhood of the Stickney crater, where they reach their greatest width, while the opposite hemisphere of Phobos is devoid of them, strongly suggests that their origin is tied to that of the crater; they may be surface manifestations of underlying fractures caused by the violent impact that created the crater. We also distinguish chains of small craters parallel to the equator; these formations are regarded as stemming from debris that was first ejected into space during a huge collision (the escape velocity on Phobos is only 54 kilometers [34 miles] per hour; a good thrower could toss an object directly into space from the ground!), then fell back to the surface after having orbited around Mars in an orbit close to that of the satellite. This process, which supposedly created the secondary craters, also accounts for the presence on the surface of Phobos of a compact dust layer comparable to what was revealed on the Moon. Pounded by meteorites (which we can assume are very numerous in the neighborhood of Mars, owing to the proximity of the asteroid belt), the soil of Phobos must certainly have undergone removal of dust particles, which, even though very small, would behave in a manner similar to the debris mentioned earlier.

No striations comparable to those of Phobos appear on Deimos, perhaps because of the absence of craters as large as Stickney (the satellite does not seem to have craters more than 3 kilometers in diameter). In all the photos taken by the Viking orbiters, the surface of Deimos appears clearly less marred than that of Phobos. However, the crater density on the two objects is about the same. Close-ups of Deimos, taken in 1977 by the *Viking 2* orbiter with a resolution of 2 to 3 meters (1.2 to 1..9 feet), did away with this apparent contradiction: They show that Deimos is indeed saturated with craters, but that most of them, unlike those on Phobos, are partially filled with a uniform, light-colored matter. Thus the craters on Deimos are as abundant as on Phobos but tend to be obliterated. Furthermore, we observe on the surface of Deimos a layer of various-sized debris not found on Phobos. This phenomenon might be explained by the fact that the two satellites have different mechanical properties and that meteoritic impacts on Deimos generate more debris likely to remain on the ground than on Phobos.

The origin of Phobos and of Deimos remains controversial. The synchronous rotation of the two satellites and their surface appearance testify to their ancient origin. Moreover, it seems that they have a common origin, because of the identical direction of their orbit around Mars and their similar photometric characteristics. But did they form by the accretion of matter expelled by Mars at the time of its formation and remained in orbit around the planet, or are they asteroids that were captured by the Martian gravitational field? The average density of the two objects, deduced from the mass and volume that can be ascribed to them from the data furnished by the spacecraft that flew over them, tends rather to support the second hypothesis. This average density does appear rather close to that of carbonaceous chondrites (2.3 gm/cm^3), a variety of meteorites related to most of the asteroids orbiting between Mars and Jupiter. Perhaps Phobos and Deimos are two fragments of a single asteroid that broke up during a passage near Mars or during a collision with another minor planet. But if they are captured asteroids, it remains to be explained why they orbit essentially in the equatorial plane of Mars and not in some other plane. The most plausible hypothesis is that the two satellites were captured a long time ago, at a time as much as 3 to 4 billion years ago, and that, since then, in accordance with the laws of mechanics, their orbit has become nearly circular, with a nearly zero inclination relative to the Martian equator.

12. Jupiter

Planets of the same type as the Earth, and their satellites, which have been discussed in the preceding chapters, represent only a very small fraction of the total planetary mass of the solar system. More than 99.5 percent of the planetary matter lies beyond the orbit of Mars. This region is dominated by a family of four planets having very different characteristics from those of the telluric planets: Their dimensions are much larger, their radii ranging between 3.7 and 11.2 times that of the Earth; their average density, on the other hand, is much lower, varying from 0.7 to 1.7 gm/cm³. Finally, they have an extremely rapid rotation, with periods ranging from 10 to 16 hours, causing their very pronounced flattening. The largest and most massive of these giant planets is also the closest: Jupiter.

Jupiter: a world of hydrogen, prototype of the giant planets. Photograph taken by Pioneer 11 *on December 2, 1973, from a distance of about 2.6 million kilometers (1.6 million miles). The visible details are cloud formations; the small black disc is the shadow of Io, one of the planet's principal satellites. NASA Photo.*

The Orbit

Jupiter completes its nearly circular orbit in 11 years, 10 months 17 days at an average distance from the Sun of 778.3 million kilometers (483.7 million miles) (or 5.2 times the average distance of the Earth from the Sun). Because of its remoteness, Jupiter receives 27 times less solar energy, per unit surface area, than the Earth. At its perihelion, Jupiter comes within 740 million kilometers (460 million miles) of the Sun; at the aphelion, it lies 816 million kilometers (507 million miles) away. The inclination of the plane of its orbit relative to that of the Earth's orbit is only 1° 18′ 28″; hence the orbit barely deviates from the ecliptic.

Size and Mass

The ancients had the right idea in giving this planet the name of the ruler of Olympus. It is indeed by far the largest planet in the solar system. Its dimensions and mass now are known with fair precision; visits by spacecraft to this planet have made it possible to reduce the uncertainty surrounding previous estimates made from Earth. Its equatorial diameter is 142,796 kilometers (88,748 miles), or 11.2 times that of the Earth, and its polar diameter is 133,540 kilometers (82,996 miles); the difference between these two measurements indicates a flattening of 6.2 percent. Its mass is 318.95 times that of the Earth and nearly 2.5 times the sum of the masses of all the other planets together. Its average density compared with water is only 1.31, which in view of its volume (1,338 times that of the Earth), implies a structure very different from that of the Earth, which has a density four times greater.

The great red spot and neighboring regions, photographed by Voyager 2 *on July 3, 1979, from a distance of 600,000 kilometers (372,902 miles). We can clearly see the whirlpool structure of the spot, as well as a vast zone of turbulence to the west (at left). NASA-Jet Propulsion Laboratory Photo.*

Nomenclature of Jupiter's bands and zones.

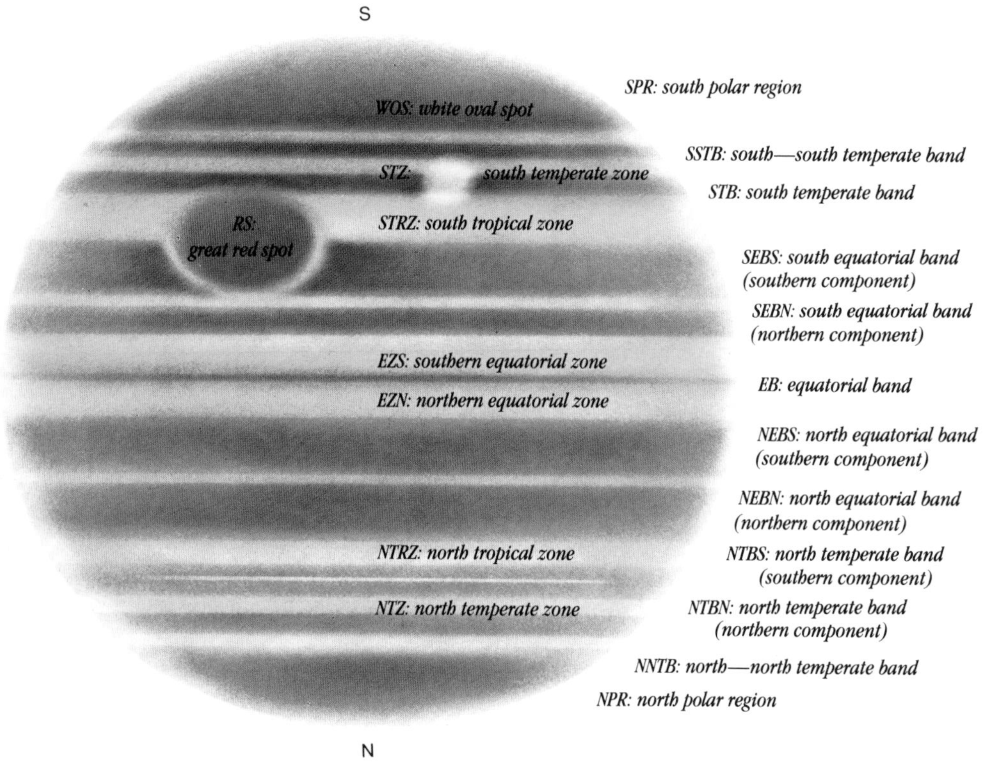

S
WOS: white oval spot
STZ: south temperate zone
RS: great red spot
STRZ: south tropical zone
EZS: southern equatorial zone
EZN: northern equatorial zone
NTRZ: north tropical zone
NTZ: north temperate zone
SPR: south polar region
SSTB: south—south temperate band
STB: south temperate band
SEBS: south equatorial band (southern component)
SEBN: south equatorial band (northern component)
EB: equatorial band
NEBS: north equatorial band (southern component)
NEBN: north equatorial band (northern component)
NTBS: north temperate band (southern component)
NTBN: north temperate band (northern component)
NNTB: north—north temperate band
NPR: north polar region
N

Telescopic Appearance

Seen through a telescope, Jupiter's disk—the apparent diameter of which reaches 48 arc seconds during the most favorable oppositions—has a very characteristic striped appearance. Light and dark parallel bands alternate on both sides of the equator. The light bands usually are denoted as zones, the term "band" being reserved for the dark bands (also called belts). In them we see knots and filaments, which often represent an exchange of matter from one region to another. The more or less rapid evolution of these features reveals the existence of currents that may be rising, falling, lateral or eddies. We also note the presence of structures that appear to be permanent but that change their contour, color or contrast over time; the most famous of these is the great red spot in the Southern Hemisphere.

Rotation

All of the visible characteristics on Jupiter's surface belong in fact to cloudy layers; their period of rotation has been precisely measured. It varies with latitude (we are clearly in the presence of a fluid), but instead of gradually increasing from the equator to the poles, as on the Sun, it changes suddenly, making it necessary to establish several systems of coordinates. The equatorial zone, which extends up to latitudes of $+10°$, forms what is called system I, with the rest of the planet constituting system II. Neither of the two rotates in a uniform manner, but standard rotation periods have been assigned to them:

- 9 hours, 50 minutes, 30.003 seconds (or 9.8416675 hours) for system I
- 9 hours, 55 minutes, 49.062 seconds (or 9.9277950 hours) for system II

Thus we can determine the longitude of a visible detail on the planet by noting the time at which it crosses the central meridian and consulting tables that give the longitude of this meridian in the two systems at regular intervals (system I being used in the equatorial zone and system II everywhere else). If the detail rotates exactly with the standard period, it retains the same longitude at each crossing of the central meridian; if not, its longitude increases or decreases from one crossing to the next, according to whether it rotates slower or faster than the standard period. In general, the observed details do not rotate exactly with the standard period, but their real rotation period may easily be calculated from their gradual change in longitude. Finally, the study of radio-frequency emissions from Jupiter has led to the establish-

The Great Red Spot

Discovered in 1664 by the Briton Robert Hooke and observed soon after by Jean Dominique Cassini, this strange oval formation, of variable size, is visible in Jupiter's Southern Hemisphere, where it extends over 28,000 to 40,000 kilometers (17,402 to 24,860 miles) in longitude and 13,000 kilometers (8,080 miles) in latitude. Bordered on the south by the south temperate band and on the north by a "cavity" in the south equatorial band, it appears to be completely immersed in the brightest zone, the south tropical zone. Its period of rotation is not strictly the same as the planet's at the same latitude, and therefore it undergoes a kind of irregular longitudinal drift, as though it were "floating" in the Jovian atmosphere. Thus, after having oscillated slightly around an average position from 1880 to 1910, it completed three revolutions around the planet in the forward direction between 1910 and 1942, then a retrograde revolution between 1942 and 1962. This overall motion is supplemented by small oscillations with a period of exactly 90 days. Moreover, it is a center of intense activity, as shown by its extremely fluid appearance: vortices develop within it, and it exchanges matter with exterior regions.

Finally, we note its cyclical evolution. Its color, noticed for the first time in 1872 by Lord Rosse, may be more or less intense and may vary from pink to brown, passing through an entire range of intermediate colors. Very prominent in some periods, as in 1962–63 and 1973–74, sometimes it becomes so pale, on the other hand, that it disappears completely, as in 1665, 1875, 1927 and 1936, though its site remains marked by the "cavity" it forms in the south equatorial band.

It was long thought that what was involved was a solid island floating on a dense at-

mosphere (a hypothesis offered by G. Hough in 1881). But we now know that no solid substance can have a density low enough to float on Jupiter's upper atmosphere. According to another, more recent suggestion, the spot is the top of a Taylor column—i.e., a stationary fluid column produced by the interaction of an atmospheric current and a topographical irregularity, or else from a local irregularity in the magnetic field, but this theory encounters many objections.

Now that we have high-resolution photographs of this spot, thanks to the space probes, it has become clear that what is involved is a gigantic atmospheric vortex, as the American Gerard Kuiper first suggested. This giant hurricane originated at the boundary between opposing horizontal currents circulating north and south of the south tropical zone. The phenomenon, then, is closely linked to the particular zone one is observing. This hypothesis explains the existence of smaller spots having otherwise more or less similar characteristics and that have been seen in other bright regions of Jupiter. However, these spots often are ephemeral, and the permanent nature of the great red spot remains to be explained.

Measurements taken in infrared have revealed that the top of the spot is at a temperature of −146°C, lower by 2°C than that of the surrounding clouds; thus it would emerge at some 8 kilometers (5 miles) above the cloud layer.

As for the color of this strange formation, it could be due to the presence of red phosphorus, possibly formed by photolysis of the phosphine discovered in Jupiter's high-contrast infrared spectra. Alternatively, it may be produced by organic compounds formed in the very special medium that is Jupiter's atmosphere.

ment of system III, which has a rotation period of 9 hours, 55 minutes, 29.75 seconds, corresponding to that of the planet's magnetic field (see page 178).

Atmosphere

The composition of the atmosphere has been determined by spectroscopic measurements. The first observations of absorption bands in the spectrum of Jupiter were made in 1863 by the Italian Angelo Secchi, but it was not until 1932 that the Americans R. Wildt and T. Dunham showed that those spectral bands were due to the presence of methane (CH_4) and ammonia (NH_3). Although these bands are the dominant characteristics of Jupiter's optical spectrum, methane and ammonia are not the principal constituents of the Jovian atmosphere. The major constituent is in fact molecular hydrogen (H_2), the very faint lines of which were identified in Jupiter's spectrum in 1960 by the Americans C.C. Keiss, H.K. Keiss and C.H. Corliss. Jupiter's low average density indicated the presence of another major constituent, which could not be detected in the optical spectrum because of low atmospheric temperature: helium. This element was indeed revealed in 1973 by the *Pioneer 10* probe owing to helium's ultraviolet line at 58.4 millimeters wavelength. With 82 percent hydrogen and 17 percent helium (the other elements account for only 1 percent), Jupiter's atmosphere has a composition very close to that commonly supposed for the nebula from which the solar system is believed to have emerged.

Other molecules have also been detected in small quantities: ethane (C_2H_6), acetylene (C_2H_2) and phosphine (PH_3). *Voyager 1* uncovered the presence of water vapor (in small quantities), which previously had been uncertain. Finally, rare isotopes exist in trace quantities, particularly deuterium (heavy hydrogen) and carbon-13. It is probable that other molecules exist that can not be detected as yet because of the low intensity of their spectral lines.

The visible clouds—or at least their upper layers—are made up of particles of solid ammonia, analogous in this to the terrestrial cirrus clouds made up of ice crystals. This layer seems relatively tenuous, since it allows an underlying layer of thicker clouds to show through, near the center of Jupiter's disk. The composition of this underlying layer is poorly understood but its principal constituent probably is ammonium hydrosulphide (NH_4SH). At still lower levels there may be a layer of clouds made up of ice crystals, and a zone of water droplets.

Various suggestions have been made to explain the color of the clouds. The planet's dominant yellowish color might be due to the presence of ammonium sulphide within the cloudy layer of ammonium hydrosulphide. The presence of phosphine is perhaps at the origin of the formation of phosphorus, which would explain the reddish coloring of some cloudy formations.

Study of the propagation of the radio signals from Voyager probes through Jupiter's atmosphere has made it possible to draw up a vertical temperature profile. Thus it has been possible to locate the level of the tropopause—i.e., the top of the troposphere (containing the cloudy layers). There one finds a pressure of 100 millibars and a temperature of $-160°C$. If we take the 20 bars pressure level as zero altitude (that is the pressure of the deepest layers we have been able to analyze) and assume an adiabatic temperature variation, we calculate that the tropopause should be at 160 kilometers (99 miles) altitude. In addition, space probes have demonstrated a warming in the stratosphere; the temperature there increases with altitude, rather than remaining roughly constant, as in the Earth's stratosphere.

Solar energy that is not reflected by the clouds is absorbed by the atmosphere. If the quantity of absorbed energy alone were reemitted, Jupiter would have a real temperature of 105°K ($-168°C$)—i.e., it would radiate the same quantity of energy as a black body at 105°K. However, measurements show that the real temperature is actually 125°K ($-148°C$), implying that Jupiter radiates more than twice as much energy as it absorbs from the Sun.

Hence, there exists an internal energy source, and its presence has major consequences for the structure of the atmosphere. One of the great contributions of the space missions was precisely to foster a better understanding of the processes of atmospheric circulation.

Spacecraft measurements have revealed that the dark and sunlit hemispheres of Jupiter both have the same temperature, implying that the solar energy absorbed into the atmosphere is transported to the dark side by the planet's rotation in a time sufficiently short that it is not reemitted by the atmosphere. Likewise, it seems that there is no temperature difference between the polar and the equatorial regions: Jupiter's atmosphere is a huge heat reservoir.

In addition, direct measurements from spacecraft have shown that the temperature of the upper layers of the zones is lower (by about 9°K) than that of the upper layers of the bands and that the altitude of the zones is greater by about 20 kilometers (12.4 miles) than that of the bands.

Unlike the Earth's atmosphere, the movements of which are controlled from outside, by incident

At left, Jupiter photographed by Voyager 1 *on January 24, 1979, from a distance of 40 million kilometers (24.9 million miles). Photo: NASA-Jet Propulsion Laboratory.*
At right, Jupiter photographed by Voyager 2 *on May 9, 1979, from a distance of 46.3 million kilometers (28.8 million miles). Photo: NASA-Jet Propulsion Laboratory.*
A comparison of these two photographs, taken at a 3½-month interval, makes it possible to assess the importance of the changes taking place in Jupiter's atmosphere, despite the existence of some permanent structures. We can note, in particular, important changes around the great red spot: An oval white spot, southwest of the red spot, shifted 60 degrees to the east during the period between the two photos, enabling another cloud formation, located directly to the west (left) of the white spot in the picture taken by Voyager 1, *to position itself just below the red spot. At the very bottom of the photo taken by* Voyager 1, *we see the satellite Ganymede. Photo: NASA-Jet Propulsion Laboratory.*

solar radiation, Jupiter's atmosphere is governed from inside, by the radiation issuing from the interior of the planet itself. This internal heat sets up convection currents inside the atmosphere as it rises. The lighter-colored zones should be regarded as regions where matter rises, heated from below in this way. On the other hand, the darker bands are trenches into which the cooled matter falls. According to the explanation generally accepted nowadays, these bands and zones are, to a certain extent, the equivalent of cyclones and anticyclones in the Earth's atmosphere. Their shape appears very different, however, for Jupiter's very rapid rotation generates an extremely strong Coriolis force, which stretches the cells of rising and falling matter lengthwise, giving them the shape of toroidal systems surrounding the planet like veritable belts and leading to the planet's well-known striped appearance.

At latitudes greater than 50 degrees, however, the parallel band structure disappears. There one finds turbulent formations resembling terrestrial cyclones in that they are more or less circular and of variable size, the difference being that they are stationary and can remain for several weeks without any noticeable structural changes.

In addition to the vertical circulation we've described, there is a horizontal circulation. Rising matter, when it reaches the top of a zone, spreads out in all directions, but its movements toward the north or south are transformed into flows in the east–west directions as a result of the Coriolis force. Each zone is thus characterized by currents circulating in opposite directions in its northern and southern halves. The differences in velocity among these currents may generate winds of 600 kilometers (373 miles) per hour. This results in the formation of enormous whirlwinds.

The width of the zones is inversely proportional to the intensity of the Coriolis force, which explains why the widths decrease as we move away from the equator.

The American Galileo mission, which is supposed to be launched in 1986 and which will carry a probe designed to penetrate Jupiter's atmosphere, certainly will make possible many other discoveries.

Internal Structure

Jupiter's low average density, which is only 1.3 times that of water, implies a very different composition from that of the telluric planets. It reflects a structure based on hydrogen and helium, those two elements being the only ones likely to have a sufficiently low density at the temperature and pressure conditions that probably exist within the planet.

In a liquid or a solid state, hydrogen and helium can mix only in certain proportions. Therefore, it is believed that Jupiter's interior contains a series of well-differentiated layers of matter, some rich in hydrogen, others rich in helium.

According to some astronomers, Jupiter formed in the same way as a star. In the nebula out of which the Sun formed, there may have been a second region where the density was sufficient for matter to condense under the effect of its own gravitational field; but as the total mass of this second condensation was too small, the temperature could not become high enough to initiate thermonuclear reactions; Jupiter's present mass is only about 0.001 times that of the Sun, whereas the critical mass required to set off thermonuclear reactions within a planet is estimated at 6 percent of the solar mass. In order for Jupiter to become a star, it would have been necessary for its mass to be 60 times greater than it is.

According to another theory, Jupiter possesses a high-density core with a composition similar to that of the telluric planets. This core was the first part of the planet to form; it then gravitationally captured the interplanetary gas in its vicinity, eventually giving the planet the appearance familiar to us. As the captured gas came from the solar nebula, it would have the same chemical composi-

Jupiter's north polar region, seen from a distance of 1 million kilometers (807,955 miles), by the Pioneer 11 *probe in 1974. It shows a granular structure consisting of convection cells, which contrast with the regularly alternating dark bands and light areas of the equatorial region. Photo: NASA.*

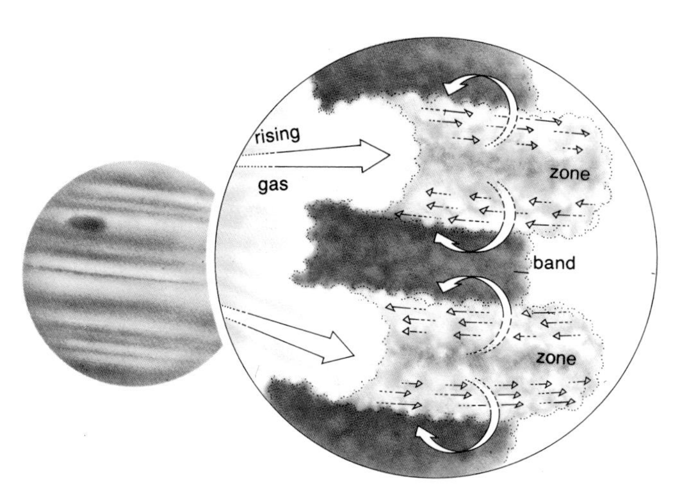

Atmospheric circulation around Jupiter. Drawing based on a NASA document.

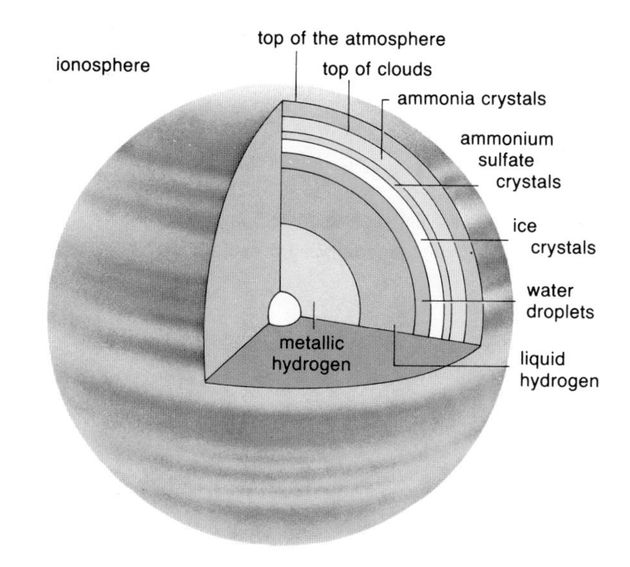

Model of Jupiter's atmosphere and internal structure.

tion as the Sun—i.e., it would be made up almost entirely of hydrogen and helium.

In one of the theoretical models, Jupiter's interior is envisioned to contain, aside from a possible dense nucleus with a mass eight times that of the Earth, exclusively hydrogen-rich layers: peripheral layers of molecular hydrogen, gaseous in the atmosphere, liquid underneath, and deep layers of metallic hydrogen—a phase in which hydrogen acquires the properties of a metal, its atoms no longer being grouped in molecules but arrayed in a gridlike structure within which electrons move about freely. Another model predicts the existence of a helium-rich nucleus surrounded by a layer rich in metallic hydrogen, and an atmosphere of molecular hydrogen. Within each layer, energy would be transported to the outside by convection, while at the border between two layers, it would be transported by conduction.

The internal temperature profile, calculated by the American Hubbard from data gathered during the *Pioneer 10* and *11* missions, indicates 2,000°C at a distance of 1,000 kilometers below the top of the clouds (it is thought that this is the level of the gas-liquid transition); 5,500°C at 3,000 kilometers' (1,865 miles') depth, a pressure of 90,000 atmospheres; and 11,000°C at 25,000 kilometers (15,538 miles) (at the level where hydrogen would change to a metallic state), the pressure then reaching 3 million atmospheres. As for the central temperature, it is thought to be about 30,000°C, the correspondiing pressure being 100 million atmospheres.

Any theoretical model of Jupiter's interior should account, moreover, for the planet's internal heat source, revealed by measurements of its luminosity in the infrared range; they have indeed shown that Jupiter radiates about 2.5 times more energy than it receives from the Sun. The first hypothesis envisioned was that this energy is of gravitational origin and results from a slow contraction of the planet (a contraction of 1 millimeter per year could explain the excess luminosity observed); but now it is accepted that it instead stems from the gravitational separation of hydrogen and helium. On the other hand, it is very unlikely that a mere cooling through radiation of the heat acquired during Jupiter's formation contributes to energy production in a significant way.

Radio-Frequency Radiation

In 1955, the American astronomers B. Burke and K.L. Franklin, who were working on a radio-frequency map of the sky on a frequency of 22.2 MHz (corresponding to a wavelength of 13.5 meters), accidentally discovered that Jupiter is a powerful source of radio waves.

Later, more elaborate studies of the planet's radio waves showed that there were several types of emission. At wavelengths shorter than 7 centimeters, the emission is mainly of thermal origin and comes from the body of the planet. At longer wavelengths, on the other hand, it is mainly nonthermal, and two components can be distinguished: one centimetric and decimetric, at wavelengths of less than 3 meters; the other decametric, appearing at wavelengths greater than 7.5 meters.

The decametric radiation—the first to be discovered—was detected at frequencies ranging from 450 kHz to 39.5 MHz—i.e., at wavelengths ranging between 670 and 7.5 meters. It is emitted in a sporadic fashion in the form of brief, intense bursts, each of which occurs only in a narrow frequency band, or of a series of many bursts (storms) lasting from several minutes to several hours. It has been shown that this radiation comes from many sources of very small dimensions near the planet, at well-defined longitudes and high latitudes. One of its fundamental characteristics, discovered in 1964 by the astronomer Bigg, is that it is modulated by Io, one of Jupiter's four main satellites and the closest to the planet. The intensity of the bursts varies according to Io's position in its orbit.

The centimetric and decimetric waves, lying between 5 centimeters and 3 meters, have very different properties. Contrary to the preceding type, they are strongly linearly polarized and show very little circular polarization. Moreover, the radiated power is practically constant; it comes from a region much larger than the planetary disk and varies in synchronization with Jupiter's rotation.

To interpret these observations, astronomers were led to assume that Jupiter has an intense magnetic field and belts of energetic particles; the centimetric and decimetric composition of the radio flux thus can be attributed, in particular, to the synchrotron radiation of electrons moving at relativistic speeds and spiraling around the lines of force of the magnetic field.

Magnetic Field and Magnetosphere

The measurements taken in Jupiter's vicinity by the Pioneer and Voyager spacecraft made it possible to confirm these hypotheses. Jupiter has an intrinsic magnetic field that may be described as a magnetic dipole. The axis of this dipole is inclined by about 11 degrees relative to the planet's axis of rotation, so that any phenomenon tied to the field shows a variation according to the planet's rotation (this applies, in particular, to radio-wave emissions; the period of rotation, deduced from the study of these emissions, does in fact correspond to the period of rotation of the magnetic field). The polarity is opposite to that observed for the Earth's field: The magnetic South Pole is near the geographic North Pole.

Moreover, the center of the dipole does not exactly coincide with Jupiter's geometric center but lies about 700 kilometers (435 miles) to the north, and 7,100 kilometers (4,413 miles) away from the axis of rotation in a direction parallel to the equator. This peculiarity explains why different field intensities have been observed at the equator and at the poles: At the altitude of the cloud cover, the intensity of the field reaches 14.8 gauss at the North Pole and 11.8 gauss at the South Pole but is only about 4.2 gauss (about 10 times that of the terrestrial field) at the equator. It is assumed that Jupiter's magnetic field, like that of the Earth, is generated by a dynamo mechanism within the liquid portion of the planet's interior.

Jupiter's magnetosphere—the region of space where the planet's magnetic field dominates that of the Sun—is very extended and may be divided into three zones. The internal magnetosphere, which extends out to about 20 Jovian radii from the planet, displays a behavior that is practically analogous to that of the terrestrial magnetosphere. The dipole field predominates in this region; nevertheless, significant perturbations due to the satellites Europa and Ganymede, which orbit in this region, have been observed.

At greater distances from Jupiter, the field's shape becomes more complicated. In the middle magnetosphere, between 20 and 60 Jovian radii from the planet, there exists a disk-shaped region in which the magnetic field is nearly constant and measures about 10^{-8} tesla. The field traps a low-energy plasma in which electric currents circulate, forming what is called a current sheath that rotates with the planet. As a result of the inclination of Jupiter's magnetic axis, this sheath is distorted: On one side, it is above the equator; on the other side, beneath it. Beyond this region, the field is weaker and more similar to a dipole.

Finally, in the outer magnetosphere, the field is very irregular. The distance between Jupiter and

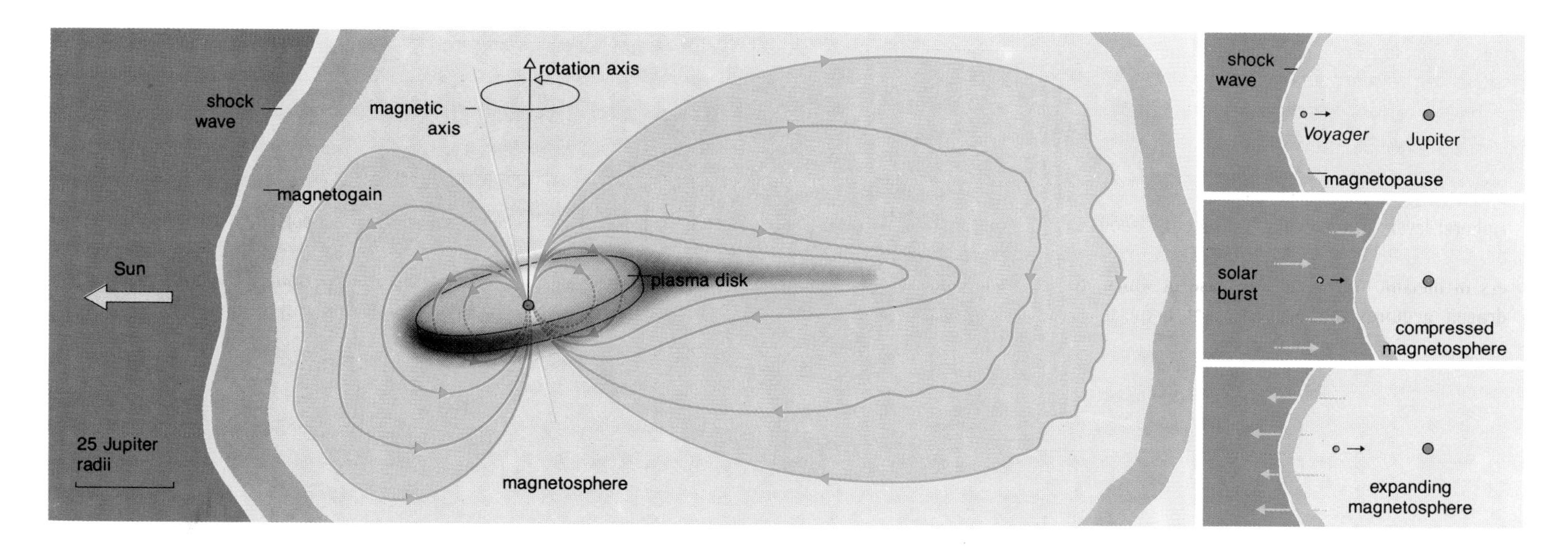

Jupiter's magnetosphere. In spiral lines along the lines of force of the magnetic field, the charged particles trapped in the radiation belts manage to acquire considerable energy.

the boundary of its magnetosphere in the direction of the Sun can vary considerably, moving between 60 and 100 Jovian radii in a short time. To explain this phenomenon, we assume that the magnetosphere's rapid rotation tends constantly to push the particles trapped within it toward the exterior, thus tending to push the boundaries of its outer region farther and farther from the planet; but strong gusts of solar wind, falling on the magnetosphere, compress it and compel it to retract. As soon as the wind pressure subsides, the expansion movement takes the upper hand, and so on.

On the side opposite the Sun, the field takes the shape of a long magnetic tail, like that of the Earth. However, it appears much more extensive than that of our planet, since space probes have detected it near the orbit of Saturn; hence, it extends over at least 600 million kilometers (373 million miles).

Around Jupiter, as around the Earth, we find two radiation belts where the particles trapped by the magnetic field (mainly electrons and protons) accumulate, but these particles have much greater energy than in the terrestrial Van Allen belts.

The interior belt, which is included in the internal magnetosphere, is by far the more intense. It forms a kind of doughnut around Jupiter, but the radiation is very intense only in a region that coincides with the magnetic equatorial plane. As one approaches the planet, the intensity appears to undergo a tenfold increase every 210,000 kilometers (130,516 miles). For the electrons, the intensity reaches a maximum at 64,000 kilometers (39,776 miles) above the cloud cover, and for the protons, 18,000 kilometers (11,187 miles), inside of which it decreases. Compared with the peaks of intensity observed in the terrestrial belts, that of the electrons near Jupiter is 10,000 times greater; that of the protons, several thousand times greater. Among the four main satellites of Jupiter, three—Io, Europa and Ganymede—are immersed in this belt. As they plow through the radiation belts, they absorb some of the energetic particles it contains and thus reduce the intensity of the radiation in the neighborhood of the planet by 10 to 100 times.

The outer belt extends into the middle and outer magnetospheres. Its average thickness is 700,000 kilometers (435,053 miles) but, as for the inner belt, the zone of intense radiation is reduced to a disk parallel to the equator. The particles they contain probably are not completely trapped; but as they escape continuously, other particles are captured from the interplanetary medium.

Much closer to Jupiter, an ionosphere has been discovered composed of several ionized layers rising up to 3,000 kilometers (1,865 miles) above the clouds. The Voyager probes detected auroral phenomena in them in the polar regions. As in the Earth's upper atmosphere, these are stimulated by the massive arrival of charged particles channeled by the lines of force of the magnetic field.

The Ring and the Satellites

Among the discoveries with which the Voyager space probes can be credited is the detection of a thin ring of matter around Jupiter. Located in the planet's equatorial plane, it extends over a width of about 6,500 kilometers (4,040 miles), but its thickness does not exceed 1 kilometer. Its outermost edge lies about 57,000 kilometers (35,426 miles) above the top of the clouds of the Jovian atmosphere. It probably is made up of a mixture of dust and ice.

This discovery, after that of the rings of Uranus (see "Uranus," page 196), shows that Saturn is not the only planet with a system of rings and that, on the contrary, the formation of rings of matter

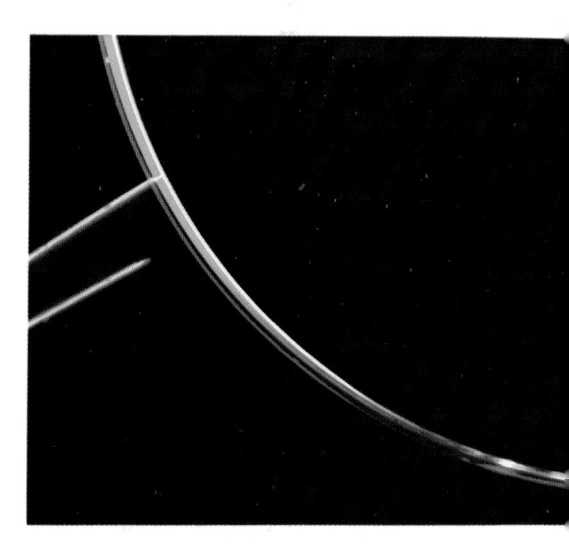

Jupiter's ring. Voyager 2 *was 1.45 million kilometers (901,181 miles) from Jupiter and about 2 degrees below the plane of the ring in the planet's shadow when it took this picture with orange and violet filters. The ring appears orange. Its lower portion is hidden near the edge of Jupiter by the planet's shadow. As a result of the spacecraft motion during the taking of the picture, Jupiter's bright limb takes the shape of concentric circles (colored white, blue and red). Photo: NASA-Jet Propulsion Laboratory.*

probably was a stage in the evolution of all the giant planets.

Moreover, the space missions have made possible a better understanding of Jupiter's entourage of satellites. We now know of 16 satellites orbiting the planet. All are small objects, with the exception of the four satellites discovered by Galileo in 1610 and termed, for that reason, the Galilean satellites.

Jupiter's System

Jupiter and the Galilean satellites—Io, Europa, Ganymede and Callisto—comprise a veritable small-scale replica of the solar system, and studying it can improve our understanding of the history of that system. Not only do they have planetary dimensions (the largest, Ganymede, is bigger than Mercury), but also, like the planets around the Sun, they all orbit around Jupiter in the same direction, following quasicircular orbits that are close to a single plane. Moreover, their distances from the planet conform to a law comparable to that of Titius-Bode: From one satellite to the next the distance from Jupiter increases by a factor of 1.66. Finally, the farther they are from Jupiter, the lower their density, exactly the same as for the planets and the sun—probably because, in both cases, the bodies closest to the central object had their lighter chemical elements gradually evaporate into space as a result of the heat given off by that object.

Thus the physical study of these objects makes an important contribution to knowledge of all the planets.

IO: INTENSE VOLCANIC ACTIVITY

Io—which takes its name from the princess whom Jupiter loved and turned into a heifer to protect her from Juno's jealousy—lies some 420,000 kilometers (261,032 miles) from Jupiter. Particularly striking photographs of it were taken in 1979 by *Voyager 1,* which came within only 18,170 kilometers (11,293 miles) of it. The surface of this object, 3,640 kilometers (2,263 miles) in diameter, appears dominated by volcanic features, between which lie vast plains separated by escarpments. Some 100 volcanic chimneys more than 25 kilometers (15.5 miles) in diameter have been noted, some of them surrounded by radial lava flows. The most spectacular discovery, however, was of erupting volcanoes spewing plumes of gas (sulphur compounds, particularly sulphur dioxide) up to altitudes of 280 kilometers (174 miles), which implies an ejection velocity of about 1,000 meters (3,281 feet) per second. This volcanism must evolve rapidly, since one of the eight active volcanoes photographed by *Voyager 1* in March 1979 appeared dormant five months later, during the passage of *Voyager 2.* According to all evidence, the volcanic activity ensures a continual remodeling of the surface: It shows no impact crater more than 1 kilometer in diameter, which testifies to its extreme youth—1 million years, at most. Furthermore, the bright colors of the soil are explained by the presence of a large range of

Jupiter's system. A montage made from photographs taken by Voyager 1 *in March 1979 showing the planet surrounded by its four main satellites: Io, Europa, Ganymede and Callisto. Dimensions and distances are not to scale. Photri Photo.*

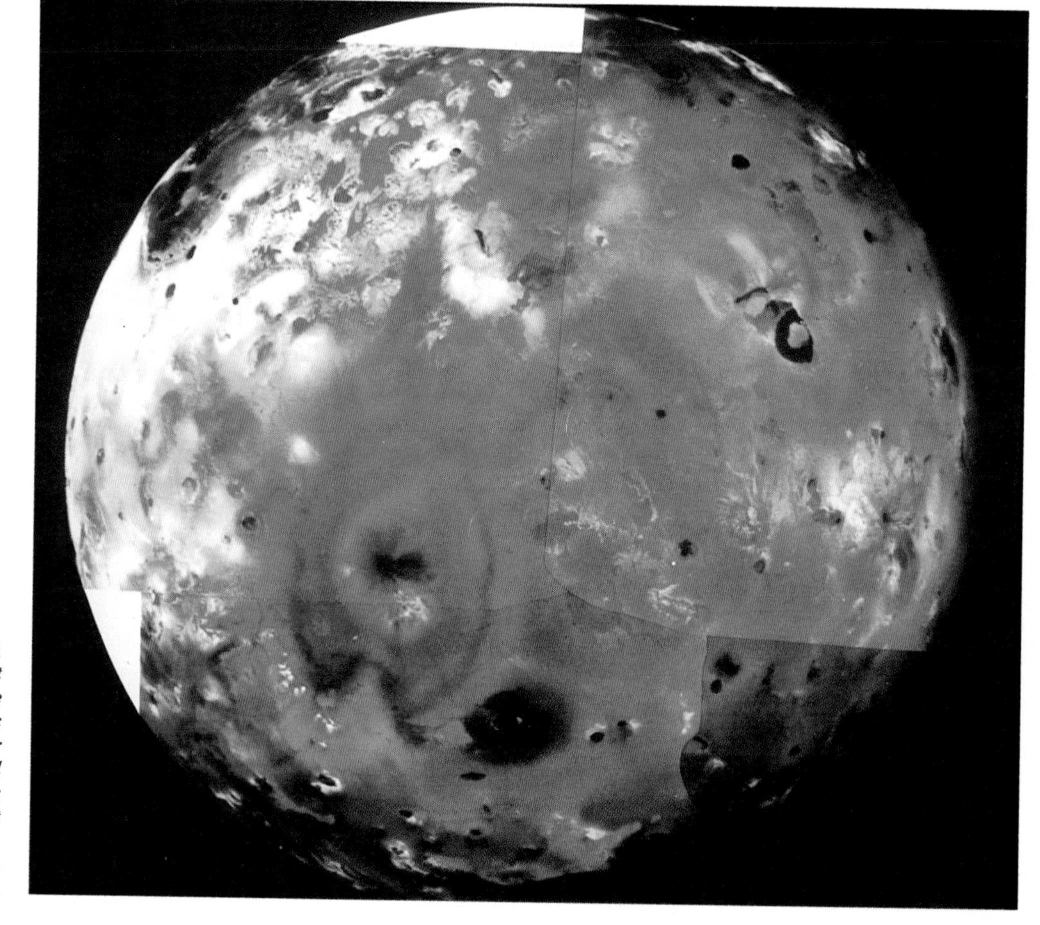

The surface of Io photographed by Voyager 1 *on March 5, 1979, from a distance of about 380,000 kilometers (236,172 miles). The black spots mostly correspond to volcanic craters; on the other hand, no impact crater can be seen, which indicates a continual renewal of the surface. Deposits of various sulfur compounds form a palette of bright colors in which red and orange predominate. Resolution is about 8 kilometers (5 miles). NASA Photo.*

Volcanic eruption on Io photographed by Voyager 1 *from a distance of 490,000 kilometers (304,537 miles). Ejected matter rises to about 160 kilometers (99 miles) in altitude. P.P.P.-IPS Photo.*

sulphur compounds of volcanic origin, which are stable in the range of temperatures prevailing on the satellite's surface (60° to 120°K).

Along Io's orbital trajectory, *Voyager 1* detected a plasma torus and found it to be rich in sulfur and ionized hydrogen, probably stemming also from volcanic discharges. Another feature of Io, which observations from Earth had led us to suspect, was thus confirmed by *Voyager 1*. The satellite resides inside Jupiter's radiation belts, and its surface is linked to the planet's ionosphere by a closed electric circuit. Inside a tube of magnetic flux flows a current with an intensity of 5 million amperes. The process responsible for the production of this torrent of energy remains unexplained. A ring of neutral sodium was also detected near the satellite's orbit and is considered to be made up of atoms ripped from its surface by the charged particles bombarding it.

With an average density of 3.5, slightly greater than that of the Moon, Io is regarded as a rocky world where silicates predominate. But its volcanic activity suggests the presence of molten matter at great depths. In particular, to explain the volcanic plumes and the existence of white clouds that appear to emerge in places from cracks in the crust, some experts have suggested the presence, several hundred meters below the surface, of pockets of liquid sulfur dioxide comparable to the aquifers of subterranean water observed beneath the surface of the Earth. It is out of the question that such a small object could have continued to be heated to this day by the disintegra-

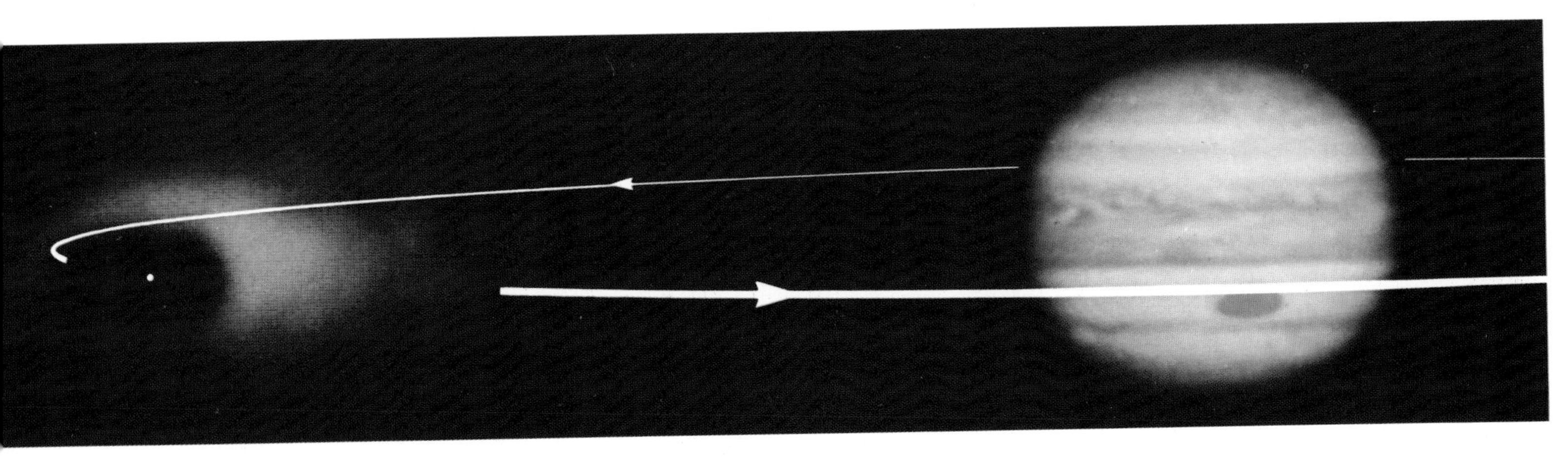

The sodium cloud surrounding Io. This photograph was actually produced by combining two monochromatic images taken with a three-hour time lapse in the light emitted by the D_1 and D_2 bands of sodium (at 5,986 Å and 5,890 Å wavelength), using the Coudé spectrograph of the 61-centimeter (24-inch) telescope of Table Mountain Observatory, in the United States, on February 19, 1977. The spectrograph-lit had been replaced by a glass plate containing a small aluminum disk designed to prevent the light reflected by Io's surface from entering the instrument (we see the shadow of this occulting disk in the photograph). The detector used was a vidicon television tube. After recording in a numerical form (by intensity level), the picture was processed to eliminate background contributions. We note that the cloud (diffuse orange spot) is very long and remains concentrated around Io's orbital plane; its shape resembles a banana. We also note that it is denser in front of the satellite than behind it. As a reference, a photograph of Jupiter has been superimposed on the document, and Io's orbit has been drawn in (white line), as well as the satellite's position (white dot) on this orbit when the pictures were taken. Photo: Jet Propulsion Laboratory.

tion of radioactive elements. According to the American Stanton Peale (who, in an article published several days before the *Voyager 1* flyby, predicted just such volcanic activity on the satellite), this internal heating is maintained by tidal phenomena of large magnitude resulting from the combined gravitational pull of Europa, Ganymede and Jupiter.

EUROPA: A WORLD OF ICE

At 671,000 kilometers (417,029 miles) from Jupiter lies the second Galilean satellite, Europa, which has a totally different appearance. Its name recalls the sister of Cadmos, whom Jupiter, changed into a bull, carried on his back as far as Crete. Its surface is largely covered by ice, as could be foreseen from its high reflectance (64 percent) and its infrared spectrum, and it appears devoid of any relief apart from a series of ridges and mounds, the altitude of which does not seem to exceed 100 meters (328 feet). But it does show a network of interwining dark lines, localized, for the most part, in a region extending over 40 degrees latitude on either side of the equator; these formations, several dozen kilometers wide and sometimes extending over several thousand kilometers, probably are fractures in the surface layer of ice, though it cannot yet be determined if they are deep cracks or mere fissures. The object may have some internal activity (maintained, as with Io, by tidal phenomena, but on a much smaller scale).

The near-total absence of impact craters (only three, about 20 kilometers in diameter, have been identified) might be explained by the existence of a dynamic ice formation process: The reforming of ice on the surface would gradually wear away the scars left by the fall of meteorites.

Given its average density (3.0 gm/cm^3), scarcely lower than that of the Moon, it is believed that the satellite, which has a diameter measuring 3,130 kilometers (1,945 miles), has an essentially rocky structure, with an ice crust no more than 100 kilometers thick. According to measurements in the infrared taken by *Voyager 2*, the temperature on the surface is about 90° K (−183°C) near the terminator, which, by extrapolation, yields a maximum temperature of 125°K (−148°C) on the sunlit hemisphere.

GANYMEDE: LAND OF CONTRASTS

The most voluminous of Jupiter's moons, Ganymede appears to be a land of contrasts. It bears the name of a Trojan prince, the handsomest of the mortals, whom Jupiter transformed into an eagle and carried off to be the gods' cupbearer. On the surface of this world, 5,276 kilometers (3,279 miles) in diameter, are juxtaposed terrains of very different ages, which seem to reflect a complex geological history. Zones riddled with impact craters, probably hollowed out more than 4 billion years ago, in a period when the planets of the emerging solar system were subjected to intense bombardment by very large meteorites, lie side by side with apparently much younger regions (in which crater density is substantially lower) traversed by long parallel streaks that no doubt represent fractures. Here and there one can also see transverse faults that greatly resemble the transforming faults that cut the Earth's midocean ridges at right angles. A major characteristic, however, is the absence of notable relief: Ganymede possesses neither high mountains (no peak is

The surface of Europa photographed by Voyager 2 *on July 9, 1979, from a distance of 241,000 kilometers (149,782 miles). It appears to be crisscrossed by a complex network of linear structures that might be filled-in fractures in the satellite's crust of ice; the very small number of impact craters suggests that this crust still is being renewed today. P.P.P.-IPS Photo.*

Ganymede photographed by Voyager 1 *on March 4, 1979, from a distance of 2.6 million kilometers (1.6 million miles). The surface of this satellite, made up of a mixture of ice and rocks, is reminiscent of that of the Moon: The dark regions resemble the lunar seas; the light regions seem to be impact craters. The latter have a starlike structure, being surrounded by long, bright rays like the younger lunar craters. P.P.P.-IPS Photo.*

higher than 1 kilometer in altitude) nor deep basins. Perhaps on Ganymede, as on the Earth, there exists intense tectonic activity. The structural models of the satellite predict the existence of an icy crust, 75 kilometers (47 miles) thick at its maximum, enveloping a mantle of liquid or frozen water and a core rich in silicates; given its average density (1.9 gm/cm³), it is believed that about 50 percent of the satellite is water. According to some experts, the streaks mark the boundaries of tectonic plates; they would correspond to lines along which underlying matter emerges, or to the subduction of planets. The American astronomer E. Shoemaker has suggested that the faults are caused by changes in phase of ice coming from deep within the satellite; this ice, initially under very high pressure, would go from one phase to another, increasing in volume and thus producing the observed shearing phenomena.

Like Io, Ganymede is surrounded by a very tenuous atmosphere; its density would be less than one-billionth of that of the terrestrial atmosphere at ground level.

CALLISTO: A FROZEN SURFACE SATURATED WITH CRATERS

The outermost Galilean satellite, Callisto, takes its name from the daughter of the king of Acadia, with whom Zeus fell in love; he changed her into a she-bear and Artemis killed her in a hunt. With a period of 16.69 days, it orbits at a distance of 1.8 million kilometers (1.1 million miles) from Jupiter, or a distance representing about 14 times the planet's diameter.

With a diameter of 4,840 kilometers, it is only slightly smaller than Ganymede, and it is presumed that the two objects have a similar structure, although the proportion of water must be higher on Callisto, which has a lower density (1.8 gm/cm³). Its surface is nonetheless very different from that of Ganymede. Made up, it is thought, of "dirty" ice (ice mixed with various impurities) and rocks, it appears much darker (its reflectance is only 17 percent) and almost entirely riddled with craters (at least on the hemisphere facing Jupiter, the only one photographed). The most remarkable structure revealed by the Voyager probes is a bright circular region about 600 kilometers in diameter, located 10 degrees north of the equator, surrounded by a series of concentric rings, separated by 50 to 200 kilometers (31 to 124 miles) and extending over a total radius of about 1,500 kilometers (932 miles). In the central part of this group of rings, the craters are three times less numerous than anywhere else on Callisto; evidently the preexisting craters were erased within a radius of about 300 kilometers (186 miles). It is presumed that this strange formation resulted from the impact of an enormous meteorite while the crust was still soft; the oscillations and stresses produced by this event fractured the crust and gave rise to the system of concentric rings. Later the ice layer hardened, solidifying to great depths.

The surface of Callisto, despite its resemblance to those of the Moon and Mercury in terms of the abundance of craters, differs from them in various ways, particularly in terms of the absence of mountains and large craters with a diameter greater than 150 kilometers (93 miles) and the shallowness of the medium-sized craters. It is probable that ice flows gradually obliterated the craters.

On the dark hemisphere, *Voyager 2* revealed a temperature of -118°C.

The surface of Callisto photographed by Voyager 2 *on July 7, 1979, from a distance of 2.3 million kilometers (1.4 million miles). It appears saturated with meteoritic craters, which testifies to its age. P.P.P.-IPS Photo.*

The Other Satellites

For a long time, Io was considered to be the satellite closest to Jupiter. But in 1892, the American Barnard discovered a fifth satellite, orbiting only some 110,000 kilometers (68,365 miles) from the highest clouds of Jupiter. It was given the name Amalthea, the goat that nourished the infant Zeus, and one of whose horns became the horn of plenty.

The pictures taken by *Voyager 1* show that Amalthea is a reddish asteroid, with an elongated shape, measuring about 155 by 270 kilometers (96 by 168 miles), with its long axis pointing toward Jupiter and its perpendicular short axis lying in its orbital plane. This object is very dark; its reflectance is only 5 percent, on average; however, at certain locations, it reaches levels three times higher. Infrared measurements taken by *Voyager 1* seem to indicate that Amalthea is hotter than it should be if all it did was absorb and reemit the radiation coming from the Sun and Jupiter. This excess heat is perhaps due to the electric currents flowing along the lines of force of Jupiter's magnetic field, or to bombardment by particles trapped in the planet's inner radiation belt. From the irregular shape of the satellite, we conclude that its internal rigidity is great and that it contains few volatile substances.

On the Voyager photographs, two other satellites were identified, much smaller and even closer to Jupiter. Both orbit in the immediate vicinity of the ring discovered around the planet. The first one, provisionally named 1979 J1, orbits in 7 hours, 8 minutes at about 58,000 kilometers (36,047 miles) from the top of Jupiter's clouds; the second, 1979 J3, completes a very similar orbit in 7 hours, 4 minutes, 30 seconds. Both have a diameter of about 40 kilometers (25 miles).

A third satellite, 1979 J2, was also spotted, at 151,000 kilometers (93,847 miles) from the top of the clouds, between the orbits of Amalthea and Io. The period of revolution of 1979 J2 is 16 hours, 16 minutes, and its diameter seems to be between 70 and 80 kilometers (44 and 50 miles).

The family of Jupiter's known satellites also includes eight other members. Very small, much farther away from the planet and divided into two very different groups, these are most certainly captured asteroids.

Jupiter and the Amateur

The number and diversity of details that can be observed on Jupiter using either a reflecting or refracting telescope, even one of low power, make this planet—which also offers the advantage of shining all night for long periods—one of the most interesting for the amateur.

A magnification of 30 to 40 times is sufficient to see the pronounced flattening of the disk. With a reflecting telescope having a 60-millimeter (2½-inch) aperture, you begin to see the parallel bands of clouds at the equator. The great red spot appears in instruments of at least 80 millimeters (3 inches). With a telescope of 150 millimeters (6 inches) and magnifications of 150 to 200 times, a detailed study of the cloud formations becomes possible: In them we see many irregularities, particularly knots and filaments evolving relatively quickly, which reveal an extremely turbulent atmospheric circulation. By concentrating the observation on a characteristic spot, it is easy to measure approximately the planet's period of rotation. To study the changes in structure of the cloudy bands, it is best to draw the planet as it appears in the eyepiece. To do this, use special templates that exactly match the elliptical shape of Jupiter's globe. First, place the details in latitude, particularly the bands, which can be carefully outlined, allowing for their width. Then position the details in longitude, being careful first to locate those closest to the planet's western edge (right edge when the picture is turned over), for these are the ones that disappear first as a

Close-up of Callisto taken by Voyager 1 *on March 6, 1979, from a distance of 350,000 kilometers (217,526 miles). The resolution is 7 kilometers (4.4 miles). The dominant formation is a large circular basin, about 600 kilometers (373 miles) in diameter, surrounded by concentric rings. This structure is reminiscent of the Mare Orientalis on the Moon, or the Caloris basin on Mercury, but its chief peculiarity is the absence of high relief. It is assumed that this is the result of the impact of a huge meteorite in the period when the satellite's crust still was soft. P.P.P.-IPS Photo.*

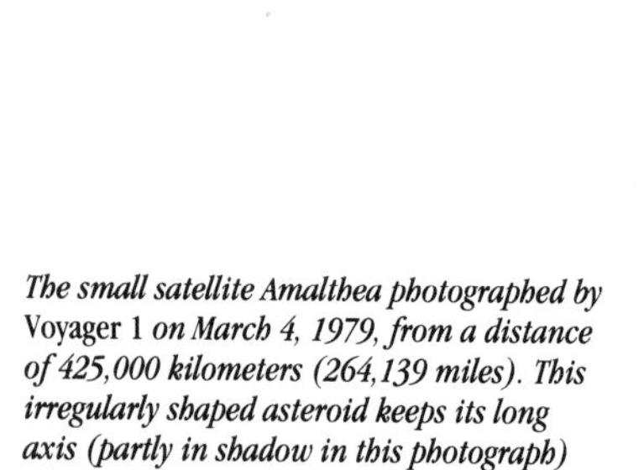

The small satellite Amalthea photographed by Voyager 1 *on March 4, 1979, from a distance of 425,000 kilometers (264,139 miles). This irregularly shaped asteroid keeps its long axis (partly in shadow in this photograph) always pointed toward Jupiter. Its surface is dark and has a characteristic reddish tinge. P.P.P.-IPS Photo.*

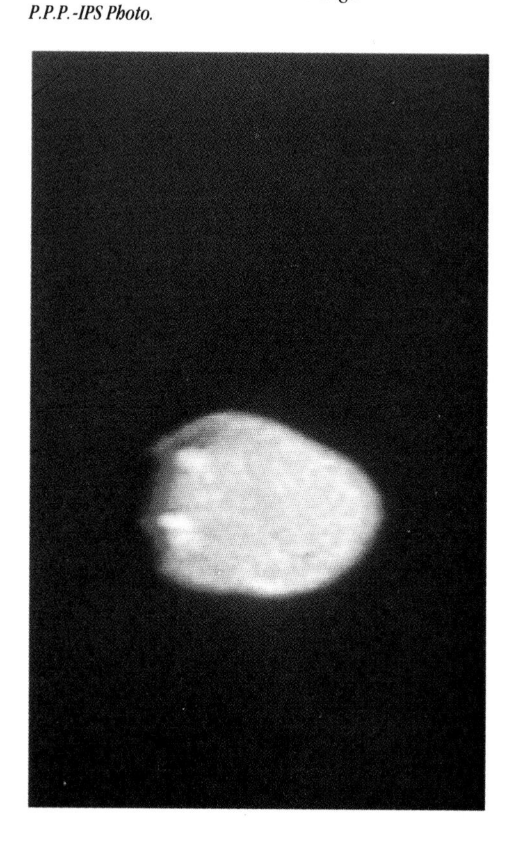

The first group includes the satellites Himalia and Elara, discovered by the American Perrine in 1904 and 1905, respectively; the satellite Lysithea, discovered in 1938 by the American Nicholson; and the satellite Leda, spotted in 1974 by the American Kowal. All four orbit at some 11.5 million kilometers (7.1 million miles) from Jupiter, in periods close to 250 days, in orbits inclined by about 28 degrees from that of the planet. Such a similiarity in their orbital parameters suggests that they have a common origin. The diameter of Himalia is estimated to be 170 kilometers (106 miles); that of Elara, 80 kilometers (50 miles); Lysithea, 20 kilometers (12.4 miles); and Leda, only 10 kilometers (6.2 miles).

The second group is at nearly twice this distance from Jupiter and includes the satellites Pasiphae (pointed out by the Briton Melotte in 1908), Sinope, Carme and Aranke (discovered by the American Nicholson in 1914, and for the last two, 1938). These four small objects, with a diameter ranging from 10 to 30 kilometers, revolve in about 700 days around neighboring orbits inclined to Jupiter's by about 150 degrees, their revolution thus occurring in a retrograde direction. They must have been captured when Jupiter was at the aphelion of its orbit (while the satellites of the first group must have been captured at the perihelion).

result of rotation. Finally, you can make out the light and dark zones, indicating the different shades by finer or coarser cross-hatching (when the drawing is recopied, you can use a series of shadings and erasures).

Apart from Jupiter itself, its four principal satellites, Io, Europa, Ganymede and Callisto (or, more simply, I, II, III and IV), visible with even the least powerful pair of binoculars, provide the amateur with fascinating observations. Orbiting in a plane very close to Jupiter's equator, these four satellites always appear practically aligned. But, very often, one of them is missing. Several phenomena may occur during their orbit around Jupiter: eclipse (disappearance into the planet's shadow), occultation (disappearance behind the planet) or transit of the planet's disk (a phenomenon accompanied by the satellite's shadow passing across Jupiter's disk). These phenomena, which recur at each revolution of satellites I, II and III (because of perspective, satellite IV—whose orbit is much larger—usually avoids Jupiter's disk and its shadow), can easily be followed with a telescope having an 80-millimeter aperture, except for the passages of the satellites across Jupiter's disk, which require an instrument of at least 100 millimeters (4 inches). The satellites' ballet is particularly interesting to follow when Jupiter is in an oblique position (quadrature) relative to the Earth and the Sun, for then we can observe the largest number of phenomena.

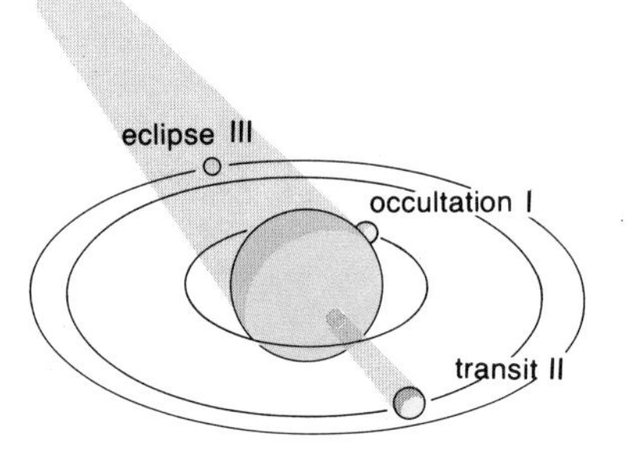

13. Saturn

Beyond Jupiter lies Saturn—the most distant of the planets known since antiquity. It is another giant world that has many features in common with Jupiter, but with sufficient differences to make it a special link in the chain that may describe the evolution of the planets. Saturn's main peculiarity is that it is surrounded by a vast system of rings. This fact, more than three centuries after its discovery, still greatly intrigues the experts.

The flyby of Saturn in 1980 by the American probe Voyager 1 *opened up a new era in our knowledge of this fascinating planet. Three photographs, taken on October 18, 1980—in green, violet and ultraviolet—were combined to produce this false-color image in which we clearly see the banded structure of the atmosphere.*

Orbit

On the average, Saturn lies 1,425 million kilometers (886 million miles) from the Sun. At its perihelion, it approaches the Sun to within 1,350 million kilometers (839 million miles); while at aphelion, it recedes to a distance of 1,509 million kilometers (938 million miles). The planet inscribes its orbit in 29 years, 167 days. It reappears at the same position relative to the Sun and Earth every 378 days, which means that it can be observed under good conditions for several months each year. From one year to the next, its oppositions recur with 13 days' delay. It remains an average of 2.5 years in each of the zodiacal constellations, and deviates only slightly from the ecliptic, with which the plane of its orbit makes an angle of only 2°30'.

Dimensions and Mass

By its dimensions Saturn is the second largest planet in the solar system after Jupiter. Its equatorial diameter is 120,660 kilometers (74,991 miles) (9.41 times that of the Earth), but its polar diameter is only 109,050 kilometers (67,775 miles). Due to its rapid rotation (see following material), the planet shows a substantial flattening (0.0912), greater than that of Jupiter.

Its volume is about 800 times that of the Earth for a mass that is only 95 times greater; this planet has the peculiarity—unique in the solar system—of having a lower average density (0.69 gm/cm^3) than that of water, which suggests that it is made up essentially of hydrogen and helium, with, however, a structure fairly different from Jupiter's.

Rotation

Saturn's rotation period is more difficult to determine precisely than Jupiter's since visible features on the disk are rarer and more blurred. Numerous measurements made from Earth and confirmed by spacecraft that have flown over the planet have shown that this rotation is differential, as with Jupiter, and that it occurs over a period varying from 10 hours, 14 minutes at the equator to 10 hours, 40 minutes at the poles. The rotation takes place around an axis inclined by 26°44' to the plane of Saturn's orbit (the planet, therefore, tilts its two poles alternately toward the Sun and Earth).

Atmosphere

Like Jupiter, Saturn has a disk striped with parallel bands, alternately light and dark, but with less contrast. The surface phenomena observed from Earth are less intense. Ephemeral features and seasonal variations in the planet's appearance nevertheless reveal some level of activity. In several cases it was possible to follow the evolution of bright white spots for several days. We also note a seasonal phenomenon: There is a very bright band that circles the planet in the equatorial zone underneath the rings; this band is not stationary but crosses the equator and switches hemispheres every 15 years, always remaining on the bright side of the rings. The rings probably act as reflectors, maintaining a special condition in the atmosphere that is responsible for this feature.

Like Jupiter, Saturn is surrounded by a very dense atmosphere, and what we can observe of the planet corresponds in fact to the upper part of this atmosphere, in which long, thin clouds circulate parallel to the equator because of the planet's rapid rotation. The characteristics of this atmosphere were studied in detail by *Voyager 1* and *Voyager 2*, which flew by the planet in 1980 at 124,000 kilometers' (77,067 miles') altitude and in 1981 at 101,000 kilometers (62,772 miles). The low-contrast pictures taken by *Voyager 1* had suggested the presence of a thick layer of mist above the clouds. But the better-quality photographs transmitted by *Voyager 2* invalidated that hypothesis and revealed many similarities with Jupiter's atmosphere.

In terms of its composition, Saturn's atmosphere also appears very comparable to Jupiter's and consists essentially of hydrogen and helium in a proportion that nevertheless differs from the one observed in the Jovian atmosphere (see page 190); the spectrum of light reflected by the planet also reveals the presence of methane and ammonia, which probably are the major components of the clouds.

At the top of the clouds, the temperature varies from −181°C to −187°C, with the lowest temperatures occurring near the center of the equatorial zone. It thus appears that Saturn, like Jupiter, possesses an internal heat source, radiating about three times more energy than it receives from the Sun.

The temperature profile is of the same type as in Jupiter's atmosphere; the temperatures, however, are lower owing to the planet's greater distance from the Sun. Assuming that the main constituent of the clouds in the upper atmosphere is ammonia crystals in both cases, the temperature difference explains the hazier appearance of Saturn's disk. Around Jupiter, conditions are such that the ammonia crystals can form only in a narrow strip of altitude; in the much colder atmosphere of Saturn, they can exist over a much wider range of altitude, and in practice, we see a dense mist. In addition, phosphine crystals might form more readily in Saturn's atmosphere than in Jupiter's.

Atmospheric circulation is similar to Jupiter's, with east–west horizontal currents in opposite directions. But while the strongest winds are observed at the periphery of the bands on Jupiter, there is no analogous correlation on Saturn; instead they undergo curious undulations. Their velocity gradually decreases with latitude. The strongest winds, which reach 1,800 kilometers (1,119 miles) per hour at the equator, are four to five times faster than on Jupiter, but they practically disappear at about 40 degrees' latitude. As on Jupiter, the polar atmosphere is different from that in the equatorial regions and is characterized by great optical depth, apparently resulting from the fact that the general cloud level is lower than it is directly over the equator.

Generally speaking, we see fewer convection cells on Saturn than on Jupiter. However, large oval spots—resembling the famous red spot of Jupiter—were photographed by the Voyager probes in both the Northern and Southern hemispheres at approximately the same latitudes (50°–55°). Like the red spot of Jupiter, they are colder than the surrounding atmosphere and may correspond to whirlwinds caused by long-lasting cyclonic activity.

Auroral discharges were observed by Voyager in the polar regions; the probe also recorded auroral-type discharges in the ultraviolet near the limb. Moreover, radio emissions characteristic of lightning discharges were recorded, but the pictures of the dark hemisphere of Saturn do not reveal the presence of any lightning; the discharges may come from Saturn's rings.

Finally, *Voyager 1* revealed that Saturn is surrounded by a vast cloud of neutral hydrogen,

[C]lose-up of a region of the Northern [H]emisphere taken by Voyager 1 *on [N]ovember 5, 1980, from a distance of 9 [m]illion kilometers (5.6 million miles). [T]he picture was computer-processed and [t]he colors were reinforced to bring out [t]he visible details, the smallest of which [m]easure about 175 kilometers (109 [m]iles). In particular, in the brown band [w]e see many cloud formations, which [r]eflect heavy turbulence. P.P.P. Photo.*

View of a portion of the rings produced by making a composite of photographs taken in different colors by Voyager 1. *In particular, the contrasts within the brilliant outer rings A and B are softened through the addition of a photograph taken in ultraviolet, while the Cassini division and the c ring—usually very dark—stand out in blue. The faintly visible colored formations within ring B are artifacts related to the length of the exposure. P.P.P. Photo.*

Saturn and its rings photographed by Voyager 1 *on October 18, 1980, from a distance of 34 million kilometers (21 million miles). This extraordinary image reveals many formerly unsuspected details in the system of rings: the presence of matter within the Cassini division, a division in the crepe ring, etc. We also note the shadow of the rings, which stands out with particular sharpness on the planetary disk. P.P.P. Photo.*

The spot mentioned below is more visible in this picture taken by Voyager 1 *on November 6, 1980. Situated at about 55 degrees south latitude, this structure measures some 12,000 kilometers (7,458 miles) in its largest dimension. It probably is a cyclonic formation comparable to those we see in Jupiter's atmosphere. However, none had ever been detected before in Saturn's atmosphere. P.P.P. Photo.*

Close-up photograph of a region of the Southern Hemisphere taken by Voyager 1 *on November 6, 1980, from a distance of 8 million kilometers (5 million miles). The picture was processed by computer, and the colors were retouched to bring out the band structure of the atmosphere. The small black disk visible at upper left is the shadow of the satellite Dione. Below, at right, near the limb, we see a brown oval formation, a miniature replica of Jupiter's big red spot. P.P.P. Photo.*

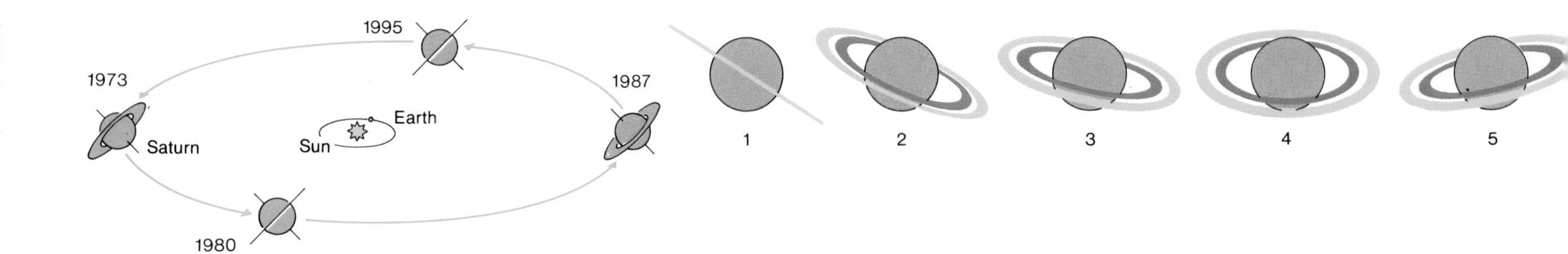

Seen from the Earth, Saturn's rings may have a very different appearance, depending on the season. At regular intervals, the Earth crosses their plane, and they are seen edge-on in profile (1). They then appear to open (2, 3), reach a maximum tilt (4) and then gradually close up again (5, 6, 7). The Earth may cross the rings' plane from north to south or from south to north. When Saturn's South Pole is tilted toward the Sun, we see the southern side of the rings, masking a part of the planet's Northern Hemisphere (1 to 7). Then it is the rings' northern side that shows (8 to 13). The visibility of the northern or southern side lasts about 15 years. During one of Saturn's revolutions around the Sun, the rings successively take on each of the different aspects shown in this illustration. Their next maximum tilt (north side) will occur in 1987.

Model of Saturn's Internal Structure

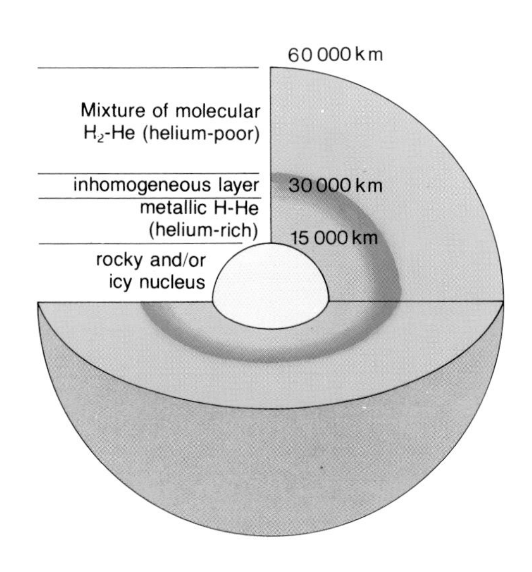

which extends between the orbits of its satellites Rhea and Titan. This cloud is thought to be continuously replenished by the hydrogen escaping from Titan's upper atmosphere.

Magnetic Field

As was supposed, Saturn has a magnetic field. It was one of the principal discoveries that can be credited to *Pioneer 11* in 1979. Weak by comparison with the Jovian field, the intensity of this field at the level of the clouds at the equator is similar to that of the field on the surface of the Earth: 0.2 gauss (compared with 0.3 for the Earth). This field may be represented as a dipole, its axis nearly coinciding with the planet's axis of rotation, unlike what we observe on Jupiter and the Earth: At most, the magnetic center appears slightly displaced (by less than 1,200 kilometers [746 miles]) from the planet's geometric center.

The intensity of the field is sufficient to create a magnetosphere, the dimensions of which are nearly one-third those of Jupiter's magnetosphere. The magnetospheric shock wave created by the solar wind is situated, on average, 1.8 million kilometers (1.1 million miles) from Saturn, while the magnetopause is some 500,000 kilometers (310,752 miles) closer to the planet, but the dimensions of the magnetosphere undergo large fluctuations depending on the intensity of the solar wind.

Within the magnetosphere, *Pioneer 11* counted, per square centimeter per second, up to 30,000 protons with energy exceeding 35 MeV, and 3 million electrons with energy greater than 3.4 MeV. The fluxes of high-energy particles trapped in the magnetosphere are strongly affected by the presence of the rings and satellites. The study of the radio-frequency radiation reveals that the magnetosphere has a rotation period of 10 hours, 39 minutes, 26 seconds.

Schematic illustration of Saturn's rings and the orbits of the planet's closest satellites. (The observer is assumed to be directly over Saturn's North Pole.)

Internal Structure

Saturn's low average density indicates that the planet consists mainly of hydrogen and helium. However, the data collected by *Voyager 1* indicate that, in the atmosphere, helium accounts for only 11 percent of the mass, as opposed to 19 percent in Jupiter's atmosphere. It therefore appears that a differentiation of hydrogen and helium has occurred, with the heavier helium becoming concentrated toward the planet's core. This phenomenon is a source of heat that may explain the planet's internal radiation (which, as with Jupiter, would not stem from a slow gravitational contraction). In any case, it shows that we were wrong formerly to have regarded the giant planets' atmospheres as merely a residual of the protosolar nebula that had undergone practically no evolution since the birth of the solar system.

Based on the data supplied by space exploration, a model of Saturn's structure has been proposed. The atmosphere, consisting of a dense fog dispersed in a mixture of gaseous hydrogen and helium, would be about 1,000 kilometers (622 miles) thick. Below that, most of the plane would be made up of hydrogen, liquid at first, and then, at about 30,000 kilometers' (18,645 miles') depth where the pressure becomes sufficient, in a metallic phase (the free electrons would then behave as they do in a conducting metal). Finally, the planet's core would have a radius of about 12,000 kilometers (7,458 miles) and a mass equal to 25 times that of the Earth; it would contain a liquid layer (water, methane, ammonia) surrounding a rocky core (magnesium oxide, silica, iron sulfide, etc.).

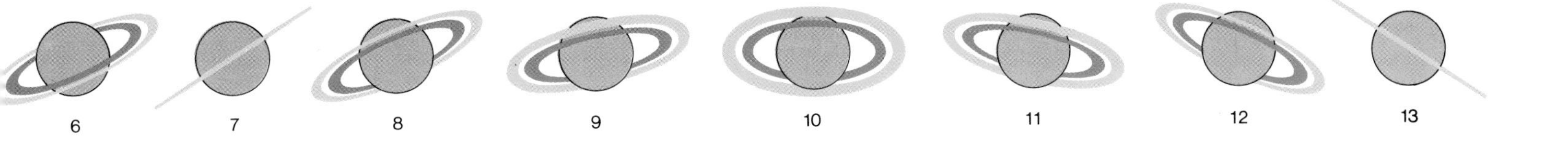

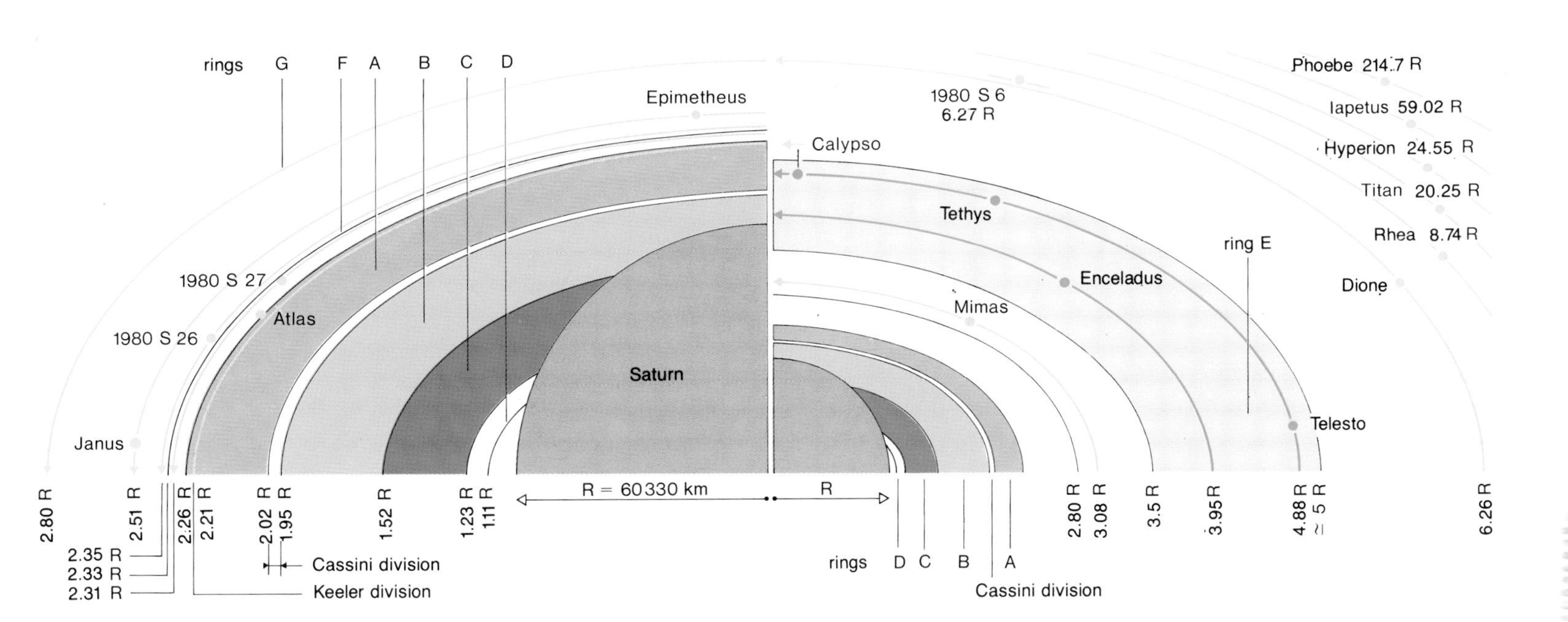

The Rings

THE FIRST ASTRONOMERS WHO observed Saturn through a telescope were intrigued by its strange appearance. Galileo in 1610 perceived two bright spots on either side of the planet. He saw them as "two servants helping old, slow Saturn on its way." By 1612, these two companions had disappeared. Galileo saw them again in 1616 but did not manage to recognize their true nature.

For some time, Saturn remained the "triplet" planet. Hevelius, Scheiner, Fontana, Riccioli and Gassendi drew the planet adorned with handles, without having the slightest notion of the structure of those formations. It was only in 1659 that Huygens solved the riddle in these terms: "Saturn is surrounded by a thin ring not adhering to the planet at any point and inclined on the ecliptic."

Later observations revealed that there is, in fact, a series of concentric rings (see following material). They are exactly in Saturn's equatorial plane; hence, during the planet's revolution, they appear at various inclinations to a terrestrial observer, the tilt depending on the epoch, and their faces are alternately illuminated by the Sun: the north side for 15 years, 9 months, the south side for 13 years, 8 months. They are so thin that they become invisible when the Sun's light hits them edge-on, or when the Earth crosses their plane, as in 1979–80.

In 1675, the French astronomer Jean Dominique Cassini discovered that the bright ring revealed by Huygens actually is divided into two fairly equal parts by a dark zone that now bears his name (Cassini division). In 1850, the American William Bond spotted a new ring, closer to the planet than the previous ones, much darker (hence the name "pancake ring" that is often given to it) and semitransparent (stars can be seen through it). Up through the end of the 1960s, it was believed that there were three rings separated by two divisions—i.e., from the outside to the inside: ring A (outer and inner radii: 136,200 to 121,000 kilometers [84,649 to 75,202 miles] from the center of Saturn); the Cassini division (width: 4,500 kilometers [2,797 miles]); ring B, by far the brightest (117,500 to 92,200 kilometers [73,027 to 57,303 miles]); the Lyot division (width: 4,200 kilometers [2,610 miles]); and ring C (89,300 to 73,000 kilometers [55,500 to 45,370 miles]); in addition, an internal division (the Encke division) separates two parts of slightly different brightness within ring A.

In 1969, the Frenchman Pierre Guérin detected on photographs taken at the Pic du Midi Observatory (by masking Saturn's disk) an additional interior ring, ring D, which is very faint. This picture of the rings has been substantially modified since 1979 by a series of discoveries. First, in September 1979, *Pioneer 11* came within 20,000 kilometers (12,430 miles) of Saturn and with its photopolarimeter was able to take an entire series of pictures of the planet and its rings. While these pictures did not make it possible to see ring D, they did confirm the existence of the Lyot division, which remained highly controversial and led to the discovery of a new outer ring, ring F. This very narrow ring extends over a width of less than 1,000 kilometers, at an average distance of 141,000 kilometers (87,632 miles) from the center of Saturn, and is separated from ring A by a

An outer ring of Saturn (the E ring) photographed on March 22, 1980 by A. Dollfus and S. Brunier with the 1-meter (39-inch) telescope at the Pic du Midi Observatory in France. A device derived from the coronagraph principle reduces the light scattered around the planetary disk, itself partially masked by an absorbing blade. Photo by kind permission of A. Dollfus.

division 4,000 kilometers (2,486 miles) wide known as the Pioneer division. Moreover, as a result of the intense campaign of observations carried out from Earth during a favorable period at the end of 1979, when the rings appeared edge-on, the Frenchman Audouin Dollfus and his assistant Serge Brunier discovered an even more distant outer ring, ring E, which extends out to more than 300,000 kilometers (186,451 miles) from Saturn's center. The existence of this ring later was confirmed by other teams.

The most spectacular discoveries, however, were made in November 1980 and August 1981 by the two Voyager probes during their close flyby of Saturn. *Voyager 1*, in crossing the planet's shadow, was able to confirm the existence of inner ring D (which begins some 7,000 kilometers [4,351 miles] from the top of the atmospheric clouds) and ring E. But above all, the pictures taken by the high-contrast television cameras on the two spacecraft revealed that the structure of the rings is much more complex than had been imagined. Each of the previously identified rings actually is made up of a series of narrower component rings, with the total number of rings reaching several thousand. Even the Cassini division—formerly considered to be devoid of matter—appears in fact to be filled with dark-colored dust divided among at least 20 thin rings. This dust does not sufficiently reflect solar light to be observable from Earth; however, these rings can be seen very clearly in Voyager photographs taken with the Sun behind them, as a result of the scattering of light. Many irregularities also can be noticed within the rings: variations in thickness or brightness, nonuniformity in the distribution of matter, etc. Within ring B, long, dark, radial structures are evident in the Voyager photographs. They appear especially in the part of the ring that has just emerged from the planet's shadow and persist for several hours without losing their identity, whereas the differential rotation between the inner and outer edges of the ring (see following material) should quickly destroy them. These radial corrugations rotate along with Saturn's magnetic field, which suggests that they are tied to electromagnetic phenomena. Experts believe they might be caused by small particles that have acquired an electrostatic charge as a result of their long stay in the magnetosphere, and are broken off from the ring and held in suspension above it by electric repulsion. As for ring F, the *Voyager 1* pictures show it to appear braided—i.e., composed of three narrow filaments more or less intertwined, but this peculiarity does not appear on the *Voyager 2* photographs; in addition, the ring has accumulations of matter in some places, apparently tied to the gravitational perturbations induced by the presence of two small satellites, one orbiting just beyond the ring's outer fringe, and the other just inside its inner fringe.

All of these peculiarities have provoked a review of the theory accepted until recently to explain the characteristics of the ring system. We know that the rings are made up of a multitude of small bodies rotating independently on concentric orbits. Cassini assumed this in 1705, and theoretical mechanics supplied the proof. In addition, this could be verified after 1895 by taking spectra in which the Doppler-Fizeau effect shows that the orbital velocity of the various parts of the rings continually decreases toward the outside according to the laws of planetary motion—i.e., inversely as the square root of the distance to the center. If the rings were made up of a single, continuous body, the velocities would be proportional to the radius.

Finally, the fact that the rings become invisible when viewed edge-on proves they are extremely thin. According to the most recent estimates, their thickness does not exceed 2 kilometers (1.2 miles).

There are two competing hypotheses to explain the origin of the rings. Some experts see them as arising from the breakup of one or more satellites

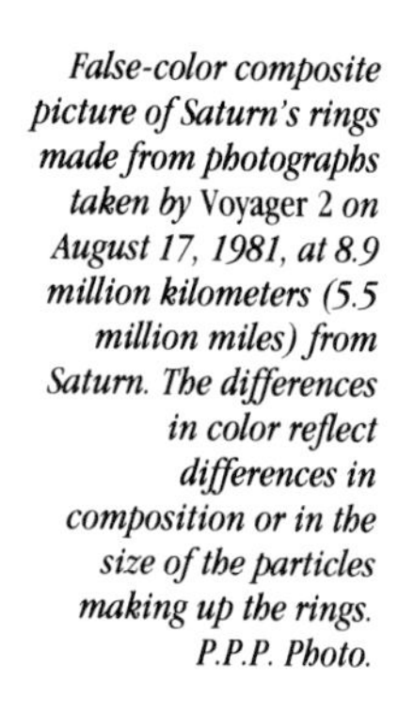

False-color composite picture of Saturn's rings made from photographs taken by Voyager 2 *on August 17, 1981, at 8.9 million kilometers (5.5 million miles) from Saturn. The differences in color reflect differences in composition or in the size of the particles making up the rings. P.P.P. Photo.*

Saturn, its rings and two of its satellites, Tethys and Dione (at top), photographed by Voyager 1 *on November 3, 1980, from a distance of 13 million kilometers (8.1 million miles). The rings' shadow and that of Tethys are projected onto the planetary disk. Note the transparency of the Cassini division. P.P.P. Photo.*

as a result of the tidal forces to which they were subjected, because they were orbiting at a distance from Saturn of less than 2.44 times the planet's equatorial radius—i.e., within the Roche limit. For others, the ring material is a residual of the formation of the planet itself, since all of the planets, at the time they condensed, must have been surrounded by a ring of dust and gases. The data gathered up to now are not yet sufficient to choose between these two hypotheses. Moreover, it is not out of the question that the inner rings may have resulted from fragmentation, while the outer rings, located beyond the Roche limit, would be residuals of the accretion epoch. Whatever the case, now that astronomers are more familiar with their idiosyncracies, they view Saturn's rings as a valuable source of information on the dynamical processes at work in the universe.

The stability of the system is ensured by the gravitation of Saturn and its satellites. However, the peculiarities detected in the Voyager photographs suggest that Saturn's magnetic field might also play some role in shaping the fine ring structure. The divisions between the rings visible from Earth represent zones where the density of matter is substantially lower than at neighboring positions. According to the traditional explanation, the bodies orbiting there cannot remain in that position for long, for their periods of revolution would bear a simple integer ratio to those of Saturn's principal satellites. Thus the Cassini division corresponds to a region where the period of revolution is equal to one-half that of Mimas, one-third that of Enceladus and one-fourth that of Tethys. This phenomenon of gravitational resonance is analogous to what is observed in the asteroid belt under Jupiter's influence. However, as mentioned earlier, the void in the divisions between the rings appears very relative in the Voyager pictures; this makes new hypotheses necessary.

The rings differ essentially in the density of matter and dimensions of the bodies they contain. Their characteristics were somewhat clarified during the Voyager missions through experiments that made it possible to study the manner in which they scatter sunlight or radio waves emitted by the spacecraft. Analysis of the data gathered has made it possible to establish that the bodies making up the rings have widely varying dimensions, ranging from a few microns to several meters. While rings D, E and F seem to contain very small particles, ring A and ring C are characterized, rather, by a predominance of large bodies. These differences suggest that the rings may not have formed all at the same time. Spectrophotometry and infrared radiometry suggest that the bodies present in the rings are covered with—or even made up of—ice.

The Satellites

In addition to its rings, Saturn is surrounded by a group of satellites, the largest of which probably are made up of a mixture of rocks and ice and that appear to have arisen from a process of accretion out of a primitive nebula, like the planets of the solar system.

Until the first flyby of Saturn by a spacecraft, in 1979, it was known with certainty that there were nine satellites (see table, page 67), which were named after the titans and giants of mythology. From the closest to the farthest from the planet they are Mimas, Enceladus, Tethys, Dione, Rhea, Titan, Hyperion, Iapetus and Phoebė. The existence of two small bodies closer to Saturn than Mimas and discovered by some astronomers in 1966, remained controversial. Finally, Themis—a satellite whose discovery was announced by the American W. Pickering in 1900—usually was cited only for the record, as it had never been seen again.

With the space missions and intensive Earth-based observations of Saturn in 1979 and 1980, during the favorable period when the rings appeared edge-on, the family of known satellites around Saturn more than doubled. The existence of the two objects discovered in 1966 now is firmly established. These small satellites have the peculiarity of being coorbital: The difference between their average orbital radii is less than the sum of their diameters. Thus their orbital velocities are very similar but not equal: The inner satellite slowly catches up with the outer satellite in its orbital motion. When they approach one another, gravitational attraction changes their angular momentum. The momentum of the inner satellite increases, and it gains a "higher" orbit where its orbital velocity is smaller; the momentum of the outer satellite decreases, and it is displaced onto a "lower" orbit, where its orbital velocity is greater. Thus the two satellites change places. This *pas de deux* is repeated every four years.

Some of the other recently discovered satellites are "Lagrangian" satellites: They move on the same orbit as a larger satellite, staying 60° ahead of or behind it, at points known as the "Lagrangian points" (points of dynamic stability, studied for the first time in 1772 by the mathematician Joseph Lagrange). Thus, for example, two such satellites were identified on the orbit of Tethys and two on the orbit of Dione.

Finally, a third category of small satellites is made up of objects spotted on the boundary of some rings. The gravitational fields of these objects are thought to confine the ring particles within narrow limits, and for this reason, these satellites are called "shepherd satellites." One of them seems to shape the very sharp outer edge of ring A; two others are on either side of ring F.

Still, one of the major results of the Voyager missions is the revelation of the physiognomy of the known satellites, about which little was previously known (apart from their orbital characteristics). All of them, except for Titan, appear only as bright points in a telescope.

Titan. The close flyby (at a distance of only 6,500 kilometers [4,040 miles]) of the largest satellite, Titan, was one of the major objectives of the *Voyager 1* mission. This planet, which is more than 5,000 kilometers (3,108 miles) in diameter, has especially attracted the attention of researchers because, of all the planetary satellites in the solar system, it is the only one endowed with a dense atmosphere. The conditions in the atmosphere have remained very close to what they probably were on the planets soon after their formation, and it is thought that the chemical reactions within it may generate some of the organic molecules considered to be the precursors of life on Earth.

The pictures of this satellite supplied by *Voyager 1* unfortunately turned out to be very disappointing. They showed that Titan, like Venus, is surrounded by a thick haze that makes it impossible to observe even the most prominent detail on its surface. This layer of orange clouds, moreover, does not have any special structure that would make it possible to determine its period of rotation. The most we can see on high-contrast photographs is that the satellite's Northern Hemisphere is darker and redder than its Southern Hemisphere. This difference in hue—as yet poorly understood—is ascribed to a seasonal phenomenon. It might be an indication of a sizable thermal gradient in Titan's atmosphere, which thus would harbor extremely strong winds and would rotate much more rapidly than the satellite itself.

The study of Titan by the scientific instruments of *Voyager 1* was, however, extremely fruitful and provided a completely different picture of the object. In particular, observations in the infrared and ultraviolet, as well as analysis of the radio signals emitted by the spacecraft just before its occultation by Titan, made it possible to establish the chemical composition and thermal profile of the satellite's atmosphere.

This atmosphere turns out to consist of nitrogen in a proportion of 82 to 94 percent. Methane—far from being a major component, as some experts had averred on the basis of an erroneous interpretation of spectra obtained from Earth—comprises only about 6 percent. Other hydrocarbons (propane, ethane, ethylene, acetylene, etc.) and various more complex compounds (such as hydrocyanic acid) also are present, but in trace quantities, and molecular hydrogen has an abundance of only 0.2 percent. These minor constituents probably derive from nitrogen and methane by a series of chemical reactions initiated by the Sun's ultraviolet radiation and by the charged particles in Saturn's magnetosphere. Finally, the presence of a substantial quantity of argon (12 percent) seems probable. The temperature—close to −100°C in the upper layers of the atmosphere—drops to about −180°C on the surface, where the pressure is estimated at about 1.6 bar (1.6 times the average pressure on Earth at sea level).

The satellite itself—with a density of 1.95 gm/cm³—probably consists of rocks and ice in nearly equal proportions. According to the *Voyager 1* measurements, it has no notable magnetic field, which indicates that it has no conducting metallic core. Its internal structure probably is very similar to that of Jupiter's two main satellites, Ganymede and Callisto, with a rocky core surrounded by a thick ice mantle. Titan is a planet more or less of the same type as Earth (with a dense nitrogen atmosphere) plunged into a deep freeze.

The other satellites. The other satellites of Saturn with which we now are familiar are miniplanets with masses of 1 to 10 percent of those of Jupiter's major satellites. They are ice-covered worlds whose surfaces often appear riddled with numerous craters, probably scars from the violent meteoritic bombardment to which the planets of the solar system were subjected early in their history and that remained intact in the absence of erosion and intense geological activity.

Mimas, which lies about 125,000 kilometers (77,688 miles) from Saturn's edge, turns out to be smaller than presumed, the estimate of its diameter being reduced from 500 to 390 kilometers (311 to 242 miles). Its front side contains a large crater with a diameter one-third that of the satellite itself, and with a central peak 9 kilometers (5.6 miles) high, and on the opposite side of the planet, there are large fissures that were probably created by the shaking following the impact that created the large crater. Evidently this satellite was struck at one time by an asteroid that nearly split it apart.

Dione, on the other hand, lying some 317,000 kilometers (197,017 miles) from Saturn, appears much larger than observations from Earth suggested; its diameter is 1,120 kilometers (696 miles). It shows a certain resemblance to Ganymede. Dione's surface has many craters, reaching up to 100 kilometers (62 miles) in diameter. Many of them have a central peak like lunar craters, and bright trails can be seen to emanate from some of them, interpreted by experts as ice flows. One of the remarkable features of this satellite is that it shows a bright and a dark side, the former containing plains filled with craters, while the latter has a network of bright trails in the center of which is a large basin—a probable vestige of a large impact crater.

Rhea, whose orbital radius is 497,000 kilometers (308,888 miles) and whose diameter is 1,530 kilometers (951 miles) could be photographed in some detail, since *Voyager 1* came within only 59,000 kilometers (36,669 miles) of it. Its surface—very similar to that of Dione—strangely resembles that of the Moon or Mercury, despite a different structure: At its relatively great distance from the Sun, the ice is so hard that it reacts like rock to the impact of meteorites. Saturated with craters, it combines bright and darker regions, the former appearing to be richer in

Saturn and the Amateur

For the amateur astronomer, Saturn is the jewel of the sky. Its rings constitute one of the most beautiful sights it is possible to see. The rings, which appeared edge-on in 1980, thus becoming invisible, now are gradually opening out, letting us discover their northern side.

When they are at maximum inclination, they can be detected with a telescope having a magnifying power of only about 40. But it is only in at least a 60-millimeter (2.5-inch) reflector, having a magnification of 100, that they stand out really well. An instrument with an 80-millimeter (3-inch) lens makes it possible to note the difference in brightness between the A and B rings and, under the best conditions, to see the Cassini division separating them at the "anses." This becomes very noticeable with an instrument that has a 100-millimeter (4-inch) aperture. We can then also see the C ring, and detect the rings' shadow on the planet, as well as the planet's shadow on certain portions of the rings.

It is only with telescopes of at least 150-millimeter (6-inch) apertures that we can see details on the planet: cloudy bands or spots.

larger craters than the latter, as if the impacts of two separate groups of objects, of different dimensions, had helped to give it its present appearance.

Tethys, lying 235,000 kilometers (146,053 miles) from Saturn, also appears to be strongly cratered, and has a bright and a dark hemisphere. Its diameter is 1,050 kilometers (653 miles). Photographs of its surface by *Voyager 1* show two remarkable features: a large circular structure, which might be a mountain range, and opposite that, a huge fault 750 to 800 kilometers (466 to 497 miles) long and 60 kilometers (37 miles) wide.

Enceladus, 178,000 kilometers (110,628 miles) from Saturn, could be observed only from a great distance by *Voyager 1,* but *Voyager 2* came within 87,000 kilometers (54,071 miles) of it. Of all the known satellites of Saturn, its surface has the highest reflectivity (nearly 100 percent) and appears to be by far the most tectonically active. In particular, there are large plains practically devoid of impact craters but merely sprinkled with hills and fractures, which are no more than a few hundred million years old. The youth of these formations implies a periodic remodeling of the surface as a result of the melting of ice trapped inside the body. The satellite's dimensions are too small to enable the disintegration of radioactive nuclei to account for the observed geological phenomena. Rather, its internal warming is considered to result from tidal interactions with Saturn and Dione.

Hyperion, whose orbital radius is close to 1.5 million kilometers (932,256 miles) could also be photographed by *Voyager 2.* This is an irregular asteroid, about 410 by 220 kilometers (255 by 137 miles). It was expected that an object of this size would be spherical, and it is quite possible that it is no more than a fragment of a larger body destroyed in a collision.

Even stranger is Iapetus, which lies more than 3.5 million kilometers (2.2 million miles) from Saturn. The photographs taken of it by *Voyager 2* confirmed a surprising peculiarity of this satellite, noticed in the 17th century by Jean Dominique Cassini and still unexplained: Its trailing hemisphere (the one that is behind relative to the direction of orbital motion) is very bright, with a reflectivity of about 50 percent, while its front hemisphere is very dark, with a reflectivity of only 3 to 5 percent. To explain this phenomenon, the hypothesis has been advanced that particles of dark matter are broken off from the surface of Saturn's outermost satellite, Phoebe, by the impact of micrometeorites, and then swept up by Iapetus after spiraling through space, leading to a pollution of the leading hemisphere of Iapetus; but the well-marked, complex boundary between light and dark regions, and the presence, on the trailing hemisphere of Iapetus, of a series of craters with hidden bottoms, rather suggests that the dark matter comes from the interior of the satellite.

Finally, some 13 million kilometers (8.1 million miles) from Saturn lies Phoebe. Photographs taken of it by *Voyager 2* reveal an approximately spherical body 220 kilometers (137 miles) in diameter, with a very dark, reddish surface, and rotating on its axis in about 9 hours. Its orbital characteristics seem to indicate that it is an asteroid that was captured by Saturn's gravitational field.

Dione, photographed by Voyager 1 *on November 12, 1980, from a distance of 162,000 kilometers (100,684 miles). Its surface bears a certain resemblance to that of the Moon or Mercury. The largest visible crater (at bottom) is nearly 100 kilometers (62 miles) in diameter and has a large central peak. On its right side, near the terminator, we see a number of winding valleys. P.P.P. Photo.*

Rhea, photographed by Voyager 1 *on November 12, 1980, from a distance of some 130,000 kilometers (80,796 miles). The resolution is at 2.5 kilometers (1.6 miles). The abundance of impact craters is evidence of the great age of the surface. However, other regions of Rhea turn out to have fewer large craters, which leads some to thinking that the satellite may have undergone some internal activity causing a remolding of its surface. The brilliant structures visible on the rims of some craters (bottom left) may represent recent deposits of ice or condensed volatile materials on steep slopes.*

The large satellite Titan is visible in smaller instruments (a reflector with a 40- to 50-millimeter [2-inch] diameter); Iapetus and Rhea become visible with an 80-millimeter reflector; Tethys and Dione with a small telescope. The other satellites remain inaccessible with amateur instruments.

14. Uranus

As William Herschel—a German organist who had immigrated to England and spent his spare time on astronomy—was studying the sky with a 16-centimeter (6-inch) telescope on March 13, 1781, he noticed a small luminous spot of appreciable apparent diameter in the constellation of Gemini that could not be a star.

Soon he recognized that the object was in motion relative to the nearby stars, but he did not hypothesize that a new planet was involved. He believed he was dealing with a comet—despite its lack of a tail—and announced its discovery at the Royal Society on April 26. Different astronomers then attempted—although in vain—to represent the orbit of the faint body as a parabola or as a very elongated ellipse; calculation showed that this orbit was in fact nearly circular, with a radius 19 times greater than that of the Earth's orbit, and that it corresponded perfectly to the prediction of the Titius-Bode law, published several years earlier. At that point, it was necessary to bow to the evidence: The new body was indeed a planet. This discovery, which suddenly pushed the boundaries of the solar system beyond Saturn's orbit, won considerable acclaim.

Herschel proposed to call the planet he had just discovered Georgium Sidus, George III's planet (after the then king of England); Lalande suggested rather giving it the name of its discoverer, Herschel, and this name was adopted for some time, but the German astronomer Bode, loyal to the tradition of mythological names adhered to for the other planets, had suggested the name of Saturn's father, Uranus, and this name finally prevailed. With Jupiter, Saturn and Uranus, the son, father and forebear thus follow one another in the solar system, in order of descendance.

This picture of Uranus was compiled from images returned on January 17, 1986 by Voyager 2 *when the spacecraft was 9.1 million kilometers (5.7 million miles) from the planet. The picture has been processed to show the planet in true color, as human eyes would see it from the vantage point of the spacecraft. The blue-green color results from the absorption of red light by methane gas in Uranus' deep, cold and remarkably clear atmosphere. The darker shadings at the upper right of the disk correspond to the day-night boundary on the planet. Beyond this boundary lies the hidden northern hemisphere of Uranus that remains in total darkness as the planet rotates. NASA/Jet Propulsion Laboratory.*

Rings

Prior to the *Voyager 2* encounter, the most significant modern discovery about Uranus occurred in 1977. On March 10, 1977, Uranus crossed in front of a small star in the constellation of Libra, SAO 158 687 (a star bearing the number 158 687 in the large catalog published by the Smithsonian Astrophysical Observatory, which gives the coordinates of 258,997 stars). Several teams of astronomers were assigned to follow the phenomenon. Occultations of stars provide information on the extent, density and composition of the occulting planet's atmosphere: Just before the star is hidden, the radiation it emits toward Earth skims the edge of the occulting planet's disk, the atmosphere of which changes the light that reaches the Earth. The diameter and position of the occulting planet may also be precisely measured at such times.

Contrary to all expectations, the star's brightness underwent several sudden, short reductions in intensity during the 40 minutes that preceded the occultation. After the star moved behind Uranus and reappeared, a new series of occultations, symmetrical to the preceding series, was again recorded.

It quickly became clear that these results could not be explained except by invoking the existence around Uranus of a series of thin concentric rings of matter. The first configuration envisioned contained five rings, respectively designated as α, β, γ, δ and ϵ, from the closest to the farthest from the planet. After three further occultations, observed on December 23, 1977, April 4, 1978,

Orbit

Uranus lies, on average, 2,875 million kilometers (1,787 million miles) from the Sun—hence, at a distance approximately twice that of Saturn. Owing to the eccentricity of its orbit, this distance falls to 2,742 million kilometers (1,704 million miles) at perihelion and reaches 3,008 million kilometers (1,869 million miles) at aphelion. The last time the planet reached its shortest distance from the Sun was in 1967.

Uranus requires no less than 84 years to complete its revolution, and since its discovery, it has returned only twice—in 1865 and in 1949—to the point on its orbit at which it was discovered. However, the observations we have of it extend in fact over more than three revolutions, since several astronomers saw it before Herschel and took it for a star. The first of these observations, credited to the British astronomer John Flamsteed, dates from 1690; from 1750 to 1769, Le Monnier had Uranus in the field of his reflector 12 times without recognizing its true nature!

Dimensions and Mass

Under favorable observing conditions, Uranus—which has a magnitude of 5.7—can be seen with the naked eye, using a good star atlas. In a telescope, the planet appears as a small, faintly luminous greenish disk, very dark around the edges, and with a diameter of less than 4 arc seconds, which makes measurement of its real dimensions difficult. The most recent calculations, based on observation of occultations of stars, imply a diameter of 50,800 kilometers (31,572 miles).

Its mass, on the other hand, is well known as a result of the study of the motion of its satellites: It is nearly 14.6 times that of the Earth.

Internal Structure and Rotation

Before the recent *Voyager 2* encounter, scientists knew comparatively little about the planet Uranus. Quite possibly its most interesting feature, unique in the solar system, is that its axis of rotation lies almost on the plane of its orbit, with which it makes an angle of only 8 degrees. Furthermore, its rotation is retrograde, as established by the Frenchman Henri Deslandres in 1902 using measurements with a spectrograph.

This arrangement makes us see—at the time of the Uranian solstices—one of the planet's poles nearly at the center of the disk, which then appears perfectly circular because its circumference corresponds to the equator; such a situation occurred in 1946 and recurred in 1985. It is only toward the equinoxes (1923, 1965) that we can observe the equatorial regions of Uranus, and its flattening becomes apparent. At those times, some observers have been able, under certain circumstances, and with powerful instruments, to perceive faint parallel bands on the disk, recalling those of Jupiter and Saturn.

The origin of the tilt of the Uranian axis of rotation is totally unknown, but it has been calculated that a collision of the planet with a body having a mass 10 times smaller would have been sufficient to cause it. The arrangement of the planet's satellites leads us in any case to assume a very ancient event (see page 199).

Like the other massive planets of the solar system, Uranus has a low density, approximately 1.2. Before the *Voyager 2* encounter, theorists believed that the planet possessed a rocky core about 16,000 km (9,944 mi) in diameter, surrounded by a mantle of hot water containing ammonia and methane which extended out some two-thirds of the planet's radius. Surrounding this rocky core and Uranian "ocean" was an atmosphere, composed mostly of hydrogen and helium, some 10,000 km (6,215 mi) thick.

Voyager 2 data helped support this planetary model and also indicated that Uranus possesses a significant magnetic field. While the presence of a magnetic field did not greatly surprise scientists, its unusual tilt, offset from Uranus' axis of rotation by some 60 degrees, was totally unexpected. Furthermore, preliminary analysis of *Voyager 2* data has indicated that Uranus' magnetic center may be offset from its physical center by perhaps 8,050 km (5,000 mi)—with the shift being toward its shadowed (north) pole. This unusual situation makes Uranus resemble what astronomers call an "oblique rotator." Uranus' tilted magnetic field (the largest offset detected so far in the solar system) produces an interesting effect on the Uranian magnetotail ("downwind" portion of its magnetosphere). As Uranus rotates, it exerts subtle, screw-like turns in the lobes of its magnetic field, inducing a corkscrew pattern in the magnetotail. Scientists can use gyrations of a planet's magnetotail to establish the rotation rate of its core. For example, during the *Voyager 2* encounters with both Jupiter and Saturn, scientists were able to determine internal rotation rates by observing each planet's magnetic field activity. Similarly, using *Voyager 2* data, scientists have been able to observe the period of "wobble" in the Uranian magnetic field, which they then assume is related to the spin rate of the planet's solid interior. For example, charged particles locked in magnetic field lines release telltale bursts of radio wave energy as they encircle Uranus. By observing these periodic radio emissions, scientists have now established a rotation rate of 17.3 hours for Uranus.

Voyager 2's detection of the Uranian magnetic field encouraged scientists to postulate that there is a dynamo mechanism active within the planet's interior. However, this magnetic field appears to be too strong to originate solely in Uranus' rocky core. Therefore, based on preliminary *Voyager 2*

Small particles are distributed throughout Uranus' ring system, although most ring particles appear to be "chunks of matter" larger than one meter (3 ft) across. This Voyager 2 *image was taken while the spacecraft was at a distance of 236,000 km (147,000 mi) from Uranus. This image has a resolution of about 33 km (20 mi). Because of its unique geometry (the highest phase angle at which* Voyager 2 *imaged the Uranian rings) and its long exposure (96 seconds), this image permitted scientists to see lanes of fine dust particles not visible from other viewing angles. The image also shows the nine previously known rings plus relatively bright dust lanes not previously observed. However, the long exposure also produced streaks caused by trailed stars, as well as a noticeable nonuniform smear. NASA/Jet Propulsion Laboratory.*

and April 10, 1978, as well as more thorough analysis of the initial recordings, four additional rings were brought to light: chi (χ) [also designated ring 6], iota (ι) [also designated ring 5], and phi (ϕ) [also designated ring 4] and η. Because different groups were involved in these discoveries, the Uranian ring system is most commonly designated with a mixed Greek letter and numerical system—which is by no means standardized by the international astronomical community. The most common [pre-*Voyager 2*] designations for the nine rings of Uranus are as follows: (going from the innermost ring outward) ring 6, ring 5, ring 4, alpha (α), beta (β), eta (η), gamma (γ), delta (δ), and epsilon (ϵ).

The *Voyager 2* encounter not only confirmed the existence of these nine rings, but preliminary data analysis has shown a faint 10th ring (initially called 1986U1R) about 3 km (1.9 mi) wide and located between the outermost delta and the epsilon rings, some 49,900 km (31,000 mi) from the planet's center. Using *Voyager 2* imagery, scientists have also located a faint, broad band of material situated about 1,600 km (1,000 mi) inside ring 6. In addition, preliminary interpretation of *Voyager 2*'s photopolarimeter data has indicated the presence of three more closely spaced narrow rings (each some 152 meters [500 ft] wide) well outside the epsilon ring.

Using *Voyager 2* data, scientists also learned that the epsilon ring contains very few particles smaller than one meter (3 ft) across. Prior to the *Voyager 2* encounter, scientists had also speculated that "shepherd" satellites were responsible for keeping Uranus' rings so narrow. But preliminary inspection of *Voyager 2* data indicates that only two of Uranus' newly discovered 10 satellites appear to be performing shepherding duties. As scientists thoroughly evaluate the *Voyager 2* data, a much more precise and thorough understanding of the dark Uranian ring system will emerge.

data, some scientists are now speculating that the Uranian magnetic dynamo is actually located farther out in the planet's "ion-rich, electrified" ocean of water, ammonia and methane at great pressure and several thousand degrees Kelvin.

The sunside portion of the Uranian magnetosphere appears to extend out about 459,000 km (285,000 mi) or about 18 planetary radii (18 R_U). All the Uranian rings and most of the planet's satellites lie inside its magnetosphere. *Voyager 2* instruments detected energetic electrons and protons everywhere inside Uranus' magnetosphere, but heavier ions from elements such as helium, carbon or oxygen were only rarely detected within the magnetosphere.

Atmosphere

The bland, aquamarine face of Uranus more closely resembles a deep, slowly varying terrestrial ocean than a frigidly cold gaseous envelope. The distinctive bluish color of Uranus' atmosphere comes from methane's absorption of red (sun) light.

Prior to the *Voyager 2* encounter, spectral analysis had revealed the existence of only two gases: methane, which is manifested as strong absorption bands, the most intense of which was detected in 1869 by the Italian Angelo Secchi, and molecular hydrogen, discovered in 1949 by the Americans Gerard Kuiper and Gerhard Herzberg. However, it was presumed that the atmosphere of Uranus also contained helium and ammonia. Models also predicted the existence of different cloudy layers: At a level where the pressure would be about 1 atmosphere, there would be a layer of methane; deeper within, at a level corresponding to a pressure 10 times greater, there would be clouds made up of ammonia crystals. The pressure on the "surface" might be about 100 atmospheres.

Voyager 2 measurements have now provided important information about Uranus' atmosphere. For example, its stellar occultation data indicate that Uranus possesses a very thin upper atmosphere of hydrogen atoms and hydrogen molecules some 4,000 km (2,484 mi) above its "surface." This rarified region of essentially pure hydrogen gas had a sunlit side temperature of approximately 750 Kelvin and a surprisingly "hotter" darkside temperature of about 1,000 degrees Kelvin. Other *Voyager 2* instruments detected what seems to be a thick layer of clouds, assumed to contain methane ice crystals, at about the 1.6 atmosphere pressure level. *Voyager 2* temperature profile data indicate that atmospheric temperatures drop sharply as one goes deeper, reaching a frigid minimum of about 51 degrees Kelvin at the 0.1 atmosphere level. Beyond that point, the temperature begins to rise with depth, due to increasing ambient pressure. As somewhat of a surprise to scientists, *Voyager 2* data indicated that the temperatures in Uranus' atmosphere were quite constant from equator to pole—with the Sun-pointing south pole at 65 degrees Kelvin actually being a bit warmer than the equator, which was at 64 degrees Kelvin. In addition, a "cold collar" region where temperatures dropped to about 62 degrees Kelvin was detected at approximately 20 and 40 degrees (south) lattitude.

Prior to the *Voyager 2* encounter, Earth-based infrared measurements had indicated that the Uranian atmosphere might contain as much as 40 percent helium. But initial interpretation of *Voyager 2* measurements indicates that Uranus' helium abundance lies somewhere between 10 and 15 percent, a range of values more compatible with values established for Jupiter and Saturn.

Careful processing of *Voyager 2* imagery revealed a brownish haze or smog concentration over Uranus' south (sunlit) pole. Some scientists now hypothesize that this "polar smog" results from photochemical dissociation of methane (CH_4) into molecules of acetylene (C_2H_2). While *Voyager 2* instruments detected only trace amounts (i.e., parts per million) of hydrocarbons such as acetylene in Uranus' atmosphere, these quantities are considered possibly sufficient to cause this brownish haze over Uranus' south pole region.

Further processing of *Voyager 2* images also showed four faint but distinct clouds deep in the planet's atmosphere between 25 and 50 degrees lattitude. Scientists were able to track these clouds and to determine that the winds on Uranus were *prograde* (in the direction of rotation) and possessed high velocities on the order of 100 meters per second (328 ft/sec). These clouds further appeared to rotate around the planet in lattitude-dependent periods ranging from 16.2 hours to 16.9 hours. These cloud rotation rates are much faster than the planet's currently assumed rotation rate of 17.3 hours.

The discovery of "electroglow" in Uranus' atmosphere by the *Voyager 2* spacecraft provided scientists yet another surprise. Electroglow is an ultraviolet (UV) emission believed to be caused in Uranus' upper atmosphere when electrons collide with high altitude hydrogen molecules. This as yet not completely understood UV emission phenomenon appears to occur only on the daylight (sunlit) side of the planet. Weaker electroglow measurements were made by the *Voyager 2* when it encountered both Jupiter and Saturn.

While passing through the Uranian system, the *Voyager 2* also observed a weak ultraviolet aurora on the planet's dark (night) side. This auroral UV emission occurs when energetic electrons from the magnetosphere cascade into the Uranian atmosphere.

The Satellites

The first major discovery resulting from the *Voyager 2* encounter with Uranus concerned its family of moons. Ten new satellites, ranging in size from about 16 km (10 mi) to 160 km (100 mi) in diameter, were discovered during December 1985 and January 1986. The two innermost of these new Uranian satellites, called 1986U7 and 1986U8, are located on either side of the epsilon

Voyager 2 *image of the Uranian satellite Ariel taken on January 24, 1986. NASA/Jet Propulsion Laboratory.*

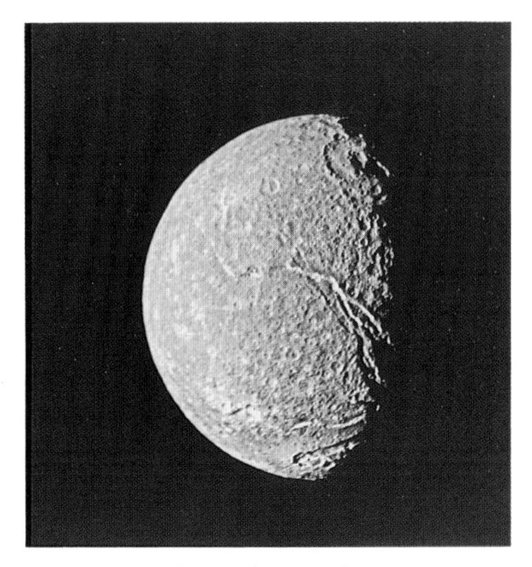

Voyager 2 *image of Uranus' largest satellite, Titania. Many impact craters of many sizes pockmark its surface. NASA/Jet Propulsion Laboratory.*

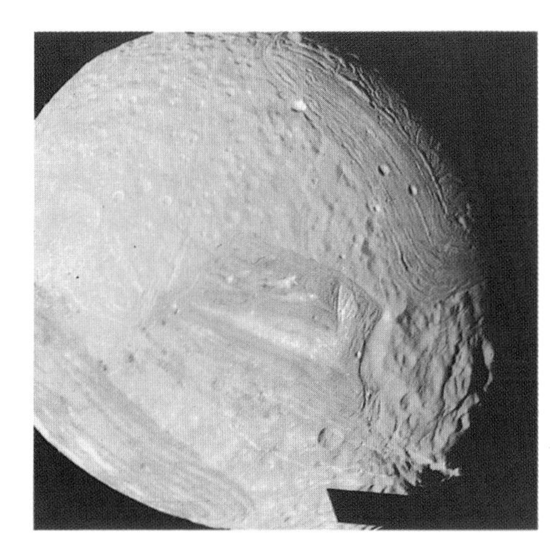

Uranus' innermost major satellite, Miranda, as photographed by Voyager 2 *on January 24, 1986. The spacecraft flew closer to Miranda than to any other body it has visited. Features as small as 560 meters (1,840 feet) are visible. NASA/Jet Propulsion Laboratory.*

ring and perform what astronomers call shepherding duties for that ring. The remaining eight new satellites orbit Uranus in the region between its rings and Miranda, the smallest of the previously known "major" Uranian moons.

Before *Voyager 2* passed through the Uranian system, astonomers had discovered five satellites: Miranda, Ariel, Umbriel, Titania and Oberon. Since these moons range in size from approximately 500 km (300 mi) diameter for Miranda to about 1,600 km (1,000 mi) diameter for Oberon and Titania, these moons have been called "intermediate-sized." This means that, in general, they are just large enough for their own gravitational attraction to pull them into spherical shape, yet not quite large enough to generate a sufficient amount of internal heat energy necessary for driving major geologic activity such as fault creation. However, scientists also feel that these intermediate-sized satellites might be able to generate enough internal heat (perhaps through radioactive decay processes) to stimulate minor amounts of geological activity. In any case, *Voyager 2* provided scientists with their first real opportunity to study the celestial objects encircling Uranus.

Voyager 2 data have confirmed that these moons are denser than the icy satellites of Saturn, suggesting perhaps that the Uranian family of satellites are worlds of half ice and half rock. With surface temperatures typically at 80 degrees Kelvin and reflectivities on the order of only about 20 percent, scientists are now speculating that ice may have mixed with significant quantities of methane during the planets' formation and then years of radiation exposure within Uranus' magnetosphere may have caused this methane-water ice mixture to darken as other organic compounds were formed.

Voyager 2 imagery showed that the outermost satellite, Oberon, was heavily cratered, with bright rays emanating from the craters. Scientists were surprised, however, to find dark patches on the floor of several craters.

Titania, Uranus' largest moon, has an extensively cratered ancient surface. The most prominent features are fault valleys that stretch to nearly 1,500 km (about 1,000 mi) in length and some 75 km (45 mi) in width.

Ariel is densely pitted with craters that are typically 5 to 10 km (3 to 6 mi) across. Numerous valleys and fault scraps crisscross Ariel's crater-scarred terrain. These valleys may have formed over down-dropped fault blocks (i.e., graben). A preliminary investigation of the *Voyager 2* imagery indicates that extensive faulting has occurred as a result of expansion and stretching of Ariel's crust. However, scientists do not yet know whether Ariel's sinuous scarps and valleys were formed by faulting or by the flow of fluids.

Voyager 2 imagery of Umbriel depicts the 1,200 km (750 mi) diameter Uranian satellite as an essentially featureless, inert, dark and ancient world. Umbriel's surface appears to be uniformly saturated with large craters, some 100 to 200 km (60 to 125 mi) across. After the *Voyager 2* encounter, scientists are now busy developing hypotheses as to why its surface is so much darker that the other Uranian satellites.

The smallest of Uranus' major satellites, Miranda, offered scientists perhaps their most unexpected surprises from the *Voyager 2* encounter. The 500 km (300 mi) diameter Miranda exhibited an incredibly diverse landscape. On Miranda, scientists noted that ridges and valleys of one geologic province are cut off against the boundary of the next province. This moon possesses very large scarps or cliffs, ranging from 0.5 to 5.0 km (0.3 to 3 mi) in height, which is higher than the walls of the Grand Canyon on Earth. Because Miranda looks like a collection of the strangest geologic forms found anywhere in the solar system, scientists are now postulating an equally unusual theory to explain this most unusual world. This first, very tentative, hypothesis speculates that at sometime in its early history Miranda was broken up or jumbled by an impact of some type. Then, as time passed, chunks of the original moon, both ice-side out and rock-core side out, came together again to form this most intriguing "patchwork of geologic forms."

Because of the *Voyager 2* encounter with Uranus in January, 1986, scientists now have an extensive data base to help them more fully understand the bland, aquamarine-colored seventh planet and its system of rings and satellites.

Satellite	Year Discovered/Discoverer	Diameter [km]*	Radius of Orbit [km]*	Period [days]
Oberon	1787, Herschel	1,554	584,000	13.463
Titania	1787, Herschel	1,594	436,000	8.706
Umbriel	1851, Lassell	1,191	267,000	4.144
Ariel	1851, Lassell	1,167	191,000	2.520
Miranda	1948, Kuiper	500	129,000	1.414
1985U1	1985, Voyager 2	160	86,000	0.762
1986U5	1986, Voyager 2	50	75,000	0.622
1986U4	1986, Voyager 2	50	70,000	0.558
1986U1	1986, Voyager 2	100	66,000	0.513
1986U2	1986, Voyager 2	80	64,400	0.493
1986U6	1986, Voyager 2	50	62,700	0.475
1986U3	1986, Voyager 2	80	61,800	0.463
1986U9	1986, Voyager 2	50	59,100	0.433
1986U8	1986, Voyager 2	25	53,300	0.372
1986U7	1986, Voyager 2	15	49,300	0.330

* approximate values

Table showing the most current information available about the 15 known satellites of Uranus. NASA/Jet Propulsion Laboratory.

15. Neptune

In the mid-nineteenth century, careful study of Uranus's motion revealed the existence of an even more distant planet with very similar characteristics.

In 1821, the mathematician and astronomer Alexis Bouvard was appointed by the French Bureau of Longitudes to compile tables of the motion of the three large planets known at the time: Jupiter, Saturn and Uranus. For the first two, he observed that the calculated theoretical positions matched the observed positions perfectly; but for Uranus, he found inexplicable deviations from the orbit predicted from the old observations. (Before being recognized as a planet in 1781, the body had been seen about 20 times, beginning in 1690, but was assumed to be a star, as we saw in the previous chapter.) In making up his tables, Bouvard decided to rely only on observations subsequent to 1781, "leaving to future ages the task of explaining whether the difficulty in reconciling the two systems results in fact from the inaccuracy of the old observations, or whether it depends on some strange, unperceived force that may have acted on the planet."

Facts supplied an answer before too long. In 1830, Uranus's position in the sky differed by 20 arc seconds from the one predicted by Bouvard's tables. The discrepancy, which gradually increased, exceeded 2 arc minutes in 1845. Most astronomers at the time considered the existence of a body more distant than Uranus that perturbed its motion without, however, having an appreciable influence on Jupiter and Saturn.

It was a Cambridge student, John Couch Adams, who first began research into this mysterious body. In 1845, after two years of calculations, he was able to demonstrate that the observed anomalies could be explained by the presence of a new planet, whose mass and orbit he defined. He communicated this result to Sir George Airy, who unfortunately did not take it seriously. At the same time, in France, where Adams's name and work remained completely unknown, Urbain Le Verrier, a lecturer in astronomy at the École Polytechnique, began similar research under the direction of Arago. In a memorandum to the Academy of Sciences dated August 31, 1846, Le Verrier in turn announced the existence of a new planet, giving its mass and orbit. Then, on September 18, he communicated the coordinates of this planet to the German astronomer Galle in Berlin. The letter reached its destination on September 23. That very evening, Galle aimed his reflector at the point in the constellation of Aquarius indicated by his correspondent and discovered the announced planet, only 52 arc minutes (less than twice the Moon's apparent diameter) from the indicated position. The new planet was called Neptune.

This discovery, which was a remarkable success for celestial mechanics, had considerable repercussions. In Great Britain it aroused a degree of bitterness: In July 1846, the astronomer Challis at the Greenwich Observatory, using Adams's data, had observed the planet several times without identifying it. In fact, Neptune may have been discovered much earlier: Michel Lefrançois de Lalande, nephew of the famous astronomer Jérôme Lefrançois Lalande, observed it on two occasions —May 8 and 10, 1795—with the great quarter-circle refractor of the École Militaire, but he thought it was a star and ascribed its slight displacement in the sky from one observation to the next, relative to nearby stars, to an error. Galileo himself, according to the American astronomer Charles T. Kowal and his Canadian colleague Stillman Drake, who examined his observation notebooks, may have seen the planet twice, on December 28, 1612, and on January 28, 1613, also assuming it to be a star, while it was in the immediate vicinity of Jupiter in the sky.

Le Verrier discovers Neptune by calculation. Engraving of the period. EDMA Archives Photo.

Orbital Characteristics

Neptune traces an immense, nearly circular orbit around the Sun, with a radius of some 4.5 billion kilometers (2.8 billion miles), (slightly more than 30 times the average distance of the Earth from the Sun), and inclined by 1°47′ with respect to the ecliptic.

Having a velocity of only 5.4 kilometers (3.4 miles) per second, the planet takes no less than 164 years, 280 days to complete this circuit. Since its discovery, it has not yet made a complete revolution! Thus it appears to drag itself through the sky with extreme slowness, remaining visible for years in the same constellation.

The orbits calculated by Adams and by Le Verrier—while they differ little from one another—diverge rather substantially, on the whole, from Neptune's real orbit. But they are in essential agreement with it near their perihelion. As luck would have it, from 1840 to 1850 Neptune was precisely in that portion of its orbit—that was the reason why the planet could be discovered practically at the point in the sky assigned to it by the calculations.

Physical Characteristics

Invisible to the naked eye, Neptune appears in the telescope as a body of magnitude 8. A magnification of at least 300 is necessary to perceive the greenish-blue disk, whose apparent diameter never exceeds 2.9 arc seconds. Its real dimensions are very close to those of Uranus, and given the imprecisions in the measurements, it was long difficult to know which of the planets was larger.

According to the most recent determinations, Neptune would have an equatorial diameter of 48,600 kilometers (30,205 miles), 4.3 percent less than that of Uranus; Neptune's polar diameter would measure about 1,300 kilometers (808 miles) less.

Le Verrier had estimated the mass of the planets responsible for the perturbations of Uranus to be 32 times that of the Earth, and Adams, 45 times. In fact, since Neptune orbits closer to Uranus than predicted by the calculations, its mass is appreciably lower; it is only 17.2 times that of the Earth. From this we infer an average density of 2.25, suggesting a structure very close to that of Uranus.

Even in the largest instruments, it is rare and difficult to be able to observe details on Neptune's surface. Irregular and poorly contrasted features sometimes have been seen on the planet's tiny disk, but parallel band structures, as on Jupiter or Saturn, never have been noted. In any case, the few details observed have not made it possible to measure the period of rotation. This, however, has been determined by other methods. Since the spectroscopic measurements taken in 1928 at the Lick Observatory by the Americans Moore and Menzel, it is accepted that Neptune rotates with a 15.8-hour period in a forward direction around an axis that is inclined by about 29 degrees relative to its orbital plane. According to recent (spectroscopic and photometric) determinations, it might be slower (18.2 hours according to M. Belton, L. Wallace and S. Howard in 1980).

Fairly similar to Uranus in size, mass and period of rotation, Neptune nonetheless differs in the composition of its atmosphere. At most, the center-edge darkening of the planet's disk—more pronounced than for Uranus—may be an indication of a thicker atmosphere.

At the beginning of the 20th century, the American V. M. Slipher had noted the similarity in the optical spectra reflected by the two planets, Neptune's being distinguished only by its much more intense absorption bands. Since then, methane and molecular hydrogen have been identified around Neptune, as around Uranus. But the presence of helium—and, deep down, of a layer of ammonia clouds—is not excluded. At high altitudes, there may also be argon clouds.

Like Uranus, Neptune has an albedo of 0.5. The greenish-blue appearance of the two planets is explained by the fact that their atmosphere reflects more blue light than red light, the latter being selectively absorbed by methane.

Neptune receives 900 times less energy from the Sun than the Earth does per unit surface area. Assuming that the solar light it absorbs is its only energy source, one calculates that its effective temperature must be about −228°C. This value does not differ greatly from those indicated by direct measurements. However, the planet is an intense source of infrared radiation and seems to give off a bit more energy than it receives; it may undergo some internal warming as a result of the tidal forces created by its massive satellite Triton. When all is said and done, in view of their many similar traits and the physical peculiarities that distinguish them from Jupiter or Saturn, Uranus and Neptune are regarded by certain specialists, such as the American astronomers Alastair G. W. Cameron, William Hubbard and James Alfred Van Allen, as the prototypes of a subcategory of giant planets.

The Satellites: Triton and Nereid

Two satellites have been identified around Neptune: Triton, discovered on October 10, 1846 (less than a month after the planet itself) by the British astronomer William Lassell, and Nereid, spotted on May 1, 1949, by the American astronomer Gerard P. Kuiper. Each satellite has special peculiarities. Triton, of magnitude 14, is more massive than the Moon. Triton's diameter is poorly known but is estimated to be about 4,400 kilometers (2,735 miles), making it one of the largest satellites in the solar system. In 5 days, 21 hours, at only 355,000 kilometers (220,634 miles) from Neptune, it completes a nearly circular orbit whose plane makes an angle of about 10 degrees with the planet's equatorial plane.

In the solar system, all the inner satellites rotate in the same direction as their planet, and—except for the Moon—are situated nearly in its equatorial plane. Triton, therefore, is a strange exception. By its retrograde motion and the inclination of its orbit, it suggests a captured asteroid, while on the other hand, its circular orbit suggests that it indeed formed where we observe it today and that it always has been a satellite of Neptune.

To explain such anomalies, Lyttleton and Kuiper suggested that the planet Pluto—the orbit of which also has surprising features (see page 203)—was originally a satellite of Neptune like Triton. The two objects thus would originally have made perfectly normal orbits around Neptune in the forward direction. But they would have passed so close to one another that as a result of enormous mutual perturbations, Pluto would have become independent, while Triton's orbit would have undergone substantial changes. In 1978, D. P. Cruikshank detected a light methane atmosphere on Triton.

The second satellite, Nereid, of magnitude 19, is much less massive: Its diameter is no more than 300 kilometers (186 miles). It revolves around Neptune, in a forward direction, in about 360 days, on an unusually elongated orbit, inclined 28 degrees on the planet's equatorial plane. Of all the satellites in the solar system, it is the one with the most elliptical orbit: During one revolution, it comes to within only 140,000 kilometers (87,011 miles) of Neptune before moving to a distance of 9.5 million kilometers (5.9 million miles). It may perhaps be a captured asteroid.

16.
Pluto

In 1915, two American astronomers, Percival Lowell and William H. Pickering suggested independently of one another the existence of an unknown planet beyond Neptune's orbit. Their suggestions were based on faint, unexplained perturbations in the motions of Uranus and Neptune.

As Le Verrier had done for Neptune some 70 years earlier, Lowell indicated the region of the sky where, in his opinion, the body he called "planet X" should be located. In 1919, the astronomer Milton Humason began a photographic search at the Mount Wilson Observatory, but this remained futile. It was only on February 18, 1930, that a young assistant at the Lowell Observatory in Flagstaff, Arizona—Clyde W. Tombaugh—discovered the anticipated planet by examining photographs taken several weeks earlier. The planet, which was in the constellation of Gemini about 5 degrees from the predicted position, received the name of Pluto, the first two letters of which are the initials of Percival Lowell.

By comparing these two photographs, taken on January 23, 1930 (at left) and January 29, 1930 (at right), the American Clyde W. Tombaugh discovered Pluto by noticing its motion among the stars. Lowell Observatory Photo.

ticular, estimates of its mass and diameter have had to be revised several times.

Percival Lowell had assigned his "planet X" a mass equal to seven times that of the Earth, and William Pickering estimated the mass of his "planet O" to be double that of the Earth. Pluto's apparent brightness turned out in fact to be appreciably less than predicted, and three possibilities had to be imagined for the planet: a poorly reflecting surface; an exceptionally dark limb; or a very small diameter (implying a density relative to water of nearly 40!).

In 1950, Gerard P. Kuiper attempted a first determination of the diameter. Based on visual observations in the 5-meter (200-inch) telescope on Mount Palomar that had just been placed in service, he attributed to Pluto an apparent diameter of 0.23 arc seconds (or the diameter of a quarter seen at 21 kilometers) (13 miles), corresponding to a real diameter of 5,900 kilometers (3,667 miles). However, such measurements, based on poorly resolved images, are subject to large systematic errors.

The following year, W. J. Eckert, D. Brouwer and G. M. Clemence were led to determine the mass of the planets farthest from the Sun in order to do the first dynamical simulation of the outer solar system, using an electronic calculator. They attributed to Pluto a mass nearly 1.1 times that of the Earth. With the diameter measured by Kuiper, this meant the planet had a density as much as 60 times that of water—a totally unreasonable figure.

During the night of April 28–29, 1965, Pluto happened to pass a star at a very small angular distance, which, however, was not occulted. The measurements taken at that time made it possible to establish that its diameter is less than 6,800 kilometers (4,226 miles)—a result compatible with Kuiper's. Moreover, in the following years, a painstaking reexamination was begun of the perturbations in the motions of Uranus and Neptune, to try again to determine Pluto's mass: This was first estimated, in 1968, to be slightly less than 0.2 times that of the Earth, then in 1971, slightly more than 0.1 times—this last figure implying for Pluto an average density of about 8 gm/cm^3, which was tolerable though still quite high. It was not until the end of the 1970s that the riddle of the planet's mass and diameter finally was solved.

First, in 1976, study of the spectrum of light reflected by Pluto in the near infrared revealed that the planet very likely is covered largely by frozen methane. This implies that its surface has a higher reflectivity than was previously thought (an albedo equal to at least 0.4) and hence that its diameter is smaller than previous estimates implied. Then, in 1979, it was possible to establish by "speckle interferometry" (see "Instruments for and Techniques of Exploring the Universe," page 25), hence by direct determination, that Pluto's diameter ranges from 3,000 to 3,500 kilometers (1,865 to 2,175 miles). The discovery in 1978 of the satellite Charon (see later in this chapter), moreover, made it finally possible to determine the planet's mass with acceptable precision (through Kepler's third law—see "The Solar System," page 68): It is only about 0.002 times that of the Earth, or approximately one-quarter that of the Moon. Pluto, then, has a mass much too low to perturb the orbital motions of Uranus and Neptune in an appreciable way, and we must now admit that chance played a large role in its discovery: It was the result of systematic search but not of precise mathematical prediction.

Recent determinations of Pluto's diameter and mass lead finally to attributing to it an average density relative to water of 0.5 to 1—comparable to that of the other distant planets in the solar system.

We know only very little of Pluto's surface composition; its feeble intensity makes study of the spectrum and the polarization of solar light it reflects extremely difficult. It seems—as has been said—that it is largely covered by methane, but some measurements also suggest the presence of silicates on its surface.

Different models of internal structure have been envisioned. The one now considered to be most valid predicts the existence, at Pluto's core, of a rocky nucleus about 1,200 kilometers (746 miles) in diameter, the mass of which represents approximately 25 percent of the total mass; this core would be surrounded by a thick ice mantle, itself topped by a crust of frozen methane 50 kilometers (31 miles) thick, representing only 3 percent of the total mass. Other models contain larger methane layers, up to 700 kilometers (435 miles) thick.

Unlike most of the other planets, which have only apparent variations in intensity resulting from these planets' changing place relative to the Earth and Sun, Pluto has an intrinsic variation in brightness over a period of nearly 6.39 days. This fact, discovered in 1955 by Robert H. Hardie, is ascribed to the planet's rotation on its axis. The most recent measurements make it possible to set the

Orbital Characteristics

Even today, Pluto's orbit still is not known with as much precision as those of the planets closer to Earth. Even so, it is clearly distinguished from the other planets by a very strong inclination on the plane of the ecliptic (more than 17 degrees) and by an exceptionally large eccentricity (0.246). As a result of these particularities, Pluto deviates up to more than 2 billion kilometers (1.24 billion miles) from the plane of the ecliptic, and its distance to the Sun varies by a substantial amount in the course of its revolution, which takes about 248.5 years. At its aphelion, which it last reached on July 12, 1866, the planet was 7,400 million kilometers (4,599 million miles) from the Sun, but at its perihelion, where it will arrive on September 12, 1989, this distance falls to 4,425 million kilometers (2,750 million miles). Over a portion of its orbit, Pluto thus comes as close to the Sun as Neptune. In fact, since January 23, 1979, Neptune has become the most distant planet, and Pluto will not regain its usual place as the last planet in the solar system until March 15, 1999. Nevertheless, as a result of their different orbital inclinations on the ecliptic, Neptune's orbit and that of Pluto do not cross: When it comes inside of Neptune's orbit, Pluto is more than 1 billion kilometers (622 million miles) above it, and the distance between the two planets never drops below 2.5 billion kilometers (1.6 billion miles). Thus there is no risk of collision between them.

Physical Characteristics

From the physical standpoint, little is yet known about Pluto more than half a century after its discovery. Its distance makes it extremely difficult to study, and the planet retains a semistellar appearance even in the most powerful telescopes. In par-

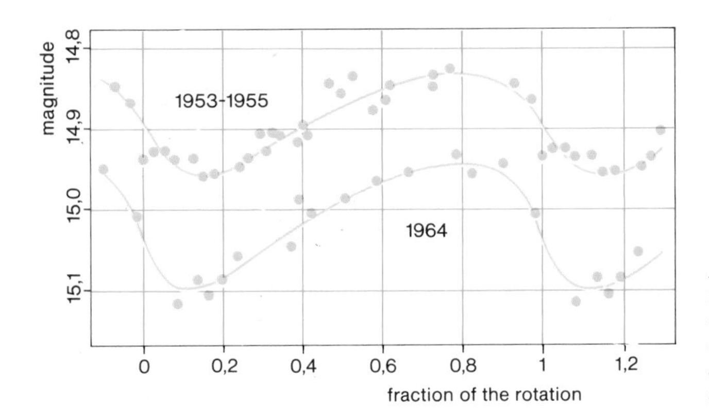

Photoelectric observations of Pluto's magnitude, showing the periodic variations in brightness over time, owing to the planet's rotation on its axis. From Robert Hardie, 1953–55 and 1964.

period at 6.38673 days, with an uncertainty of no more than 1 minute. This periodic fluctuation in brightness appears as a slow increase followed by a more rapid decrease, which may be explained by assuming that regions of different reflectivity are juxtaposed on Pluto's surface. Moreover, since its discovery, the amplitude of its variations has grown from 0.11 to 0.22 magnitudes. Finally, we note that Pluto is gradually darkening: In 20 years, its average magnitude, calculated by taking into account variations in its distance from the Earth and Sun, has increased by 0.20. These phenomena suggest that Pluto's appearance, seen from the Earth, is slowly changing while the planet completes its orbit, thus implying that its axis of rotation is not perpendicular to the plane of its orbit.

Origin

Pluto appears to be a strange world indeed. Traditionally it has been classified as one of the major planets in the solar system. But its small size and the peculiarities of its orbit suggest that it should be placed, rather, among the minor planets: It would then be the largest and most distant asteroid we know. Could it be the main representative of a second asteroid belt, situated beyond Neptune? In any case, it is unlikely that Pluto formed under the same conditions as the major planets of the solar system.

Some people see it as one of the many bodies that formed on the fringe of the primitive solar nebula during the period of contraction: Thus it would be a parent of the comets, perhaps even a giant cometary nucleus.

However, because of its size, mass and density, Pluto greatly resembles certain satellites of Saturn, which suggests that Pluto may originally have been tied to one of the large distant planets.

According to a hypothesis put forward by G. Kuiper and R. Lyttleton—who polled many experts—Pluto is a former satellite of Neptune that managed to escape the planet's gravity, perhaps after colliding with another body. This would explain the fact that Pluto is in gravitational resonance with Neptune—Pluto making two revolutions while Neptune makes three—and the astonishing current peculiarities of Neptune's system: Of the two satellites orbiting around it, the larger, Triton, is the only example in the solar system of a satellite close to its planet that has a retrograde motion; the smaller satellite, Nereid, has an unusually elongated orbit, which is exceeded in eccentricity only by some of the cometary orbits.

Planet X: Myth or Reality?

As recent measurements have revealed, Pluto's mass—even with its satellite added—appears much too low to perturb the motions of Uranus and Neptune appreciably. Inasmuch as such perturbations are quite real and cannot be explained by the known planets, the question arises as to what is causing them. One hypothesis, often put forward in the past few years, invokes the presence of a large planet beyond Pluto that remains to be discovered—planet X, so called because of its hypothetical nature but also because it would be the 10th large planet of the solar system. In short, this would be the planet imagined by Lowell and Pickering, which Pluto cannot claim to represent.

In 1972, D. Rawlins and M. Hammerton found that the perturbations of Neptune's motion could be explained by the existence of a planet having 2 to 5 times the Earth's mass and situated at a distance of 50 to 100 a.u. (astronomical units) from the Sun. For their part, Van Flandern and Harrington refer to such a planet to explain the origin of Pluto and its satellite: In passing close to Neptune, this planet would have torn Pluto away from Neptune's gravity, while the tidal forces tore away the satellite.

Finally, to account for the perturbations affecting the trajectory of Halley's comet, J. Brady in 1972 put forward the hypothesis of a planet beyond Pluto with a mass equal to 280 times that of the Earth and circulating at 60 astronomical units from the Sun; but shortly afterward, this theory was refuted by P. Goldreich and W. Ward: The existence of such a massive object making the predicted orbit is incompatible with the present configuration of the solar system.

In any case, the search for a possible planet beyond Pluto is an extremely tricky venture, for, so far from the Sun, its intensity and apparent motion relative to the stars could only be very weak.

As of now, systematic observations of the sky carried out earlier with a view to discovering Pluto enable us to state that there is no planet of Neptune's size near the plane of the ecliptic less than 270 astronomical units from the Sun. While it cannot be totally excluded, the existence of a large planet beyond Pluto seems very unlikely. It appears more plausible to attribute the still unexplained perturbations in the motions of Uranus and Neptune to a massive comet cloud within the confines of the solar system.

One Satellite: Charon

On June 22, 1978, the American astronomer James W. Christy was carefully examining enlargements of photographs of Pluto taken a few weeks earlier with the 155-centimeter (61-inch) astrometric telescope of the U.S. Naval Observatory (with the aim of determining its orbital parameters more precisely). He noticed that the planet's image showed a protuberance with a regularly varying orientation, making a complete rotation in a bit more than six days. From this he deduced the existence of a satellite, too close to Pluto to appear separated from it. This discovery was confirmed several days later by the examination of new photographs as well as plates dating from 1965 and 1970.

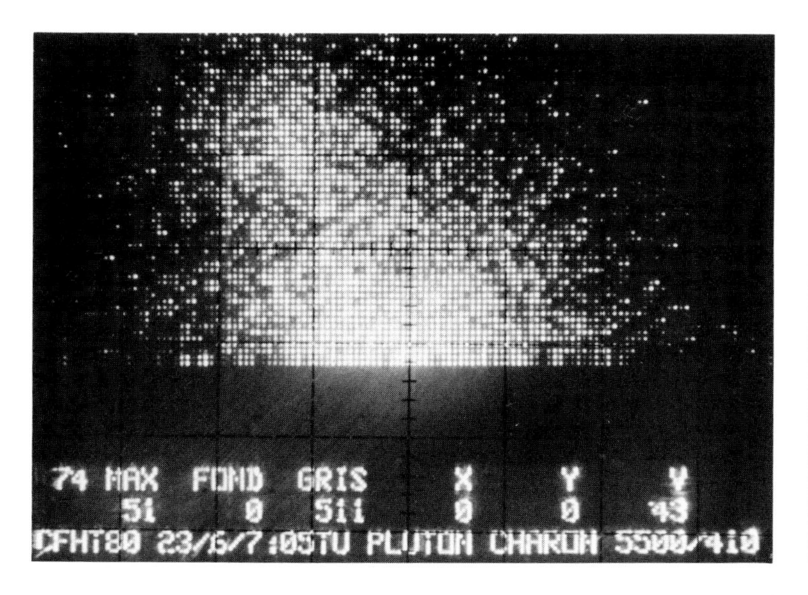

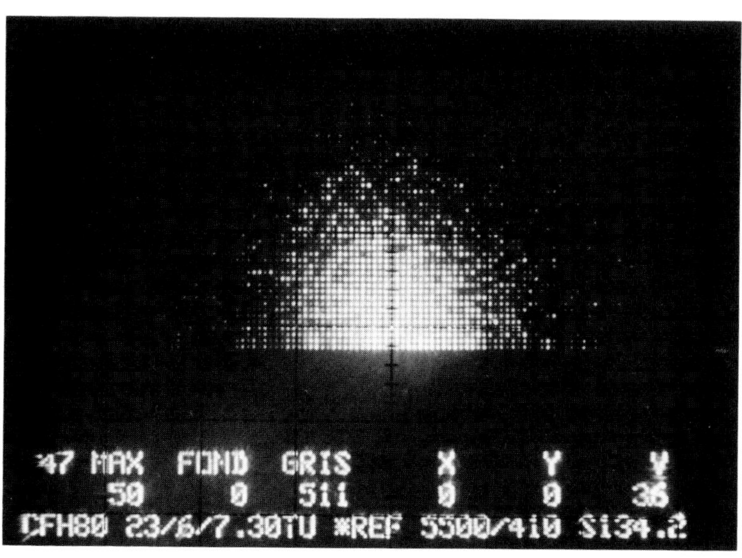

The first resolution of the Pluto-Charon system. Speckle interferometry and A. Labeyrie's photon-counting camera made possible the first resolution of the Pluto-Charon system, with the 3.60-meter (142-inch) French-Canadian telescope in Hawaii. We see here the average of the system's greatly enlarged autocorrelations; these pictures are recorded on videotape at the television scan rate: 20 milliseconds per image. The central spot is the sum of the autocorrelation of Pluto's disk with that of Charon's disk. The secondary peak, clearly separated, is the correlation of Pluto's disk with that of Charon. The distance between the two peaks, and the direction of the axis connecting them, respectively measure the angular distance between Charon and Pluto, and their position angle, to within 180 degrees on the celestial sphere. The picture at right represents the average autocorrelation of the images of an unresolved star in Pluto's field. This autocorrelation serves as a reference. From the pictures, taken in June 1980, we conclude that the magnitudes at the average opposition of Pluto and Charon are 15.3 and 16.9, respectively; the respective diameters are 4,000 and 2,000 kilometers (2,486 and 1,243 miles), respectively; the distance between the two bodies is 19,000 kilometers (11,809 miles); the albedos are roughly equal to 0.2; the total mass of the system is 2.10^{25} grams; the densities probably are identical and close to 0.5. Thus Charon appears more as a companion than as a satellite of Pluto. This result supports the notion of a common origin for the two bodies. Photo by D. Bonneau, R. Foy, published in Astronomy and Astrophysics.

Officially named Charon (the ferryman of Greek mythology who transported the souls of the dead in his barge and took them to Hades, Pluto's realm), this satellite makes a nearly circular orbit with a radius of about 20,000 kilometers (12,430 miles) inclined 65 degrees to the plane of Pluto's orbit. In fact, Charon orbits in a retrograde direction (that is, in a clockwise direction for an observer at the North Pole of the ecliptic), so the inclination is considered equal to 115° (180° − 65° = 115°). Its period of revolution is strictly equal to the period of the planet's rotation on its axis—this phenomenon being ascribed to an effect of gravitational resonance. It is probable that the two objects are in total synchronization, the plane of the satellite's orbit coinciding with the plane of the planet's equator. Charon, then, always would be at the vertical of the same point on Pluto's surface, which certainly must cause a permanent tidal effect inside the planet.

A new indication of the existence of this satellite was obtained on April 6, 1980. As he was studying Pluto with the 1-meter (39-inch) telescope of the Sutherland Observatory in South Africa, the astronomer Alistair R. Walker noted the disappearance, close to the planet, for about 50 seconds, of the light from a star of magnitude 12. It seems that this occultation may be ascribed to Charon, which then would have a diameter of at least 1,200 kilometers (746 miles), its mass, moreover, being estimated at about 10 percent of Pluto's. This now makes it necessary to consider the Pluto-Charon system as a double planet.

Pluto and its satellite. In this negative, taken on July 2, 1978 with the 1.55-meter (61-inch) astrometric telescope of the U.S. Naval Observatory, a small outcropping at the top of Pluto's image reveals the presence of Charon, the planet's satellite. Photo by kind permission of J.W. Christy, U.S. Naval Observatory.

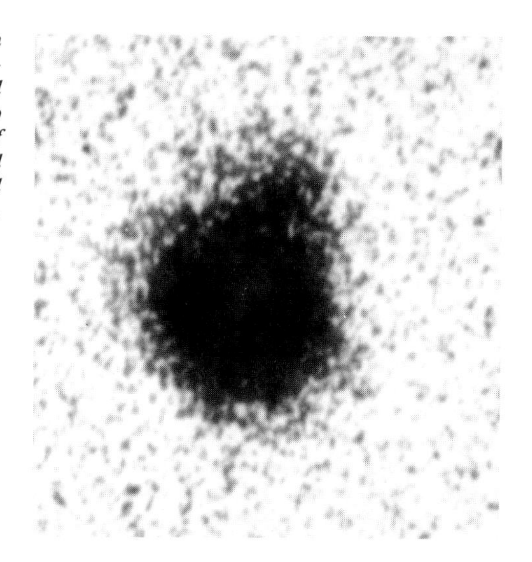

17.

The Asteroids

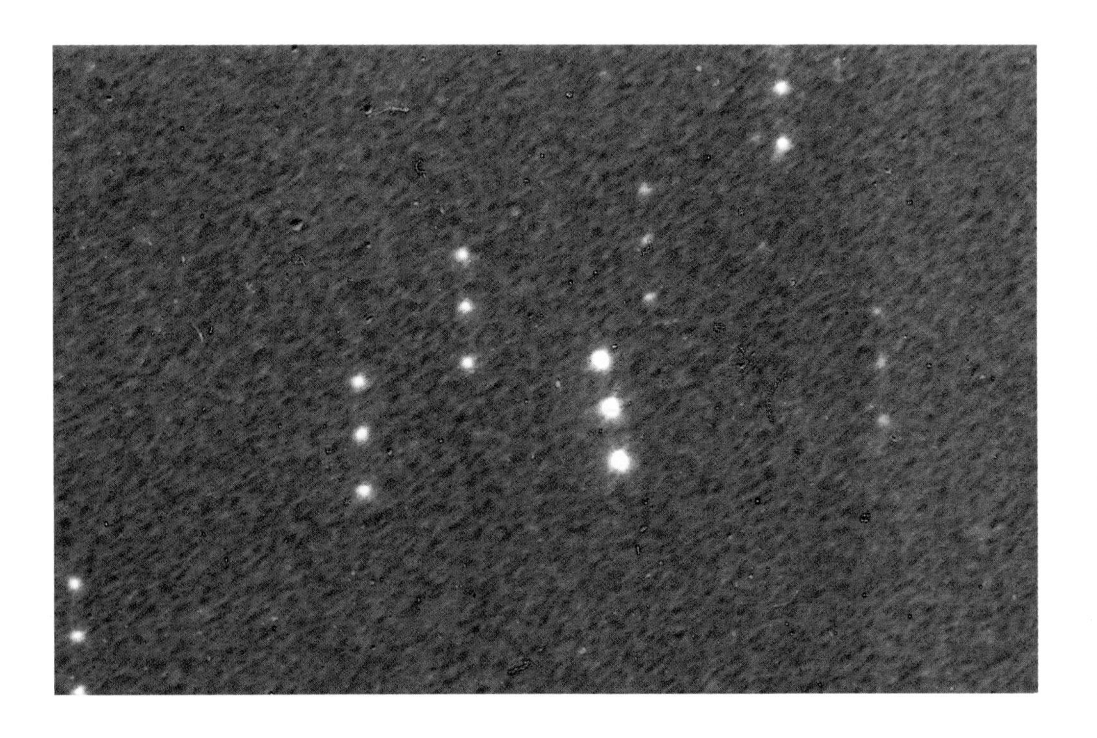

Between Mars, the last of the telluric planets, and Jupiter, the first of the giants, lies a huge gap that cuts the solar system in two. In the second half of the 18th century, publication of the numerical relationship known as the "Titius-Bode law" led astronomers to contemplate the presence of an as-yet-unknown planet in this empty space.

Asteroid 433 Eros moving among the stars. To reveal the minor planet's motion relative to the stars, a single photographic plate placed at the focus of an astrograph was used to take three exposures of two minutes each, three minutes apart. Each star in the field appears as a group of three dots lined up parallel with the edges of the photograph. The asteroid can be seen at first glance, as its proper motion causes a different alignment of the dots representing it. Photo by kind permission of Bernard Milet, Nice Observatory, France.

Thus it was that at the urging of Bode, director of the Berlin Observatory, and of Baron von Zach, a Hungarian amateur astronomer, a group of observers was constituted with the goal of systematically searching for the mysterious planet.

Someone got there before this "sky police," however. On January 1, 1801, at Palermo, Father Giuseppe Piazzi, who was observing stars in the constellation of Taurus and noting their exact positions, was surprised to discover an object not appearing on any map. On the following nights, he observed the object's motion among the neighboring stars and concluded at first that it was a comet. But the calculations done by Gauss to determine its orbit proved that it was in fact a new planet (named Ceres), orbiting at an average distance of 414 million kilometers (257 million miles) from the Sun—almost exactly at the position predicted by the Titius-Bode law—with a period of 1,680 days.

The gap finally appeared to be filled in. However, on March 28, 1802, an astronomer from Bremen, Olbers, perceived in the constellation of Virgo a "star" of magnitude 7 that was not on the map he was using. It was another planet, orbiting at the same average distance from the Sun as Ceres but tracing an orbit, in 1,686 days, of unusually great eccentricity (0.23) and inclination to the ecliptic (35 degrees). Then, on September 1, 1804, at the Lilienthal Observatory near Bremen, Harding discovered Juno, which orbits in 1,593 days at approximately 399 million kilometers (248 million miles) from the Sun.

At that point, it was necessary to bow to the evidence: The gap between Mars and Jupiter was not occupied by a single object but by a group of minor planets.

On March 29, 1807, Olbers, looking in the constellation of Virgo, further discovered Vesta—with an average distance of only 353 million kilometers

(219 million miles) from the Sun and a period of revolution of 1,325 days.

Then the years passed. As there did not seem to be any more asteroids, the "sky police" discontinued its searches. But in 1845, an amateur astronomer from Berlin, Hencke, discovered the fifth minor planet, Astrea. This unexpected success, at a time when the harvest seemed complete, reinspired observers, who, from then on, threw themselves excitedly into the search for new asteroids, aided in this by the publication of highly detailed celestial atlases.

Since 1848, not a year has gone by without a new discovery. In 1868, the number of known and identified asteroids reached 100; in 1879, 200; in 1890, 300; in 1900, 449.

Up until that time, a purely visual search method was used, requiring much patience and perseverance. One spent one's time going over the star atlases very scrupulously, looking through the eyepiece of a reflecting or refracting telescope until one found an unidentified bright point in the sky, which was revealed through daily observation to change position relative to the neighboring stars. The photographic method, introduced in 1891 by Max Wolf at the Heidelberg Observatory and by Auguste Charlois at the Nice Observatory, soon was adopted and, being much safer and quicker, made it possible to multiply discoveries. It consists of photographing a region of the sky with a long exposure (about 1 hour) at the focus of a telescope. If an asteroid is in the field, then as a result of its own motion it is recorded on film as a small linear trail and thus can be distinguished from the stars, which appear as points. A similar method consists of examining two short-exposure photographs of the same region of the sky, taken some time apart, in a blink or stereo comparator. In the latter, all the stars appear on the same plane, while an asteroid seems to stand out in front owing to its change of position from one photograph to the other.

Nomenclature

When a supposedly new asteroid is discovered in a photograph, it is given a provisional designation, consisting of two letters placed after the thousandth of the current year. Objects discovered between January 1 and 15 are designated respectively, in the order of their discovery, by the letters AA, AB, AC; those discovered between January 16 and 31 by BA, BB, BC; and so on, up to those discovered between December 16 and 31, which are designated by YA, YB, YC (the letter J is not used).

After the discovery, an attempt is made to determine the orbital characteristics of the observed object and to verify that it is indeed a new minor planet. If so, it is entered into the official list only

Orbits of a few unusual asteroids.

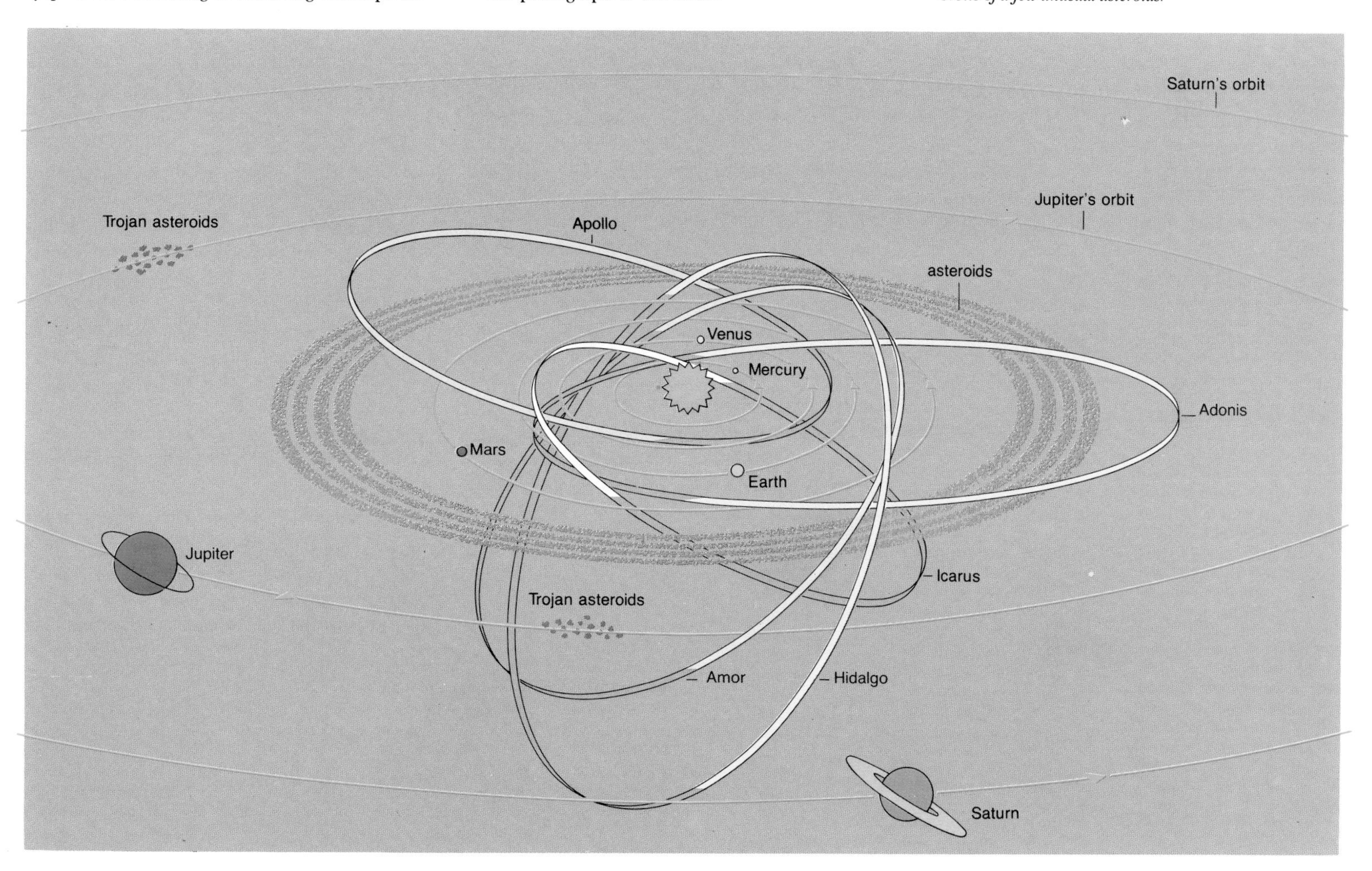

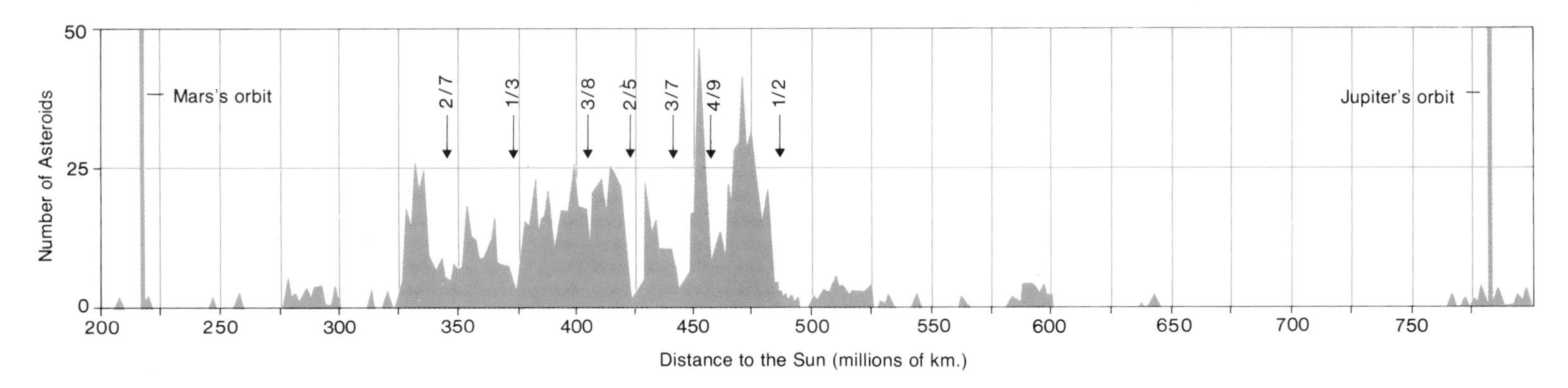

Distribution of the asteroids in interplanetary space as a function of their distance from the Sun. The gaps indicated by arrows correspond to regions where the asteroids are in gravitational resonance with Jupiter, and their period of revolution around the Sun (noted above the arrows) is in a simple ratio to that of the planet.

after it is once again observed during another opposition. Then it is given a definitive final name, comprised of a reference number followed by a name chosen by the observer.

At present, some 2,300 asteroids have been inventoried in this manner, and about 1,000 have been given a provisional designation. But thousands probably remain to be discovered. Estimates made after two systematic searches for asteroids—at the McDonald Observatory in 1950–52 and at Mount Palomar in 1960—have led to the estimate of more than 500,000 minor planets brighter than the photographic magnitude 21.2 during their average opposition. Moreover, the number of asteroids with a diameter greater than 1 kilometer (.62 mile) is estimated at about 400,000.

Orbits

Most of the asteroids are concentrated in a ring between the orbits of Mars and Jupiter at a distance from the Sun ranging from 320 to 495 million kilometers (199 to 308 million miles). But even within this ring, they are not evenly distributed. Some regions, known as Kirkwood gaps (after the American astronomer who first discovered them in 1866), appear practically empty because, according to the most generally accepted interpretation, they correspond to gravitational resonance zones where Jupiter's gravity prevents asteroids from remaining. These gaps are seen, in particular, at distances from the Sun of about 374, 422, 443 and 491 million kilometers (232,262, 275 and 305 million miles), corresponding to orbits in which the period of revolution would be in a simple ratio with Jupiter's period of revolution (1/3, 2/5, 3/7 and 1/2, respectively).

Some asteroids, however, diverge appreciably from the zone in which most of their counterparts remain. Their highly elliptical orbits enable them periodically to approach the Earth, Venus and even Mercury. Those that can graze the Earth in this way are called Earth-grazers, Earth-grazing asteroids, or, more simply, EGAs (see box). Among them is Hermes, a rock less than 1 kilometer in diameter that, in 1937, came within only 780,000 kilometers (484,773 miles) of the Earth, about twice the distance of the Moon. Some EGAs even have the distinctive feature of orbiting inside the Earth's orbit and thus have a period of revolution of less than 1 year.

The Trojan asteroids (see box) are other particular examples of minor planets revolving outside the asteroid belt.

The farthest known asteroid from the Sun is Chiron, discovered in 1977, with an orbit between those of Saturn and Uranus (see box). Another remarkable object is 944 Hidalgo, which is the asteroid with the most eccentric known orbit (0.66): At perihelion it comes within 300 million kilometers (186 million miles) of the Sun, but at aphelion it lies 1,450 million kilometers (901 million miles) away, nearly reaching Saturn's orbit. It also is the only asteroid known to come within less than 150 million kilometers (93 million miles) of Jupiter.

Furthermore, there can be no doubt that Mars's two small satellites—Phobos and Deimos—are former asteroids that were captured by the planet's gravity. The same hypothesis has been advanced for Jupiter's outer satellites, for Phoebe in Saturn's system and for Nereid in Neptune's. Generally

Asteroids That Can Graze the Earth

We know of several dozen asteroids with very eccentric orbits capable of coming quite close to the Earth. In fact, such objects probably are very numerous (60,000 having diameters that exceed 100 meters [328 feet], according to some estimates), but their small size makes it difficult to identify them. According to their orbital characteristics, we often distinguish:

1. objects of the Apollo type, with a distance from the Sun, at perihelion, of less than 1 astronomical unit (i.e., the Earth's average distance from the Sun), which, owing to that fact, regularly come within the Earth's orbit
2. objects of the Amor type, with a perihelion slightly outside the Earth's orbit, at a distance from the Sun ranging from 1 to 1.38 astronomical units, which, as a result, can graze the Earth at certain favorable times.

In reality, this distinction is arbitrary. Some asteroids can move from one group to the other under the combined effect of the large planets.

The origin of these minor planets still is in dispute, but as new ones are discovered, a twofold solution seems to be emerging. It seems that 75 to 80 percent of them have a *planetary* origin and emerged from the breakup of asteroids of the principal belt that initially did not come within the orbit of Mars. The 20 to 25 percent remaining would have a *cometary* origin: They would be the residues of the nuclei of former comets that had lost all their volatile materials.

The Trojan Asteroids

The Trojan asteroids—so named because they bear the name of heroes of the Trojan War—are a famous example of asteroids not belonging to the principal belt. They are concentrated along Jupiter's orbit, around two points, one preceding the giant planet by 60 degrees, the other following it 60 degrees behind, each group thus forming a triangle with Jupiter and the Sun that always is appreciably equilateral. This remarkable arrangement was predicted in a purely theoretical way in 1772 by the French mathematician Lagrange. It corresponds to a particular solution of the three-body problem (study of the relative motions of three mutually attracted bodies in accordance with Newton's law): Calculation shows that the points in the neighborhood of which the Trojan asteroids are concentrated (the Lagrangian points) are dynamically stable. In fact, each of the asteroids in question oscillates in a complex fashion in a zone on either side of these points.

A systematic photographic study of the two regions harboring the Trojan asteroids showed that there exist about 700 asteroids brighter, on average, than magnitude 20.9 at their opposition, in the group preceding Jupiter, and about half as many in the group that follows Jupiter. But at present we really know of only 22 Trojan asteroids, the first (588 Achilles) having been discovered on February 22, 1906, at the Heidelberg Observatory, and the last four in 1971 at Leiden in the Netherlands. Thirteen of them, situated in front of Jupiter, form the Achilles group; these are: Achilles, Hector, Nestor, Agamemnon, Ulysses, Ajax, Diomedes, Menelaeus, Antiloch, Odysseus, Telemon (the names are those of the Greek heroes of the *Iliad*, except for Hector, which was named too hastily) and two asteroids that as yet have only a temporary name. The other nine, behind Jupiter in its orbit, form the Patrocles group. These are: Patrocles, Priam, Aeneas, Anchises, Troilus (the names refer to Trojan heroes, except for Patrocles, which, like Hector, appears to be held hostage among its enemies) and four as yet unnamed minor planets.

Some Asteroids with Unusual Orbits

Order Number and Name	Year of Discovery	Type[1]	Distance to the Sun (millions of km.) Perihelion	Aphelion	Inclination of the Orbit to the Ecliptic (degrees)	Eccentricity of the Orbit	Sidereal period of Revolution (years)	Shortest Possible Distance to the Earth[2] (millions of km.)	Estimated Diameter (km.)	Presumed Origin[3]	Remarks
1221 Amor	1932	Am	162	413	11.91	0.436	2.67	16.0	0.8	P	
433 Eros	1898	Am	169.5	267	10.83	0.223	1.76	22.3	27 × 16	P	
1862 Apollo	1932	Ap	97	343	6.36	0.560	1.78	3.7	2.1		
2101 Adonis	1936	Ap	66	493.5	1.42	0.764	2.56	2.1	0.3	C	came within 2 million km. of the Earth on February 7, 1936
1937 UB (Hermes)	1937	Ap	92	398	6.22	0.624	2.10	0.3	0.8		came within less than 800,000 km. of the Earth on October 30, 1937
1685 Toro	1948	Ap	115	294	9.37	0.436	1.60	7.4	3		this asteroid's motion is in resonance either with that of the Earth or with that of Venus
1566 Icarus	1949	Ap	28	294.5	22.95	0.827	1.12	5.5	1.6	C	came within 6.3 million km. of the Earth on June 14, 1968; its perihelion lies within Mercury's orbit
1620 Geographos	1951	Ap	124	248.5	13.32	0.335	1.39	4.7	3	P	
1973 EA	1973	Ap	93	438	39.85	0.650	2.37	0.4	1.3		
2060 Aten	1976	Ap	118	171	19.05	0.183	0.95		1.3		circulates mainly between Venus and the Earth; first asteroid known to have a period of revolution of less than 1 year
2100 Rá-Shalom	1978	Ap	70	179.5	16	0.439			3	C	

1. Am = Amor; Ap = Apollo.
2. It varies more or less rapidly over time as a result of the perturbations imposed on the asteroid's motion by the major planets.
3. C = cometary; P = planetary

speaking, all the known asteroids have direct orbits like the planets, but their orbits, with an average eccentricity of 0.14 and an average inclination of 7 degrees to the ecliptic, are more elliptical and more inclined than those of the planets.

Masses and Dimensions

Determining the masses and dimensions of the asteroids is an important part of their study.

Theoretically, we can hope to deduce the mass of a minor planet from the gravitational perturbations it causes in the motion of another. In fact, such perturbations generally are too weak to be measurable, and the method could be applied to only three major asteroids: Ceres (1.17×10^{21} kg.), Pallas (2.6×10^{20} kg.) and Vesta (2.4×10^{20} kg.). For the others, it is necessary to know their diameter and to estimate their average density in order to estimate their mass.

Determining an asteroid's diameter requires prior knowledge of the object's brightness and distance. But brightness can be measured fairly easily with photometers, and distance can be determined without difficulty once we know the parameters of the orbit.

Assuming that these two quantities are known, there was, until the recent past, only one method for determining the diameter of a minor planet: observing it visually in a telescope and measuring its apparent diameter. But most of the asteroids appear as points, even in the largest instruments,

Chiron: Asteroid or Former Satellite of Saturn?

On October 18, 1977, the American astronomer C. T. Kowal, using the 48-inch Schmidt telescope at Mount Palomar Observatory, discovered an object of magnitude 18 in the constellation Aries that orbited the Sun, but the exact nature of which remained mysterious.

Rediscovered in old photographs, some of which went back to 1895, this object, which had been given the name Chiron, circulates between the orbits of Saturn and Uranus, in a region of the solar system where no asteroids had yet been observed. Its orbit shows a very marked eccentricity: 0.379. At perihelion, it comes within 1.3 billion kilometers (807,955 million miles) of the Sun, thus crossing Saturn's orbit, but at aphelion Chiron moves to 2.8 billion kilometers (1.7 billion miles) away from it. Moreover, Chiron is a minor planet with an estimated diameter of 150 to 600 kilometers, (93 to 373 miles), thus possessing dimensions similar to those of the major asteroids. If we assume that we are dealing with a body made up entirely of ice, with a high albedo, its diameter must be close to the lower end of this range. However, the existence of "all-ice" asteroids, predicted by some theorists, remains fairly problematical, and it seems more plausible that we are dealing with a type C carbonaceous asteroid, its diameter thus situated at the top of the range of possible values, between 500 kilometers and 600 kilometers (311 miles). Many astronomers think that Chiron might be the largest member of a second ring of minor planets, this time orbiting between Saturn and Uranus. Nothing, in fact, prevents asteroids from being present in the farthest regions of the solar system; but when we are concerned with very faint objects having extremely slow apparent motion, their identification proves to be particularly difficult. Such is undoubtedly the reason why Chiron remains a special case at present.

According to some specialists, Chiron is a former satellite of Saturn. It has been calculated that Chiron's period of revolution is 50.682 years, which thus is in a simple ratio (5/9) with that of Saturn, and can be explained by a gravitational resonance effect.

It also has been suggested that Chiron may be a cometary nucleus, but the estimates of its dimensions eliminate that hypothesis: Typically such nuclei less arc than 10 kilometers (6.2 miles) in diameter and cannot exceed 100 kilometers (62 miles) at the very most.

Perhaps the riddle of Chiron's nature may be resolved when the object next returns to its perihelion, in 1996. Then the planet will appear about 40 times brighter than at the time of its discovery, which makes physical study of it easier.

and this method could be used only for the four most voluminous. Even so, there was a fairly large margin of uncertainty, the angular diameter of the observed images reaching at most a few tenths of an arc second.

Since 1970, however, two new, indirect techniques have been used successfully. Each involves the albedo of the asteroid being studied and illustrates the fact that one can indirectly determine the diameter of an asteroid (presumed to be spherical) if one knows its brightness, distance and albedo. The first of these techniques consists of measuring the polarization of light reflected by the asteroid at different phase angles; analysis of meteorites in the laboratory has in fact shown that the darker the surface (the smaller the albedo) the more pronounced the polarization. The second is based on measuring the asteroid's brightness in both the visible and the infrared ranges, which makes it possible to compare the quantity of incident solar radiation reflected by the object to the quantity it absorbs and reemits in the form of heat.

These two methods, totally independent of one another, yield entirely compatible results. Already they have made it possible to estimate the dimensions of some 200 minor planets, with diameters ranging from about 1,000 kilometers (622 miles) (Ceres) to less than 1 kilometer; only eight have a diameter greater than 300 kilometers (see table), and there are hardly more than a hundred with a diameter that reaches or goes beyond 100 kilometers.

The 10 Major Asteroids

Order Number and Name	Year of Discovery	Average Distance from the Sun (millions of km.)	Period of Revolution (years)	Diameter (km.)	Type
1 Ceres	1801	414	4.60	1,000	C
2 Pallas	1802	414	4.61	610	U
4 Vesta	1807	353	3.63	540	U
10 Hygeia	1849	471	5.59	450	C
31 Euphrosyne		562.5	5.61	370	C
704 Interamnia		458	5.34	350	C
511 Davida		477	5.70	320	C
65 Cybele		511.5	6.33	310	C
52 Europa		464	5.45	290	C
451 Patientia		458	5.35	280	C

Note: This table does not include the asteroid Chiron, the diameter of which is still uncertain.

The occultation of a star by a minor planet is a rare phenomenon, but its observation also makes it possible to determine precisely the dimensions of the asteroid in question. Applied in 1977 to the minor planet 532 Herculina, it not only made it possible to assign it a diameter of almost 220 kilometers but also revealed above all that that minor planet has a satellite about 50 kilometers in diameter orbiting at about 1,000 kilometers from the main object: Herculina thus appears as the first known double asteroid. Since then, careful study of the photometric curves recorded during occultations of stars has made it possible to imagine other possible cases of binary asteroids, particularly Pallas, Juno, Eros, Hebe, Antigone and Melpomene.

Rotation Period: Light Curve

Most of the asteroids have an irregular shape and/or a reflectivity that is not strictly uniform at all points on their surface, so that the amount of light they reflect toward Earth varies with the phase of their rotation. Consequently, we can generally deduce an asteroid's period of rotation from observation of its light curve, and this has actually been done in about 50 cases: the measured periods range from 2.273 hours for 1566 Icarus to 18.813 hours for 532 Herculis. Some objects have even longer periods, the record being held by 393 Lampetia (38.7 hours) and 128 Nemesis (39.0 hours).

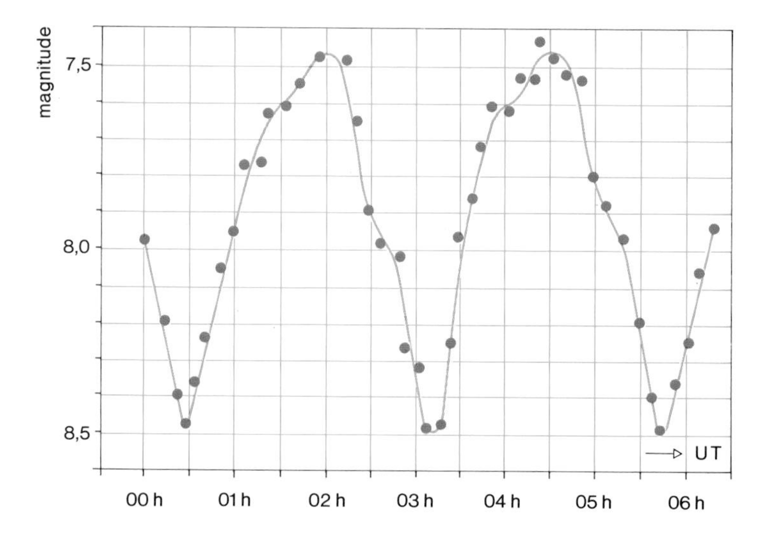

However, in some cases, the light curves can be difficult to interpret, as the example of Vesta illustrates particularly well. Astronomers have long noted that Vesta's brightness varies by about 0.15 magnitude and reaches a maximum every 5 hours, 20 minutes. There are two possible interpretations of this phenomenon: Either Vesta is a nearly spherical body with a rotation period of 5 hours, 20 minutes, or it is an elongated object with a rotation period twice as long. A detailed analysis of photometric observations of the asteroid, done in 1967 by Tom Gehrels, tended to indicate that the correct rotation period was the shorter of the two. But in 1971, R. C. Taylor did a new series of observations that clearly show that the successive maximum luminosities are alternately more and less intense and that the real period is 10 hours, 41 minutes. These observations were carried out at a time when Vesta was showing its Southern Hemisphere to the Earth. When its Northern Hemisphere faces us, the two maxima of differing amplitude that occur during each rotation period become imperceptibly different as a result of different variations of albedo in the two hemispheres. We can account for the observations by assuming that Vesta is a flattened spheroid, one of whose three axes is at most 15 percent longer than the other two. Likewise, examination of the Trojan asteroid Hector's light curve (the brightness of which varies by a factor of 3.1 during its rotation period of 6.9 hours) testifies to the peculiar shape of that small body. According to the Americans William K. Hartmann and Dale Cruikshank, that elongated object, about 150 by 300 kilometers, is made up of two asteroids touching one another. It would then present an example of the process of accretion by collision that governed the formation of the planets.

The characteristics of other asteroids should be revealed in years to come with the increase in observations of these objects, particularly in the infrared.

Surface Composition

Many techniques (photometry, spectrophotometry, polarimetry, infrared radiometry, etc.) now are being employed to study the physical characteristics of asteroids. One of the major results obtained in this field since the beginning of the 1970s is the discovery of six major physical types of asteroids:

- type C (47 percent of the asteroids studied), which includes objects with carbonaceous, hence very dark surfaces (albedo: 0.02 to 0.06), with reflecting properties similar to those of carbonaceous meteorites
- type S (35 percent of the asteroids studied), which includes objects with a silicate surface (albedo: 0.10 to 0.22), with a reflection spectrum similar to that of the silicate-rich stony meteorites
- type M (3 percent of the asteroids studied), which includes objects rich in metals (albedo: 0.08 to 0.15)
- type R (1 percent of the asteroids studied), which includes objects similar to ordinary, iron-poor chondrites (albedo: 0.20 to 0.30)
- type E, which includes a few rare objects having surfaces in which iron is completely absent (albedo: 0.30 to 0.38)
- the Vesta type, which includes a few rare objects having a surface corresponding to that of some basaltic chondrites (albedo nearly 0.25)

Light curve of the minor planet Eros. This curve, established by Alain Figer, is the result of combining 219 visual measurements taken from January 12 to 25, 1975, by seven observers, and relating them to a reference period, January 14, 1975, from 0:24 to 5:40 UT.

Types C and S alone account for more than 80 percent of the asteroids studied. The proportion of type C objects increases with distance from the Sun: at 300 million kilometers from the Sun, they constitute 50 percent of the total number of asteroids, but beyond 450 million kilometers, their proportion reaches 95 percent. This suggests that the outer region of the asteroid belt contains a large population of very dim objects remaining to be discovered.

Finally, about 13 percent of the asteroids studied do not fall into any of the six preceding categories and provisionally are designated U (unclassified). They can probably be added in the future to the other known categories of meteorites.

Three asteroids—433 Eros, 1566 Icarus and 1685 Toro—were detected by radar during their passage through the Earth's vicinity. Eros, in particular, could be studied in 1975 under good conditions at 3.8 centimeters wavelength, the measurements revealing that its surface is rough. In addition, based on the width of the radar spectrum and the period of rotation, it was possible to determine the asteroid's dimensions independently of, but in agreement with, the results supplied by the polarimetric and photometric measurements; the different methods show that Eros is an elongated object with its longest axis measuring about 36 kilometers and the shortest 15 kilometers.

Origin

In spite of the recent progress in knowledge about asteroids, their origin still is under investigation.

In the past century, the German astronomer Olbers hypothesized that these small bodies were the debris of a large planet that had exploded for unknown reasons. This theory still finds supporters. In particular, it was adopted in 1972 by the American T. Ovenden, who thought that the explosion occurred 16 million years ago (hence in a recent period, on an astronomical scale) and would have involved a planet having a mass 90 times that of the Earth.

On the other hand, according to another hypothesis, offered in 1964 by the Swede H. Alfvén, asteroids are materials in the process of accretion.

However, the majority of experts regard the minor planets rather as condensations of the original solar nebula that did not succeed in becoming agglomerated into a single body owing to their insufficient total mass (estimated at 10^{-9} solar mass, or only 1 percent of the mass of Mercury) and/or, as the American Kuiper suggested in 1950, because of the existence of a gravitational instability (due to Jupiter's gravitational attraction) in the region of the solar system concerned.

Space Probe Exploration Project

The flyby of certain asteroids by space probes should make it possible in the near future to obtain new information to answer the many questions these minor planets still raise.

A group of French and European scientists, assisted by engineers from the French CNRS (National Center for Space Studies), worked in 1978–79 to try to see whether Europe, with its own space resources, might carry out such a mission. This initial study showed that, with an improved version of the Ariane rocket and a launching from the Guyanese space center, a satellite with a total mass of 500 to 700 kilograms (1,102 to 1,543 pounds) might be sent into the asteroid belt before the end of the decade. By combining Earth-based guidance and on-board navigation using a camera to photograph the sky against a backdrop of stars, it is possible to estimate that the altitude at which the flyby of a given asteroid would take place might be less than 1,000 kilometers. Leaving aside the possibility of a landing on some asteroid because of the large relative velocity (2 kilometers per second or more) that it would be necessary to attain, it is possible to envision the flyby of several minor planets, which would make it possible to collect information on objects of different types. A useful scientific payload of 45 to 70 kilograms (99 to 154 pounds), essentially based on imaging, radar observation and infrared spectrophotometry, would make it possible to have a thorough description of each object flown over.

18. Comets

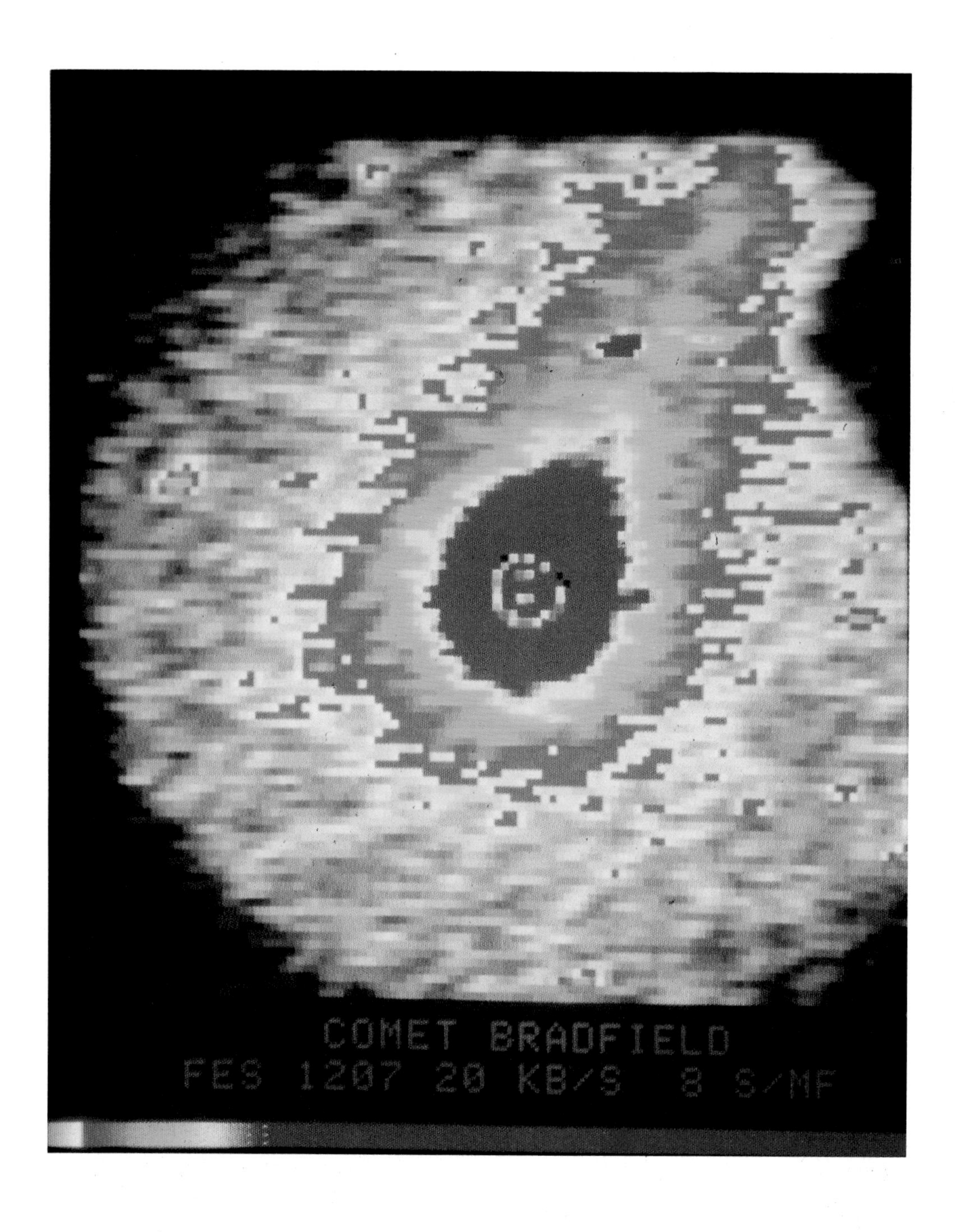

Observed since antiquity because of their occasionally spectacular displays, comets often have aroused terror. Their sudden and unpredictable apparitions, their strange appearance, their changing form, and their surprising motion among the stars explain why they have been regarded for centuries as messengers of the gods, harbingers of catastrophes. It was only in the 18th century, when the fundamental problem of their motion had been solved, that they lost most of their mystery. Nonetheless, there remains great interest in the scientific study of comets.

Comet Bradfield (1979 I), photographed in ultraviolet light by the IUE *satellite in January 1980. An appropriate color coding reveals the structure of the head. ESA Photo.*

Progress made within the past few decades in understanding the physical and chemical characteristics of comets has led researchers to regard them as vestiges of the primitive solar system; therefore, by studying them, specialists hope to find some additional pieces of the still very incomplete puzzle of how our planetary system originated and evolved.

Nomenclature

The rapid growth in the number of known comets has prompted the International Astronomical Union to catalogue them according to precise rules.

At its discovery, a comet receives a temporary designation given by the year of discovery, followed by a lower-case letter indicating the order in which it was discovered: a for the first comet found in the year, b for the second, etc. Later, when its orbit has been calculated and the date of its perihelion passage has been determined, it receives a definitive designation composed of the year of its perihelion passage followed by the Roman numeral indicating the chronological order of the passage in that year. Thus, for example, comet 1940a (the first comet discovered in 1940) became 1939VIII (the eighth comet to reach perihelion in 1939), while, inversely, 1969i became 1970II.

These rules apply not only to new comets but also to periodic comets, which, at each return, receive a new designation.

Each new comet also is given the name of the observer who discovered it. In the case of almost simultaneous observations, the names of the first two or at most three discoverers are applied. Thus comet 1969IX also is called Tago-Sato-Kosaka.

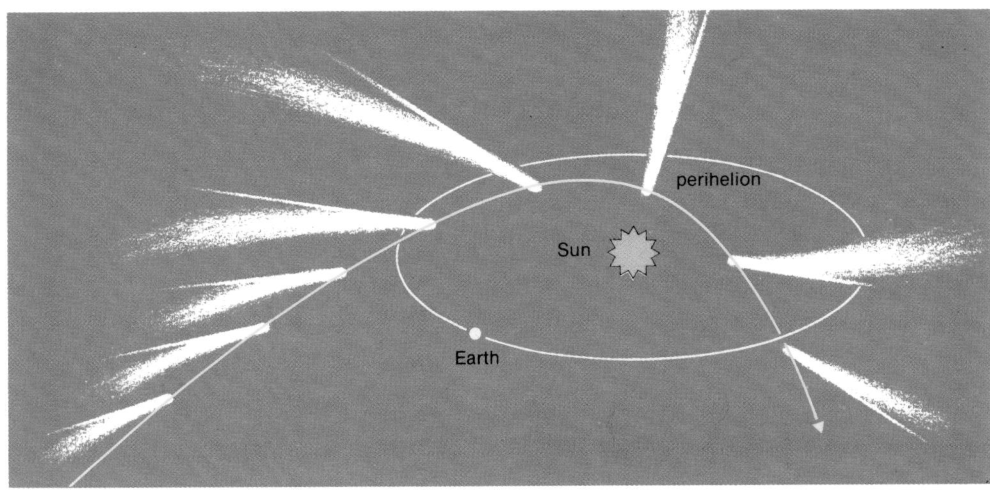

Evolution of a comet on its orbit. The tail, always opposite the Sun, may precede the head.

Comets having a period less than 200 years are additionally designated by the symbol P/ in front of the discoverer's name (or names). If the same observer discovers several periodic comets in succession, each one then carries an ordering number at the end of its name; for example, P/Tempel 1 and P/Tempel 2 refer to two periodic comets discovered in 1873 by Tempel.

When a comet has been lost for a long time, the name of the observer who rediscovers it is sometimes added to the names of the original discoverers.

Finally, some comets are given the name of the astronomer who first calculated their orbit; this is the case for Halley's comet (see box, page 214); for Encke's comet (see box, page 216); and for Lexell's comet (1770I), which was discovered by Messier.

Cometary Orbits

By April 1, 1977, a total of 922 appearances of comets had been observed with sufficient precision to calculate their orbits. They correspond to 641 different comets, among which 108 are short-period (less than 200 years) and 533 are long-period.

Among the 533 observed long-period comets, 283 of the orbits are considered as being parabolic, 165 as elliptical and 85 as hyperbolic. It must be emphasized, however, that the perfectly parabolic orbit represents a particular idealized case that is extremely improbable and that orbits are classed in this category only when they cannot be distinguished from a parabola given the uncertainties of the measurements, even though they may in reality be extremely elongated ellipses or only slightly opened hyperbolae.

The parameter that specifies the nature of a cometary orbit is its eccentricity. The eccentricity of an ellipse is less than 1 (the circle, a particular case of an ellipse, has an eccentricity of 0), that of a parabola is equal to 1 and that of a hyperbola is greater than 1.

The cometary orbit having the largest known eccentricity is that of the comet Sandage (1972IX);

Comets and the Death of Famous People

Of the gratuitous superstitions concerning comets, one of the most popular in history has been the view of these bodies as omens of the death of famous people.

Thus comets "announced" the death of Emperor Constantine (336); Attila (453); Emperor Valentinian III (455); Merovius (577); Chilperic (584); Maurice, the Emperor of Byzantium (602); Mohammed (632); Louis the Debonair;* (837); Emperor Louis II (875); Boleslas I, king of Poland (1024); Robert II, king of France (1033); Casimir, king of Poland (1058); Henry I, king of France (1069); Pope Alexander III (1181); Richard I, king of England (1198); Philippe Auguste (1223); Emperor Frederick (1250); Pope Innocent IV (1254); Pope Urban IV (1264); Jean-Galeas Visconti, duke of Milan (1402); Charles the Foolhardy (1476); Philippe the Handsome, father of Charles Quint (1505); François II, king of France (1560); etc.

In antiquity, chroniclers did not hesitate to change—sometimes by several years—the date of a comet's appearance to make it coincide with the death of a famous person.

This form of superstition did not begin to die out until the 17th century. Thus Henri IV—to whom it was reported that astrologers predicted his death because a comet had just been observed—aptly replied: "One of these days, their prediction will come true, and the single occasion when the prediction was fulfilled will be better remembered than all of the other occasions when it was not."

Likewise, on January 2, 1681, Mme de Sévigné wrote to her cousin, the comte de Bussy: "We have here a comet, which is very long; it is the finest tail that can possibly be seen. All of the great persons are alarmed, and firmly believe that Heaven, intent on their ruin, is giving warnings of it through this comet. They say that, Cardinal Mazarin having been given up by his physicians, his courtiers felt it necessary to honor his dying moments with a wondrous event, and told him that a great comet was appearing and causing fear. He had the strength to ridicule them, and told them quite amusingly that the comet was doing him too great an honor. In truth, we ought to say as much about it as he, and human pride does itself too much honor to think that there are great doings among the stars when we must die."

Nevertheless, we still find traces of this superstition in the 19th century. In 1821, on St. Helena, a dying Napoleon (he died on May 5) was struck, it is said, by the appearance of the comet discovered on January 21 by Nicollet, in Paris, and that became visible in the Southern Hemisphere in April. Later, in 1861, some people saw in the appearance of a large comet an omen of the death of Pope Pius IX.

* Louis the Debonair actually died in 840, three years after the appearance of the comet that was supposed to announce his death.

the value of its eccentricity is 1.006288. Most hyperbolic orbits, however, have an eccentricity less than 1.001. Even tiny gravitational perturbations can suffice to transform a very elongated elliptical orbit into one that is slightly hyperbolic. And we know that the planets, particularly Jupiter and Saturn, are capable of causing significant perturbations to the orbits of comets that pass through their neighborhoods—a whole family of comets that have been captured by Jupiter has been identified. The gravitational attraction of this giant planet had transformed their orbits into short-period ellipses (see box, page 216: Encke's comet). When the past trajectories of those comets that currently have hyperbolic orbits are calculated, one finds that in all cases where the calculation can be done with sufficient precision, the orbit prior to entering the zone of influence of the giant planets was a very elongated ellipse. This result is of great significance—it implies that comets originated with the solar system. Initially they all gravitate around the Sun like planets, although in much more distant orbits, and only after their first perihelion passage do some of them, as a result of perturbations to their motion, come to follow a parabolic or hyperbolic trajectory that permits them to escape the Sun's attraction.

Cometary orbits are also characterized by their inclination to the ecliptic plane, which, in contrast to planetary orbits, can take any value from 0° to 180°. When the inclination exceeds 90°, the comet's motion is retrograde.

It should be noted that there is no known comet having a period between 18 and 27 years. This gap is due to a gravitational resonance phenomenon with Jupiter and Saturn analogous to that discovered for asteroids.

As mentioned earlier, hundreds of comets have been observed from antiquity through recent times. But these are only the brightest comets and those that pass sufficiently close to the Sun and the Earth. We only know of those comets having a minimum distance from the Sun of 300 million kilometers (186 million miles) or less. Allowing for the existence of those having a minimum distance of perhaps up to 600 times larger, one can then estimate the total number of comets as 100 billion. Kepler probably was right when he wrote in 1613, "The sky is full of comets, as the sea is full of fish."

Halley's comet in May 1910. Below, at right, shines the planet Venus. Photo: Lowell Observatory, Flagstaff, Arizona.

Mass

The study of perturbations exerted by comets upon the paths of planets is the only direct means one has for determining comets' mass. However, to date these perturbations always have been too small to be revealed by measurement. In particular, the passage of comet Lexell 2.4 million kilometers (1.5 million miles) from the Earth in 1770 did not noticeably change the length of the year. Similarly, the motions of Jupiter's satellites were not affected in any appreciable manner by the passage of comet P/Brooks 2 through the Jovian system in 1886. Only upper limits to the masses of comets can be deduced from these negative results. For comet Lexell, one finds less than 6 million billion tons (6.10^{15} t.), which is a millionth of the mass of the Earth.

Various positions of Halley's comet on its orbit.

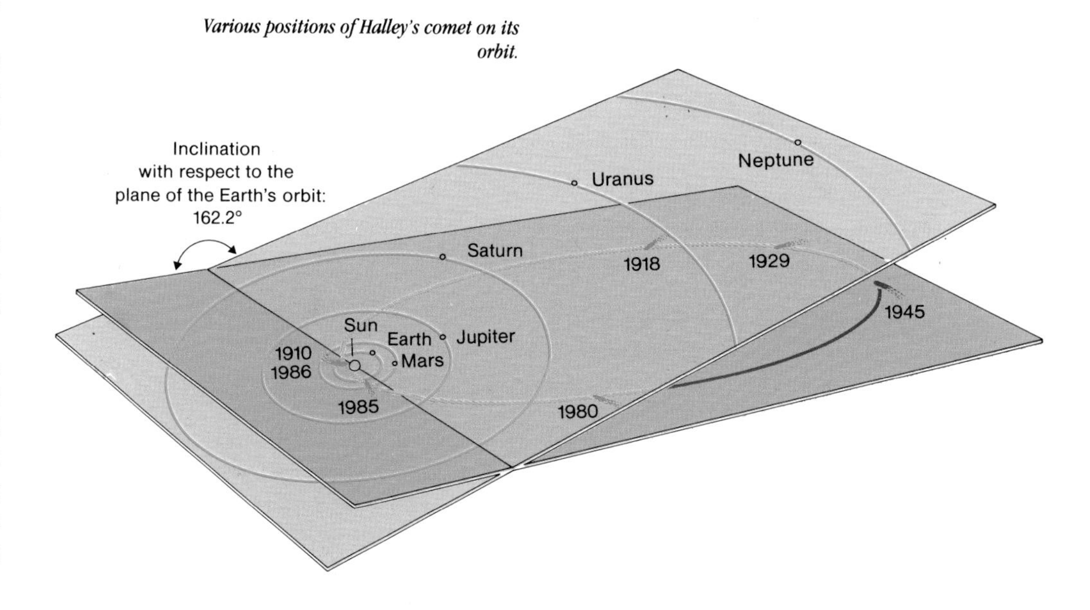

Halley's Comet

This comet—without doubt the most famous in history—now bears the name of the English astronomer Edmund Halley (1656–1742), who, in 1705, first calculated its orbital elements, recognized its periodic character and predicted its return.

Halley had undertaken to calculate the orbits of 24 comets for which a sufficient number of observations existed, taking into account planetary attraction and using the method developed by Newton. Thus he was led to think that the comet observed by Apian in 1531, which Kepler and Logomontanus described in 1607 and which he himself had seen in 1682, was the same object, and thus he predicted it would return at the end of 1758 or the beginning of 1759 (the uncertainty having to do with the difficulty of precisely estimating, at that time, the perturbatory influence of the large planets Jupiter and Saturn on the comet's orbit).

In June 1758, Lalande and Mme. Nicole Lepaute (wife of a famous clockmaker) decided to calculate the exact date of the comet's return to perihelion, using algebraic formulas devised by Clairaut. These calculations, completed in November, showed that the comet would be delayed 518 days by Jupiter's influence and 100 days by Saturn's; in other words, its revolution would be longer by one year, eight months than its previous revolution, and it passage through perihelion would occur toward the middle of April 1759, give or take a month.

Rarely had a scientific prediction aroused keener curiosity from one end of Europe to the other. The comet faithfully kept the date: On December 25, 1758, it was sighted by amateur astronomer J. G. Palitzsch, a farmer of the Dresden area, and its passage through perihelion occurred on March 12, 1759, barely one month before the announced date (a more precise prediction would have been possible if the masses of Jupiter and Saturn had been known with more accuracy at the time). The event caused a considerable stir: For the first time, a comet's return was observed after having been predicted according to the laws of celestial mechanics, and it was shown that comets—at least some of them—orbit around the Sun like planets, but follow much more elongated orbits.

The comet was seen again in 1835 and in 1910. Its appearance was particularly spectacular: After its passage through perihelion, on April 20, 1910, it became clearly visible to the naked eye, in the morning before dawn; it had a magnificent tail, which on May 20 reached a record length of 140 degrees; in addition, it passed exactly between the Sun and the Earth on May 18, and our planet then passed through its tail.

The orbit of Halley's comet now is fully

Adoration of the Magi *by Giotto. Above the manger, the star that guided the Magi shown in the likeness of the comet that appeared in 1301. Scala Photo.*

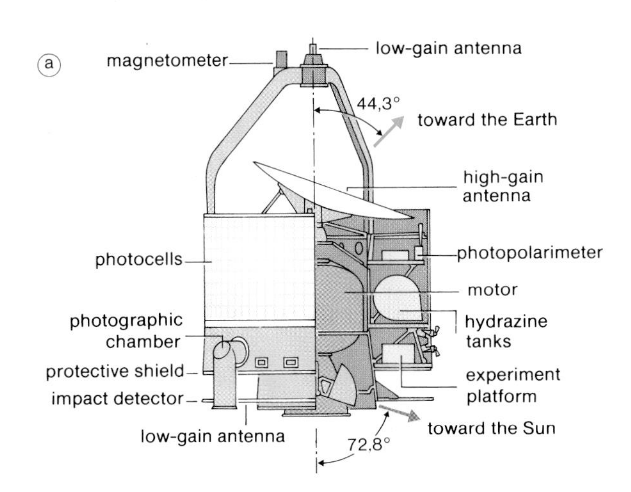

The European Spacecraft Giotto, *which made the closest approach to Halley's Comet (its name derives from the painting by Giotto in 1301*—The Adoration of the Magi—*which shows the comet). Launched on July 2, 1985, this probe reached the comet on March 13, 1986, and flew past it with a relative velocity of 68 kilometers per second, at a distance of less than 500 kilometers. It carried a dozen scientific instruments having a total mass of 53 kilograms, including a camera, mass spectrometers, an impact detector, a plasma analyzer, a photopolarimeter, etc.*

a. Outer appearance of the probe. From the ESA bulletin, with kind permission of the ESA.

b. Comparison of trajectories of the Earth, Halley's comet (projected onto the ecliptic) and the probe. On the three orbits, intervals of about one month are marked off. From an ESA document.

c. Geometry of the probe's meeting with the comet. From an ESA document.

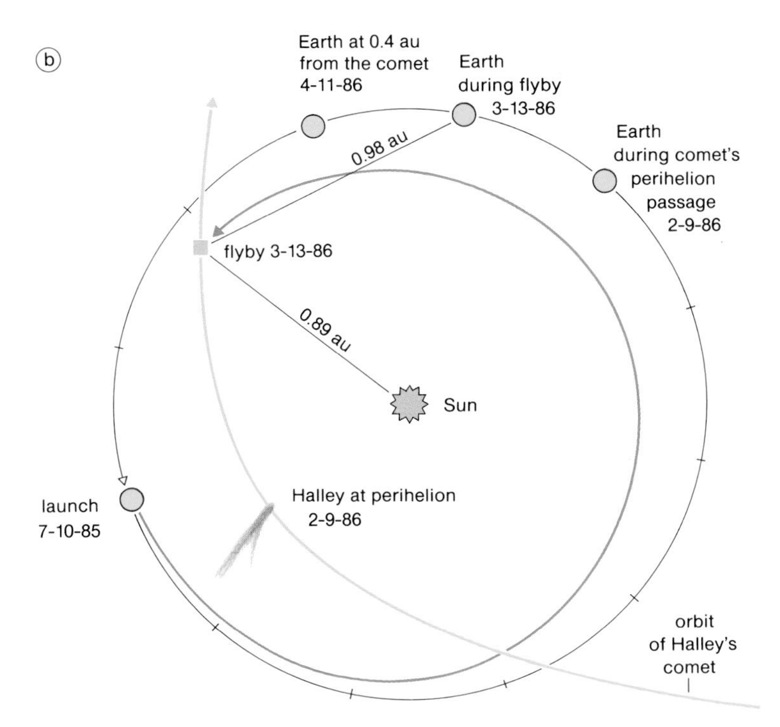

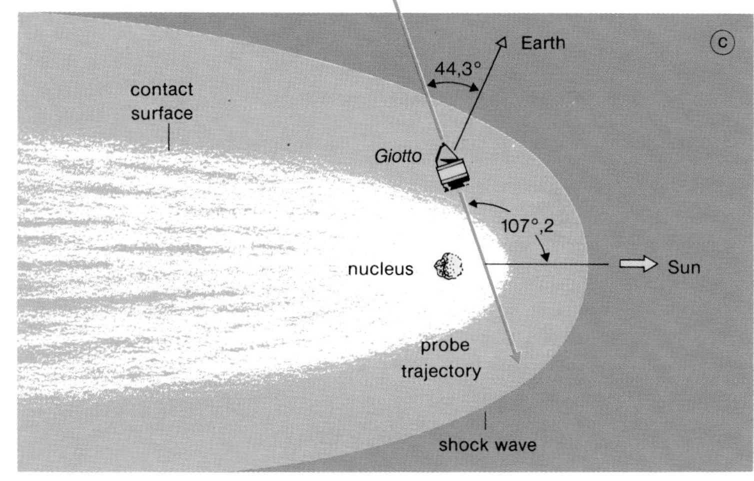

determined. Its aphelion lies 5,300 million kilometers (3,294 million miles) from the Sun, slightly beyond the orbit of Neptune, and its perihelion lies at 88 million kilometers (55 million miles), between the orbits of Mercury and Venus. Its average period is 76 years, but it may vary between 74.7 and 79.1 years, as a result of perturbations of the giant planets, depending on whether it passes close to one or the other.

By calculating the dates of the comet's previous passages through perihelion, it has been possible to link it to several comets mentioned in ancient chronicles.

Its most ancient confirmed observation goes back to 240 B.C.; it may have been observed in 1057 B.C., but its identification with a comet mentioned in Chinese annals of that year remains doubtful.

One of its still famous appearances was in 1456. Historians of the time described it as great and terrible, with a very long tail resembling an undulating flame. Appearing three years after the capture of Constantinople by the Turks, which had thrown the Christian world into a panic, its appearance no doubt increased the terrors of war. But—contrary to a persistent legend—nothing indicates that Pope Callistus III saw in it a sign of the invasion of Christianity by the Turks and ordered public prayers to chase away the perturbing body. In fact, no papal document makes reference to the comet. The sole purpose of the prayers and processions ordered by the bull solemnly promulgated by Callistus III in 1456 was to obtain God's help in delivering the Christians from the peril of the Turks.

At the present time, the comet is leaving us again, as it decelerates out of the inner solar system. It passed through perihelion on February 9, 1986, and should remain discernible in telescopes until well into 1987. The comet was dim and unspectacular during this passage because the circumstances of its apparition were quite unfavorable. Its maximum brightness, which it achieved in early April 1986, was about the same as a 4th magnitude star, that is, somewhat less than the brightness of Polaris. In other words, it could be seen with the naked eye, but only with difficulty, and only from good sites. Its closest passage to Earth occurred on April 11, 1986, when it passed by at a distance of 63 million km (39 million miles). Several spacecraft were launched toward Halley's comet in order to study the chemical composition and the physical properties of the coma, to photograph the nucleus, and to determine the composition of the dust. In July 1985, the European Space Agency launched *Giotto*, a 950-kg probe which came within 500 km (312 miles) of the comet's nucleus on March 13, 1986. The U.S.S.R. launched two probes, *Vega 1* and *2*, in December 1984. These two-ton spacecraft, which had 85-kg instrument platforms, first swung around Venus before encountering the comet on March 6 and 9, 1986 respectively, both at a minimum distance of 10,000 km (6,000 miles). Finally, Japan launched two 138-kg spacecraft: *Sakigake*, which flew by the comet at a distance of 7,000,000 km (4,350 miles) on March 11, 1986, and *Planet-A*, which came within 200,000 km on March 8, 1986.

The study of the mutual perturbations of the fragments of a comet that ruptured, comet Biéla, has given a value 10 times smaller: 600,000 billion tons (6.10^{14} t.). Finally, more recently, the study of the progressive separation of the two fragments of the nucleus of comet Wirtanen led the astronomer Elizabeth Roemer to attribute to this comet a mass slightly greater than 100 billion tons, which is only one sixty-millionth of the mass of the Earth. According to the astronomer Richter, cometary masses lie between 100 million tons (10^8 t.) and 100,000 billion tons (10^{14} t.); therefore they always are less than a hundred millionth of the Earth's mass.

Structure

Comets can exhibit a variety of different appearances. When they reach their maximum development in the neighborhood of the Sun, however, they generally show a diffuse circular or oblong nebulosity (the *coma*), out of which extends a trail of luminosity, sometimes very long, pointing away from the Sun (the *tail*). In the center of the coma, a bright spot can be seen with a telescope (the *nucleus;* the coma and nucleus together are called the *head*). In addition, observations made with the help of artificial satellites since 1969 have established that comets are surrounded by a vast halo of hydrogen.

THE NUCLEUS

The only permanent part of a comet is its nucleus. It is from the nucleus that the coma and tail arise in the neighborhood of the Sun. There is no doubt about its existence, even though none has ever been directly observed. The most that can at times be seen in the middle of the coma is a kind of faint star. Its exact nature still is unknown, but hypotheses relating to its structure all conform to two basic models. Up until about 1950, most specialists imagined it as a tenuous swarm of small, largely separated solid particles. But this "sand bank" model, for which the American Lyttleton was one of the most ardent proponents, no longer receives much support today. The preferred model is the "dirty snowball," proposed in 1950 by the American Whipple, who describes a cometary nucleus as a conglomerate of rocks, dust, ice and frozen gas. This model explains a larger number of observations, in particular the fact that comets can approach within a very small distance of the Sun without being totally vaporized, as well as the secular acceleration or deceleration of some of them, such as Encke's comet (see box). Furthermore, it accounts particularly well for the formation of the coma and tail as the Sun is approached in terms of an outgassing phenomenon caused by solar radiation.

Recent observations of several comets, notably at radio wavelengths, confirm that water must be a major constituent of cometary nuclei; however, their composition remains quantitatively very uncertain.

Assuming, like Whipple, that cometary nuclei are solid, spherical bodies that scatter sunlight, one can deduce their diameter from their apparent brightness. Elizabeth Roemer, in 1965, thus estimated photometrically the nuclear radius of 29 comets for two extreme values of albedo encountered in the solar system: 0.02 and 0.70. The results obtained differ according to the type of comet. For periodic comets with elliptical orbits, the radius was found to be less than or equal to 10 kilometers (6.2 miles). For comets with quasi-parabolic orbits, it reaches a minimum of 65 kilometers (40 miles). These extremely small values explain the pointlike appearance of cometary nuclei in telescopes: To have a measurable size with the best ground-based telescopes, the nucleus of a comet at a distance of 1 a.u. (astronomical unit) from the earth must have a diameter greater than 700 kilometers (435 miles)! Still, these estimates probably exceed the true size: The visible nucleus corresponds not merely to the solid nucleus but also to the cluster that the nucleus forms with a halo of small ice crystals that increase the apparent luminosity, especially near the Sun; the nucleus itself would have a diameter ranging only from 0.5 to 5 kilometers (0.3 to 3 miles).

THE COMA

When it comes sufficiently close to the Sun (at a distance of about 600 million kilometers [373 million miles]), the cometary nucleus begins to sublimate—i.e., to pass directly from a solid to a gaseous state under the effect of heat. Thus it liberates gas and dust, which form around it a more or less circular nebulosity, the coma. At that point the comet becomes visible, as the dust escaping from the nucleus scatters sunlight, while

Encke's comet photographed on October 22, 1947, at the Meudon Observatory, France. Photo by Charles Bertaud, kindly contributed by him.

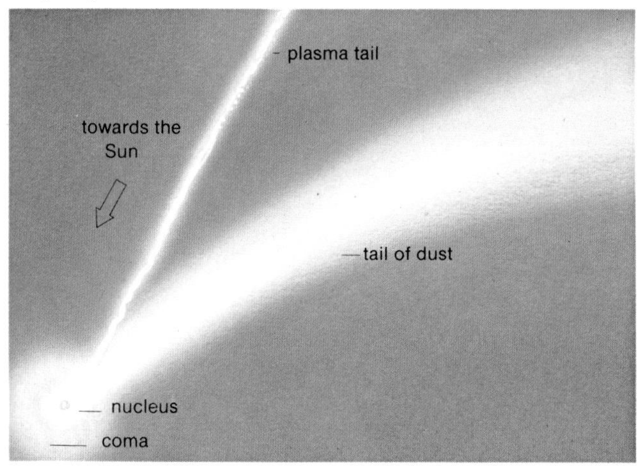

Diagram of a comet.

Encke's Comet

In 1818, the German astronomer Johann Encke managed to show that the four comets discovered, respectively, on January 17, 1786, by Pierre Méchain in Paris; on November 17, 1795, by Caroline Herschel in Slough, England; and on October 20, 1805, and November 26, 1818, by Jean-Louis Pons in Marseille were in fact a single body. According to custom, this comet has been named, not after its first discoverer, but after the astronomer who recognized its periodic nature, studied its motion and determined its orbit.

Of all the comets we know of, this one has the shortest period of revolution (3.3 years) and therefore has been studied the most. At perihelion, it comes to within 0.34 a.u. (51 million kilometers or 32 million miles) of the Sun; at aphelion, the comet moves to a distance of 4.1 a.u. (615 million kilometers or 382 million miles). Since 1818, it has been observed at each of its returns, except in 1944. in 1975, it was photographed while it was close to its aphelion; thus it is observable throughout its orbit. Its appearance never is spectacular; generally invisible to the naked eye, it appears in telescopes as a dull glow, sometimes accompanied by a faint tail. Its spectrum is especially remarkable because of the extreme faintness of its continuum component (which is the sign of a very low dust content) and by the intensity of emissions from the triatomic molecule of carbon (C_3).

In 1838, Encke announced a systematic decrease in the period of revolution, the comet returning to perihelion each time 2.5 hours in advance of the moment predicted by calculation. For a long time, attempts were made to explain this phenomenon as the result of a resisting medium, but this hypothesis ran up against the fact that most of the other comets with a short period did not show a similar acceleration; some even had an increasing period. The explanation now considered most likely is the one proposed in

the gas absorbs the sunlight and reemits it at different wavelengths. When it reaches its greatest extent, about 200 to 300 million kilometers (124 to 186 million miles) from the Sun, the coma has, in general, a diameter between 50,000 and 250,000 kilometers (31,075 and 155,376 miles). That of the great comet of 1811 reached a second diameter, still unequaled, 1 million kilometers (621,504 miles). For some comets, a decrease in the diameter of the head can be noticed as it nears perihelion.

Concentric halos can at times be seen in the coma of comets. Such halos are believed to result as gas is ejected from surface regions of the nucleus over which the ice sublimates particularly rapidly and that are repeatedly exposed to sunlight by the rotation of the nucleus. The most striking example of this is Donati's comet, observed in 1858, which showed halos appearing for three weeks in regular succession with clocklike precision. From measurements of halo diameter and from theoretical data on the expansion velocity of the gas as a function of distance from the Sun, one can calculate the time at which the regions emitting gas had faced the Sun. This method permitted the astronomer Fred Whipple to determine the rotation period of several comets (4.6 hours for Donati's comet, 6.5 hours for Encke's comet, etc.).

Arend-Roland comet (1957 III) photographed on April 29, 1957, with the 1.20-meter (48-inch) telescope of the Mount Palomar Observatory. Photo: Mount Wilson and Mount Palomar observatories.

THE TAIL

The most spectacular part of a comet is, in general, that which appears last: the tail. This is nothing but an emanation from the coma. The fact that it always appears to point in the direction opposite the Sun suggests that the Sun exerts a repelling force on the matter composing it, as Kepler first hypothesized at the beginning of the 17th century.

About 100 years ago, the Russian astronomer Bredikhin undertook a morphological study that led him to distinguish three types of cometary tails, differing primarily in their curvature. Since then, this classification has been reduced to only two types:

- Type I tails, narrow and linear. They are not rigorously directed away from the Sun but lie in general at an angle of a few degrees from that direction. They show an outstanding filamentary structure, with very narrow rings in which bright condensations can at times be seen moving away from the heads. Their length can be considerable, sometimes greater than the distance between the Earth and Sun (320 million kilometers [199 million miles] in the case of the great comet of 1843).
- Type II tails, larger, more diffuse and curved (their concavity being turned away from the direction of the comet's motion). They show little structure and generally are shorter than type I tails.

Spectroscopic studies show that these two types of tails are of a fundamentally different nature. Type II tails have a spectrum that is identical to that of sunlight; therefore they merely reflect sunlight. From this it can be deduced that they are composed of dust. From more complete observations it has been established that the dust particles have dimensions between about 0.25 and 5 thousandths of a millimeter. A large fraction of these particles must be formed from metallic silicates, since the silicate "signature" has been detected in the infrared reflection spectra of recent bright comets (Bennett, Kohoutek, Bradfield, etc.). Moreover, a number of metals were identified in the spectrum of comet Ikeya Seki when it grazed the Sun in 1965; it is thought that these are the metals of silicate grains vaporized by heat. Type I tails, on the other hand, are formed of ionized gas: carbon monoxide $CO+$ (the $+$ sign indicates an ion carrying a positive electric charge), nitrogen N_2+, carbon dioxide CO_2+, water vapor H_2O+, etc.

In fact, most comets possess both a dust tail and an ion (or plasma) tail, but they only appear separated if the Earth happens to be in the direction perpendicular to the plane of the comet's orbit, as was the case, for example, for comet

1949 by the American astronomer Fred Whipple, stating that variations in the period are related to rotation of the nucleus (assumed to consist of ice and dust). As a result of that rotation, a given area on the surface will be alternately illuminated or in shadow. The sublimation of the ice, and the degassification that results, take place from the "morning" until sometime during the "day." If the direction of rotation is contrary to that of the orbital motion, the thrust exerted by the reaction of the nucleus to the ejected gases reduces the comet's kinetic energy and brings the direction of the motion back toward the interior, causing a shortening of the orbit and a consequent reduction of the period: That is what would occur for Encke's comet. On the other hand, for a comet in which the nucleus rotates in the same direction as the orbital motion, we would see an increase in the period.

However, for more than a century, the increase in the motion of Encke's comet has steadily diminished. At present, its period is reduced by no more than a few minutes from one passage to the perihelion to the next.

This phenomenon may perhaps be explained by the fact that the quantity of gas released by the nucleus is gradually thinning out. According to Whipple and Sekanina, however, it results rather from a change in direction, in space, of the axis of rotation of that nucleus and from the corresponding change in geometry of the forces of reaction. Indeed, the nucleus may not be spherical, but elliptical and flattened at the poles, which would involve a precessional motion analogous to that affecting the axis of the terrestrial poles, but much more rapid: According to Whipple and Sekanina, the direction of the axis of Encke's comet thus would have turned more than 100 degrees in 191 years.

Mrkos just after August 15, 1957. On the other hand, if the Earth lies in the orbital plane, the two tails are superimposed, preventing one from distinguishing them. Naturally, all intermediate cases can arise.

The effects of perspective play a large role in other ways as well in determining the forms of cometary tails. When it is approximately directed toward the Earth, the tail of a comet is much reduced in apparent length and can appear more foreshortened than another that is seen in profile, even though the other is smaller in reality. Thus, although the great comet of 1861 was extended over 118 degrees in the sky, it had a tail that was in fact three times smaller than the comet of 1843, which only stretched over 68 degrees. In the same way, a comet's tail, whatever its curvature, appears almost linear if its plane makes a very small angle with the observer's line of sight; for the true form to be revealed, this angle must be somewhat large. Finally, when it points toward the Earth, the tail of a comet seems to spread out like a fan; this phenomenon is enhanced if the end of the tail is near our planet (in the same way that the sides of a road seem to open out around an observer standing in the road). The effect of perspective also can cause the appearance of a short and narrow luminous extension from the nucleus toward the Sun, short in opposition to the principal tail; this anomalous tail (or antitail) is due to light scattered by dust lying in the comet's orbital plane and that moves away from the Sun along an apparently retrograde trajectory.

The formation and the characteristics of the two types of cometary tails result in the following process: When the comet approaches the Sun, ices on the nucleus sublimate because they absorb the heat and light of the Sun. The liberated gas carries dust and small ice particles with it into space. These ice particles vaporize in a spherical halo some hundreds of kilometers across. The dust, on the other hand, is swept in all directions out to several tens of thousands of kilometers before ultimately being pushed outward from the Sun by solar radiation pressure; it then forms the dust tail, the curvature of which can be explained by the fact that the comet's nucleus moves along its orbit while the dust in the tail falls into a somewhat different orbit under the influence of the Sun's radiation pressure.

As for the gas that initially swept the dust from the nucleus, it is barely influenced by solar light, for it is practically transparent. Hence it continues its expansion around the nucleus for a long time, producing the coma. As a result of various factors,

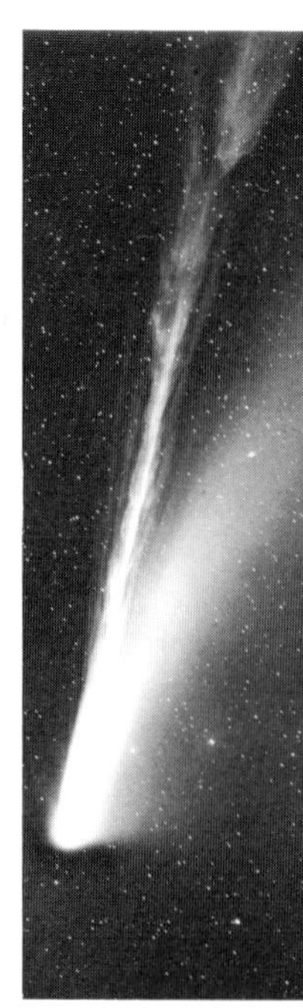

Mrkos's comet (1957) photographed on August 22, 24, 26 and 27, 1957, with the 1.20-meter (48-inch) Schmidt telescope of the Mount Palomar Observatory. We clearly see the gas tail (narrow and straight) and the dust tail (wide and diffuse). Note the evolution of the gas tail. Photo: Mount Wilson and Mount Palomar observatories.

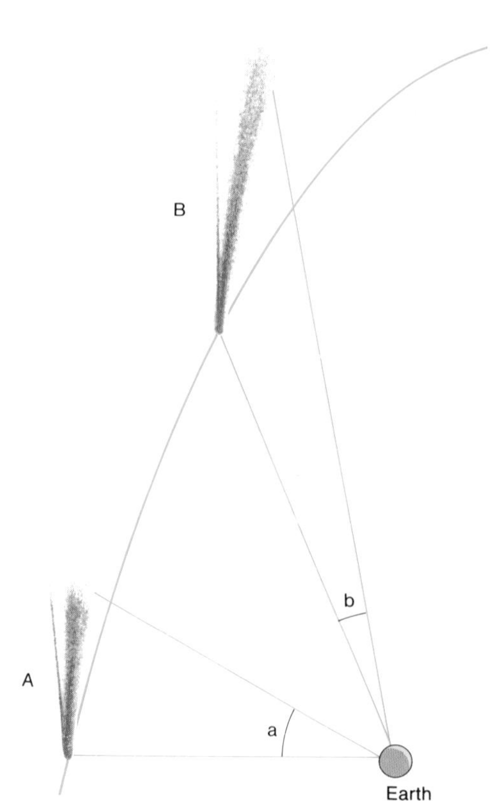

The apparent length of cometary tails depends on their direction relative to the Earth.

Comet-Hunting: An Activity Accessible to Amateurs

Because it does not require powerful instruments, the discovery of comets is not the sole prerogative of professional astronomers. Quite the contrary—since the 18th century, several amateurs have distinguished themselves by discovering a number of comets after systematic searches for such bodies using modest means. Such was the case, in particular, of Charles Messier (1730–1817), a clerk at the French Navy's map repository, who discovered 13 comets and thus deserved to be nicknamed the "Comet Ferret" by Louis XV; today Messier remains known mainly for his catalogue of stellar clusters, nebulae and galaxies, but it was specifically to aid his search that he began counting all the nebulae in the sky liable to be confused with comets. Another famous comet-hunter, Jean-Louis Pons (1761–1831), began as a custodian at Marseille Observatory, but through the lessons given him by the various directors of that institution, he went on to achieve a brilliant career as an astronomer, ending as director of the Florence Observatory. He alone discovered 37 comets, which earned him the nickname "Comet Magnet." We may also mention the German Wilhelm Tempel (1811–89), a lithographer, who discovered 17 comets, and the American Edward Emerson Barnard (1857–1923), who discovered 19. The discovery of comets by amateurs is less common nowadays, however, for comets now are frequently detected by powerful photographic devices, deployed in observatories, while the comets still are not very luminous and thus are inaccessible to amateurs' instruments. However, some amateurs still are distinguishing themselves in this field, such as the Australian William Bradfield, who identified 10 comets between 1972 and 1980.

The amateur wishing to begin a systematic search for comets must be thoroughly familiar with the sky in order not to have to consult an atlas whenever he observes a nebula. He must also be endowed with patience, for according to various estimates by experts, a discovery can occur only after an average of 150 to 200 hours of observing the sky.

particularly solar ultraviolet light, the various molecules present in the gas become dissociated and ionized. Near the nucleus, and up to distances of about 10,000 kilometers (6,215 miles) from it, the density is sufficient for collisions to occur between the molecules and the products of their dissociation and ionization; these collisions generate chemical reactions. Generally these are slow; some, however, are very fast—those that exchange electrical charges between ions and molecules. They transform the water vapor and other parent molecules given off by the nucleus (the nature of which remains largely undetermined) into the series of ions and radicals that can be observed by spectroscopy. At a greater distance from the nucleus, between about 10,000 and 1 million kilometers, there are no more collisions owing to the rarefaction of the gas, but the radicals are again photodissociated by solar ultraviolet radiation until they produce atoms. As a result, the resonance lines of carbon, oxygen and hydrogen are observed out to very great distances from the nucleus, in the ultraviolet, by rockets or artificial satellites.

When the comet encounters the solar wind, the lines of force of the magnetic field accompanying the solar wind fold around the comet's head like an umbrella; then they form a kind of magnetic bottle with a single opening directly opposite the flow of incoming solar particles. The cometary ions that strike this magnetic wall follow its contours and escape out the back, forming the plasma tail. This tail is straight (for it moves quickly) and narrow (for the magnetic bottle's neck is not wide); often it contains knots and spirals showing the turbulence of the interaction between the plasma coming from the comet and the solar plasma.

Origin

Where do comets come from? Any theory of their origin must first and foremost explain the great predominance of very elongated ellipses among their orbits, before the planetary perturbations come into play. It must also account for the nearly haphazard distribution of inclinations to the ecliptic of these semiparabolic orbits. Finally, it must lead to a model in conformity with the knowledge we have regarding the structure and physical properties of comets.

According to some early theories, comets were of planetary origin. In 1814, the mathematician Lagrange, starting from the existence of the family of comets tied to Jupiter, hypothesized that comets were bodies ejected by the large planets. This idea was adopted in 1874 by the Englishman Proctor, in 1890 by the Frenchman Tisserand and in 1925 by the Englishman Crommelin, who similarly ascribed a planetary origin to the group of comets that graze the Sun. From 1930 to 1950, the Soviet Vsekhsvyatsky upheld a variant of that theory: Comets were ejected, not by the planets themselves (which would require considerable velocity: at least 42 kilometers [26 miles] per second from Saturn and 67 kilometers [42 miles] per second from Jupiter), but by their satellites. This hypothesis, however, is not very satisfactory. In particular, it does not account for the existence of parabolic cometary orbits. According to another theory, proposed in 1901 by the Englishman Chamberlain, comets resulted from the fragmentation of small asteroids passing near the large planets. But it is difficult to explain in this way the existence of comets with orbits that are strongly inclined to the ecliptic and that never come into the vicinity of any planet.

A second group of theories ascribes an interstellar origin to comets. Laplace, in 1813, considered comets as small condensations of interstellar matter that are captured by the Sun when they enter its sphere of influence. But the original cometary orbits should then be hyperbolic, which does not agree with the observations. Moreover, since the Sun moves relative to the nearby stars at a velocity of about 20 kilometers (12.4 miles) per second in a certain direction (the apex), we should—if Laplace's theory were correct—observe more comets coming from that direction than from any other, since the Sun is always rushing to meet them at high velocity; in the other direction, on the other hand, there would be few comets fast enough to catch up with the solar system. However, there does not appear to be any special direction from which comets predominantly arrive.

Laplace's theory was revised and improved in about 1950 by the Englishman Lyttleton. According to him, when the Sun goes through a cloud of interstellar dust, some of it might accumulate into discrete condensations as a result of solar attraction. Tens of thousands of such condensations would be captured each time the Sun traverses a nebula. But this hypothesis led to the sand-bank model for comets (see page 216), which has now been abandoned.

Today comets are regarded rather as residues of the primitive solar nebula coming from the outer reaches of the solar system.

The most widely accepted theory, developed since 1950 by the Dutch astronomer Oort, is based on a statistical study of about 50 of the best-known semiparabolic cometary orbits, corrected for planetary perturbations; this study reveals a remarkable predominance of orbits in which the aphelion is at a distance from the Sun ranging from 40,000 to 150,000 a.u. (0.6 and 2.3 light-years). According to Oort, this region, far from the inner solar system, would constitute a vast reservoir of as many as 100 billion comets (the total mass of which, however, would be no more than one-tenth that of the Earth!). As a result of the perturbations created by nearby stars, comets would gradually emerge from this reservoir. Some would finally be expelled from the solar system and would remain forever invisible to us; on the other hand, others would be sent toward the Sun and would constitute the "new" comets we observe.

The formation of the comet cloud, however, remains to be explained. It probably did not form directly at its present distance from the Sun, in an area where the primitive nebula was necessarily very tenuous. Rather, comets probably formed, at the same time as the planets, in the outer region of the planetary system, from which they were then pushed into stable but much more distant orbits as a result of planetary perturbations.

According to the American Whipple, the planets Uranus and Neptune might have formed out of the agglomeration of a large number of comets. And some specialists even think that a ring of comets with a total mass approaching that of the Earth remains beyond Neptune, at distances of about 40 to 50 a.u. from the Sun.

Here is some advice that will be useful to those who wish to try their luck:

1. Choose an open site protected from background light (even at the horizon).
2. Use 7 × 50 or 10 × 50 binoculars, or a reflector equipped with an eyepiece providing a modest magnification (20 to 40 ×), to take advantage of the maximum field of view and luminosity.
3. Scan the sky systematically in horizontal or vertical bands. Observe mainly toward the west just after sunset, or toward the east just before dawn, since comets reach their greatest intensity in the Sun's neighborhood.
4. if you observe a diffuse, dully luminous object, consult a celestial atlas to make sure it is not a stellar cluster, nebula or galaxy. Repeat the observation on the following day, or whenever possible, to see if the object has moved with respect to the neighboring stars. If so, and if the object still has a nebulous appearance, there is a good chance that it is indeed a comet. In that case, note its coordinates (even approximately), evaluate its brightness, note its appearance and report these observations as soon as possible to the closest observatory, or to the Central Bureau for Astronomical Telegrams, Smithsonian Astrophysical Observatory, Cambridge, MA 02138, U.S.A.

Those who lack the time or the patience to go comet-hunting may be satisfied with observing those discovered by others. First study the head; see if the nucleus is visible, and carefully examine the halo around it, particularly its shape: If it lengthens in a certain direction, this may be a sign of a rupture of the nucleus in the direction of lengthening. Next, observe the tail; its length may be evaluated by following it as far as the limit of perception, using the visible stars in the instrument's field as reference points. Follow the body's evolution from day to day, especially at the time it reaches its shortest distance from the Sun.

Such notions remain highly speculative, however, and the problem of the comets' origin is far from being solved.

Fate of a Comet

We know much more about the comets' ultimate fate, at least those that are not ejected from the solar system by planetary perturbations.

Comets that come within about 750 million kilometers (466 million miles) of the Sun are sufficiently warm to lose a substantial quantity of matter by evaporation, which is the main source of interplanetary gas and dust. It is believed that a comet may lose about 5 percent of its mass each time it reaches perihelion. Periodic comets also undergo a gradual disintegration. The dust they release gradually disperses along the cometary orbit, giving rise to swarms of meteorites (see "Meteorites and the Zodiacal Cloud," page 225).

After their reserves of volatile materials have been exhausted, it is likely that cometary nuclei continue to orbit around the Sun like very faint asteroids and thus are very difficult to detect. This hypothesis is reinforced by the discovery of several dozen asteroids, of the Apollo or Amor type (see page 209), of which the diameter resembles that of cometary nuclei (about a kilometer) and for which the very elongated orbit is much more similar to those of comets than to the orbits of ordinary minor planets. Moreover, we know of two comets—P/Arend-Rigaux (discovered in 1951) and P/Neujmin (discovered in 1913)—that generally have a perfectly stellar appearance; their motion does not seem to be affected by any gravitational perturbations, which leads to regarding them as transitional objects between comets and asteroids. Likewise, comets P/Encke, P/Tempel 2 and P/Reinmuth 2 seem to be approaching the end of their active life, although they are at a less advanced stage of evolution than the two previous ones.

The slow evolution of comets sometimes is accelerated by the fragmentation of their nucleus. We know of about 20 examples of comets with a fragmented nucleus. The most famous instance is that of comet Biéla. It was discovered on February 27, 1826, at Josephstadt, in Bohemia, by the Austrian officer Wilhelm von Biéla, and independently 10 days later in Marseilles, by the astronomer Gambart. Calculation of its orbit made it possible to recognize its periodic nature and to identify it with two comets observed earlier—one in 1772, the other in 1805. Its period being 6.7 years, it should have reappeared in 1778, 1785, 1792, 1813 and 1819, but it was not observed during those years, probably because of its low brightness. Spotted again in 1832, it again remained invisible in 1839 because its apparent position was very close to the Sun. Calculations by the Italian Santini set the date of its next perihelion passage at February 11, 1846. On November 28, 1845, the comet was seen simultaneously by De Vico in Rome and Encke in Berlin. After December 19, it appeared as a slightly elongated, pear-shaped nebulosity, without anyone paying much attention to this phenomenon; but on the 29th, Herrick and Bradley in New Haven noted with surprise that it had split in two. Possessing a short tail, each of its components—one of which clearly was brighter than the other—remained visible for several weeks, gradually moving apart from one another. The twins reappeared in September 1852, their distance from one another then being 2.4 million kilometers (1.5 million miles). At the next return, in May 1859, they were not found again, for the same reason as in 1839. Astronomers strongly hoped to see them again in 1866, as their position in the sky then would supposedly be favorable—but in vain. A new attempt, in 1872, also failed. But on November 27, 1872, a magnificent meteor shower occurred. The number of meteors increased for several hours, and at the peak of the shower there were several hundreds of them per minute. Their total number was estimated at 160,000. All of these meteors seemed to emanate from the vicinity of the star γ Andromeda. It could be shown that the meteorites responsible for the phenomenon were following the same orbit as the Biéla comet; their position on that orbit corresponded to that which the comet must have occupied 12 weeks earlier. This shower of meteorites seemed to be the product of the comet's final disintegration. But when past records were examined, a trace was found of meteor showers occurring in November in 1741, 1798, 1830 and 1838—hence prior to the disappearance of the comet. The meteorites then were nearly 500 million kilometers (311 million miles) ***ahead*** of the comet, while those responsible for the shower of falling stars in 1872 were situated 300 million kilometers (186 million miles) ***behind***. The meteor shower associated with the comet thus extended over a distance of at least 800 million kilometers (497 million miles). On November 27, 1885, a spectacular meteor shower again took place, nearly as intense as the one in 1872. Since that date, no more meteorites associated with the Biéla comet have appeared; perturbations probably have affected their orbit in such a way that they no longer cross the Earth's orbit. As for the comet itself, it has never been seen again; it probably disintegrated in space.

More recently, in 1965, the nucleus of comet Ikeya-Seki (1965VIII) also broke in two, sometime after the comet passed within a mere 470,000 kilometers (292,107 miles) of the Sun's visible surface. Finally, in 1976, the nucleus of comet

Comet West (1976 VI) photographed on March 7, 1976, at the cometary research observatory of the New Mexico Institute of Mining and Technology and the Goddard Space Flight Center of NASA. Discovered by the astronomer Richard M. West on September 24, 1975, with the Schmidt telescope of the ESO (Chile), this comet passed through perihelion on February 25, 1976, and remained visible to the naked eye for about one month. It was the subject of numerous observations, which led in particular to the identification in its spectrum of emission bands of the ionized water molecule and, at a frequency of 1667 MHz (radio range), a band caused by the hydroxy (OH) radical. We observe, moreover, the spectacular rupture of its nucleus into several large fragments. NASA Photo.

The region of Tunguska devastated by the cataclysm of June 30, 1908. Photo: Palais de la Découverte, Paris.

The Mysterious Cataclysm of Tunguska

On June 30, 1908, shortly after 7:00 A.M., the Tunguska valley in central Siberia, 800 kilometers (497 miles) northwest of Lake Baikal, was the scene of a strange cataclysm. The closest witnesses, peasants from the village of Vanarava, saw an enormous ball of fire, much brighter than the Sun, cross the sky from the southeast to the northwest, followed by a long, dark trail. Then the object exploded in a blinding flash, and intense heat overcame the witnesses, who were lifted into the air. The deflagration was heard up to more than 1,000 kilometers (622 miles) away, the shock wave was perceived in the state of Washington and Java and, during the next few nights, the night sky in Europe, Asia and America showed an unusual brightness that persisted for more than two months.

In 1927, a scientific expedition, headed by the Soviet mineralogist Leonard Kulik, visited the site. The specialists then were convinced that the phenomenon had been caused by the fall of a huge meteorite. They actually discovered, in the region of presumed impact, an area about 60 kilometers (37 miles) in diameter where the taiga was devastated and the ground strewn with fallen trees, their branches torn off and burned to ashes. But they found no meteoritic fragment, nor any impact crater capable of verifying the fall. The most they saw were some 200 small, shallow depressions filled with water, which turned out to be ruts.

Since then, other teams of researchers have visited the site, and the devastated area has been minutely explored. Elaborate studies were carried out in particular from 1956 to 1962 by the Geochemistry Institute of the USSR Academy of Sciences. It was discovered that the trees covering the ground are lying radially around an epicenter that has been located, to within 500 meters (1,641 feet), at the center of a small region characterized by its lesser devastation. Moreover, a multitude of small spheres of metal and silicates has been found in the soil. The absence of a crater rules out the hypothesis of an explosion on the ground or in its immediate neighborhood. On the other hand, the large number of fallen trees testifies to the violence of the shock wave created by the projectile.

West (1976VI) was seen to break up into four fragments—an unusual phenomenon that had been seen only twice before: in 1882 with the comet 1882II and in 1889 with comet P/Brooks 2.

Sometimes the breakup occurs without an apparent cause, as with comet Biéla. Perhaps at such times it results from a collision of the comet with a large meteorite (which would explain why we generally see such an occurrence in the neighborhood of the ecliptic), unless it is due to a phenomenon of internal pressure, whereby volatile matter contained in solid pockets vaporizes under the effect of solar radiation until it exerts sufficient pressure on the overlying crust to make it explode.

Most often, however, the breakup occurs near perihelion. It can then be explained—particularly if the comet has grazed the Sun (as with the 1882II and 1965VIII comets)—by a tidal effect tied to the difference in the gravitational forces acting on the side of the nucleus that faces the Sun and on the opposite side. A similar phenomenon, but this time due to Jupiter's gravity, would explain the breakup of the nucleus of comet P/Brooks 2 in 1889 after it passed within a short distance of Jupiter.

From the extent and configuration of the devastated zone, it has been concluded that the explosion must have occurred at about 8 kilometers (5 miles) in altitude, and the released energy must have been on the order of 10 million billion joules, or the equivalent of a nuclear bomb of several megatons.

Many hypotheses have been offered: arrival in the atmosphere of a block of antimatter (C. Cowan, C. Atlury and W. Libby, 1965); collision between the Earth and a black microhole (A. Jackson and M. Ryan, 1973); explosion of an extraterrestrial spaceship in flight (J. Baxter and T. Atkins); explosion of a nuclear bomb (F. Zigel, 1978), etc.

However, the most generally accepted hypothesis nowadays is the one proposed in 1930 by the American Fred Whipple, according to which a small cometary nucleus, having lost all its volatile materials, entered the atmosphere and completely disintegrated at a low altitude, owing to the compression of air layers.

This explanation has the merit of accounting for most of the characteristics of the phenomenon—in particular, the devastating explosion; the very strong deflagration,; the blast of wind accompanying the sudden release of heat; the dispersion in the atmosphere of dust, diffusing solar light; and the absence of a crater on the ground (the low density of cometary matter enabling its pulverization in the atmosphere).

The principal objection that has been raised is that the cometary nucleus referred to was not observed before it crossed the atmosphere. But the Soviet astronomer V. Fesenkov calculated in 1966 that the object was coming from the Sun's direction, at a speed greater than 100,000 kilometers (62,150 miles) per hour; thus its detection was impossible during the nights prior to its arrival in the atmosphere. Perhaps the projectile was a fragment of Encke's comet, as was suggested in 1978 by the Czechoslovakian astronomer Lubor Kresak. Despite the research that has been going on for more than half a century, the Tunguska cataclysm still retains a share of its mystery.

19. Meteorites and the Zodiacal Cloud

As long as astronomers could observe the sky only with their eyes, the only bodies in the solar system they could describe were the Sun, the nearby planets, and interplanetary dust appearing in the form of zodiacal light.

A shower of falling stars. The Leonids are meteorites stemming from Tempel's comet that create a periodic shower, sometimes visible from Earth about November 16. At the time this photograph was taken, in 1966, rates much higher than 1,000 per hour were reached. Photo © 1974 Charles Capet, Hansen Planetarium, Salt Lake City, Utah.

These fragments of solid interplanetary matter now are being studied by elaborate techniques, and we are discovering that the interplanetary medium is anything but a perfect vacuum; it represents a danger to space probes and a nuisance for astronomical observation of weak sources, but it also is a way of studying the evolution of the solar system.

Visual Observation of Meteorites or Shooting Stars

On a fine clear night without the Moon, it is common to see the sky streaked by ephemeral trails of light, or occasional meteors, poetically referred to as shooting stars. Before midnight we can see an average of five per hour, while after 2:00 A.M. the average rises to 15 per hour.

On certain exceptional nights, the rate is so high that we can speak of a shower or even a rain of shooting stars: The meteors all seem to come from the same point in the heavens, known as the radiant. Thus, every year, the night of August 12 is characterized by an hourly rate of about 50 meteors, apparently issuing from the constellation of Perseus and thus named Perseids.

Two observers located at two relatively distant stations simultaneously might see the same meteor appear; but the path of the light trail does seem to them to occur in the same direction relative to the stars. This is an effect of perspective. Calculations show that meteors appear in the Earth's upper atmosphere, at altitudes ranging from 110 to 70 kilometers (68 to 44 miles), and that they move along at speeds between 11 (6.8 miles) per second (the escape velocity of the Earth) and 74 kilometers (46 miles) per second (the maximum observable velocity from the Earth for a body belonging to the solar system).

The phenomenon, then, results from the heating and vaporization, in the dense layers of the atmosphere, of fragments of solid matter of interplanetary origin; these fragments are called

*Some parts of this contribution were taken from an article ("Lumière zodiacale et poussières interplanétaires") published by the author in *l'Astronomie,* bulletin of the French Astronomical Society, November 1979.

Hourly evolution of the rate of visual meteors. On average, more meteors are visible in the morning than in the evening; this is simply because, in the Earth's motion on its orbit, the morning region is facing forward.

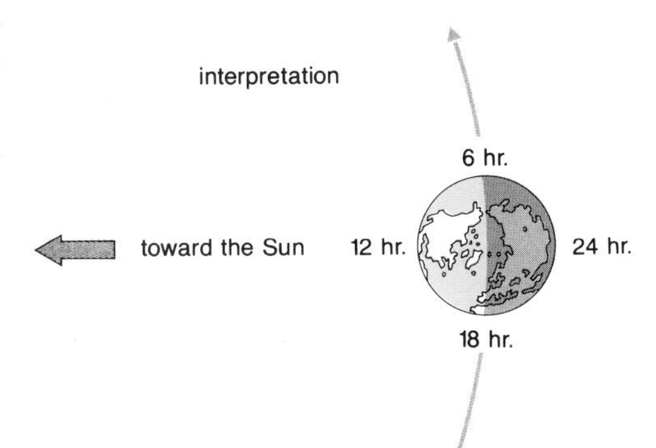

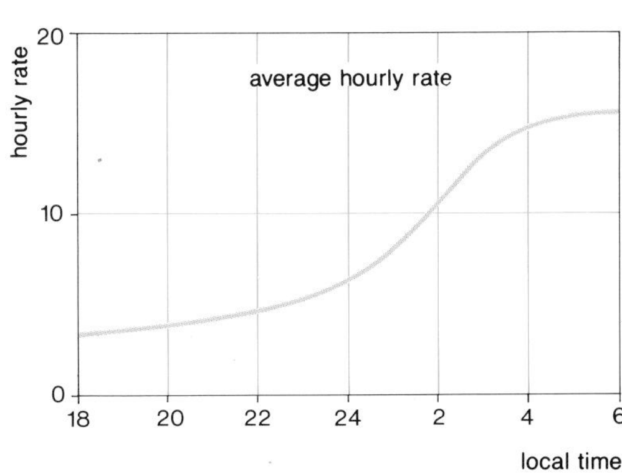

meteoroids, or, if they reach the Earth, meteorites. The fact that more meteors are visible in the morning than in the evening is easy to interpret: The morning region, directed forward in the Earth's motion on its orbit, encounters more meteoroids than the evening region, like the windshield of a forward-moving car that receives more rain than the rear windshield.

Reconstruction of the Saint Severin meteorite, done by the Center for Weak Radioactivity from molds of the recovered pieces. Photo: Center for Weak Radioactivity

Meteorites

Scientific study of meteorites is based on analysis of samples gathered from the ground, on the monitoring of visual meteors by networks of super-Schmidt cameras and on the diurnal detection by radio telescopes of the ionization of the atmosphere along the luminous trail. Formation of the sporadic E layer of the ionosphere, and the appearance of luminous nocturnal clouds at about 80 kilometers (50 miles) in altitude, probably are also caused by meteorites. Within the Earth's orbit, there are many more fragments of very low mass than of average or high mass; the number of meteoroids increases in inverse proportion to their mass, following a logarithmic curve.

The smallest meteoroids—hence the most numerous—are slowed by the Earth's atmosphere but not destroyed. The heat they absorb as a result of friction in the dense atmospheric layers is insufficient to raise their temperature to the boiling point. Their mass does not exceed 10^{-7} g, and

The Earth and Meteorites

According to an already old estimate by the American astronomer C. C. Wylie, an area equal to that of France would receive, on average:

1. six meteorites with an initial mass (at their entry into the atmosphere) of nearly 5 kilograms (11 pounds) each year
2. a meteorite with an initial mass greater than 3 t every 20 years
3. a meteorite with an initial mass greater than 250 t every 50,000 years

According to other estimates, the Earth would be struck every 10 million years on average by a projectile liable to produce a crater 12 kilometers (7.5 miles) in diameter.

The frequency of such encounters was much higher in the early solar system, particularly during the first 700 million years. At that time, the solar system still was very crowded, and planetary surfaces were literally pummeled by a large number of objects stemming from the breakup of asteroids in collisions: It has been calculated that such objects represent nearly 5 percent of the mass of the Earth's crust, bounded by the Mohorovicic discontinuity at a depth of 30 kilometers (19 miles).

While the Moon, Mercury and Mars have retained vestiges of the multiple impacts that formerly scarred their surfaces, erosion on Earth has gradually erased most of the craters dug by very large meteorites during the first stages of planetary evolution, and only the largest structures, dating from less than 1 billion years ago, still can be recognized, particularly in the ancient formations of the Canadian shield and Siberia. Modern research has led to the incontrovertible identification of craters of meteoritic origin—confirmed by the presence of meteoritic fragments at the site (see table)—and of other craters with a *probable* meteoritic origin (for they show the effects of unquestionable metamorphic impact, although no meteorite has been found in them).

Known Meteor Craters on the Earth's Surface (according to R. A. F. Grieve and P. B. Robertson, 1978)

Site	Latitude	Longitude	Number of Craters	Diameter of the Main Crater (m)	Year of Discovery
Meteor Crater, Barringer, Arizona (U.S.)	35° 02′ N.	111° 01′ O.	1	1200	1891
Wolf Creek (Australia)	19° 10′ S.	127° 47′ E.	1	850	1937
Boxhole (Australia)	22° 37′ S.	135° 12′ E.	1	185	1937
Odessa, Texas (U.S.)	31° 48′ N.	102° 30′ O.	3	168	1921
Henbury (Australia)	24° 34′ S.	133° 10′ E.	14	150	1931
Kaalijärvi, Estonia (USSR)	58° 24′ N.	22° 40′ E.	7	110	1928
Morasko (Poland)	52° 29′ N.	16° 54′ E.	7	100	
Wabar (Saudi Arabia)	21° 30′ N.	50° 28′ E.	2	97	1932
Campo del Cielo (Argentina)	27° 38′ S.	61° 42′ O.	20	90	1923
Sobolev, Siberia (USSR)	46° 18′ N.	137° 52′ E.	1	51	
Sikhote Alin, Siberia (USSR)	46° 07′ N.	134° 40′ O.	122	26.5	1947
Dalgaranga (Australia)	27° 43′ S.	117° 05′ E.	1	21	1923
Haviland, Kansas (U.S.)	37° 37′ N.	99° 05′ O.	1	11	1933

The certain meteoritic craters are young formations, no older than the Pleistocene, and relatively small: The largest one is Arizona's famous Meteor Crater, which measures 1,200 meters (3,937 feet) in diameter, and is estimated to be 24,000 years old.

The probable meteoritic craters are older and larger. Many exceed 10 kilometers (6.21 miles) in diameter, and their age ranges between 100 million and 500 million years; two of them—in Sudbury, Ontario, Canada, and in Vredefort, South Africa—reach 140 kilometers (87 miles) in diameter and are thought to be nearly 2 billion years old. These fossil craters commonly are referred to nowadays as astroblemes (from the Greek, "star wounds"), a term proposed in 1960 by R. S. Dietz.

The Antarctic is a large "reservoir" of small meteorites. After their fall, they are transported by ice flows to the edges of the continent and are concentrated in well-defined areas, the relief of which forms a barrier for the ice. Since the discovery by a Japanese team in 1969 of meteorites near Mount Yamato, more than 1,000 meteoritic fragments have been gathered in the Antarctic. Their study holds much interest, for most of them are primitive specimens that have undergone practically no change since their fall.

their dimensions are about several dozen micrometers (1 μm = 10^{-6} m); we will describe later in further detail these micrometeorites or interplanetary dust particles that make up the zodiacal cloud.

The meteorites with a mass ranging from 10^{-7} g. to 1 kg. are destroyed in their passage through the atmosphere. Those with a mass below 10^{-2} g. are detected by the ionization they produce; those with a mass above 10^{-2} g. are revealed by the light trails they produce.

The largest meteorites, those with a mass greater than 1 kg., reach the ground at a frequency of about 2 to 10 per day. Typically broken up and burned to some degree, they have lost about four-fifths of their mass in their traversal of the atmosphere. The existence of these bodies that fall from the sky was not recognized until the beginning of the 19th century through the work of the physicist Biot. They produce impact craters on the ground similar to those with which the Moon and Mercury (whose surfaces are neither protected by an atmosphere nor altered by perpetual erosion) are riddled. Some large meteorites of iron-nickel composition (siderites) or of iron/silicate composition (siderolites) are certainly asteroidal debris, created by collisional ablation. But most are of a stony nature (aërolites, made up essentially of silicates) and seem to have an interstellar or cometary origin, followed by disintegration of the comet's nucleus after repeated passages through perihelion.

The importance of the cometary origin of meteorites is confirmed by the regularity of certain meteor showers (see table): the Quadrantids, Lyrids, Perseids, Orionids, Taurids, Geminids and Ursids, in particular. Every year at the same time, the Earth encounters meteoritic showers consisting of meteorites that follow common elliptical paths (hence, through perspective, the appearance of a radiant point); the orbital parameters of these showers are very close to those of certain known comets—e.g., Giacobini's, Encke's or Halley's.

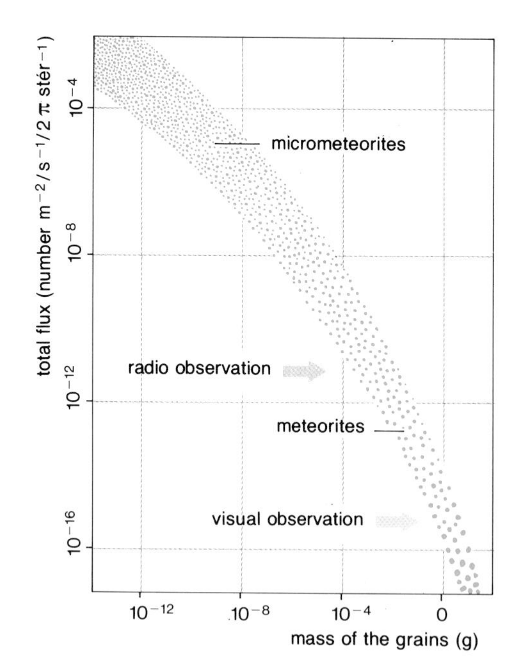

Distribution by mass of meteorites reaching the Earth. The total flux of interplanetary grains having a mass lower than a given mass is shown as a function of the mass of those grains (logarithmic scales). The number of meteorites rapidly decreases when their mass increases.

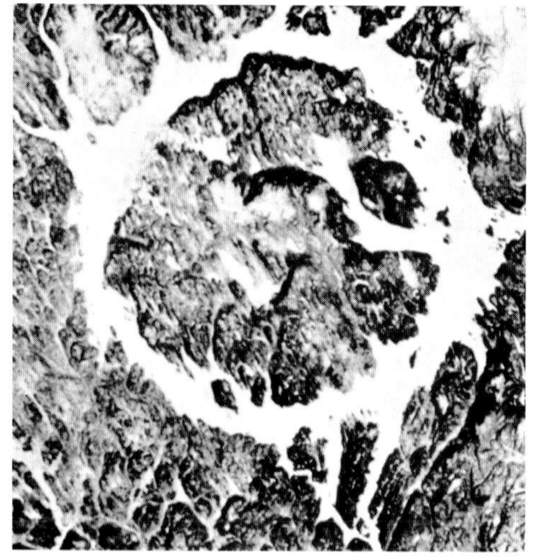

Impact crater in Quebec. The region of Lake Manicouagan in Quebec shows, over 150 kilometers (93 miles) of its diameter, a structure of concentric rings that geologists attribute to a very ancient meteoritic impact. The ring-shaped lake (diameters: interior, 55 kilometers [34 miles]; exterior, 70 kilometers [44 miles]), still frozen in this photograph taken by the Landsat *satellite on April 20, 1974, occupies the central region of the depression. Many impact craters with smaller dimensions (between 20 meters [66 feet] and 3 kilometers [1.9 miles]) have been identified, particularly in Australia, North America and Siberia. NASA Photo.*

In sum, the Earth—in its daily and annual motion—encounters meteorites of various sizes. These are solid fragments present everywhere in the solar system and probably, like it, 4.6 billion years old.

Visual Observation of Zodiacal Light

In the absence of all background light and on a clear, moonless night, we can sometimes detect, shortly after sunset or before dawn, the phenomenon known as zodiacal light: A portion of a very elongated ellipse appears in the sky. Its long axis is nearly coincident with the ecliptic on the celestial sphere; it seems to be centered on the Sun, which then is about 20 degrees below the horizon.

The Different Types of Meteorites

According to composition, there are three main types of meteorites:

1. *aeroliths*, or stony meteorites, consisting mainly of silicates; aeroliths are again divided into two categories: *chondrites* (the more numerous), so called because they contain very characteristic granules, chondrules (spheroidal aggregates of olivine and pyroxin, with a diameter of nearly 1 mm and that are not found in terrestrial rocks); and *achondrites*, which do not have such granules
2. *siderites*, consisting chiefly of iron and nickel (the proportion of the two metals is variable, although iron is the more abundant element)
3. *sideroliths* or *lithosiderites*, consisting of approximately half iron and nickel and half silicates.

According to a statistic established by B. Mason, aeroliths represent 61 percent of the total number of meteorites collected, the siderites 35 percent and sideroliths 4 percent. Aeroliths are mostly collected immediately after their fall, while, later on, it is mainly only siderites and sideroliths that are gathered, because they are more easily recognized and resist erosion better.

Finally, we should note the existence of a special variety of meteorite, tectites. These are glassy rocks with rounded shapes, black or dark green in color, very rich in silica (75 percent), containing a small amount of aluminum (10 to 15 percent) but low in iron, magnesium and sodium and thus resembling volcanic rocks of the obsidian type, although they differ from them in having a complete lack of crystals.

Unlike ordinary meteorites, they are found only in certain regions of the Earth's surface (about 10 in all), including Australia, Czechoslovakia and Indochina. The most an-

The Largest Known Meteorites

Site	Mass	Year of Discovery
Hoba (South Africa)	60	1920
Ahnighito, Cape York (Greenland)	36	1895
Chingo (China)	30	?
Bacubirito (Mexico)	27	1863
Mbosi (Tanganyika)	25	1930
Armanty (Mongolia)	20	?
Agpalilik (Greenland)	17	1963
Willamette (Oregon, U.S.)	15	1902
Chapaderos (Mexico)	14	1852
Otumpa (Argentina)	13.6	1783
Mundrabilla (Australia)	12	1966
Morito (Mexico)	11	1600

Characteristics of the Principal Known Meteor Showers

Name	Comet of Origin	Approximate Date of the Maximum	Period of Activity	Average Coordinates of the Radiant Point in Degrees (1950) α	δ	Average Hourly Rate	Notes
Quadrantids	Kozik-Peltier	Jan. 3	1/1 → 1/4	232	50	40	
Lyrids	Thatcher	Apr. 21	4/19 → 4/24	274	34	12	
η Aquarids	Halley	May 4	5/1 → 5/8	336	0	20	
Arietids		June 7	5/29 → 6/19	45	23	60	observed by radio techniques during the day
ζ Perseids		June 9	6/1 → 6/17	62	24	40	same as above
β Taurids	Encke	June 28	6/24 → 7/2	87	20	30	same as above
δ Aquarids S		July 29	7/21 → 8/15	339	− 17	20	Southern Hemisphere
δ Aquarids		July 29	7/15 → 8/18	339	0	10	
α Capricornids	Mrkos	Aug. 1	7/15 → 8/20	308	− 10	5	
ι Aquarids S		Aug. 5	7/15 → 8/25	338	− 15	10	Southern Hemisphere
ι Aquarids		Aug. 5	7/15 → 8/25	331	− 6	10	
Perseids	Swift-Tuttle	Aug. 12	7/22 → 8/18	46	58	50	
χ Cygnids		Aug. 20	8/18 → 8/22	290	55	5	
Giacobinids (or Draconids)	Giacobini-Zinner	Oct. 9	10/9 → 10/10	262	54	1,000	periodic shower—period approx. 6 yrs.
Orionids	Halley	Oct. 20	10/18 → 10/26	95	15	25	Southern Hemisphere
Taurids S	Encke	Nov. 5	09/15 → 12/15	53	14	15	
Taurids	Encke	Nov. 10	10/15 → 11/30	57	22	5	
Leonids	Tempel	Nov. 16	11/14 → 11/20	152	22	10,000	periodic shower—period approx. 33 yrs.
Geminids		Dec. 13	12/7 → 12/15	113	32	50	
Ursids	Tuttle	Dec. 22	12/17 → 12/24	217	80	15	

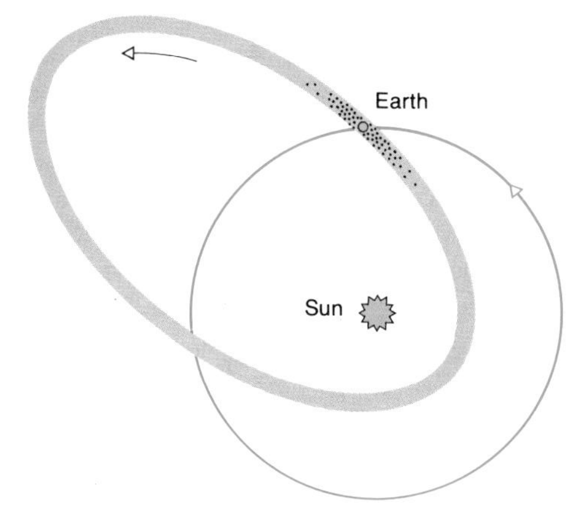

relatively young periodic shower

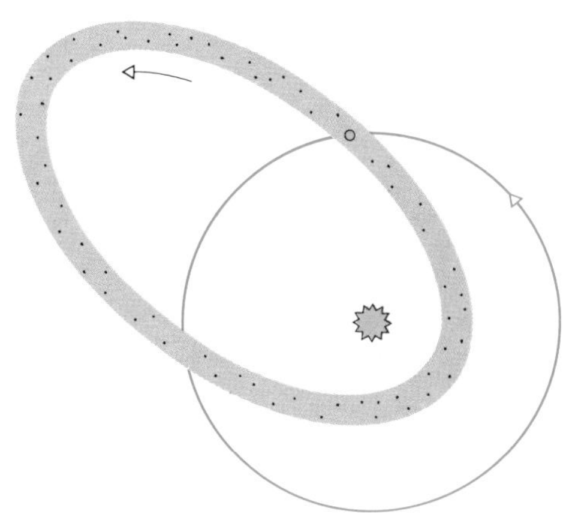

older permanent shower

Periodic and permanent showers. The meteorites of a permanent shower are evenly distributed around their orbit and may produce visual meteors every year at the same time. In the periodic, younger showers, the meteorites are close together and may produce, with a periodicity of several years, very spectacular showers of falling stars.

The intensity of the zodiacal light is at a maximum along the path of the ecliptic; its name, of course, comes from the fact that essentially it is visible to the naked eye only in the directions of the constellations of the zodiac; however, detectors more sensitive than the eye reveal that at lower brightness it covers the entire celestial sphere.

The zodiacal light also becomes more intense toward the horizon, hence in the direction of the Sun. When arising from regions near the Sun, this light merges with that of the Sun's F corona (the dust corona, as opposed to the K or plasma corona); it is no longer directly observable, owing to the strong luminosity of the daytime sky; these regions can be studied only outside the diffusing terrestrial atmosphere or during a solar eclipse.

Thus, on Earth, visual observation is possible only when it is completely dark and the plane of the ecliptic rises high above the horizon. Observation is easier near the equator, where the long axis of the ellipse always is fairly close to the vertical.

The origin of zodiacal light was explained in 1672 by Cassini. Quite simply, it involves solar light that is scattered by small dust grains distributed throughout the solar system. Hence it is a phenomenon comparable to the fact that suspended dust in a dark room shines when it is hit by a ray of sunlight. Moreover, there is no sub-

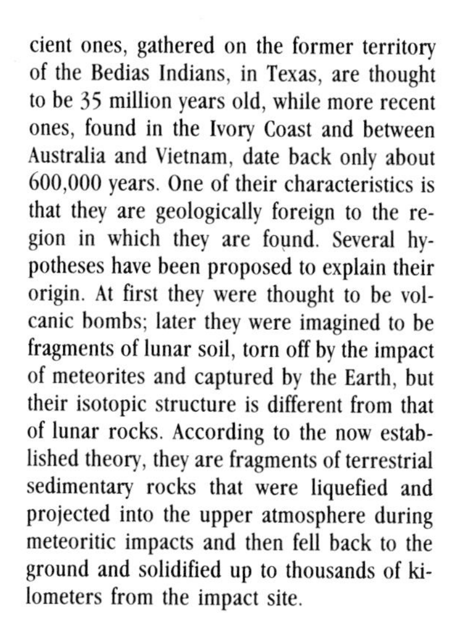
cient ones, gathered on the former territory of the Bedias Indians, in Texas, are thought to be 35 million years old, while more recent ones, found in the Ivory Coast and between Australia and Vietnam, date back only about 600,000 years. One of their characteristics is that they are geologically foreign to the region in which they are found. Several hypotheses have been proposed to explain their origin. At first they were thought to be volcanic bombs; later they were imagined to be fragments of lunar soil, torn off by the impact of meteorites and captured by the Earth, but their isotopic structure is different from that of lunar rocks. According to the now established theory, they are fragments of terrestrial sedimentary rocks that were liquefied and projected into the upper atmosphere during meteoritic impacts and then fell back to the ground and solidified up to thousands of kilometers from the impact site.

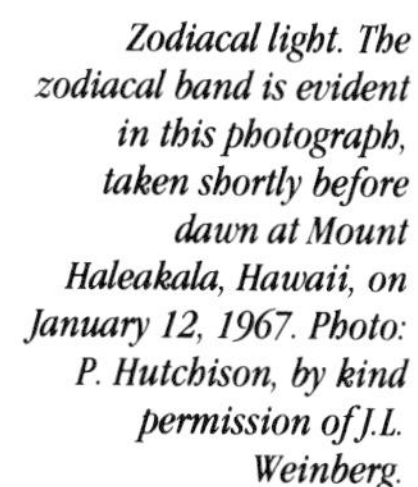
Zodiacal light. The zodiacal band is evident in this photograph, taken shortly before dawn at Mount Haleakala, Hawaii, on January 12, 1967. Photo: P. Hutchison, by kind permission of J.L. Weinberg.

stantive difference between micrometeorites and grains of interplanetary dust; together they make up a vast cloud at least 600 million kilometers long—the zodiacal cloud.

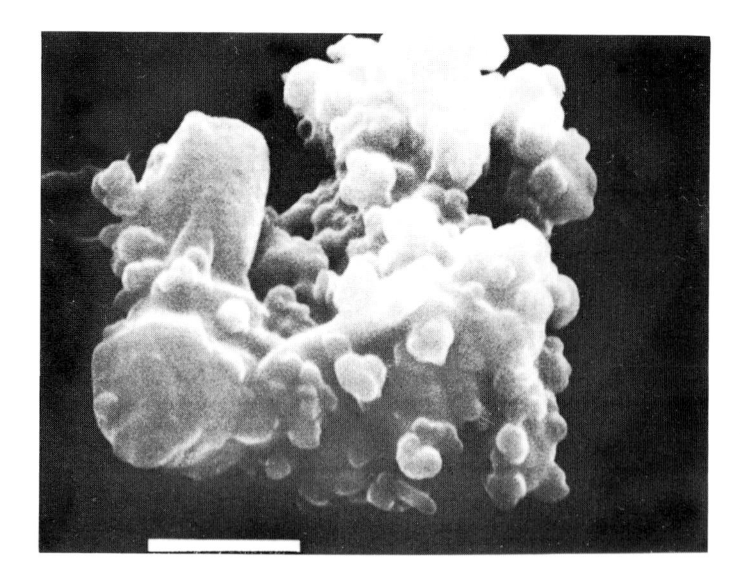

Micrometeorite gathered in the stratosphere. Dust grain collected in the stratosphere, probably of interplanetary and perhaps of cometary origin. The horizontal bar represents 1 µm. Photo: D.E. Brownlee.

Methods of Studying the Zodiacal Cloud

Although it is several hundred times denser than the solar wind (which, together with it, makes up the interplanetary medium), the zodiacal cloud is an extremely rarefied medium; its total mass, estimated from local or integral studies that will be described later, probably is much less than 10^{-10} times the total mass of the planets. At 1 a.u. from the Sun in the plane of the ecliptic—i.e., at the distance of the Earth's orbit—its density is nearly 10^{-23} g./cm^3. As most of the interplanetary grains have radii ranging from 1 to 100 µm, this numerical value means in essence that, near the Earth, there is hardly more than 1 micrometeorite in a cube of space 200 meters on each side; in other words, one could expect a daily flux on our planet of about 1 grain of a dozen µm on a collecting surface of 1 m^2. Under such conditions, methods of study are understandably difficult.

In addition to the large meteorites mentioned earlier, micrometeorites have been sought on Earth or in the Earth's atmosphere. To avoid contamination by terrestrial dust, samples have been taken from ice floes, from sedimentary layers or from the stratosphere. Some black magnetic spherules of extraterrestrial origin have been found in marine sediments. During stratospheric flights in an airplane or balloon, a group from Brownlee (University of Washington in Seattle) collected fine specimens; these are porous aggregates chiefly constituted by the accretion of grains of magnesium, calcium, iron, silicon, nickel and cobalt. But no space probe has yet been able to take samples of interplanetary grains far from the Earth and bring them back for study in our laboratories; thus it is necessary, for complete understanding of the zodiacal cloud, to rely on indirect local methods (impact study) or integral methods (study of scattered solar light).

The impact craters produced by micrometeorites as well as the residual matter left after collision, have been studied on targets transported by manned satellites (Gemini, Skylab) and (thanks to the Apollo missions) on lunar rocks exposed to interplanetary dust for billions of years. Of course, it has not yet been possible to analyze targets on Earth, from distant regions of the zodiacal cloud. But a limited study, likely to confirm the information deduced from observations of zodiacal light, could be carried out aboard space probes such as *Pioneer 10* or *11* and *Helios 1* or *2*, which retransmitted via telemetry data concerning the impact of dust on microphones; charged electrostatic detectors; or thin, ionized layers.

Complete and precise photopolarimetry of the zodiacal light has been conducted from ground stations and, since the end of the 1960s, from balloons, rockets, artificial satellites or distant space probes. The zodiacal light perceived in a certain direction is the sum (or, in mathematical terms, the integral) of light locally scattered in the direction of the observer by all the interplanetary grains present along the line of sight; the luminance of this locally scattered light depends on solar illumination, on the local density of the cloud and on the physical properties of the dust. Thus we go from integral information to local information by comparing the observed luminance to those that would be implied by theoretical models of the zodiacal cloud, or, more rigorously, by taking the inverse of the luminance integral.

Ground-based observations are durable and relatively inexpensive; but they are limited in wavelength by atmospheric opacity in the ultraviolet and infrared, and they are difficult to analyze owing to the presence of background light emission (of atmospheric or interstellar origin) that is absorbed and scattered by the Earth's atmosphere. They are done with small-field telescopes using filters to screen out the spectral lines of atmospheric luminescence; the atmospheric continuum is estimated by geometric reductions or is based on color differences. Preferably, instruments should be installed at high altitude (to minimize absorption and atmospheric scattering) and at low latitude (to approximate observational conditions at the equator).

Hawaii, Tenerife and the Pic du Midi are very good sites; quality observations have been made there for several years by the groups of Weinberg (Space Astronomy Laboratory, Gainesville, Florida), Dumont (Bordeaux Observatory) and Robley (Observatory of the Pic du Midi and Toulouse).

Observations in space are of threefold interest: It becomes possible to detect rays emitted at any wavelength, without limitation by atmospheric absorption; problems of contamination by light from cities or by atmospheric luminescence no longer exist; nor do problems of poor weather conditions. Finally, observations can be made at various heliocentric distances (and, as we will see farther on, soon can be made outside the plane of the ecliptic). But it is certain that, owing to the high cost of experiments in space, there are very stringent limits on the duration, volume, weight or electrical energy.

The American satellites *OSO 2, OSO 5* and *OAO 2*, the French satellites *D2-A Tournesol* and *D2-B Aura* and the orbital stations *Skylab* and *Salyut 6* have contributed to the study of the zodiacal light. Finally, the American *Pioneer 10* and *11* probes and the German probes *Helios 1* and *2,*

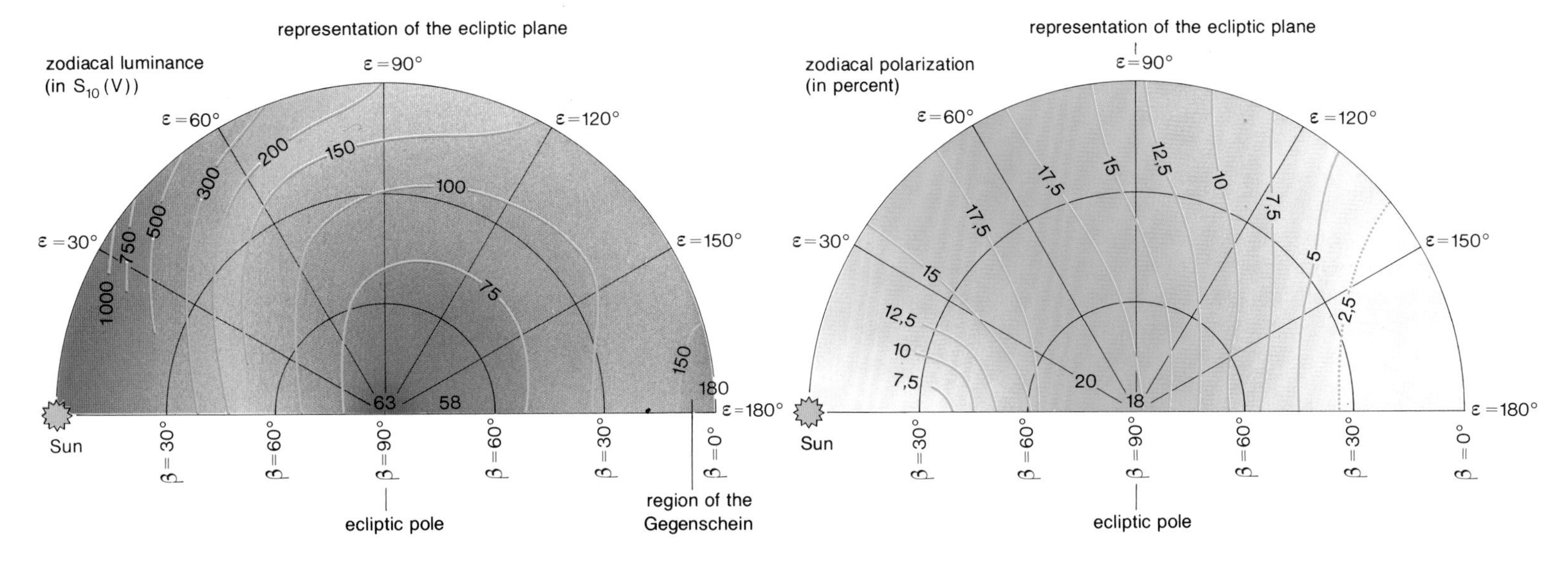

Luminance and polarization of the zodiacal light. Zodiacal light, which can be detected from Earth or in orbit around the Earth, increases when β and ε diminish. The polarization is a maximum at ε = 60° and decreases toward both larger and smaller values of ε. From R. Dumont and F. Sánchez-Martínez, with the kind permission of R. Dumont.

mentioned earlier, have made zodiacal measurements outside and inside the Earth's orbit, respectively.

Interplanetary Dust

From the midultraviolet to the midinfrared, the spectrum of zodiacal light reproduces the solar spectrum rather faithfully, which firmly establishes that what is involved is solar light scattered by interplanetary dust. In the far infrared, the grains' own thermal emission is superimposed upon it. Many uncertainties remain as to the scattering efficiency of the grains in the far ultraviolet, at less than 250 nm.

The figure shows the variation of luminance and polarization with ecliptic latitude β (angular height above the plane of the ecliptic) and ecliptic longitude ε (angle between the Sun and the projection of the line of sight onto the plane of the ecliptic). The unit used for luminance, specific to this type of work, is the S_{10} (V), (or the equivalent number of solar-type stars of visual magnitude 10 per square degree. In the region around the pole of the ecliptic, the luminance reaches a minimum of about 60 S_{10}(V), or, in other words, of about 70 stars of magnitude 28 per square arc second, which unquestionably places a limit on the performance of large space telescopes. In the direction opposite the Sun, a slight intensification known as the *gegenschein* can be seen. This intensification, discovered by Humboldt at the turn of the 19th century, and that can be observed in Europe in autumn in the constellation Pisces, was long attributed to a dust cloud tied to the Earth. Observations by the *Pioneer 10* and *11* probes refuted this hypothesis by proving that the phenomenon, still detectable far from the Earth, must be tied to the backscattering properties of interplanetary grains. As for the *polarization* of zodiacal light, it is far from negligible, since it reaches a maximum of about 20 percent; it may be that a slight negative polarization exists around the antisolar direction.

The density distribution in the plane of the ecliptic may be measured using ground-based observations in directions of small elongation or on board the Helios and Pioneer probes: a model of the cloud showing a decrease of $1/R^{1.3}$ as the heliocentric distance R increases is in agreement with observations, at least up to 2 a.u. from the

Study, from Salyut 6, *of the zodiacal light. At 300 kilometers (186 miles) in altitude, the emission of ionospheric regions D and E is visible, as is the internal zodiacal light. The colored isophotes were obtained from the photograph taken in space on March 10, 1978, by processing with the microdensitometer of the I.A.P. Photo: G. Gretchko and S. Koutchmy, with the kind permission of S. Koutchmy.*

Sun; farther out the zodiacal light becomes too weak to be conveniently measured, which proves, moreover, that the asteroid belt does not correspond to an appreciable increase in the density of dust.

The question of variations in zodiacal light—that is, fluctuations in the zodiacal cloud—has been the subject of numerous controversies, since the variations in question (a few dozen $S_{10}(V)$) sometimes are below the threshold of sensitivity of the experiments. There exists an apparent variation in the luminance at high or intermediate average ecliptic latitude; it stems simply from the Earth's annual motion relative to the symmetry plane of the zodiacal cloud. Observations carried out at Tenerife and aboard the *D2-A* satellite yield an inclination of this plane to the ecliptic of nearly 1.5 degrees with the ecliptic longitude of the intersections being about 96 degrees, which confirms that at the distance of the Earth, the symmetry plane of the zodiacal light corresponds quite well with the invariant plane of the solar system (a plane perpendicular to the total angular momentum of the Sun and planets). Observations simultaneously carried out in Hawaii and on the Helios probes tend to prove that, within the Earth's orbit, the inclination of the symmetry plane is more pronounced and that it may merge with Venus's orbital plane.

Overall, the zodiacal cloud is a stable phenomenon, and it does seem that no variation in luminance greater than 10 percent has been caused by extreme fluctuations of solar activity, except perhaps for the *Gegenschein*. However, the discovery by *D2-A* of small highly localized short-period variations implies that there exist within the zodiacal cloud inhomogeneities of about several dozen million kilometers in size. The time of year and direction in which these inhomogeneities are observed are correlated with the passage of the Earth through swarms of meteoroids that form a network of elliptical toroids within the zodiacal cloud. This result, then, means that among the different possible sources of interplanetary dust (the primitive solar nebula, asteroids, interstellar clouds and comets), comets contribute appreciably to the zodiacal cloud.

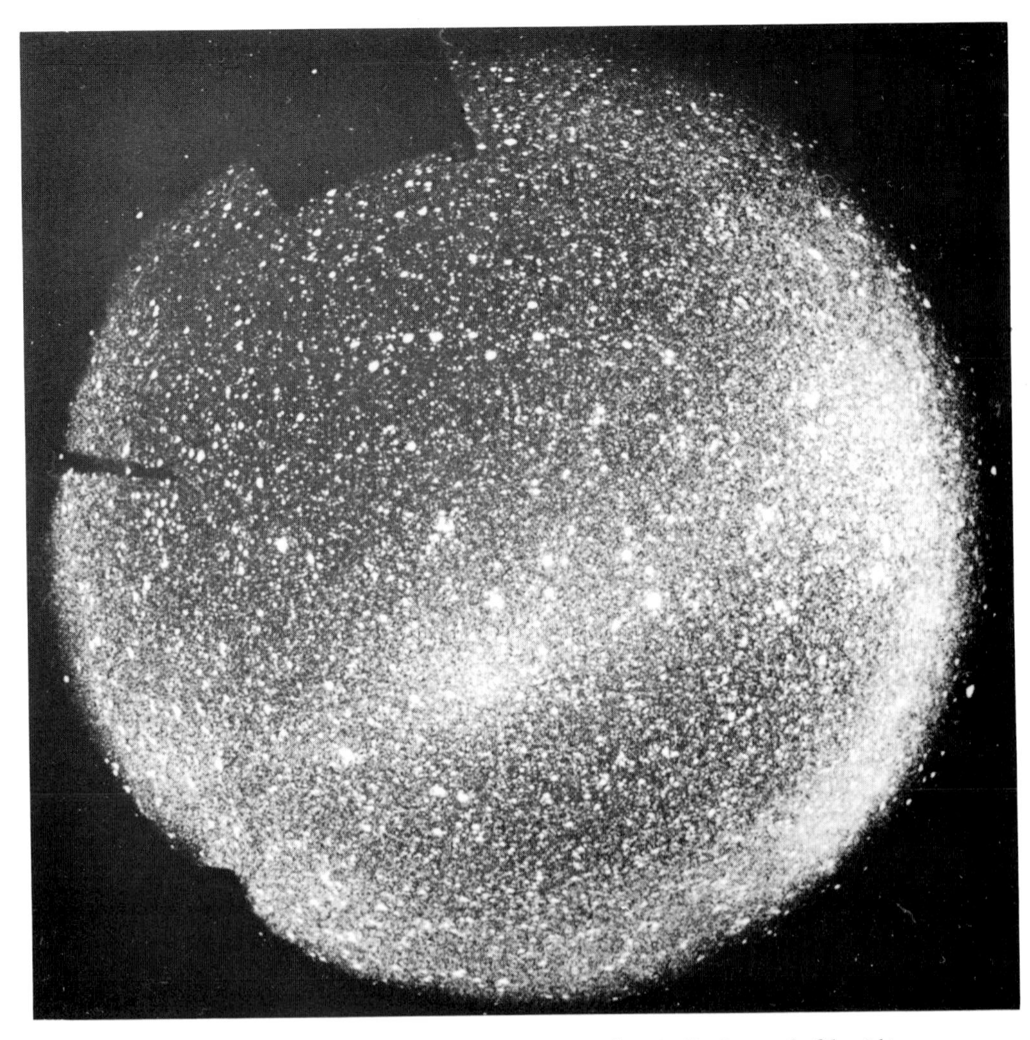

Gegenschein. *In this photograph of the night sky taken on March 12, 1972, in Ogasawara, Japan, we see: the Milky Way on the right; the trace of the ecliptic plane, from lower left to upper right; the* Gegenschein *is a little below the center. Photo: A. Miyashita, with the kind permission of H. Tanabé.*

Evolution of Meteoroids and Interplanetary Dust

Local or integral observations and theoretical studies make it possible to consider the evolution of interplanetary dust in a global manner and to grasp the fundamental underlying physical phenomena.

While most of the grains (with the exception of some, known as β meteoroids, which have hyperbolic trajectories) seem to follow elliptical orbits around the Sun, like most of the planets, all are subject to forces that modify their orbit, owing to their low mass. The Poynting-Robertson effect, or the loss of angular momentum by reflection of solar light, makes them spiral gradually closer to the Sun, while the pressure of solar radiation continually pushes the particles of submicron size toward the exterior of the solar system. The presence of very small dust particles in the zodiacal cloud, and the loss of meteoroids through impact on planets (while a collecting surface of 1

Trajectories of several space probes. In the neighborhood of the ecliptic, Helios 1 *and* 2 *and* Pioneer 10 *and* 11 *studied the zodiacal cloud, at less and at more than 1 au, respectively. The ISPM probe will bring a third dimension to knowledge of local interplanetary matter.*

m² receives hardly more than a grain of a few dozen micrometers per year in the neighborhood of the Earth's orbit, our planet, in its entirety, receives several thousand tons of interplanetary matter per year!) imply that there is a constant addition of dust to the globally stable zodiacal cloud. Its evolution can be represented as follows:

Comets, the nuclei of which are composed of matter from the initial protosolar system or from red giants, break up (as did Biéla's comet, for example, in 1845); their dust, mixed with ice that sublimates during the passage through perihelion, escapes as a swarm of meteoroids. Under the influence of gravitational forces and collisions, the orbits of individual meteoroids evolve differently: The diameter of the swarm increases; its density decreases; the dust is distributed gradually along the trajectory, and the swarm, which, if it reached the Earth, would be a periodic shower (like the Giacobinids or Leonids), gradually becomes a permanent shower visible every year at the same time. Over time, the meteoroids of cometary origin, and also probably the fragments produced by collisions between asteroids, as well as the dust from interstellar clouds, are slowly diluted in the zodiacal cloud. The size of these grains, as they approach the sun as a result of the Poynting-Robertson effect, gradually decreases under the combined influence of collisions and erosion caused by the solar wind. At the end of their life (about 10,000 years on the average), interplanetary grains too close to the Sun are sublimated or pushed out by radiation pressure.

Prospects for Future Study

Observations of the zodiacal light were made in 1982 aboard the American Space Shuttle (photopolarimeter) and during the flight of the French experimental package on board the Soviet orbiting station *Salyut 7* (PIRAMIG high-sensitivity polarimetric camera). Later the Space Shuttle should, through the LDEF (long-duration exposure facility), place recoverable targets in orbit at a few hundred kilometers' altitude for several consecutive months; thus it will be possible to learn more about the average density and chemical composition of the micrometeorites.

In 1985, a probe was launched toward Jupiter and will use that planet's gravitational field to leave the plane of the ecliptic and fly over the Sun's poles (see page 111). This project, known as ISPM (International Solar Polar Mission), would have made it possible to observe the zodiacal light, among other things; but as NASA rejected participation of the European Space Agency in this mission, we will have to wait for a future date to determine the distribution of interplanetary dust outside the ecliptic.

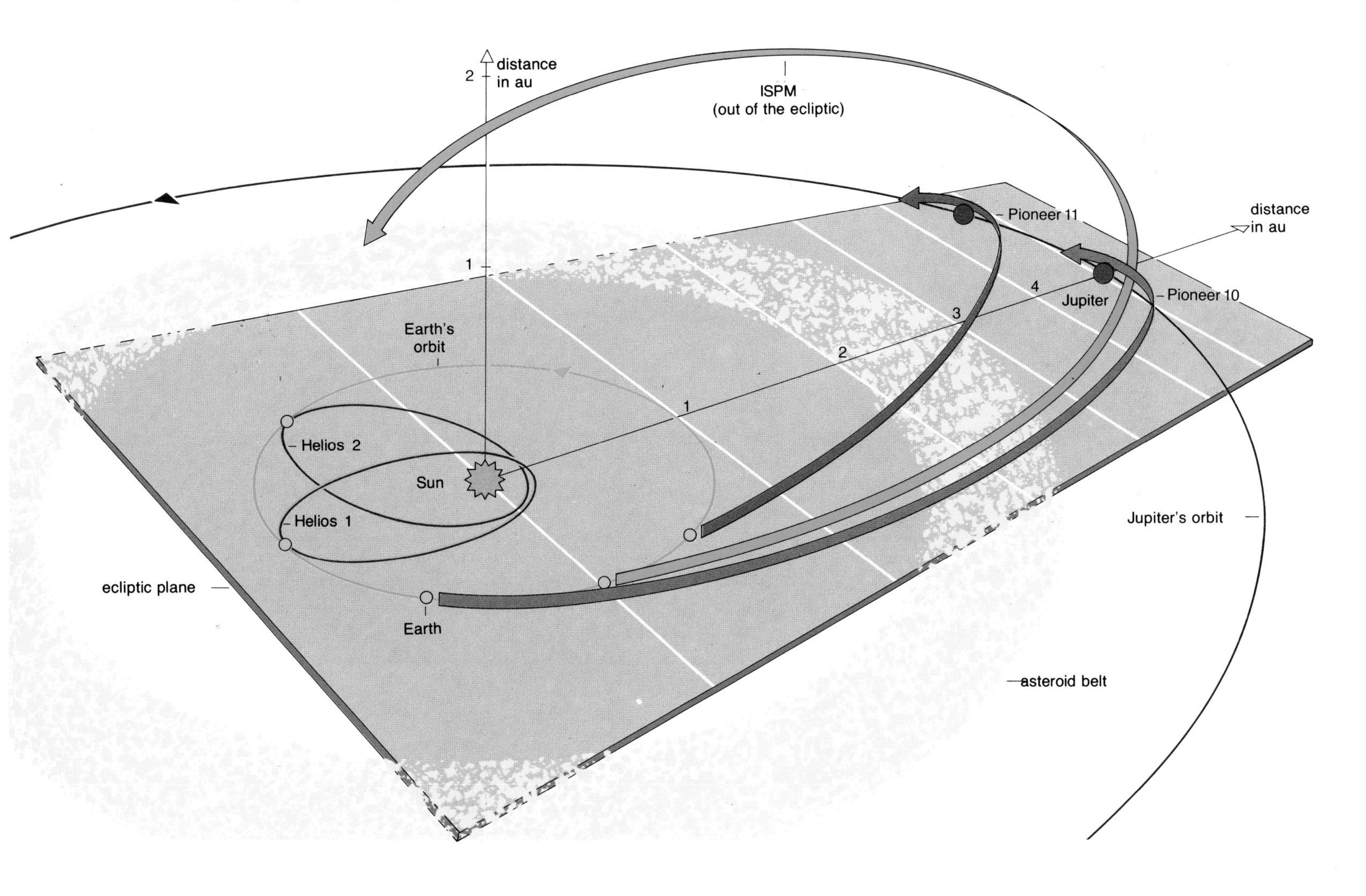

Finally, in 1986, Halley's comet—which last appeared in 1910—again passed in proximity to the Sun. This active comet is ideal for a space mission; in March 1986, a European probe (*Giotto*), two Soviet probes (*Vega*) and a Japanese probe (*Planet A*) flew by it (see box, page 214).

An even more precise evaluation of the zodiacal light will become a strict necessity for future astronomical observations carried out with large orbiting telescopes, for which the zodiacal light will be the only residual contaminant.

However, the goal of all studies of meteoroids or interplanetary dust is knowledge of the origin and evolution of the solar system, as well as knowledge of interstellar dust, which seems to play an essential role in astrophysical processes such as the formation of stars or the synthesis of molecules.

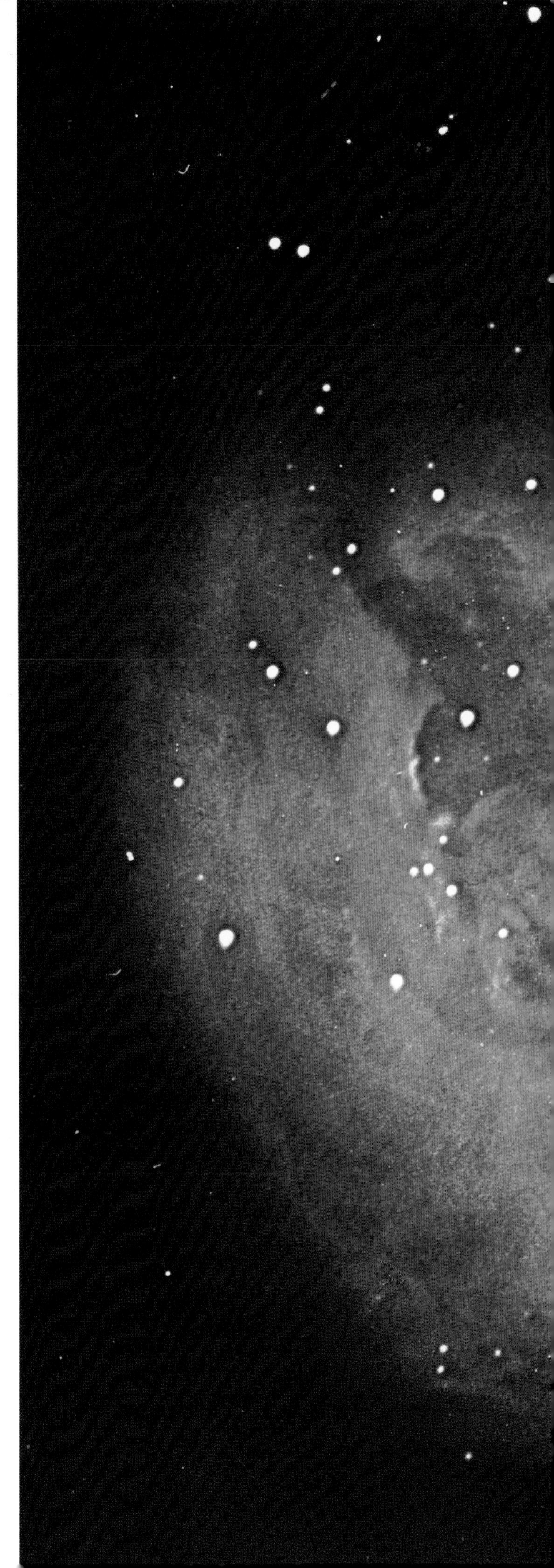

Situated at a distance of about 1,600 light-years and visible to the naked eye as a faint nebulosity in the constellation Orion, this vast hydrogen cloud, lit up by the hot stars it contains, is a veritable breeding ground for stars. USIS Photo.

III

The Realm of the Stars

20.

The Stars in the Sky

While the solar system is the realm of space most accessible to our investigations, it nonetheless represents an extremely small portion of the universe. To grasp ϕ the extent of the field astronomy proposes to explore, one has only to glance at the sky on a dark night and think that each of the twinkling points one sees there is a star more or less similar to the Sun. Moreover, the few thousand stars visible to the naked eye represent only a tiny sampling of the stellar populations accessible to present-day instruments!

Parallel with the progress in observational techniques, the extreme diversity of the sidereal universe has been revealed. There are both immense and tiny stars, hot and cold stars, young and old stars, stable and "variable" stars, single and "multiple" stars; some are isolated, others grouped in more or less compact clusters. In the space between stars, huge gas clouds of varied shapes—nebulae—sometimes are found. The astronomer's first job is to catalogue all the varied objects he discovers. With the extension of observations to the entire spectrum through the development of space astronomy, many previously unsuspected objects have been identified, proving that this task that began with the invention of the telescope is far from finished.

Matter ejected during the explosion of a star—which, it is estimated, occurred some 12,000 years ago—slowly dispersed in space to form this complex nebula, which can be seen in the constellation Vela in the Southern Hemisphere. Photo © Anglo-Australian Observatory.

But observation is not an end in itself. All the phenomena revealed among the sea of stars must be understood, or interpreted, and their genesis explained, their implications foreseen.

Thus, since the advent of astrophysics, experts have tried to develop a theory of stellar evolution. While we are now able to account, more or less, for the evolution of different types of stars, it would be useless to imagine that the job is finished. Many unknowns persist, particularly with regard to the processes of formation, and conceptions regarding the final stages of stellar evolution remain largely speculative. Moreover, the endless crop of new data leads to continual refinement of the models when it does not challenge them.

This study of the processes of stellar evolution is itself only one of the elements essential to the understanding of the galaxies, vast systems of gas, dust and dynamically isolated stars that actually represent the units with which the universe is populated. Among these galaxies, we will, of course, first describe in detail the one in which the solar system is located—*the* galaxy—before turning our attention to the other similar systems and the structure of the universe in its broad outlines.

Constellations

From the earliest civilizations, observers of the sky, to find and name more easily the stars they saw, imagined them to be grouped according to the more or less arbitrary figures they form in the sky. This gave rise to the constellations, which were given the names of heroes, animals or objects associated with myths and legends, which varied from one country to another.

Current nomenclature is based largely on Greek mythology. It seems it was Aratos, a physician and poet at the court of the Macedonian ruler Antigonos Gonatas, in the third century B.C., who thought of giving the various constellations names taken from Greek mythology. The map of the celestial Northern Hemisphere is based on the one established in the second century by Ptolemy, who noted 48 constellations. The southern constellations, on the other hand, are of much more recent origin, as astronomers could not observe the sky of the Southern Hemisphere until much later. It was mainly Bayer and Hevelius, in the 17th century, and Lalande and La Caille, in the 18th century, who named them. Most were given names of birds or scientific instruments: the Peacock, the Toucan, the Microscope, the Octant, the Sextant, the Telescope, etc.

The boundaries of the constellations remained imprecise for a long time. Difficulties arose when it was necessary to catalogue the many low-brightness stars identified through telescopes, especially since some astronomers took it upon themselves to name new celestial groupings as they wished, overlapping with the old ones. At the end of the 19th century, 108 constellations had thus been counted, though their boundaries were not unanimously acknowledged. Thus, beginning in 1922, the International Astronomical Union undertook a revision of the constellations. According to a suggestion by the Belgian astronomer E. Delporte, it was decided in 1927 to substitute arcs of parallels and meridians for the former imaginary boundaries. Since then, the entire sky has been divided into 88 constellations (see table, pages 236 to 237), each containing, in addition to the grouping of bright stars that initially gave it its name, a conventionally delimited region of the sky.

Apparent rotation of the stars around the celestial pole; the circumpolar stars trace full circles above the horizon. Photo: Observatory of Haute-Provence, France.

The Celestial Sphere: Systems of Coordinates

Observation shows that during the night, the celestial sphere seems to rotate as a unit in a counterclockwise direction around a point that, in the Northern Hemisphere, is very close to the north star. In North America and Europe, the constellations closest to the north star always remain visible; those farther away from the pole rise in the east and set in the west. This is merely an apparent motion. In reality, it is the Earth that rotates, from west to east, in 23 hours, 56 minutes, around an axis that goes through its poles (see "The Earth," page 128). But as the stars are very distant, we may roughly consider them as being situated all at the same distance, attached to an imaginary sphere with a very great radius—the celestial sphere—with the Earth at its center, rotating in 23 hours, 56 minutes around an axis that coincides with the Earth's polar axis.

The polar axis of the celestial sphere is that axis formed by extending the line connecting the Earth's poles toward the north and south; it intersects the celestial sphere at two points, called the celestial poles. The plane perpendicular to that axis and going through the center of the celestial sphere is ϕ ϕ none other than the plane of the Earth's equator; it cuts the celestial sphere in a great circle called the celestial equator. To locate the position of a point on the Earth's surface precisely, the geographic coordinates—latitude and longitude—were invented. The equator is taken as the plane of reference. The Earth's surface is thought of as a network of circles, some parallel to the equator (parallels, which except for the equator, are small circles), the others going through the poles (meridians, which are always great circles). Thus each point on the Earth's surface is at the intersection of a parallel and a meridian. Its position is perfectly determined when we know the angular distance between the parallel it is on and the equator (its latitude) on the one hand, and that between the meridian going through that point and a meridian arbitrarily chosen as a reference—in practice, that which goes through the old Greenwich Observatory in England (its longitude).

To determine the position of bodies on the celestial sphere, astronomers have invented similar systems.

ALTITUDE-AZIMUTH COORDINATES

The plane of reference of this system is the plane parallel to the horizon that passes through

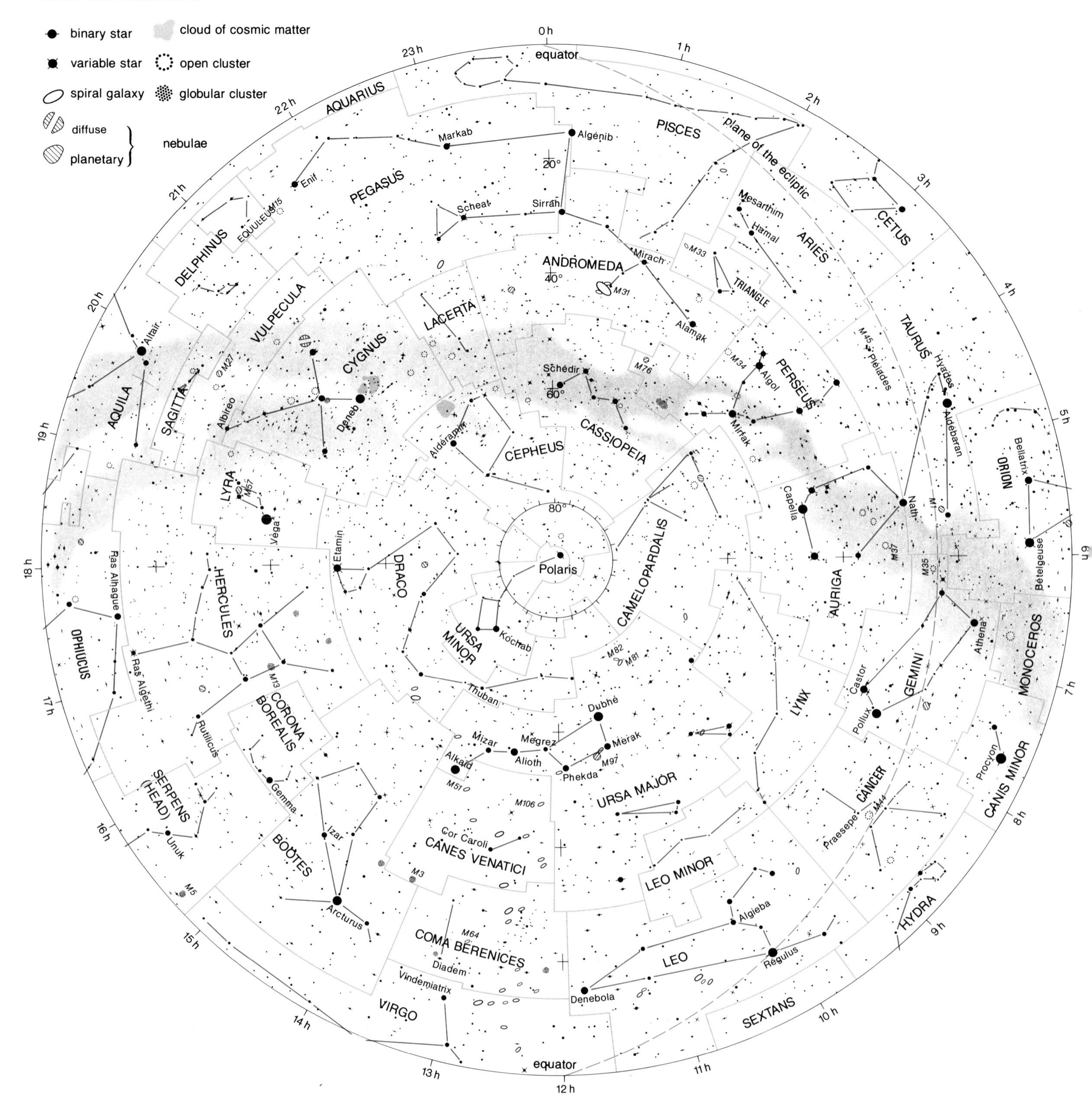

Naming the Stars

At one time, a star was given a name that referred to its position in the mythological figure identified with the constellation to which the star belonged.

Moreover, the brightest stars (see table, page 240) were given names—chiefly by the Arabs in the Middle Ages—that have been preserved by tradition: Sirius, Rigel, Aldebaran and so on.

In 1603, the German astronomer Bayer got the idea of introducing a simple, rational nomenclature using the letters of the Greek alphabet:

α alpha	η êta	ν nu	τ tau
β bêta	θ thêta	ξ ksi	υ upsilon
γ gamma	ι iota	ο omicron	φ phi
δ delta	κ kappa	π pi	χ khi
ϵ epsilon	λ lambda	ρ rhô	ψ psi
ζ zêta	μ mu	σ sigma	ω oméga

In each constellation, the brightest star is designated by α, the second brightest by β, then γ and so on. This nomenclature, now universally adopted, has a few exceptions established by custom: Thus the brightest star of Ursa Major is ϵ, not α. When the Greek alphabet is exhausted, the Latin alphabet is used, and then numbers. Even so, we manage to designate in this way only the stars that are visible to the naked eye. The stars of lesser brightness, revealed by telescopes, are designated solely by their order number in reference catalogues. As the constellations are designated officially by their Latin names (comprehensible throughout the world), a star's official name can be found by placing after the letter designating it the genitive of the Latin name of the constellation to which it belongs.

SOUTHERN HEMISPHERE

The stars are represented by dots whose size is proportional to the stars' brightness (magnitudes from 6.5 to 1).

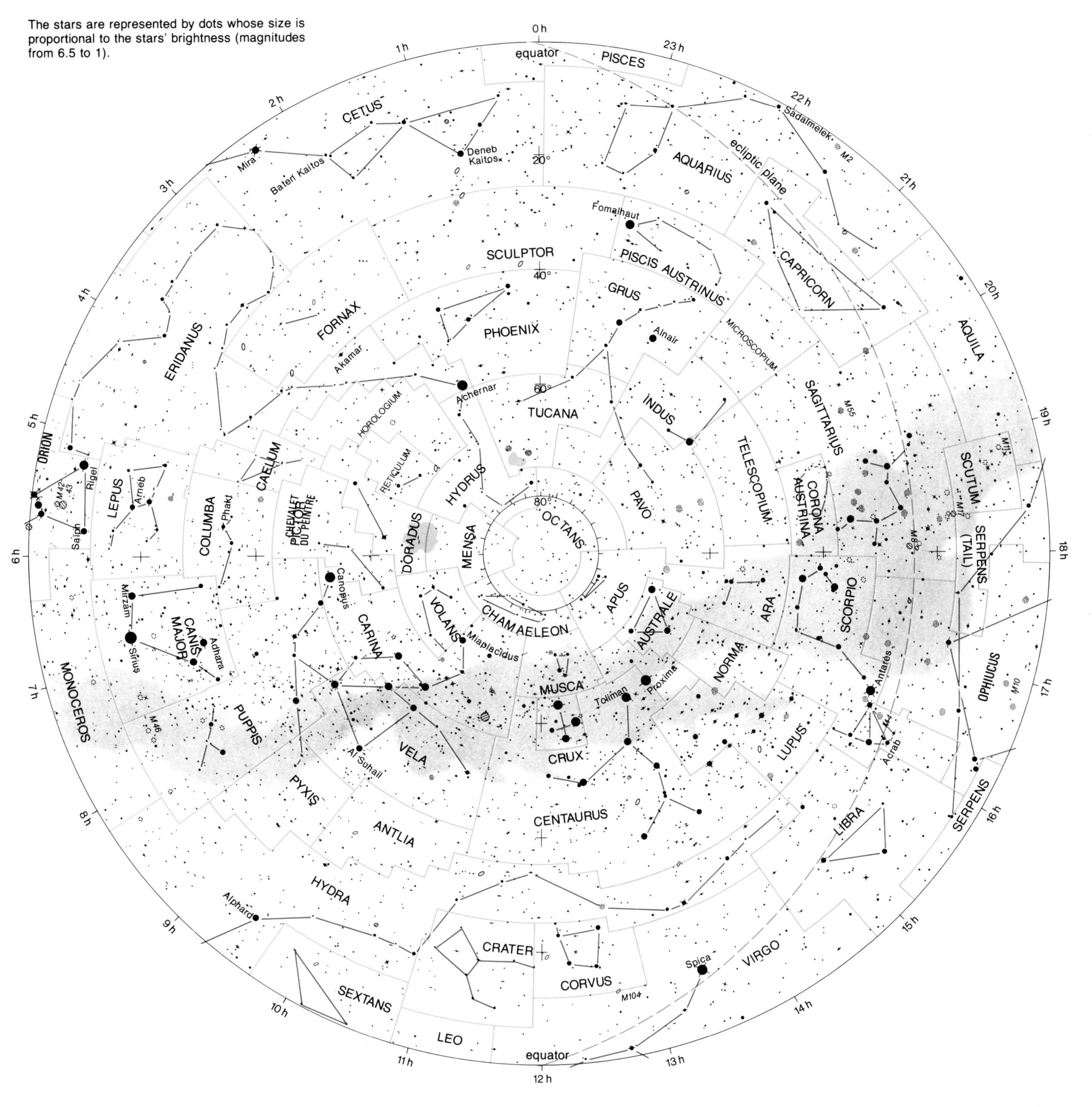

Celestial Atlases

Replacing maps carved on wood, the first celestial atlases appeared in the 17th century with the development of printing.

One of the most famous is *Uranometria,* published in 1603 by the German Johann Bayer. The constellations are designated in it by Greek letters according to a system of nomenclature still used.

In the *Coelum stellarum christianum,* published by the German Julius Schiller, mythological names are replaced by names of Christian origin; thus the twelve signs of the zodiac are replaced by the twelve apostles.

The atlas of the Pole Hevelius, which appeared in 1690, met with a success comparable to that of Bayer.

The Englishman John Flamsteed's *Atlas coelestis,* published in 1729, gives the exact positions of the stars for the first time. One of the last great illustrated atlases was the German Johann Elert Bode's *Uranographie,* published in 1801.

Since the end of the 19th century, photographic atlases have replaced illustrated atlases.

In 1887, at the prompting of Admiral Ernest Mouchez, director of the Paris Observatory, an international effort was undertaken to produce a map of the sky; it consisted of photographing the stars down to the 14th magnitude, then publishing the coordinates for stars brighter than the 11th magnitude. The catalogue was completed, but not the map.

The most recent of the great celestial atlases is the *Palomar Sky Atlas,* consisting of photographs taken in 1950 with the Schmidt telescope at the Mount Palomar Observatory, which has a 1.22-meter (48-inch) aperture; it includes about 500 million stars down to magnitude 21.

the place of observation. The straight line perpendicular to that plane (the vertical) and passing through the observer contains the two poles of the system: the zenith, above the horizon, and the nadir, below it. Any point on the celestial sphere is found by its altitude—the angular distance separating it from the horizon—and its azimuth, the angle it makes with the southern direction. The altitude is measured from the horizontal plane, from 0° to + 90° above it, and from 0° to − 90° below it. The zenith angle, which is the complementary angle (90°—altitude), also is used. The azimuth is counted from 0° to 360°, in the retrograde direction. This system of coordinates is employed by instruments that have an altazimuth mount—i.e., movable about horizontal and vertical axes. But in this system, both coordinates of a body change continuously owing to the Earth's rotation. Thus sometimes it is preferable to use a different system, called the equatorial coordinate system, which we now are going to describe.

zenith
celestial sphere
North
South
horizon
nadir
z = zenith distance from star S
h = altitude a = azimuth

Horizontal Coordinates

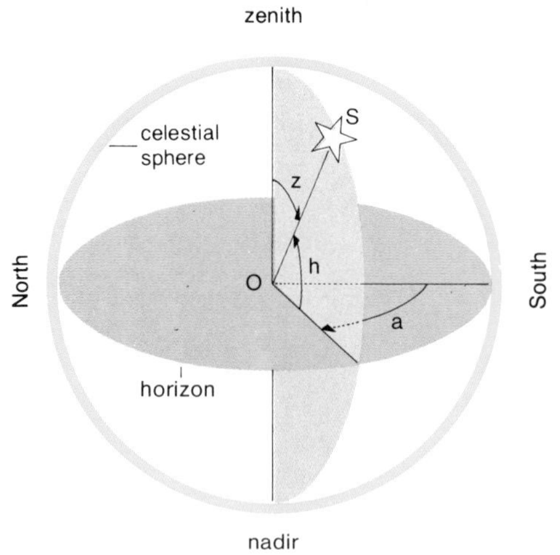

EQUATORIAL COORDINATES

This system, which takes the celestial equator as the plane of reference, is the exact replica of the geographic coordinate system. However, the great circles passing through the poles are called hour circles, the term meridian (more precisely, observer's meridian) being reserved for the great circle that passes through the poles and the zenith.

The Constellations

Latin Name (and genitive ending)	Official Abbreviation	English Name	Approximate Boundaries		Length (in square degrees)	Number of Stars Brighter Than Magnitude 6	Principal Star(s)
			Right Ascension	Declination			
Andromeda (-ae)	And	Andromeda	22 hr., 56 min. to 2 hr., 36 min.	+ 21.4° to + 52.9°	722	100	Alpheratz or Sirrah, Mirach, Almak
Antlia (-ae)	Ant	Air Pump	9 hr., 25 min. to 11 hr., 3 min.	− 24.3° to − 40.1°	239	20	
Apus (-odis)	Aps	Bird of Paradise	13 hr., 45 min. to 18 hr., 17 min.	− 67.5° to − 82.9°	206	20	
Aquarius (-ii)	Aqr	Water Bearer	20 hr., 36 min. to 23 hr., 54 min.	+ 03.1° to − 25.2°	980	90	
Aquila (-ae)	Aql	Eagle	18 hr., 38 min. to 20 hr., 36 min.	− 11.9° to + 18.6°	652	70	Alatïr
Ara (-ae)	Ara	Altar	16 hr., 31 min. to 18 hr., 6 min.	− 45.5° to − 67.6°	237	30	
Aries (-tis)	Ari	Ram	1 hr., 44 min. to 3 hr., 27 min.	+ 10.2° to + 30.9°	441	50	
Auriga (-ae)	Aur	Charioteer	4 hr., 35 min. to 7 hr., 27 min.	+ 27.9° to + 56.1°	657	90	Capella
Bootes (-is)	Boo	Herdsman	13 hr., 33 min. to 15 hr., 47 min.	+ 07.6° to + 55.2°	907	90	Arcturus
Caelum (-i)	Cae	Graving Tool	4 hr., 18 min. to 5 hr., 3 min.	− 27.1° to − 48.8°	125	10	
Camelopardalis (-)	Cam	Giraffe	3 hr., 11 min. to 14 hr., 25 min.	+ 52.8° to + 85.1°	757	50	
Cancer (-cri)	Cnc	Cancer (or Crab)	7 hr., 53 min. to 9 hr., 19 min.	+ 06.8° to + 33.3°	506	60	
Canes (-um) Venatici (orum)	CVn	Hunting Dogs	12 hr., 4 min. to 14 hr., 5 min.	+ 28.0° to + 52.7°	465	30	Cor Caroli
Canis (-) Major (-is)	CMa	Big Dog	6 hr., 9 min. to 7 hr., 26 min.	− 11.0° to − 33.2°	380	80	Sirius, Adhara, Mirzam, Wezen
Canis (-) Minor (-is)	CMi	Little Dog	7 hr., 4 min. to 21 hr., 57 min.	− 00.1° to + 13.2°	183	20	Procyon
Capricornus (-i)	Cap	Sea Goat	20 hr., 4 min. to 21 hr., 57 min.	− 08.7° to − 27.8°	414	50	
Carina (-ae)	Car	Keel	6 hr., 2 min. to 11 hr., 18 min.	− 50.9° to − 75.2°	494	110	Canopus
Cassiopeia (-ae)	Cas	Cassiopeia	22 hr., 56 min. to 3 hr., 36 min.	+ 46.4° to + 77.5°	598	90	Schedir, Caph, Tsih
Centaurus (i)	Cen	Centaur	11 hr., 3 min. to 14 hr., 59 min.	− 29.9° to − 64.5°	1,060	150	Rigil kentarus
Cepheus (-i)	Cep	Cepheus	20 hr., 1 min. to 8 hr., 30 min.	+ 53.1° to + 88.5°	588	60	Alderamin
Cetus (-i)	Cet	Sea Monster (or Whale)	23 hr., 55 min. to 3 hr., 21 min.	− 25.2° to + 10.2°	1,231	100	Diphda, Mira
Chamaeleon (-ontis)	Cha	Chameleon	7 hr., 32 min. to 13 hr., 48 min.	− 75.2° to − 82.8°	132	20	
Circinus (-i)	Cir	Compass	13 hr., 35 min. to 15 hr., 26 min.	− 54.3° to − 70.4°	93	20	
Columba (-ae)	Col	Dove	5 hr., 3 min. to 6 hr., 28 min.	− 27.2° to − 43.0°	270	40	
Coma (-ae) Berenices	Com	Berenice's Hair	11 hr., 57 min. to 13 hr., 33 min.	+ 13.8° to + 33.7°	386	50	
Corona (-ae) Australis	CrA	Southern Crown	17 hr., 55 min. to 19 hr., 15 min.	− 37.0° to − 45.6°	128	25	
Corona (-ae) Borealis	CrB	Northern Crown	15 hr., 14 min. to 16 hr., 22 min.	+ 25.8° to + 39.8°	179	20	Margarita
Corvus (-i)	CrV	Crow	11 hr., 54 min. to 12 hr., 54 min.	− 11.3° to − 24.9°	184	15	
Crater (-is)	Crt	Cup	10 hr., 48 min. to 11 hr., 54 min.	− 06.5° to − 24.9°	282	20	
Crux (-cis)	Cru	Southern Cross	11 hr., 53 min. to 12 hr., 55 min.	− 55.5° to − 64.5°	68	30	
Cygnus (-i)	Cyg	Swan	19 hr., 7 min. to 22 hr., 1 min.	+ 27.7° to + 61.2°	804	150	Deneb
Delphinus (-i)	Del	Porpoise	20 hr., 13 min. to 21 hr., 6 min.	+ 02.2° to + 20.8°	189	30	
Dorado (-us)	Dor	Swordfish	3 hr., 52 min. to 6 hr., 36 min.	− 48.8° to − 70.1°	179	20	
Draco (-nis)	Dra	Dragon	9 hr., 18 min. to 21 hr.	+ 47.7° to + 86.0°	1,083	80	Etamin
Equuleus (-i)	Equ	Little Horse	20 hr., 54 min. to 21 hr., 23 min.	+ 02.2° to + 12.9°	72	10	
Eridanus (-i)	Eri	River	1 hr., 22 min. to 5 hr., 9 min.	+ 00.1° to − 58.1°	1,138	100	
Fornax (-acis)	For	Furnace	1 hr., 44 min. to 3 hr., 48 min.	− 24.0° to − 39.8°	398	35	
Gemini (-orum)	Gem	Twins	5 hr., 57 min. to 8 hr., 6 min.	+ 10.0° to + 35.4°	514	70	Pollux, Castor
Grus (-is)	Gru	Crane	21 hr., 25 min. to 23 hr., 25 min.	− 36.6° to − 56.6°	366	30	Alnaïr
Hercules (-is)	Her	Hercules	15 hr., 47 min. to 18 hr., 45 min.	+ 03.9° to + 51.3°	1,225	140	
Horologium (-ii)	Hor	Clock	2 hr., 12 min. to 4 hr., 18 min.	− 39.8° to − 67.2°	249	20	
Hydra (-ae)	Hya	Sea Serpent	8 hr., 8 min. to 14 hr., 58 min.	+ 06.8° to − 35.3°	1,303	130	Alphard
Hydrus (-i)	Hyi	Water Snake	2 min. to 4 hr., 33 min.	− 58.1° to − 82.1°	243	20	
Indus (-i)	Ind	Indian	20 hr., 25 min. to 23 hr., 25 min.	− 45.4° to − 74.7°	294	20	

Equatorial Coordinates

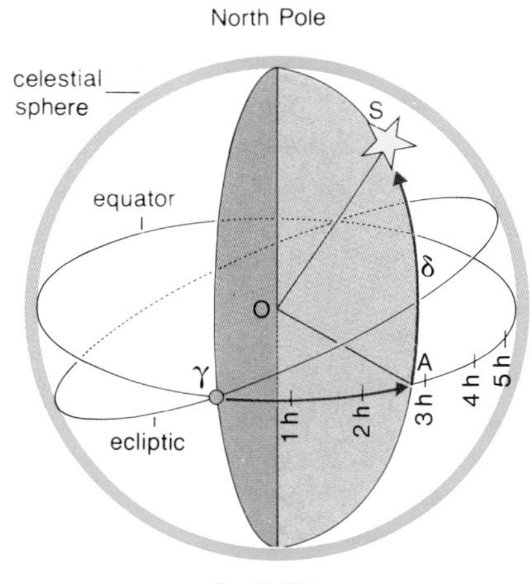

$\widehat{\gamma A}$ = Right ascension of stars
δ = declination

Any point on the celestial sphere is specified by its right ascension and its declination. The right ascension (α) of a point is the angular distance between an hour circle arbitrarily taken as a reference—in practice, the one that goes through the point in the sky where the Sun is located at the vernal equinox, and the hour circle of the given point; its declination (δ) is the angular distance between the celestial equator and the parallel going through that point. The right ascension is measured in hours, minutes and seconds, from 0 to 24 hours, in a direct sense; the hour is considered here as an angle unit of 15°; the convention adopted is such that the right ascension of bodies passing through the meridian increases with time. The declination is counted in degrees and minutes from 0° to +90° to the north, and from 0° to −90° to the south.

Through the choice of time units, the right ascension of the observer's meridian has the same value as sidereal time.

The hour circle passing through a given body and the meridian of the place of observation cut the celestial equator at two points, between which the angular distance (counted from 0 to ± 12 hours, less often from 0 to 24 hours, in the retrograde direction) is called the body's hour angle.

The system of equatorial coordinates offers the advantage of being independent of the place of observation. On the other hand, as it involves the equator and the terrestrial poles, it is susceptible

The Constellations (cont'd.)

Latin Name (and genitive ending)	Official Abbreviation	English Name	Approximate Boundaries: Right Ascension	Approximate Boundaries: Declination	Length (in square degrees)	Number of Stars Brighter Than Magnitude 6	Principal Star(s)
Lacerta *(-ae)*	Lac	Lizard	21 hr., 55 min. to 22 hr., 56 min.	+ 34.9° to + 56.8°	201	35	
Leo *(-nis)*	Leo	Lion	9 hr., 18 min. to 11 hr., 56 min.	− 06.4° to + 33.3°	947	70	Regulus, Denebola
Leo *(-nis)* Minor *(-is)*	LMi	Little Lion	9 hr., 19 min. to 11 hr., 4 min.	+ 23.1° to + 41.7°	232	20	
Lepus *(-oris)*	Lep	Hare	4 hr., 54 min. to 6 hr., 9 min.	− 11.0° to − 27.1°	290	40	
Libra *(-ae)*	Lib	Balance	14 hr., 18 min. to 15 hr., 59 min.	− 00.3° to − 29.9°	538	50	
Lupus *(-i)*	Lup	Wolf	14 hr., 13 min. to 16 hr., 5 min.	− 29.8° to − 55.3°	334	70	
Lynx *(-cis)*	Lyn	Lynx	6 hr., 13 min. to 9 hr., 40 min.	+ 33.4° to + 62.0°	545	60	
Lyra *(-ae)*	Lyr	Lyre	18 hr., 12 min. to 19 hr., 26 min.	+ 25.6° to + 47.7°	286	45	Vega
Mensa *(-ae)*	Men	Mountain	3 hr., 20 min. to 7 hr., 36 min.	− 69.9° to − 85.0°	153	15	
Microscopium *(-ii)*	Mic	Microscope	20 hr., 25 min. to 21 hr., 25 min.	− 27.7° to − 45.4°	210	20	
Monoceros *(-otis)*	Mon	Unicorn	5 hr., 54 min. to 8 hr., 8 min.	− 11.0° to + 11.9°	482	85	
Musca *(-ae)*	Mus	Fly	11 hr., 17 min. to 13 hr., 46 min.	− 64.5° to − 75.2°	138	30	
Norma *(-ae)*	Nor	Carpenter's Level	15 hr., 25 min. to 16 hr., 31 min.	− 42.2° to − 60.2°	165	20	
Octans *(-tis)*	Oct	Octant	00 hr., 00 min. to 24 hr., 00 min.	− 74.7° to − 90.0°	291	35	
Ophiuchus *(-i)*	Oph	Serpent Holder	15 hr., 58 min. to 18 hr., 42 min.	+ 14.3° to − 30.1°	948	100	
Orion *(-is)*	Ori	Orion, the Hunter	4 hr., 41 min. to 6 hr., 23 min.	− 11.0° to + 23.0°	594	120	Rigel, Betelgeuse, Bellatrix, Alnilam, Alnitak, Saiph, Mintaka
Pavo *(-nis)*	Pav	Peacock	17 hr., 37 min. to 21 hr., 30 min.	− 56.8° to − 75.0°	378	45	
Pegasus *(-i)*	Peg	Pegasus	21 hr., 6 min. to 13 min.	+ 02.2° to + 36.3°	1,121	100	Markab, Scheat
Perseus *(-i)*	Per	Perseus	1 hr., 26 min. to 4 hr., 46 min.	+ 30.9° to + 58.9°	615	90	Algenib
Phoenix *(-cis)*	Phe	Phoenix	23 hr., 24 min. to 2 hr., 24 min.	− 39.8° to − 58.2°	469	40	Algol, Mirfak
Pictor *(-is)*	Pic	Easel	4 hr., 32 min. to 6 hr., 51 min.	− 53.1° to − 64.1°	247	30	
Pisces *(-ium)*	Psc	Fish	22 hr., 49 min. to 2 hr., 4 min.	− 06.6° to + 33.4°	889	75	
Piscis (-) Austrinus *(-i)*	PsA	Southern Fish	21 hr., 25 min. to 23 hr., 4 min.	− 25.2° to − 36.7°	245	25	Fomalhaut
Puppis (-)	Pup	Stern	6 hr., 2 min. to 8 hr., 26 min.	− 11.0° to − 50.8°	673	140	
Pyxis *(-idis)*	Pyx	Compass	8 hr., 26 min. to 9 hr., 26 min.	− 17.3° to − 37.0°	221	25	
Reticulum *(-i)*	Ret	Net	3 hr., 14 min. to 4 hr., 35 min.	− 53.0° to − 67.3°	114	15	
Sagitta *(-ae)*	Sge	Arrow	18 hr., 56 min. to 20 hr., 18 min.	+ 16.0° to + 21.4°	80	20	
Sagittarius *(-ii)*	Sgt	The Archer	17 hr., 41 min. to 20 hr., 25 min.	− 11.8° to − 45.4°	867	115	Kaus australis, Nunki
Scorpius *(-ii)*	Sco	Scorpion	15 hr., 44 min. to 17 hr., 55 min.	− 08.1° to − 45.6°	497	100	Antares, Schaula, Dschubba, Acrab
Sculptor *(-i)*	Scl	Sculptor's Tools	23 hr., 4 min. to 1 hr., 44 min.	− 25.2° to − 39.8°	475	30	
Scutum *(-i)*	Sct	Shield	18 hr., 18 min. to 18 hr., 56 min.	− 04.0° to − 16.0°	109	20	
Serpens *(-tis)*	Ser	Serpent	15 hr., 8 min. to 18 hr., 56 min.	+ 25.7° to − 16.0°	637	60	
Sextans *(-tis)*	Sex	Sextant	9 hr., 39 min. to 10 hr., 49 min.	+ 06.6° to − 11.3°	314	25	
Taurus *(-i)*	Tau	Bull	3 hr., 20 min. to 5 hr., 58 min.	+ 00.1° to + 30.9°	797	125	Aldebaran, El Nath
Telescopium *(-ii)*	Tel	Telescope	18 hr., 6 min. to 20 hr., 26 min.	− 45.4° to − 56.9°	252	30	
Triangulum *(-i)*	Tri	Triangle	1 hr., 29 min. to 2 hr., 48 min.	+ 25.4° to + 37.0°	132	15	
Australe *(-is)*	TrA	Southern Triangle	14 hr., 50 min. to 17 hr., 9 min.	− 60.3° to − 70.3°	110	20	
Tucana *(-ae)*	Tuc	Toucan	22 hr., 5 min. to 1 hr., 22 min.	− 56.7° to − 75.7°	295	25	
Ursa *(-ae)* Major *(-is)*	UMa	Big Bear	8 hr., 5 min. to 14 hr., 27 min.	+ 28.8° to + 73.3°	1,280	125	Alioth, Alkaid, Dubhe, Merak, Phecda
Ursa *(-ae)* Minor *(-is)*	UMi	Little Bear	00 hr., 00 min. to 24 hr., 00 min.	+ 65.6° to + 90.0°	256	20	Polaris, Kochab
Vela *(-orum)*	Vel	Sails	8 hr., 2 min. to 11 hr., 24 min.	− 37.0° to − 57.0°	500	110	
Virgo *(-inis)*	Vir	Virgin	11 hr., 35 min. to 15 hr., 8 min.	+ 14.6° to − 22.2°	1,294	95	Spike
Volans *(-tis)*	Vol	Flying Fish	6 hr., 35 min. to 9 hr., 2 min.	− 64.2° to − 75.0°	141	20	
Vulpecula *(-ae)*	Vul	Little Fox	18 hr., 56 min. to 21 hr., 28 min.	+ 19.5° to + 19.4°	268	45	

to the precession of the Earth's axis (see "The Earth," page 128), the vernal equinox shifting slowly on the celestial equator, which it circuits in a period of nearly 26,000 years. Thus it is essential to specify the date or era to which the measurements refer. The epochs 1950.0 or 2000.0 frequently are used, meaning that the indicated coordinates refer to January 1, 1950, or to January 1, 2000. The annual variations of the right ascension and the declination are approximately given by these formulae:

$\Delta\alpha = 3.074 + 1.336 \sin\alpha \tan\delta$ (in seconds of time)

$\Delta\delta = 20.041 \cos\alpha$ (in seconds of arc)

GALACTIC COORDINATES

To study the structure of the galaxy, we have invented a special system of coordinates for which the plane of reference is the plane of symmetry of the galaxy (the galactic plane).

In this system, a body's position is defined by its galactic latitude (b), which is its angular height above the galactic plane (counted from 0° to 90°, positive toward the north and negative toward the south), and by its galactic longitude (l), which is the angular distance (counted from 0° to 360° in the direct sense) between the half-plane going through the body and through the poles of the galaxy, and a half-plane arbitrarily chosen as a reference. Initially, the ascending node of the galactic plane (the intersection of that plane when it goes from the Northern Hemisphere to the Southern Hemisphere with the celestial equator) had been chosen as the origin of longitude. More recently, the origin has been set in the direction of the center of the galaxy. The galactic coordinates expressed in the old system are designated by index I; those expressed in the new system by index II or by no index. The origin of longitude in the new system has as equatorial coordinates (referring to epoch 1950.0): α = 17 h 42.4 min and $\delta = 28°55'$. As for the equatorial coordinates of the galactic north pole, they are (also referring to epoch 1950.0):

α = 12 h 49 min and $\delta = +27°24'$

Stellar Brightness: Apparent and Absolute Magnitudes

A mere glance at the sky is enough to note that not all stars are equally bright.

In the second century B.C., Hipparchus, author of the first stellar catalogue that has come down to us, had the idea of classifying the stars in six "sizes," by decreasing order of brightness: The 1st magnitude was assigned to the brightest stars, the 2nd to those that appeared slightly less luminous, and so on, up to the 6th, which characterized the stars just barely perceptible to the naked eye. Over the entire sky, there are about 20 stars of 1st magnitude, 50 of 2nd magnitude, 150 of 3rd magnitude, 450 of 4th magnitude, 1,350 of 5th magnitude, and 4,000 of 6th magnitude, making about 6,000 stars visible to the naked eye.

Since the invention of optical instruments, the discovery of stars dimmer than those visible to the naked eye has led astronomers to extend and refine the scale of magnitudes.

A body's apparent magnitude, which characterizes its brightness when seen from the Earth, is defined by the relationship established in 1856 by Norman Pogson: $m = -2.512 \log E + \text{constant}$, where m is the body's apparent magnitude and E its luminosity. We are more apt to use the relative definition: $m = m_0 = 2.5 \log E_0/E$, in which m and m_0 denote the magnitudes of two bodies of which the respective luminosities are E and E_0. By conventionally assigning a given magnitude to a few stars used as a reference, we can determine the magnitude of all the others by comparison with the former.

The brighter a body, the lower the value of its magnitude. The scale of magnitudes has been set up so that there is a brightness ratio of 100 between stars of magnitude 1 and those of magnitude 6. A difference of one magnitude between two bodies corresponds to a brightness ratio of 2.512; a difference of n magnitudes corresponds to a brightness ratio of $(2.512)^n$. Thus a star of magnitude 1 is about 2.5 times brighter than a star of magnitude 2; 6.4 times brighter than a star of magnitude 3; 16 times brighter than a star of magnitude 4 and 40 times brighter than a star of magnitude 5.

The scale of sizes empirically set up by Hipparchus corresponds approximately to the scale of magnitudes as a result of a property of the eye by which the variations in brightness that the eye can perceive follow a logarithmic progression: "Sensation varies with the logarithm of excitation" (Weber-Fechner law).

However, some stars recognized by Hipparchus as being of 1st magnitude are in reality even brighter, which leads, on the modern scale, to their having a 0 or negative magnitude. Thus Canopus, in the constellation Caring, has an apparent magnitude of −0.7; Sirius, the brightest star in the sky, in the constellation Canus Major, has a magnitude of −1.4. On the same scale, the full Moon has a magnitude of −12.5; and the Sun, −26.7.

The naked eye can detect stars up to magnitude 6; with a pair of binoculars, one can reach magnitude 9; with a 75-millimeter (3-inch) amateur refractor, magnitude 11; with a 200-millimeter (8-inch) reflector, magnitude 13.5. Under the best conditions, the 5.08-meter (200-inch) telescope of the Mount Palomar Observatory makes it possible to observe stars of magnitude 20.6 and to photograph celestial objects of magnitude 23.5. NASA's future space telescope should make it possible to reach magnitude 29.

The determination of magnitudes relies on precise measurements of the intensity of stellar radiation. We speak of visual magnitudes when these measurements are taken visually; of photographic magnitudes when they are taken on ordinary photographic plates; of photovisual magnitudes when they are taken on special photographic plates sensitive to the same wavelengths as the eye (orthochromatic plates sensitive to yellow radiation, used with a filter opaque to other radiation); of photoelectric magnitudes when the measurements are taken with photoelectric photometers. Such specificity regarding the kind of detector used is essential, for different detectors do not react in the same way to the radiation that strikes them; they have different spectral sensitivities. In particular, measurements of magnitudes using photographic plates require that the plate's calibration curve be known beforehand. On the other hand, measurements by photoelectric photometry avoid that drawback.

The use of appropriate filters makes it possible to determine the magnitude of a star in any color range and over spectral ranges from a few hundred angstroms down to only a few angstroms. As a star's energy distribution varies according to wavelength, it is essential to specify the spectral region in which the measurement was taken.

In the most common photometric system, called the UBV system, magnitude is determined successively in the ultraviolet (U), the blue (B) and the yellow (V—for visible), with color filters that select radiation around 360, 420 and 540 nm ($1 \text{ nm} = 10^{-9}$ m) wavelength. Other standard photometric systems should also be mentioned, such as the R (red), I, J, K, L, M and N (infrared) systems, centered respectively on 700 nm, 900

Light-year and Parsec

To express stellar distances, the kilometer is much too small a unit; the astronomical unit (a.u.) itself, used within the solar system (see page 62), also is too small.

The *light-year*, (l.y.) is used; commonly it represents the distance light travels in a vacuum (and, practically speaking, in space) in one year.

Nevertheless, astronomers generally prefer to use another slightly larger unit, the parsec (pc), which represents the distance from the Earth of a star for which the annual parallax would be 1 arc minute (i.e., the distance from which the semi-major axis of the Earth's orbit would subtend an angle of 1 arc minute). Very great distances are expressed using two multiples of that unit: the kiloparsec (kpc.), which equals 1,000 parsecs, and the megaparsec (Mpc.), which equals 1 million parsecs. The distance in parsecs of a star of parallax π, expressed in seconds of arc, equals $1/\pi$.

Among the light-year, the parsec, the astronomical unit and the kilometer, the following correspondences exist:

1 l.y. = 9.4605×10^{12} kms. = $6{,}324 \times 10^4$ a.u. = 0.3066 pc.

1 pc. = 3.0856×10^{13} kms. = 2.0626×10^5 a.u. = 3.2616 l.y.

ities along the line of sight (radial velocities), from proper motion and from the direction of motion, we can deduce the distance of these stars. This method made it possible, e.g., to determine the distance of the Hyades cluster (120 light-years). It is applicable out to about 300 light-years.

Some stars, concentrated in a relatively small volume, form a homogeneous group of which the average distance can be statistically determined, based on the fact that the group's velocity can be broken down into an average velocity and an individual random velocity for each star. Thus we can estimate the distance of more remote stars than allowed by the previously mentioned methods, but with less precision. This method, called the method of statistical parallaxes, has a range of about 1,000 light-years.

Another important technique is the method of spectroscopic parallax. The characteristics of a star's spectrum make it possible to place it on the Hertzsprung-Russell diagram by its spectral type and luminosity. Its absolute magnitude M then can be deduced from this. Moreover, if we know its apparent magnitude m, we can deduce its parallax π by the relationship mentioned earlier.

$$M - m = 5 - 5 \log d; \text{ for } d = 1/\pi, \text{ hence}$$

$$\log \pi = \frac{M - m - 1}{5}$$

This method is applicable out to 300,000 light-years—i.e., beyond our galaxy.

Finally, the photometric parallax method is based on photometric measurements of variable stars, such as Cepheids or RR Lyrae stars, in which the period of brightness variation is tied to luminosity, or of explosive events (novae and supernovae) for which the absolute magnitude can be deduced from the apparent magnitude at maximum brightness. As before, the relationships among the apparent magnitude, absolute magnitude and distance lead to determination of the parallax. This method, which involves highly luminous stars, hence those perceptible at great distances, has a much more extensive range than the prior ones: Variable stars permit distance estimates up to 7 million light-years, and supernovae up to 50 million light-years. Thus this is a very valuable method for estimating the distances to the closest galaxies.

However, trigonometric parallax, though it can be used only for the stars close to the Sun, continues to play a fundamental role, for it is the only method that can determine a star's distance without making any hypothesis about its physical nature; it is the basic method for calibrating all the other methods that permit a much deeper penetration into space.

Sizes of the Stars

The stars are much too far away to measure their diameters directly using telescopes. Thus several indirect methods have been developed. The most important one, historically, is based on analysis of their spectra, their total luminosity and temperature being tied to their surface area, hence to their radius, by a simple mathematical relationship.

There is also a method that can be applied to binary stars showing eclipses; first we determine the diameter of the orbit traveled by these stars; then, from their light curve, we determine the fraction of the orbit during which a star is hidden by the eclipse; from this we then can deduce the diameters of the two stars.

A recent method that seems promising but that has as yet been applied only to a few nearby stars, consists of measuring the star's angular diameter by a technique of optical interferometry (speckle interferometry), giving a speckled image of the diffraction spot. Developed in France by Antoine Labeyrie, this method has been applied to Betelgeuse, Aldebaran, Sirius and Vega, among others.

The largest stars, called supergiants, have a diameter that may exceed 1,000 times that of the Sun. Thus the largest known star, a red supergiant in the constellation of Auriga, has 2,700 times the diameter of the Sun and therefore is 20 million times more voluminous. If it were placed in the Sun's position, it would encircle all planets out to Saturn.

On the other hand, there are stars much smaller than the Sun. Thus, white dwarfs have planetary dimensions: In Canus Major, the companion of Sirius is one of the largest white dwarfs but measures only 10,400 kilometers (6,464 miles) in diameter—making it slightly smaller than the Earth. But there exist even smaller stars, neutron stars, which manifest themselves as pulsars (see chapter following page 258); their diameter is only 20 to 30 kilometers (12.4 to 18.6 miles).

Comparative dimensions of stars

The Largest Stars

Name	Constellation	Diameter (expressed in solar diameters)
ε Aurigae	Auriga	2,700
VV Cephei	Cepheus	1,200
Rosalgheti	Hercules	800
Antares	Scorpio	500
Mira Ceti	Cetus	460
Betelgeuse	Orion	400

Stellar Masses

Stellar masses can be determined directly only for the components of binary systems (see box, page 244), and then only in the most favorable cases. Visual binaries—i.e., those resolvable in a telescope—have orbits that can be directly observed; they trace ellipses against the background of distant stars. If we know the linear dimensions of the orbits and the individual motions of the two components, we can deduce the masses of the two stars using Kepler's third law. In practice this method could be applied only to a few dozen binary systems. If a piece of information is missing (for example, if one of the stars is not bright enough to be observed), we can then determine the ratio and, ultimately, the sum of the masses of the components but not their individual mass.

Spectroscopic binary stars cannot be resolved with a telescope, but the wavelength shift by the Doppler effect of the lines of their spectra reveals

Hipparcos: First Satellite of Fundamental Astronomy

Postional astronomy should make great strides in a few years as a result of the *HIPPARCOS* satellite (its name is an acronym for *h*igh-*p*recision *par*allax *co*llecting *s*atellite, but above all evokes the name of the famous Greek astronomer Hipparchus, considered to be the founder of positional astronomy). This satellite, which was created by a decision of the European Space Agency in 1980, will be placed by the European rocket Ariane in a geostationary orbit in about mid-1986 and should remain in operation for at least 2½ years. It will scan the entire sky continuously and systematically, using a telescope that can measure with great precision the angular distance separating two stars in fields about 70 degrees apart. At the conclusion of the mission, the digital reduction of several millions of such measurements will produce, for each of the 100,000-odd stars in the observation schedule, a determination of its positional coordinates (right ascension and declination) within merely 0.002 arc second of its trigonometric parallax, and of the two components of its proper motion. This will represent major progress: At present, stellar positions in the best of cases are known only to within 0.04 arc second; the parallaxes of the brightest stars have been determined only to within 0.010 arc second and in the measurement of proper motions traditional methods would yield a precision comparable to what is hoped for with the satellite only after 50 years of observation, using the best data available. Moreover, the systematic errors will be less than 0.001 arc second, while now they reach 0.005 arc second for parallaxes and 0.04 arc second for stellar positions.

All the fields of astronomy and astrophysics will be affected, directly or indirectly, by this influx of homogenous and very precise data—in particular, determination of the distance scale in the universe, which makes it possible, through successive calibrations, to go from the stars close to the Sun to the most distant stars in our galaxy, then to the brightest stars of the galaxies close to our own, then to the more distant galaxies, where we cannot identify the stars individually; as well as all the studies of galactic structure and evolution (distribution of the different types of stars, correlation between that distribution and the average age of these types of stars, location of the spiral arms, etc.) and various problems pertaining to stellar physics (tests of theories of internal structure and evolution of the stars, determination of stellar masses and diameters, calibration of the mass-luminosity and period-luminosity relationships, etc.).

Distance to the Stars

People have known since antiquity that the stars were much farther away than the Sun, Moon or planets. After the invention of optical instruments in the 17th century, when it became obvious that the stars were not attached to a sphere but that the universe had depth, the problem of determining stellar distances was brought sharply into focus. It was not until 1838 that the German Friedrich Bessel was able to measure the first stellar distance, that of the star 61 Cygni, by the method of trigonometric parallax. This method is analogous to the one used by surveyors on the Earth's surface to determine the distance of an inaccessible point. It consists of a triangulation, using the diameter of the Earth's orbit as a base. Because of the Earth's motion around the Sun, the closest stars trace a small ellipse on the sky in relation to the distant stars. In one year, the apparent position of a star varies by an angle 2p, of which half, p, is defined as the star's parallax. In principle, it is sufficient to observe the star at a six-month interval—i.e., knowing the radius of the Earth's orbit, from two opposite positions on the Earth's orbit—in order to determine its parallax. In fact, parallaxes are very small angles, always less than 1° (the angle at which one sees a thickness of 1 millimeter from a distance of 200 meters), and determining them requires observations spread out over several years. In the past, observations were visual. Today photography is used, and measurements have become 100 times more accurate.

The stars closest to the solar system. Most are small, dim stars; only Sirius, Procyon and Alpha Centauri are more luminous than the Sun.

Principle of measuring the distance of a star by the trigonometric parallax method.

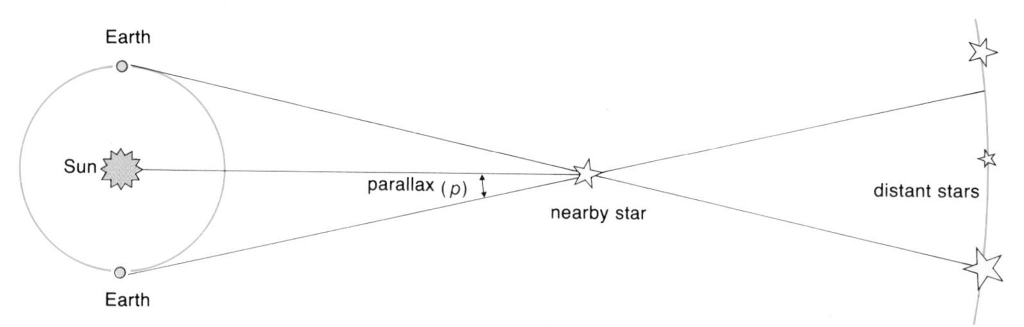

This method is at the origin of the basic astronomical unit of distance known as the parsec (see box, page 238).

The parallax found by Bessel for the star 61 Cygni was 0.35 arc second; according to modern measurements, it is 0.29 arc second, implying a distance of 11.1 light-years for the star. Shortly after Bessel's announcement of his result, two other stellar parallaxes were found by the same method: that of the star α Centauri by the British astronomer T. Henderson, at the Cape of Good Hope, and that of the star Vega by the Russian astronomer W. Struve at Dorpat, Estonia.

We now know that there are about 40 stars less than 15 light-years away, and only 11 at less than 10 light-years. The closest, Proxima Centauri, is 4.22 light-years away (see box, page 243).

The trigonometric parallax method has been applied to some 10,000 stars having distances of up to 40 parsecs. Beyond that, results are too imprecise, and it is necessary to resort to other methods.

Some groups of stars have a common motion and make up what is called a moving group. The motions of all of the stars belonging to such a group are roughly parallel, but as a result of perspective, their motions seem to converge on a point in the sky. From the measurement of veloc-

The 20 Closest Stars

Name	Constellation	Distance (in light years)	Apparent Visual Magnitude m_v	Absolute Visual Magnitude m_v	Spectral Type	Mass (relative to the Sun)	Radius (relative to the Sun)
Proxima Centauri	Centaurus	4.22	11.05	15.45	M5	0.1	
α *Centauri A*	Centaurus	4.35	−0.01	4.3	G2	1.1	1.23
α *Centauri B*	Centaurus	4.35	1.33	5.69	K5	0.89	0.87
Bernard's star	Ophiucus	5.90	9.54	13.25	M5		
Wolf 359	Leo	7.60	13.53	16.68	M8		
Lalande 21185	Ursa Major	8.12	7.50	10.49	M2	0.35	
Sirius A	Ursa Major	8.64	−1.45	1.41	A1	2.31	1.8
Sirius B	Ursa Major	8.64	8.68	11.56	wd*	0.98	0.022
UV *Ceti A*	Cetus	8.87	12.45	15.27	M5	0.044	
UV *Ceti B*	Cetus	8.87	12.95	15.8	M6	0.035	
Ross 154	Sagittarius	9.45	10.6	13.3	M4		
Ross 248	Andromeda	10.27	12.29	14.8	M6		
ϵ *Eridani*	Eridanus	10.76	3.73	6.13	K2		0.98
Luyten 789-6	Aquarius	10.76	12.18	14.60	M7		
Ross 128	Virgo	10.82	11.10	13.50	M5		
61 *Cygni* A	Cygnus	11.08	5.22	7.58	K5	0.63	
61 *Cygni* B	Cygnus	11.08	6.03	8.39	K7		
ϵ *Indi*	Indus	11.21	4.68	7.00	K5		
Procyon A	Canis Major	11.41	0.35	2.65	F5	1.77	1.7
Procyon B	Canis Major	11.41	10.7	13.0	wd*	0.63	0.01

wd = white dwarf

Some proper motions, such as that of Sirius, display a sinusoidal character, betraying the existence around the star in question, of a massive satellite that is too dim to be observed directly. The measurement of proper motions also makes it possible to discover star streams—i.e., groups of stars moving together in space—and to determine their distance (see following material).

A star's proper motion reflects only one of the components of its overall motion: that which is perpendicular to the direction of observation. A star that moves in space along the line of sight (either approaching or receding from the Sun) has a zero proper motion and appears immobile in the sky.

To know a star's real motion, it is essential, therefore, also to know its component along the direction of observation. This velocity component, known as the radial velocity, may be determined by measuring the wavelength shift, through the Doppler-Fizeau effect, of its spectral lines toward the blue or the red. The stars moving away from us have a positive radial velocity; those approaching us have a negative radial velocity.

By combining information about a star's distance with its proper motion, we get the value of its tangential velocity—i.e., the value of the component of its velocity projected on the plane perpendicular to the line of sight. Knowing its radial velocity, we can then determine its space velocity relative to the Sun, which reflects its true motion through space. In the Sun's vicinity most of the stars have a space velocity of less than 50 kilometers (31 miles) per second. The Sun itself moves in relation to the stars. Discovered by Herschel in 1783, this motion takes place, at a speed of 20 kilometers (12.4 miles) per second, toward the direction of a point called the apex, which is between stars ν and o of the constellation Hercules.

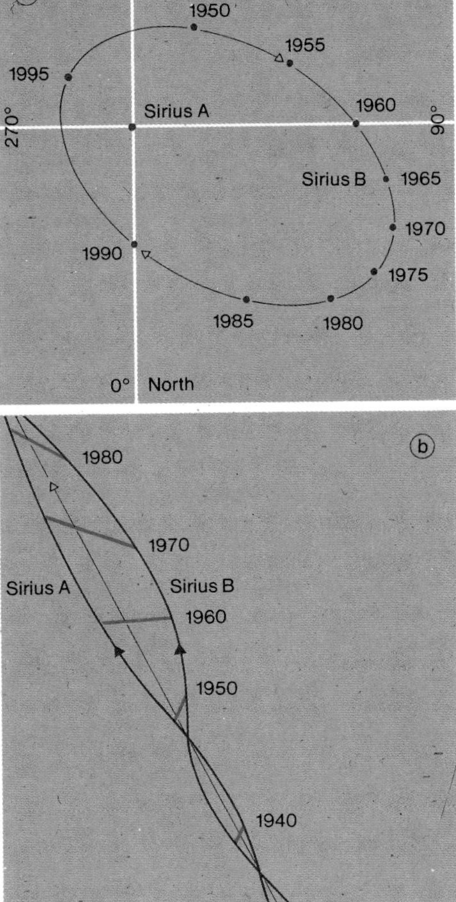

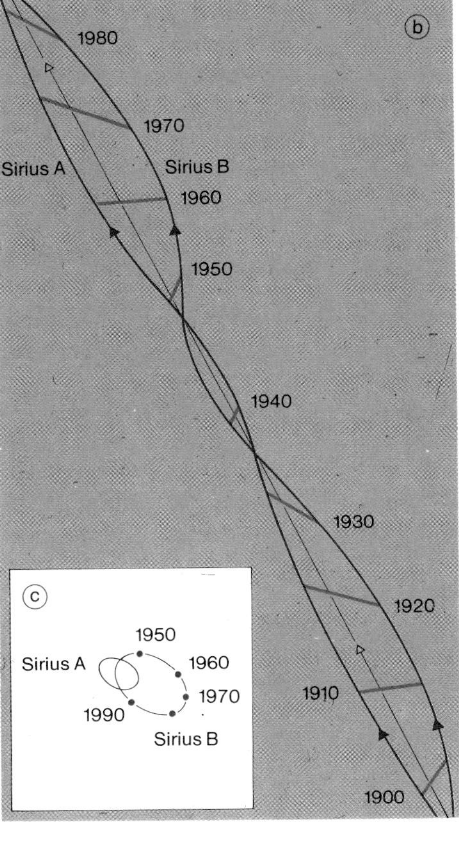

Motion of Sirius and its companion. In 1834, the German Friedrich W. Bessel observed irregularities in the proper motion of Sirius and attributed them to the presence of a low-brightness companion (the two components orbiting around their center of gravity and forming what is called a binary system, or double star). Discovered in 1862 by the American Alvan Clark, this companion (Sirius B) is a white dwarf 10,000 times less luminous than the primary component and with a radius estimated at only 16,000 kilometers (9,944 miles), with a mass equal to that of the Sun.
a. apparent orbit of Sirius B around Sirius A (primary component)
b. proper motions of Sirius A and Sirius B in the sky (the trajectory of the center of gravity is a straight line)
c. apparent orbits of Sirius A and Sirius B around their center of gravity G (projection onto the sky of the real orbits inclined 43 degrees)

Components of a star's motion in space.
S_r = radial speed
S_t = tangential speed
S = spatial speed
μ = specific motion of the star E
Knowing that $S^2 = S_r^2 + S_t^2$, one can find S by determining S_r (by measuring the shift in the star's spectral lines by Doppler-Fizeau effect) and S_t (linked to two measurable dimensions—the parallax and the motion itself).

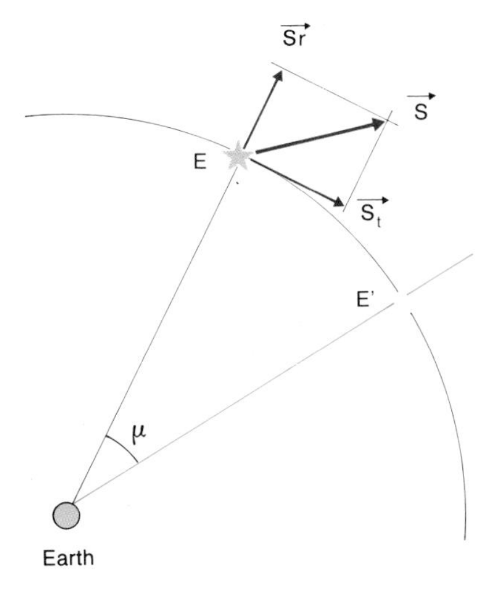

The 20 Brightest Stars

Name: Common	Name: Official	Constellation	Apparent Visual Magnitude m_v	Absolute Visual Magnitude M_v	Spectral Class	Distance (in light years)
Sirius	α CMa	Canis Major	− 1.45	+ 1.41	A1	8.64
Canopus	α Car	Carina	− 0.73	+ 0.16	F0	190
Rigel Centarus	α Cen	Centaurus	− 0.10	+ 4.3	G2	4.37
Arcturus	α Boo	Centaurus	− 0.06	− 0.2	K2	36
Vega	α Lyr	Bootes	+ 0.04	+ 0.5	A0	26.5
Capella	α Aur	Lyra	+ 0.08	− 0.6	G8	45
Rigel	β Ori	Charioteer	+ 0.11	− 7.0	B8	660
Procyon	α CMi	Auriga	+ 0.35	+ 2.65	F5	11.41
Achernar	α Eri	Canis Minor	+ 0.48	− 2.2	B5	130
Agena	β Cen	Eridanus	+ 0.60	− 5.0	B1	390
Altair	α Aql	Aquila	+ 0.77	+ 2.3	A7	16.1
Betelgeuse	α Ori	Taurus	+ 0.80*	− 6.0*	M2	650
Aldebaran	α Tau	Crux	+ 0.85	− 0.7	K5	68
Acrux	α Cru	Virgo	+ 0.9	− 3.5	B2	260
Spike	α Vir	Scorpius	+ 0.96	− 3.4	B1	260
Antares	α Sco	Gemini	+ 1.0	− 4.7	M1	425
Pollux	β Gem	Piscis	+ 1.15	+ 0.95	A0	36
Fomalhaut	α PsA	Austrinus	+ 1.16	+ 0.08	A3	23
Deneb	α Cyg	Cygnus	+ 1.25	− 7.3	A2	1,600
Mimosa	β Cru	Crux	+ 1.26	− 4.7	B0	490

*average value (variable star)

nm, 1.25μm, 2.2μm, 3.4μm, 5μm and 10.2μm, the last five wavelengths having been chosen to coincide with the atmosphere's narrow windows of transparency to celestial infrared radiation.

Whatever the method used and spectral range involved, the measurements must be corrected for absorption due to interstellar matter.

The difference between the apparent magnitudes of a single body measured at two different wavelengths is a parameter for objectively characterizing the body's color: It is called the color index.

The color index traditionally adopted is the excess of the photographic magnitude over the visual magnitude. It ranges from -0.3 for the whitest and hottest stars (O and B spectral types, surface temperature greater than or equal to 20,000° K) to $+1.6$ for the reddest and coldest (K and M spectral types, surface temperature near 3,000° K). By convention, the zero of this scale corresponds to the star Vega (spectral type A, surface temperature 11,000° K).

Other indices also are used, based on a different or a more complex system of wavelengths. Thus, in the UBV system, the indices U-B and B-V generally are considered.

A body's apparent magnitude depends not only on its real brightness but also on its distance. To compare the brightness of bodies, a conventional magnitude has been defined, absolute magnitude: This, for a given body, corresponds to the apparent magnitude it would have if it were situated at 10 parsecs (or 32.6 ly). The effect of distance then is clearly revealed. Thus the star Sirius, which seems to be the brightest in the sky with an apparent magnitude of -1.45, is in fact substantially less bright than Rigel (apparent magnitude 0.11), their absolute magnitudes being, respectively, $+1.4$ and -7.0. The intrinsic brightnesses of stars may vary considerably, their absolute magnitudes ranging from -9 for the brightest (800,000 times brighter than the Sun) to $+17$ for the dimmest (100,000 times less bright than the Sun).

Knowing a star's apparent and intrinsic brightnesses, one can calculate its distance, since the brightness falls as the square of distance. Given a body of absolute magnitude M, apparent magnitude m and distance d, expressed in parsecs, there exists this relationship: $M - m = 5 - 5 \log d$. This formula makes it possible either to calculate the absolute magnitude of a star for which the distance has been directly determined, or to estimate the distance of a body for which the absolute magnitude has been estimated. It plays a fundamental role in determining the distances of remote bodies. The quantity $m - M$ is called the distance modulus.

Proper Motion; Radial Velocity

For intervals of time limited to a few centuries, the figures that the constellations form in the sky seem immutable, but an observer who lived several tens of thousands of years would realize that they were gradually changing shape. Indeed, the stars composing them are slowly shifting in relation to one another. It was the British astronomer Edmund Halley who first discovered, in 1718, this shifting of the stars on the celestial sphere, called proper motion, by noticing that the bright stars Sirius, Arcturus, Procyon and Aldebaran occupied a slightly different position from the ones listed in Ptolemy's catalogue, compiled 16 centuries earlier.

Since then the proper motions of more than 300,000 stars have been measured: The measurements are made either from observations with a meridian transit instrument, or by comparing photographs taken at different epochs. The star with the largest proper motion is Barnard's star in the constellation of Ophiuchus (see box, page 316), which shifts by 10.31″ per year, and thus, in 180 years, covers a distance in the sky equivalent to the full Moon's apparent diameter; next comes Kapteyn's star, in the southern constellation of the Painter, with an annual displacement of 8.76 arc seconds. These two stars are invisible to the naked eye. Most stars, owing to their distance, have an annual proper motion of less than 1 arc second.

100,000 years ago

today

100,000 years from now

The shape of Ursa Major at different epochs. From top to bottom: 100,000 years ago; today; 100,000 years from now. The arrows show the direction in which the proper motion of each of the stars takes place.

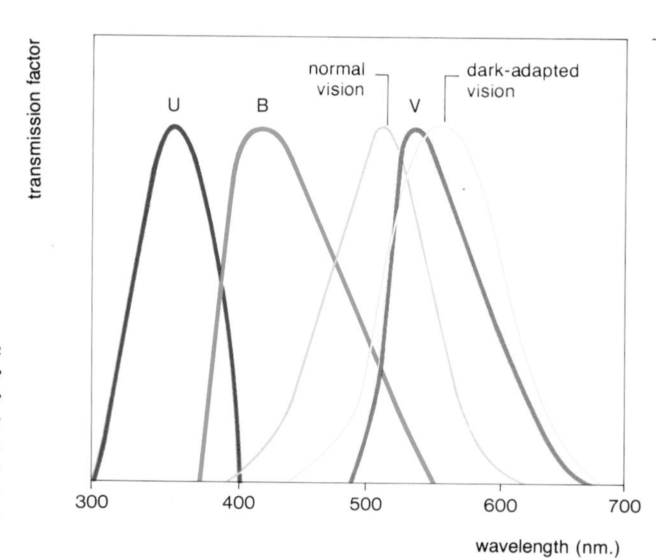

Response curves of the filters used in the UBV photometric system. These curves show how the intensity of the light transmitted by the filters varies according to wavelength. In comparison, the eye's response curve for normal vision and that of the eye for dark-adapted vision also are shown.

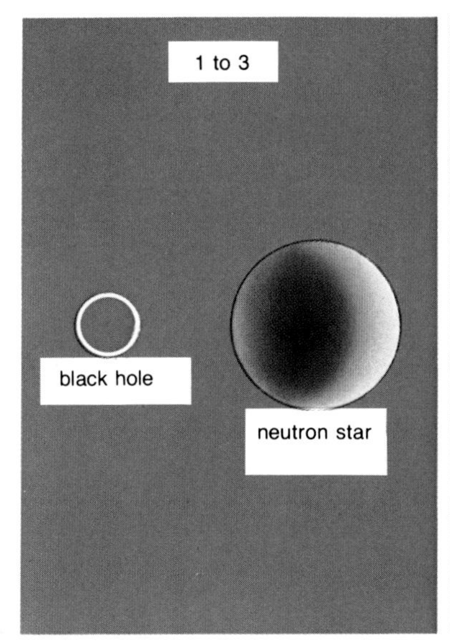

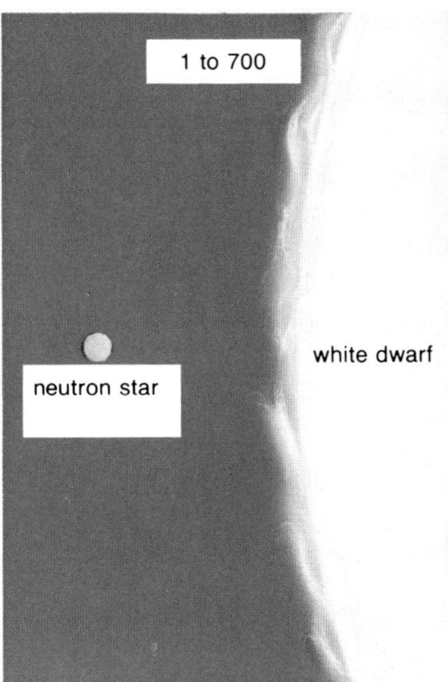

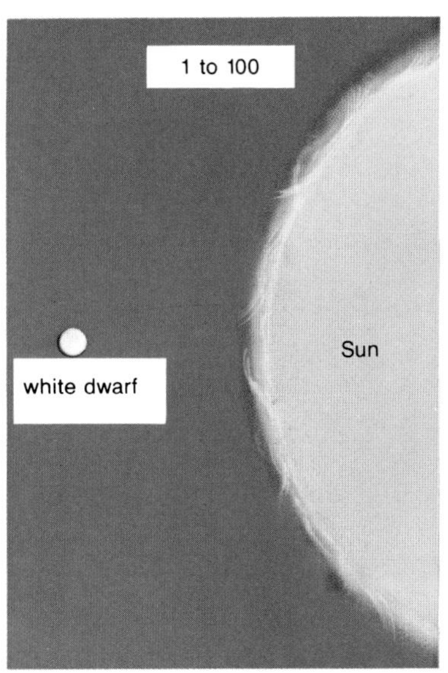

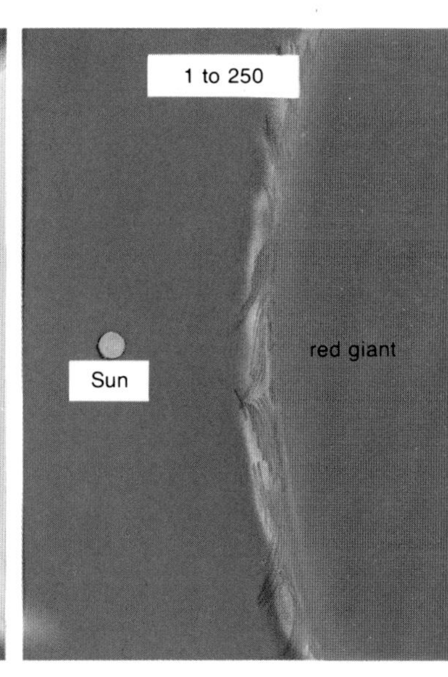

their orbital motion. In general, the inclination of the orbital plane relative to the line of sight is unknown, and we can determine only a lower limit to their masses. However, for the binaries with eclipses, the orbital plane approximately coincides with the line of sight, and the masses of the two components can be determined with fair accuracy.

A third method has been applied to a few white dwarfs, but it cannot be used for ordinary stars. The white dwarfs are so dense that their spectral lines undergo a red shift of gravitational origin. Relative to its usual position in the spectrum, each line suffers a shift in wavelength $\Delta\lambda$ such that: $\frac{\Delta\lambda}{\lambda} = \frac{GM}{Rc^2}$, where λ denotes the line's usual wavelength, M the star's mass, R its radius, G the gravitational constant and c the speed of light. Thus we can determine the mass of a white dwarf when we are able to estimate its radius from its luminosity.

Some white dwarfs turn out to be components of close binary systems (for instance, Sirius B and Procyon B), so that their mass can also be determined by the method applied to double stars.

The results obtained show that the masses of white dwarfs range from 0.5 to 1.4 times the Sun's mass ($M_\odot$).

With the preceding methods, we have been able to determine the masses of only about 200 stars until now. However, these constitute a representative sample of all stellar types, from giants to white dwarfs. The values obtained are spread over a relatively narrow range, from 0.05 to 50 $M_\odot$, approximately.

There also exists an indirect method of determining stellar masses. In 1924, the British astrophysicist Eddington showed that a star's luminosity varies according to its mass, the least luminous stars being the least massive and the most luminous ones the most massive. There are, in fact, different relationships according to the star's spectral type: For stars having a mass in the range between 0.3 and 20 times that of the Sun, the luminosity varies approximately as the fourth power of the mass. This mass-luminosity relationship makes it possible to calculate a star's mass based on its absolute magnitude. White dwarfs and red giants do not, however, conform to this rule.

Star Densities

Once the radius and mass of a star are known, determining its average density is merely a matter of calculation. The densities of stars on the main sequence of the Hertzsprung-Russell diagram (see "Structure and Evolution of the Stars," page 246) range from 0.05 to 4 times that of water. Remember that the Sun, which belongs in that category, has an average density of 1.4. The giant stars are much less dense; their average densities, ranging from 0.03 to 0.00003, are comparable to that of gases on Earth. Antares, a red giant with a diameter up to 500 times that of the Sun, has an average density of almost 0.0002. The supergiants have even lower densities. The density of a star like Betelgeuse is only one-ten-millionth that of water, which corresponds to the density prevailing in a relatively high vacuum, where the pressure does not exceed 0.2 millimeter of mercury.

On the other hand, there are some extremely dense stars: white dwarfs and neutron stars. At the center of a white dwarf, 1 cm³ may weigh up to 100 tons; at the center of a neutron star, up to 100 million tons. The existence of even denser objects, black holes (see chapter following page 000), is strongly suspected.

Temperature and Chemical Composition of the Stars

Observation shows that stars have a variety of colors: Sirius is white, Vega bluish, Aldebaran and

The Closest Stellar System: Alpha Centauri

Our closest stellar neighbor, the Alpha Centauri (or α Centauri) system, consists of three stars. The A and B components make up the lovely visual double star α Centauri strictly speaking, the double nature of which was discovered in 1689 by the astronomer Richard in Pondichéry. The C component is a red dwarf of magnitude 11, situated 2°11′ from the A-B pair and having an analogous proper motion. The astronomer Innes, having found a parallax for star C slightly greater than that of α Centauri, suggested calling it Proxima Centauri.

In 1978, K. W. Kamper and A. J. Wesselink published a new study of this famous triple system, based on photographs taken from 1897 to 1971, all prepared in a uniform manner. The parallaxes obtained are as follows:

α Centauri: 0.750″ (to within 0.005″)
Proxima: 0.772″ (to within 0.004″)

This results in a distance of 1.295 parsecs (4.22 l.y.) for Proxima Centauri and 1.333 parsecs (4.35 l.y.) for the A-B pair. The Kamper and Wesselink study confirms, moreover, the nearly identical proper motions of Alpha Centauri and Proxima Centauri: 3.692″ and 3.847″ per year, respectively.

The masses of the A and B components are, respectively, 1.10 and 0.91 times the Sun's mass. The mass of Proxima Centauri, estimated according to its luminosity, is 0.12 times that of the Sun. The distance between the A-B couple and Proxima Centauri, projected on the sky, is 10,500 astronomical units (a.u.); the real distance must be about 13,000 a.u., or some 430 times the Sun's distance from Neptune.

Antares reddish. This is related to the fact that their surface temperatures are different. Just as a heated iron bar goes from dark red to orange, yellow and finally to white as the temperature increases, a white star is hotter than a yellow star, which is hotter than a red star. In fact, a star's color is the result of a mixture of luminous radiations that can be broken down by a prism into a spectrum. The study of stellar spectra is the basis of astrophysics. Among other things, it makes it possible to know what physical conditions exist in the outer layers of stars as well as the chemical elements present and their relative abundances.

These concepts will be more fully developed in the following chapter.

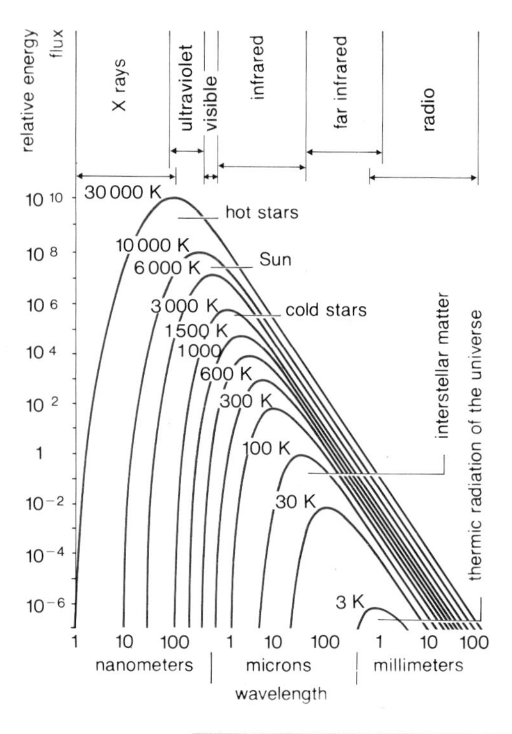

Each of the curves in this diagram indicates how the energy emitted by a black body at a given temperature is distributed over the electromagnetic spectrum. It reflects approximately the distribution of the energy emitted by a star of the same surface temperature. The hotter the star, the more energy is emitted at short wavelengths.

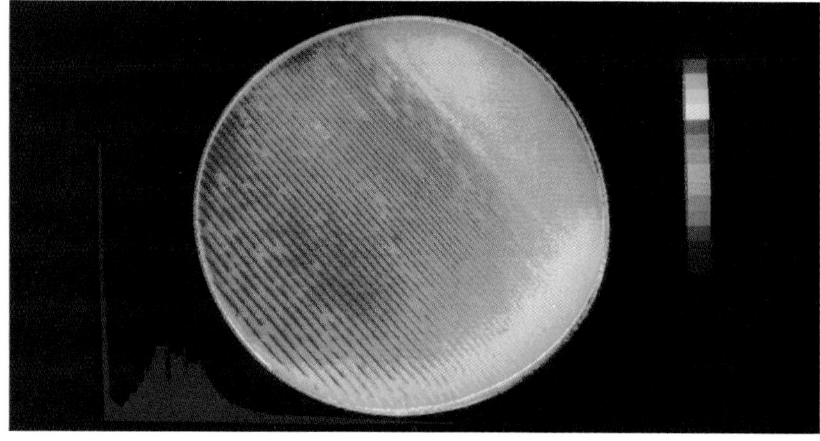

A high-dispersion spectrum in the ultraviolet (between 1,150 Å and 2,000 Å wavelength) of a hot star, taken by the IUE satellite. The colored bar at right represents the scale of luminous intensities. Photo: J. M. Chourgnoz-ESA.

Double Stars

When two stars appear extremely close in the sky, they form a double star.

Sometimes their proximity results simply from the effect of perspective, the components actually being at very different distances from the Earth; in that case, we speak of optical doubles. But, more often, especially when doubling turns out to be difficult to verify, the components are linked by mutual gravitation, and each one orbits around the system's center of gravity, forming a physical double star (or binary). When the components can be separated by visual observation in a telescope, we speak of visual double stars. Binaries that can be separated by an interferometer also are placed in that category.

Other binaries, called spectroscopic, are revealed by a periodic doubling of their spectral lines, owing to the Doppler-Fizeau effect. Their components are much closer together than those of the visual binaries, and as a result of their mutual gravitation, they can exchange matter (in the case where they have very different masses, there is a transfer of matter from the less massive to the more massive star).

A third category of double stars, called photometric or eclipsing binaries, reveal themselves through a periodic variation in brightness owing to the mutual occultations of the two components passing alternately behind one another, each one then being more or less completely hidden, depending on the inclination of the orbit and the diameters of the two stars.

Finally, certain stars have an invisible companion, either of a planetary nature, or simply not bright enough to be visible. Each of the components then traces a Keplerian orbit around the system's center of mass, and the perturbed motion of the primary star may reveal the presence of the hidden companion. As this involves the result of positional measurements, such pairs are described as astrometric.

Origin of Double Stars

The origin of double stars remains mysterious, for we have never seen a pair form, and the variety of their appearances cannot—as in the case of single stars—be interpreted as representing an evolution. Four basic hypotheses have been forwarded on this point.

First is the capture hypothesis. However, a star passing close to another can be captured only if it is somewhat slowed down by the medium; otherwise, the result is a hyperbolic, not a closed, orbit. In any event, it is difficult to explain by this hypothesis the close pairs and only slightly eccentric orbits, which are not at all a minority among the known pairs.

The second hypothesis involves the fissioning of a rapidly rotating star. This idea runs up against mechanical objections and could generate neither widely separated pairs nor multiple systems. However, a category of binaries—of the W Ursa Major type—might be accounted for by this hypothesis, and various studies since 1960 have examined this question. There again, fissioning could not be a general origin of double stars.

The third hypothesis considers an interstellar cloud in its initial state, where two neighboring nuclei might condense separately. This is the only hypothesis that might account for the formation of multiple systems, and there are no fundamental theoretical objections to it. On the other hand, it does not explain the closer couples well, for we cannot imagine two nuclei capable of forming stars if they are very close at the outset. For that it is necessary to envision a process that would enable the two stars to come together once they are formed (contrary to what happens with fissioning), and none has been found.

A fourth theory that has been proposed recently is a kind of extension of the previous one. It states that stars in a cluster might combine into several systems containing a few dozen objects, each of which would be unstable and ultimately would form a multiple star centered around a close binary. Certain components then would escape from the system.

In fact, no hypothesis is satisfactory by itself. In addition, all the studies that try to predict the distribution of semi major axes, eccentricities, double stars by type, etc., starting from a chosen hypothesis, and then to compare the result with reality, run up against a major obstacle: The gap between our data and the facts as a result of the various, very important selection effects in data.

21. Structure and Evolution of the Stars

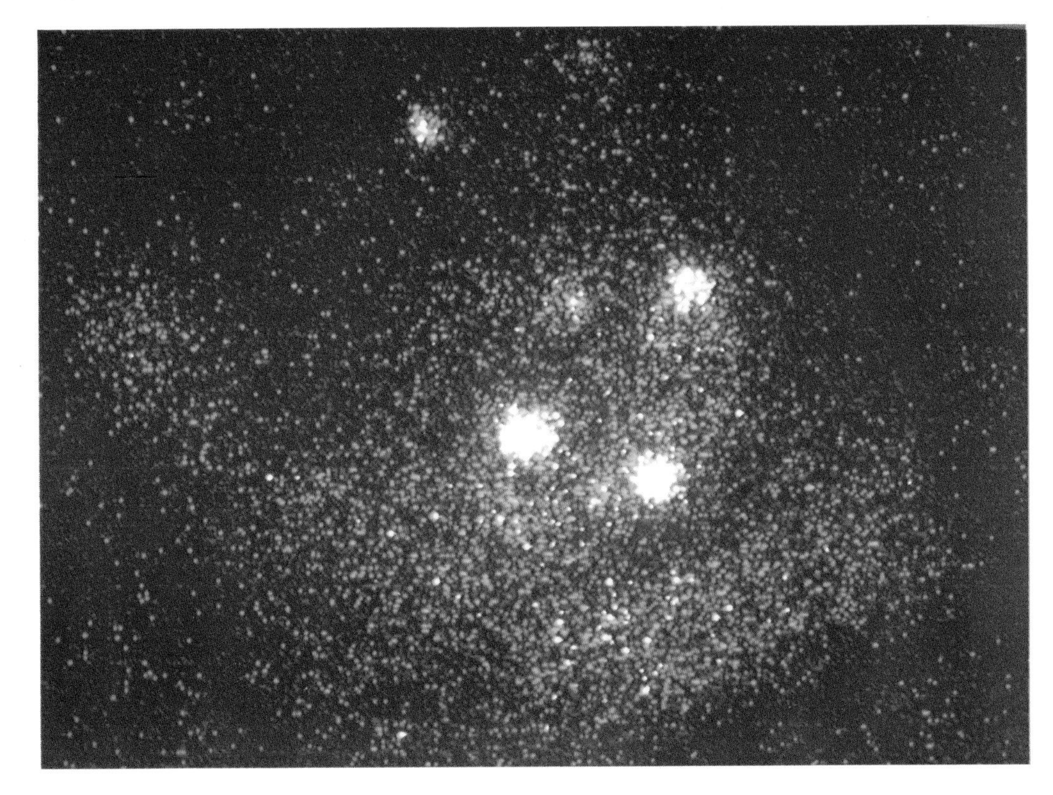

Taken by the American satellite HEAO 2 *in the X-ray range, this picture shows, within the* η *Carinae nebula, several intense X-ray sources corresponding to young stars. This discovery was a surprise to astronomers, who did not know that at this stage in their development, stars can be X-ray emitters. P.P.P.-IPS Photo.*

It's a demonstrated fact: The stars are changing, evolving. In this chapter we will describe the successive phases in the life of a star like the Sun, from its formation to its death.

We will discuss phenomena pertaining both to its internal structure and to its external appearance, particularly when seen from Earth. Then, for each evolutionary phase, we will show the effect of the star's mass on its evolution and, more briefly, the effect of its chemical composition.

This will lead us to present the so-called Hertzsprung-Russell diagram, which is a remarkable synthesis of all the evolutionary phases of all types of stars.

Birth of a Star

THE AGE OF THE SOLAR system is about 4.6 billion years. The universe is much older; it was formed probably about 10 billion to 12 billion years ago. The Sun, then, was born well after the origin of the universe. Most of the stars in the Milky Way are in this category.

Stars are formed from clouds of interstellar matter contracting inward on themselves. Let us consider such a cloud in its prestellar phase—

i.e., shortly before it becomes a star. Its diameter is about the same as that of the solar system; it is sufficiently dense to be opaque. The cloud contracts because of its own mass, and the gravitational energy recovered in the contraction is converted into heat—that is how the cloud's temperature gets to be a few hundred degrees absolute (or Kelvins, K), which still is low for a future star.

Having reached this stage of contraction, the cloud suffers a catastrophic event: The temperature becomes high enough so that part of the contraction energy no longer serves to "heat up" the cloud but to ionize certain atoms and dissociate the cloud's molecules (particularly H_2). The molecules previously absorbed much of the radiated energy; their disappearance makes the cloud much less opaque, and then it radiates much more energy toward the exterior; it becomes luminous, it is transformed into a star. But this energy loss creates a net imbalance between the gravitational force that produces the contraction and the gaseous pressure that normally restrains it to a considerable extent: The cloud collapses on itself, virtually at the speed of free fall, a collapse that has the effect of raising the gaseous pressure. This catastrophe ends when the gravitational force and the gaseous pressure are once again in equilibrium. It lasts about six months. When it ends, the star has a diameter about 100 times greater than that of our Sun—, i.e., it fills Mercury's orbit. The temperature at its surface is nearly 4,000° K, and it radiates 100 times more energy than the Sun.

Since it lasts such a short time, this phenomenon has a very low probability of being observed. However, by chance, the sudden appearance of a new star was observed in 1936: It was FU Orionis in the huge cloud of gas and dust making up the Orion nebula. The characteristics of this star are those of the category of T Tauri variables, named after the first of such stars to have been studied in detail by Joy in 1942.

The most obvious of these characteristics for the observer is variability. The magnitude of the T Tauri stars varies in a wholly irregular manner. First, there is no apparent period; the fluctuations may occur over a few hours or be nonexistent for years. Second, the amplitude of these brightness variations varies in an unpredictable way (but can reach several magnitudes). The origin of these fluctuations is not really known at present. They might be due to the appearance and disappearance of active eruptive centers carried by the star's own rotational motion, which probably is very rapid.

The other characteristics of T Tauri stars are deduced from examination and study of their spectra. A T Tauri spectrum is immediately recognizable by its very numerous and very intense emission lines. It resembles a spectrum of the solar chromosphere. The lines are greatly broadened, although it is not yet possible to know whether the broadening results from the Doppler-Fizeau effect due to the star's rotation, or if it is due to large-scale motions in the stellar atmosphere. In addition, these lines are systematically shifted toward the red—a sign that the emitting matter is collapsing onto the star.

Finally, the spectra of T Tauri stars show intense lithium lines, indicating that that element is more than 1,000 times more abundant there than in the Sun. We will see that this seemingly innocuous detail is, in fact, one of the few pieces of information we possess regarding the internal structure of these stars.

Once the protostellar cloud is fully collapsed, the new star thus born continues to contract more slowly. At about 4,000° K, the opacity of the star's atmosphere begins to increase rapidly with temperature—the luminosity decreases strongly, at a nearly constant temperature. Simultaneously, the contraction makes the central temperature increase, the transfer of thermal energy from the center to the surface being achieved by convection. When the central temperature reaches the critical threshold where the thermonuclear reactions of hydrogen fusion can begin, the flux of energy thus produced balances the gravitational force: Temperature and luminosity increase slightly until the star reaches the most stable and longest phase of its existence, that of a dwarf star. It took the Sun 50 million years to go through the evolutionary sequence just described.

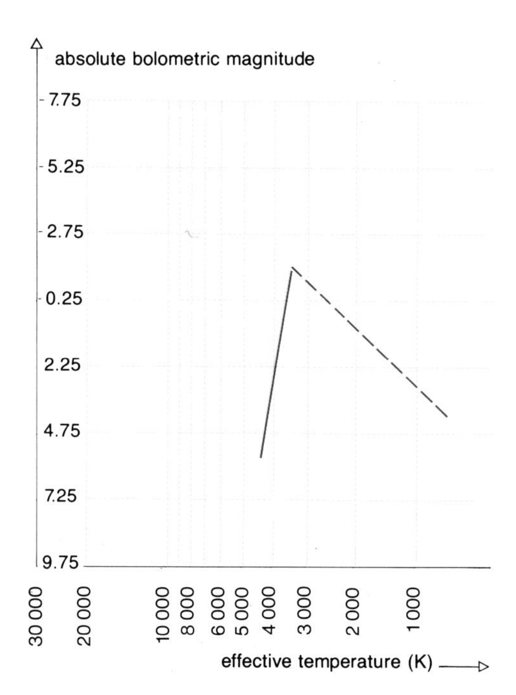

Cold phase of the collapse of the protostar in the Hertzsprung-Russell diagram: absolute bolometric magnitude vs. logarithm of temperature. The rapid increase in the opacity of the atmosphere of the star makes the luminosity drop. On the facing page, the contraction slowly continues, increasing luminosity and temperature until thermonuclear reactions begin at the center of the star. The thickness of the lines qualitatively indicates the length of the evolution of stars of different masses: The thicker a line is, the slower the evolution.

The formation of a star:
a. A cloud of interstellar matter contracts under the effect of its own mass.
b. The cloud heats up, toward 1,000° to 1,500°K.
c. While the temperature still is rising, the cloud's molecules break up and therefore do not absorb any more radiation; the cloud radiates a great deal and collapses.
d. The gas pressure equilibrates the gravitational force.

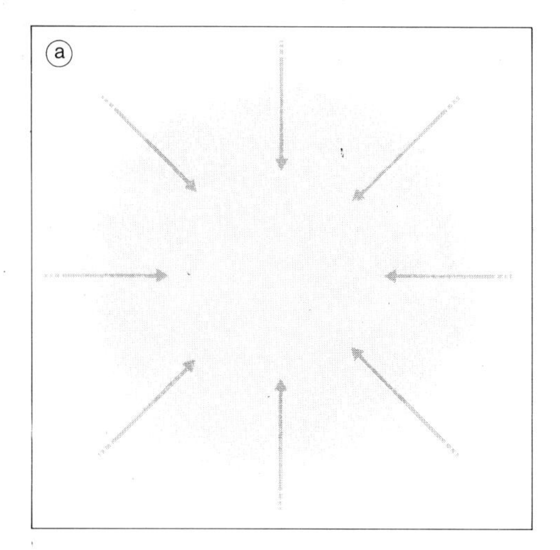

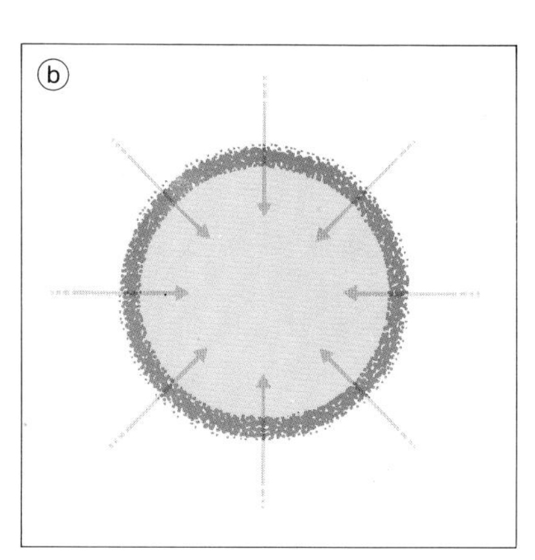

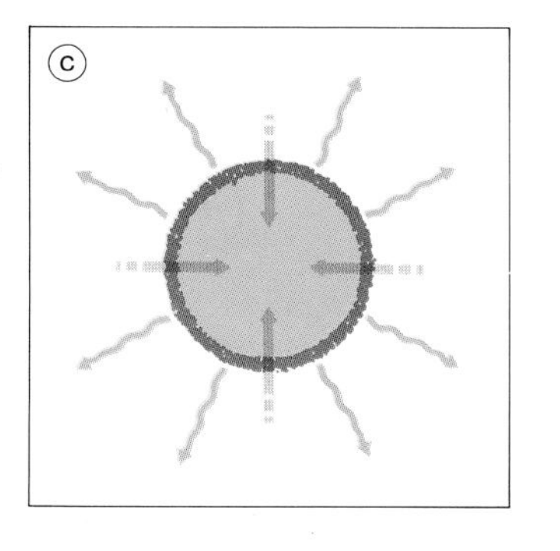

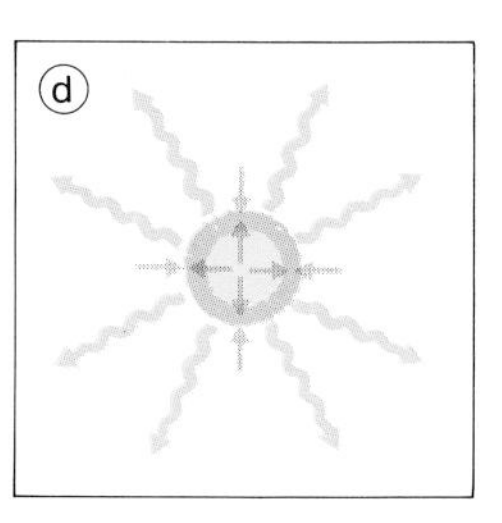

scale factor: (a) = 200 (b) = 30 (c) = 10 (d) = 1

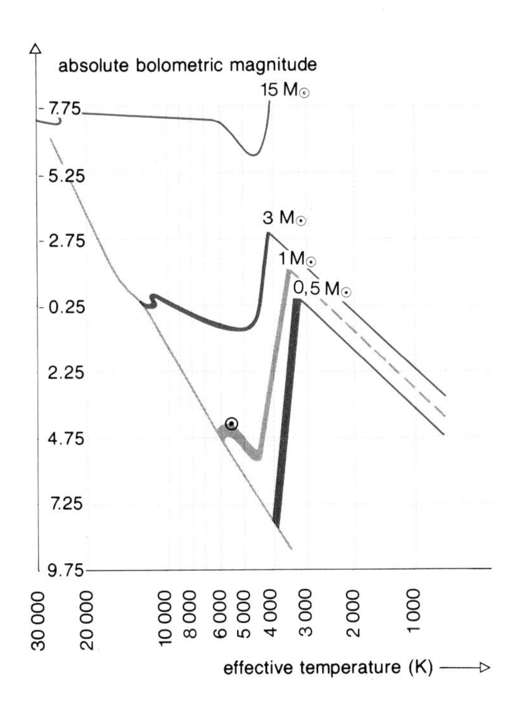

The Main Sequence

WHAT HAPPENS TO PROTOSTELLAR clouds with masses different from the Sun's? The speed of a star's formation depends on the initial mass of the interstellar material of which it will be constituted.

The more massive the cloud of interstellar matter, the slower the contraction that culminates in a protostar; the most massive stars are the last ones formed in a star cluster! On the other hand, the final phase of contraction—the collapse of the protostar under its own weight—more rapid as the stellar mass gets larger. The same goes for the continuing contraction after the collapse, up to the ignition of the hydrogen at the center: a star 10 times more massive than the Sun (i.e., having 10 solar masses) will no longer take 50 million years, like the Sun, but only 100,000 years to go through those two phases of evolution, whereas a star of 0.5 solar mass will require 100 million years.

Moreover, at each point in this evolution, a massive star always will be more luminous than a less massive star. Also, during the period leading up to the ignition of hydrogen, a massive star is much hotter than a low-mass star: The massive dwarf stars are hot (up to 20,000° to 50,000° K at the surface); the low-mass stars are "cold" (2,000° to 3,000° K).

The group of stars burning hydrogen at their center make up the main sequence: They form a strip in the "bolometric-magnitude temperature" diagram established by the Danish astronomer Hertzsprung in 1905 and, independently, by the American astronomer Russell in 1914; it is better known in the form absolute visual magnitude versus spectral type.

Spectral Classification

The color of a star roughly reflects its surface temperature: A hot star like Rigel is blue; the Sun is yellow; a very "cold" star like Antares is red. This means that the hotter the star is, the richer it is in short-wavelength radiation—blue or even ultraviolet. The breakdown of stellar light by a spectrograph reveals the lines of atoms and molecules present in the stellar atmosphere; usually these are absorption lines, but sometimes emission lines are present as well.

The line spectrum varies considerably depending on the temperature of the stellar atmosphere; this has made it possible to classify stars by their spectral types. The spectral classifications are empirical. The most common systems of qualitative classification rest on the classification developed at Harvard at the beginning of the century. The spectral types of 225,000 stars were determined in this system, and the listing of these constitutes the Henry Draper catalogue.

The spectral types are named, from the hot to the cold types, by the following letters of the alphabet: O, B, A, F, G, K, M. (You can remember this sequence with the help of the following sentence: Oh, Be A Fine Girl, Kiss Me!) Each type is divided into 10 subtypes, from 0 to 9. It is the intensity ratios of certain lines as well as the mere presence or absence of certain lines that constitute the classification criteria. The Harvard system developed by the American astronomers Morgan and Keenan uses essentially the following spectral characteristics:

Spectral Type	Absolute Magnitude	Criteria	Typical Star	Temperature (K°)
O	− 6 to − 4	Neutral and ionized helium: He I and He II; twice-ionized carbon: CIII; thrice-ionized silicon: Si IV. After 05, classification based on the ratio He I 4471/He II 4541..	λ *Orionis* (08)	30,000
B	− 4 to 0	Weak He II, disappears at B5; intense He I, maximum at B2. Lines of O II and N II; H becomes most intense.	ε *Orionis* (B0)	28,000–10,000
A	0.5 to 2.5	He I disappears; H very intense, maximum from A0 to A3. Ca II, Fe II, Cr II and Ti II, Fe I, Cr I increase from A0 to A9. Classification based on the ratio Ca II 3934 Hδ. For many stars, strong peculiarities (for example, very intense bands of Si or Eu).	Vega (AD)	10,000–7,500
F	2.6 to 4.3	Spectrum dominated by many lines of neutral, single ionized metals; the line of Ca I at 4,226 Å is the most intense. H rapidly decreases from F0 to F9. Ca II increases and becomes very intense. Classification based on the ratio Hγ 4341/Ca I 4226.	α *Persei* (F5)	7,500–6,000
G	4.4 to 5.8	The lines of neutral metals dominate the spectrum. Classification based on the ratio Hγ 4341/Fe I 4325. Molecular bands of CN and CH appear.	Sun	6,000–5,000
K	5.9 to 8.9	The lines of neutral metals are still most intense. Classification based on the ratio of Ca I 4226/Fe I 4290 +; band of CH at 4300. CN and CH increase. TiO appears at K5.	Arcturus (K2 giant)	5,000–3,500
M	9 to 16	Still very many metallic bands. TiO dominates the spectrum. Very many other molecules present.	Antares (M1 giant)	3,500
The rarer types of stars should be added to this list. They are, first, the Wolf-Rayet stars, at least as hot as those of type O . . .				
W	− 6 to − 4	He II in emission.		
WC WN	− 5.3 − 4.7	CII, CIII, CIV, OII, OIII, OIV, OV, NIII, NIV, NV.		23,000 38,000
S	− 1	No TiO, or very little; very abundant ZrO; organic molecules.	2,500	
C (carbon star)	− 1 ?	No oxide bands, or very few.		
R N		CH, C_2 CN, C_2		

No star of the S or C type is a dwarf; these types are mentioned here only for the sake of showing all the spectral types in a single table.

Different systems of quantitative classification have been suggested. All are more or less derived from the three-dimensional classification of the two French astronomers Barbier and Chalonge. This classification is essentially based on measurement of the Balmer discontinuity and on the distribution of the stars' continuous energy. The hydrogen lines seen in the visible spectrum are the lines of the Balmer series: The more one approaches the blue, the closer together they are (i.e., their order number in the series increases). At wavelengths shorter than 3650 Å, the lines merge: We no longer have absorption lines but a continuous absorption. This discontinuity, which occurs at 3,650 Å, is the Balmer discontinuity. The contrast of the continuum absorption on both sides of that discontinuity, its apparent wavelength and finally the distribution of continuum energy along the visible spectrum make it possible to distinguish the stars according to their temperature, luminosity and chemical composition.

Effects of Chemical Composition

Up to now, we have considered stars of the same chemical composition as the Sun. Now the position of the main sequence in the Hertzsprung-Russell diagram depends on the stars' chemical composition. Thus, dwarf stars containing substantially fewer metals than the Sun will lie on a less luminous main sequence than that of stars of solar chemical composition. In fact, for a given mass, a paucity of elements other than H and He reduces the opacity of stellar matter (for there is less absorption of energy through ionization of these elements); this produces a greater temperature and luminosity, such that, at constant temperature, the luminosity of metal-poor stars is less than that of normal stars. A certain number of stars is deficient in metals by a factor of about 100 relative to the Sun; the main sequence of these stars is less luminous by about 1 magnitude relative to that of the stars of solar chemical composition. It is because of this shift toward low luminosities that the metal-poor dwarf stars are described as subdwarfs, and their main sequence as the subdwarf main sequence. This term in no way prejudges the diameter of the subdwarfs relative to the dwarfs—at a given temperature, the diameter of a dwarf is 1.6 times that of a subdwarf. But as the mass of the two stars differs in that case, the comparison makes little sense.

A change in the abundance of hydrogen, the principal constituent of stars (90 percent of the atoms) relative to that of helium, the second constituent (nearly 10 percent of the atoms), also changes the position of the main sequence. In a normal star, at the time when nuclear reactions are beginning at the center, hydrogen represents about 71 percent of the star's mass, helium 27 percent and other elements 2 percent. If, in another star, the abundance of hydrogen is less, the average molecular weight of the stellar matter is greater and the number of particles per gram is smaller; to maintain its equilibrium, the star has to increase its pressure through contraction. Thus its central temperature increases, which speeds up the nuclear reactions and implies an increase in surface luminosity and temperature. There again, the main sequence slides toward lower luminosities along a line of constant temperature (and thus of inconstant mass).

Mass-Luminosity Relationship

Along the main sequence, therefore, mass and luminosity are not two independent parameters. In particular, the study of double stars makes it possible to estimate quantitatively the relationship between these two quantities. A single expression cannot correctly represent this relationship everywhere on the main sequence. For hot stars, of types O to K2 approximately, luminosity is proportional to the fourth power of a star's mass, which is written:

$$L/L_\odot = (M/M_\odot)^4$$

where L and $L_\odot$ respectively denote the luminosity of the star and the Sun, and M and $M_\odot$ their mass. For later types, from K5 to M5, the dependency of the two parameters M and L is more complex; we have:

$$L/L_\odot = 0.6\ (M/M_\odot)^2$$

These two expressions make up the mass-luminosity relationship, which is fairly well explained by the models of internal structure and the calculations of the stars' energy sources.

The Internal Structure of Dwarf Stars

The fact that two different expressions describe the mass-luminosity relationship implies that the internal structure and energy source of dwarf stars do not follow a single model. The internal structure of cold stars may be validly represented by that of the Sun, which has already been described. We briefly recall that structure here to understand better the evolution of the cold dwarf stars beyond the main sequence.

The solar core represents about half the Sun's mass in 1.5 percent of its volume (one-fourth of the radius). The central temperature is about 12 million Kelvins; 99 percent of all solar energy is produced in the core, mainly by the "proton-proton" nuclear reaction. Remember that the final result of this reaction is the conversion of four hydrogen atoms, or protons (H^1), into one helium atom (He^4) and radiated energy, according to these formulae:

$$H^1 + H^1 \rightarrow H^2 + e^+ + \nu$$

(H^2, deuterium; e^+, positron; ν, neutrino)

$$H^2 + H^1 \rightarrow He^3 + \gamma \text{ (radiation)}$$

$$He^3 + He^3 \rightarrow He^4 + He^4 + H^1 + H^1$$

Out to a distance equal to 85 percent of the Sun's radius, energy is transported mainly by radiation. Beyond that, and as far out as the photosphere, the energy transfer is achieved by convection. At that point, the temperature is sufficiently low for the nuclei of heavy elements to capture electrons, which leads to a sudden increase in the opacity of matter through which the solar energy must travel; a convection instability is established, and through that convection the energy makes its way to the surface.

The internal structure of hot stars is the reverse of this. The higher temperature of the core causes its energy to flow out of it by convection. On the other hand, the heavy elements remain too highly ionized to enable a convective zone to be created above the core; energy is transported by radiation as far as the star's surface. The central temperature, higher than 15 million Kelvins, enables nuclear reactions of the CNO cycle to take place, as follows:

$$C^{12} + H^1 \rightarrow N^{13} + \gamma$$

$$N^{13} \rightarrow C^{13} + e^+ + \nu$$

$$C^{13} + H^1 \rightarrow N^{14} + \gamma$$

$$N^{14} + H^1 \rightarrow O^{15} + \gamma$$

$$O^{15} \rightarrow N^{15} + e^+ + \nu$$

$$N^{15} + H^1 \rightarrow C^{12} + He^4$$

As in the case of the proton-proton chain, the overall result of this chemistry is the combustion of four hydrogen atoms to produce one helium atom and, of course, energy. The atoms of carbon, nitrogen and oxygen have served only as relays, or catalysts, in this reaction; there is neither more nor less before or after the reaction. However, it is important to note that there are at least two known variants of this cycle:

$$N^{15} + H^{1} \rightarrow O^{16} + \gamma$$
$$O^{16} + H^{1} \rightarrow F^{17} + \gamma$$
$$F^{17} \rightarrow O^{17} + e^{+} + \nu$$
$$O^{17} + H^{1} \rightarrow N^{14} + He^{4}$$

and

$$O^{17} + H^{1} \rightarrow F^{18} + \gamma$$
$$F^{18} \rightarrow O^{18} + e^{+} + \nu$$
$$O^{18} + H^{1} \rightarrow N^{15} + He^{4}$$

The second branch merely recycles the N^{15} from which the first branch in fact began; on the other hand, the first branch produces an atom of N^{14}, whereas it in fact began, via the principal cycle, with an atom of C^{12}. Thus the result is a change in the abundance ratio between carbon and nitrogen; farther on, we will see the effect of this in the star's evolution.

Life on the Main Sequence

This evolutionary phase is particularly stable. Thus the Sun will remain in a dwarf state for about 10 billion years. But it is not an inert phase, for the central hydrogen is gradually being converted into helium; at present, the Sun has already burned up half of its central hydrogen.

The transformation of hydrogen into helium involves a decrease in the number of particles per gram in the star's central region (since helium is heavier than the hydrogen it replaces). Consequently, the pressure drops and the gravitational force dominates the pressure forces; this leads to a compression of this central region. The density therefore increases until the forces of pressure are again in balance with the gravitational force. The increase in density is, of course, accompanied by an increase in temperature. The output of nuclear reactions increases very rapidly with temperature; thus the star's luminosity increases (at the end of this process, the star's outer layers expand slightly, leading to a slight drop in its surface tempera-

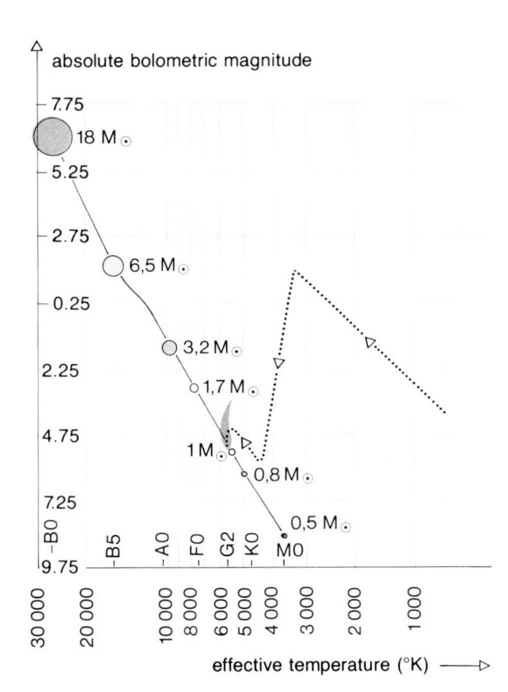

Hydrogen combustion at the center of the star causes its nucleus to contract, bringing a rise in temperature; the rate of nuclear reactions accelerates, and the star's luminosity increases (in pink).
The stars on the main sequence. The mass of each star relative to the Sun's mass is written next to the circle that represents it.

ture). The Sun now is brighter by half a magnitude than at its arrival on the main sequence 5 billion years ago.

The more massive a star is, the higher its central temperature, and consequently the faster the hydrogen combustion reactions and the faster the central hydrogen is exhausted. Hence, the more massive a star is, the shorter its stay on the main sequence for a star of 15 solar masses, of spectral type approximately BO, the time it remains on the main sequence is only 10 million years! For an AO star, like Vega or Sirius, of about three solar masses, this time is only 250 million years. It exceeds 4 billion years for type F5, such as Procyon, which has a mass 25 percent greater than our Sun.

Stars less massive than the Sun remain dwarfs for a period longer than the age of the galaxy; such a star, born at the same time as the galaxy, with a chemical composition similar to that of the Sun, still would be a dwarf today; a star of 0.7 solar mass would remain a dwarf for 20 billion to 25 billion years: It would practically be a fossil!

The Case of the Least Massive Stars

We saw that nuclear reactions are more rapid as the mass, and consequently the stars' central

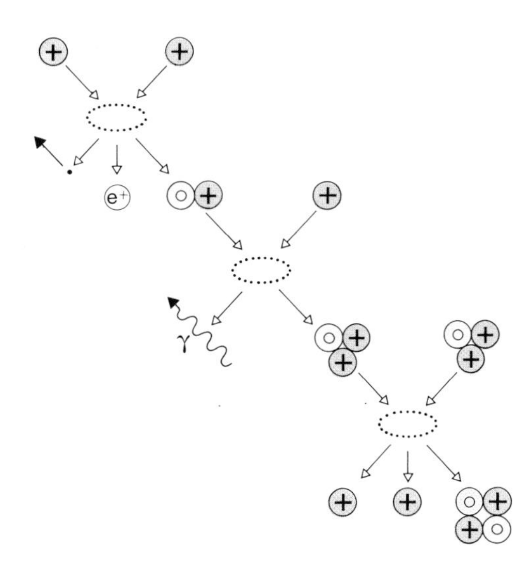

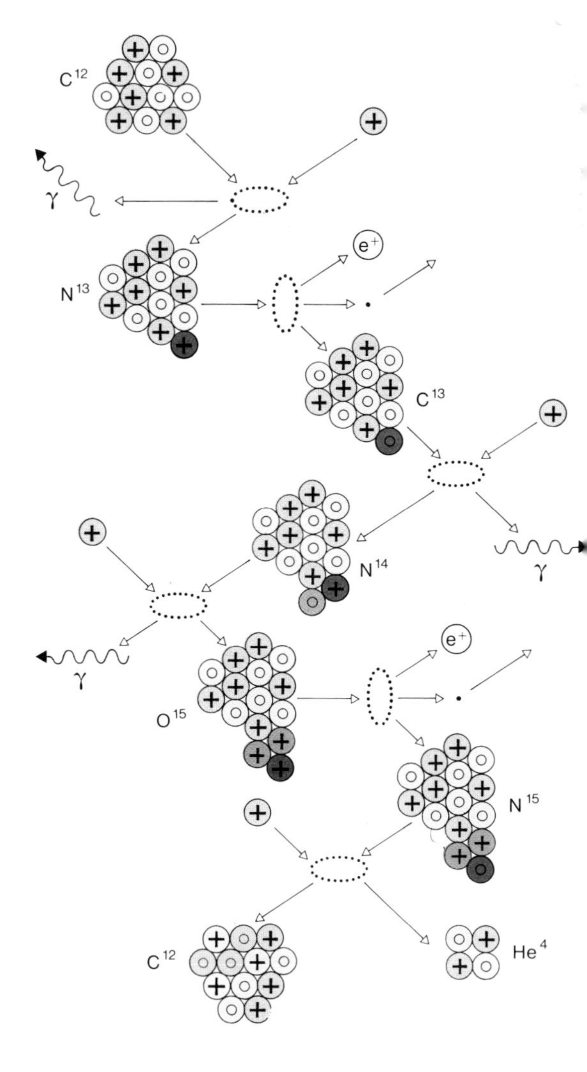

The proton-proton cycle of hydrogen-burning. The basic elements we find in all of the nuclear reactions described here are: the proton, or hydrogen nucleus, with a positive electric charge, designated by + ; the neutron, with practically the same mass but with zero charge, designated by o ; the electron, 1,840 times less massive, with a negative charge, designated by e^- and its antiparticle, hence with a positive charge, designated by e^+; the neutrino, a particle without mass and without a charge, designated by ·→ —its only energy is that associated with its rotation, a motion on its own axis, and its interaction with matter is minimal (a neutrino may cross 10 planets the size of Earth without being absorbed); the photons of very short wavelengths are γ rays, designated by γ→
The CNO cycle of hydrogen-burning. At each stage in the cycle, the particle of the heavy nucleus affected by the last nuclear reaction is shown in red. The additions of carbon-12 to the initial nucleus are colored differently: We see that carbon serves only as a support to the gradual fusion of the two protons and two neutrons of the helium nucleus that is ultimately produced.

temperature, increase. The converse is true, of course, and there exists a critical mass beneath which the central temperature never becomes high enough to initiate the hydrogen combustion in the proton-proton chain. We should say from the start that the different theories of these stars' internal structure do not agree on that critical mass: 0.25 solar mass, 0.1 or even less? Substantial progress in convection theory is essential before this threshold can be made more precise; indeed, below about 0.4 solar mass, a star is entirely convective from the center to the surface (owing to its low temperature, which makes the opacity very great throughout the star).

Nevertheless, stars of very low mass exist. The source of the energy they radiate is gravitational contraction: They are in permanent slow contraction, the heat given off by that contraction being transported by convection to the star's surface, where it is radiated away.

Is there a lower limit to a star's mass? Here is a problem that may not have a solution; it is not clear, in fact, that there is a distinct boundary between stars of low mass, and giant planets with a structure similar to Jupiter's. Moreover, it has often been suggested that Jupiter is a microstar, aborted for lack of sufficient mass. In fact, it radiates about twice as much energy as it receives from the Sun, the source of that energy difference being, according to some, its inward contraction.

The Case of the Most Massive Stars

On the other hand, there is certainly an upper limit to a star's mass. But there again, our knowledge of physics is not advanced enough to be precise. Beyond a certain mass, the central temperature and density become so high that the radiation and gas pressure overwhelm the gravitational force stemming from the star's mass; therefore it is not stable, and it explodes.

What nuclear reactions can occur in the star during such a catastrophe? What is the critical mass? At least 30 solar masses, perhaps 60, even 100. But can such monsters even form? It is not in fact clear that a cloud of interstellar matter with such a mass would be sufficiently homogeneous to condense into a single star. It is more likely that density inhomogeneities would lead to a fragmentation into several protostellar clouds, the progenitors of a multiple stellar system, like the Trapezium in Orion, or even a cluster of stars.

Giant Stars

LET US NOW GO BACK to the description of the phenomena that occur in the heart of a star of the same mass as the Sun, phenomena that make up its evolution; it is the fate our Sun will suffer in about 5 billion years. We will next see how the star's mass and, briefly, its chemical composition, affect that evolution.

The Sun's Evolution Beyond the Main Sequence

What, then, will happen to the Sun when it has burned up all the hydrogen at its center? Gradually, a spherical shell surrounding the core, still rich in unburned hydrogen, will become the site of hydrogen burning; the role of the CNO cycle will become more and more important relative to that of the proton-proton chain. Thus the star's energy no longer comes from its core but from a thick shell around it.

This shell will move away from the center of the star as hydrogen burning proceeds, in order, so to speak, to find fresh hydrogen in the stellar atmosphere (the zone that extends from the star's surface to the hydrogen-burning layer). As the shell expands, it becomes thinner. Temperature and density increase, which increases the efficiency of the nuclear reactions.

But this increase only partially compensates for the reduction in the shell's thickness. Hence, the production of nuclear energy diminishes. Moreover, this energy production is explosive, so that matter is expelled toward the atmosphere, causing it to expand rapidly. Since it is expanding, the atmosphere cools. Half of the energy necessary for the expansion comes from the energy production in the hydrogen shell undergoing combustion; the other half comes from the thermal energy of the atmosphere itself, which therefore cools more rapidly. The nuclear energy is thus transformed into mechanical energy and is no longer radiated away; in addition to its surface cooling, the star's luminosity decreases.

As the stellar atmosphere cools, it becomes more opaque to radiation, for more free electrons can then recombine with atoms heavier than helium, and these atoms then absorb more radiation by ionization. Hence it becomes more and more difficult for the energy to travel in the form of radiation; when the surface temperature falls to

Hydrogen burns in a layer around the inert helium nucleus. This combustion causes an expansion of the outer layers (the atmosphere) of the star; the star appears colder; it is a subgiant. Beginning at about 5,000°K, the opacity of the atmosphere rapidly decreases; the star's luminosity increases, helped along by the development of an intense convective zone.

about 5,000° K, convection becomes the main mode of energy transfer between the hydrogen-burning shell and the star's surface.

Just at the star's surface, above the convective zone, the principal source of opacity is the negative ion of hydrogen, H^-. This H^- ion is the combination of a hydrogen atom and a very close free electron. Their binding energy is very low: 0.75 eV. (Free electrons in the atmospheres of cold stars come from the ionization of atoms having low ionization potentials: iron, magnesium and silicon between about 6,000° K and 4,500° K; sodium, calcium and potassium below that.) This opacity decreases when the temperature also decreases. The star's outer layers thus become more transparent, which is reflected in an increase in the star's luminosity. Moreover, in the convective zone, the energy used for the expansion of the atmosphere is of exclusively thermal origin.

Hence, as the convective zone enlarges, the portion of nuclear energy used in expanding it decreases; as a result, the share of nuclear energy that can be radiated toward the exterior increases. This phenomenon also contributes to increasing the star's luminosity.

While the hydrogen-burning shell grows thinner, the central temperature and density of the nucleus rise rapidly. The density becomes so great that the pressure from the degeneracy of electrons assumes a predominant role relative to the gas pressure. As a result of this appearance of degeneracy, the conductivity of the electrons is constantly increasing: The star's nucleus becomes isothermal—i.e., it has the same temperature at the center as at the periphery. When this high conductivity is established, it is, of course, accompanied by a drop in the central temperature (and a rise in the peripheral temperature). Then the entire nucleus continues to heat up, being at the same temperature everywhere, until the temperature reaches about 100 million Kelvins (remember that the Sun's current central temperature is only about 12 million Kelvins!).

At this point in its evolution, the Sun will be a red giant of spectral type K3 or K5 approximately, with a surface temperature of 3,500° K (instead of the current 5,780° K). Despite this much lower temperature, it will radiate much more energy: 400 times more than it does now! in the Earth's sky, it will be a monster; its diameter will be 55 times greater than now; the Sun's apparent diameter will be nearly 30 degrees!

Helium-Burning at the Core

When the temperature at the center, where all the hydrogen is replaced by helium, reaches 100 million degrees Kelvin, a new nuclear reaction will take place: that of helium burning. Three helium nuclei will fuse to produce an atom of carbon, and, of course, energy. After this reaction, a carbon atom can fuse with another helium atom, producing an oxygen atom, and more energy:

$3\ He^4 \rightarrow C^{12} + \gamma$

$C^{12} + He^4 \rightarrow O^{16} + \gamma$

Since the stellar core composed of helium is isothermal, this reaction begins simultaneously at all points in the core, not just at the center: It is an explosive reaction. Because of the great electron conductivity, the energy released remains trapped in the isothermal core, creating a sharp rise in temperature and therefore of gas pressure. The temperature increase stops when the gas pressure is again greater than the degeneracy pressure; the core ceases to be degenerate. Subsequent to that moment, the pressure again becomes proportional to temperature, and the core expands because of the high temperature and thus cools.

The dominant source by far of the star's energy still is hydrogen burning in the shell surrounding the helium (95 percent of the energy at this stage of evolution).

To maintain an equilibrium between the amount of energy produced and that transferred to the exterior, density and temperature must remain high in the hydrogen-burning shell, in spite of the reduction in density and temperature that should result from the expansion of the nucleus. Equilibrium then is maintained by the contraction of the stellar atmosphere above the shell: This contraction heats and compresses the matter around the shell, thus making it possible to maintain a high yield of the nuclear reactions of the CNO cycle. Since the temperature is increasing in the atmosphere, the opacity decreases on the average (because the atoms become more ionized and can absorb less radiation); as a result, convection disappears. On the other hand, throughout this phase of helium combustion at the star's center, the core never ceases to expand.

The central reaction of helium combustion supplies a greater and greater share of the star's total energy—up to 35 percent. The fusion reactions of carbon with helium, $C^{12} + He^4 \rightarrow O^{16} + \gamma$, and oxygen O^{18} with helium, $O^{18} + He^4 \rightarrow Ne^{22}$, become more important.

At this stage in its evolution, the Sun thus will become a variable star of the RR Lyrae type. Indeed, when the stellar atmosphere is ϕ in rapid contraction, the gas pressure does not instantly balance the gravitational force: In its contraction motion, the star, in its momentum, contracts more than the gas pressure allows; then the gas pressure becomes greater than the gravitational force, and the star expands, again going beyond the point of equilibrium in the opposite direction, then it contracts again and so on. Hence the star oscillates. Normally such oscillations decay very rapidly, owing to damping, and the star stabilizes. But variations in temperature and pressure in the "helium ionization zone" may keep the oscillations going.

Let us now return to the star's evolution toward higher temperatures. This evolution ceases when, as a result of the decrease in the central density, the gas pressure becomes insufficient to balance the gravitational force. Then the core contracts and, conversely, the atmosphere expands. This evolutionary phase is comparable to the star's evolution right after it leaves the main sequence. The star cools at nearly constant luminosity.

When nearly all of the helium is converted into carbon, oxygen and neon, at the center, the entire star, core and atmosphere, contracts to maintain the gas pressure despite the decrease in the number of particles. The star heats up, and at the periphery of the core, the temperature, pressure and abundance of helium are sufficient for the helium-burning to continue. This helium-burning shell is initially thick, owing to the convective nature of the core, then gradually becomes thinner. When it becomes very thin and lies far from the center of the star, the part of the star above this shell is pushed outward and expands; it then cools, so that physical conditions in the helium

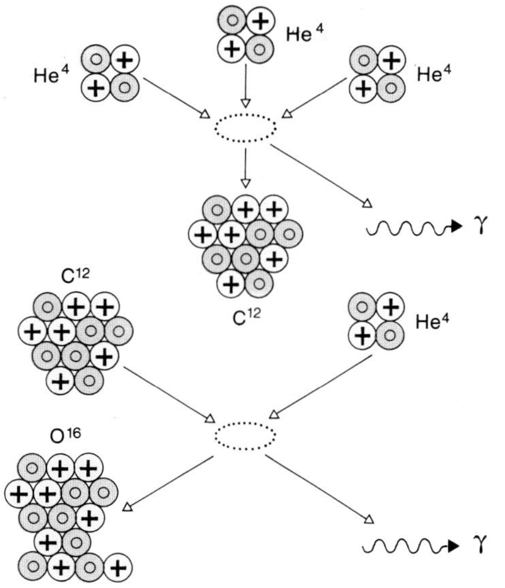

Fusion reaction of three helium nuclei to produce an atom of carbon. The second reaction—synthesis of a nucleus of oxygen by fusion of a nucleus of carbon and a nucleus of helium—does not affect all the nuclei of carbon. The end result of these reactions is that the relative abundances of carbon and oxygen remain practically unchanged.

Degeneracy

Heisenberg's uncertainty principle states that the position and quantity of motion (the product of mass times velocity) of a particle cannot be precisely determined simultaneously; the product of the uncertainties regarding the position and quantity of motion is greater than a constant equal to $6{\cdot}6 \times 10^{-34}$ joule per second.

The position of an electron subjected to the influence of an atomic nucleus is very well defined (the force to which it is subjected is large). Owing to the uncertainty principle, its quantity of motion, however, is ill defined: The electron has a continuous motion around the position it occupies. This motion exerts pressure on the surrounding environment, exactly as the thermal agitation of the particles of a gas exerts its pressure. This pressure is called the pressure of electron degeneracy. This pressure, since it is nonthermal in origin, is, of course, independent of temperature. It depends only on density: The greater the density, the closer the electrons of the gas will be to one another, and the more each of them will feel the degeneracy pressure of the others. It is, however, only at very high densities that degeneracy pressure becomes comparable or superior to gas pressure: We then say that the matter is degenerate.

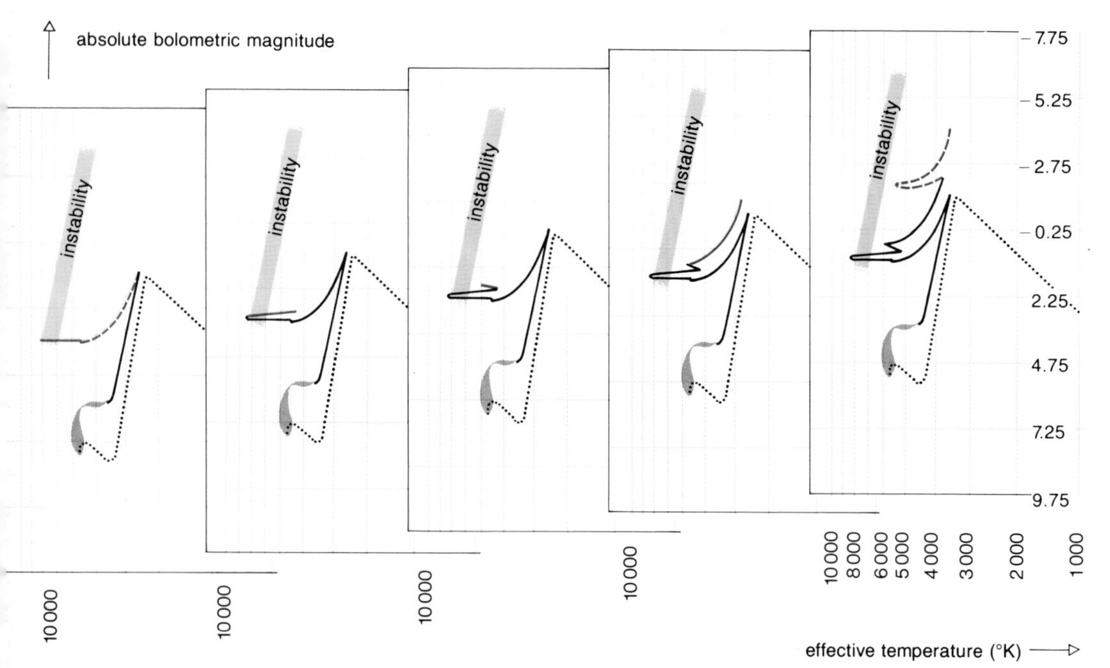

The explosive burning of helium begins at the center of the star. In a very short time, the star travels a path that might be the dotted part (at left). When the burning is no longer explosive, the nucleus expands; the star slowly contracts and heats up. When it crosses the instability strip, it is a variable star of the RR Lyrae type. There comes a time when the central pressure of the expanding nucleus no longer equilibrates the gravitational force. Then the nucleus begins to contract, while the atmosphere expands; the star cools (at the center). When nearly all of the central helium has been burned, the entire star contracts to compensate for the decrease in gas pressure at the center. The atmosphere again cools and the star becomes more luminous, owing to the rapid decrease in opacity in the atmosphere (at right). When the central temperature reaches about 1 billion degrees, carbon-burning begins in the nucleus. The star's evolution, at this stage, remains uncertain.

shell are no longer favorable for the helium-burning reaction. The star's sole energy source then comes from helium-burning in a thin shell around the carbon and oxygen nucleus.

The subsequent evolution is parallel to that of a star of which the energy source comes from hydrogen burning in a shell surrounding the helium core. The oxygen core continues its contraction and heats up rapidly, while the increasing absorption in the cooling atmosphere tends to lower the star's luminosity. But absorption in the stellar photosphere rapidly decreases so that the star's luminosity rapidly rises. The large energy loss resulting from the escape of neutrinos produced during nuclear reactions acts as a brake on that evolution.

When the central temperature reaches 1 billion degrees Kelvin, the carbon atoms begin to fuse according to this reaction:

$$C^{12} + C^{12} \rightarrow Ne^{23} + H^{1} \rightarrow Ne^{20} + He^{4}$$

At this high temperature, the protons and helium nuclei (α particles) fuse rapidly according to these principal reactions:

$$Na^{23} + H^{1} \rightarrow Mg^{24} + \gamma$$

and

$$Ne^{20} + He^{4} \rightarrow Mg^{24} + \gamma$$

The final result of these reactions is the formation of many atoms lying near magnesium in the periodic table.

The burning of carbon is explosive: In the degenerate core, the energy is trapped, causing a very rapid temperature increase, which ceases when the degeneracy is lifted—i.e., when the gas pressure again exceeds the degeneracy pressure. The core and surrounding layers then expand and cool to the point where the temperature and pressure conditions within the helium-burning shell no longer allow that reaction to take place; it is practically extinguished. However, the stellar atmosphere contracts, heating and compressing the former hydrogen-burning shell; the combustion of hydrogen then can begin again. The net effect on the star's luminosity is not certain; the star would make a new loop toward high temperatures in the Hertzsprung-Russell diagram.

The subsequent evolution is even more uncertain. Either the burning of carbon or that of helium is unstable. This instability is reflected in rapid increases in the star's temperature and luminosity; the period of these instabilities is about several thousand years, each of the bursts being more violent than the previous one. The variable stars of the Mira Ceti type would represent this stage of stellar evolution, although the reason for the variability of these stars is not yet understood. In this phase, the Sun will be very cold: 1,800° to 2,500° K, depending on whether it is at its minimum or its maximum luminosity. Its diameter will be colossal: about 300 times its current diameter—it will engulf the orbits of Mercury, Venus, the Earth and practically Mars. Our planet, then, will have disappeared from the map of the cosmos shortly before this era: It will be the true end of the terrestrial world, which will evaporate in the solar inferno. Long before that, of course, all life will have disappeared from the Earth's surface because of the torrid heat that will prevail there.

A more violent burst would appear in the form of a nova, a veritable stellar explosion. At such a time, the star radiates as much energy in a few months as the Sun in 10,000 years! Part of the stellar atmosphere is ejected, giving rise to a planetary nebula, a cloud of gas excited by the radiation of the rest of the star: Its nucleus of degenerate matter still is very hot. Thus the central star of the Helix nebula has a temperature of about 100,000°K and is in the process of cooling.

What is the duration of the evolutionary phase we have just described after the Sun's departure from the main sequence? The phase of expansion of the atmosphere before the rapid increase in luminosity will last about 200 million years, while the phase of high luminosity will last at least 1 billion years. After the ignition of helium at the

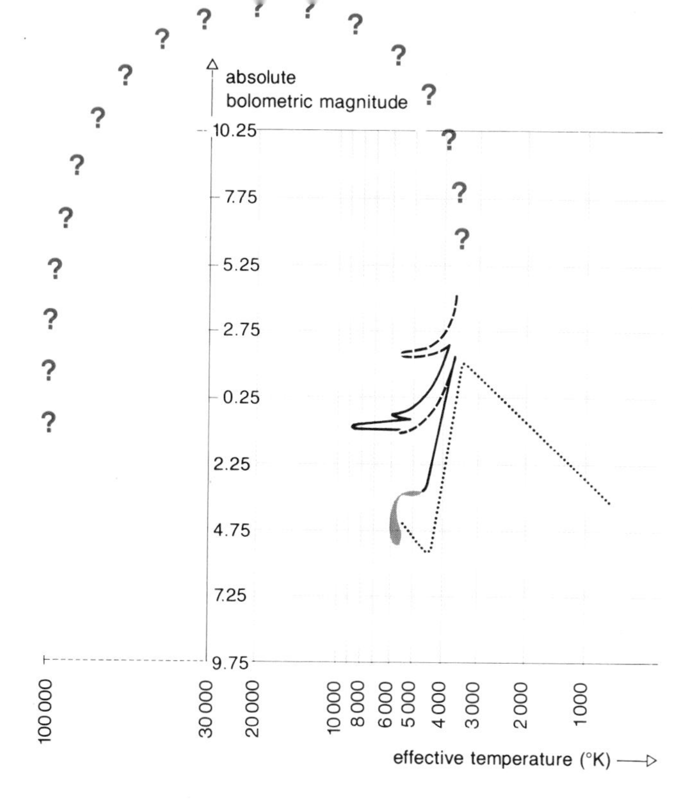

Beginning a new loop in the evolutionary diagram, the star becomes violently unstable and ejects a portion of its atmosphere; it forms a planetary nebula surrounding the very hot nucleus of degenerate matter. There again, the evolution is uncertain, and the dotted line is merely indicative.

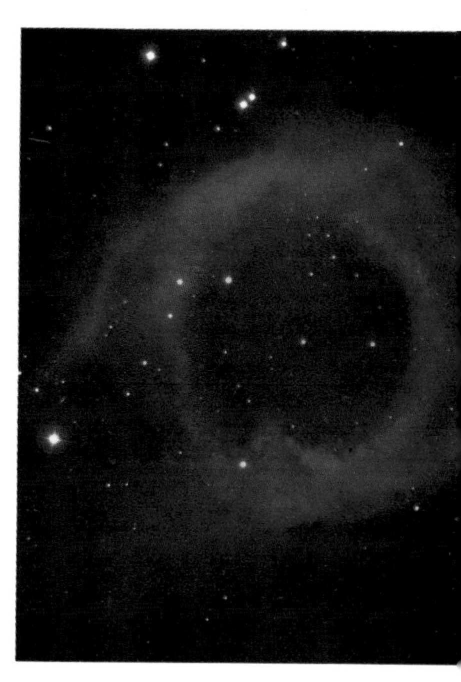

The Helix nebula (a planetary nebula) is comprised of the gas ejected by a star that exploded as a nova. The very hot object that remains at the center of the nebula is what remains of that star: Its nucleus consists of degenerate matter. The nebula, of magnitude 6.8, is easily observable with instruments having an aperture of at least 100 millimeters (4 inches). Observation of the nucleus is more difficult, since its photographic magnitude is 13.4. Its temperature being about 100,000°K, it is even fainter in the visual. Photo: California Institute of Technology and Carnegie Institution of Washington.

core, evolution speeds up: Perhaps there are 100 million years until the nova stage, perhaps less. The current state of our knowledge of the internal structure of the highly evolved stars does not permit us to be more precise.

The Evolution of Stars with Different Masses

BETWEEN 1 AND 1.1 SOLAR MASSES

Although the problem has not been studied very thoroughly, it seems that the evolution of stars less massive than the Sun does not differ very much from the pattern described above.

If a star's mass is between 1 and 1.1 solar masses, its evolution will be identical to the Sun's. This mass interval corresponds to the interval of spectral type F8-G2 on the main sequence. For stars slightly more massive than 1.1 solar masses, evolution following the main sequence is different from the Sun's; for stars more massive than 2.25 solar masses, evolution in the giant phase also is different.

EVOLUTION AFTER THE MAIN SEQUENCE

Let us go back, then, to the state of a star more massive than 1.1 solar masses when it is at the end of its stay on the main sequence. The difference in evolution that appears above 1.1 solar masses is tied to the emergence of the central convective zone.

We saw that, at the end of the hydrogen-burning phase, the central region contracts and thus heats up. This contraction accelerates until the mass of hydrogen reaches about 5 percent of the mass of the core. At that moment, the entire star, not just its core, begins to contract: In the stellar atmosphere, the gravitational energy is converted into heat and transferred by convection into the regions around the core and then, by radiation, beyond that. Thus the star heats up and becomes more luminous, until the hydrogen represents no more than 1 percent of the core mass.

At that point the energy released by the hydrogen-burning reactions has decreased to the point where it is no longer sufficient to compensate for the energy loss resulting from the difference in temperature between the center and the edge of the core. This insufficiency causes a very rapid contraction and a cooling of the nucleus. Simultaneously, the outer layer of the nucleus, still rich in hydrogen, of course, also collapses; its density and temperature increase so that the hydrogen can be burned according to the reactions of the CNO cycle. The reaction is particularly explosive; matter is expelled on both sides of the shell where the new burning takes place. The expulsion of the atmosphere toward the exterior causes an increase in the star's radius; as in the case of the Sun, part of the nuclear energy thus is converted, no longer into radiated energy, but into mechanical energy. The star's luminosity decreases slightly. Very quickly, this dissipation of energy into a mechanical form becomes negligible, so that the star's luminosity again slowly increases as the shell where the hydrogen is burning moves away from the core; the ashes of that burning accumulate at the core, its mass increases and the contraction continues. The evolution subsequently conforms to the pattern of solar evolution.

EVOLUTION OF GIANTS HAVING MORE THAN 2.25 SOLAR MASSES

If a star's mass is greater than 2.25 solar masses, an important difference occurs in its evolution. When the hydrogen burns in a layer around the helium core, the latter never becomes degenerate. Because of the larger mass, the central temperature also is higher, and therefore the gas pressure always is much greater than the degeneracy pressure. The evolution at the moment when the helium ignites therefore is different from what we saw in the case of a degenerate nucleus.

There again, the helium-burning reaction is explosive: The amount of energy produced by the reaction is greater than can be transferred to the surface of the star. The helium core therefore expands upon ignition; this expansion will alter the temperature and density in the hydrogen-burning shell, so that this combustion supplies much less energy. The star's luminosity plummets; the stellar atmosphere contracts, and consequently the surface temperature rises; the subsequent evolution is then similar to what will happen to the Sun.

Stars more massive than the Sun pass through the instability strip due to the helium ionization zone with a greater luminosity; therefore their light curve is different and has a longer period: They become variables of the cepheid type.

The more massive a star is, the more luminous it is at a given stage of its evolution, and the more rapid that evolution is. We saw that the Sun took 1 billion years to climb up the asymptotic red-giant branch. This time is reduced to 200 million years for a star of 1.5 solar masses (650 million years from the main sequence to helium ignition at the center), to 4 million years for a star of 3 solar masses (20 million years after the main sequence), and to only 60,000 years for a star of 9 solar masses (300,000 years after the main sequence).

Beyond the point of helium ignition, the evolutionary time varies from 70 million years for a star of 3 solar masses to 4 million years for a star of 9 solar masses. A star of 15 solar masses (of type B0 on the main sequence), which remains on the main sequence for 10 million years, goes through the evolutionary phase just described in 1.5 million years.

Soon after leaving the main sequence, stars are classified as subgiants (luminosity class IV), then as giants (luminosity class III) and then as bright giants (class II). The very massive stars rapidly become supergiants (class I), first blue and then red.

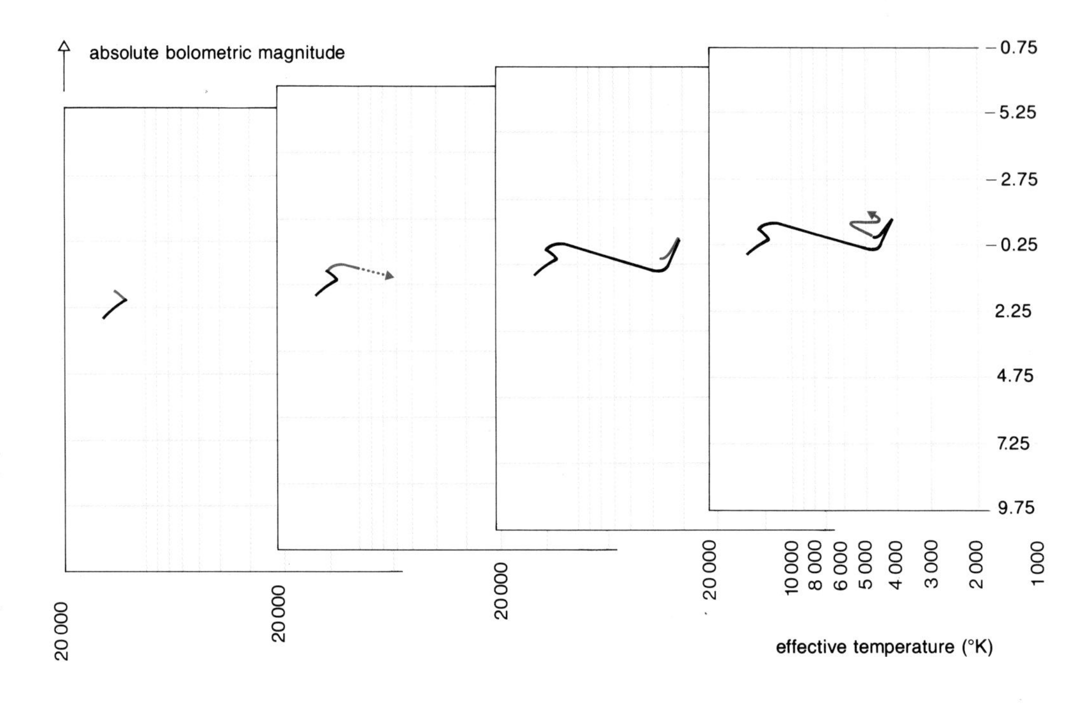

The Final Evolution of Very Massive Giants

The very massive stars do not become novae at the end of their giant phase; rather, they become supernovae—i.e., they experience a truly catastrophic event. Their luminosity is very high for a very short time; they reach absolute magnitude −16! In a few dozen days, the star radiates as much energy as the Sun in 10 billion years! Here is the process that leads to this explosion.

After having burned carbon at its center, a massive star then burns the oxygen that was produced during the combustion of the helium at the center. This burning process occurs through the fusion of a helium atom with a carbon atom. This reaction cannot begin until the central temperature reaches 2 billion degrees. Then it takes place according to these reactions:

$$O^{16} + O^{16} \rightarrow Si^{31} + n$$

where n is a neutron, or

$$O^{16} + O^{16} \rightarrow P^{31} + H^{1}$$

or

$$O^{16} + O^{16} \rightarrow Si^{28} + He^{4}$$

The neutrons, protons and helium nuclei produced are immediately captured by the heavy atomic nuclei present, including those of magnesium and other nuclei of similar atomic mass that were formed during carbon burning. The outcome of these oxygen-burning reactions is the production of many atoms of atomic mass close to that of silicon.

Again, through contraction of the star's core, the central temperature continues to rise after the oxygen-burning phase. When the temperature reaches 3 billion to 4 billion degrees, a fraction of the atoms of silicon and other heavy elements are gradually dissociated into light particles (protons, neutrons and helium nuclei). These light particles are immediately captured by the remaining silicon atoms, thus making heavier atoms, which of course also dissociate, leaving silicon in particular. These reactions reach an equilibrium, at the end of which the most abundant element happens to be iron, Fe^{56}.

The stellar core continues to evolve, either by reaching very high temperatures (7 billion degrees) or very high densities. If the temperature is very high, all the heavy nuclei at the center of the star are destroyed according to reactions of this type:

$$Fe^{56} \rightarrow 13\ He^{4} + 4\ \mathbf{n}$$

These reactions are endothermic—i.e., they absorb energy. The result is a drop in temperature at the center of the star. Finally, the gravitational force is no longer at equilibrium, and the star collapses inward.

If, on the other hand, the central temperature does not reach such high values but the density increases considerably, the electrons are captured by the atomic nuclei. Since it is the electrons that contribute the most to the pressure at the center of the star, the decrease in the number of electrons entails a decrease in pressure. If this decrease occurs rapidly, the star becomes unstable and collapses.

The evolution of a star more massive than 1.1 solar masses (broken down here, from left to right, into four phases) is different from that of the Sun at its arrival on the main sequence. When less than 5 percent of the original hydrogen remains in the stellar nucleus, the entire star contracts and heats up (here, evolution of a star of 3 solar masses). The hydrogen then ignites in a shell around the nucleus. Next comes a drop in luminosity, then a slow increase in luminosity as the shell moves away from the center. The star's nucleus, made up of helium, contracts and heats. When the combustion of helium begins at the center of the star, the nucleus expands, so that the temperature and pressure conditions no longer permit hydrogen-burning in the shell surrounding the nucleus; the star's luminosity abruptly drops. The star's mantle contracts and heats. The subsequent evolution resembles that of the Sun.

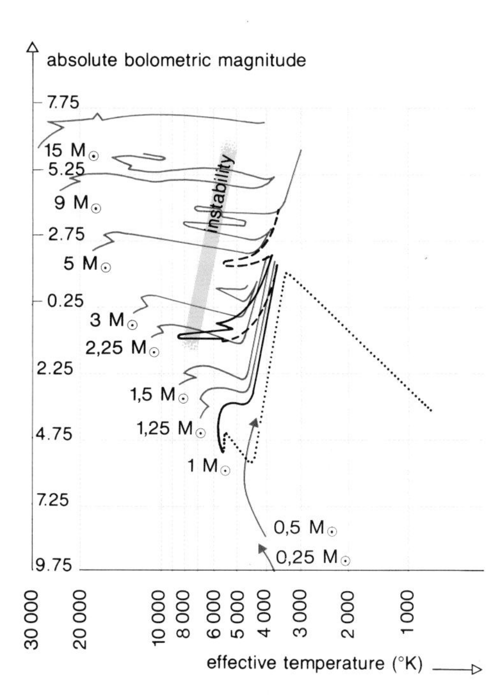

Evolutionary diagram of stars of different masses.

In both cases, the huge rise in temperature that accompanies the gravitational collapse transforms most of the matter at the center of the star into neutrons according to this reaction:

$$H^1 + e^- \rightarrow n + \delta$$

When the neutrons are sufficiently compressed, the pressure due to these neutrons resists the collapse and interrupts it. At these very high temperatures, the rate of formation of neutrino-antineutrino pairs is very high. These neutrinos diffuse very rapidly toward the outside of the stellar core and are partially absorbed by the star's outer layers. Thus these layers heat up considerably, so that the gas pressure in them becomes substantially greater than the gravitational force: The star's atmosphere is literally blown away. This is the supernova explosion. The ejection of matter may take place at speeds as high as 10,000 kilometers (6,215 miles) per second.

This process contributes greatly to regenerating the interstellar medium; during the explosion, the layers of carbon, oxygen and silicon burn briefly and incompletely as far as the latter is concerned. The matter expelled into interstellar space thus is enriched in elements ranging principally from silicon to iron-peak elements.

We will continue studying the core of the supernova after briefly showing the influence of chemical composition on the evolutionary paths of the giants, as well as how observation supports the theory of evolution we have described.

Effect of Chemical Composition

The effect of chemical composition is much more complex with regard to the giants than the dwarfs, since several types of nuclear reactions take place in the giants. The more deficient in metals a star is, the bluer will be the asymptotic branch through which it travels: Its entire evolutionary path is shifted toward higher temperatures. Likewise, the more a star is deficient in metals, the more luminous its horizontal branch is—i.e., the more luminous it is when it burns helium in its center.

In the stars that are deficient in metals, the helium-burning phase, which is on the Hertzsprung-Russell diagram at the blue end of the horizontal branch, is unstable: We find variable stars of the RR Lyrae type.

Observational Aspects of Stellar Evolution

SEVERAL OBSERVATIONAL TESTS are possible at different stages in the theoretical evolution we've described. The basic tool for the observational study of stellar evolution is the Hertzsprung-Russell diagram of star clusters, both the open and the globular clusters.

The very young open clusters, like h and χ Persei, which are less than 10 million years old, make it possible to determine the zero-age main sequence—i.e., the locations of stars of different masses that have just begun to burn hydrogen at their center. The comparison between the main sequence of theoretical age zero and an observed sequence serves to adjust the opacities used in the calculation of the radiative transfer within the stars.

The older open clusters, such as M67 or NGC 188, serve to probe the evolution beyond the main sequence: The shape of theoretical evolutionary paths and the isochronic curves must follow that of the observed diagram. For example, the position of the asymptotic giant branch also is very sensitive to many parameters of the models of internal structure, particularly those that govern convection. In such a comparison, an important problem is the chemical composition of the stars of the observed cluster: We saw that the evolutionary paths are sensitive to the chemical composition adopted for calculating them. The chemical composition of cluster stars can be determined by photometry—e.g., for a given star, the difference between the magnitude in the blue (B: 4,430 Å) and the visible (V: 5,550 Å), termed B-V, is a function of the chemical composition. The more abundant the metals are, the more intense the photospheric absorption lines are; these lines are more numerous and more intense in the blue than

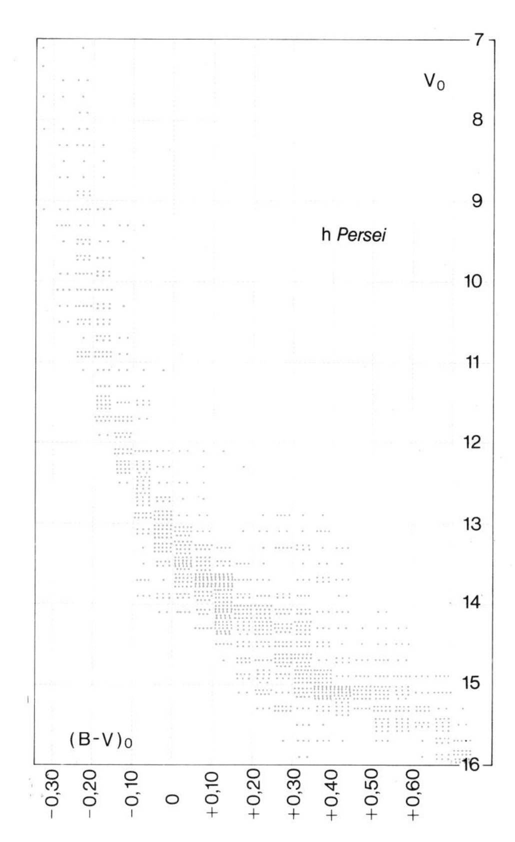

Hertzsprung-Russell diagram of the very young open cluster h Persei. Except for the most massive among them, the stars in this cluster have not had time to evolve on the main sequence; therefore they define the shape of the zero-age main sequence. The less massive stars still are at the end of contraction, even before reaching the main sequence.

in the visible. The two colors B and V therefore differ in their sensitivity to chemical composition. Of two stars at the same temperature, the one that has the highest B-V color index will be richer in metals.

A much more precise determination of chemical composition is possible through detailed analysis of high-dispersion stellar spectra (between 1 and 15 Å/mm.): In this case, several hundred absorption lines are analyzed independently of one another using theoretical stellar atmosphere models—i.e., models of the distribution of temperature, pressure and opacity as a function of depth in the stellar atmosphere.

The density of stars at each evolutionary stage in the Hertzsprung-Russell diagram of clusters makes it possible to estimate the duration of the stage: The more rapid a stage is, the less time a star remains in it, and thus the fewer stars there will be in that stage at the time of observation. Theoretical evolutionary paths should predict speeds of evolution compatible with those deduced from the distribution of stars on the HR diagrams of observed clusters.

Comparisons between theory and observation should take into account observational difficulties. Clusters, particularly globular clusters, are very far from the Sun, so the observed stars are very faint. As a result, photometry is subject to errors that may be considerable. Determining the distances of the clusters is tricky, owing to the observational dispersion of the stars either on the main sequence or on the horizontal giants branch, which serves as a reference, and also because of interstellar extinction. Consequently, the absolute magnitudes, or luminosities, are likely to be tainted with systematic errors, which then affects the theoretical models that have been adjusted to fit these observed cluster diagrams. Also subject to error is the temperature scale that is used to establish the abscissa of the observed diagrams.

The open cluster M 67. About 50 faint stars are visible within it with an 80-millimeter (3-inch) refractor; several hundred with a 300-millimeter (12-inch) reflector. M 67 is one of the oldest known open clusters. Photo: Haute-Provence Observatory, France, of the C.N.R.S.

Hertzsprung-Russell diagram of the old open cluster M 67. All of the stars more massive than about 1 solar mass already have left the main sequence; the position of the blue extremity of the main sequence thus makes it possible to assign an age to the cluster.

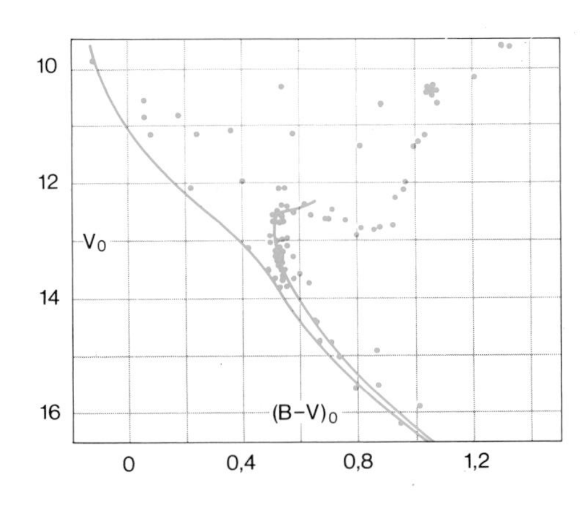

But the chief test of any theory of evolution is the Sun. The models of internal structure that are made to evolve must be capable of reproducing the Sun's characteristics: temperature, luminosity and age. Often parameters applied to calculation of convection are empirically adjusted to be able to reproduce these solar parameters.

h and χ Persei, twin open clusters perfectly visible to the naked eye, since they are of 4th magnitude. While a dozen stars of magnitudes 6 to 7 can be brought into focus with binoculars in each of the two clusters, this number reaches 100 in a 150-millimeter (6-inch) reflecting telescope. In the study of stellar evolution, h and χ Persei are of great interest, for they are among the youngest known clusters: They are about 10 million years old. Photo: Haute-Provence Observatory, France, of the C.N.R.S.

The Death of Stars: An Endless Process

White Dwarfs

Let us return to the evolution of a star like the Sun at the moment it is a nova, then a nucleus surrounded by a planetary nebula. After several tens of thousands of years, the planetary nebula dissipates into the interstellar medium. The luminosity of the star's nucleus rapidly plummets:

It becomes a hot white dwarf of the DO type, with a temperature of about 20,000° K. It is mainly degeneracy pressure of electrons that keeps the gravitational force in equilibrium. The density is considerable, since the mass of white dwarfs is about 1 solar mass, while their radius is comparable to the Earth's; on average, 1 cm³ of a white dwarf weighs some 500 kilograms (1,102 pounds); in fact, the density varies from 0 at the surface (by definition) to about 16 t./cm³ at the center. If the outer layers of a white dwarf are gaseous, the nucleus is solid.

Because of the white dwarfs' very high density, the pressure in the atmosphere of these stars is enormous; the spectral lines are greatly broadened by collisions. The appearance of the spectrum depends, on the one hand, on the star's temperature and, on the other hand, on the diffusion phenomena in and under the atmosphere. The spectrum may show only hydrogen lines, or even no spectral line. The cold white dwarfs show either a continuous spectrum, or a few molecular lines (CH) and some metallic lines that usually are very intense in other cold stars.

A white dwarf isolated in space has no other source of energy except thermal energy: All of the energy it radiates comes from its continual cooling. Since the degeneracy pressure is independent of temperature, this cooling produces no decrease in the star's pressure, as would happen with a non-degenerate star; the star is not destabilized. A white dwarf, then, evolves with decreasing temperature and luminosity. This evolution slows down as it advances—it lasts several billion years. But, in fact, we should not speak of the duration of evolution, for this phase has no end.

Many white dwarfs are part of a double system. The most famous of the white dwarfs, discovered in 1862, is Sirius B, the companion of Sirius. If the pair is sufficiently close, there may be an accretion phenomenon: Matter from the nondegenerate component is drawn toward the white dwarf. Two processes are possible: Either the white dwarf "collects" the matter from a stellar wind given off by the primary component, or the matter is drawn in directly from the normal primary star to the white dwarf. The matter collected in this way is compressed in its fall to the white dwarf, because of the very strong gravitational field on its surface; this matter heats up greatly and radiates; finally, the original gravitational energy of this matter is converted into radiated energy. The white dwarf then appears much brighter: It becomes a novoid. Some believe that the companion to Mira, presumed to be a white dwarf, is made luminous in this way.

The density of this matter, when it reaches the surface of the white dwarf, is not less than 10 kg./cm³. This density increases with the quantity of matter thus accreted, until it passes the threshold for hydrogen ignition. The accreted mass must then be about 1/10,000 of the Sun's mass (some red giants eject such a quantity of matter in only a few years!). The explosive combustion of the accreted hydrogen then takes place on the surface of the white dwarf. Then the increase in the star's luminosity is fantastic: This is a different type of nova. In the case of white dwarfs, the accretion would not take place continuously, with matter falling in large but discontinuous lumps onto the white dwarfs and thereby causing small, explosive, hydrogen-burning events.

The mass of a white dwarf cannot be arbitrary. The greater the mass, the greater the gravitational force. Thus the degeneracy pressure also has to be greater to balance it. This can be achieved only through an increase in density—i.,e., through a smaller radius. Therefore we arrive at this unusual result: The more massive a white dwarf is, the smaller its radius. Now, according to what we have seen of the principle of degeneracy, if the radius of a white dwarf is smaller, the free space for an

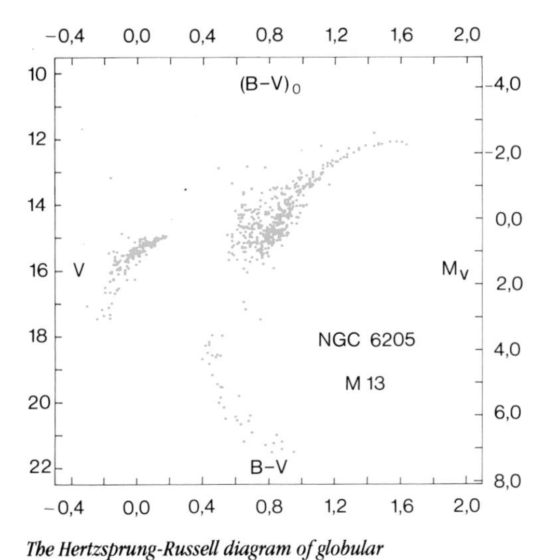

The Hertzsprung-Russell diagram of globular clusters is very characteristic: a main sequence populated only in its cold part, a clearly indicated asymptotic branch, and a horizontal branch where the stars burn their helium. The RR Lyrae variables are located toward the hot extremity of this horizontal branch. Such diagrams (here, that of M 13) serve to test the theory of the evolution of stars that are very deficient in metals, as well as the helium content of such stars; the luminosity of the horizontal branch is sensitive to the original abundance of helium. The B-V color index serves as an indicator of the stars' temperature: The hotter a star is, the smaller its B-V index. Position on the Hertzsprung-Russell diagram of the close and the brightest stars.

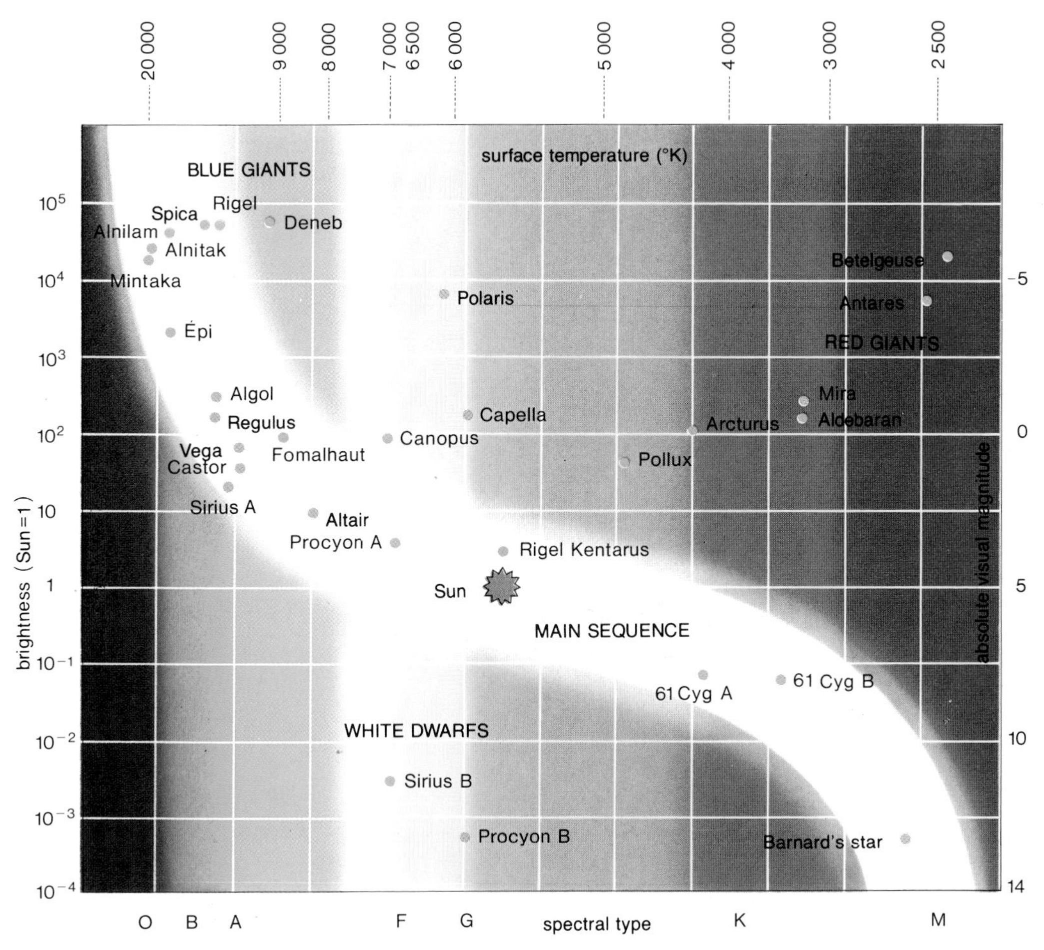

electron is smaller. Hence its quantity of motion must be greater according to Heisenberg's uncertainty principle; therefore its speed increases. When the electrons become relativistic—i.e., approaching the speed of light—the pressure of the degenerate gas of relativistic electrons becomes dominant. Now, this pressure is much less effective than when the electrons were not relativistic; thus the star cannot be stable. The threshold of stability is about 1.4 solar masses (the Chandrasekhar limit).

What happens to a white dwarf for which the mass, through accretion, goes beyond that fatal threshold? The answer really is unknown. It probably implodes, with the large quantity of gravitational energy thus recovered serving, in part, to blow away the star's outer layers. The core becomes even denser, perhaps a neutron star. The star must look like a nova, perhaps even a supernova.

Neutron Stars

We saw that a massive enough star—about 4 solar masses or more—ended its evolution in the superluminous part of the Hertzsprung-Russell diagram by exploding to produce a supernova. The star's atmosphere is completely ejected into the interstellar medium; on the other hand, its core, after implosion, remains, made up almost exclusively of neutrons, and perhaps a tiny fraction of electrons (0.5 percent).

Neutrons are to a neutron star what electrons are to a white dwarf. It is the degeneracy pressure of the neutrons that equilibrates the gravitational force. The radius of a neutron star is incredibly small: a neutron star of 1 solar mass has a radius of 10 kilometers (6.2 miles); its density is about 500 million tons per cubic centimeter! If the Earth had such a density, its diameter would be no more than 30 meters (98 feet); Mount Everest would be no taller than 2 centimeters, and an average person (if there could be any on its surface!) would measure no more than 10 μm (1/100 millimeter!). In fact, it is difficult to speak of a diameter for neutron stars, since their enormous density distorts the space around them, so that the notion of length depends on the distance to the star's center; it is preferable to speak of circumference: A walker on the surface of a neutron star of 1 solar mass has to travel only 60 kilometers (37 miles) to complete the circuit (still, that is a lot if you are only 10 μm tall).

The structure of a neutron star is fairly simple, and not without analogy to that of a planet. The atmosphere, if there is one, is very thin, for it is immensely compressed by the star's colossal gravitational field—at most a few centimeters in thickness (the thickness of the solar photosphere is a few hundred kilometers, and that of a variable of the Mira Ceti type, a few astronomical units!). The star's surface is a solid crust, with a crystalline structure, a thickness of about 1 kilometer (.62 mile), and a very high temperature: about 1 million degrees on the surface. The star's interior is fluid, with perhaps zero viscosity and zero electrical resistance.

Like white dwarfs, neutron stars cannot exceed a certain critical mass; the greater the mass of these stars, the smaller their radius. Like the white dwarfs' electrons, the neutrons of neutron stars take on greater and greater velocities, until they become relativistic. The degeneracy pressure of the neutrons then decreases, which has the effect of making the star collapse inward. Calculating the critical mass at which that collapse occurs is complex, owing to the nuclear reactions that may take place at that time, and also owing to the role of the flux of elementary particles that are emitted during collisions between neutrons. It is estimated that the mass of a neutron star cannot exceed 3 or 4 solar masses.

The existence of neutron stars was theoretically predicted in 1935. But they did not enter into the "world of astrophysics" until 1967, when radioastronomers at Cambridge discovered the first pulsar: a source giving off a radio wave with a duration of 50 milliseconds every 1.33730 seconds! Since then, more than 400 pulsars have been discovered. Often they are associated with sources of X rays varying with the same period. In the central region of the Crab nebula, the debris of the supernova of 1054, a pulsar was discovered and optically identified with a variable star of magnitude 16.

Pulsars: Neutron Stars in Rapid Rotation

The regularity of the period of pulsars is remarkable and can only be related to a rotational motion. Explanation of this period by stellar pulsations is not satisfactory—the possible periods of pulsation of white dwarfs are too large. On the other hand, the pulsational periods calculated for neutron stars are much too short. Only rotation, then, can account for the regularity of the observed phenomenon, and the rotating object must be a neutron star, for only stars of this type are small enough to be able to rotate so quickly—one rotation per second, or nearly—without instantly disintegrating (the maximum rotational period of a white dwarf is about 10 seconds).

The rapid rotation of pulsars results from contraction of the core of the star that gave rise to it. This core was rotating like the star itself before it exploded as a supernova; but since the core collapses, its diameter of course decreases and its rotation accelerates, just as an ice skater spins faster if he gathers his arms close to his body.

The magnetic field of the presupernova's stellar core is heavily compressed during the implosion of the degenerate core. The result is that the dipolar magnetic field of a neutron star is on the order of 100 million teslas (T), (while the Sun's global field is 1/10,000 T). This field would be 10,000 times greater still at the moment the neutron star appears. As this dipolar field is rotating, it gives off electromagnetic waves. Its field lines spin along with the star like a solid body, so that their tangential speed grows in proportion to their distance from the star, without, however, exceeding the speed of light, c. At the distance from the star where their speed approaches c, the shape of the field lines changes. They tend to wrap around the star in a spiral. At a great distance from the neutron star, the magnetic field, accompanied by an electric field, is identical to an electromagnetic wave.

Interpretation of the radio-frequency, X or visible rays is not yet clear, particularly because of the role the pulsar's magnetosphere plays. Many theories have been offered, based on the idea that a current of charged particles is emitted by the neutron star; these particles are strongly accelerated by the intense magnetic field up to relativistic speeds, probably along the axis of the magnetic poles. These particles then radiate, according to different processes. Even if, in the reference system of an emitting relativistic particle, radiation is emitted in all directions, in any other system, and particularly in any terrestrial reference system, it is confined within a narrow beam parallel to the direction of motion of the relativistic particle—i.e., following the axis of the magnetic poles. This beam is carried around by the rotation motion of the neutron star, which means that it sweeps through space like the beam of a flashlight in the dark. If the Earth is in this beam, the astronomer will see radio, X or visible pulses.

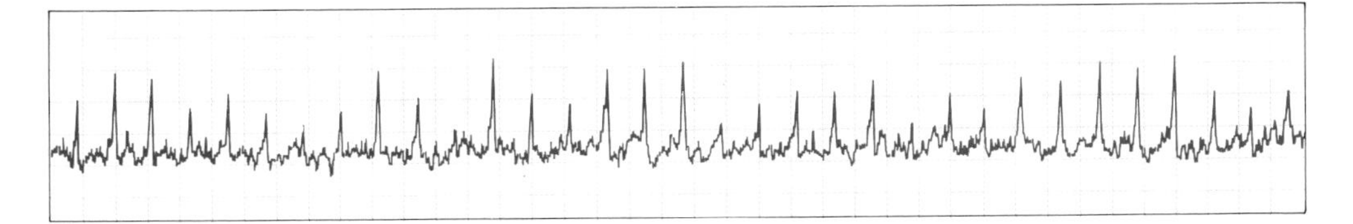

The radio signal emitted by a very regular pulsar. Here we see the periodic flashes of pulsar CPO 328 observed at 21 centimeters wavelength with the radio telescope of Nançay, France; the pulses last 7 milliseconds and return with a period of 0.714518603 second. They probably are emitted by particles situated above the magnetic poles and are carried by the very rapid rotational motion of the neutron star on its axis.

In 1054, a new star appeared in the sky. It was brighter than Venus and was visible in broad daylight: This was a supernova, mentioned in the Chinese chronicles of the time. Today there remains of that surpernova, first, ejected matter—comprising the Crab nebula, above—and second, a very dense star, the Crab pulsar. This combination of a pulsar with a supernova remnant is a confirmation of the theory of the origin— and of the very existence—of neutron stars. Photo: California Institute of Technology and Carnegie Institution of Washington.

This ejection of particles and their subsequent radiation require energy that, as it cannot be produced here by a nuclear reaction, is made up of the kinetic energy resulting from the star's ultra-rapid rotation. As the neutron star radiates, its rotation slows.

This slowing causes the final peculiarity of neutron stars that we will discuss here: starquakes. A star in rapid rotation is not perfectly spherical; it is flattened at the poles and bulging at the equator. In particular, the outer crust of a neutron star is strained. When the pulsar's speed of rotation slows, the centrifugal force decreases, and the star should gradually become less flattened. Now the crust of a neutron star is solid; thus it does not deform gradually but breaks when its internal stresses (resulting from the disequilibrium between the preponderant gravitational force and the decreasing centrifugal force) become too strong. Then we observe a starquake. The effect of this quake is to change the distribution of the star's matter, which changes the pulsar's speed of rotation and hence its period. Such discontinuous variations in period have been observed for several pulsars. The changes in the profile of the radio pulses accompanying these quakes are valuable means of investigating the internal structure of neutron stars, just as the analysis of terrestrial seismic waves gives us information about the Earth's internal structure.

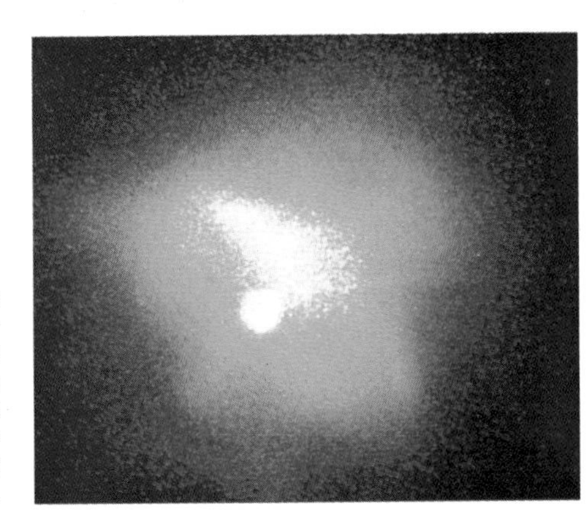

In the X-ray range, the Crab nebula has a totally different appearance. The pulsar at the center appears as an extremely bright object. The picture (taken by the American satellite HEAO 2*) also reveals a complex energy transfer from the pulsar to the surrounding medium. P.P.P.-IPS Photo.*

Black Holes

The evolution of very massive stars (exceeding about 8 solar masses) beyond the supernova phase leads, very probably, to a core too massive to become a neutron star; in the most massive neutron stars, the neutrons assume relativistic speeds, so that the degeneracy pressure becomes relativistic itself. This relativistic pressure increases very slowly with the star's mass, so that, if the core of a presupernova is appreciably more massive than the critical mass at which the degeneracy pressure becomes relativistic, this pressure cannot equilibrate the gravitational force (which increases in proportion to the mass). The star, then, can only collapse on itself; no pressure can then prevent this catastrophic plunge. The entire mass of the core of the presupernova is concentrated at the central point, where the density becomes infinite! Such an object is a genuine hole in the universe.

Since the density is infinite at the center, the escape velocity from the center point is itself infinite; therefore it would be necessary to accelerate a particle located. e.g, at this center point to a speed greater than the speed of light (300,000 kilometers [186,451 miles] per second) in order for it to leave this hole once and for all. That, of course, is impossible. Thus nothing that falls into the hole can ever come out of it; in particular, no radiation can escape and reach us, hence the name "black hole" or "occluded star" given to such an object. It is remarkable that, in 1796, Laplace foresaw the existence of bodies too massive and dense for any radiation to escape.

If no particle can escape from the black hole itself, we may wonder from what distance from the black hole it is possible to return. It is well known, and easily proved, that the farther one is from a body, the easier it is to free oneself from its gravity. Thus, for each black hole, there is a distance beyond which the escape speed becomes less than the speed of light. This distance is called the "black-hole radius." This does not refer to the radius of the central mass, since it is a point, but to the radius of the region of space surrounding it, which is completely cut off from the rest of the universe. A space vessel that had the dangerous mission of closely studying a black hole could not return to Earth if it came close enough to the black hole to cross that boundary. We will see farther on what happens to any matter that crosses it.

The radius of a black hole is very small and obviously dependent on its mass: It is 3 kilometers (1.9 miles) for a black hole of 1 solar mass, and 10^{-24} millimeter (i.e., zero point twenty-three zeroes followed by 1) for a black hole of 1 kilogram, or a hundred-thousandth-billionth of the dimension of an atom! Such a hole would pass matter without hindrance. On the other hand, a black hole of 50 solar masses has a radius of about 160 kilometers (99 miles). But, even more than for a neutron star, it is preferable to speak here of circumference rather than radius; the circumference of a black hole of 3 solar masses is about 60 kilometers (37 miles), while that of a black hole of 50 solar masses is 1,000 kilometers (622 miles).

In fact, the problem of a particle escaping from a black hole does not arise, since there is nothing inside a black hole, anyway. The particles themselves are reduced to a zero diameter; they collapse on themselves ϕ upon reaching the center of the black hole. Since there are no particles in a black hole, there is no problem of interaction between particles! Thus the center region of a black hole appears as a simple object to the physicist. It is a singularity in space of nonzero mass—i.e., something infinitely more simple than an atom, even one of hydrogen.

Different Possible Types

There is no theoretical limit on the mass of a black hole in either direction. The theory of stellar evolution predicts the existence of black holes of several solar masses, but two other theories predict the existence of other types of black holes.

On the one hand, the cosmology of the "big bang," of the initial explosion that is thought to have produced our current universe, says that small black holes may have been formed during this explosion. But the black holes lighter than 10^{16} g. (about the mass of Lake Geneva) should slowly give off a small amount of light and particles, through complex processes arising on the surface of their horizon; in this way they should lose some of their mass and disappear in the form of a violent eruption of γ rays: these are mini-black holes. Mini-black holes less massive than 10^{15} g. would already be dead. Those that still would be observable now would have masses between 10^{15} and 10^{16} g. Current physical conditions no longer permit the creation of such black holes.

On the other hand, n-body theory (theory of gravitational interactions between a total of n distinct bodies) predicts the possible formation of massive black holes, up to several hundred solar masses, at the center of globular clusters. Through collisions or very close encounters between stars, a tight double system should form at the center of the cluster, and would evolve into a central black hole; subsequent collisions of stars with the central black hole would only make it heavier. A similar process might take place in the nucleus of our galaxy. This time, it would involve a gigantic black hole: Its mass would be about a million solar masses; its circumference, about 10 million kilometers (6.2 million miles). We will not discuss these two types of black holes any farther here, since they do not result from the evolution of a star.

The shape of the horizon of a black hole is not necessarily spherical. All stars are rotating; the Sun rotates on its axis in about 25 days, which corresponds to a rotation speed at the equator of about 2 kilometers (1.2 miles) per second (four times more than on Earth). Some hot stars have equatorial rotation speeds of more than 300 kilometers (186 miles) per second! Likewise, the core of a star collapsing to form a supernova is rotating while it collapses inward. Consequently, the black hole that results from that collapse is rotating. The effect of this rotation is to flatten the horizon of the black hole at the poles, just as Jupiter, which rotates in 10 hours, appears clearly flattened even in the smallest telescope. The equatorial radius of the horizon is not affected by the rotation, while the polar regions are pushed inward. However, if the black hole rotates very rapidly, the shape of its horizon no longer can be described in our Euclidean space.

The effects of the distortion of space around a black hole often are compared to what happens in a whirlwind, with the center of the whirlwind representing the black hole. A simpler representation, because it has a single dimension, might be the following:

Let us imagine a canoe being paddled between Lakes Erie and Ontario, upstream from Niagara Falls. At a great distance from the falls, in spite of the current, the paddler can maneuver as he wishes, to go where he wishes. He can even keep the canoe motionless by resisting the current. If the vessel approaches the falls, it reaches an area where it can no longer move against the current because the force of the current has increased. The paddler can then only move his canoe sideways to get away from the falls. For a black hole, the equivalent of this boundary between the areas of total freedom and those of "controlled" freedom of the photons is called the "static limit." And the equivalent of the limit beyond which it is no longer possible to "climb back up the current," even sideways, is ϕ none other than the horizon of the black hole. The region of space stretching from the static limit to the horizon is called the ergosphere.

Myth or Reality?

The major problem of research in the area of black holes is, at the present time, to prove by observation that they are not merely the products of theoretical speculations. To do that, it is necessary to discover at least one. But since black holes are invisible, how could we confirm by observation that they exist?

If a black hole is isolated in interstellar space, it is practically impossible to detect its presence unless one is right next to it. Two potential tracers of its presence have been suggested by theoreticians—one very difficult to observe, the other very unlikely.

The first way makes use of the interaction of the black hole with the interstellar matter within which it may be found. This matter falls into the black hole and is heavily compressed in its fall; as a result, it heats up and gives off radiation. Thus we could detect a black hole by this emission. The spectral energy distribution of this radiation would be very different from that of a star. Unfortunately, the density of interstellar matter is not sufficient for this radiation to be substantial: The absolute magnitude (about 16) that the black hole would reach would make it extremely difficult to detect.

The other possibility for detecting an isolated black hole rests on the "gravitational lens" effect. Because of its density and mass, a black hole distorts the space around it, so that it deflects light rays, somewhat like a lens. If, then, a star lies on the line of sight of a black hole, but beyond the black hole relative to the Earth, part of the radiation given off by that star will be focused in the direction of the Earth. For the terrestrial observer, the star will appear brighter, and angularly larger, during its passage behind the black hole. We can well imagine that an Earth-black hole-star alignment is very rare. But, in addition, the focusing effect will be greater, hence more observable, if the occulted star is far from the black hole. This greatly reduces the probability of such an alignment as well as the probability that the occulted star would be sufficiently bright to be known.

Thus observers are not concentrating their efforts on isolated black holes. If, on the other hand, the black hole is a component of a binary system, the situation is more favorable.

Like any member of a binary system, a black hole should affect the motion of its companion. The orbit of this companion, in particular, depends on the mass of each of the two objects in the system. The motions of many stars have been studied in efforts to determine whether they are members of a double system in which the second member is invisible. The mass of this invisible companion deduced from the motion of the primary in no case makes it possible to state that what we are faced with is a black hole.

A more promising case of a binary is that of the very tight binaries, with periods of revolution ranging from several hours to several dozen hours. The prototype of the tight binaries is well known to amateur astronomers, since it involves Algol, or β Persei, the period of which is 69 hours (but neither of Algol's two components is a black hole!). In the case of a sufficiently tight pair, the strong gravitational field resulting from the high mass of the black hole may cause matter from the outer layers of the primary to be drawn toward it. The matter thus lost from the primary arrives near the black hole when it has already moved appreciably along its orbit; the matter thus "misses" the black hole; instead of falling into it, it passes behind it and becomes a satellite, so to speak, around the black hole, forming a disk of gas that girdles it. This disk is called the "accretion disk." The enormous gravity present near the surface of the black hole compresses this gas so intensely that it heats up very rapidly. In addition, the internal layers of the accretion disk spin much faster around a black hole than the outer layers; this results in intense friction in this accretion disk, which contributes to heating the gas to temperatures of about 1 million degrees. Up to 40 percent of the mass of the gas that falls in a spiral around the black hole would be converted into radiation! This very energetic radiation is emitted mainly in the X-ray range.

A Serious Candidate

The X-ray telescope of the *Uhuru* satellite, launched in 1970, made possible the discovery of many X-ray sources. Some can be identified as neutron stars surrounded by an accretion disk and forming a binary with a "normal" star. Hercules X-1 (the first X-ray source discovered in the constellation of Hercules) gives off an X-ray pulse every 1.24 seconds; it is eclipsed behind the primary component—the variable star HZ Hercules—for 5.8 hours every 1.7 days. The cause of the X-ray pulse is the fall and compression of gas from the accretion disk onto the magnetic poles of a neutron star; the gas is brought there, still at temperatures greater than 1 million degrees, and as a result the magnetic poles give off X rays. The rotation of the neutron star brings these poles alternately into view and out of view of a terrestrial observer.

At least one of the X-ray sources discovered by *Uhuru* probably is associated with a black hole. The pattern observed in Hercules X-1 cannot apply to the first X-ray source discovered in Cygnus, Cygnus X-1. Through its radio radiation, investigators have been able to identify Cygnus X-1 with the supergiant HDE 226868, a spectroscopic binary with a single-lined spectrum. The variations in the radial velocity of this star imply that the mass of Cygnus X-1 is at least 6 and probably 10 solar masses. Thus Cygnus X-1 is too massive to be a neutron star. Probably it is a black hole, the first black hole known.

Only more elaborate observations will be able to confirm (or invalidate) this hypothesis. In particular, the theory predicts that hot spots could form on the accretion disk, with a life-span of several periods of revolution. These hot spots would be responsible for the X-ray bursts observed on Cygnus X-1. The fine structure of these bursts should reveal the rotation period of the disk at the distance from the black hole where the spot is located; the flux of radiation is not constant, for the Doppler-Fizeau shift makes the X rays apparently more intense when, in the course of its rotation, the spot approaches the Earth, and less intense when it moves away. For a black hole of 8 solar masses, as Cygnus X-1 might be, these variations should have periods of from 3.6 to 0.6 millisecond, depending on whether the rotation of the black hole itself is 0 or maximal.

The search for these variations on the X-ray bursts probably will lead to the least questionable identification of an X-ray source in a binary system with a black hole.

Thus three outcomes are possible for a star, depending on its mass: a white dwarf, which gets its energy from its slow cooling; a neutron star, which gets its energy from the gradual slowing of its rotation speed; and a black hole, which gets its energy from the action of its gravitational field on the matter surrounding it. None of these three processes has a reason to stop once and for all, so even if the energy flux it emits can decrease indefinitely until it becomes minute, a star never dies.

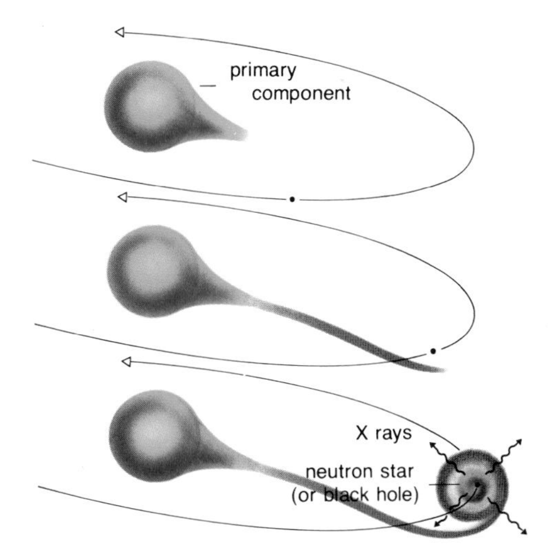

Diagram of a close binary, of which the secondary component is a neutron star or a black hole. The strong gravitational field of the secondary object attracts matter from the top layers of the primary object, through a tidal effect. This matter winds around the secondary object in a disk; there, the friction it undergoes heats it to temperatures of about 1 million degrees. The disk therefore radiates in the X-ray range. In the case of a neutron star, it is perhaps the very dropping of matter onto the magnetic poles of the neutron star that causes the principal heating and hence the principal source of X rays.

22. The Interstellar Medium

The study of the interstellar medium occupies an important place in astrophysics. In this medium, rich in everything that lies between the stars, there exist very special physical conditions that we are unable to simulate in our laboratories on Earth but that we nonetheless understand relatively well. This medium plays a fundamental role in the problem of star formation and the formation and evolution of our galaxy. That is why, in the following pages, we will try to present the various constituents of this medium, even though such a breakdown may seem artificial at times.

In the region of the star ρ Ophiuchi (above, shrouded in a nebula by blue reflection) is a large cloud of dust and gas masking stars behind it. This dark nebula, which appears very vast owing to its relative proximity (about 700 light-years), is an important source of molecular emission (more than 20 molecules have been detected in it) and an intense source of γ radiation. The bright star visible below, at right, surrounded by a red and yellow nebula, is Antares. To the right of it, we see the globular cluster M 4. Photo © 1979 Anglo-Australian Telescope Board.

Interstellar Matter

OBSERVATION OF DARK OR bright nebulae in our galaxy and study of their optical spectra have shown that the matter of the universe is not contained only in stars but that a certain quantity of gas and dust exists in the galaxies. This extremely dilute matter is between the stars—hence its name, interstellar matter. It constitutes a substantial proportion of the mass of many galaxies. While elliptical galaxies contain very little interstellar matter, spiral galaxies comparable to ours have several hundredths of their mass in this form, while certain irregular galaxies, such as the Magellanic clouds, contain nearly 40 percent. Some irregular dwarf galaxies seem to be made up almost exclusively of gas.

The importance of interstellar matter is considerable—first, the stars and planets that move within it were formed at various epochs, and some are being formed at present from this matter. As stars evolve, they lose mass, either peacefully (supergiants stars, red giants, planetary nebulae) or violently (eruptive novae and supernovae). The Sun loses part of its mass in the form of the solar wind, and some magnetic storms are due to the bombardment of our planet by particles emitted by the Sun. This matter ejected by the stars ends up in the interstellar medium and mixes with the gas and dust already there. Its composition is different from the original interstellar matter since it depends upon the evolution of the star from which it was expelled, and we note that it has been enriched with heavy elements (carbon, nitrogen), while initially it was made up only of hydrogen and helium (the big bang theory). In addition, a study of isotope abundance ratios (carbon-12 and carbon-13, for example) shows which types of stars have contributed to the enrichment of the interstellar medium with heavy elements.

Physicists and chemists who study interstellar matter find a wealth of subjects. The physical conditions there are highly varied and cannot be achieved in a terrestrial laboratory. The density generally is low, from one to 1 billion (10^9) atoms per cubic centimeter (the best "vacuums" created on Earth contain several million billion—10^{15}—atoms per cubic centimeter). The temperature also can vary, from 3 degrees to 10 million (10^7) degrees absolute (K). More typical values would be a density of 10 to 10,000 atoms/cm^3 and a temperature of 10° K to 10,000° K. The ionization rate (ratio of the number of ions to the total number of atoms) varies greatly, depending on whether one is close to a star or far away from it, and its measurement is one of the most interesting problems in the dense clouds where practically no radiation can penetrate. Hence it is not surprising to find, under such conditions, highly varied phenomena that are difficult to re-create in a laboratory, such as the forbidden emission lines in gaseous nebulae, and radio emission from molecular masers in the vicinity of newly forming stars, to which we will return in detail later. The chemistry of the interstellar medium also is rich and very different from that of a laboratory: More than 50 molecules have been identified in the former, in general by their emission spectra in the millimeter-wavelength domain. They form either by catalysis on grains of dust, or by gas-phase chemistry involving highly reactive ions, such as HCO^+ or H_3^+, of which the equivalent does not exist naturally on Earth.

Interstellar Dust

Closely mingled with interstellar gas, interstellar dust represents about 1 percent of the mass of the gas in our galaxy. It manifests itself to the observer in various ways.

First, it dims the light from the stars behind it. On the one hand, it absorbs starlight; on the other hand, it scatters part of it in directions that differ from the direction of incidence. Some dust clouds are visible on photographs of the sky as zones in which stars are absent. This idea of dust interposed between the stars and the observer was not obvious, and many astronomers spoke of a region of the sky in which stars were missing. The Sombrero galaxy appears divided in two by such a band of absorption; in fact, a photograph shows that the dust forms a disk flatter than that of the stars; this disk, however, is far from homogeneous, and it has a microstructure. We note within it dark nebulae of large dimension, as well as small nebulae appearing as dark globules within bright nebulae, called Bok globules. The dust, then, is divided among clouds of greatly varying dimensions, from 0.03 to several dozen parsecs. Interstellar extinction is a considerable nuisance for the optical astronomer, for it prevents him from seeing and studying stars more than 2 or 3 kpc away in the plane of the galaxy. Thus the center of our galaxy is completely inaccessible to optical studies.

For a star not affected by interstellar extinction, the relationship between distance d, apparent magnitude m, and absolute magnitude M is written: $m = M + 5 \log d - 5$, d being expressed in parsecs.

Interstellar extinction A_v increases the magnitude, which then becomes, for a star showing extinction A_v in magnitudes, $m = M + 5 \log d - 5 + A_v$. However, it is necessary to specify the wavelength at which the extinction is measured; in the UBV system, the extinction A_v in color V of average wavelength 5500 Å is about 1.5 mag-

The Lagoon nebula (M 8) in the constellation Sagittarius. This vast hydrogen and dust cloud contains several variable stars of the T Tauri type, considered to be very young, and dark globules that are thought to be stars in formation. Photo: Kitt Peak National Observatory.

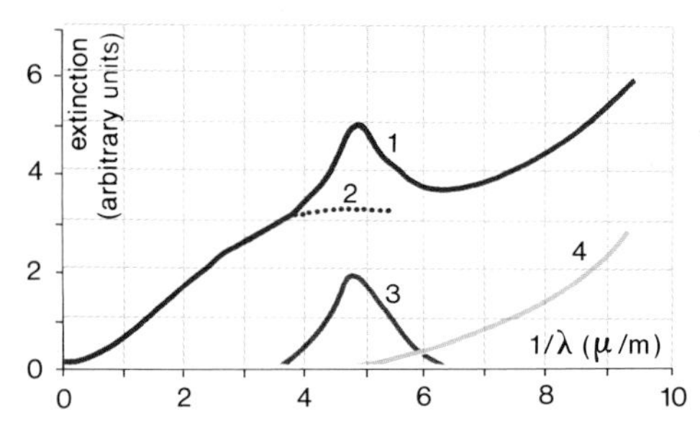

Schematic curve of interstellar extinction (1). Curve (2) is due to ordinary grains, while the small grains produce an extinction peak at 4.6 μm (3), perhaps caused by graphite, and an extinction at very short wavelengths (4). From M. Greenberg, 1978.

nitude/kpc. A direct measurement of extinction is difficult, for it is necessary to determine both M (based on spectroscopic analyses) and d (by a nonphotometric method). Fortunately, there are other indirect means of measuring extinction—e.g., by studying neutral gas or the infrared radiation from the dust.

Interstellar extinction is selective—different wavelengths are not affected in the same way. Weak in the infrared, extinction increases greatly when the wavelength decreases, and becomes very important in the ultraviolet (that is what explains the observational difficulties of the Copernicus satellite below 1500 Å). The light received from the stars thus appears redder than if extinction did not exist; this is interstellar reddening.

Moreover, interstellar dust scatters part of the light it receives. That is what explains the fact that we can see luminous emission from dust clouds illuminated laterally or frontally by a star close to the cloud. These are reflection nebulae, which should properly be called scattering nebulae. The nebula surrounding the Pleiades is a famous example of this.

The dust is not oriented randomly in interstellar clouds. Several authors have suggested that the dust is aligned owing to the existence of an interstellar magnetic field and that this alignment creates a polarization of the light received from stars. The degree of linear polarization reaches several percent, and a very weak component of circular polarization recently has been discovered.

Finally, interstellar dust gives off thermal radiation in the far infrared, owing to its fairly low temperature. This radiation, ranging essentially between 60 microns and 1 millimeter in wavelength, is accessible only to space vehicles, airplanes, balloons or rockets, and occasionally from the ground (at about 350 μm; the atmospheric temperature at a high mountain site permits direct observations of the denser clouds). As extinction is low at these wavelengths, it is possible to map clouds of interstellar matter throughout the galaxy by their thermal infrared emission. A comparative study of these maps and of the molecular emission of the dust clouds makes it possible to determine the distribution of these clouds in our galaxy and to define their physical parameters.

The Physics of Interstellar Grains

The theory of the absorption, emission and scattering of electromagnetic waves is already an old topic, but it is undergoing a revival of interest owing to recent observations in the infrared and ultraviolet.

Specialists are trying to reproduce the law of extinction and the albedo based on models of grains, the properties of which are known *a priori;* they are guided by the fact that the grains must be formed from the most abundant elements (carbon, nitrogen, oxygen, silicon and iron) and cannot have a total mass greater than about 1 percent of the mass of interstellar matter (the rest being gas, the mass of which is determined separately), for they are made up only of the heavy elements just mentioned, and their abundance relative to hydrogen and helium—the essential constituents of the mass of the galaxy—is known.

• *Dimensions of interstellar dust grains* — To account for the strong extinction observed in the ultraviolet, it is necessary for there to be very small grains, with a radius of about 100 angstroms (Å). But we cannot say exactly what the composition of these grains is. Some authors think that the extinction maximum at around 2200 Å wavelength is due to particles of graphite or silicates that possess a characteristic absorption at this wavelength. It is difficult to say more, for the mass of these grains represents only a small fraction of the total mass.

Extinction in the visible is easily explained by grains having a radius 10 times larger than the preceding ones. Excellent agreement with observation is achieved with grains containing a nucleus of silicates of radius 50 Å covered by an ice mantle that brings the radius to about 1,000 Å.

It is quite tempting to think that the small grains are nuclei on which materials making up the ice mantle condensed.

The grains quickly reach thermal equilibrium and, if we know the radiation field, we can calculate their temperature, which we find to be 10° to 15° K for the average interstellar field. We will see that this temperature can be measured indirectly by study of interstellar molecules, and directly from the thermal radiation of the dust at various wavelengths (from 10 μm to 1 mm., usually). Most of the observations were made from a balloon or a rocket, for this wavelength region is scarcely accessible from the ground. Unexpected results were obtained: In many interstellar clouds, the temperature of the grains is closer to 30° K, rather than 10° to 15° K, and therefore there certainly is an internal heat source in the cloud. To heat up these clouds takes a power 100,000 to 1,000,000 times that of the Sun; only young, massive stars in the cloud or in close proximity to it can deliver such power. In some cases, we can directly detect these stars, either because they are optically visible outside the cloud, or when their extinction shows they are inside the cloud or behind it, or when they are measured as relatively pointlike infrared or radio-astronomical sources. The infrared emission from these objects corresponds to the emission of the star itself or to that of the dust close to the star, which is very hot and which radiates in the near or medium infrared.

An original determination of the type of star buried in the cloud can be made by studying the relative dimensions of the ionized regions in which we observe the recombination lines of hydrogen, carbon and sulfur, as we will see later.

Recent observations of a large part of the galactic emission in the far infrared make it possible, moreover, to state that the galaxy gives off as much energy in the form of thermal emission from the dust in the far infrared as in all the other wavelength regions combined. Thus we can estimate that half of the energy radiated by stars is absorbed by dust, which reemits it in the far infrared.

• *Nature and origin of interstellar dust*— Only very recently has substantial progress been made in the subject of dust grains. Our total knowledge comes from observations in the infrared. First, a wide absorption band at about 9.7 μm wavelength was found in the spectrum of many infrared sources heavily affected by interstellar absorption—i.e., those that must be behind a large quantity of dust. Another, narrower band is found at 3.1 μm. These bands are due to silicates and ice, respectively; they have been observed in the laboratory in these materials, and the existence of these constituents in interstellar grains has been demonstrated. Solid ammonia, NH_3, also can contribute to the band at 3.1 μm, and we will see that its presence in interstellar grains is a certainty. The two bands have variable intensities from one object to another: It is clear that the ratio of the quantity of ice to the quantity of silicates varies from one place to another. In some cases, the dust is hot enough for the silicate band to appear in emission (it has to be about 300° K); such a situation occurs in circumstellar envelopes or in regions where the dust is very close to hot stars.

In 1978, new observations supplied fundamental information: Wide emission lines were discovered in the near and medium infrared spectra of a fairly large number of dusty objects: planetary nebulae, gaseous nebulae, and even nuclei of galaxies. These lines were not immediately identified, and the initial explanations were unconvincing. In fact, these lines are characteristic of the radiation emitted by many molecules contained in the grains. When these grains are close to those of a hot star—a source of ultraviolet light—the molecules they contain absorb some of that light and turn up in excited vibrational states: Their subsequent deexcitation leads to the emission of the observed wide lines, the equivalent of which has been reproduced in the laboratory. In this way it has been possible to recognize ice (H_2O) by its emission at 3.1 μm, which is the emission equivalent of the absorption band mentioned earlier, and by other emission features (in particular that at 11.3 μm). Methane (CH_4), ammonia (NH_3), acetylene (C_2H_2) and perhaps (H_2CO), carbon monoxide (CO) and nitrogen monoxide (NO) also have been recognized.

This list is very close to that of the molecules observed in dense molecular clouds, as we will see below. This method certainly will make it possible to observe other constituents, but it seems that the most abundant molecules making up grain mantles already have been observed. Unfortunately, a measurement of the relative abundances of these various molecules cannot yet be made in the dust grains, contrary to the situation in dense molecular clouds.

These observations clearly confirm the dual nature of interstellar dust: Small silicate particles appear to form the small grains responsible for extinction in the far ultraviolet and at the same time are the condensation nuclei onto which the molecules making up the mantles come to stick. The origin of the small silicate grains is fairly easy to explain. The material constituting them is the first to condense when a very hot gas having the composition of the stars (and of interstellar matter) is cooled. When stars eject part of their envelope—which happens frequently at the end of their life, in the stages represented by red giants, Mira variables, planetary nebulae and even supernovae—small silicate grains condense there as the gas cools. Where and how are the volatile materials making up the mantles deposited? Two conflicting ideas have been suggested: According to some, this deposition takes place soon after the formation of the small silicate grains, in the next stage of the gas cooling; for others, it is in the dense interstellar clouds that the mantles grow by the accretion of interstellar molecules that happen to get stuck there. Which of these two models is the right one? In the first, a portion of the interstellar molecules can form from preexisting mantles, while in the second, these mantles form from interstellar molecules. It seems that these gas-dust exchanges can take place in both directions, so the time situation perhaps is a mixture of the two possibilities.

Diffuse Interstellar Gas

This component of the interstellar medium was discovered at the beginning of this century from optical interstellar absorption lines. In stellar spectra, we observe extremely narrow absorption lines—much narrower than stellar lines—that are relatively scarce. Often the bands are multiple, which is interpreted as the presence between the star and the observer of several clouds having different radial velocities, each of which produces a component shifted relative to the others by the Doppler effect. These observations reveal the existence of interstellar gas that is inhomogeneously distributed among the clouds. In 1951, the 21-centimeter line of atomic hydrogen—the principal constituent of this medium—was discovered. More recently, observations by rocket, balloon and satellite have supplied a vast amount of information on the composition and physics of the interstellar medium.

The 21-Centimeter Line of Atomic Hydrogen

The line at 21 centimeters (v_o = 1420.405752 MHz) corresponds to the transition between two hyperfine sublevels of the lowest energy level of atomic hydrogen. Indeed, in atomic hydrogen—as in any other atom or molecule having a nucleus of nonzero spin—the two possible relative orientations of the nuclear spin and the electronic spin correspond to a slight energy difference, and the wavelength of the associated transition is situated in the microwave region. This transition is forbidden, meaning that it has only a very low probability of occurring; a hydrogen atom left to itself in the upper level will take 10 million years to fall back to the lower level. Despite the very small probability of this event, there are so many hydrogen atoms in the interstellar medium that the corresponding line (at 21 centimeters) has an appreciable intensity. This situation, in which we

The Pleiades. The stars of this famous cluster in the constellation Taurus appear to be shrouded in nebulae that are typical examples of "diffuse clouds" in which the CN, CH and CH+ radicals were observed in 1937. Photo © 1961 by California Institute of Technology and Carnegie Institution of Washington.

observe lines that are difficult or impossible to observe in a laboratory, will recur in other contexts (H II regions, molecular clouds, etc).

A very large number of observations of the 21-centimeter line in emission have been made since its discovery, both in our galaxy and in external galaxies. Our galaxy, in particular, has been entirely mapped in the 21-centimeter line. In general, we observe that in a given direction, the line has a complex profile with several components. This profile is due to the radial component of the motion of the emitting atoms: The various components correspond to separate clouds of different radial velocities (or perhaps to the superposition of several clouds with the same radial velocity); the width of each of the components is linked to the internal motions in each cloud, of which a fairly small contribution is made by the thermal agitation of the atoms, since the temperature is not very high, as we will see farther on.

If observations in emission are valuable sources of information for studying the distribution of interstellar matter in the galaxy (this point is discussed in the following chapter), observations in both emission and absorption yield crucial information concerning the physics of the interstellar medium, since they make it possible to measure its temperature and, to some extent, its density. These observations show that the largest interstellar clouds have temperatures of about 50° to 100° K but that a certain number of clouds have temperatures that reach or exceed 1,000°K. The internal motions of the clouds reach an average of 1.7 kilometers per second, (internal-velocity dispersion); finally, interstellar clouds have random velocities relative to one another, with a dispersion of about 6 kilometers per second along a line of sight, and 10 kilometers per second in three dimensions. It is not possible to define a "standard" cloud, for the properties of interstellar clouds are extraordinarily diverse. Let us say, for the sake of argument, that their density ranges from 1 to 1,000 atoms/cm^3 and their mass from a fraction of a solar mass to several hundred solar masses.

Measurements in emission/absorption also enable us to find the distance of radio continuum sources in the following manner: In the emission spectrum observed at a position adjacent to the source, generally we see several components with different radial velocities, which can in general be at different distances. In the absorption spectrum observed in the direction of the radio source and corrected for emission, we see the absorption counterpart of certain emission components but not of others. This makes it possible to localize the source at a distance greater than that of the absorbed components but closer than that of the unabsorbed components. Thus we get the relative distances of the various spiral arms. By determining the scale of these distances using the independently measured radial velocities of the H II regions, we are finally able to establish the structure of the galaxy.

Interstellar Absorption Lines in the Visible and Ultraviolet

As we said, very narrow absorption lines of interstellar origin have been observed since the beginning of the 20th century in the spectrum of many stars. In the visible range, these bands are scarce, and only a small number of atoms (Na, K, CA) and ions (Ca^+, Fe^+, Ti^+) as well as a few molecules (CN, CH, CH^+) can be observed in this way. The ultraviolet spectrum of stars is much richer in interstellar lines, particularly in the far ultraviolet range: 912 to 1300 Å. Hundreds of lines have been observed, first by rocket probes, then by the specialized American satellite *Copernicus,* which was equipped with a 90-centimeter (35-inch) telescope and a high-resolution spectrograph. The most intense is the Lyman α line of atomic hydrogen at 1,216 Å, but we also observe the great majority of abundant atoms and ions, as well as a few molecules: H_2, CO, OH and C_2. The observed lines always correspond to resonance transitions between the lowest energy levels (or levels of very small excitation) and high levels; indeed, only the lowest levels are substantially populated, owing to the rather low temperatures prevailing in the neutral interstellar medium. Up to now it has been possible to study interstellar lines only in front of relatively nearby stars, generally at less than 1 kpc. occasionally more; the more distant stars are too faint for observations at high spectral resolution. Likewise, observations are difficult or impossible in front of stars suffering strong interstellar extinction, especially in the far ultraviolet, where that extinction is particularly strong. Thus the interstellar absorption lines make it possible to study only the diffuse interstellar clouds, the sources of 21-centimeter line emission and only near the Sun. However, they bring us essential information about the composition of the interstellar medium.

Composition and Physics of Diffuse Interstellar Clouds

As a result of the previously mentioned observations, it is now possible to get an idea of the composition of the diffuse interstellar medium. Most of the elements are less abundant than in the solar system, sometimes by a factor of more than 1,000. Even though small differences may exist between the abundance of elements in the solar system, which is 4.6 billion years old, and the current interstellar medium because of the evolution of the galaxy, they certainly do not exceed a factor of 2, and cannot be called on to account for the enormous deficiencies observed in the interstellar medium. The natural explanation for these deficiencies is that the missing fraction of the elements is found in the form of interstellar grains. Moreover, the most deficient elements (calcium, aluminum, iron, silicon, etc.) also are the most refractory, and in particular are those that form silicates easily, which we have seen are an important constituent of the grains. Even carbon, nitrogen and oxygen (the most abundant elements after hydrogen and helium) are deficient, but to a lesser degree; their abundance, unfortunately, is poorly determined, since they have very strong lines. Where and how these elements condensed onto the grains still is uncertain. The fact that they seem to be only slightly or not at all deficient in the lowest-density clouds may indicate that they condense around the nuclei of silicate grains to form the mantles of ice only in the fairly dense interstellar clouds.

Observation of interstellar lines reveals that many elements, particularly the metals, exist not only in the form of atoms but also in the form of ions: for example, Ca and Ca^+, Mg and Mg^+, C and C^+, etc. This is easily explained: The rather diffuse interstellar clouds being discussed here are easily penetrated by the ultraviolet radiation from the hot stars of the galaxy. Hydrogen is so abundant that it rapidly absorbs the photons with a wavelength shorter than 912 Å, becoming ionized near the stars. However, it lets the photons of longer wavelength pass, some of which can ionize other elements. The elements likely to become ionized are those having an ionization energy less than 13.6 electron volts (that of hydrogen); in fact, this includes all of the elements except oxygen, nitrogen and the rare gases, which thus remain as neutral atoms. When the abundance of an element in its atomic *and* its ionized form (for example, that of Ca and Ca^+, from interstellar lines in the visible), can be measured at the same time in the same cloud, we can deduce a very important quantity from it: the density of free electrons in the interstellar medium. We can, in fact, state that the number of ionizations per second of a chosen element, which is proportional to the flux of ultraviolet radiation from the stars and to the density of the atoms of that element, is equal to the number of recombinations, which depends on the densities of ions and electrons; in general, we find that there is about 1 free electron for 10,000 hydrogen atoms. That is what we except, given the abundance of the principal elements capable of being ionized, chief among which is carbon (the density of the electrons is, of course, equal to the sum of the densities of all the ions). When such an element exists in two states of ionization, the abundance of the atoms is much less than that of the ions.

The Role of Ions

Even though the ions we have been discussing are present in trace quantities, since they scarcely represent more than 10^{-4} of the total number of atoms (dominated almost totally by hydrogen and helium), they play an important role in the physics of interstellar gas, since they are its principal cooling agents. Here is how this occurs. Many ions, in particular C^+, have their fundamental energy level doubled as a result of the interaction between the orbital angular momentum and the spin of the electrons (the fine structure interaction). As a result of collisions between the free electrons of the gas and the ions, some of the ions may be carried into the uppermost fine structure level, resulting in a loss of energy by the electron that participates in the collision. This energy will subsequently be radiated by the deexcitation of the ion when it falls back down into the lowest energy level, giving off a photon in the far infrared (at 156 μm wavelength in the case of C^+). Thus the energy of the free electrons is gradually lost through this process. It can be shown that the atoms and ions are in thermal equilibrium with the electrons, so that all of the gas cools.

A Still Uncertain Heating Mechanism

The cooling is mainly due to the emission by C^+in the line at 156 μm; this line has not yet been widely observed, although the population of the two fine-structure sublevels of that ion, which produce two different absorption lines in the ultraviolet, has been observed. This cooling is significant and must be counterbalanced by an effective heating so that the temperature of the interstellar clouds (measured with the 21-centimeter line) may reach or exceed 100 degrees K. The heating mechanism still is uncertain. The energy release accompanying the ionization of this medium by ultraviolet light from the stars is insufficient. A more effective but as yet poorly understood mechanism involves ionization of the surface of the dust grains by the same ultraviolet photons—an ionization that would eject from those surfaces a substantial quantity of electrons with energies of several electron volts. The most interesting mechanism involves the molecule H_2; observation from the *Copernicus* satellite has shown that it is abundant in interstellar clouds. This molecule forms, as we will see, as a result of the meeting of two hydrogen atoms stuck to a dust grain (this is an example of catalysis); this strongly exothermic reaction may lead to ejection of the molecule from the grain with a residual kinetic energy of about 2.2 electron volts; the energy thus contributed appears sufficient to heat the medium. The H_2 molecules themselves are photodissociated by stellar ultraviolet radiation, which closes the cycle.

Other heating mechanisms are possible but perhaps more hypothetical. These include heating by ionization of interstellar gas; by low-energy cosmic rays (a few million electron volts), the existence of which has not been proved; or by soft X rays. It is interesting to note that a large number of studies have been devoted to the problem of the heating of diffuse interstellar clouds. This problem remains nearly intact, but these studies have led to considerable progress in some areas of atomic physics, particularly our knowledge of the collision mechanisms among atoms, ions and electrons.

Intercloud Gas and the Coronal Gas

Observations in the 21-centimeter line and observations with the Copernicus satellite have led to two fundamental discoveries, the ramifications of which probably have not yet been fully realized.

The first is relatively old, since it dates back about 15 years; but recent observations have contributed important new data. This is the discovery of a hot neutral medium revealed by observations in emission and absorption at 21 centimeters, as mentioned earlier. The emission at 21 centimeters is independent of the temperature of the gas—at least if the optical depth is small—while the absorption is inversely proportional to temperature. Thus if there is hot and cold gas along a line of sight toward a radio source, only the cold gas will be easily detected in absorption, while both will appear in emission if the radiotelescope is aimed adjacent to the source. Observation shows that this situation occurs frequently: In nearly all directions, there are wide components observed in emission at 21 centimeters, of which the corresponding absorption against the radio source is either not detectable, or is detectable only with difficulty. Thus, in general, only a lower limit can be obtained to the temperature of this gas, which is about about 1,000° K. In a few cases where the absorption could be detected without ambiguity, we observe temperatures of 5,000° to 10,000° K. The traditional interpretation of these observations is that the interstellar clouds are immersed in a medium that is at least partly neutral (since it is revealed by the 21-centimeter line), hot and thin (about 0.15 atoms/cm^3). Unlike the heating of the clouds, the heating of such a medium presents no problem despite its higher temperature, for its density is very low and the electron-atom collisions responsible for the cooling are rare. The soft X rays observed by various satellites (the origin of these X rays we will see later on) may be sufficient heating agents via their ionization of the atoms of the medium. It is interesting to note that the intercloud medium, thus defined, has a total mass essentially equal to that of the clouds in the neighborhood of the Sun, and it has a greater thickness than the clouds perpendicular to the galactic plane. The total mass of the diffuse clouds and the intercloud medium projected onto the galactic plane is about 7 solar masses per square parsec, divided roughly equally between each component.

New Insights

This description, which has become relatively traditional, has nevertheless run into two difficulties recently. The first is observational: Some radioastronomers have pointed out (not without justification) that the wide emission profiles of the 21-centimeter line so characteristic of the intercloud medium might be artifacts from the detection by radio telescopes of galactic emission at 21 centimeters far from the line of sight. Since the radial velocity of these regions varies with direction, the superposition of the parasitic emission of several regions may simulate a wide profile. It is necessary to make new observations with specially calibrated antennae to verify this idea. However, it is unlikely that *all* the wide emission profiles should not be real and hence that the intercloud medium does not exist. But it is not possible to be so positive about its mass and density. The other difficulty concerning the usual interpretation of the intercloud medium has to do with the observations we are now going to discuss.

The most surprising observation—made by the Copernicus satellite—probably is the discovery of absorption lines of five-times-ionized oxygen, O^{5+}, or, in astronomical notation, O VI*, at wavelengths of 1,032 and 1,038 Å in the spectra of most of the observed stars. These observations reveal the existence of a gas at a temperature of about 500,000° K, which must be very common in the vicinity of the Sun. Their interpretation is not unambiguous: The stars that serve as light sources for observation of interstellar absorption lines in the ultraviolet are all hot stars of type O (or early B); this is reasonable, for the other stars do not emit enough ultraviolet radiation to be useful. However, we now know that these stars are continuously losing mass in the form of a "wind" blowing at speeds of 1,000 or 2,000 kilometers (622 or 1,243 miles) per second. This gas ejected by the star might contain O^{5+}, and the lines of this ion might therefore be of circumstellar rather than of interstellar origin. The situation is not yet clear, but there are strong indications (based especially on the fact that the radial velocity of the O^{5+} lines is not correlated with that of the observed star) that they are of interstellar origin. If this is indeed true, and if we assume that the very hot medium thus detected is approximately in pressure equilibrium with the interstellar clouds, we find that it must fill most of the volume of the local interstellar medium, with the clouds and neutral "intercloud medium" taking up only a small part of it. This is in no way incompatible with observation. The discovery of this hot medium has clarified X-ray observations of the past several years by rockets and satellites that showed the presence throughout the celestial sphere of low-energy radiation (less than 1,000 electron volts); initially it was believed that this radiation was extragalactic, but now it has been proved that it arises in nearby regions; quite naturally it is interpreted as thermal emission (thermal *Bremsstrahlung*) of the hottest parts (a few million degrees) of the diffuse medium.

What is the origin of this very hot gas, which, although it fills up most of the volume of interstellar space, has only a negligible mass relative to the rest of the interstellar medium (its density is only about 10^{-4} particle per cubic centimeter)? The prevailing notion is that it consists of gas filling the inside of supernova remnants. Massive stars end their lives in a gigantic explosion—the supernova—which ejects matter at an enormous speed (2,000 to 20,000 kilometers) (1,243 to 12,430 miles) per second and thus exerts a profound influence on the structure of the interstellar medium. In the last phases of its evolution, a supernova remnant reaches a considerable diameter, reaching or exceeding 40 pc, and is filled with a thin, high-temperature, completely ionized gas. The remnants of supernovae that have exploded in various places finally join together and merge, forming a kind of tunnel with very irregular shape and filled with very hot gas; successive explosions easily keep the gas at a high temperature despite the cooling processes, which are not very effective in any case (see page 267) and constantly change the structure of the medium. The cavities hollowed out by the stellar winds of hot stars also make their contribution to this structure. Thus we arrive at a dynamic and no longer static conception of the interstellar medium, where the cold or warm clouds are like irregular islands forming barriers between enormous very hot and thin cavities. The fairly hot but rather neutral intercloud medium, on the other hand, forms thin parts of these walls or transition zones between clouds and cavities. Its heating poses no problem, since it is provided by the radiation from the cavities.

Although this new model is extremely appealing, it cannot be considered as firmly established. Many observations still are necessary to understand more fully the complex structure of the interstellar medium.

Dense Clouds and Interstellar Molecules

The existence of molecules in interstellar matter has been known since the 1940s; we already mentioned, in the preceding paragraph, the discovery of three molecules that, incidentally, are free radicals—CH, CH^+ and CN—through their interstellar absorption lines. However, these are the only ones that have sufficiently intense lines to be observed in the visible range, and the subject was practically abandoned after the initial discoveries. Nonetheless, during the 1950s it was realized that observation in the far ultraviolet and in the radio wave range might well be more profitable than optical observation. The first molecules discovered through their absorption lines in the vacuum ultraviolet were detected only in 1969: These were H_2 and CO, more recently followed by OH and C_2. In radio, the first interstellar molecule observed was OH in 1963, which produces four lines at 18 centimeters wavelength. In 1968, more complex molecules were discovered: H_2O, NH_3 and H_2CO. But most of the 50-odd interstellar molecules known today were observed by their rotational lines in the millimeter range, which are more intense and more numerous than in the range of decimeter and centimeter waves. Thus recent progress in this area has been and still is tied to the development of sensitive receivers. As soon as the first good millimeter receiver was placed in service by Bell Telephone Laboratories in 1969, CO was discovered through its line at 2.6 mm wavelength; then in rapid succession, other molecules. Practically speaking, all of the molecules known today have been identified accurately through the efforts made to observe the corresponding radio transitions in the laboratory, or, failing that, to calculate them with precision when the molecule is too unstable under laboratory conditions. Most of the molecules yield several observed transitions anyway, since the total number of molecular lines detected in the radio range exceeds 300.

DENSE MOLECULAR CLOUDS

Certain molecules such as H_2, CO, OH, etc., are present nearly everywhere in the interstellar medium (at least in its cold, neutral portions), although in small, variable quantities. However, they are mainly found, like the more complex molecules, in the particularly dense interstellar clouds, of which almost nothing was known before observations of radio molecular lines were made. Nevertheless, dark nebulae—the relatively starless regions mentioned earlier—had long since been observed, but the gas that should normally be associated with the dust thus observed had not been detected. On the other hand, emission from these regions in the 21-centimeter line was weak or nonexistent. A reasonable hypothesis, which was subsequently verified, is that if the 21-centimeter

* In this notation, neutral oxygen O is OI, O^+ is O II, H is HI, H^+ is H II, etc.

line is not visible, it is because hydrogen does not exist there in the form of atoms but in molecules. Unfortunately, the H_2 molecule could not be observed directly in such clouds; this molecule has lines only in the far ultraviolet, below 1,100 Å, and the dense clouds are too opaque at these wavelengths to make it possible to see them. But many other molecules, apparently much less abundant than H_2 (10^4 to 10^7 times less abundant), have been observed there, the most important being CO. It is possible to estimate indirectly the density in these clouds based on observations of CO and other molecules, and to calculate their mass by combining this estimate with their distance (based on measuring the radial velocity of the lines, as in the case of the 21-centimeter line) and dimensions. In general, the density is greater than several hundred molecules per cubic centimeter; it may attain and exceed 10^9 molecules per cubic centimeter, a much larger figure than for the diffuse clouds observed in the 21-centimeter line. The clouds' dimensions range from a tiny fraction of a parsec to 20 parsecs, and their greatly varying mass may reach 10^5 or 10^6 solar masses. Now it is possible to map these clouds—e.g., by using the line of CO at 2.6 millimeters. We note that mapping them also is possible, but much more difficult at present, by measuring the thermal emission from their dust in the far infrared (see page 273). Various partial or complete maps of the galaxy in the CO line—similar to what was done with the 21-centimeter line—have been published, and soon we will know the distribution of the molecular clouds in the galaxy, as well as that of the diffuse clouds observed in the 21-centimeter line. Already we know that the two distributions are not identical: The closer we get to the center of the galaxy, the more molecules there are, and the molecular clouds form a disk even flatter than the rest of the interstellar matter. The total mass of the gas in molecular form in the galaxy is about the same as that in atomic form; in the neighborhood of the Sun, it is about half.

CHEMISTRY OF INTERSTELLAR MOLECULES

The first two tables give the interstellar molecules detected as of June 1982. All of the polyatomic molecules, with the exception of acetylene C_2H_2 and methane CH_4, were detected only in radio, as were several diatomic molecules (CS, SO, NS, SiO and SiS). This list is far from complete, for the symmetric molecules (such as H_2, C_2, acetylene C_2H_2, ethane C_2H_4, etc.) do not have radio lines and are difficult to detect otherwise. Moreover, the frequencies of the radio lines of certain light molecules such as HC1 fall into the submillimeter range and are not yet observable with current methods, while others—such as those of MgO, FeO or HNSi, probable interstellar molecules—have not yet been measured in the laboratory and are very difficult to calculate with sufficient precision. Finally, it is certain that many complex molecules remain to be discovered.

Examination of these tables shows that the great majority of interstellar molecules are organic molecules. This is not surprising, for the atoms making up the organic molecules (hydrogen, then carbon, nitrogen and oxygen) are by far the most abundant in the universe (here we exclude helium, which is very abundant but chemically inert). Among the most abundant elements after these, sulfur and silicon form observable molecules, but not magnesium or iron. This is explained by the fact that the transition frequencies of the corresponding simple molecules are not known, as mentioned earlier.

However, interstellar chemistry has hardly any relationship with the chemistry in our laboratories, for conditions in interstellar clouds are extremely different: a very high vacuum, despite the term "dense clouds," which is altogether relative; and a very low temperature (7° to 100° K). In conclusion, collisions are rare, which would not seem at the outset to favor the formation of molecules (but they have a very long time in which to do this). However, the very rarity of collisions means that species that are unstable in the laboratory—free radicals such as CH, CN, OH, C_2H,

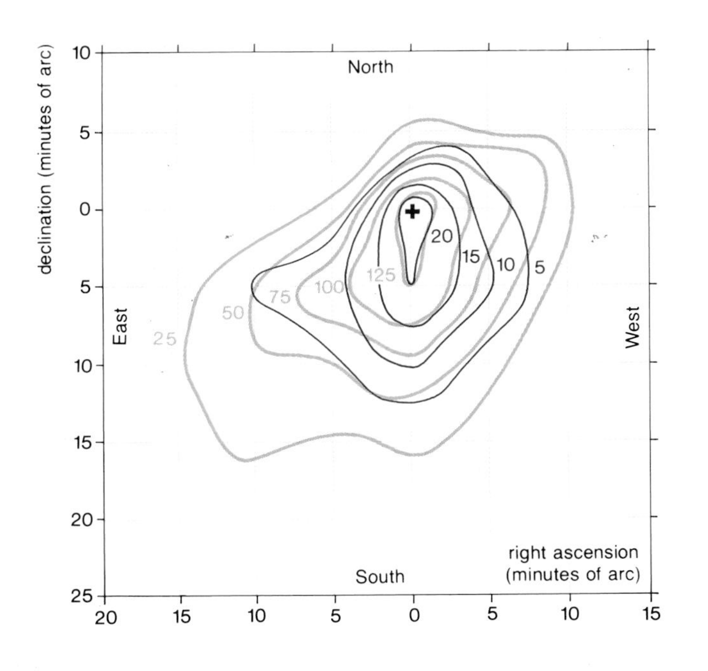

Structure of the molecular cloud associated with the ionized hydrogen region Sharpless 106. The isophotes shown in orange refer to the isotope ^{12}C of the CO molecule; those shown in red, through the ^{13}C isotope of the ^{13}CO molecule. The red cross shows the placement of the infrared protostar responsible for the excitation of this H II region.

Interstellar Molecules (February 1986)
(the number indicates the total number of atoms)

2	3	4	5	6	7	8	9	11	13
H_2	H_2O	H_2CO	HCOOH	CH_3OH	CH_3C_2H	CH_3COOH	$(CH_3)_2O$	HC_9N	$HC_{11}N$
CH	HCO	NH_3	HC_3N	NH_2CHO	CH_3CHO	CH_3C_3N	CH_3CH_2OH		
CH^+	HCO^+	HNCO	CH_2NH	CH_3CN	HC_5N		CH_3CH_2CN		
CN	CCH	H_2CS	NH_2CN	CH_3SH	CH_3NH_2		HC_7N		
CO	HCN	C_3N	H_2CCO		CH_2CHCN		CH_3C_4H		
CO^+	HNC	HNCS	C_4H						
CS	N_2H^+	$HOCO^+$	C_3H_2						
OH	H_2S	C_3H							
SO	OCS	$HCNH^+$							
NS	SO_2	C_3O							
SiO	HNO								
SiS	HCS^+								
C_2	SiC_2								
NO	H_3^+								
HCl									

etc., or molecular ions such as HCO^+ or N_2H^+—can exist there for a very long time. Moreover, given the low temperatures, no appreciable kinetic energy exists in the atoms that collide with one another, and the only possible reactions are exothermic. In addition, it can be shown that, with a few rare exceptions, reactions between neutral atoms or molecules are impossible under these conditions; for neutral atoms or molecules, to combine with one another, must in general overcome a small repulsive potential barrier, and their kinetic energy is not sufficient to achieve this (at temperatures that can be reached in the laboratory, on the other hand, this difficulty exists only rarely). Thus, in the gaseous phase, it is generally

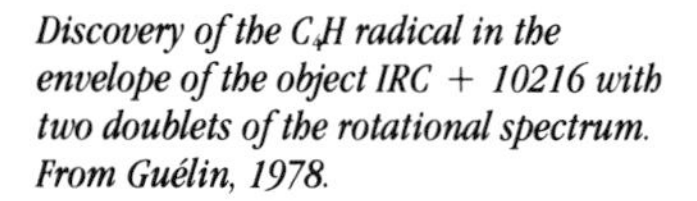
Discovery of the C_4H radical in the envelope of the object IRC + 10216 with two doublets of the rotational spectrum. From Guélin, 1978.

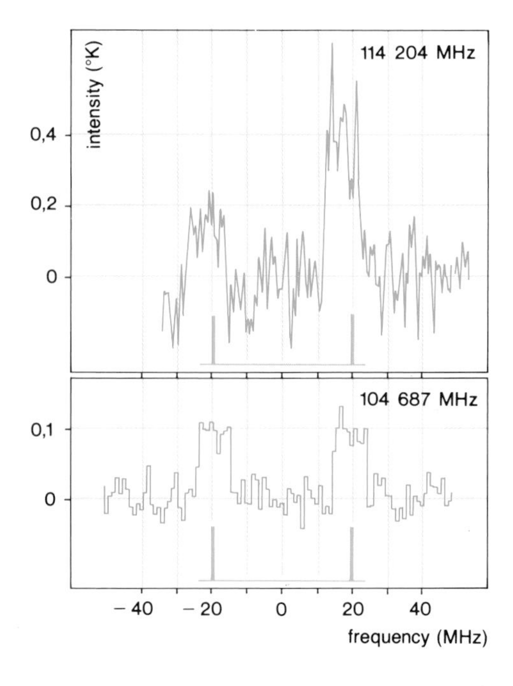

necessary for ions to be involved. How can they be produced? In diffuse clouds, ultraviolet radiation from stars can ionize atoms, but that is not the case in the dense clouds, where radiation does not penetrate owing to strong extinction by dust. Here is where cosmic rays come in—those high-energy charged particles (at least 100 million

YEAR	MOLECULE	SYMBOL	STRUCTURE	IDENTIFYING WAVELENGTH	TYPE OF SIGNAL
1937	methylidyne radical	CH	C—H	4,300 Å	optical absorption
1940	cyanogen	CN	C≡N	3,875	optical absorption
1941	methylidyne ion	CH^+	$C—H^+$	3,745 Å	optical absorption
1951	hydrogen	H	H	21 cm.	emission
1963	hydroxyl radical	OH	O—H	18 cm.	normal absorption
				18 cm.	maser emission
				18 cm.	normal emission
				6.3 cm.	maser emission
				5.0 cm.	maser emission
				2.2 cm.	maser emission
1968	ammonia	NH_3	H \ N—H / H	1.3 cm.	maser emission
1968	water vapor	H_2O	H \ O / H	1.3 cm.	maser emission
1969	formaldehyde	H_2CO	H \ C=O / H	6.2 cm.	normal absorption
					normal absorption
				6.2 cm.	anomalous
				2.1 cm.	normal absorption
				1.0 cm.	normal absorption
				2.1 mm.	normal absorption
					normal emission
1970	carbon monoxide	CO	C=O	2.6 mm.	normal emission
1970	cyanogen radical	CN	C≡N	2.6 mm.	normal emission
1970	hydrogen	H_2	H—H	1100 Å	U.V. absorption
1970	hydrocyanic acid	HCN	H—C≡N	3.4 mm.	normal emission
1970	formyl ion	HCO^+	$H—C=O^+$	3.4 mm.	normal emission
1970	cyanoacetylene	HC_3N	H—C≡C—C≡N	3.3 cm.	normal emission
1970	methyl alcohol	CH_3OH	H | H—C—O—H | H	36 cm.	normal emission
1970	formic acid	HCOOH	H—C—OH ‖ O	18 cm.	normal emission
1971	carbon monosulfide	CS	C≡S	3 mm.	normal emission
1971	methyl cyanide	CH_3CN	H | H—C—C≡N | H	2.7 mm.	normal emission
1971	isocyanic acid	HNCO	H—N=C=O	3.4 mm.	normal emission
1971	formamide	$HCONH_2$	H / N / \ H—C H ‖ O	6.5 cm.	normal emission
1971	carbonyl sulfide	OCS	O=C=S	2.7 mm.	normal emission
1971	silicon monoxide	SiO	Si=O	2.3 mm.	normal emission
1971	hydrogen isocyanide	HNC	H—N≡C	3.4 mm.	normal emission
1971	methylacetylene	CH_3C_2H	H | H—C—C≡C—H | H	3.4 mm.	normal emission
1971	acetaldehyde	CH_3COH	H | H—C—C—H | ‖ H O	30 cm.	normal emission
1971	thioformaldehyde	H_2CS	H \ C=S / H	9.5 cm.	normal emission
1972	hydrogen sulfide	H_2S	H / S \ H	1.8 mm.	normal emission
1972	sulfur methanimine	CH_2NH	H \ C=N / \ H H	5.7 cm.	normal emission
1973	sulfur monoxide	SO	S=O	3 mm.	normal emission
1974	silicon monoxide	SiO	Si=O	3.6 and 2.3 mm.	1st maser millimetric
1974	ethynyl radical	CCH	C≡C—H	3.5 mm.	normal emission
1974	dimethylether	$(CH_3)_2O$	H H | | H—C—O—C—H | | H H	3.2 mm.	normal emission
1975	cyanamide	NH_2CN	H \ N—C≡N / H	2.98 and 3.72 mm	normal emission

YEAR	MOLECULE	SYMBOL	STRUCTURE	IDENTIFYING WAVELENGTH	TYPE OF SIGNAL
1975	silicon sulfide	SiS	Si=S	2.75 and 3.30 mm.	
1975	nitrogen sulfide	NS	N≡S	2.6 mm.	
1975	diatomic carbon	C_2	C=C	10140Å	normal absorption
1975	diazenyl ion	N_2H^+	N=N—H$^+$	3.22 mm.	
1975	formyl radical	HCO	H—C=O	3.46 mm.	
1976	sulfur dioxide	SO_2	O=S=S (O=S, with S doubly bonded to lower S)	3.46 mm. and 3.58 mm.	
1977	cyanoethynyl radical	C_3N	C≡C—C≡N	3.03 mm. and 3.37 mm.	
1977	ketene	H_2C_2O	(H)(H)C=C=O	2.94 mm. 3 mm and 3.67 mm.	
1977	acetaldehyde	CH_3CHO	H—C(=O)—C(H)(H)—H	28.1 cm.	
1977	cyanodiacetylene	HC_5N	H—C≡C—C≡C—C≡N		
1977	methylamine	CH_3NH_2	H—C(H)(H)—N(H)(H)	3.48 mm. and 4.1 mm.	
1977	vinyl cyanide	CH_2—CH—CN	(H)(H)C=C(H)—C≡N	21.86 mm.	
1977	cyanohexatriyne	HC_7N	H—C≡C—C≡C—C≡C—C≡N	2.95 cm.	
1977	nitroxyl	HNO	H—N=O		
1978	butadiynyl radical	C_4H	C≡C—C≡C—H		
1978	nitrogen monoxide	NO	N=O	1.99 mm.	
1978	methyl formate	$HCOOCH_3$	H—C(=O)—O—C(H)(H)—H	18.6 cm.	
1978	ethanol	CH_3CH_2OH	H—C(H)(H)—C(H)(H)—O—H	2.8 mm. 3.3 mm and 3.5 mm.	
1978	propionitrile (ethyl cyanide)	CH_3CH_2CN	H—C(H)(H)—C(H)(H)—C≡N	2.58 mm. and 3.06 mm.	
1978	cyanooctatetrayne	HC_9N	H—C≡C—C≡C—C≡C—C≡C—C≡N		
1979	methyl mercaptan	CH_3SH	H—C(H)(H)—S—H	2.93 = 3.95 mm	normal emission
1980	thiocyanic acid	HNCS	H—N=C=S	2.13 = 3.65 mm	normal emission
1981	thioformyl ion	HCS^+	H—C=S$^+$	1.2 = 3.5 mm	normal emission
1981	carbon monoxide ion	CO^+	C≡O$^+$	2.56 mm.	
1981	formic acid ion	$HCOOH^+$	H—C—O—H$^+$		
1982	cyanopentaacetylene	$HC_{11}N$	H—C≡C—C≡C—C≡C—C≡C—C≡C—C≡N		
1981	protonated carbon dioxide	$HOCO^+$	H—O—C=O$^+$	2.34, 2.80, 3.51 mm	normal emission
1984	methylcyanoacetylene	CH_3C_3N	H—C(H)(H)—C≡C—C≡N	9.1–14.5 mm	normal emission
1984	methyldiacetylene	CH_3C_4H	H—C(H)(H)—C≡C—C≡C—H	1.23, 1.47 cm	normal emission
1984	silicon dicarbide	SiC_2	Si bonded to both C of C≡C (ring)	1.76–3.22 mm	normal emission
1984	tricarbon monoxide	C_3O	C≡C—C≡O	1.56 cm	normal emission
1985	propynylidyne	C_3H	C—C≡C—H	2.12–9.18 mm	normal emission
1985	(?)	H_3^+	triangle H, H—H$^+$	804 mm	normal emission
1985	cyclopropenylidene	C_3H_2	ring C, C=C, each lower C bonded to H	1 mm to 1.6 cm	normal emission and absorption
1985	hydrogen chloride	HCl	H—Cl	479 mm	normal emission
1985	protonated hydrogen cyanide	$HCNH^+$	H—C=N—H$^+$	1.34, 2.03, and 4.05 mm	normal emission

* The transition at 1.3 mm. of CO was observed on July 4, 1973; the one at 0.83 mm., in the summer of 1978; and the one at 120μm, in 1980.

electron volts for those propagating near the Sun) that can penetrate the densest clouds and ionize the atoms and molecules they encounter—i.e., essentially He and H_2, which are the most abundant components. The number of ions formed in this way is very small, given the rarity of the cosmic rays and their mediocre ionization capacity. Nonetheless, it is sufficient to initiate interstellar chemistry.

This chemistry now is fairly well understood in principle; with many reactions having known or recently measured rates, we can even calculate the abundance of the interesting molecules and compare them to the observed abundances, which makes it possible to verify and improve the theory. An odd and interesting feature of interstellar chemistry is that it can be different for two isotopes of a single element, even though those isotopes have similar chemical properties.

Isotopic Interstellar Molecules

H_2 HD
${}^{12}CH^+$ ${}^{13}CH^+$
${}^{12}C^{16}O$ ${}^{13}C^{16}O$ ${}^{12}C^{17}O$ ${}^{12}C^{18}O$ ${}^{13}C^{18}O$
${}^{16}OH$ ${}^{18}OH$ ${}^{17}OH$
${}^{12}C^{32}S$ ${}^{13}C^{32}S$ ${}^{12}C^{33}S$ ${}^{12}C^{34}S$
$H_2{}^{16}O$ $HD^{16}O$ $H_2{}^{18}O$
$H^{12}C^{14}N$ $D^{12}C^{14}N$ $H^{13}C^{14}N$ $H^{12}C^{15}N$
$H^{14}N^{12}C$ $D^{14}N^{12}C$
$H^{12}C^{16}O^+$ $D^{12}C^{16}O^+$ $H^{13}C^{16}O^+$
$H_2{}^{12}C^{16}O$ $H_2{}^{13}C^{16}O$ $H_2{}^{12}C^{18}O$
HC_3N $H^{13}CC_2N$ $HC^{13}CCN$ $HC_2{}^{13}CN$
${}^{14}N_2H^+$ ${}^{14}N_2D^+$
${}^{14}NH_3$ ${}^{14}NH_2D$
$H^{14}N^{12}C$ $D^{14}N^{12}C$
CH_3NH_2 CH_3NHD ${}^{29}SiO$ ${}^{30}SiO$

The accompanying list indicates detected isotopic molecules, which are very numerous. The molecules containing deuterium (D) often are nearly as abundant as the corresponding hydrogenated molecules, while the D/H abundance ratio is only about 10^{-5} (this ratio is measured directly in the diffuse clouds using the Lyman lines of these two elements). This has to do with the existence of important isotope exchange reactions such as $CH_3^+ + HD \rightarrow CH_2D^+ + H_2$

As it has been shown that reactions involving free electrons compete with this reaction, study of the isotopic deuterium-containing molecules may help to determine the abundance of free electrons in dense clouds, an abundance extremely small that is ($n_e/n_{H2} \leq 10^{-7}$ to 10^{-8}).

Even though a large part of interstellar chemistry is a gas-phase ion chemistry, as directly proved by the presence of ions such as HCO^+ and N_2H^+, this is not all. In particular, it can be established that the molecule H_2, the most abundant of all, cannot form in the gas phase and must necessarily form by the encounter between hydrogen atoms on interstellar grains (catalysis).

Rather complex interstellar molecules (cyanopolynes such as HC_9N) have been detected. A search for prebiotic molecules is being carried out by several groups. Glycine—one of the configurations of which has indeed been measured in the laboratory—has been detected by only one American team, by a single one of its transitions, and skepticism about this discovery is in order. Nothing allows us to state that life began in the interstellar medium, but we may simply note that thermodynamics orients simple primitive systems (mixtures of methane, ammonia and water) toward the formation of comparable molecules.

EXCITATION OF INTERSTELLAR MOLECULES; MOLECULAR MASERS

The physics of excitation of interstellar molecules—i.e., the processes that populate the energy levels of the molecules between which the observed transitions occur—is one of the most interesting topics of current astrophysics. Owing to the complexity of the molecular energy levels, one finds a wide variety of interesting effects that have actually been observed, including maser emission.

Interpretation of the intensities of the interstellar absorption produced in the visible or ultraviolet molecules is extremely simple: The optical depth in the line is directly proportional to the column density of the molecules in the lower level of the transition and, as in the case of atoms, we can use a curve-of-growth analysis to estimate this column density. The advantage of this method lies in making it possible to calculate the relative populations of all the corresponding energy levels from different lines; this is how the Copernicus satellite made it possible to conclude that high rotational levels of the hydrogen molecule may be substantially populated even though the medium is cold. This observation is interpreted as being due to the effect of ultraviolet radiation on the H_2 molecules, which carries some of them to very high energy levels, from which they redescend by radiative decay, populating the observed rotational levels. This situation corresponds to the diffuse clouds and not to the dense molecular clouds.

The physics of the formation of radio molecular lines is considerably more complicated and requires study of the population of the corresponding energy levels, which often are far from thermodynamic equilibrium (we have just seen the example of H_2; however, the rotational lines of this molecule are not observable).

In some cases, the medium acts as an amplifer of ambient radiation—i.e., it behaves as a maser. Several interstellar molecules display such natural maser effects, the most spectacular of which correspond to transitions of the OH molecule at 18 centimeters wavelength, and to the transition of H_2O at 22 GH_z. The interstellar maser sources are localized within dense molecular clouds, generally in the vicinity of the ionized zones (H II regions within the cloud or on its edge). Generally their lines are complex, very intense, narrow and variable over time (in the case of H_2O masers, they may appear or disappear in the space of a week). Each component of the overall line profile generally comes from a very small region, with angular dimensions far below 1 arc second (and often even below 0.001 arc second in the case of H_2O masers), these regions being grouped together. The groups of interstellar masers often (but not always) coincide with infrared sources, an indicator that the interstellar matter at that location is hot. In spite of a great many studies devoted to interstellar masers since their discovery in 1965, we still do not have an entirely convincing theory of their excitation. This is because the possible mechanisms are numerous: Population inversions may be caused by collisional transitions alone; or by radiative transitions among rotational, vibrational or electronic levels as a result of far-infrared, near-infrared or ultraviolet radiation; or by a combination of these various effects. We can even imagine molecules forming in a state corresponding to inverted populations. However, recent studies have clarified the situation in the case of OH masers. Whatever the case, it is certain that interstellar masers must correspond to particularly dense and hot regions of the interstellar medium (10^7 to 10^9 molecules per cubic centimeter and more than 400° K for the H_2O masers), and generally they are considered to have something to do with the formation of stars, which occurs through contraction of the molecular clouds; we will come back to this later.

Ionized Gaseous Nebulae (H II Regions)

Ultraviolet radiation from hot discharge ionizes the nearby gas and thus forms gaseous nebulae, also called emission nebulae or H II regions. Their luminous emission, which makes them so spectacular, comes partly from the recombination of ions and from various other processes we will examine later.

The brightest and largest of these nebulae were reported in the first catalogues of diffuse objects (*Messier Catalogue, New General Catalogue* and its supplement, the *Index Catalogue*) and often are referred to by their number in these catalogues (for example, M 42, NGC 2024 or IC 1396). More recent optical catalogues are those of Shajn and Haze (1955) and Sharpless (1959). These objects also are detectable by their radio emission in the continuum and by lines, which makes it possible to observe them in distant regions in the galaxy obscured by interstellar extinction; a recent catalogue containing the most intrinsically intense H II regions is that of Smith, Biermann and Mezger (1978). Finally, by combining the optical and radio information and observations regarding the exciting stars, Georgelin and Georgelin (1976) carried out a detailed study of the distribution of the principal H II regions in the galaxy that provides the best available description of its spiral structure. Apart from these relatively intense objects, there exists in the galaxy an infinite number of weak and relatively diffuse H II regions revealed by monochromatic photographs in the $H\alpha$ line of hydrogen (on this subject, consult the $H\alpha$ atlas of the galaxy by Sivan, 1974) or by observations of the diffuse radio background in the continuum or in lines; it would not make much sense to try to catalogue them. Finally, we will have another opportunity to mention the compact H II regions that have been detected within certain molecular clouds.

Ionization of the Interstellar Medium

In the presence of ultraviolet radiation, atoms ionize by absorbing photons; conversely, the ions and electrons thus created recombine to form new atoms. A steady state is established in which the number of ionizations per unit of time and volume is equal to the number of recombinations.

The most frequent case is that of hydrogen. But, of course, hydrogen is not the only element ionized; all the others are ionized to some extent. However, helium, which has an ionization potential of 24.5 eV as compared with 13.6 eV for hydrogen, is ionized only if the star is hot enough, and is found concentrated toward the central regions. On the other hand, an element such as carbon, which has an ionization potential of only 11.2eV, may be ionized outside the hydrogen ionization zone by photons having energies ranging from 11.2 to 13.6eV—i.e., those that are not absorbed by hydrogen. Generally speaking, for a uniform gaseous nebula, each element is ionized in a sphere, the radius of which depends on the ionization potential. This is how oxygen, which has nearly the same ionization potential as hydrogen, is ionized into O^+ in the same volume in which hydrogen is ionized. But O^+ may itself be ionized into O^{++} in a smaller volume comparable to the one in which helium is ionized, etc.

This description is slightly academic, however, for the gaseous nebulae have, in general, anything but a uniform density, so that the ionization structure usually is not clearly visible. The more uniform planetary nebulae, which will be described later, fit the theory better. But the most important departures from this theory stem from the existence of dust within the H II region; we will discuss this in the following paragraph.

Continuum Emission from Gaseous Nebulae

Gaseous nebulae display continuous emission in all wavelength ranges, from radio to ultraviolet. This radiation is produced by a variety of processes.

At radio wavelengths it involves the process known as thermal *Bremsstrahlung*, or free-free emission. Radiation is produced by the acceleration of free electrons passing close to the ions.

In the infrared, free-free emission also occurs, but the radiation is mainly from the dust heated by the stellar radiation and situated either inside, on the edge of or outside the H II region (unfortunately, often it is impossible to know exactly where this radiation comes from). The emission is very intense and contains a large proportion, sometimes the majority of the total energy radiated by the nebula. This shows that a large fraction of the stellar photons—including, of course, ionizing far-ultraviolet photons—is absorbed by dust. This phenomenon substantially complicates the study of H II regions. In particular, the flux of photons in the Lyman continuum *is not* the flux emitted by the exciting star or stars but is the flux actually absorbed by ionization of the gas; the emitted flux may be greater by a factor that is difficult to estimate, but is perhaps about two. Thus it is dangerous to use the observations of H II regions incautiously to determine indirectly the emission properties in the ultraviolet of the exciting stars. Finally, another consequence of the existence of dust is that the H II region is diminished relative to the predictions of the simple theory.

In the visible and the ultraviolet, free-free emission still is significant. It is superimposed on the stellar light scattered by dusts in the nebula, and on a special radiation called two-photon radiation, corresponding to transitions between levels of hydrogen and helium involving an unquantified intermediate level (the sum of energies of the two photons is constant, but their individual energies are arbitrary). Finally, continuum emission arises from the recombination of ions with electrons. Since free electrons have unquantified energies, their recombination to a given level of an atom produces a continuous spectrum that is abruptly limited on the long-wavelength side. The energy of a photon at this discontinuity corresponds to the recombination of an electron of zero kinetic energy. The most interesting of these discontinuities is the Balmer discontinuity of hydrogen at 3,646 Å, which corresponds to recombinations of electrons to the atomic level $n = 2$. As the intensity of this discontinuity depends on the temperature T_e in a different way from the intensity of the recombination lines of hydrogen—which we now are going to discuss—a measurement of the nebular temperature may be deduced by comparing them. This is practically the only application of the study of the optical or ultraviolet continuum spectrum of gaseous nebulae.

Recombination lines

The recombination of ions with free electrons may occur at all the energy levels of the atom. The subsequent cascades of these energy levels toward the fundamental level produce lines known as "recombination lines." All elements produce such lines in recombining; however, the only intense ones are those of the two most abundant elements, hydrogen and helium. The lines that have been most studied in the optical are the Balmer lines of hydrogen, which correspond to transitions culminating at the $n = 2$ level. The Hα band, in the red, corresponds to the transition $n = 3 \rightarrow 2$, Hβ to $4 \rightarrow 2$ and so on. The theory of recombination is relatively simple in principle: For each level, one writes the number of radiative and collisional transitions per unit of volume and time that culminate at that level, and the number of transitions that begin from it; it is assumed that the system is independent of time—i.e., that the two numbers are equal. This produces an equation involving the population of all the relevant levels connected to the level under consideration, as well as the population of the continuum level (density of free electrons). Once such equations are written for all the other levels, the system of equations thus defined is solved to get the level populations. It is then simple to calculate the intensity of the recombination lines which simply is proportional, in the optical case, to the product of the population of the level at which the transition begins times the probability of the corresponding spontaneous emission. The intensity of the lines is, of course, proportional to the number of recombinations per unit of volume and time and therefore to the square of the electron density (in the case of hydrogen).

The ratio of intensity between the Balmer lines (the Balmer decrement) is practically independent of the physical conditions. The relative intensities of these lines are:

$I_{H\alpha}$	$I_{H\beta}$	$I_{H\delta}$	$I_{H\gamma}$
2.86	1	0.47	0.26

The observed relative intensities generally are different, for they are affected by interstellar reddening. The comparison between the observed intensities and the theoretical intensities makes it possible to estimate this reddening and subsequently to correct all the intensities of the lines and continuum for interstellar extinction. Moreover, the intensities of the recombination lines of helium and of several other elements may be used to determine their abundance relative to hydrogen.

About 15 years ago, many recombination lines of hydrogen and helium were discovered in the radio range. These lines correspond to transitions between levels of extremely high quantum numbers (roughly from $n = 50$ to $n = 200$). We denote as α lines those corresponding (as for the optical lines) to the transition between two consecutive levels—for example, 110 and 109; for the H109α line at 6 centimeters β lines corresponding to transitions of $n + 2 \rightarrow n$, etc. The physics of these lines, which essentially are nearly impossible to observe in the laboratory, is very interesting; in particular, they may display maser effects tied to population reversals. They may be used to measure the temperature of the electrons.

The abundance of helium also may be deduced from the comparison of helium and hydrogen lines. Moreover, these lines, characteristic of H II regions, may serve unambiguously to distinguish H II regions that are invisible in the optical and are detected by their radio continuum radiation from other sources of continuum radio emission (particularly supernova remnants). The possibility they offer of measuring, through their Doppler shift, the radial velocity of distant H II regions makes them a prime tool for the study of galactic structure.

Carbon and sulfur, in recombining, also produce radio recombination lines. In H II regions, generally they are camouflaged by the great intensity of the continuum, but they are clearly detectable in the direction of certain molecular clouds (the accompanying graph shows an example of this), where the existence of sources of ionization thus is revealed; these sources probably are stars of type B, whose ultraviolet radiation ionizes the surrounding carbon and sulfur, but not hydrogen, and certainly not helium, for which we observe no recombination lines.

Forbidden Lines

The most unusual property of the H II regions is their high-intensity emission lines corresponding to transitions of very low emission probability and that thereby are extremely difficult to reproduce in the laboratory. Most of the ions in gaseous nebulae produce such lines, which are somewhat incorrectly called "forbidden" lines. Forbidden bands are designated by astronomers by the name of the ion in brackets followed by the wavelength; e.g., [0 III] 4,363 Å (remember that 0I = 0, 0II = 0+, 0III = 0+ +, etc.).

The emission process of these forbidden lines is very peculiar: The upper level of the transition is populated from the lower levels through collisions with free electrons, collisions for which there are no selection rules, as there are for the radiative transitions. If the density is low, the collisional deexcitation of the upper level is very slow, and the atom has no other choice, in order to become deexcited, than to emit a photon, despite the small probability of that event.

Forbidden lines are of great importance for the physics of gaseous nebulae. Numerous and intense, they radiate more energy than the continuum and recombination lines; thus they are the principal cooling agents of gaseous nebulae. By measuring their intensity, it is possible to estimate the abundance of the corresponding ions; this is the chief source of information we possess about the abundance of elements in the universe, together with that gleaned from studies of the solar system. Finally, the intensity ratios of the lines are sensitive either to density (as with the ratios of the [0 III] 51.7 μm/88.2 μm, or [0 II] 3,727 Å/3,729 Å or to temperature (ratio of [0 III] 4,363 Å/5,007 or 4,959 Å) and allow for the best determinations of physical parameters in H II regions, knowledge of which in turn is essential for going from the intensity of the lines to the abundance of the emitting ions.

The Thermal Balance of Gaseous Nebulae

What is the process controlling the temperature of gaseous nebulae? The heating corresponds to the kinetic energy of the electrons torn from the atoms by ionization, and their energy clearly comes from the exciting star. The free electrons, in turn, essentially lose their energy in collisions with the ions, during which these ions are brought to high energies, from which they descend again, giving off forbidden lines. Therefore, as we said, the lost energy is emitted chiefly in this form. Since the intensity and number of forbidden lines increase strongly with temperature, these lines act as a thermostat that regulates the temperature of the H II region. The temperature in equilibrium depends mainly on the abundance of these ions, particularly those of 0+ and 0+ +, which are the most numerous; Given the measured abundance of oxygen, it is about 8,000°K near the Sun, climbing to 15,000°K in the outer regions of spiral galaxies, where the abundance of oxygen is less by a factor of as much as 10, and falls to 3,000° or 4,000°K in the inner regions, where oxygen is superabundant by a factor of about three.

Evolution of Stars; Relationships Between Interstellar and Circumstellar Matter

It is beyond doubt that stars form within molecular clouds by gravitational collapse of a part of the cloud, but this process still is poorly understood. The existence of condensations within the

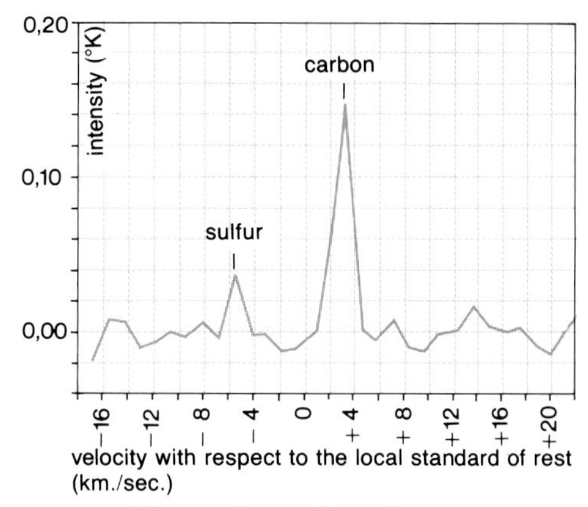

Detection of the recombination lines of carbon and sulfur in the direction of the molecular cloud ρ Ophiuchi. From Cesarsky, 1975.

cloud is known only through an indirect interpretation of the observations of molecular lines, but the limited resolution of the radio telescopes used for that purpose has not yet made it possible to reveal them directly. However, as we have mentioned, the molecular maser sources probably are small, dense and hot regions. Can they be identified as protostars? That is possible—but not certain—in the case of certain H_2O masers, which would then be the first observed manifestations of star formation. Moreover, in the infrared we have observed small, bright sources within certain molecular clouds, which are due to the thermal emission from hot dust and which sometimes coincide with H_2O masers. Here again, the interpretation is ambiguous: What may be involved is either part of a cloud in final contraction, in the protostellar phase or, on the other hand, a region of a cloud heated by an already formed star.

When a star is formed and arrives on the main sequence, its effects on the remaining interstellar matter around it become clearly observable, at least when it is hot enough for that. If an O star is involved, it ionizes a portion of the still dense cloud around it, thus forming a compact H II region. Many compact H II regions are now known through radio astronomy (they are only rarely visible in the optical, owing to extinction by what remains of the cloud). They are always coincident with infrared emission from the dust inside the H II region or enveloping it and often are surrounded by OH or H_2O masers, which must represent small condensations in the envelope of the H II region. Subsequently, the strong internal pressure of the H II region and the dynamical effects of the matter ejected by the star (stellar wind) will dissipate the cloud; the H II region will grow and generally will become optically visible. But some part of the molecular cloud nearly always remains, within which new stars can form (the molecular complex near ρ Ophiuchi is a typical example of this).

The less hot and less massive stars do not produce such spectacular effects unless they are very luminous (supergiants), in which case they heat the neighboring dust in a detectable way. We said earlier that the presence of relatively hot B stars in molecular clouds had been demonstrated by the existence of radio recombination lines of carbon and sulfur, ionized by their ultraviolet radiation (we may perhaps also find weak compact H II regions). The colder stars of the main sequence are barely detectable, with current methods, in molecular clouds. However, it appears that all stars, at the moment of their birth, eject gaseous matter that produces dynamical effects on the surrounding interstellar gas, effects that often are visible. The phenomenon of ejection and its consequences are directly visible in the spectrum of the T Tauri stars and the nebulae surrounding them. The mysterious Herbig-Haro objects—small, bright, often variable nebulae—probably represent the emission of gas excited by the stellar wind given off by a young, invisible star (T Tauri stars and Herbig-Haro objects often are accompanied by infrared sources and molecular masers).

The Omega nebula (M 17). Located in the constellation Sagittarius, near the border with the constellation Scutum, it may be seen with binoculars. It is a diffuse nebula that, like many other objects of this type, shows an intertwining of bright and dark regions. Photri Photo.

Stellar Winds and their Effect on the Interstellar Medium

We now know, as a result of observations by the *Copernicus* satellite, that all stars of mass greater than about 15 times the Sun's mass continually lose matter, which is ejected at the rate of 10^{-6} to 10^{-5} solar mass per year and at a speed of up to 2,000 kilometers per second. Such a stellar wind produces considerable dynamical effects on the interstellar medium, the importance of which we are only beginning to understand. In particular, it is possible that the characteristic aspect of certain "bubble-shaped" H II regions—which correspond to dense, more or less spherical shells practically empty of gas in their interiors—is due to winds from the central exciting stars. Stellar winds from 0 stars also contribute in an appreciable way to the renewal of interstellar matter.

Another category of stars that eject matter into the interstellar medium consists of the giants and red supergiants. Optical observation of such ejections is difficult, but recently radio observation came to the rescue with regard to the most evolved of these objects, which are long-period red variables (the Miras) or similar objects (carbon stars, etc.). The circumstellar envelopes of these objects constitute an important source of molecular emission, particularly OH maser emission, which is being actively studied. Study of such emission makes it possible to show that relatively cold matter, very probably full of dust, is ejected at a relatively low speed (a few kilometers per second to several dozen kilometers per second). The dynamical effect on the surrounding interstellar gas is small, but these ejections are important sources of interstellar dust.

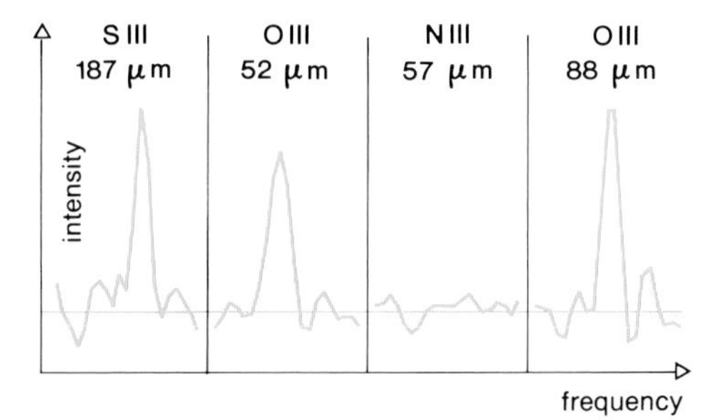

Fine structure lines detected in the M 17 nebula by Baluteau et al. (1980). These lines make it possible to measure several physical parameters of the ionized hydrogen region from which they emerged.

The Death of Stars: Planetary Nebulae and Supernovae

After the red-giant stage, stellar evolution accelerates, and it rapidly culminates in the death of the star. This occurs in a different fashion depending on whether its mass is less or greater than a threshold of about four solar masses. The lower-mass stars, after having gone through the Mira stage, gradually eject their whole envelope, exposing a carbon core that is initially very hot. The ultraviolet radiation from this core ionizes the ejected matter, which then acts as a H II region: We have a planetary nebula. The dimensions of these objects are small (a few seconds to a few minutes of arc), but their brightness makes them easily observable. A large number of them were included in the *New General Catalogue* and the *Index Catalogue* and are designated by NGC or IC numbers; about 1,000 of them are known altogether.

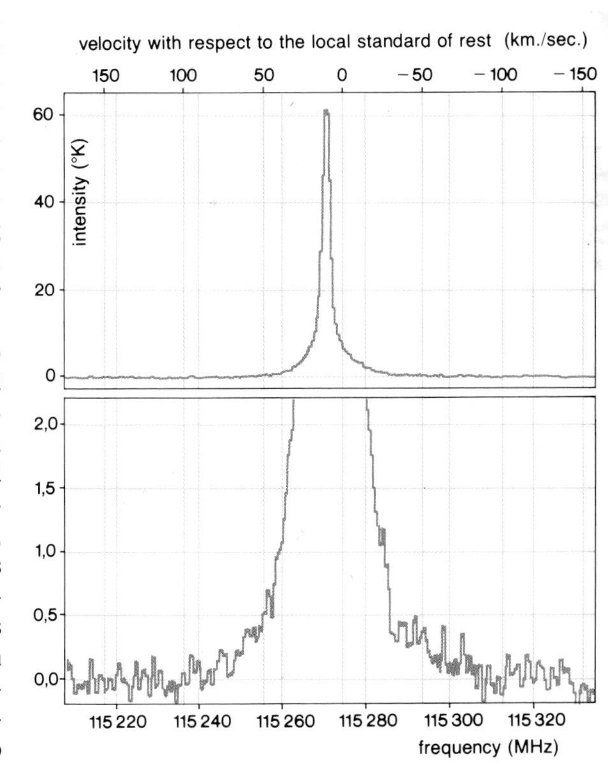

Emission spectrum of carbon monoxide, CO, in the direction of the Orion molecular complex. The lower part represents the expanded-scale spectrum of this emission. The observed widening at the base of the spectrum corresponds to a stellar wind of 100 kilometers (62 miles) per second and makes it possible to determine the quantity of matter expelled at each moment by the active star.

The physics of planetary nebulae is entirely similar to that of H II regions, the only difference being that, owing to the frequently higher temperature of the central star of a planetary nebula (up to 100,000°K), which therefore furnishes a harder ultraviolet radiation, certain elements often are in higher states of ionization (in particular, we observe He^{++} by its recombination lines). The relative simplicity of these objects means that the ionization structure predicted by the Strömgren theory sometimes is quite evident. The planetary nebulae are expanding with speeds of about 20 kilometers (12.4 miles per second), and their lifespan should not exceed a few times 10^4 years. Recently radio astronomers have discovered that dense molecular envelopes exist around several planetary nebulae in which we observe the emission from various molecules, particularly CO. Thus the total mass of these objects may be substantially greater than the ionized mass alone (about half of a solar mass); this is in agreement with the predictions of the theory of stellar evolution. Finally, molecular observations around certain "cold" stars have revealed the (predictable) existence of intermediate objects between Miras and planetary nebulae (e.g., the objects CRL 2688 and IRC + 10216, which have been studied extensively). Planetary nebulae are of considerable importance in the evolution of the galaxy, since they are the principal source of interstellar matter ejected by the stars, matter they have enriched with certain elements, particularly ^{13}C and ^{14}N formed in the envelope, and ^{12}C from a part of the core ejected along with the envelope. Dust (visible because of its infrared emission) is also ejected by planetary nebulae.

The massive stars end their life in a catastrophic manner (forming a supernova and a pulsar), ejecting nearly all their mass at high velocity into the interstellar medium. The ejected matter and pulsar are particularly visible, and have been well studied in the case of the Crab nebula, the remains of a supernova that exploded in 1054. The effects of the ejected matter on the surrounding gas are, of course, enormous. They may be summarized by saying that the ejected spherical envelope captures the matter it encounters, which gradually slows its expansion. After 10,000 years, the diameter is about 50 parsecs and the rate of expansion is only about 100 kilometers (62 miles) per second, as compared with 20,000 kilometers per second at the time of explosion. The gas is brought to considerable temperatures by compression, and it emits at practically all wavelengths, from X rays to radio waves. In particular, the envelope is made visible by its optical emission lines; in this way, about 30 supernova remnants have been optically detected in our galaxy. Supernova remnants are extremely effective in stirring and mixing up the interstellar medium, which they further enrich with synthesized heavy elements before and during the explosion. The most spectacular effect of supernova remnants is, however, their probable structure of tunnels filled with a very hot diluted gas, which form the largest part of the volume of the interstellar medium. This is no doubt one of the chief areas toward which future research on the interstellar medium will be directed.

Supernova remnants are intense sources of X rays. This picture, taken by the American satellite HEAO 2 *(Einstein Observatory) in the X-ray range, reveals a rapidly expanding nebula in the constellation Cassiopeia, already known as an intense radio source (Cassiopeia A); it is thought to be the vestige of a supernova that exploded unnoticed around 1660. P.P.P.-IPS Photo.*

The Interstellar Magnetic Field and Cosmic Rays

Most of the energy present in forms other than the thermal agitation of gas or dust particles consists of the magnetic and kinetic energies of the extremely rapid particles that make up cosmic rays.

In addition, both the magnetic field and the rapid particles have interactions with the gas that affect it in extremely important ways. The interstellar medium, always weakly ionized, actually is a plasma that tends to "freeze" the lines of force of the magnetic field it contains. Thus the presence of the field is important for the dynamics of the gas: The gas moves preferentially along the lines of force. Cosmic rays, for their part, contribute to the partial ionization of the medium and provide a kinetic energy that, when converted into thermal energy, is decisive in heating the gas. Finally, acceleration of the electrons of cosmic rays in the magnetic field causes the emission of synchrotron radiation, which enables us to detect the presence of extremely rapid particles and a magnetic field as far away as the most distant galaxies and the quasars.

The Interstellar Magnetic Field

Measurement of such a weak magnetic field (a few microgauss, or a few millionths of the Earth's magnetic field) at such great distances calls for rather complicated methods, such as measuring the polarization of stellar light, the Faraday rotation of radiation from radio sources, and the Zeeman effect.

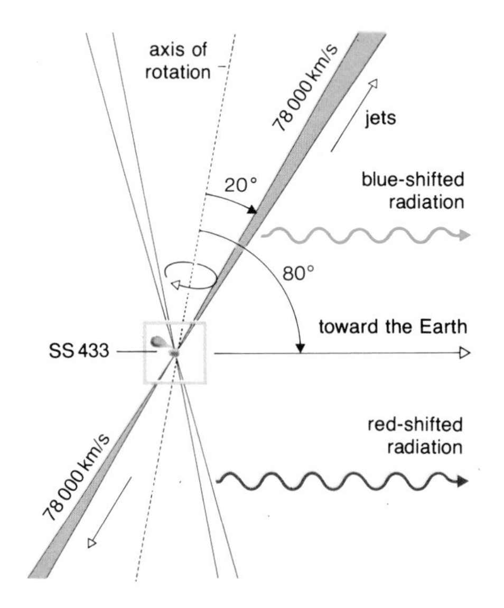

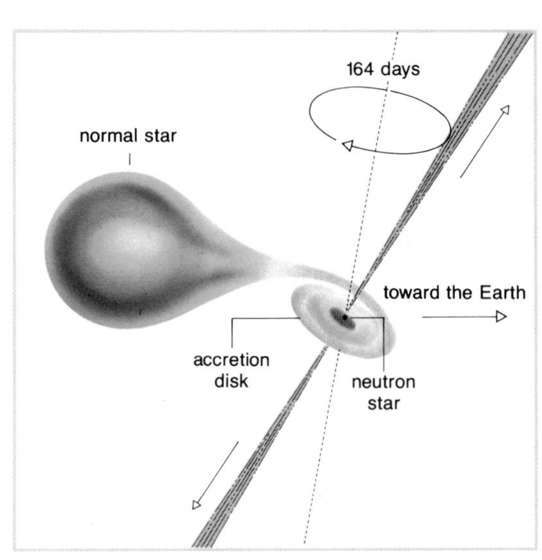

Binary Pulsars

Pulsars do not emit only in the radio range. Two pulsars—one at the center of the Crab nebula, the other in the Veil nebula—have been associated with stars that pulsate in visible radiation with the same period as at radio wavelengths, For the Crab pulsar, we also observe X-ray and gamma-ray pulses.

Objects very similar to radio pulsars have been discovered by observing the sky in the X-ray range with telescopes aboard satellites. Such objects emit X-ray pulses with a very regular periodicity of about 1 second to several hundred seconds. But these "X-ray pulsars" do not seem to emit radio waves and therefore would have emission mechanisms different from those of radio pulsars.

While the majority of X-ray pulsars belong to binary systems (the pulsar's own period being modulated by its stellar partner, there were only three known binary radio pulsars at the end of 1980. One of them, P 1913 + 16 (this number indicating its equatorial coordinates on the celestial sphere), is particularly remarkable because its period of revolution is only 7 hours, 45 minutes, though it can reach several dozen to several hundred days for the other binary pulsars. The precision with which we have been able to measure this period and its slow variations has made it possible to test in a few years several predictions of the general theory of relativity, and the results obtained are in excellent agreement with the theoretical pre-

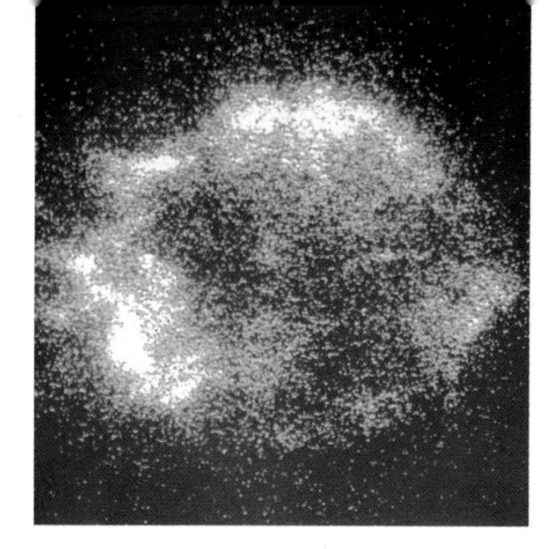

Polarization of Stellar Light

In 1949, two American astronomers, W. A. Hiltner and J. S. Hall, discovered independently of one another that the light from many stars has a high degree of linear polarization. (While natural light is represented by an electric field vector that has a randomly distributed direction in the plane perpendicular to the direction of propagation, in linearly polarized light the electric field vector has a special direction called the "direction of polarization.") In addition, the stars with the highest polarization rays were precisely those with reddened radiation characteristic of strong absorption by dust. Hiltner and Hall concluded from this that the phenomenon of polarization probably was due to the presence of interstellar dust along the path that the light had traveled from the star to the telescope, rather than to an internal mechanism of the star itself. An interpretation was offered by Davis and Greenstein: The dust grains are elongated in shape and are aligned by the magnetic field, so that their long axis is perpendicular to the field; thus the light polarized perpendicularly to the field is weaker than the light polarized parallel to the field. Thus unpolarized incident light finally appears after crossing the dust cloud as polarized parallel to the magnetic field. Observation of the direction of polarization thus gives the projection of the magnetic field direction on the plane of the sky; after observing a number of stars, we can draw a map of the magnetic field in the galaxy. Such a task was accomplished by Mathewson and Ford.

FARADAY ROTATION FROM RADIO SOURCES

Radio emission from most radio sources are due to the synchrotron effect. This involves the continuous-emission spectrum of relativistic electrons accelerated in the intense magnetic field of the sources. Initially this emission is linearly polarized (perpendicular to the magnetic field); but the propagation of these waves through interstellar regions where there exists a magnetic field with a component parallel to the direction of propagation, and a certain density of free electrons, leads to a rotation of the direction of polarization. This is known as the Faraday effect. The polarization vector of a wave propagating in the direction of the magnetic field rotates in a counterclockwise direction; it rotates in a clockwise direction if the magnetic-field direction is opposed to the direction of propagation of the wave. The angle of rotation is proportional to the magnetic field and to the square of the electron density of the medium, as well as to the square of the wavelength. Thus measuring the polarization at various wavelengths makes it possible to find the product Bn_e^2L (the rotation measure). But in general the electron density n_e is unknown.

However, there is a case in which the measurement of n_e is simple: It involves pulsars. Their radiation consists of pulses emitted with a well-defined period of repetition; the speed of propagation of the pulses through the interstellar medium depends on the square of the electron density and the square of the wavelength (dispersion). Thus the study of relative arrival times of the pulses at different wavelengths yields the product n_e^2L (pulsar dispersion measure). The rotation measure/dispersion measure ratio then gives the component of the galactic magnetic field along the line of sight. This field has an intensity of 3 microgauss.

THE ZEEMAN EFFECT

This is in fact the only direct measurement of the interstellar magnetic field. The Zeeman effect appears when an electromagnetic spectral line is emitted in the presence of a static magnetic field (see "The Sun", page 91). Usually the line is split into three components by the magnetic field: One has an unchanged frequency; the other two are shifted, one toward higher frequency, the other toward low frequency.

Measurement of the frequency gap between the two shifted components makes it possible to measure the magnetic field in the region where the atoms radiate. This is how it is possible to measure fields of some 20 microgauss, by studying the Zeeman components of the 21-centimeter line of neutral hydrogen in the Perseus arm, while fields of a few milligauss have been measured using the OH radical in circumstellar envelopes.

Map of the galaxy's radio continuum emission, prepared with the aid of the Effelsberg radio telescope, by Wielebinski, Haslam and Beck (1981). Each color specifies an intensity level. The asymmetrical emission (near the top, center), known as the north galactic spur, might correspond to a supernova remnant. Photo kindly contributed by P. Encrenaz.

Comparison of Interstellar Energy Densities

Radiation from stars	0.7×10^{-12} erg/cm.^{3}
Turbulent gas motions	0.5×10^{-12} erg/cm.^{3}
Cosmological microwave background at 2.8°K	0.4×10^{-12} erg/cm.^{3}
Cosmic rays	1.6×10^{-12} erg/cm.^{3}
Magnetic field	1.5×10^{-12} erg/cm.^{3}

dictions.

A faint star in the constellation of Aquila (The Eagle), SS 433, associated with a radio source, seems to be, in a different way, an exceptional binary pulsar.

The optical spectrum of this star shows intense emission lines of neutral hydrogen and helium, shifted toward the red and the blue by amounts that, interpreted as a Doppler effect, imply radial velocities varying in a sinusoidal fashion from −30,000 kilometers (18,645 miles) per second (approach) to +50,000 kilometers (31,075 miles) per second (moving away), with a period of 164 days. In addition to these "mobile" lines, the spectrum also shows "stationary" emission lines—i.e., without strong shifts.

The simplest model envisioned to account for the characteristics of this amazing object is described on page 276. The mobile emission lines would be produced by two jets of particles emitted at a velocity of 78,000 kilometers (48,477 miles) per second by a dense central object. The observed velocity variations would be due to precession of the axis containing the particles. The axis of precession would make a 20-degree angle with the axis of the particles and an 80-degree angle with the direction of observation. A study of the stationary emission lines has further revealed a sinusoidal variation in their radial velocity, with an amplitude of 80 kilometers (50 miles) per second and a period of 13 days. These values are very typical of those that characterize certain binary systems. Therefore SS 433 would belong to such a system, the pulsar's period being modulated by its revolution in 13 days around an invisible normal star.

Cosmic Rays

Cosmic rays consist of ionized nuclei (protons and heavier nuclei) and electrons. At a given energy E, electrons represent about 1 percent of the total number of cosmic rays. It seems that the most energetic cosmic rays (10^{18} to 10^{20} eV) that reach the Earth come from galaxies external to our own. Cosmic rays with energy lower than 1 GeV (10^9 eV) are heavily affected by the Sun's magnetic field. The international solar polar mission (see box, page 91) should give us information about this in a few years.

One of the oldest means of studying cosmic rays consists of analyzing the traces they leave in samples of material of different thicknesses. Likewise, analysis of the flaws they cause in meteorites and lunar dust enables us to infer the composition of the incident cosmic ray.

Ions moving at very close to the speed of light strongly interact with the interstellar magnetic field and give off synchrotron radiation (nonthermal rays from the galaxy and supernova remnants). The radio source Cassiopeia A is the most intense in the sky at centimeter wavelengths; it emits synchrotron radiation due to the relativistic electrons spiraling in its magnetic field. The intensity of the emitted radiation decreases over time (by about 1 percent per year), which has made it possible to determine the age of the radio source (about 300 years).

The interactions of cosmic rays with the gas of the interstellar medium is of fundamental importance in studying γ radiation, nuclear spallation reactions and the ionization of molecular clouds.

γ radiation. The collisions of protons with the atoms or molecules of hydrogen give rise to radiation that cannot be observed by balloons or satellites.

γ radiation is a tracer of the dense regions of our galaxy; thus the excess emission observed by the *SAS 2* and *COS B* satellites between 3 and 5 kiloparsecs from the center of our galaxy coincides with the concentration of molecular clouds detected through rotational transitions of the CO molecule.

Nuclear spallation reactions. The only means of forming certain light nuclei results from the fragmentation of heavy nuclei by the impact of cosmic-ray protons. Despite their very low abundance, such nuclei play a fundamental role in understanding the nucleosynthetic history of our galaxy.

Ionization. In the dense molecular clouds, stellar ultraviolet radiation does not penetrate; only energetic cosmic rays penetrate deeply and ionize the most abundant elements (hydrogen H, carbon C). Although there is a very low abundance of electrons in these clouds ($n_e/n_H = 10^{-7}$ to 10^{-8}), all of interstellar chemistry rests on these ions being capable of interacting rapidly among themselves, despite the low temperatures prevailing in the medium in question.

The distribution of cosmic rays also is different from that of interstellar matter; the disk within which cosmic rays propagate is much thicker than the gas disk, and some authors even think that the radio halos of certain galaxies (NGC 891) is due to synchrotron radiation from cosmic-ray electrons forming a much thicker system than the gas of the disk.

The origin of cosmic rays has long been controversial. The most likely source of these rays is the supernovae. The scattering of cosmic rays by the irregularities of the magnetic field have rapidly made it isotropic, except perhaps around the remnants of the younger supernovae (Cas A, Crab nebula).

Other possible sources are pulsars and eruptive stars.

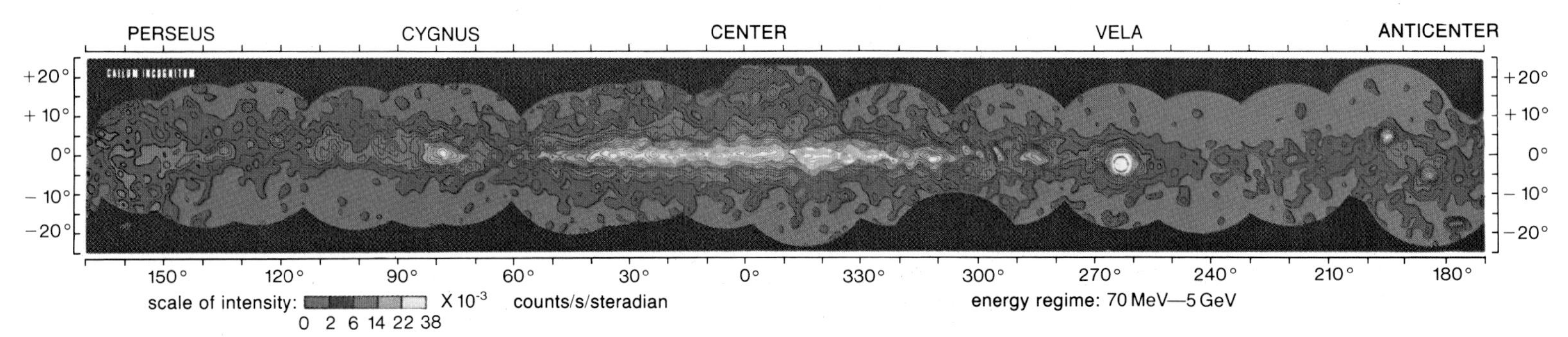

Map of the galaxy's gamma-ray emission, prepared with the help of the European satellite COS-B. *Most of the observed sources are inside the galaxy; one of them is associated with the molecule complex of* ρ *Ophiuchi. ESA Photo.*

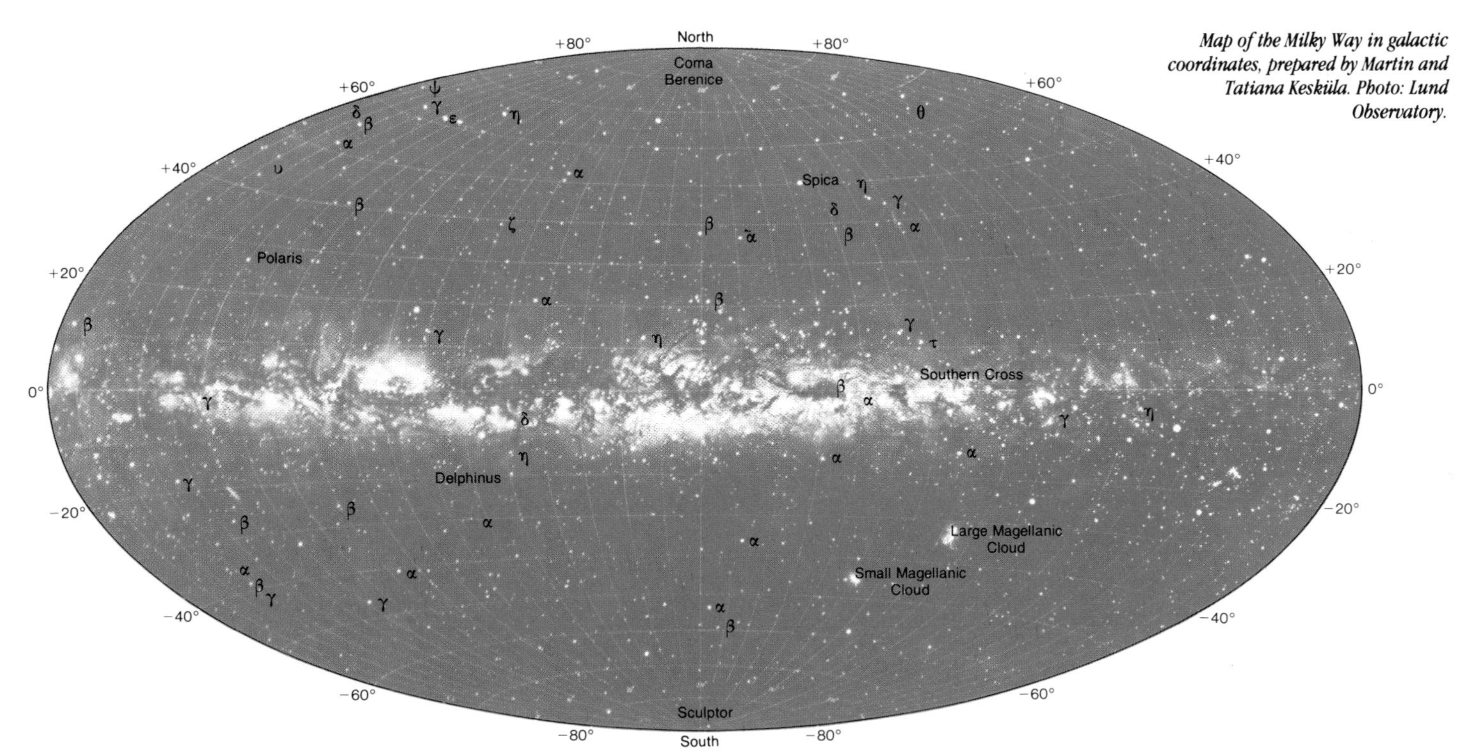

Map of the Milky Way in galactic coordinates, prepared by Martin and Tatiana Keskküla. Photo: Lund Observatory.

23. Structure and Evolution of the Galaxy

Gazing at the night sky, with no background light, an observer can discern a rather wide whitish trail on the dark backdrop sprinkled with bright points; the Milky Way. Seen in an edge-on view, it is the gigantic concentration of stars, dust and gas to which the solar system belongs—our galaxy.

Comparison with exterior galaxies, and direct observation, agree in describing our galaxy as a very flat system, thickening in the central regions; in the plane, its structure is a spiral made up of many dense arms—studded with bright stars—which all join together in a luminous, dense nucleus at the center; a view of the galaxy in profile would show two luminous disks with a dark, rather fine layer of dust sandwiched between them. The accompanying illustration represents the galaxy in these two projections and shows the position the Sun occupies in it; it is situated nearly in the galactic plane, in one of the spiral arms, but 30,000 light-years from the galactic center, which explains the asymmetry of the Milky Way seen from the Earth: The white strip mentioned above is the

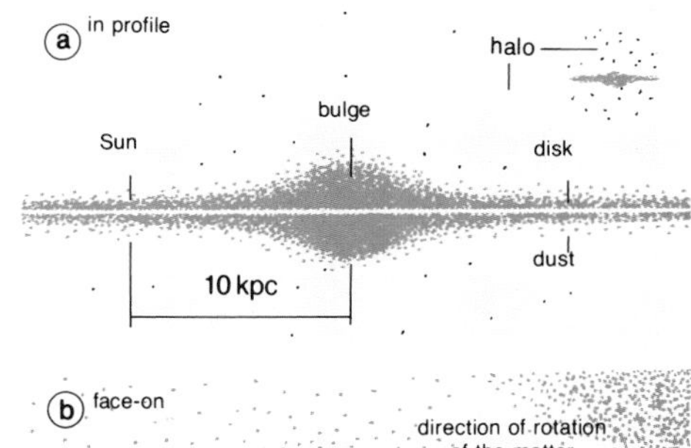

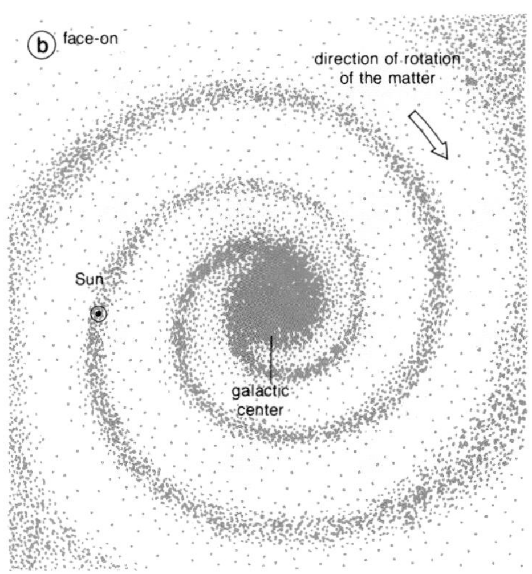

Schematic representation of the galaxy. (a) seen in profile—the globular clusters (black dots) form a nearly spherical halo around the galactic center; dust is concentrated in the galactic plane, (b) seen face-on—the galaxy is a spiral of the S^b type, all the matter of which rotates around the center in the same direction as the winding of the spiral arms.

trace of the galactic disk in the sky. The disk measures about 100,000 light-years in diameter and 1,000 light-years in thickness. The galaxy has a halo of stars and groups of stars called globular clusters symmetrically distributed in a concentric sphere at the galactic nucleus. The total number of stars in the entire galaxy is estimated at 100 billion (an estimate based on the evaluation of the galaxy's total mass and the average mass of a star). In the halo, the globular clusters, numbering about 200, may contain from several thousand to a million stars (100,000 on average). These are the largest concentrations in the galaxy. Their spherically symmetric distribution around the center of the galaxy enables us to localize it and to determine the Sun's off-center position. This center is in the direction of the constellation of Sagittarius (more precisely, the galactic nucleus is located at a very intense radio source, Sagittarius A: a = 17 hours, 42.4 minutes; o = −28°55.0' [1950.0]).

Characteristics of the Galaxy

diameter of the disk:	~100,000 light-years
diameter of the spherical halo:	~150,000 light-years
total mass:	140 billion solar masses
perpendicular distance between the spiral arms:	5,200 light-years
thickness of a spiral arm:	2,000 light-years
Sun's distance to the center of the galaxy:	30,000 light-years
speed of the Sun's rotation:	220 km./sec.
presumed age:	12 billion years

THE GALAXY'S SYSTEMS OF COORDINATES

The galactic longitudes and latitudes l^{II} and b^{II} are the systems of galactic coordinates used by the International Astronomical Union:

pole: α = 12 hr., 49. min., δ = + 27°24.′(1950.0)
galactic center: (l^{II} = 0, b^{II} = 0):
α = 17 hr., 42.4 min., δ = − 28°55.′(1950.0)

The galaxy also is composed of a gaseous interstellar medium, neutral or ionized, sometimes concentrated into dense gas clouds made up of closely mingled atoms, molecules and dust. It is the changes made by this medium in the emission spectra of the stars behind it that make it possible to detect and analyze it.

All of the matter—gas, dust and stars—participates in a rotation around a central axis perpendicular to the galactic plane. Thus the centrifugal force caused by the rotation balances out the galactic plane. Thus the centrifugal force caused by the rotation balances out the gravitational force, which tends to draw all the matter toward the center. In particular, the Sun rotates around the galactic center at a speed of 220 kilometers (137 miles) per second and makes a revolution in 300 million years; since its birth, the Sun has made about 15 revolutions. The galaxy does not rotate as a solid body—the angular velocity of its matter varies with the distance from the center; this is differential rotation. The matter far from the center is slower and therefore makes fewer rotations in a given amount of time. Differential rotation thus tends to wind the spiral arms more tightly around the galactic center.

Methods of Observation

For a long time, the only means of observing our galaxy was the optical telescope. However, visible light—corresponding to wavelengths of 0.4 to 0.8 μm—is strongly affected by dust, which reddens or even completely obscures it if there is a great quantity of dust, as, e.g., in the galactic plane. This is due to the dimensions of the dust grains, solid particles a few tenths of a micrometer in diameter, about the same as the wavelength of visible light. The region of the galaxy that can be optically studied, therefore, is limited to a sphere with a radius of about 10,000 light-years, centered on the Sun. Observations of external galaxies, by comparison, provide indications about the large-scale optical aspect of our galaxy. On the other hand, not all radiation with wavelengths larger than 1 micron are stopped by dust or interstellar matter. Thus radio-frequency waves emitted chiefly by interstellar gas make it possible to gain information on the large-scale structure of the galaxy; in particular, the spiral structure can be observed at the 21-centimeter wavelength emitted by neutral atomic hydrogen. The galactic center, inaccessible by optical methods, has been revealed by its radio and infrared emissions.

Recent techniques for detecting more energetic photons than those in the visible also have contributed their share: ultraviolet rays, X rays and gamma rays. In particular, the accelerated gas falling into a black hole gives off many X rays, and the interaction of cosmic rays with interstellar matter produces gamma rays.

Owing to the improvement and diversification of observational methods, great progress has been made in the past 30 years in knowledge of the structure of our galaxy. This chapter is primarily concerned with the description of this structure, as it has been observed or as we can envision it today.

But observational data also inform us indirectly about the evolution of the galaxy. Thus study of the stars it contains, combined with study of their motions and positions, enables us to understand the stages of its inward contraction at the time of its birth, while radio-astronomical study of the interstellar gas it contains leads to theories of the formation and stability of its spiral structure. This evolution will be described in the final part, and the theoretical models that attempt to retrace it will be presented in summary form.

The Stars in the Galaxy

EVER SINCE THE OBSERVATION of spiral-shaped galaxies in the universe, astronomers have tried to demonstrate such a structure in our own galaxy. Unfortunately, all the laborious counting of stars has been in vain: The spiral arms do not owe their luminosity predominantly to a greater density of stars but to a relatively small number of very luminous, very young stars. It was the classification of stars in populations, proposed by the American Baade in the 1940s, that enabled astronomers to reveal two spiral arms in our galaxy in 1951; first, in the vicinity of the Sun (the range of optical telescopes being limited by the dimming due to dust); then, several months later, throughout the galaxy, based on observations of the 21-centimeter line of hydrogen.

Stellar Populations

Baade's classification was rather crude; it distinguished only two categories of stars: population I, of which the brightest stars are blue, very hot and thus young; and population II, of which the brightest stars are cold, red, large and older. Population I is found mainly in the spiral arms of the galaxy but never in the regions between arms; population II, mainly toward the center and in the globular clusters. This classification should be compared to the Hertzsprung-Russell diagram (see "Structure and Evolution of the Stars," page 255). The stars in population I belong to the main sequence; the brightest ones have a temperature of 30,000° K and an intrinsic luminosity equal to 10,000 times that of the Sun. The bright stars of population II are grouped along the red-giant branch.

Since then, the classification has been refined. The stars are grouped in many more categories, according to their temperature and luminosity, and there is an ongoing evolution from one category to another. During a conference in Rome in 1957, astronomers agreed on a classification into five populations, known since then as the Vatican conference classification, which takes into account temperature, luminosity, age, chemical composition and the kinematics of the stars (see table).

Age and Chemical Composition of the Stars

Why are population I stars found mainly around the spiral arms of the galaxy, and never, or almost never, in the interarm regions? Star formation takes place predominantly in the spiral arms, as will be explained later, and the stars of population I, since they are very young, have not yet had time to move very far from their place of formation. In addition, a star's intrinsic luminosity is a very rapidly increasing function of its mass, so that the massive stars are extremely bright; but they also exhaust their resources much faster than smaller stars and very quickly cease to shine, the hottest ones in a few million years (by way of comparison, the age of the Sun—an average star—is 5 billion years). This phenomenon explains the strong luminosity of the spiral arms (which makes the structure of the spiral galaxies particularly visible). In effect, the brightest stars, which burn out rapidly, practically never leave the spiral arms, the place of their formation.

The basic component of the stars and the interstellar medium is hydrogen, but heavier elements are slowly synthesized within the stars; their chemical composition—and especially their heavy-element abundances—depend on when they were formed, on their age. When a star has used up all its nuclear fuel in the formation of heavier and heavier elements (see "Structure and Evolution of the Stars," page 254), it becomes a supernova, disintegrating in a huge explosion, or else a nova, undergoing a series of smaller explosions. In this way, through injections from the stars, heavy elements appear in the interstellar medium, so that the stars now old, which were born at the beginning of the galaxy, at a time when it contained only hydrogen, are poorer in heavy elements than the young stars, which now are forming from an interstellar medium enriched by the accumulation of ejecta from all the previous stars.

The Motions of Stars

The motions of stars are directly related to their age of formation, and their study makes classification much easier. Like all matter, stars share in the rotation of the galaxy, but they also have their own motion, both in the plane of the galaxy and perpendicular to it. These perpendicular motions are periodic oscillatory motions, in which the restoring force is the gravity exerted by the galactic plane. Their maximal velocity in that direction, reached during the crossing of the plane, is proportional to the maximal height reached on either side of the plane, where the velocity goes through zero and the star retraces its path. For a given star, the speed perpendicular to the galactic plane depends on the phase of its cycle of oscillation at which it is observed. But statistically, in a group of stars, the greater the speed, the greater the height reached above the plane. It has been possible to show in this way that globular clusters have quasispherical motions around the galactic center; the maximum height they can reach above the plane is about the same as the radius of the galaxy, and their speed perpendicular to the plane is about 100 kilometers (62 miles) per second. On the other hand, the brightest and hottest stars are more or less confined to the plane of the galaxy, with a perpendicular velocity of 5 kilometers (3.1 miles) per second. Between these two extremes there is a constant gradation in the populations of stars, which is related to their chemical composition.

This phenomenon is very interesting, for it supports the most commonly accepted ideas regarding the evolution of the galaxy. At the beginning, there existed only a cloud of gas, equivalent in mass to about 100 billion stars. Because of gravitational instability, this gas cloud contracted inward and fragmented into smaller and smaller clouds; these clouds, through condensation, simultaneously produced stars or groups of stars, while the galaxy continued to contract. At the beginning, the initial cloud had a tiny speed of rotation on its own axis. But in contracting, it spun faster and faster to maintain its angular momentum, like a ballet dancer whose rotation accelerates when she folds her arms against her body. The rotational motion is organized around a particular axis, determined by the slight initial movements, and the contraction takes place essentially along that axis, with the gravitational force gradually equilibrating the centrifugal force. The shape of the galaxy, more or less spherical at the outset, flattens more and more, becoming the nearly flat disk observed today. The stars born during this evolution preserve, in their trajectories and dynamics, the characteristics of the galaxy at the time of their formation. Once formed, they have very little interaction with their environment, undergoing only an insignificant braking owing to infrequent collisions. This is how the older stars—those of the globular clusters—show a quasi-spherical arrangement and trajectory around the galactic center, a reminder of the structure of the primordial cloud; while the brighter and younger stars of population I have small motions on either side of the galactic plane, because they have recently been formed from a cloud now compressed and shrunken to the galactic disk.

Stellar Populations

	Population I		Old Population	Population II	
	Extreme	Less Young	(disk)	Intermediate	Extreme (halo)
environment	gas, dust, spiral arms	open clusters	galactic center	globular clusters	
stars	supergiants rich in heavy elements	Sun, giants,	novae, dwarfs, poor in heavy elements	large-velocity dispersion	subdwarfs, very poor in heavy elements
average height above the plane (in pc.)	120	160	400	700	2,000
average velocity perpendicular to the plane (in km./sec.)	8	10	16	25	75
distribution	dispersed in aggregates			homogeneous	
average age in billions of years	< 0.1	0.1–1.5	1.5–5	5–6	> 6
abundance of heavy elements relative to H	0.04	0.02	0.01	0.004	0.001

The Interstellar Medium

BETWEEN THE STARS OF our galaxy, the existence of vast clouds of gas and dust has long been known owing to the dimming of starlight for which they are responsible. Two centuries ago, William Herschel already spoke of "holes in the sky." Today the densest of these clouds are called "molecular clouds" because about 50 molecules have been observed within them.

The structure of the interstellar medium, as it is known in our time, is very diversified. The main component of the gas clouds is hydrogen; it may occur in its neutral atomic form, H I, in its ionized form, H II; or in its molecular form, H_2. The latter is found especially in the dense molecular clouds and remained undetected for a long time; it can be observed directly only by its ultraviolet emission, and indirectly by the collisional excitation of the molecules of carbon monoxide, detected at millimeter wavelengths.

Ionized hydrogen may be observed both by its optical and by its radio emission. Optically, it involves the recombination lines of hydrogen, produced when a free electron is captured by the proton and cascades down the levels of the atom. In the radio, recombination lines also may be detected, but concerning much higher energy levels of the atom; usually what is involved is continuum thermal emission resulting from the mere braking of electrons in their encounters with protons. The ionized hydrogen regions, or H II regions, are found in the neighborhood of the very hot, massive and young stars and are ionized by their ultraviolet radiation; these are tracers of the zones where stars are formed. Optically, H II regions are easily detected only in the vicinity of the Sun, with a high resolution (however, the interference methods of the Marseilles Observatory have made it possible to extend such observation up to 10 kiloparsecs), while radio observations provide the large-scale distribution of ionized hydrogen in the galaxy. This distribution reflects the galaxy's spiral structure.

Neutral atomic hydrogen—very common in the galaxy—is observed in radio waves by its emission at 21 centimeters' wavelength. What is involved is a transition between two hyperfine levels of the atom; the probability of spontaneous emission is very small for this transition: An atom of hydrogen emits a photon of 21-centimeters' wavelength every 10 million years. This low probability is compensated for by the very large quantities of hydrogen capable of emitting such radiation. Observation of the 21-centimeter line, which began in 1951, was one of the first triumphs of radio astronomy because of the importance of its results.

After gas, the most abundant constituent of the interstellar medium is dust, closely mingled with gas clouds and about which little is known up to now. The mass of dust should represent about 1 percent that of the gas. The dust grains are solids, with dimensions of about a micron, and thus obscure the optical wavelengths. They are believed to consist essentially of silicates and ice. Indeed, in the infrared, deep absorption bands are observed at 10 μm, a band characteristic of the silicates. These dust grains may act as catalyzers of reactions to form molecules, particularly molecules of hydrogen H_2.

Finally, cosmic rays—very-high-energy atomic particles (especially protons)—traverse the galaxy at very great speed and mostly interact with its magnetic field. Particles with an energy of up to 10^{20} eV have been detected. The interaction of cosmic rays with molecular clouds produces gamma rays.

Atomic Hydrogen in the Galaxy

Neutral hydrogen, present everywhere, represents about 5 percent of the total mass of the galaxy (with the stars making up most of the rest). As a rough approximation, the gas has an average temperature of 100° K and a density ranging from 1 to 10 atoms/cm³ within the spiral arms and 0.1 atom/cm³ in interarm regions. More refined observations, and theoretical considerations, lead to a "two component" model: One consists of cold gas clouds with a density of 10 to 100 atoms/cm³ and a temperature of at least 100° K; the other component is a more diffuse, very hot gas, with a temperature reaching 10,000° K but with a density of only 0.1 atom/cm³. This thin, hot medium maintains a pressure around the denser, colder clouds that ensures the overall equilibrium. The spiral arms contain two components; on the other hand, the cold clouds are nearly absent in the interarm regions.

In the 1950s and 1960s, observations of neutral hydrogen supplied a map of the galaxy that until very recently was considered to be representative of the interstellar medium. Today it is known that the distribution of molecular hydrogen is very different; it even appears that atomic hydrogen is an exception among all the so-called population I constituents of the galaxy; those young constituents combined with stars of population I—the gas or dense molecular clouds necessary for the formation of stars, the ionized gas testifying to the presence of young stars with ionizing radiation, the vestiges of very massive stars that have evolved very rapidly, supernova remnants, pulsars—all have a comparable radial distribution, while atomic hydrogen extends twice as far from the center of the galaxy (see page 287).

By what means can we reconstruct the map of the galaxy in neutral hydrogen H I, as an outside observer sees it, and do so using emission received at a particular point near the galactic plane, the Sun? In other words, how can we know the distance of each emitting source, and separate the various sources of emission coming from the same direction? All of this is possible, for the emission of atomic hydrogen corresponds to a spectral line and thus shifts in frequency as a result of the Doppler effect, making it possible to determine the speed of the source relative to the observer. This speed, combined with a theoretical model of the galaxy's rotation curve, then determines the distance of the source. Observation shows that the distribution of hydrogen is anything but uniform; several concentrations appear in succession on a single line of sight. Analysis of these spectra is possible only if the various contributions do not overlap, which generally is the case: Each H I cloud shows a velocity dispersion that is less than the difference in velocity between two clouds. The typical velocity dispersion for a concentration of hydrogen is 10 kilometers (6.2 miles) per second. This dispersion is not thermal in origin (a temperature of 100° K corresponds to a dispersion of 0.9 kilometer [0.6 mile] per second); it is due to turbulence motions. Several ambiguities persist, however, in determining the distances of all the concentrations of hydrogen.

In the final analysis, the large-scale spiral structure of hydrogen in our galaxy cannot be estab-

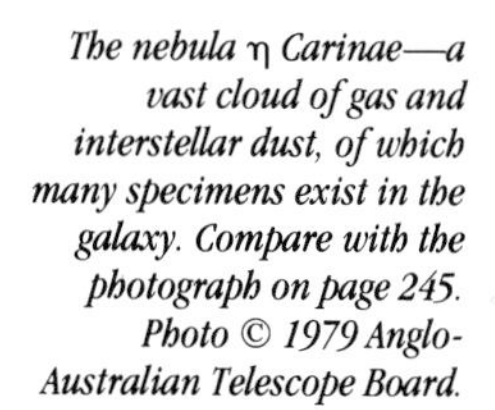

The nebula η Carinae—a vast cloud of gas and interstellar dust, of which many specimens exist in the galaxy. Compare with the photograph on page 245. Photo © 1979 Anglo-Australian Telescope Board.

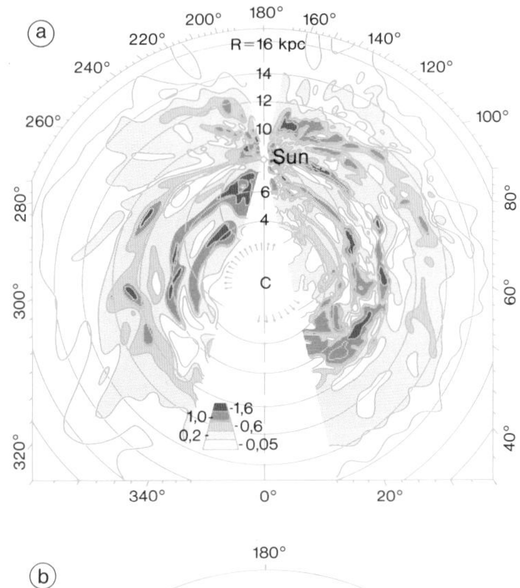

Distribution of neutral hydrogen in the galaxy, based on profiles observed in the 21-centimeter line. The longitudes close to 0 degree or 180 degrees are not shown, for the radial velocity of hydrogen relative to the sun is zero in those directions, and it is impossible to deduce the distance of the corresponding sources from them. This map shows multiple condensations, implying a spiral structure with tightly wound arms. However, radio astronomers now prefer to confine themselves to raw observations, considering the conversion to distance to be too subject to risk.

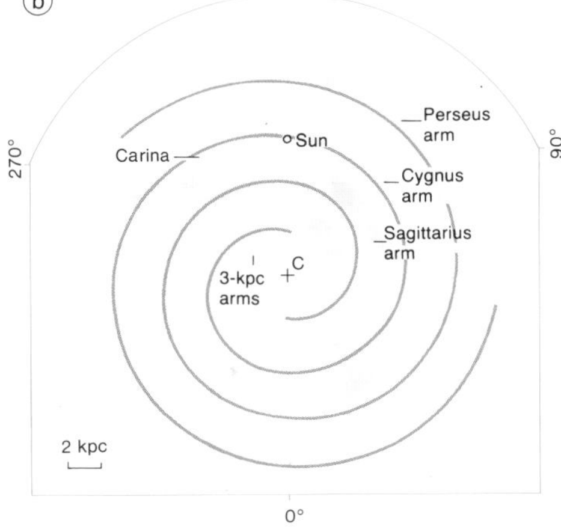

Sketch of the spiral structure superimposed on the distribution of hydrogen. The closest arms that can be optically identified are also seen in the radio range: Perseus, Cygnus, Sagittarius.

model that involves the determination of little-known distances.

The galactic center is a very special region, where the profiles of neutral hydrogen differ from those in other regions of the galaxy. These profiles reveal, in particular, a regular emission source between the galactic longitudes 1 = 335° and 1 = 4°, in which the radial velocity varies almost linearly with longitude between −50 and −200 kilometers per second. This emission feature is interpreted as a spiral arm situated about 3 kiloparsecs from the center and that is believed to be expanding away from the galactic center at a speed of about 50 kilometers per second, unlike other spiral arms. This arm is seen in absorption in front of the galactic center, which proves its presence

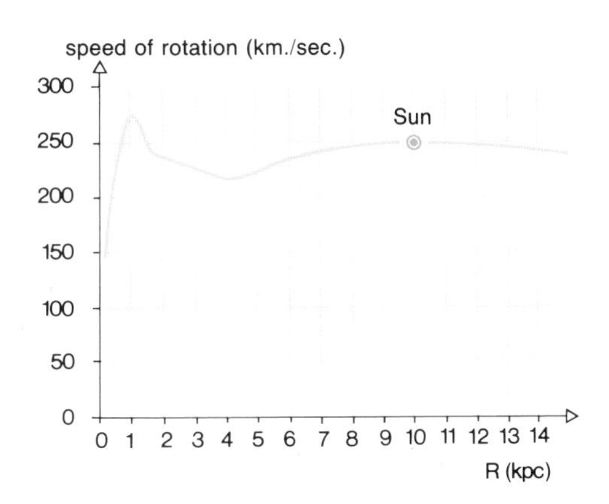

Rotation curve of the galaxy, experimentally determined. Note that at distances from the center of 1 to 15 kiloparsecs, the linear velocity remains nearly constant (very close to 250 kilometers [155 miles] per second), implying a constantly decreasing angular velocity for an increasing distance from the center; this is differential rotation. A comparison with other spiral galaxies suggests a linear velocity that is still nearly constant beyond 15 kiloparsecs. It is only in the neighborhood of its center, at radii less than 1 kiloparsec, that the galaxy rotates as a solid body.

lished clearly and directly, especially in the regions closer to the center than the Sun. On the other hand, many parameters of the spiral have been obtained: inclination, separation of the arms, contrast of arm-interarm density, etc. If we confine ourselves to kinematic data, the spiral arms can be clearly identified and isolated, and their characteristics are even in excellent agreement with the analogous data inferred from optical observations in the vicinity of the Sun. The use of raw data appears much more satisfactory today than the reconstruction of those data in terms of a

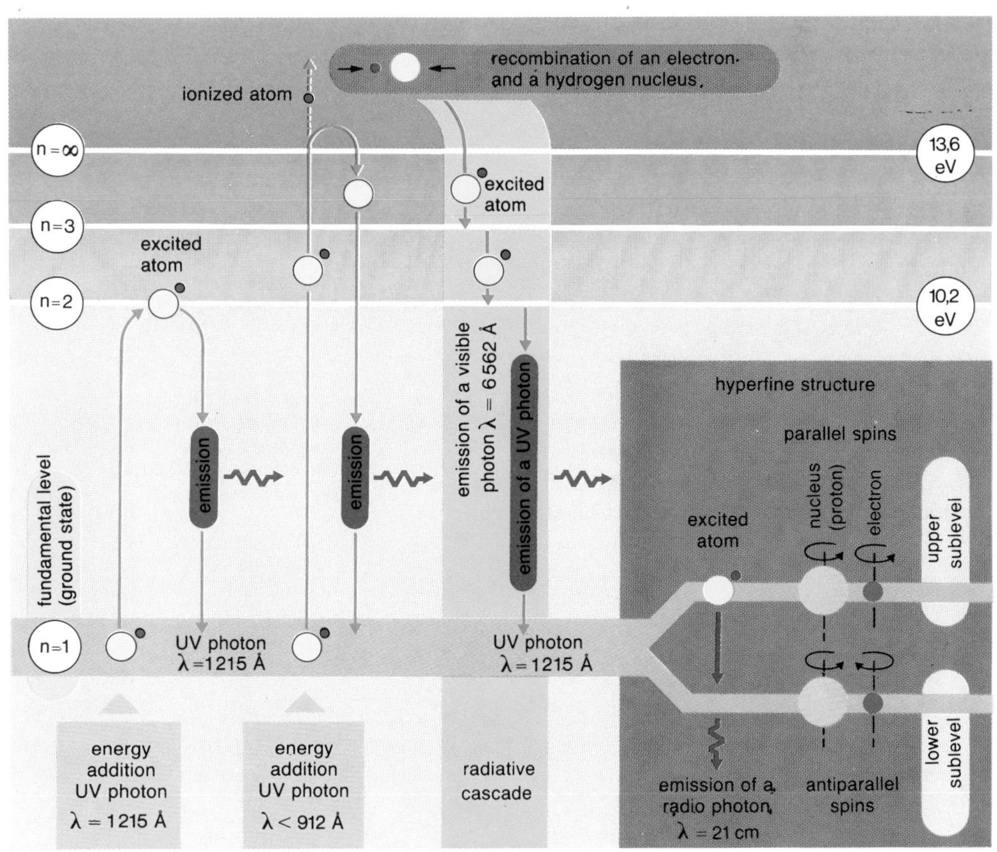

Energy levels of the hydrogen atom. In its most stable state, the hydrogen atom (neutral hydrogen) is in a level of minimum energy (ground state level). Addition of energy can bring it to a more energetic state (excited level); then it returns spontaneously to the ground state level while emitting radiation. An extreme case occurs when the energy provided is sufficient to tear the electron away from the atom (ionization). The electron may again be captured subsequently by a proton, to form another excited atom, which, in returning to its stable condition, wil emit light (recombination lines). In fact, the fundamental level includes two very slightly different sublevels, tied to the electron's direction of rotation. In going from the more to the less energetic, the atom emits a photon at 21-centimeter wavelength: This line of neutral hydrogen is an effective means of studying interstellar gas.

between the Sun and the galactic center. A similar emission source is observed between longitudes 0° and 10°, with radial velocity ranging from +50 to +200 kilometers per second; but this hydrogen feature, unlike the arm at 3 kiloparsecs, does not absorb the intense radio emission from the galactic center; thus it is interpreted as an arm situated on the opposite side of the galactic center relative to the Sun and moving away from the center at a speed of 135 kilometers per second. Its position is more difficult to estimate but probably is about 2.5 kpc. from the center. Finally, the profiles also show an arm at strong negative velocities between 356° and 0° longitude, which seems to have a counterpart at positive velocities between 0° and 4° longitude. This emission, according to Rougoor and Oort, would be due to a central disk of hydrogen of 300 parsecs in radius and to a ring between 500 and 590 parsecs from the center.

A number of hydrogen clouds at high radial velocity, known as "high-velocity clouds," have been detected at high galactic latitude. The problem of their nature has not yet been solved. Are they tied to the galaxy and rotating rapidly around its center? Or are they extragalactic, and do they collide with the galaxy? Or, again, are they the traces of a collision between our galaxy and the Magellanic Clouds?

A Distorted Disk

The variation of the hydrogen emission with galactic latitude b shows that neutral hydrogen is distributed in a very thin, nearly flat disk. The thickness of this disk, determined between the two points at which the emission intensity falls to half of its maximum value, is reduced in the central parts of the galaxy to 120 parsecs and increases toward the exterior.

The disk of neutral hydrogen is flat only at first glance. In reality, the plane is distorted, or warped. This distortion is symmetrical relative to the center and consists of an elevation of the edge of the disk, on one side, and a lowering on the other. The height of the deviation for a given radius is a nearly sinusoidal function of longitude. This distortion appears only for rather large-radii R, starting from the position of the Sun, approximately (R = 10 kiloparsecs), and reaches, for R = 15 kiloparsecs, a height of 1 kiloparsec above the galaxy's midplane. The inclination of the hydrogen plane is not peculiar to our galaxy but also is found in a large number of other spiral galaxies.

Several explanations have been offered for this phenomenon:

1. The warping could be primordial oscillations of the plane of the galaxy (Lynden-Bell, 1965)—i.e., oscillations created at the time the galaxy was formed. This hypothesis has been placed in doubt by the Hunter and Toomre study (1969): Any oscillation of the plane of the galaxy is rather rapidly dissipated by differential precession; its life-span does not exceed 100 million to 1 billion years, yet the age of the galaxy is more than 10 times greater than this.
2. An oscillatory mode of the plane of the galaxy would be unstable and could increase at the expense of the potential or kinetic energy instead of disappearing. However, an analytical study of all oscillatory modes, done by Hunter and Toomre, reveals no unstable modes.
3. The oscillation of the plane is caused by tidal interaction with nearby galaxies. For our galaxy, interaction with the Large Magellanic Cloud would result in distortion of the plane. The Large Magellanic Cloud is now too far from our galaxy for the present interaction to be sufficient, but it is very probable that its past trajectory brought it close to us. If this occurred in the fairly recent past (less than 100 million years ago), so that the oscillation has not had time to die out, the sinusoidal distortion observed today might have been created by that passage.

This explanation—a likely one as far as our galaxy is concerned—is not, however, universal; a number of galaxies with a distorted hydrogen plane are completely isolated in the sky; they have no companion close enough to have had a tidal interaction with them in a sufficiently recent past. That is why other hypotheses have been proposed: The galaxy moving at a certain speed relative to the intergalactic medium might distort the plane of the galaxy by exerting asymmetrical pressure on its halo (Kahn and Woltjer). The accretion of intergalactic gas by the galaxy, if it occurred asymmetrically relative to the galactic plane, might disequilibrate it.

In conclusion, the problem of the origin of the warping of the hydrogen plane has not yet been solved, especially not in a general way.

The Theory of Density Waves

We described our galaxy as a spiral galaxy, without, however, having an explanation for that kind of structure. It is not only observations in the radio range that have confirmed the spiral structure on a large scale, but also comparison with other galaxies similar to ours, such as M 81, a galaxy of the Sb type. In the Hubble classification, our galaxy was placed in the category of spiral galaxies of the Sb type.

How was this spiral structure formed, and how is it maintained? The problem might appear simple at first glance; according to the rotation curve, the outer parts rotate less rapidly than the regions immediately inside, as a result of differential rotation. A lack of homogeneity of density or luminosity in the galaxy's disk at the beginning is sufficient for differential rotation to form a spiral; e.g., if there were a linear arrangement of rela-

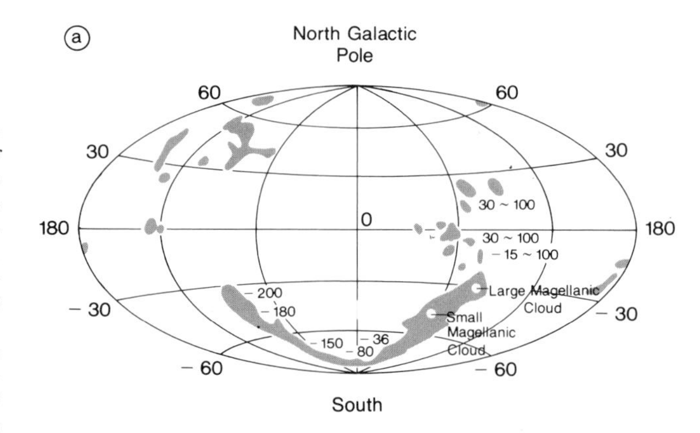

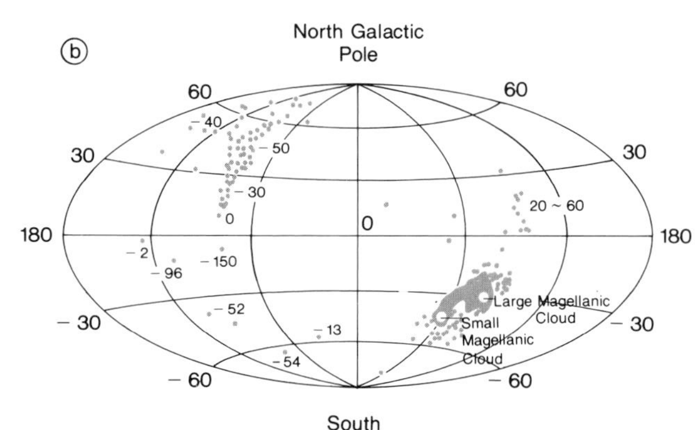

The "Magellanic stream" is a trail of a gas, forming a ring of neutral hydrogen clouds nearly 90 degrees from the plane of the galaxy. This "current" is the vestige of the interaction between the Magellanic clouds and our galaxy. (a) Observed neutral hydrogen clouds. (b) Computer simulation by Fujimoto and Sofue, 1976 (the hydrogen clouds are represented by points of negligible mass; the numbers represent the radial velocities in kilometers per second.

A spiral galaxy of the same type as our own: M 81, in Ursa Major. This is how the galaxy would appear to us if we could observe it from the exterior. Photo: Hale Observatories.

tively luminous matter, the outer extremity of that line would lag behind the inner one; in a short time, the ray would curve and would wind around itself, becoming a spiral. But that is exactly the problem: The winding would occur very quickly, in a few hundred million years, whereas the age of the galaxy is estimated to be 100 times greater. Hence, the spiral structure formed according to this principle would last only a hundredth of the life of the galaxy. It would be very unlikely that we would be present just in time to observe it, and above all, there would not be so many spiral galaxies in the sky. (The rotation curves and life-spans are analogous from one spiral galaxy to another.)

This winding problem thus eliminates the possibility, for the observed spiral arms, of their being arms of matter sharing in the rotation of the galaxy. An alternative was proposed in the 1960s by the Swedish astronomer Bertil Lindblad: The spiral arms would be density waves moving at a constant angular velocity different from that of matter; at a given instant, matter collects in an arm, resulting in a greater luminosity, and then leaves it, giving way to a continuous flux of rotating matter. The phenomenon might be compared to a swell on the sea: The matter (here the molecules of water, the motion of which is revealed by a floating cork) participates in the wave as it passes but is not pulled along by it; the cork remains in place, while the wave moves (to get an equivalent pattern with galactic waves, it is necessary to place oneself in the rotating frame of reference in which matter is stationary).

The origin of this wave may be a slight distortion in the initial system, which is propagated by gravity and, aided by rotation, forms a spiral-shaped perturbation. That is what the simple computer-simulated models constructed by Lindblad, using a very small number of particles, show. What happens chronologically? The gas and stars rotate together around the center of the galaxy. But their trajectories are not perfect circles; they make small excursions around an average trajectory. It is not surprising, therefore, that a small concentration of matter should occur rather frequently at the expense of the neighboring regions. These inhomogeneities produce perturbations of the gravitational potential. If the sum of the perturbations cancels out, or has an effect contrary to the concentrations that gave rise to it, there are no particular consequences. On the other hand, whenever the perturbations can be organized, they augment themselves. This results in a global perturbation, or wave, that is propagated in the disk of the galaxy. In this way, spiral arms may be formed that are veritable potential wells, where matter has a tendency to be concentrated and thus to accumulate momentarily.

Lindblad's idea was developed and quantified by Lin and his collaborators. Their approach is hydrodynamic and statistical: They treat the galaxy as a fluid of stars. Beginning with simplifying hypotheses (a spiral structure wound around itself, a constant angular velocity for that structure, a spiral perturbation already present in the potential), they calculate the perturbation that results in the distribution of matter. If that calculated perturbation is sufficient to maintain, by its gravitation, the initial potential perturbation, the starting hypothesis is justified *a posteriori*: The spiral wave can exist. The result is positive only when certain conditions are fulfilled between the velocity and the shape of the spiral.

A Fertile Hypothesis

This theory turns out to be very rich in consequences and applications. One remark can be made from the start: While the spiral perturbation is generated principally by the stars, it is the distribution of gas that is most affected by it. Most of the mass—hence of the gravitional potential—is supplied by stars; the gas represents only a small percentage of the total mass of the galaxy. Moreover, the stars show large random velocities around their average motion; they have a substantial velocity dispersion. This is certainly due to the absence of collisions between stars, the stellar medium being very diluted. On the other hand, the velocity dispersion of the gaseous medium is much smaller; collisions and viscosity brake the random motions. This results in a greater coherence of the gas, which thus responds better to perturbations of potential than the stars. Thus it is the gas that is primarily accumulated in the potential wells that are the spiral arms; the stars, moving at large velocities in all directions, are only too capable of leaving them. Why, then, do the spiral arms seem so bright to us? The answer is that the formation of new stars is provoked by gas traveling through the spiral arms. When the gas encounters the spiral wave, it is held back and accumulates, but then the pressure increases: Each cloud undergoes compression as it passes through an arm, which breaks its equilibrium with the extend medium and unleashes a gravitational instability; the cloud contracts more and more, and may go to the point of forming a star, which will radiate. If the star is massive enough, it radiates intensely, in a short period of time, and its radiation is predominantly ultraviolet, which can ionize the surrounding hydrogen gas. This ionized gas (H II) then emits, in recombining, a strong line, which makes the ionized region around the star (H II region) very bright. These luminous phenomena have a fairly short life-span relative to the time it takes matter to rotate around the galaxy, and particularly to leave the spiral arm; that is why light seems to come predominantly from the spiral arm.

When we diagram the various stages matter goes through in crossing an arm, we note that the theory might be tested by several observations:

1. The gas is first compressed in the wave over a long enough time to enable us to detect a dark strip of clouds and dust in the concave side of the spiral arm (in our galaxy, the wave does not rotate as fast as matter; matter enters through the con-

cave part of the spiral arm; the arms "lag behind," as is the case, moreover, in the majority of galaxies). It is easy to verify that, quite often, dust is present in the concave edges of the spiral arms; in addition, dense molecular clouds may be detected around the spiral arms by virtue of the carbon monoxide line in the radio range; such experimental verification is very difficult to do in our galaxy, where we do not always know where to "place" the source of emission, but it has been done in a number of external galaxies.

2. The gas is then accelerated to supersonic speeds during its passage through maximum compression; the density and pressure of the gas are then high in a very narrow zone, and the compression is accompanied by a shock wave. The formation of stars and of H II regions is accelerated by this, and, following the dark strip along the concave side of the arm, there comes the bright zone; the stronger the shock wave, the thinner it is.

3. Subsequently, interstellar gas that has not been transformed into stars undergoes an expansion and forms—after the two narrow zones just described—a wider strip of neutral hydrogen, detected at 21 centimeters. Observations clearly show the narrowness of the dark and bright zones (especially in external galaxies), but until now it has been difficult to place neutral hydrogen relative to the H II regions or the dense molecular clouds. The spatial resolution of H I observations at 21 centimeters rarely is sufficient to permit direct discrimination. An indirect measurement would be possible, however, as a result of the small differences in velocity expected during the passage of a spiral arm or any concentration of matter: These are "currents," so to speak; the rotation speed of matter is diminished outside the arm and increased on the inside relative to the average speed of rotation. These small differences in velocity are rather tiny, about 5 to 10 kilometers (3.1 to 6.2 miles) per second. Up to now, the velocity resolution of the observations has not been sufficient to make it possible to demonstrate a difference in velocity between neutral hydrogen and carbon monoxide or ionized hydrogen, for example; more sophisticated observations perhaps will make it possible in the near future.

The Distribution of Molecular Hydrogen

Only recently, radio observations revealed the importance of molecular hydrogen. It was necessary to wait for millimetric observations (these were developed in the 1970s) to be able to observe the rotation line of carbon monoxide (at 2.6 millimeters) and thus indirectly obtain the abundance of the H_2 molecule. The fact is, since that molecule is symmetrical (H-H), it cannot radiate a molecular rotation spectrum. It is directly observable only through its ultraviolet lines; it is to be noted that this wavelength domain, also explored fairly recently, does not make it possible, owing to absorption and extinction, to obtain the distribution of molecular hydrogen throughout the galaxy.

Observation of carbon monoxide, CO, makes it possible to infer the masses of regions of molecular hydrogen, for a proportional relationship has been established between the CO emission and the column density of H_2:

To emit a rotational line, carbon monoxide must be excited by collisions with hydrogen molecules; the denser the medium is in molecular hydrogen, the more radiation emitted by CO. This proportional relationship was empirically determined in molecular clouds near the Sun, for which we know both the density of H_2 (as a result of measurements in ultraviolet) and the intensity of the CO emission. The CO/H_2 abundance ratio thus is estimated at approximately 6×10^{-5} to 10^{-4}.

Carbon monoxide was observed in 1976 throughout the galaxy by Gordon and Burton. They took a sampling of the galactic plane every 2 degrees and sometimes 0.5 degree in longitude. The radial velocity-longitude diagram they deduced from this is strangely devoid of negative velocities (unlike the equivalent diagram in neutral hydrogen, for example); we can infer from it, with the help of a model of the galaxy's rotational field, that very little carbon monoxide is observed outside a disk of radius R_o, representing the distance of the Sun from the galactic center. By establishing an average value for the abundance of carbon monoxide at each galactic radius, we can construct the radial distribution of CO and therefore that of H_2. It is remarkable that the bulk of the interstellar gas at the center of the galaxy is in the form of molecular hydrogen. In the disk, CO emission is essentially present only between radii of 4 and 8 kiloparsecs, with a maximum at about 5.5 kiloparsecs. At 4 kiloparsecs, the intensity suddenly drops; on the other hand, beyond 8 kiloparsecs, the decrease is gentler.

An attempt was made to explain this radial distribution with the density wave theory: Carbon monoxide usually is associated with dense molecular clouds formed from neutral hydrogen through compression and contraction during the passage through a spiral arm; the result is that the CO emission should be abundant only in regions where spiral waves exist. Now, in density-wave theory, the 4-kiloparsec radius corresponds to the inner Lindblad resonance, inside of which waves cannot propagate. This explanation is seductive, but then no carbon monoxide should exist at all at the center; as this is not the case, an alternative has been suggested: an insufficient abundance of gas between 2 and 4 kiloparsecs (the neutral hydrogen also is underabundant in this region) in order to explain the absence of molecular cloud formation.

The gently sloping decrease beyond 8 kiloparsec' galactic radius is, on the other hand, well interpreted by the density wave theory. Indeed, the more effective and substantial the formation of molecular clouds the greater the compression of the gas during its journey through the wave. This compression may be quantified by the difference in speed between the rotating gas and the wave, at least the projection of that difference perpendicular to the spiral arm. One can specify the curve of the variations, as a function of galactic radius, from the rotation speed of matter V(R) and the angular velocity of the wave Ω_{wave} and thus represent the compression factor as:

$$C = [V(R) - \Omega_{wave} XR] \text{ S sin } i;$$

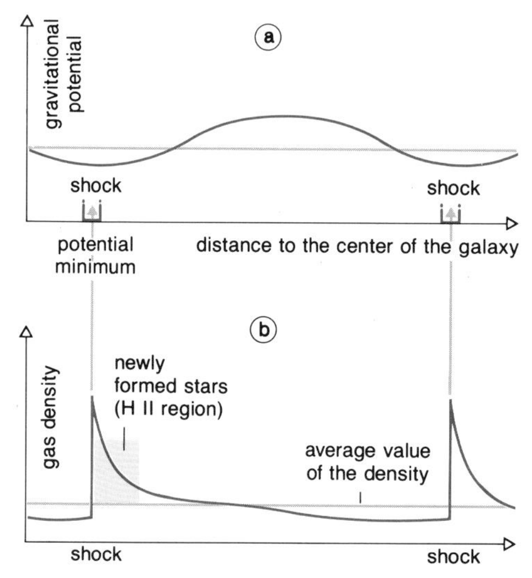

Shock waves in the interstellar gas of the spiral arms. The perturbed distribution of the stars generates gravitational potential wells (a). In entering one of these wells, the gas is compressed and creates a shock. Compression increases the rate of star formation, and the ultraviolet radiation from the massive stars formed causes ionization of the gas (b).

Radial distribution in the galaxy of population tracers: carbon monoxide (CO), molecular hydrogen (H_2)—which is deduced from the preceding— neutral hydrogen (H I), ionized hydrogen (H II) and supernova remnants. Only atomic hydrogen, H I, extends very far toward the exterior.

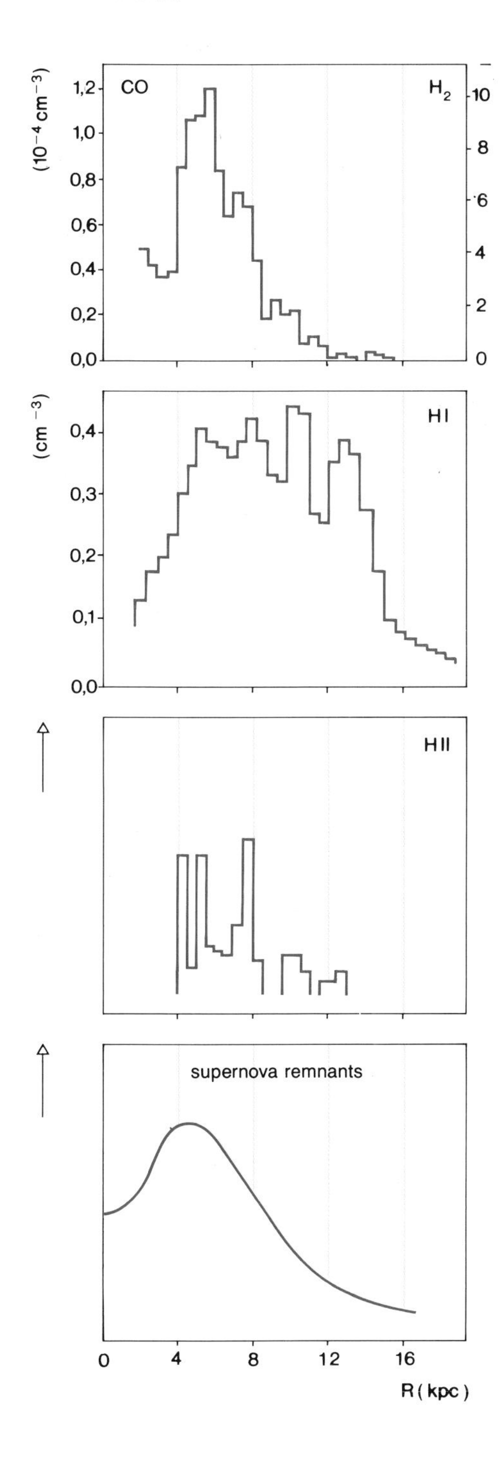

i being the inclination angle of the spiral relative to the tangent to a circle centered on the galaxy. The compression reaches its maximum at about 5.5 kiloparsecs and decreases gently, disappearing at about 10 kiloparsecs. This radius is the radius of corotation, where matter rotates at the same speed as the wave and hence does not undergo compression. This decrease is quite similar to the gently sloping decrease in the abundance of carbon monoxide and of molecular hydrogen.

A more accurate correlation is provided by the theory between the positions of the spiral arms and that of the CO emission. There does seem to be such a correlation in our galaxy (in spite of the difficulty of estimating distances within it); on the other hand, in Andromeda, our closest neighbor and an Sb spiral galaxy like ours, carbon monoxide and the dense clouds have been observed in correlation with the arms of neutral hydrogen, as expected. The confirmation of this result in many spiral galaxies could clarify the spiral structure of our own galaxy.

The Galactic Center

THE GALACTIC CENTER, THE NUCLEUS, is an exceptional region in the galaxy. It is optically invisible, hidden by thick clouds of dust, and has been revealed only recently by its radio, infrared and X-ray emissions. These observations show that the nucleus is the site of gigantic explosions; the radio and infrared emissions are very strong there, and a flux of matter is ejected from the center at high velocity. These phenomena qualitatively resemble those occurring in Seyfert galaxies, spiral galaxies with a tiny active nucleus surrounded by gas that they have ejected at amazing speeds (see page 301). The source of the energy supplied by these nuclei remains mysterious. Our galaxy, by comparison, nevertheless possesses a relatively calm nucleus, since it has been able to attain only a hundredth of the brightness of a Seyfert galaxy, even in the past, when it was brighter than it is today.

Observations

Observations in the 21-centimeter line of neutral hydrogen have revealed the ejection of large quantities of gas (see page 284)—10^5 to 10^6 solar masses—at speeds of several hundred kilometers per second (the "3-kiloparsec arms" have a dynamic energy of 10^{46} Joules).

Radio observations at several wavelengths, particularly centimeter wavelengths, show the great complexity of the galactic center; large masses of ionized gas (H II) have been detected by their radio continuum radiation and the recombination lines of hydrogen and helium. At the center of symmetry of the system is an H II region with extremely wide lines, corresponding to internal motions of more than 200 kilometers (124 miles) per second. However, this region is very compact, since its diameter does not exceed 2 pc. It is at the center of a powerful radio source, Sagittarius A, 12 pc. in diameter, the radiation from which is not entirely thermal (like that of H II regions) but synchrotron emission; it consists of continuum emission by very high energy electrons, which spiral in a magnetic field. This emission testifies to the past explosion of supernovae, sources of relativistic electrons. Other supernova remnants have been detected by synchrotron emission in the galactic center, with Sagittarius A remaining the most powerful source.

More recently, radio astronomers have detected many interstellar molecules in the galactic center, with lines that fall in the millimeter range. These molecules, very abundant in comparison to the rest of the galaxy, indicate the presence of many very dense clouds, representing a considerable mass. Some of these clouds form a nearly complete ring around the galactic center, with a radius of about 300 parsecs. The ring is believed to be expanding outward from the center at a rate of 100 kilometers per second. Most of the dense clouds are associated with H II regions: Young, hot and ionizing stars recently were formed within some of the molecular clouds. The densest of the clouds contains as many as 100 million particles per cubic centimeter, or about 100,000 times the density of the average molecular cloud in the galactic disk.

Infrared emission at 2.2 μm, detected for the first time in 1968 by Becklin and Neugebauer, indicates a very great abundance of cold stars in the galactic center; the density—about 10 million times greater than in the solar vicinity—is as much as a million stars per cubic parsec. From 4 to 20 μm, infrared observations have made it possible to detect the dust emission surrounding very hot, very massive stars (50 to 100 solar masses). In the far infrared, at about 100 μm, the heated dust also radiates enormous quantities of energy. The X-ray detectors aboard the artificial satellite *Uhuru* revealed an extended source coinciding with the 100 μm source. It may be thermal emission from very hot gas at the galactic center.

Hypotheses

The explosions at the galactic center have been computer-simulated by Sanders, Wrixon and Prendergast. Several simplifying hypotheses have been put forward: The stars, constituting the majority of the mass, practically determine the potential by themselves, while the gas is the only thing affected by the explosion and leaves the potential practically unchanged; consequently the numerical model has only to follow the motions of the gas in a constant potential. The gaseous component is schematically represented by a disk 100 parsecs in thickness, rotating around the center. The explosion is simulated by placing a large concentration of hot gas at the center of the disk. The hot gas immediately begins its expansion in all directions. Experiments show that the majority of the gas leaves the plane of the galaxy and escapes completely in 10^7 years. The gas remaining in the disk is swept up and forms an expanding ring 3 kiloparsecs from the center after about 10^7 years. In the simulation, no more gas remains between the ring and the galactic center and, in addition, the gas of the ring has cooled and is no longer ionized; it could emit only the 21-centimeter line of neutral hydrogen. However, the ring still is rotating.

The gas forming the ring was previously near the center of the galaxy, and its rotation was rapid, but through conservation of angular momentum, its rotation slows considerably through expansion (as a dancer's rotation slows when he extends his arms). In the slowdown, the centrifugal force decreases faster than the gravitational force of attraction toward the center, and the ring then stops expanding and comes back, accelerating toward the center. The radius at which it stops depends on the initial energy of the explosion. To observe an arm at 3 kpc. expanding at 50 kilometers (31 miles) per second, an explosion of 3×10^{51} J is required, and a mass of more than 100 million solar masses must be ejected. Governed alternately by centrifugal force and by gravitational force, the gas ring oscillates. These oscillations are damped by dissipation of energy (viscosity, collisions), and a position of equilibrium is reached in a few hundred million years. The various arms of gas—of neutral hydrogen far from the center, and molecular hydrogen in proximity to the nucleus—each represent a ring ejected by an explosion (the ring of molecular clouds representing the most recent explosion). Therefore it can be concluded that an explosion occurs about every 500 million years, after which time the motions of the gas have stopped; the total energy necessary for all these explosions is then about the same as the estimated energy of the explosions observed in Seyfert galaxies. Therefore the interpretation proposed is plausible, but the origin of the energy of the explosions still must be explained. The most acceptable hypothesis consists of assuming the presence of the center of a supermassive star of 1 million solar masses in which the thermonuclear reactions would be violent enough to produce the predicted explosions. If that supermassive star were formed by collisions of stars or by accretion of gas, we should then observe a large concentration of mass in the nucleus. Yet the infrared observations that have enabled us to estimate the density of the stars, and the observations at 21 centimeters of neutral hydrogen, which make it possible to derive the total mass from the velocity of the gas, do not reveal the existence of a sufficient concentration of mass in the nucleus. Therefore we must assume that the last explosion used up nearly all the mass of the nucleus and that the mass is reconstituted between each explosion by a flux of gas coming either from the galaxy itself or from outside. The problem of the astounding activity of the galactic center still is far from resolution.

Evolution of the Galaxy

THROUGHOUT THIS CHAPTER WE have glimpsed a few ideas regarding the evolution of the galaxy. The hypothesis of a gradual contraction and flattening of a vast cloud of gas is supported by the study of stellar populations. The stages in the formation of the galaxy thus can be diagrammed. The older stars, containing the least quantity of heavy elements, have the most spherical overall structure; their velocity dispersion and the eccentricity of their orbits are the greatest. The younger and younger populations formed flatter and flatter systems, up to population I, which is confined to the galactic disk, and which has the highest abundance of heavy elements (the stars of population I were formed from gas already enriched with heavy elements that the older stars synthesized and expelled into the interstellar medium). The outlines of the evolution of the galaxy are clear. But can we account for that evolution in greater detail?

The attempts at theoretical interpretation have dwelt essentially on two areas: chemical evolution and dynamical evolution. This separation, of course, was introduced only as a means of simplifying a very complex problem, for the interconnection between dynamics and chemistry is beyond doubt: The stars' orbits are tied to their chemical composition; moreover, the large-scale motions of the gas mix the matter and homogenize the chemical composition of the galaxy. If it incorporated a flux of intergalactic gas, as has often been suggested, that addition of new gas devoid of heavy elements would perturb its chemical evolution.

Chemical Evolution

Let us first review the facts of observation on the distribution of the abundances of elements in the galaxy, which every model explains. The abundance, relative to hydrogen, of heavy elements other than helium (which was formed at the beginning of the universe, before the formation of the galaxy) we will call Z. The stars of the halo

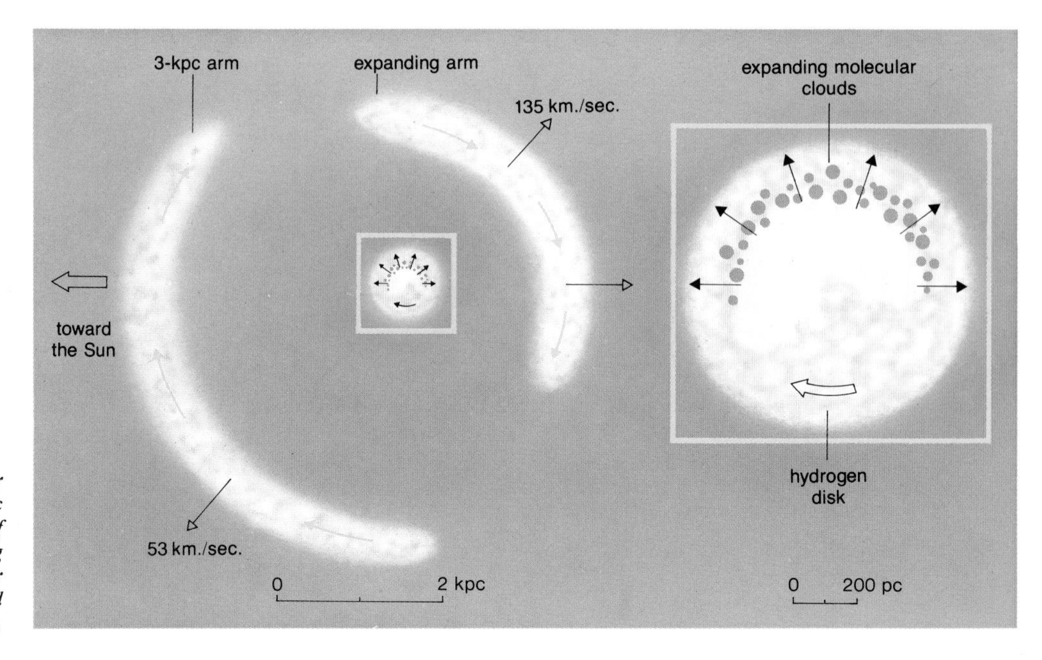

Schematic representation of the interstellar medium in the neighborhood of the galactic center, with the two expanding arms of neutral hydrogen, the hydrogen disk rotating near the nucleus and the ring of molecular clouds also expanding. From Sanders and Wrixon.

have the smallest Z, unlike the stars of the disk (where Z is high); but this is only a rough law: Z may vary slightly for stars of the same age. Gradients of abundance as a function of galactic radius have been observed (as well as in other spiral galaxies of the same type): Z grows as we approach the galactic center. Finally, the abundance ratios of isotopes of the same element differ within molecular clouds relative to the usual values in the solar system; this difference indicates the variations in chemical composition of the interstellar medium since the formation of the solar system (4.6 billion years).

How do we construct a model of the galaxy's chemical evolution? We must first take as a hypothesis a function representing the rate of star formation, depending on the mass of the star formed, the place and time of its formation and then estimate, according to the star's mass, the quantity of heavy elements the star will expel into the interstellar medium, and how long after its formation. The sum of the additions made by all the stars should represent the observed abundances, or at least be compatible with them.

The distribution of the stars' masses at their birth may be obtained partly by the luminosity function on the main sequence. For the massive stars evolving very rapidly (mass greater than two solar masses), the rate of star formation has not had time to vary since their birth. On the other hand, for the stars that should live longer than the present age of the galaxy (mass lower than or equal to 1 solar mass), the luminosity function on the main sequence represents only the distribution of past mass, corresponding to the period of the birth of these stars. Moreover, these two categories of stars have a completely different role in the chemical evolution of the galaxy: The massive stars frequently enrich the interstellar medium with heavy elements, while the others immobilize matter without contributing to nucleosynthesis. To put it simply, the models use a power law for the mass distribution of the stars at their birth, whatever their category (the Salpeter law).

The rate of star formation should also depend on the local gas density. Schmidt's law—which assumes a rate varying as the square of the density of the gas—seems rather well verified in our galaxy.

Solution of the equations then is made easier by the generally accepted hypothesis of "instantaneous recycling": The life-span of the massive stars (those with a mass greater than 2 solar masses) is ignored and the heavy elements are assumed to enter the interstellar medium at the birth of these stars. These simple models yield good results: The chemical composition depends mainly on the gaseous content and very little on the rate of star formation or the time of evolution; the abundance of an element is found to be directly proportional to the fraction of that element that each star expels into space relative to its total mass. However, the models predict many more stars poor in heavy elements than have been observed in the vicinity of the solar system.

Dynamical Evolution

The extremely rapid development of the power of computers now makes it possible to simulate the evolution of a galaxy almost step by step: These are the *n* body problems. The gravitational interaction among a million particles has been simulated in order to represent the galaxy. In fact, each particle already represents a group of stars or gas clouds; there are 100 billion stars in the galaxy. The number of 100,000 particles is, however, statistically sufficient; the increase in the number of particles no longer changes the qualitative behavior of the whole.

The initial hypotheses of these numerical models are extremely simple: The particles are randomly distributed at all radii, but in a homogeneous way, in a thick, rotating disk. The velocities and velocity dispersions are chosen in such a way that the disk is in equilibrium. The adopted radial density distribution is what is currently observed in the galaxy.

The result is not so simple. The simulated galaxy is unstable; the particles group together toward the center, forming a bar parallel to a diameter of the disk and heat up rapidly—i.e., the velocity dispersion rapidly becomes much greater than the real dispersion observed in the galaxy. Many trials to prevent the particles from heating up and forming concentrations have been futile. The solution, however, seems to have been found in the existence of a spherical halo of concentric matter at the nucleus of the galaxy; this halo, which must be at least twice as massive as the disk, exerts stabilizing gravitational forces on the particles of the disk and prevents their heating up. In addition, in the model with a halo, a spiral structure appears in the disk. This structure appears to be stable, at least during a dozen rotations of the galaxy, the time limit of the simulation. Does such a massive halo exist around the galaxy? The mass of visible stars and of globular clusters, the distribution of which is practically spherical around the galaxy, is anything but sufficient; it does not reach one-tenth of the mass of the disk. Could the matter be invisible, made up of burned-out stars or gas, under physical conditions that do not allow it to radiate? This possibility should be compared with the fact that, in many galaxies, the gas rotation curve makes it possible to infer a mass for the galaxy far greater than the mass of visible matter.

The evolution and even the structure of this complex system of stars and gas known as the Milky Way still pose serious problems for astronomers.

Numerical simulation of the explosions in the galactic center. From Sanders, Wrixon and Prendergast. (a) The hot gas begins expanding in all directions. (b) the expanding gas sweeps up and compresses the gas present in the disk, creating a rotating ring at about 3 kiloparsecs from the center; the centrifugal force no longer equilibrates the gravitation, and the ring is pulled back toward the center. (c) The ring contracts and its rotation increases. (d) The ring again moves out to 3 kiloparsecs from the center, at a speed of about 50 kilometers (31 miles) per second.

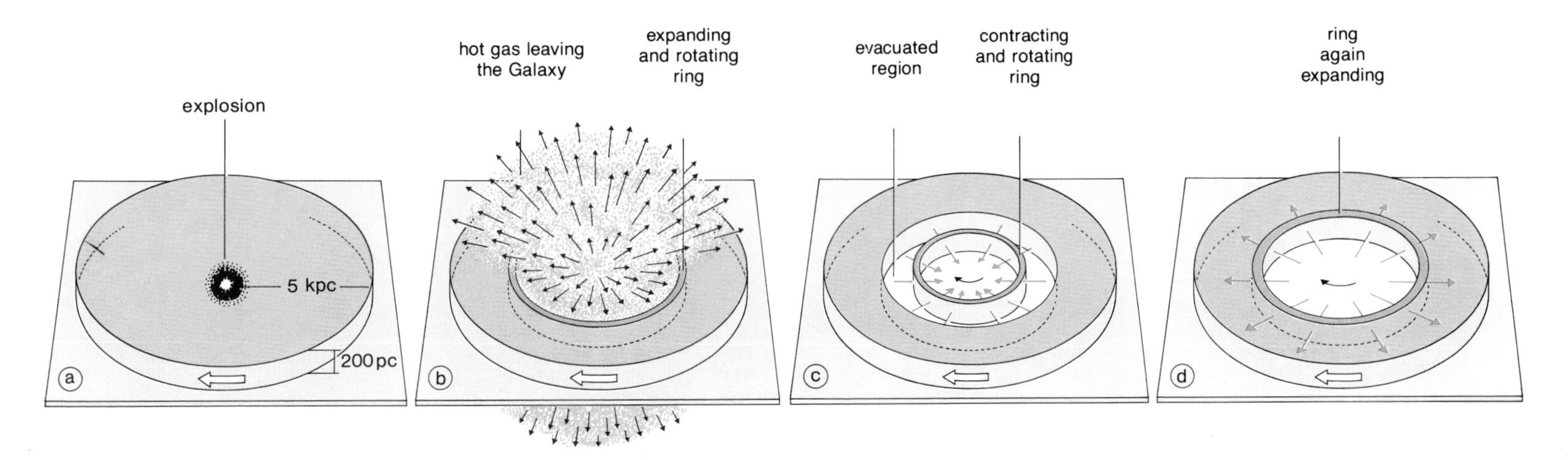

Radio galaxy NGC 5128. This giant elliptical galaxy, the equatorial plane of which appears to be streaked by a wide dust band, is 13 million light-years away. It coincides with the center of an extremely powerful and very extensive radio source, Centaurus A, and with an intense source of X rays; its central regions are presumed to be the site of very violent phenomena that are the source of the enormous energy released. Photo © 1979, Anglo-Australian Telescope Board.

IV
Toward Infinity

24. Galaxies and the Intergalactic Medium

A typical spiral galaxy, NGC 2997. The central bulge and spiral arms emanating from it are particularly visible. Photo: © 1979 Anglo-Australian Telescope Board.

The study of galaxies is relatively recent. It was made possible by the development of large instruments and sensitive detectors. Galaxies are located at such great distances that we receive only faint luminous energy from them despite the intensity of their intrinsic luminosity.

Properties of Galaxies

Classification

The first system of classification for galaxies was introduced by Edwin Hubble in 1925, using the collection of photographs he obtained at Mount Wilson. The basic outlines of this classification scheme still are used. There are three main classes of galaxies: elliptical, spiral and irregular.

Elliptical galaxies have a very regular structure, with a luminous intensity that progressively decreases away from the central regions, disappearing into the background of the sky. In general, they reveal no structure except a bright nucleus. Their apparent shape is generally that of an ellipse, which may range from a circle (EO) to a very elongated oval shape (E7). It is believed that their appearance does not necessarily correspond to the full reality, owing to the effects of projection, and that their real shape is that of a more or less flattened, revolving ellipsoid. Although it is impossible to decide if a particular elliptical galaxy really is spheroidal or, on the other hand, if it is a flattened system seen perpendicular to its plane of symmetry, there are far too many spheroidal galaxies to enable us to conclude that all of them are flattened systems. In fact, an analysis of the distribution of these objects shows that the observations agree with the hypothesis that the flattenings of elliptical galaxies have a random distribution. Among them we often see globular clusters; on the other hand, they contain only very little dust or young stars. The most luminous galaxies known are elliptical, but dwarf elliptical galaxies also exist, such as the companions of Andromeda.

The spiral galaxies clearly show their spiral-arms structure when their disk is perpendicular to the line of sight—i.e., in the plane of the sky. Seen in profile, they appear in the shape of a central spheroidal bulge and a flat disk, with a dark strip running across it. They are subdivided into two categories: the normal (S), which often have two arms emerging from two opposite sides of the nucleus; and the barred (SB), which have a bar running across the nucleus with the spiral branches starting from the ends of it. The central regions are the brightest, often the only ones visible in the telescope, while the arms appear only in a photograph.

Within the spiral category there are three divisions—Sa, Sb and Sc. The size of the central bulge and the regularity and compactness of the arms decrease from Sa to Sc. On the other hand, the proportion of bright nebulae and of very blue stars increases from Sa to Sc. These bright regions and young stars often are concentrated in the spiral arms.

Irregular galaxies have no definite structure, nucleus, arms or geometric shape. Among them we find a great deal of dust, very blue stars and bright nebulae. An example is the Small Magellanic Cloud.

The great majority of galaxies fall into one of these categories. There are, however, several exceptions (about 2 percent, according to G. de Vaucouleurs). In particular, Hubble himself was compelled to introduce the category of lenticular galaxies. They differ from the ellipticals in the sense that they consist of a large central condensation and a flattened disk. However, they have no arms, unlike spiral galaxies, and usually seem to be devoid of dust. There is general agreement that lenticulars belong to a category of galaxies midway between ellipticals and spirals. According to a study by G. de Vaucouleurs dealing with a fairly homogeneous sample of 1,500 galaxies, 13 percent are elliptical, 21 percent lenticular, 61 percent spiral, 3 percent irregular and 2 percent special objects.

Current classifications—particularly those of A. Sandage and G. de Vaucouleurs—are more detailed, showing transitions between the Hubble categories, and taking into account finer structures, such as an internal ring around the nucleus or, on the other hand, an external ring surrounding the entire galaxy. These modern classifications also have introduced new types of spirals, Sd and Sm, extending the classification beyond the Sc spirals.

Stellar Content of Galaxies

We can see individual stars only in a very small number of galaxies—those that are sufficiently close to us. Even in those favorable circumstances, we can discern only the brightest stars: a star of average luminosity, such as the Sun, situated in the Andromeda galaxy, is below the threshold of detection of the instruments we currently possess. It was with the 2.5-meter telescope of the Mount Wilson Observatory that Hubble for the first time succeeded, in 1923, in resolving individual stars in the irregular galaxy NGC 6822, and the arms of the large spiral galaxies M 31 and M 33.

Study of the small number of close galaxies that have been resolved into stars has, however, shown that the different types of galaxies have differing stellar contents. A photograph taken through a blue filter can select the very luminous, blue, supergiant stars of spectral types O and B. A photograph taken through a red filter, on the other hand, reveals the brightest of the red stars. This technique enabled the American astronomer W. Baade to show, in about 1943, that there exist in the Andromeda galaxy and its companions two types of stellar populations that are not distributed in the same way as in the galaxy. Population I is characterized essentially by giant and supergiant blue stars of spectral types o and B and by a few red giant stars of spectral type K or M. The objects of population I are found in the arms of the spiral galaxies and in the irregular galaxies.

By superimposing the results of measurements of carbon monoxide (where the + 's indicate regions where that molecule has been detected, the o's regions where attempts at detection have been fruitless upon an isophotal map of the neutral hydrogen emission in the southern arm of the Andromeda galaxy (M31), we observe a spatial coincidence between the emission maxima of those two species, which enables us to state that stars form in the spiral arms. From F. Combes, 1980.

North

East

The stars of population II are essentially red giants of spectral types G5 to K5, planetary nebulae and probably dwarf stars populating the lower part of the main sequence of the HR diagram (but the latter are not luminous enough for us to be able to detect them even in nearby galaxies). The stars of population II are found mainly in elliptical and lenticular galaxies, in the bulges of spiral galaxies and in globular clusters. In addition, we observe globular clusters at the periphery of galaxies. Hubble was the first to notice their presence in the vicinity of the Andromeda galaxy. They are

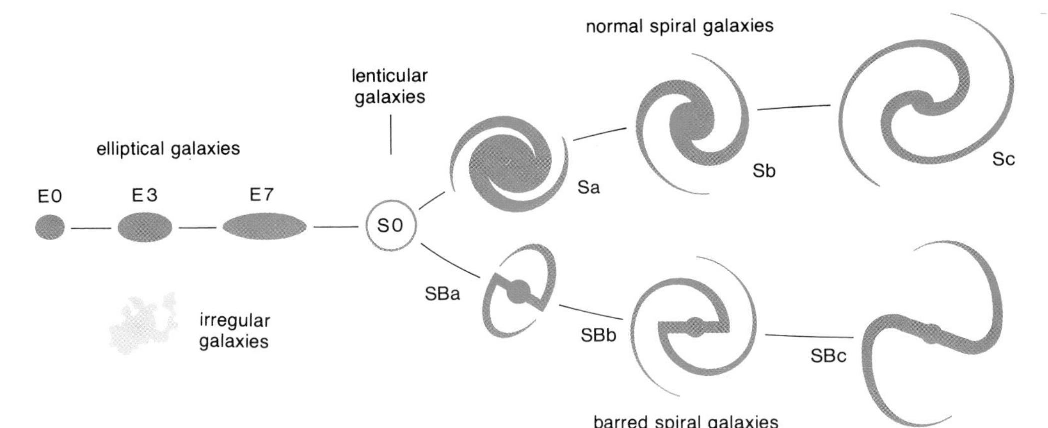

The Hubble Classification Scheme for Galaxies

as numerous and as luminous as those of our own galaxy.

These two populations are present in the spiral and irregular galaxies, in variable proportions; it seems that ellipticals incorporate only the stars of population II. The relative proportion of stars of population II compared to those of population I increases fairly regularly as we go through the sequence of types, from ellipticals to irregulars, including the Sa, Sb and Sc spirals.

In a spiral galaxy, the stars of population I, concentrated in the arms, draw our attention by their bright, spectacular appearance. They contribute, however, only a small percentage to the total luminosity of the galaxy and are superimposed on the stars of population II, regularly distributed over the entire disk.

Detailed study of these two populations of stars in our galaxy has shown that the real situation is more complex than this simplifying concept of two populations of stars. A more detailed analysis of the colors and spectra of galaxies shows that age and chemical composition also are important parameters.

Groups and Clusters of Galaxies

GALAXIES ARE NOT DISTRIBUTED uniformly in the sky. This was discovered in the 19th century by W. Herschel, then by his son John. It is due first of all to our position in the disk of the galaxy, where the distribution of absorbing matter is demonstrated by the presence of an avoidance zone at low galactic latitude, in the directions in which we observe very bright galaxies beyond our own.

However, if we take that effect into account and study the distribution of galaxies from the standpoint of direction and distance, we see that they tend to be grouped into associations of various sizes. They are rarely isolated and generally are in pairs, triplets or small groups. For example, the two Magellanic Clouds constitute a pair of galaxies joined in a looser triplet to our galaxy, with that group itself belonging to the local group. There exist much larger groupings, called clusters, which are themselves combined into superclusters. This distribution suggests the existence of a universe organized into a hierarchy in which the galaxies are distributed in groups and clusters that are combined into superclusters, and so on . . . perhaps. In fact, we know of no groupings of superclusters. We will see at the end of this chapter how this situation may, however, be compatible with the hypothesis of a uniform large-scale distribution of matter in the universe.

The Local Group

Our galaxy belongs to a group (the "local group") of about 20 galaxies occupying an ellipsoidal volume having a maximum extent of more than 2 million parsecs (2 megaparsecs, or 2 Mpc.). Our galaxy and the Andromeda galaxy are the largest galaxies in the group, and each is at one end of this diameter. Together with M 33, in the constellation Triangulum, the Large Megallanic Cloud and Maffei 2—an object made invisible by galactic obscuration and discovered through radio and infrared astronomy—these are the only spiral galaxies. The Small Magellanic Cloud and two objects of smaller dimensions are irregular galaxies. All the other systems of the local group are dwarf elliptical galaxies, except Maffei 1, which is a giant elliptical. There may be other members we have not yet discovered because they are hidden, like Maffei 1 and Maffei 2, by the band of galactic absorption. In addition, the boundaries of the local group are poorly defined, largely because distances are poorly known, and authors do not always agree on the membership of peripheral galaxies.

The galaxies in the local group are not uniformly distributed but tend to concentrate in two subgroups. The first is centered around Andromeda and its companions; the second around our galaxy and the Magellanic Clouds.

We have been able to measure the radial velocities of several of the galaxies of the local group. The Andromeda galaxy is moving toward us at a speed of 275 kilometers (171 miles) per second, while the Large Magellanic Cloud is moving away at a speed of 270 kilometers (168 miles)

Determination of Extragalactic Distances

The decisive discovery that put an end to the great debate of 1920–25 about the existence of galaxies outside our own is Hubble's determination of the distance of the Andromeda galaxy, situating it well outside our galaxy.

The great distances of galaxies cannot be measured directly, as is possible for certain stars using geometric methods such as the method of trigonometric parallaxes. We can use only indirect methods. Just as in our galaxy, these methods are based on fundamental relationships, or criteria of distance, established and calibrated with bodies of known distance. These criteria are essentially of two types: They link either luminosity, L, or geometric dimension, D, of a well-defined object to a parameter, P, that may be determined by observation without there being a need to know the distance. Observation of parameter P makes it possible to determine luminosity (or diameter) from the preceding relationship. Then we can determine distance, r, from measurement of the apparent brightness $E = L/4\pi r^2$ or from the apparent diameter α ($\tan\alpha = D/r$).

Certain of these criteria are based on calibration relationships established in our galaxy; for this reason, they are called "primary criteria." They use the variable stars of the RR Lyrae type, the cepheids and the novae. The variable stars of the same type as the star RR Lyrae all have the same average luminosity. Variable stars of the same type as the star δ Cephei have a luminosity that varies periodically over time. The period of the variations is proportional to the star's average luminosity; measurement of the period of the variations in brightness thus makes it possible to determine the star's average luminosity. Finally, novae all have about the same luminosity at their maximum brightness; more precisely, the rapidity of the decrease in their luminosity is linked to their maximum luminosity. It is sufficient, therefore, to measure the average apparent brightness of a RR Lyrae star or of a cepheid, and the maximum apparent brightness of a nova, to deduce their distance from the luminosities determined by the criteria described.

The use of these primary indicators requires observation of individual stars in the galaxies and determination of their apparent brightness. Although these stars are luminous, they can be detected only in the galaxies closest to us, those for which the distance does not exceed about 4 Mpc. Thus we have been able to observe Cepheid brightness variations in only six nearby galaxies.

This first extragalactic step is fundamental in constructing a more and more accurate scale of distances in the universe. The nearby galaxies for which the distance could be determined using primary distance indicators establish new distance criteria, which we will call "secondary" indicators because they are no longer directly calibrated in our galaxy. These criteria use, for example, the luminosity of globular clusters or blue supergiant stars, the diameters of the brightest nebulae (or H II regions). Study of the individual luminosities of globular clusters in a galaxy of known distance shows that the most luminous among them always have about the same luminosity. The same is true for blue supergiant stars. We can therefore determine the luminosity of the brightest globular clusters in a galaxy, or that of the blue supergiant stars. In a galaxy of known distance, in which we observe globular clusters or blue stars, it is then sufficient to measure the apparent brightness of the most brilliant objects to which that luminosity is attributed: The value of the distance results. With regard to the large bright nebulae, we are inclined to give them a maximal linear dimension; the comparison between that linear dimension and the apparent dimension of the largest of them that are measured in a galaxy of unknown distance then enables us to find the distance.

These methods rest on an implicit hypothesis: The objects used to establish the distance criteria (globular clusters, blue stars, brilliant nebulae) must have the same properties in all galaxies. It is not easy to verify the validity of this hypothesis; we suspect, for example, that a difference in chemical composition might mean a difference in luminosity. We have also noticed that the maximum dimensions of the brilliant nebulae increase with the luminosity of the galaxies to which they belong; the exact relationship that links this linear dimension to the apparent dimension that is measured in a galaxy,

The Principal Galaxies of the Local Group

Object (constellation)	Type*	Distance (light-years)	Radial velocity** relative to the Sun (km./sec.)	Diameter (light-years)	Mass (M.)	Apparent Visual Magnitude	Absolute Magnitude
Our galaxy	*Sb*		0	100,000	1.4×10^{11}		− 20.2
Large Magellanic Cloud (Dorado)	SB*c*	160,000	+ 270	23,000	10^{10}	0.1	− 18.7
Small Magellanic Cloud (Tucana)	I	200,000	+ 168	10,000	2×10^{9}	2.4	− 16.7
Draco	E	220,000		1,000	10^{5}		− 8.5
Ursa Minor	E	220,000		1,000	10^{5}		− 9
Sculptor	E	280,000		2,300	3.2×10^{6}	7	− 12
Fornax	E	550,000	+ 40	5,400	2×10^{7}	7	− 13
Leo I	E 4	750,000		2,000	4×10^{6}		− 11
Leo II	E 1	750,000		1,000	10^{6}		− 9.5
NGC 6822 (Sagittarius)	I	1,600,000	− 40	7,500	3.2×10^{8}	8.6	− 15.6
Wolf-Lundmark	E 5	1,600,000	− 80	5,000		11.1	− 13.3
NGC 205 (Andromeda)	E 5	2,100,000	− 240	7,800	7.9×10^{9}	8.2	− 16.3
NGC 221 = M 32 (Andromeda)	E 2	2,150,000	− 210	2,300	3.2×10^{9}	8.2	− 16.3
NGC 147	E 4	2,150,000	− 250	4,600	10^{9}	9.6	− 14.8
NGC 185	E 0	2,150,000	− 300	3,250	10^{9}	9.4	− 15.2
NGC 224 = M 31 (Andromeda)	*Sb*	2,200,000	− 275	160,000	3.2×10^{11}	3.5	− 21.1
IC 1613 (Cetus)	I	2,400,000	− 240	10,000	2.5×10^{8}	9.6	− 14.8
NGC 598 = M 33 (Triangulum)	*Sc*	2,700,000	− 280	46,000	7.9×10^{9}	5.7	− 18.8
LGS 3 (Pisces)	I	2,700,000	− 280	1,600	15×10^{6}	21	− 9.0
Maffei I	E	3,300,000	+ 17		2×10^{11}	11	− 20

* S = ordinary spiral; SB = barred spiral; E = elliptical; I = irregular.
** A positive velocity indicates a galaxy that is moving away; a negative velocity indicates an approaching galaxy.

The Large Magellanic Cloud. Visible in the sky of the Southern Hemisphere, this small galaxy, only 170,000 light-years away, is a satellite of ours. Many individual stars can be observed within it. At left we see the famous Tarantula nebula: About 800 light-years across, it is the largest diffuse nebula known; this vast cloud of ionized hydrogen is a region rich in young stars. Photo: P.P.P.-IPS.

per second. These velocities are mainly due to the Sun's motion around the center of the galaxy and to the galaxy's motion within the local group. These motions lead to a motion of the galaxies relative to the Sun. Analysis of the radial velocities of galaxies relative to the Sun, such as we determine using the Doppler-Fizeau effect, shows that the Sun is rotating around the galactic center at a speed of 250 kilometers (155 miles) per second, and that our galaxy is moving within the local group at a speed of about 200 kilometers (124 miles) per second. Study of the motions of the different galaxies of the local group leads to estimating its total mass at about 5×10^{11} solar masses.

Nearby Groups of Galaxies

As we move away from the local group, we find other similar groups of galaxies, encompassing several dozen galaxies, in the neighborhood of the local group—e.g., in Sculptor, Ursa Major, Leo, Fornax, etc. The dimensions of these groups are measured in megaparsecs. A few examples are given in the accompanying table.

The galaxies of a single group tend to show similarities of morphological type. For example, the small, loose groups generally are made up of spiral galaxies and a small minority of lenticular or elliptical galaxies, except for the dwarf ellipticals, which very often are satellites of spirals (as, for example, the satellites of Andromeda). The denser and richer groups, on the other hand, seem to be dominated by giant elliptical and lenticular galaxies. This segregation probably is related to the physical conditions that existed at the moment of the galaxies' formation. But, for the

Group Name	Distance (Mpc)	Diameter (degrees)	Diameter (Mpc)	Radial Velocity (km/sec.)
Sculptor	2.4	25 × 20	1.0 × 0.8	+ 142
Messier 81	2.5	40 × 20	1.8 × 0.9	+ 160
Canes I	3.8	28 × 14	1.9 × 0.9	+ 342
Messier 101	4.6	23 × 16	1.8 × 1.3	+ 508
NGC 2841	6.0	15 × 7	1.6 × 0.8	+ 589
NGC 1023	6.3	20 × 10	2.2 × 1.1	+ 566
NGC 2997	7.6	14 × 8	1.9 × 1.1	+ 534
Messier 66	7.6	7 × 4	1.0 × 0.6	+ 592
Canes II	8.0	22 × 12	3.0 × 1.6	+ 747
Messier 96	8.3	11 × 7	1.6 × 1.0	+ 741

to its distance and to the luminosity of the galaxy (which also involves the apparent brightness of the galaxy and the distance) no longer depends uniquely on distance. Thus the reduced relationships among the apparent dimension, the apparent brightness and the distance no longer provide a very sensitive distance criterion.

We can also understand why such criteria cannot be primary criteria. For that, we would have to be able to calibrate them directly in our galaxy; but the absorption of light by interstellar dust prevents us from seeing farther than the near neighborhood of the Sun. As these criteria refer to the most luminous or largest objects in an entire galaxy, it is highly unlikely that examples of such objects would be situated precisely in our neighborhood and therefore would be accessible to us.

In the final analysis, the secondary indicators make it possible to determine the distances of galaxies out to about 25 Mpc. These galaxies in turn provide other criteria, called "tertiary" indicators. These indicators are based on certain global or large-scale structural properties of the galaxies (such as their diameter and luminosity) or on radio properties and no longer on the observation of stars or of individual structures. They make it possible to reach galaxies at distances up to about 100 Mpc. Beyond that distance we no longer have a good distance criterion, and we use Hubble's law (see following material).

The fragility of this method of determining distances should be noted. It rests on criteria that we assume to be universal without having been able to prove it. Moreover, any defect in the calibration set up in one step would have repercussions on all the following steps. For example, if the primary criteria are wrong and provide distances that are systematically too great, this overestimate also will affect the secondary criteria, since they are calibrated according to distances of galaxies that have been overestimated by the primary criteria; the tertiary criteria are affected in the same way.

Two schools have developed over the past few years. The first consists of A. Sandage and G. Tammann. They have tried to come up with what is for them the best primary criterion (average period-luminosity relationship of the cepheids), then the best secondary criterion (linear diameter of the largest bright nebulae) and finally the best tertiary criterion (intrinsic luminosity of a galaxy). The second school, led by S. Van den Bergh and G. de Vaucouleurs, prefers to use at each step a large number of different criteria; the convergence of the results obtained provides support *a posteriori* for the validity of each of the criteria used. This involves work that is extremely tricky but fundamental to the entire field of extragalactic studies and cosmology.

Natural Selection). These complex objects, capable of transforming information, of reproducing, and of evolving by natural selection include not only those organisms that may be termed "living" according to the previous definition, but also, at the very limit, complex molecules capable of reproducing, such as DNA.

Another important problem: Does extraterrestrial life appear in a form analogous to what we know on Earth? It seems unlikely that extraterrestrial organisms would resemble living species on Earth (particularly in view of the diversity of the latter). On the other hand, we can assume that extraterrestrial life, like life on Earth, is based on carbon chemistry. There are several justifications for this hypothesis. Organic cosmic chemistry (see further) which is an indicator of chemical evolution and which has clearly been demonstrated in the past several years, suggests that carbon plays a major role. Some authors have envisioned the possibility of living systems based on other elements, such as silicon. Nothing completely rules out this hypothesis. However, it appears to have little justification, since the chemical properties of silicon are very different from those of carbon and, above all, do not offer such a variety of possibilities for molecular structures (i.e., the complexity needed for the storage of information). Other authors, on the other hand, claim that life is a physical and chemical phenomenon similar, for example, to crystalline growth, and that the laws governing this phenomenon should be considered universal—biochemistry would be the same throughout the universe, as physics or chemistry apparently are. However, it should be noted that on Earth, slight variations in environmental conditions have caused substantial variations in the living world, to the point of altering the biochemical parameters of some organisms. However, it remains probable that some biochemical processes are universal.

All of the scientists involved in this problem agree that it is reasonable to seek in the rest of the universe life forms that are essentially fairly close to those we know on Earth.

Where to Look?

Our knowledge of the processes that led to the emergence of life on Earth—although very incomplete—enables us to postulate certain conditions necessary for the formation and development of living organisms on a planet.

First, in view of the major role played by the planetary surface in the first stages of chemical evolution, a celestial body with the potential to give rise to living systems must be a planet. This planet must have an atmosphere—probably of the reducing type—and must get enough energy from its sun to permit the formation of atmospheric precursors. The energy, however, must not be so great that these precursors cannot evolve later on toward the formation of more complex compounds. Another indispensable requirement is the presence of water in a liquid state. It is necessary not only for prebiotic reactions but also for the later survival of living species. Finally, the planet's temperature must be neither too low (or the rates of reactions set in motion would be too slow) nor too high (or the organic structures would be destroyed). In fact, the sum total of these physical and chemical conditions lead to the definition of a zone surrounding the central star of a given planetary system where the development and propagation of life are possible—a ***biothermal zone.***

To define such a zone one must know the limiting value of the different parameters involved. Many studies of life under extreme conditions have been done. They show that some species live in volcanoes where the temperature approaches 100°C and that, in principle, there is no lower-temperature limit: Cold does not destroy life. However, most terrestrial organisms carry on life activities between 0° and 50°C. Other factors besides temperature, such as the value of the pressure or the intensity of the radiation flux, also play an important role. It would seem, however, that the temperature factor is one of the most important, chiefly because it determines the presence or absence on the planet of liquid water, the crucial role of which in the emergence and continuity of life has been demonstrated by all the studies conducted. Finding life first requires finding a planet within the biothermal zone of a planetary system.

A Direct Approach to Exobiology: Space Exploration

Existence of a Cosmochemistry of Carbon

Up to 1969, the only extraterrestrial materials we possessed were meteorites. The most interesting ones, as far as extraterrestrial life is concerned, are the carbonaceous chondrites, for they are rich in carbon compounds.

This is true of the famous Orgueil meteorite, in which some researchers thought they saw included or fossilized structures characteristic of one-celled organisms, which in fact turned out in some cases to come from a contamination of terrestrial origin. Thus it was thought afterward that the amino acids positively identified in that meteorite also resulted from contamination. It should be noted, however, that doubt persists with regard to many of the organic elements found in Orgueil. It does not seem possible at present to determine whether they are due to a terrestrial biological contamination, whether they are structures of an extraterrestrial biological origin or whether they are structures having a purely mineral origin. In the 1970s, many thorough analyses of organic compounds were done, particularly by the Kvenvolden team, on chondrites gathered immediately after their fall, such as those named Murchison, Murray or, still more recently, Allende. The analyses showed the presence of many amino acids, several of which are absent from living organisms. Better yet: The amino acids identified do not show the optical asymmetry of the corresponding constituents of biological origin. Similar experiments conducted by Ponnamperuma on the Orgueil meteorite have since led to the same result; these compounds, therefore, are indeed of extraterrestrial origin.

Other categories of biochemical compounds, such as the pyrimidic bases, also have been identified in these chondrites. All of the results obtained clearly establish the existence of an extraterrestrial organic chemistry, at least in the solar system. Since the development of radio astronomy, many organic molecules have been identified in very remote regions of space. At present, the list includes about 60 compounds, but it is getting longer every year. Some of these compounds, such as hydrocyamic acid or cyanoacetylene, are important precursors involved in prebiotic syntheses.

Many of the molecules identified correspond to volatile intermediaries discovered during simulation experiments in the laboratory. It should also be noted that the amino acids detected in meteorites, whether of the biological kind or not, are mostly the same as those identified in experiments such as Miller's. This suggests that the processes described above regarding the Earth's chemical evolution are taking place in many parts of the universe. There is a cosmochemistry, which, like biochemistry, is based mainly on carbon chemistry.

Life in the Solar System?

For centuries, human beings "populated" the planets of the solar system with various humanoids: Moon men, Martians, Venusians and even Saturnians. In 1896, people even believed they finally had proof of the existence of extraterrestrials, with Percival Lowell's famous description of the Martian canals. Of course, his observations did not necessarily presuppose the presence of anthropomorphic creatures on the planet Mars, but nonetheless they signified that a relatively evolved civilization had lived or was still living on our neighbor. We now know that Lowell's conclusions were false. In fact, for several decades it has been assumed that no intelligent life comparable to ours inhabits the planets of the solar system outside Earth. Our mastery of radio emission and reception for the last 50 years would have enabled us to detect any such civilizations (in view of their relative proximity to Earth).

On the other hand, so long as human beings have not explored—either directly or indirectly, through remote-controlled devices—the various planets orbiting the Sun, we cannot positively rule out *a priori* the idea that these planets harbor life, at least in the form of microscopic organisms. However, our knowledge of the physical and chemical conditions necessary for the development and maintenance of living organisms on a planet, added to the results of observations of the different planets in the solar system, should enable us to choose from among them those that seem capable of supporting life. Thus it is generally acknowledged that the biothermal zone, necessary for the survival of any terrestrial-type organism, is bounded by our two neighboring planets, Venus and Mars.

Mercury, which is too close to the Sun, has a very high temperature and has no atmosphere. These two reasons suffice to make the absence of any life on that planet very probable. On the other hand, the outer planets have extremely low temperatures, thwarting the development of life.

Nevertheless, the first two giant planets, Jupiter and Saturn, seem to possess lower layers where the temperature might be more favorable to life. Moreover, the atmospheres of those two planets have a relatively complex organic chemistry, to the point where the origin of Jupiter's famous red spot sometimes has been attributed to chromophoric biological molecules. Finally, the appreciable abundance of ammonia in the Jovian atmosphere is not incompatible with the development of living organisms. Studies done by Dr. Siegel, in particular, show that many terrestrial organisms may develop in such environments. It seems unlikely, however, that Jupiter and Saturn are capable of supporting life. Because of their composition and chemistry, their atmospheres are of great interest to our discussion. In particular, they show one of the first stages of development in the course of chemical evolution. Finally, it should be noted that some of these planets' satellites (mainly Titan, which has an atmosphere) sometimes have been mentioned as potential habitats for living organisms, despite probably very low temperature conditions.

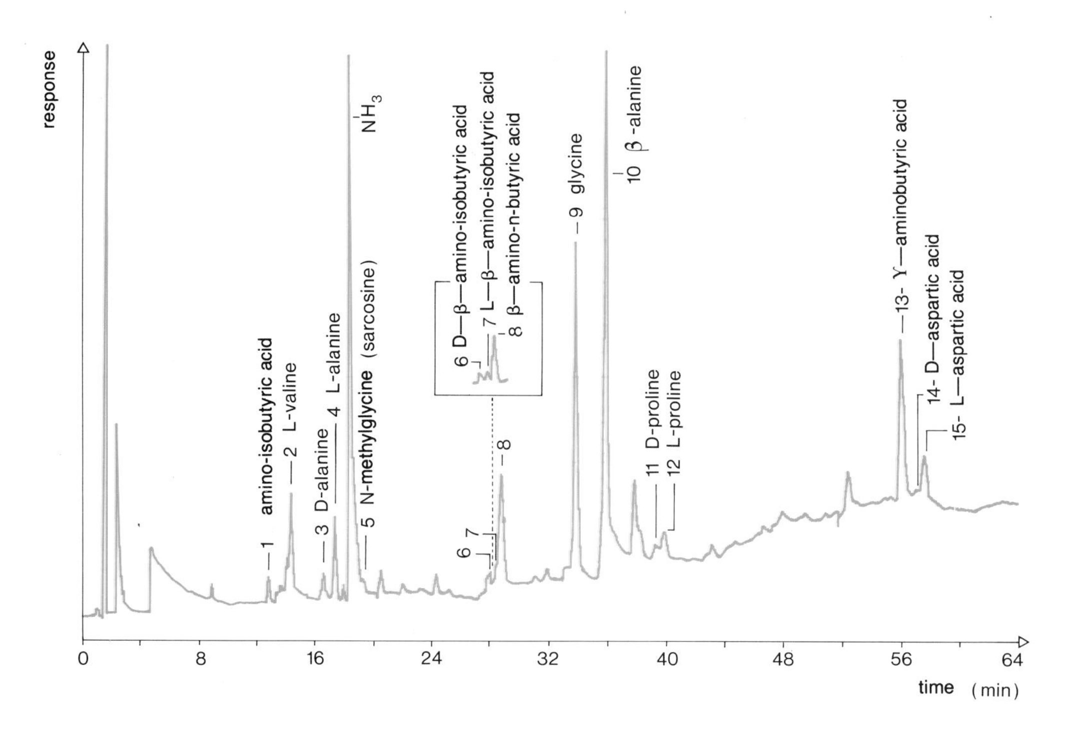

At the inner limit of the biothermal zone, Venus is far from being a planet hospitable to life, owing to its high surface temperature (about 480°C) and pressure (about 100 bars). This last point is not a fundamental obstacle to the survival of organisms (we know of living species in the Earth's oceans at pressures greater than 1,000 bars). However, we can state, owing to the temperature on the ground, that there is no living organism on the surface of Venus, at least similar to those we know. On the other hand, within the planet's atmosphere is a zone where the temperature is compatible with the emergence of life, which has led some persons to suggest that this zone might be inhabited. However, the carbon dioxide content and the presence of sulfuric acid clouds make that eventuality highly unlikely.

At the center of the biothermal zone is not only the Earth, but also the Moon. Well before man's first historic step on our natural satellite, we knew that it had no atmosphere. It seemed entirely unlikely, therefore, that life could have developed

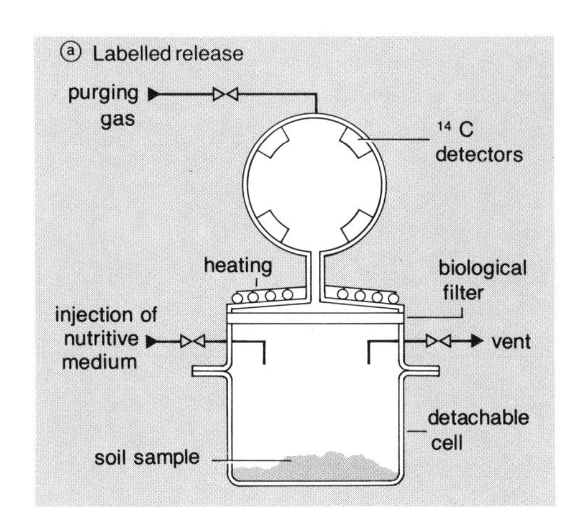

The Biological Program of the Viking Mission

Three experiments were tried aboard each of the two Viking probes to detect any Martian life. Two of them were supposed to measure the metabolic activity of living organisms within the Martian soil.

The first experiment (a), called *labeled release*, was designed to estimate the soil's capacity to "digest" a food and transform it into gaseous products. A soil sample was placed in the presence of a typical nutritive medium. The organic products making up that medium were marked with radioactive carbon-14. Following the possible assimilation of carbon-14, simple gaseous compounds that might then be released were then identified by radioactivity counts.

Two main results were obtained. First, a rapid decomposition of the organic com-

Analysis by gas chromatography of the amino acids contained in the Orgueil meteorite. Except for glycine, the amino acids show a center of asymmetry: They cannot be superimposed on their mirror images. For an amino acid of a given formula, we can therefore predict a priori *two components. These components, called enantiomers, both exist and possess different physical properties. According to conventional representation, one is called D and the other L. In general, the artificial synthesis of amino acids leads to the formation of equal quantities of L and D. This is called a racemic mixture. The presence in this chromatogram of pairs of peaks having equal areas for each type of amino acid shows that we are dealing with a racemic mixture and that the compounds detected do not stem from a biological contamination on Earth. This enables us to conclude that these compounds are of extraterrestrial origin. Document kindly provided by C. Ponnamperuma.*

and been maintained there. It should be noted in this regard that NASA still did not want to run the risk of contaminating Earth's population with possible lunar microbes and decided to place living or dead objects coming back from the Moon in quarantine. Analysis of a large number of samples brought back from our neighbor quickly provided proof that no life (nor chance of life) exists on its surface. Furthermore, no biochemical compound was identified in the lunar rocks. Since then we have been able to explain the absence of such molecules: The solar wind, made up of high-energy protons impinging upon the Moon's surface, is capable of breaking down any biochemical molecule, even those buried in the soil.

With regard to exobiology, on-site exploration of the Moon has not been totally devoid of results. Among other things, it has permitted the development of space technology and in particular may be considered as the beginning of exploration of the other bodies of the solar system, Venus and especially Mars.

Mars's Surprises

Mars was long held to be the most promising planet in terms of the presence of extraterrestrial living organisms. Indeed, Mars has an atmosphere, even though it is very thin, and boasts "relatively" clement temperatures (although they vary between −100°C and +70°C). Moreover, examination of Mars's surface seems to show that liquid water must have been present on the planet in relatively recent geological times. These various observations, together with the planet's proximity to the Earth, made it the first planet to visit after the Moon. Under the influence of exobiologists such as Klein, Young and especially Carl Sagan, the United States became involved in a vast project of exploration of the planet Mars: the Viking mission.

Launched in August 1975, two successive probes reached the Martian soil on July 20 and September 3, 1976: *Viking 1* in the Chryse basin, *Viking 2* on Utopia Planitia. Each of these two devices was extremely complex and carried an automatic laboratory, making it possible to look for any living organisms on Mars's surface.

The first very important observation was the discovery of molecular nitrogen, N_2, in the atmosphere of the red planet in a nonnegligible quantity (2 to 3 percent). Until then it had been thought that N_2 was completely absent from the Martian environment, which was an annoying problem, since nitrogen, N, is indeed one of the elements necessary for the formation of basic biological molecules.

For several months, three types of experiments designed to detect living systems were carried out on several occasions at the two sites selected for each of the two Viking probes. Many criticisms have been made of the choice of these experiments. In particular, they have been accused of being too sophisticated, difficult to carry out and susceptible to giving misleading answers, since they are not really specific to life. Indeed, it should be admitted that these biological experiments definitely provided ambiguous results that are difficult to interpret. Of the three experiments, two yielded positive results, while the third (gas exchange) gave a negative response. For many persons, the conclusion of these experiments is that there are no living organisms on Mars. However, some exobiologists do not share that opinion and feel that the Viking biological program did not prove the absence of life on our neighbor: Martian organisms, as Sagan suggests, may be fundamentally different from terrestrial ones, which might explain some of the surprising results of the biological experiments. Or else the two sites chosen were not appropriate: It would have been preferable to do these experiments in the old Martian valleys or at the poles, places that have contained or still contain water. Doubts probably should be dispelled in a few years, with a new American program to explore Mars that will use a mobile laboratory—either the Rover, an all-terrain vehicle that should make it possible to study the old Martian valleys, or the more modest but very ingenious Maboule project proposed by the French professor Blamont.

After the complex and unexpected results of the Viking biological experiments, studies done in the laboratory seem to show that the experiments may be interpreted on a purely chemical basis. That probably is the reason why the majority of scientists concerned with problems of exobiology think that the conclusion of the biological program of the Viking mission is that life is absent from Mars. This thesis is further supported by the fact that on Mars's surface is a very substantial flux of ultraviolet radiation incompatible with the survival of species—at least of the terrestrial type.

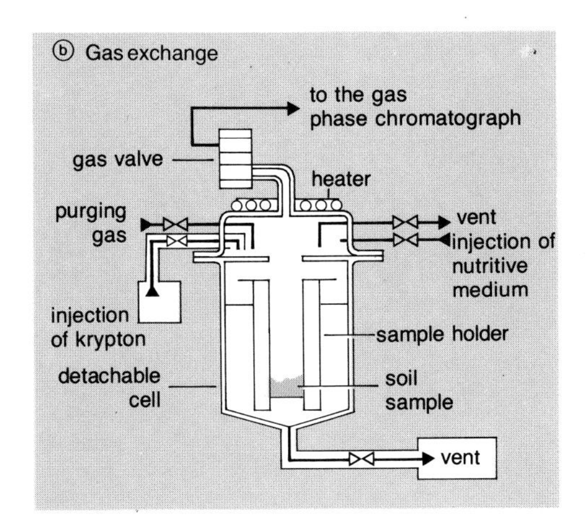

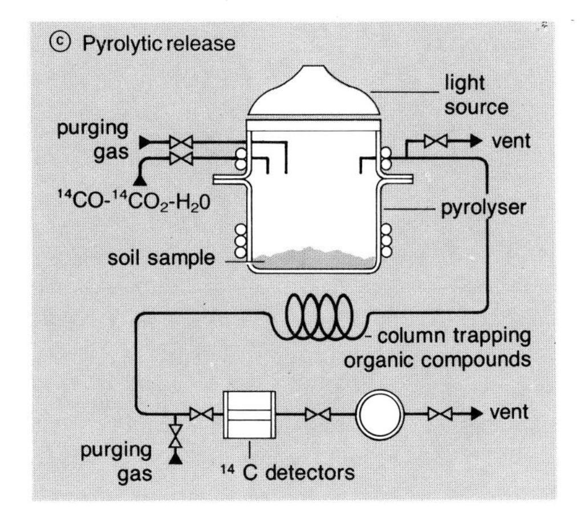

pounds of the nutritive medium was observed for all the soil samples. Finally, the "sterilization" of the samples by heating to 50°C eliminated that property.

In the second experiment (b), called *gas exchange*, the sample was also placed in the presence of a nutritive solution, but the gases given off after "digestion" were analyzed by chromatography. The analyses showed a rapid release of oxygen when the Martian soil came in contact with water. Incubation in the presence of a nutritive medium caused a slow release of CO_2. On the other hand, no gas exchange attributable to living systems was observed.

The third experiment (c), called *pyrolytic release*, was designed to reveal any activity involving photosynthesis of carbon compounds. A Martian soil sample, placed in a reaction chamber containing a Martian atmosphere simulated with CO_2 marked with ^{14}C, was exposed to light rays similar to those reaching Mars. After several days of irradiation, the sample was separated from the gaseous phase, then heated to 600°C to pyrolyze any organic compounds formed by photosynthesis.

Analysis of the radiactivity of the pyrolytic gases made it possible to estimate the photosynthetic activity. Through the trials it became apparent that small quantities of organic compounds were synthesized from CO-CO_2 mixtures. This synthesis, fostered by light, is inhibited by water. (See the accompanying diagram.)

This flux is such that it is even capable of destroying organic compounds, as John Oro recently demonstrated, which may also explain the absence of such compounds on Mars's surface.

After the Viking mission and the negative results of its biological program, it appears more and more likely that Earth is the only planet in the solar system to harbor life. Therefore we must look farther out for extraterrestrial life.

Life Outside the Solar System?

Are there other planetary systems in the universe? It must be admitted that up to now, we have no direct experimental proof of the presence of planets outside the solar system. These planets would be too dark to be seen easily from Earth. However, several arguments work in favor of the existence of other planetary systems.

Since it has been possible to measure the speed of the stars' rotation about their axes, we have noted that the majority of them move relatively slowly and that they seem to have lost part of their angular momentum (the value of which is related to mass and rotation speed among other things). The main explanation of this phenomenon amounts to assuming that these stars have lost matter that formed a planetary system. Modern theories of the formation of stars predict the existence of planets like those of our solar system, as well as more voluminous ones. Examination of the stars closest to the Sun has shown the possible existence of the latter. For example, it has been discovered that Barnard's star, the Sun's closest neighbor after Proxima Centauri and Alpha Centauri, might be surrounded by two planets comparable in mass to Jupiter. Perhaps in a few years the Space Telescope will make it possible to confirm this result and to observe planets around other nearby stars.

Since the large planets predicted by the theory were discovered (despite the extreme difficulties in detecting them), we may assume that small ones also exist. In fact, most astronomers accept that planetary systems correspond to a normal result of stellar evolution and that they probably are very numerous in the universe. It is generally estimated that there may be about 1 billion of them in our own galaxy (or about 10^{10} planets), and the figure of 10^{19} planets in the entire universe frequently is mentioned. How many of them are capable of supporting life?

Let us assume that, as is the case of our solar system, one planet out of 10 meets the conditions necessary for the development of life; we must then figure the probability of living systems appearing on such a planet. We saw that this point still was far from being clarified. Let us simply recall that there might be 10^{18} planets in the universe capable of harboring life.

Admitting that the Earth is the only planet with living systems therefore would mean that the probability of life appearing on a planet meeting the necessary physical and chemical conditions is not much larger than 1 in 10^{18}. Assuming, on the other hand, that chemical evolution on the same planet necessarily leads to life means accepting that there are 10^{18} planets in the universe on which life is, was or will be present. Taking the most favorable case, it appears probable that one of the stars close to the Sun, like Barnard's star, has such a planet in its planetary entourage. But that star is 6 light-years from Earth.

If it took only about 1½ years for the Voyager probes to travel the approximately 10^9 kilometers between the Earth and Jupiter, it would take the same type of space vehicle at least 10,000 years to reach Barnard's star (hence, twice as long for a round trip). Thus it seems clear that with the space technology we currently possess, exploration missions outside the solar system, and particularly the search for living organisms in the planetary systems close to ours, cannot be envisioned in the coming decades, even if we confine ourselves to the closest stars.

The Search for Intelligent Extraterrestrial Life

Speculations About Extraterrestrial Civilizations

Despite the preceding conclusions, there still remains the possibility of knowing in the near future if life exists elsewhere. Indeed, if we assume that there are billions of planets in the universe that harbor living organisms, it is not unreasonable to suppose that on some of them, life has evolved, as it has on Earth, to the stage of intelligent life. Some authors, moreover, even go to the point of hypothesizing that such evolution is systematic. Why not try, then, to make contact with such intelligent extraterrestrial societies? Detection of such civilizations would not only answer the question "Does life exist elsewhere?" but also might greatly influence our technological and scientific capacities.

Some researchers have done probability calculations to estimate the number of civilizations in the universe, their life-span and their minimal distance from the Earth. Most of these calculations are based on the Drake-Sagan formula.

The life-span of a civilization is an important unknown. Von Hoerner went deeper into this question by looking for the possible causes for the disappearance of a technically advanced civilization. He came up with several scenarios: the civilization disappears in a crisis that destroys all life on the planet; the civilization disappears through extinction of the beings that compose it; the civilization disappears as a result of physical and mental degeneration; the civilization loses its interest in science and technology. Van Hoerner believes that there is no chance of a civilization lasting indefinitely. In putting forth the hypothesis that our planetary system and our civilization represent a universal average, and using an analysis of the evolution of our civilization, he was led to propose a value of 5,000 to 10,000 years for the average life-span of a technically advanced civilization. The Soviet astronomer Shklovsky adds to the dangers threatening a civilization that of the overproduction of information. However, taking into account the development of possible robot

Barnard's Star

The name "Barnard's star" refers to a small star in the constellation of Ophiucus; the star was discovered in 1916 by the American astronomer Edward Emerson Barndard. It is a red dwarf, of spectral type M5 and of visual magnitude 9.5. Situated 5.9 light-years from Earth, it is the closest star to the solar system after the triple system of Alpha Centauri. But it is known primarily as the star having the greatest proper motion: Its annual displacement on the celestial sphere is 10.31 arc seconds. It is currently moving toward us at a radial velocity of 108 kilometers (67 miles) per second, and its distance is decreasing by 0.036 light-year per century. In some 10,000 years, its distance will be no more than 3.85 light-years; it will then become the closest star to the solar system. Beginning in 1937, the Dutch-born American astronomer Peter van de Kamp closely studied the perturbations of its motion at the Sproul Observatory in Swarthmore, Pennsylvania. In so doing, he came to the conclusion that this star possesses two planets, one having a mass 0.8 times that of Jupiter and with a period of revolution of 11.7 years; the other, smaller, having a mass 0.4 times that of Jupiter and a period of revolution close to 20 years. These results still are controversial.

Nearby Stars That May Possess Planets

Name	Constellation	Distance (light-years)	Remarks
Barnard's Star	Ophiucus	5.9	2 planets M_1 = 0.8; P_1 = 11.7 years M_2 = 0.4; P_2 = 20 years
Lalande 21185	Ursa Major	8.2	1 large planet M = 30; P = 420 days
Luyten 726-8	Cetus	7.9	2 planets M_1 = 1.1; P_1 uncertain M_2 = 1.4; P_2 uncertain
Ross 248	Andromeda	10.3	1 planet M uncertain; P = 8 years
ε *Eridani*	Eridani	10.7	1 planet M between 6 and 50; P = 25 years
61 Cygni	Cygnus	11.1	1 planet M = 1.6; P = 5 years
BD + 5°1668	Canus Minoris	12.3	1 planet M = 60; P = 7 years
BD + 20°2465	Leo	15.50	1 planet M = 30; P uncertain

* M = mass of the presumed planet relative to Jupiter
P = period of revolution of the presumed planet

The Drake-Sagan Formula

$$N = R \times fg \times fp \times ne \times fl \times fi \times fa \times L$$

Frank Drake and Carl Sagan have proposed the above mathematical formula for estimating the number N of technologically advanced civilizations in our galaxy:

R represents the average rate of star formation in the galaxy; fg the percentage of stars similar to the Sun; fp the percentage of those stars possessing a planetary system; ne the number of planets of the terrestrial type in the habitable zone (ecosphere) of that system; fl the fraction of those planets where life has appeared; fi the fraction of those latter planets where life has evolved toward intelligence; fa the fraction of intelligent civilizations that have developed advanced technologies; and L the average life-span of a civilization.

The chief problem, of course, lies in choosing values for these different parameters. Thus we can easily understand that, according to the authors, the estimate of, e.g., the number of technological civilizations present in our galaxy varies from 0 to 10^9 (mainly owing to the very wide range of values that can be assigned to the parameters fl, fi and fa).

civilizations (suggested by Fred Hoyle in particular), Shklovsky estimates that the life-span of a civilization may last from 100,000 to 1 million years; in any case, its longevity could not be greater than 1 billion years.

Thus we again see considerable variations in the estimates, which is not so surprising, for we should not forget that we are in an entirely speculative domain here. The same is true, of course, with regard to estimating the distance that separates us from the nearest extraterrestrial community with advanced technology: from 10 to several thousand light-years, depending on the authors. In all objectivity, it must be admitted that we are not capable, with the data and means we currently possess, of determining either the probable number of technologically advanced civilizations nor their average life-span or distance from Earth. As Orgel remarks, the estimates made in this regard "depend more on the temperament of their authors than on pertinent information."

If, however, we take the position adopted by the optimists, we should bear in mind that the universe and our galaxy itself may possess a multitude of planets inhabited by intelligent civilizations, some of which are at a stage of technological development superior to ours. The Soviet Kardashev, who proposed classifying galactic civilizations according to the quantity of energy they use for communication, goes so far as to assume the existence of civilizations that devote the energy of an entire galaxy to it. The signals emitted by such civilizations would be detectable by us, wherever their point of origin in the universe. Without going as far as such a hypothesis, we may also imagine that such civilizations have been emitting radio signals for thousands of years. Detection of such emission would indicate to us the existence of extraterrestrial civilizations.

Certain authors, such as Hart and Papagiannis, using the Drake-Sagan formula, come to the conclusion that if there are other intelligent inhabitants of our galaxy that have reached an advanced technological level, they should also be present in our own solar system. At a conference in 1977, Papagiannis even was inclined to suppose that one of the places best suited for such a space colony, after a long interstellar voyage to the Sun, would be the asteroid belt. Some authors have been inclined to assume that, in that case, our planet itself now is being observed by those extraterrestrial colonizers!

The Problem of Unidentified Flying Objects (UFO's)

Is the answer to the question "Does life exist elsewhere?" related to the problem of UFO's? It seems difficult to approach a topic here that has already caused so much ink to flow. Literature—good or bad—already is swarming with papers, most of them pseudoscientific, that deal with mysterious visits to our planet, in the past or present. It seems indispensable, however, to bring up the question of UFO's and to summarize the points that seem to us the most important for our discussion.

Official investigations have been conducted—mainly in the United States and the USSR—to study these phenomena. In the United States, the semi-public investigation carried out for 22 years by the U.S. Air Force and then entrusted to a scientific commission, led to a report published by the Condon Commission. This report concludes that UFO's do not exist, though it admits that some of the cases studied are doubtful. It appears, on the basis of statements by some scientists, such as the astrophysicist Hynek, who studied the entire U.S. Air Force record on UFO's in detail, or the atmospheric physicist MacDonald, who had access to many entries in that record, that many cases resist any "natural" explanation. However, despite the efforts of these two eminent scientists, no irreproachable official inquiry has been launched.

The majority of the world scientific community has taken a position with regard to the problem of UFO's that is far from rational; it is satisfied merely to ignore the question. This attitude is supported by the development, since the 1950s, of an entire "flying saucer" literature simultaneous with the creation of certain pseudoscientific groups whose ramblings sometimes have resulted in discrediting the problem of UFO's.

It must be admitted that rational and objective study of this question is difficult, for one must rely on indirect information. We can, however, critically look at certain points concerning the problem of UFO's. We must ask ourselves, in particular, if this problem really exists. For all the scientists who seriously studied the entire record, the answer is yes. However, the disputed or unexplained cases do not prove the existence of extraterrestrials. At most, MacDonald considers that an explanation based on the presence of extraterrestrials is the poorest hypothesis in some cases. In view of the currently available data, it seems highly unlikely to us that the UFO's can be identified as extraterrestrial spaceships.

It appears that we are confronted by an irritating problem that is difficult to study rationally: We do not know what we are looking for, nor how to look for it. It seems essential to us, however, that this question should finally be taken up officially by scientists, in order, initially, to define its limits and its methods of study.

Listening for Extraterrestrials

If the problem of UFO's is not yet taken seriously by most scientists, including those engaged in space exploration, the existence of technologically advanced extraterrestrial communities is, on the contrary, an eventuality that they consider plausible. To convince oneself of this, it is sufficient to remember the message NASA placed, at the exobiologist Sagan's urging, aboard the *Pioneer 10* probe, launched in 1972, then *Pioneer 11,* launched in 1973. This communication was aimed at any intelligent inhabitants in our galaxy or even in more distant regions of the universe.

Better yet, two successive American probes were launched in August and September 1977: the Voyager probes. After exploring Jupiter in 1979, Saturn in 1981 and Uranus in 1986, these two spaceships gradually will leave our solar system and enter interstellar space, where they will not encounter planetary systems for millions of years. Aboard each of these probes, NASA placed a videodisk recording, a message from Earth to potential extraterrestrials. This message contains various sounds: greetings from man in 54 different languages; samples of music from different civilizations and different places; and noises supposed to be representative of our planet, from the sound of an avalanche to the song of the whale. This message also contains images that can be decoded with a video system: a message from former President Jimmy Carter; a message from former United Nations Secretary General Kurt Waldheim; a message from several United Nations delegations; and more than 100 photographs explaining our planet and its inhabitants. Even though such gestures are mainly symbolic, they nonetheless show that the possibility of the existence of intelligent extraterrestrial civilizations in now taken seriously not only by scientists but by politicians as well.

First Attempts

Yet less than 20 years ago, many people chuckled when two astrophysicists, Frank Drake and William Waltman, persuaded the director of the National Radio Astronomy Observatory at Green Bank, West Virginia, to let them use the parabolic antenna of the radio telescope for an experiment of listening to the cosmos for several months. This was Project Ozma. The idea for this project came from Professors Cocconi and Morrison, who suggested searching the universe for extraterrestrial signals emitted at radio wavelengths, using one of the radio antennas available at the time. Drake and Waltman chose the wavelength recommended by Cocconi and Morrison: 21 centimeters, the wavelength of the natural line emitted by hydrogen. As for the stars to be studied, they chose τ Ceti

The globular cluster M 13 in the constellation Hercules. Discovered in 1714 by Edmund Halley, this cluster contains several hundred thousand stars within a volume less than 200 light-years in diameter. At the center, the density of stars is 500 times greater than around the Sun. Its distance is close to 25,000 light-years. Photo: P.P.P.-IPS.

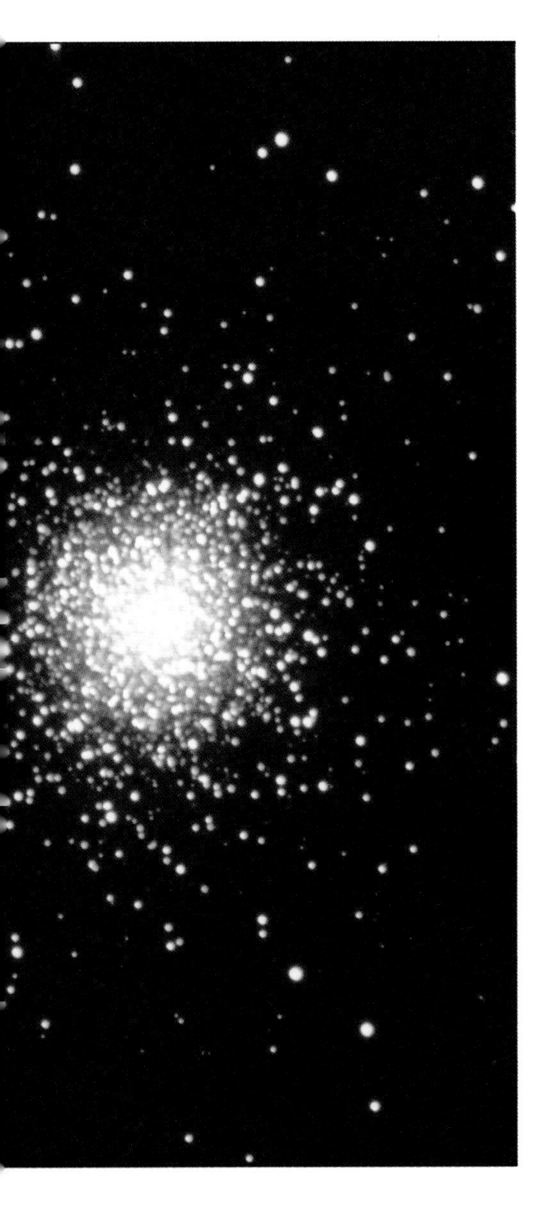

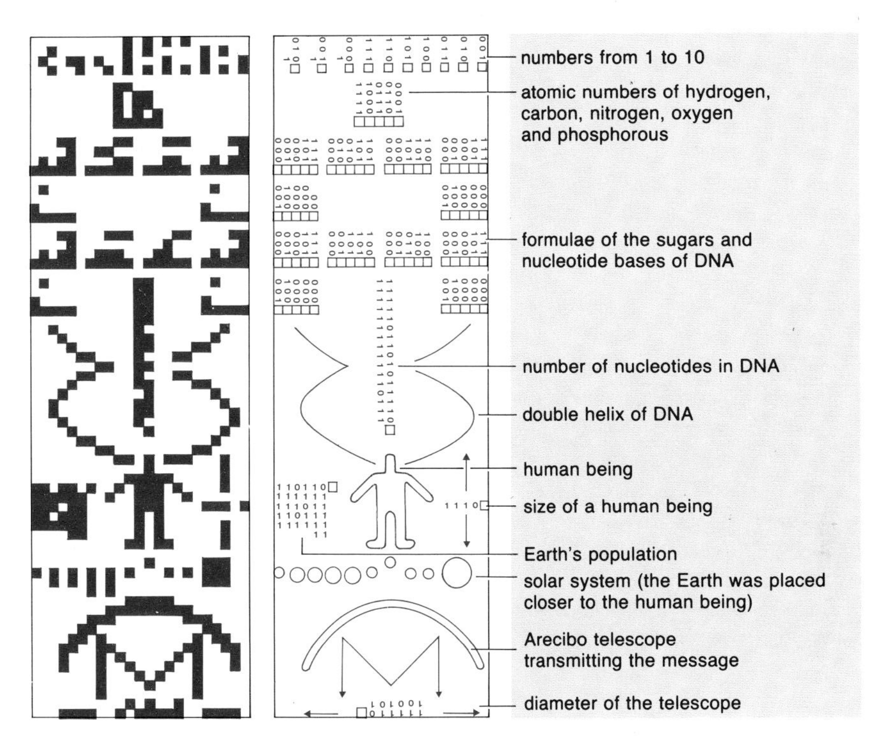

Schematic illustration of the message beamed at the globular mass M 13 for 169 seconds on November 16, 1974, at the frequency 2,380 MHz, using the radio telescope at Arecibo (see page 125). This message, composed of two kinds of pulses (1 and 0), contains 1,679 consecutive characters describing the numbers from 1 to 10; the atomic numbers of five biologically significant atoms (hydrogen, carbon, oxygen, nitrogen and phosphorus); the formulae of sugars and bases of DNA; a diagram of the double helix of DNA with an estimate of the number of its nucleotides; and drawings of a human being, the solar system, and the radio telescope used for the emission. The total number of pulses can be broken down into primary factors in only one way: 23 × 73. The illustration above, at left, is obtained by arranging the impulses in 73 lines of 23 columns and by darkening the boxes corresponding to the 1 pulses.

and ε Eridani, 11 light-years away from our Sun, their relative proximity making them accessible to the radio telescope used.

The study of the most favorable wavelength range for interstellar communications suggests that it may go from 3 centimeters to a few meters. Hydrogen is the most abundant chemical element in the universe, and we can assume that its importance has not escaped the attention of advanced civilizations. The radio-emission wavelength of atomic hydrogen—21 centimeters—thus would correspond to the universal frequency for communication between civilizations.

The experiment lasted three months, from May to July 1960, and the conclusion was that "no signal of extraterrestrial origin has been discovered during these preliminary observations." Several serious criticisms were made of Project Ozma, both of the choice of wavelength and of the choice of stars. The main criticism, a more basic one, was made by Shklovsky. He assumes that our chances of contact are much greater with civilizations more evolved than our own, which would require completely different instrumental methods of much higher performance.

A synthesis of all the principal projects and ideas concerning the search for extraterrestrial civilizations was done at the Byurakan congress in Armenia in May 1964. One year later, a Soviet astronomer, Sholomitsky, discovered a very intense radio source, CTA 102, with characteristics that appeared to correspond to those of the artificial radio sources predicted by Kardashev in his theory of supercivilizations. It took Soviet scientists several days to conclude that the source was a quasar.

A somewhat similar experience happened in 1967 to Jocelyn Bell, a student at Cambridge, when she discovered the first pulsar.

in 1971, at the instigation by NASA of the American Society for Engineering Education, a new project of listening to the cosmos was launched: Project Cyclops. On the basis of Van Hoerner's work regarding the number, life-span and distance of extraterrestrial civilizations, and on Dyson's ideas about supercivilizations, the American researchers, led by Oliver, tried to reevaluate the possibilities for interstellar listening. In view of the presumed expansion of galactic societies, going to the point of using the energy of their entire stellar system, it seems that one means of detecting advanced civilizations is to look for intermittent sources of infrared radiation.

For their part, the Soviets have more recently launched the Project Ceti (Communication with Extraterrestrial Intelligence), planned in two stages: Ceti I (1975–85) and Ceti II (1980–90). This ambitious program, which plans work over a long period, relies on a vanguard technology. It requires a constant revision of its content, which takes into account progress in detecting radio and light waves, as well as progress in theoretical studies of the interpretation of intelligent signals. Initially eight radio antennae will be permanently installed on the ground to receive potential signals from the cosmos. They will then be gradually complemented by orbiting antennae. Finally, a simultaneous attempt at reception by two sensitive radio-astronomical observatories should provide maximum receptivity.

Research in the area of detecting intelligent extraterrestrial life still is only in an embryonic stage. But now it is official. A scholarly society has been set up, Ceti, which includes scientists interested in the problem of communication with intelligent extraterrestrial systems. The first meeting of Ceti was held in 1971, in Byurakan again. Among the participants were two Nobel Prize winners. Since 1972, scientists have been meeting in the International Astronautical Academy, which set up a permanent commission on the Ceti problem. At the end of 1976, an international telecommunications commission was set up by the United States and the Soviet Union to do an in-depth study of signals that might be emitted by possible extraterrestrial civilizations. The International Consultative Committee on Radio Communications, which carried out that study, concluded that electromagnetic waves are the only practical means of detecting the existence of intelligent extraterrestrial life. It added that it should be technically possible to receive radio signals from extraterrestrial civilizations and to recognize them.

A Bottle in Space

The reception of extraterrestrial signals is not our only means of contact with possible distant civilizations. Assuming our correspondents are sufficiently evolved to have very sensitive detectors, we can also envision sending messages to inform them of our existence, probably without hope of a reply. That is what was done on November 16, 1974, from the radio telescope at Arecibo. The message, sent to the Messier 13 grouping in the constellation of Hercules, 25,000 light-years from Earth, was copied from the famous message in binary code that Frank Drake had presented several years earlier to a radio astronomy congress. He had then asked his colleagues to assume that this message had been received from a star system and to decode it.

Up to now, no extraterrestrial civilization has been detected. However, an important step has been taken, because now the problem is no longer being approached by a few isolated researchers but by an entire international scientific community.

Bibliography

General References

■ J. AUDOUZE and G. ISRAEL (eds), **The Cambridge Atlas of Astronomy** (Cambridge: *Cambridge University Press*, 1985). ■ L. BERMAN and J. EVANS, **Exploring the Cosmos**, 4th ed (Boston: *Little*, 1983; London: *Hutchinson Ed. Ltd.).* ■ B. M. FRENCH and S. P. MARAN (eds), *A Meeting With the Universe* (NASA EP-177, 1981). ■ M. FRIEDLANDER, **Astronomy: From Stonehenge To Quasars** (Englewood Cliffs: *Prentice Hall*, 1985). ■ N. HENBEST and M. MARTEN, **The New Astronomy** (Cambridge: *Cambridge University Press*, 1983). ■ R. JASTROW and M. THOMPSON, **Astronomy: Fundamentals and Frontiers,** 4th ed (New York: *Wiley*, 1984). ■ W. KAUFMANN, **Universe** (San Francisco, CA: *W. H. Freeman*, 1985). ■ P. MOORE, **The New Atlas of the Universe**, 2nd rev ed (New York: *Crown*, 1984; London: *Mitchell Beazley*, 1984). ■ I. RIDPATH and W. TIRION, **Universe Guide to Stars and Planets** (New York: *Universe*, 1985; London: *Collins*, 1984). ■ C. SAGAN, **Cosmos** (New York: *Random House*, 1980; London: *Macdonald*, 1981).

History of Astronomy

■ J. ASHBROOK, **The Astronomical Scrapbook: Skywatchers, Pioneers, and Seekers in Astronomy** (Cambridge: *Cambridge University Press*, 1985). ■ R. BERENDZEN, **Man Discovers the Galaxies** (New York: *Columbia University Press*, 1984; London: *Wm Dawson*). ■ K. BRECHER and M. FEIRTAG (eds), **Astronomy of the Ancients** (Cambridge, MA: *MIT Press*, 1979). ■ J. CORNELL and P. GORENSTEIN (eds), **Astronomy from Space: Sputnik to Space Telescope** (Cambridge, MA: *MIT Press*, 1985). ■ D. DeVORKIN, **The History of Modern Astronomy and Astrophysics** (New York: *Garland Pub*, 1985). ■ F. DURHAM and R. PURRINGTON, **Frame of the Universe** (New York: *Columbia University Press*, 1985; London: *Wm Dawson*). ■ W. T. SULLIVAN (ed), **The Early Years of Radio Astronomy: Reflections Fifty Years after Jansky's Discovery** (Cambridge: *Cambridge University Press*, 1984). ■ A. VanHELDEN, **Measuring the Universe: Cosmic Dimensions from Aristarchus to Hailey** (Chicago and London: *University of Chicago Press*, 1985).

Instruments & Techniques of Exploration of the Universe

■ M. COHEN, **In Quest of Telescopes** (Cambridge: *Cambridge University Press*, 1980). ■ J. CORNELL and J. CARR (eds), **Infinite Vistas: How the Space Telescope and Other Advances Are Revolutionizing Our Knowledge of the Universe** (New York: *Scribner*, 1985; London: *Macmillan*). ■ E. DERENIAK and D. CROWE, **Optical Radiation Detectors** (New York: *Wiley*, 1984). ■ G. FIELD and E. CHAISSON, **The Invisible Universe: Probing the Frontiers of Astrophysics** (Boston: *Birkhauser*, 1985). ■ B. LOVELL, **The Jodrell Bank Telescopes** (Oxford: *Oxford University Press*, 1985). ■ W. TUCKER and R. GIACCONI, **The X-Ray Universe** (Cambridge, MA: *Harvard University Press*, 1985).

Amateur Astronomy

■ M. COVINGTON, **Astrophotography For the Amateur** (Cambridge: *Cambridge University Press*, 1985). ■ P. DUFFETT-SMITH, **Astronomy With Your Personal Computer** (Cambridge: *Cambridge University Press*, 1985). ■ W. KALS, **Stargazer's Bible** (New York: *Doubleday*, 1980). ■ C. A. RONAN, **The Skywatcher's Handbook** (New York: *Crown*, 1985). ■ C. A. RONAN, **Amateur Astronomy** (Feltham: *Newnes*, 1984). ■ P. C. SHERROD, **A Complete Manual of Amateur Astronomy: Tools and Techniques for Astronomical Observations** (Englewood Cliffs, NJ: *Prentice-Hall*, 1981).

The Solar System

■ R. BATSON, *et al* (eds), "The Atlas of Mars," (NASA SP-438, 1979). ■ J. BEATTY and B. O. O'LEARY (eds), **The New Solar System**, 2nd ed (London: *Cambridge University Press*, 1982). ■ C. CHAPMAN, **Planets of Rock and Ice: From Mercury to the Moons of Saturn** (New York: *Scribner*, 1982; London: *Macmillan*). ■ E. CORTRIGHT (ed), "Apollo Expeditions to the Moon" (NASA SP-350, 1975). ■ M. DAVIES, et al (eds), "Atlas of Mercury" (NASA SP-423, 1978). ■ J. DUNNE and E. BURGESS, "The Voyage of Mariner 10: Mission to Venus and Mercury" (NASA SP-424, 1978). ■ E. EZELL and L. EZELL, "On Mars: Exploration Of the Red Planet 1958–1978" (NASA SP-4212, 1984). ■ R. FIMMEL, et al, "Pioneer Venus" (NASA SP-461, 1983). ■ K. FRAZIER, **Our Turbulent Sun** (Englewood Cliffs, NJ: *Prentice-Hall*, 1982). ■ D. GOLDSMITH, **Nemesis: The Death Star and Other Theories of Mass Extinction** (New York: *Walker & Co*, 1985). ■ A. L. GRAHAM, A. BEVAN, and R. HUTCHISON, **Catalogue of Meteorites,** 4th ed (Tucson: *University of Arizona Press*, 1985; London: *BM of Nat. History*). ■ W. HARTMANN, **Moons and Planets,** 2nd ed (Belmont, CA: *Wadsworth Pub*, 1983). ■ G. HUNT and P. MOORE, **Rand McNally Atlas of the Solar System** (Rand McNally, 1983; London: *Springfield Books Ltd.*). ■ D. MORRISON, "Voyages to Saturn" (NASA SP-451, 1982). ■ D. MORRISON and J. SAMZ, "Voyage to Jupiter" (NASA SP-439, 1980). ■ R. NOYES, **The Sun, Our Star** (Cambridge, MA: *Harvard University Press*, 1982). ■ B. PREISS (ed), **The Planets** (New York: *Bantam*, 1985). ■ F. WHIPPLE, **The Mystery of Comets** (Washington, DC: *Smithsonian*, 1985; Cambridge: *Cambridge University Press*, 1986).

Stars, The Galaxy

■ B. BOK and P. BOK, **The Milky Way,** 5th ed (Cambridge, MA: *Harvard University Press*, 1981). ■ D. CLARK, **Superstars: How Stellar Explosions Shape the Destiny of Our Universes** (New York: *McGraw-Hill,* 1984). ■ R. KIPPENHAHN, **One Hundred Billion Suns: The Birth, Life, and Death of the Stars** (New York: *Basic Books*, 1983; London: *Counterpoint*, 1983 and *Unwin Paperbacks*, 1985). ■ S. SHAPIRO and S. TEUKOLSKY, **Black Holes, White Dwarfs, and Neutron Stars: The Physics of Compact Objects** (New York: *Wiley*, 1983).

The Extragalactic Universe

■ J. CORNELL and A. LIGHTMAN (eds), **Revealing the Universe: Prediction and Proof in Astronomy** (Cambridge, MA: *MIT Press*, 1981). ■ D. HEESCHEN and C. WADE, **Extragalactic Radio Sources** (Reidel Holland/Kluwer Academic, 1982; USA; *Reidel Pub. Co.* 1982). ■ J. HEIDMANN, **Extragalactic Adventure: Our Strange Universe** (Cambridge: *Cambridge University Press*, 1982). ■ R. HILLIER, **Gamma Ray Astronomy** (Oxford: *Oxford University Press*, 1984). ■ P. HODGE (ed), **Universe of Galaxies** (New York and Oxford: *W.H. Freeman*, 1984). ■ W. KAUFMANN, **Galaxies and Quasars** (New York and Oxford: *W. H. Freeman*, 1979).

Cosmology

■ J. BARROW and J. SILK, **The Left Hand of Creation: Origin and Evolution of the Universe** (New York: *Basic Books*, 1983; London: *Heinemann*, 1984 and *Unwin Paperbacks*, 1985). ■ T. FERRIS, **The Red Limit: The Search for the Edge of the Universe,** rev ed (New York: *Morrow*, 1983). ■ G. GIBBONS, *et al* (eds), **The Very Early Universe** (Cambridge: *Cambridge University Press*, 1983). ■ E. HARRISON, **Cosmology: The Science of the Universe** (Cambridge: *Cambridge University Press*, 1981). ■ E. HARRISON, **Masks of the Universe** (New York: *Macmillan*, 1985). ■ W. KAUFMANN, **Relativity and Cosmology,** 2nd ed (New York: *Harper and Row*, 1977). ■ W. KAUFMANN, **Black Holes and Warped Spacetime** (New York and Oxford: *W. H. Freeman*, 1979). ■ J. SILK, **The Big Bang: The Creation and Evolution of the Universe** (New York and Oxford: *W. H. Freeman*, 1980). ■ J. TREFIL, **The Moment of Creation: Big Bang Physics from Before the First Millisecond to the Present Universe** (New York: *Scribner*, 1983; London: *Macmillan).* ■ R. WAGONER and D. GOLDSMITH, **Cosmic Horizons: Understanding the Universe** (New York and Oxford: *W. H. Freeman*, 1983). ■ S. WEINBERG, **The First Three Minutes: A Modern View of the Origin of the Universe** (New York: *Basic Books*, 1976; London: *Andre Deutsch*, 1977).

Exobiology, Interstellar Communication

■ J. ANGELO, **The Extraterrestrial Encyclopedia** (*Facts On File*, 1985; London: *Phaidon Press Ltd.).* ■ I. ASIMOV, **Extraterrestrial Civilizations** (New York: *Crown*, 1979; London: *Pan Books*, 1981). ■ J. BILLINGHAM, et al (eds), **Life in the Universe** (Cambridge, MA: *MIT Press*, 1981). ■ F. CRICK, **Life Itself** (New York: *Simon & Schuster*, 1982). ■ D. GOLDSMITH (ed), **The Quest for Extraterrestrial Life: A Book of Readings** (Mill Valley, CA: *University Science Books*, 1980). ■ D. GOLDSMITH and T. OWEN, **The Search for Life in the Universe** (Menlo Park, CA: *Benjamin/Cummings Pub*, 1980; London: *Addison Wesley*). ■ M. HART and B. ZUCKERMAN, **Extraterrestrials: Where are they?** (Elmsford, NY: *Pergamon Press*, 1982). ■ F. HOYLE, **The Intelligent Universe: A New View of Creation and Evolution** (New York: *Holt, Rinehart & Winston*, 1984; London: *Michael Joseph*, 1983). ■ R. JASTROW, **Until the Sun Dies** (New York: *W. W. Norton* 1977; London: *John Wiley & Sons*). ■ D. MILNE, et al (eds), "The Evolution of Complex and Higher Organisms" (NASA SP-478, 1985). ■ P. MORRISON, J. BILLINGHAM, and J. WOLFE (eds), "The Search for Extraterrestrial Intelligence: SETI" (NASA SP-419, 1977). ■ R. ROOD and J. TREFIL, **Are We Alone? The Possibility of Extraterrestrial Civilizations** (New York: *Scribner*, 1983; London: *Macmillan*). ■ C. SAGAN, **The Cosmic Connection** (New York: *Doubleday*, 1973; London: *Macmillan*, 1981).

index

L

M

N

R

S

T

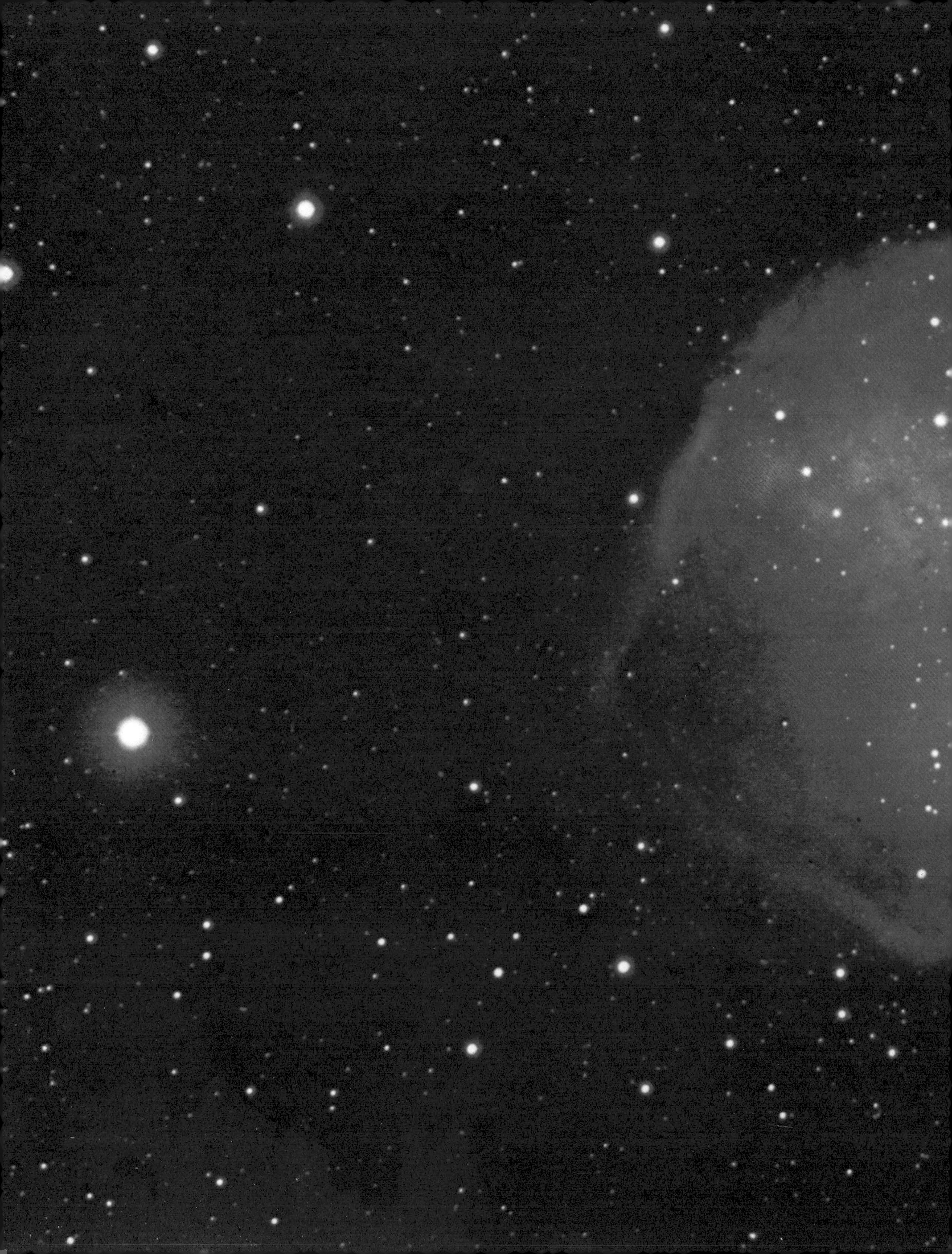